THE EDITORS

Professor T. K. Bose awarded M.Sc. (Ag.), Ph.D. degree from Calcutta University, West Bengal, India. Former Horticulturist and Secretary of the Agri-Horticultural Society of India; Project Coordinator, All India Coordinated Floriculture Improvement Project (ICAR); Professor and Head, Department of Horticulture; Dean, Faculty of Agriculture; Director of Research, Bidhan Chandra Krishi Viswavidyalaya, West Bengal; Member, State Agriculture Commission, West Bengal; Member of Scientific Advisory Committees of ICAR, CSIR (Govt. of India) and Government of West Bengal. Awarded D.Sc. (Honoris Causa) from OUAT, Bhubaneshwar, Odisha and UBKV, Coochbehar, West Bengal, India. Guided around 45 students for their Ph.D. degree, published more than 250 research papers and authored 30 books on horticultural crops.

Dr. V. A. Parthasarathy has about half a century of research experience in Horticulture particularly in the field of genetics and biotechnology. Former Director, Indian Institute of Spices Research, Kerala, India. President of National Academy of Biological Sciences. Edited/authored over 50 books, about 100 review/ invited papers and over 225 research papers.

Professor S. K. Mitra awarded M.Sc. (Ag.), Ph.D. degree from Bidhan Chandra Krishi Viswavidyalaya, West Bengal, India. Former Head, Department of Fruit Science; Dean, Faculty of Horticulture; Dean, Post Graduate Studies, Bidhan Chandra Krishi Viswavisyalaya, West Bengal, India. Guided 20 Ph.D. students, published more than 200 research papers and authored15 books on horticultural crops.

Professor B. Ghosh awarded M.Sc. (Ag.), Ph.D. degree from Bidhan Chandra Krishi Viswavidyalaya, West Bengal, India. Former Head, Department of Fruit Science, Officer in Charge of Mango database Project, DBT, Government of India. Guided 6 Ph.D. students, authored 5 books, published more than 80 research papers.

Professor I. Chakraborty awarded M.Sc. (Ag.) from Calcutta University, Ph.D. from Bidhan Chandra Krishi Viswavidyalaya, West Bengal, India. Former Head, Department of Post Harvest Technology, Bidhan Chandra Krishi Viswavisyalaya, West Bengal, India. Recipient of J.S. Pruthi award conferred by AFSTI, CFTRI, Mysore, India. Guided 7 Ph.D. students, published more than 100 research papers and written 4 books on horticultural crops.

Late Professor D. Sanyal awarded M.Sc. (Ag.), Ph.D. degree from Bidhan Chandra Krishi Viswavidyalaya, West Bengal, India. Associate Professor, NEHU, Nagaland and Professor in the Department of Horticulture, Institute of Agriculture, Visva Bharati, West Bengal. Published more than 100 research papers and written 15 books on horticultural crops including 10 volumes of "Dictionary of Horticulture".

FRUITS

Tropical and Subtropical

— *Volume 2* —

— *Senior Editor* —

V.A. Parthasarathy

ICAR – Indian Institute of Spices Research
32/482 C Narmada Nilayam
Bharathan Bazar, Chelavoor
Calicut 673571, Kerala, India

— *Editor Emeritus* —

T.K. Bose

Faculty of Horticulture
Bidhan Chandra Krishi Viswavidyalaya
West Bengal 741 252, India

— *Edited by* —

S.K. Mitra • B. Ghosh • I. Chakraborty

Faculty of Horticulture
Bidhan Chandra Krishi Viswavidyalaya
West Bengal 741 252, India

D. Sanyal

Palli Siksha Bhavan (Institute of Agriculture)
Sriniketan 731236, Visva Bharati
West Bengal

2022

Daya Publishing House®
A Division of
Astral International Pvt. Ltd.
New Delhi – 110002

Fourth Revised & Illustrated edition 2021

Cataloging in Publication Data--DK

Courtesy: D.K. Agencies (P) Ltd. <docinfo@dkagencies.com>

Fruits : tropical and subtropical / senior editor, V.A. Parthasarathy ; editor emeritus, T.K. Bose ; edited by S.K. Mitra, B. Ghosh, I. Chakraborty, D. Sanyal. -- Fourth revised & illustrated edition.

volumes cm

Includes index.

ISBN 9789354616198 (HB)

1. Tropical fruit--India. I. Parthasarathy, V. A., editor.

LCC SB359.F78 2021 | DDC 634.60954 23

Published by : **Daya Publishing House®**
A Division of
Astral International Pvt. Ltd.
– ISO 9001:2015 Certified Company –
4736/23, Ansari Road, Darya Ganj
New Delhi-110 002
Ph. 011-43549197, 23278134
E-mail: info@astralint.com
Website: www.astralint.com

From the Desk of Editor Emeritus

The undersigned had initiated the publications entitled 'Fruits: Tropical and Sub-tropical', 'Vegetable Crops' and 'Commercial Flowers' in one volume each in the 1980's in collaboration with co-editors, humbly expresses from the core of his heart that the updated revised editions of the above mentioned publications with colour illustrations in thirteen volumes which was planned in June, 2020, are being printed from October this year, a rare event in Horticultural Science has been possible by the blessing of Almighty God, with the cooperation of dedicated horticulturist as editors and contributors.

T.K. Bose

Editor Emeritus

Preface to the Fourth Edition from Senior Editor

It is nearly two decades since the third edition hit the stands. Lot of water has flown under the bridge. Many contributors to earlier editions have retired (some deceased as well). We pray the almighty to lay them in rest those who contributed to the earlier editions and since deceased. In India, the last two decades I saw a huge development in horticulture and the country has made huge strides. In terms of number of horticultural crops, the National Agricultural Research System works on over 130 horticultural crops.

India holds an eminent position in cultivation and research on Horticulture. We are the second largest producer of fruits and vegetables in the world leading in the production of mango, banana, coconut, cashew, papaya, pomegranate while our productivity of grapes, banana, cassava, peas, and papaya is the highest in the world. In the research front, we have one of the best networks of research institutes under ICAR. We have 10 Central Institutes, 6 Directorates, 7 NRCs, 13 AICRPs and 6 Network Projects / Outreach programmes. We have 62 Agricultural Universities and Under NARS; we work on about 130 horticultural crops and have developed/released about 1600 varieties of fruits, vegetables, ornamentals, spices, plantation, medicinal and aromatic crops. Our strength in our genetic resources is well conserved with about 73000 accessions being maintained at various gene banks under the control of National Bureau of Plant Genetic Resources, India.

Indian horticulture has great documentations available. Even before Linnaeus developed the hierarchical system of classification of nature, a Portuguese governor of Cochin (Kochi), Van Rheede, documented the horticultural wealth of Malabar region in 12 volumes taking about 3 decades to complete with Indian Botanists.

Van Rheede commenced his work in 1678 and completed it in 1703. History tells us the great history of Indian horticulture.

Research on perennials, particularly, fruit breeding is time consuming and tedious. Contrast to the annuals, this is life time work. We tried to rope in the best in the field to write the eleven chapters with suitable illustrations in this volume. I feel the present volume would be a valuable reference material for post graduate students, research workers, teachers and also the policy makers.

Kolkata **Senior Editor**

Preface to the Fourth Edition from Editors

The first edition of the book entitled 'Fruit Crops' published in 1985 was greatly appreciated and proved valuable to the students, teachers, researchers and extension specialists in horticulture in general and fruit science in particular, in different parts of the world. Then the book was revised twice in 1993 and 2002 with the title 'Fruits: Tropical and Subtropical' in three volumes due to its increasing demand. All the previous editions, which contained updated scientific information of all commercially grown fruit crops, were considered as valuable contribution on the subject. After about 18 years of the last revised publication in 2002, voluminous information pertaining to fruit crop improvement, production, protection and post-harvest management has been generated necessitating further revision in four volumes.

This particular edition has been enlarged by incorporation of up-to-date information on various aspects of fruit crop production, protection and improvement particularly through biotechnological interventions with meaningful illustrations, and published in four volumes. Efforts have been made to cover as much information as possible. Second volume in this edition includes 9 chapters on tropical and subtropical fruit crops covering more than 600 pages. Most of the contributors have taken care to review up-to-date research works and prepared the manuscript. The revision has been made by the editors and in few cases new contributors have been associated with some chapters.

We express our sincere thanks to all contributors who sincerely participated in revision, editing and proof-reading of the manuscript and we appreciate their commendable efforts to bring out this publication. We express our sincere thanks to senior editor Dr. V.A. Parthasarathy for thoroughly editing the manuscript and offering valuable suggestions in strengthening the same.

Apart from own collection of photographs, editors have also taken some illustrations which are in public domain from different websites. Therefore we acknowledge with enormous thanks to all State Agricultural Universities and Research Institutes under Indian Council of Agricultural Research, New Delhi, India; Public and Private Agricultural Institutes across the world for taking illustration from their websites to make this book a valuable knowledge for the readers across the world. We think images can be powerful teaching tools, as illustrations to in-class lectures, or for studying concepts outside of the classroom. We also think identification of fruit crops, their package of practices, cultivars, utilization, diseases and pests with the help of suitable illustrations in a consolidated manner will make a dent for the students and teachers of fruit science.

Our thanks are due to Shri Prateek Mittal, Director, Astral International (P) Ltd., New Delhi for his keen interest in bringing out this publication. Our special thanks go to Mr. Debashis Roy, Kolkata, West Bengal for processing the manuscript.

Kolkata **Editors**

Contents

List of Contributors

Aswath, C.

ICAR- Indian Institute of Horticultural Research, Hessarghata Lake Post, Bangalore 560089, Karnataka

Bajpai, P.N.

Department of Horticulture, CS Azad University of Agriculture and Technology, Kanpur 208002, Uttar Pradesh

Bose, T.K.

Department of Horticulture, Bidhan Chandra Krishi Viswavidyalaya, Mohanpur 741252, Nadia, West Bengal

Chakraborty, Ivi

Department of Post-harvest Technology, Faculty of Horticulture, Bidhan Chandra Krishi Viswavidyalaya, Mohanpur 741252, Nadia, West Bengal

Dhinesh Babu, K.

ICAR- National Research Centre on Pomegranate, NH-9, Solapur-Pune Highway, Kegaon 413255, Solapur, Maharashtra

George, A.P.

Maroochy Horticultural Research Station, P.O. Box No. 5083, Sunshine Coast Mail Centre, Nambour, Qld 4560, Australia

Ghosh, B.

Department of Fruit Science, Faculty of Horticulture, Bidhan Chandra Krishi Viswavidyalaya, Mohanpur 741252, Nadia, West Bengal

Ghosh, S.P.

ICAR-Central Tuber Crops Research Institute, Sreekariyam, Thiruvananthapuram, Kerala

Hanamashetti, S.I.

Kittur Rani Channamma College of Horticulture, Arabhavi, Tq: Gokak Dist: Belagavi 591218, Karnataka

Karale, A.R.

Department of Horticulture, Mahatma Phule Krishi Vidyapeeth, Rahuri, Ahmednagar 413722, Maharashtra

Karunakaran, G.

ICAR- Indian Institute of Horticultural Research, Hessarghata Lake Post, Bangalore 560089, Karnataka

Maity, S.C.

Department of Horticulture, Bidhan Chandra Krishi Viswavidyalaya, Mohanpur 741252, Nadia, West Bengal

Menzel, C.M.

Maroochy Horticultural Research Station, P.O. Box No. 5083, Sunshine Coast Mail Centre, Nambour, Qld 4560, Australia

Mitra, S.K.

Department of Fruit Science, Faculty of Horticulture, Bidhan Chandra Krishi Viswavidyalaya, Mohanpur 741252, Nadia, West Bengal

Nissen, R.J.

Maroochy Horticultural Research Station, P.O. Box No. 5083, Sunshine Coast Mail Centre, Nambour, Qld 4560, Australia

Parthasarathy, V.A.

ICAR – Indian Institute of Spices Research, 32/482 C Narmada Nilayam, Bharathan Bazar, Chelavoor, Calicut 673571, Kerala

Prakash, Jai

ICAR- Indian Institute of Horticultural Research, Hessarghata Lake Post, Bangalore 560089, Karnataka

Samaddar, H.N.

Department of Agriculture, Government of West Bengal, Kolkata 700001, West Bengal

Sankaran, M.

ICAR- Indian Institute of Horticultural Research, Hessarghata Lake Post, Bangalore 560089, Karnataka

Shukla, H.S.

Department of Horticulture, CS Azad University of Agriculture and Technology, Kanpur 208002, Uttar Pradesh

Simpson, D.R.

Maroochy Horticultural Research Station, P.O. Box No. 5083, Sunshine Coast Mail Centre, Nambour, Qld 4560, Australia

Singh, Sanjay Kumar

ICAR – National Research Centre on Litchi, Musahri, Ramna 842002, Bihar

Winks, C.W.

Maroochy Horticultural Research Station, P.O. Box No. 5083, Sunshine Coast Mail Centre, Nambour, Qld 4560, Australia

1

CITRUS

V. A. Parthasarathy, S.P. Ghosh and T.K. Bose

1.0 Introduction

Citrus is one of the most important and widely cultivated fruit crops across the globe, grown in more than 100 countries in tropical, subtropical and Mediterranean climate of the world. It is believed to have its primary centre of origin in South and South-East Asia, particularly in the region extending from Northeast India, eastward through the Malayan Archipelago to China and Japan, and southward to Australia (Mabberley, 2004). The area and production have increased many folds in the past 30 years, particularly in Japan, Brazil, Israel, Turkey and Cuba. Although, domestic consumption is the principal market for citrus fruit, the major citrus producing countries, *viz.* Spain, USA, Israel, Morocco, and South Africa export a sizeable amount of fresh fruits to the world market, in particular to European Economic Community, Russia, Canada, Saudi Arabia, Kuwait and Hong Kong. To meet the demand of frozen concentrated orange juice, the USA has to import substantial amount from Brazil (Gountur and Brown, 1983). Citrus plantations in Israel occupy more than 40,000 hectares and 80 per cent of Israel's agricultural export value lies on citrus. In Cuba, the planting is about 1,50,000 hectare, though grapefruit is mainly under cultivation. In Japan, citrus occupies the first place among fruit crops, covering nearly more than 50 per cent of the total areas under fruit crops and 2.8 per cent of the total area under cultivation of which 83 per cent is under Satsuma mandarin and 10 per cent under Natsudaidai (*C. natsudaidai*) Citrus occupies about 9 per cent of the total land under various fruits in India. The most important commercial citrus in India is the mandarin orange followed by the sweet oranges and acid limes. However, citrus production in the world is increasing faster than consumption and population which is a cause for some concern in the industry (Wolf, 1977; Myers and Grierson, 1978).

2.0 Composition and Uses

2.1.0 Composition

Different species of citrus fruits have different chemical compositions. While in the sweet group, the principal constituents of the edible portions are sugars (glucose and sucrose) and acids (primarily citric acid and little of malic acid), the fruits of acid groups contain primarily the acids in the fruit juice (Table 1). The rind of citrus fruits is rich in pectin and certain essential oils. The rind also contains certain glucosides (hesperidin in oranges, lemons and naringins in the grapefruit). Citrus fruits contain considerable amount of ascorbic acid, the vitamin C. The total soluble solids (TSS) in the fruit juice, in most of the sweet group of citrus, vary from 8-12 per cent, while the titratable acidity usually ranges from 0.5 to 1.5 per cent. Usually, for oranges and mandarins a TSS-acid ratio of 8:1 and in grapefruit the ratio of 6:1 were considered to be optimum for marketable fruits. In the lemons and limes, titratable acidity of fruit juice usually ranges from 5 to 6 per cent. The vitamin C (ascorbic acid) content in the fruit juice of different citrus species has been recorded to be variable and 25 to 85 mg per 100 ml of juice could be recorded. The nutrient composition of citrus juice on an average contains the following ingredients.

Table 1: Nutrition Information per Serving (portion)

Serving size (1 average orange)	160 g
Calories	50 kcal
Protein	1 g
Carbohydrate	13 g
Fibre, dietary	6 g
Potassium	270 mg

2.2.0 Uses

Besides world-wide demand of orange as fresh fruits, orange marmalade has long been an important product and is one of the main uses of the sour and sweet oranges. Grapefruits and other types are also used in this way. Orange, lemon, lime and grapefruit juices are bottled and canned in large scale. Lemon-barley water is prepared from the common lemon. The production of canned juice and pulp and particularly, of frozen concentrated juice has been increasing very rapidly throughout the world (Table 2). The flower, leaf and rind of citrus contain oils of good fragrance and have good commercial value. Lemon and orange oils are most important citrus oils used for flavouring purposes, followed by lime, grapefruit and tangerine oils. Other commercial products are citric acid and pectin, made primarily from cull and unmarketable fruits.

Table 2: Disposition of Citrus Crop from Leading Producing Countries (Thousand metric tonnes)

Country	*Production*		*Export*		*Processing*	
	1971-72	*1996-97*	*1971-72*	*1996-97*	*1971-72*	*1996-97*
World	45385.1	84717.0	6290.1	8027.0	14093.4	26740.0
Northern	34025.2	59010.0	5774.8	6691.0	11533.9	14777.0
Northern Hemisphere						
United States	11475	16469.3	785.1	1247.0	9048.1	11082.0
Mediterranean	11705.6	15194.4	4578.3	4834.1	1543.4	2762.8
Mediterranean Region						
Greece	668.7	1133.5	189.6	408.0	163.8	271.0
Italy	2688.6	2585.0	387.3	237.0	351.0	762.0
Spain	2745.2	4020.2	1620.4	2403.0	276.2	703.9
Israel	1609.5	1071.0	891.8	399.0	561.4	521.0
Algeria	515.5	268.0	112.6	1.5	—	—
Morocco	798.4	1152.0	563.3	532.0	95.8	67.0
Tunisia	119.7	212.0	28.4	25.0	—	—
Cyprus	245.0	259.0	184.5	168.0	16.6	70.0
Egypt	942.2	2472.1	164.5	262.0	3.0	125.9
Lebanon	315.4	—	165.6	—	—	—

Country	Production		Export		Processing	
	1971-72	1996-97	1971-72	1996-97	1971-72	1996-97
Turkey	748.2	1805.0	121.9	330.0	68.7	220.0
Former USSR	84.0	71.0	—	—	—	—
Japan	3863.4	1625.0	—	—	—	—
Cuba	162.0	586.0	47.7	95.0	—	—
Mexico	1791.6	4091.0	—	—	—	—
China	719.1	7909.5	83.1	156.0	—	—
11359.9	25707.0	515.3	1336.0	2559.5	11963.0	
Southern hemisphere						
Argentina	14.34.4	1940	40.2	298.0	245.4	7000.0
Brazil	6051.0	17693.0	53.3	91.0	1944.6	10736.0
Uruguay	73.9	—	13.4	—	1.3	—
Venezuela	235.2	—	—	—	—	—
United States	899.2	979.8	—	—	—	—
Australia	417.6	579.0	25.3	120.0	171.0	235.0
South Africa	604.3	1221.0	305.7	817.0	94.2	292.0

3.0 Origin and Distribution

It is believed that most of the species under the genus *Citrus* are native to tropical and subtropical regions of Southeast Asia, particularly, India, China and in the region between these two countries. The north-eastern region of India is considered as one of the natural homes of citrus – at least for a few species. Certain non-edible species like *C. indica, C. inchangensis, C. macroptera, C. latipes* are found to grow in wild and semi-wild state in the north-eastern region. Bhattacharya and Dutta (1956) considered them as indigenous to the area. Also different strains of citron (*C. medica*), sour pummelo (*C. megalaxycarpa*), rough lemon (*C. jambhiri*) and sour orange (*C. aurantium*) have been found to grow widely in semi-wild conditions in different north-eastern states of India (Verma and Ghosh, 1979). It is possible that since very early days, there might have been spread of different citrus species in the contiguous regions of their natural homes and there are possibilities of intercrosses, particularly between species of China and India. Chinese literature written as early as 2200 BC showed evidence of citrus cultivation in China (Ho, 1934). Hooker (1872) thought that not less than 78 species of the family Rutaceae are native of India. It is believed that citrus was introduced in Greece between 250 and 200 BC (Tolkowsky, 1938). Bonavia (1890), however, thought that Malaya is the home of lemon while pummelo is thought to be originated in the island east of Malaya (Malaysia) archipelalgo, including Fiji and Friendly islands (De Candolle, 1886) or in China (Bretschneider, 1898). Throughout the process of natural hybridization also, some new forms might have been originated. In Europe, citrus was probably introduced in the 16th century by the Portuguese. During the Ist century AD Arabians introduced citrus fruit to the east coast of Africa and it is believed that Colombus took orange

Distribution of Citrus

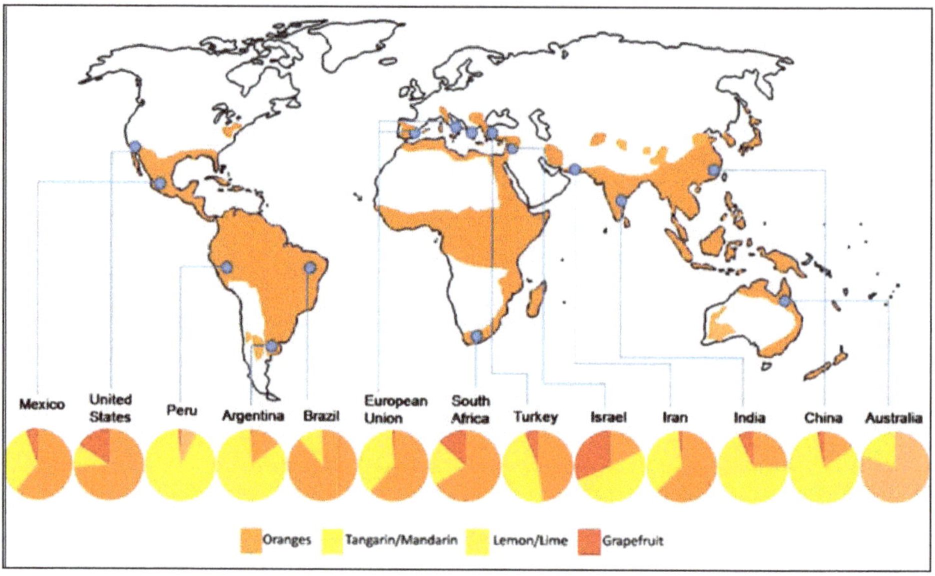

Distribution of citrus species in the world

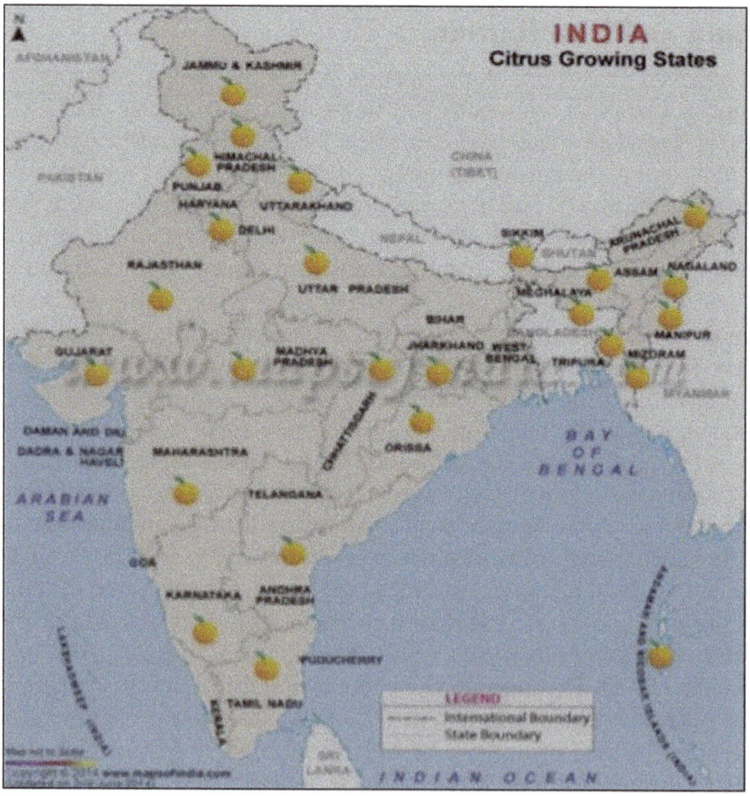

Citrus growing states in India

Uses of Citrus

Citrus fruit with slices

Juice from fruits

Juice from fruits

Canned products from fruits

Uses of Citrus

Lemon barely water from fruits

Oil from fruits

Citric acid from fruits

and lemon seeds to America (Hume, 1959). Although the introduction and spread of citrus fruits in Europe and the United States took place comparatively recently, some of them hold principal positions in the citrus industry of the world.

The major citrus producing countries of the world are USA, Spain, India, Italy, Japan, Argentina, Mexico, Brazil, Morocco, Algeria, Greece, South Africa, Australia, Israel, Egypt, Jamaica, *etc.* Sweet oranges, mandarins, grapefruits, lemons and limes are the principal citrus types and occupy important commercial positions in the citrus industry. The United States is considered to be the largest producer of citrus fruits, covering about 35-40 per cent of the world's total production and the citrus fruits have spread on commercial scale to about 30 countries of the world. In general, citrus trees are found between latitude 40oN and 40oS where conditions are not too dry, too moist or too cold.

4.0 Species and Cultivars

4.1.0 Species

There is a good deal of diversity in the views on classification of citrus, primarily, due to divergent concepts about the constitution of a valid species. The treatments of two world famous authorities on the subject – W.T. Swingle (USA) and T. Tanaka (Japan) are at two extremes. Swingle (1948) recognized only 16 species under the genus *Citrus* whereas Tanaka (1954) described as many as 144 species. Swingle's treatment was not sufficiently comprehensive and he considered most species to be of hybrid origin and rejected which were not found to occur in nature in wild form. He failed to cover many forms of horticultural importance, and many species of Japanese, Chinese and Indian origin have been denied. Swingle divided the genus *Citrus* into two subgenera, *viz., Eucitrus* having 10 species and *Papeda* having 6 species.

Tanaka's (1954) treatment, although considered more comprehensive and detailed, contained excessive number of species, some of them being of doubtful validity. In the mandarin group alone, he described 35 species, resulting in too much confusion and obviously to lesser practical utility. Tanaka divided the genus *Citrus* into two subgenera, *viz., Archicitrus* having 98 species and *Metacitrus* with 46 species.

Considering the two extreme views expressed in the treatments of Swingle and Tanaka, attempts were made from time to time to formulate an intermediate of compromise treatment with more acceptability and greater utility. Contrary to Swingle's opinion, citrus forms of hybrid origin were accepted as valid species, while species standing given to certain natural hybrids and certain cultigens by Tanaka was questionable and might be avoided. It is accepted that the characters employed for identifying a valid species of citrus should be free from the environmental influence, should be easy to discern, showing sufficient range of variation and should fall into clear-cut discontinuous categories. Certain characters which are principally employed in the identification and classification of citrus are:

(i) Leaf: Margin (entire, serrate, crenate) and petiole (degree of wing development).

(ii) Shoot: Nature of branches (angular or otherwise), coloration of new growth.

(iii) Flower: Size (small to large), solitary or clustered, colour of petals (pigmented or not), coherence of stamens (stamens free or filaments cohering in bundles), number of carpels (locules).

(iv) Fruit: Form (presence and absence of apical papilla), shape, articulation, adherence of rind (loose or tight), nature of central axis (core open or solid), colour of pulp, seed size, colour of cotyledons.

Different species of Citrus could be recognized on the basis of the following important characters (Singh and Nath, 1969):

(i) Development of petiolar wing: It is most developed in the members of a *Papeda* group, followed by the pummelo and least developed in citrons.

(ii) Nature of inflorescence: In *Fortunella*, mandarins and their hybrid flowers are mostly solitary.

(iii) Colour of the cotyledons: It is light greenish in sour lime and rough lemon, green in mandarin and deep green in *Fortunella* .

(iv) Coherence of filaments: The filaments are free in *Poncirus* and in *Papeda*, connivent into a tube in *Fortunella* and more or less united into a tube in mandarins.

(v) Number of petals: In most Papedas, pummelo, sour lime and *Citrus pennivesiculata*, the petals are four in number.

(vi) Arrangement of vesicles and presence of large opaque oil droplets in vesicles: In most of the Papedas and their hybrids like *C. pennivesiculata* and *C. assamensis* and *Jeneru tenga*, the vescicles show an opaque oil droplet, are also attached to half of the radial carpellary wall, and are often arranged in herringbone fashion.

(vii) Colour of pulp: In Papeda and sour lime it is greenish, bright orange in mandarins and their hybrids, straw coloured in lemons and pink colour in some pummelos and grapefruits.

(viii) Hairiness: The calyx is hairy in Papedas, Cephalocitrus (Tanaka), sour orange, *C. tachibana*, *C. sunki* and *C. tumida* of Tanaka. The filaments are hairy in true citrons and some of their hybrids.

(ix) Length of filaments: The stamens are markedly smaller than the pistil in pummelo. This character is also found in its hybrids.

(x) Thickness and nature of rind: The rind is thick and spongy in pummelo and very thin and smooth in sour lime.

(xi) Surface of rind: The oil glands on the surface of fruits of *Fortunella* and pummelo are markedly raised, on sweet orange moderately raised, and on grapefruit, mandarin and sour orange, they are sunken. Some Papedas and their hybrids like *C. semperflorens* and *C. pennivesiculata* are characterized by sparse deep pits on the fruit surface.

(xii) Thickness of septa: Thick septa and a large number of carpels are characteristic of the cross section of citron.

(xiii) Mamilla of the fruit: It can vary greatly, but a broad mamilla surrounded by a prominent furrow is a characteristic of the rough lemon and *C. limetta.*

(xiv) Reflexion of petals: The petals are strongly reflexed in pummelo, grapefruit and tight-skinned acidic fruits. They are boat shaped and almost straight in *Fortunella*, mandarins and mandarin-like fruits with loosely adhering rind.

(xv) Seed characters: *Fortunella* and mandarins are allied species, with smooth, short and plump seeds. Papedas generally have very large seeds as does *Poncirus*. The seeds of pummelo are large, striated and irregularly shaped. The seeds of *C. ichangensis* have a characteristic ornamentation not found in other species.

4.1.1 Major Citrus Species of Horticultural Importance

In citrus industry, quite a good number of citrus species have been exploited commercially, some for direct fruit production, some as rootstocks in propagation, *etc.* In addition, quite a large number of intergeneric and interspecific hybrids have been developed. In the intergeneric hybridization work, the related genera like *Poncirus* (trifoliate orange) and *Fortunella* (kumquat) have been extensively used in crossing programme with citrus and a few broad groups like citranges (trifoliate orange x sweet orange), citrumelos (trifoliate orange x grapefruit), citradias (trifoliate orange x sour orange), citrumuats (trifoliate orange x kumquat), limquats (kumquat x West Indian lime) have been developed. In the interspecific hybridization programme, on the other hand, a few other groups like lemonimes (lemon x lime), tangors (mandarin x sweet orange), tangelos (mandarin x grapefruit) have been developed and some of them could be found to be of horticultural value. The major citrus species and hybrids of commercial importance are:

4.1.1.1 Mandarin Group

(a) *Citrus reticulata*: A highly polyembryonic species of Chinese origin, having medium-sized upright trees, leaves lanceolate in shape with narrowly-winged petiole. Fruits medium-sized, globose, sweet in taste, segments easily separable, core open at maturity, loose skinned, orange in colour, rind thin, rind and segments easily separable, usually 10-14 segments in each fruit; seeds pointed with light green cotyledons. The important cultivars are Nagpur, Coorg, Khasi mandarins of India and Ponkan of China.

(b) *C. unshiu*: A polyembryonic species of Japanese origin, with small spreading trees, comparatively cold hardy. Midrib of the leaves is prominent and petiole narrowly winged but long. Fruits are seedless with thin rind, orange colour at maturity. The fruits are very juicy with agreeable flavour. The quality of fruit is also excellent with good blend of sugar and acidity. The famous Satsuma mandarins of Japan and Owari, Kara, Silver Hill, *etc.* are important cultivars.

(c) *C. deliciosa*: A Mediterranean origin, highly polyembryonic species having trees of medium-size with drooping growth habit. Fruits strongly compressed and yellow to light orange in colour with distinctive flavor; the fruits are juicy, rind and segments loose. Fruits seedy and cotyledons light green in colour. The Willow-leaf mandarin, Kinnow and Wilking of USA and Blinda, an Algerian selection come under the species.

(d) *C. nobilis*: A polyembryonic species, native of Indo-China, thought to be a natural tangor of ancient origin. Tree grows 4-6 m with dense top, frequently hanging branches and small leaves, petioles narrowly winged, sweet scented white blossoms at the end of the branches or in the axils. Small, flattened, orange- coloured or reddish fruit with distinctly sweet pulp; peel and segments easily separable. Oil glands are present on the skin. Fruits are juicy and of excellent quality. The Kunembo of Japan, King orange of USA, *etc.* come under the species.

(e) Other species of mandarin group namely *C. reshmi* (Spice mandarin), *C. madurensis* (Calamondin), *C. madaraspatana*, *C. tangerina* (Dancy tangerine of Florida), *etc.* are of certain specific importance.

4.1.1.2 Orange Group

(a) *C. sinensis* – sweet orange: A highly polyembryonic species of Chinese origin. Trees medium-large with blunt-pointed leaf apex and narrowly-winged petiole. Fruits subglobose to oval in shape, orange-coloured, tight skinned with solid central core. Fruit globose or oblate, light orange or reddish in colour, with smooth peel; pulp juicy and sour-sweet. Fruit is sweet and flesh colour usually orange. Seeds are with whitish cotyledons. The species is of great economic importance for its excellent quality as well as for its rootstock value to a limited extent. Quite a large number of cultivars, *viz.*, Mosambi, Malta Blood Red, Sathgudi of India, Valencia, Pineapple, Washington Navel (having a rudimentary secondary fruit embedded in the fruit apex) of the USA, Shamouti of Israel, Succari of Egypt, Dobla Fina of Spain, *etc.* are some of the commercially important cultivars.

(b) *C. aurantium* – bitter or sour orange: Highly polyembryonic and cold hardy species. Tree medium-sized, upright with dark green and distinctively scented leaves. Flowers large and very fragrant. Fruit medium-sized, orange-red skinned at maturity, central core semi-hollow and the primary oil glands in the fruit peel are beneath sunken areas. Flesh orange coloured, acidic in taste with bitter after-taste, rind separates easily. Highly seeded; seeds flattened and cotyledons white. The species was primarily used as one of the principal rootstocks and flowers are used for perfumery purpose (oil of neroli). Since the species is intolerant to certain citrus viruses like tristeza, in more recent years this has been abandoned as rootstock.

4.1.1.3 Pummelo-grapefruit Group

(a) *C. grandis* (*C. maxima*) – pummelo or shaddock: A monoembryonic species with large-sized fruits, considered to be indigenous to Malayasia and Polynesia. Spreading, round-topped, almost thornless tree. Leaves large with broadly-winged petiole. Lower surface of leaf is pubescent, particularly the main vein. Flowers very large, crowded in short axillary racemes. Fruits large-sized, subglobose to pyriform in shape, with thick and spongy rind. Fruits sweet and moderately juicy. Rind thick, smooth with large oil glands. Seeds very large, coarsely veined and white within. Two types, *viz.* white fleshed and red or pink fleshed are available in India and cultivars are named accordingly. Some of the named cultivars are Kao Pan of Thailand and Buntan of Formosa (Taiwan).

(b) *C. paradisi* – grapefruit or pummelo: A polyembryonic species probably originating in southern China and neighbouring countries, with round-topped spreading tree. Leaves large with winged petioles (less broadly winged as compared with pummelo). Unlike pummelo, leaves and new growth are non-pubescent. Fruit large but subglobose in shape. Rind surface and fruit flesh are yellowish in colour. Fruits highly juicy, sweet with bitter aftertaste; the central axis or core opens at full maturity. Seeds big but smooth-surfaced and white inside. The cultivars like Foster, Ruby, Marsh, Duncan (seedless), Thompson, *etc.* are of commercial importance.

4.1.1.4 Acid Group

(a) *C. limon* – lemon: A weakly polyembryonic species with medium-sized spreading trees. The new flushes are pigmented. Leaves not dark green and leaf margins are subserrated. Petioles medium-sized and narrowly winged. Flower buds pigmented and the pistil is densely dotted with oil glands. Fruits oval to elliptic with pointed nipple. Fruit surface smooth, light yellow and core solid; juice abundant and acidic. Seed cotyledons are white. The species is of great commercial importance and the cultivars like Eureka and Lisbon (USA), Femminello and Monachello (Italy), Bernia (Spain) are important. The so-called bush lemon types (lemon-citron) like Kaghzi Kalan, Italian Round, Assam lemon, *etc.*, of India are not true lemons. Lemon oil is one of the most important citrus oils used for flavouring purposes in soft drinks, baked foods, confectionary, *etc.* and is of good demand.

(b) *C. jambhiri* – rough lemon: A highly polyembryonic species of Indian origin (Dutta, 1958). Tree medium to large and spreading in growth habit. Leaves are light green and serrations of leaf margin are not sharp. Fruit differ distinctly in shape and colour from lemons having roundish to subglobose shape with cavity or depression surrounding the apical papilla or nipple. Fruit skin usually rough and lemon-brown in colour. Flesh yellow, core open at maturity, cotyledons light green, seeds numerous, small. Rough lemon is one of the most important citrus rootstocks of the world. The species is adapted to wide range of soil conditions but better suited to light soils. It is fairly tolerant to many citrus virus diseases, including tristeza.

(c) *C. aurantifolia* – sour lime: A highly polyembryonic, distinctive species of great commercial importance. Tree is small, bushy with small but sharp spines. Leaves small with narrowly-winged petioles. Flowers and fruits small, fruits round to oval, maturing irregularly throughout the year, greenish yellow in colour and thin skinned. Core solid at maturity, flesh greenish in colour and juice highly acidic. Seeds small, smooth and cotyledons whitish. The Kagzi lime is the most important commercial cultivar of India. The cultivar is highly susceptible to tristeza virus and bacterial canker disease.

(d) *C. limettoides* – sweet lime: A highly polyembryonic species of Indian origin. Tree medium-large and spreading in growth habit. The leaves medium-sized, petiole wing margined, leaf lamina light green and characteristically cupped (somewhat rolled). Flowers and fruits larger than acid lime and globose to ellipsoid in shape. Fruit surface smooth, greenish yellow with thin rind. Flesh pale yellow to straw colour, core semi-yellow at maturity, flavour insipid, juice sweet (flat) with little bitter aftertaste. Seeds medium in size with white cotyledons. There is no named cultivar, the 'Mithanimbu' or 'Sharbati' of India belongs to this species. It is commercially grown in countries like Egypt and has been employed as rootstock.

(e) *C. medica* – citron: A monoembryonic species, considered to be a native to India and susceptible to frost. Trees medium-large (about 3 m) spreading and irregular in growth habit. New growth pigmented, leaf margin serrated. Leaf size large and rumpled. Flowers large, petals pigmented on the lower surface and filaments generally hairy. Fruits large-sized, lemon-yellow in colour at maturity, oblong to oval in shape, in many cases style persists, semi-smooth to rough surfaced. Rind thick to very thick, hard and in some cases semi-sweet and aromatic. Pulp acidic but juice scanty. Seeds medium-sized and white inside.

(f) *C. karna* – Karna Khatta: A moderately polyembryonic species widely used as rootstock in northern India. Tree medium to large, somewhat similar to that of rough lemon. Leaves large with serrulate margin and winged petiole. Flowers large and pigmented. Fruits medium, rind surface irregular and apical papilla well developed. Rind thick and moderately adhering. Fruit surface and pulp orange coloured. Fruits juicy and juice acidic in taste. Moderately seeded and cotyledons white. The Soh Sarker of Assam belongs to this species.

(g) Other species in acid group include *C. latifolia* (Tahiti or Persian lime), *C. limetta* (Lemetta of the Mediterranean, Pani Jamir of Assam), *C. limonia* (Rangpur lime), *etc.*

4.1.2 Wild, Semi-Wild Species and Related Genera

A few citrus species of wild, semi-wild and of less economic value are found to grow widely in the hilly areas of north-eastern region of India. Bhattacharya and Dutta (1957) described 17 different species and 52 cultivars of citrus from the erstwhile Assam State alone. The north-eastern hill region is considered to be one of the sites of origin of citrus and the species available in the region are:

Species of Citrus

Citrus reticulata

Citrus unshiu ***Citrus deliciosa***

Citrus nobilis

Citrus medurensis

Species of Citrus

Citrus tangerina

Citrus sinenesis

Citrus aurantium

Citrus grandis

Species of Citrus

Citrus paradisi

Citrus limon

Citrus jambhiri

Citrus aurantifolia

Species of Citrus

Citrus limettoides

Citrus medica

Citrus karna

Citrus indica

(i) *Citrus indica* – Indian wild orange: The species has been found to occur in nature in certain parts of Assam, Nagaland and Meghalaya. Plant bushy, medium-sized and grows well in marshy areas under forest shades. Fruits inedible, small-sized with slimy juice of acidic taste and unpleasant aroma. Seeds are very big and cover the major portion of fruit.

(ii) *C. latipes* – Khasi Papeda: A species found to occur under high altitude conditions (upto 1800 m above mean sea level) and trees are of vigorous growth habit. Leaf petiole winged and broad. Fruits medium-sized, subglobose, moderately juicy and acidic. A cold tolerant species.

(iii) *C. macroptera* – Melanesian Papeda: A widely distributed species, trees vigorous. Fruits used by local tribals for medicinal purpose and in cooking. Leaf large with petiolar wing as wide and long as the lamina. Fruits medium-sized, subglobose, very juicy and juice highly acidic in taste. Two forms, namely, Sat Kara and Tith Kara, are available.

(iv) *C. ichangensis* – Ichang Papeda: Medium-sized tree with long lanceolate leaves and small fruits (4-6 cm diameter). Petiolar wing well-developed, fruits inedible. A cold hardy species.

(v) *C. assamensis* – Adajamir (*C. pennivesiculata* var. *assamensis*): A distinctive species having crushed leaf aroma similar to ginger or eucalyptus smell. Medium-sized trees with thick glossy leaves. Fruits medium-sized, almost spherical and smooth surface. The fruit is of limited home use with acidic juice. The species was described as a new indigenous species from Assam by Bhattacharyya and Dutta (1956) but it has been found to be the same as *C. pennivesiculata* (Tanaka), the Gajanimma of India.

4.1.2.1 Related Genera

(i) *Poncirus* – trifoliate orange: A distinctive genus with single species *P. trifoliata*. Plants deciduous with trifoliate leaves. Plants highly thorny and fruits inedible. Seeds polyembryonic and the species are principally used as rootstock for citrus.

(ii) *Fortunella* – Kumquat: A genus having a few species, out of which *F. margarita, F. japonica, F. crassiflora* and *F. hindsii* are important. Small shrubby trees with very small oval or round shaped orange coloured fruits. Seeds are polyembryonic. The fruits are used for preparation of candy and the plants for ornamental purpose.

4.2.0 Hybrids and Cultivars

4.2.1 Intergeneric and Intrageneric Hybrids

4.2.1.1 Intergeneric

I. Hybrids of Poncirus

(i) Citrange – A group having the parentage of trifoliate orange (*Poncirus trifoliata*) and sweet orange (*C. sinensis*). The hybrids showed intermediate characters of the parents. The leaves are mainly trifoliate but unifoliate evergreen leaves are also observed in some plants. The fruits are yellow to orange in colour and the rind is thin. The fruits are juicy, flavoured. Some of the cultivars are Troyer, Carrizo, Morton, Etonia, Rusk, Coleman, *etc.*

(ii) Citrangequat – This is the trigeneric hybrid of three different genera (*Poncirus, Citrus* and *Fortunella*).

(iii) Citrangedin – This is the trispecific hybrid of citrange with the Calamondin (*C. mitis*).

(iv) Citrangor – This hybrid is developed by back crossing of citrange and *C. sinensis.*

(v) Cicitrange – Another back cross hybrid between citrange and *Poncirus trifoliata.*

(vi) Citrumelo – This is similar to citrange, a hybrid between *Poncirus trifoliata* x *C. paradisi.*

(vii) Citrandarin – A hybrid between *Poncirus trifoliata* and *C. reticulata.*

(viii) Citremon – A hybrid between *Poncirus trifoliata* and *C. limon.*

(ix) Citradia – Very similar to citrange. A hybrid between *Poncirus trifoliata* and *C. aurantium.*

(x) Citrumquat – This is the hybrid between *Poncirus trifoliata* and *Fortunella japonica* or *F. margarita,* a very difficult hybrid to breed.

II. Hybrids of Fortunella

Procimequat – (*Fortunella japonica* x *C. aurantifolia* cv. Mexican) x *F. hindisii*

Limequat – *C. aurantifolia* x *F. japonica*

Orangequat – *C. reticulata* cv. Satsuma x *F. japonica* x *F. margarita* cv. Meiwa.

4.2.1.2 Intrageneric

(i) Tangor – mandarin x sweet orange (*C. reticulata* x *C. sinensis*) hybrids. Temple, Clementine, Monreal, Umatilla are some important cultivars, mostly monoembryonic.

(ii) Tangelo – mandarin x grapefruit (*C. reticulata* x *C. paradisi*) illustrated by cultivars like Orlando, Sampson, Minneola, Seminole, *etc.*

(iii) Lemonime – lemon x lime (*C. limon* x *C. aurantifolia*) having a cultivar like Parrine

(iv) Lemonnage (*C. limon* x *C. sinensis*)

(v Lemandarin (*C. limon* x *C. reticulata*)

4.2.2 Cultivars

4.2.2.1 Mandarin

Mandarins are commonly known as loose-skin orange. Hodgson (1967) classified mandarins into four groups – Satsuma, King, Willow-leaf (Mediterranean) and common.

Satsuma group: Satsumas are one of the best mandarins of the world and commonly grown in Japan. Some of the important cultivars of this group are Ikeda, Mikado, Owari, Silverhill and Wase. Fruits of this group are mostly seedless and are used as fresh as well as canned segments.

King group: The cultivars King of the USA and Kunembo of Japan belong to this group which is native to Indo-China region. Fruits of this group are many seeded, mature late and have strong flavour.

Willow-leaf group: This group is native to Mediterranean region. Seedless Willow leaf of Spain and Ciaculli late of Italy are two important cultivars of this group. The trees show alternate bearing tendency. Fruit juice has mild aromatic flavour and seeds are polyembryonic in nature.

Common mandarins: There are numerous cultivars in this group and some of the important cultivars are depicted below:

Nagpur Santra: The cultivation of this cultivar is centred around Nagpur region of Central India. It is reported to be the most important mandarin cultivar of India. Fruits are medium to large, subglobose, cadmium colour with smooth surface having prominent oil glands; rind adherence slight, segments 10 to 11, easily separable, juice abundant with little rag, sweet, excellent flavour, saffron yellow in colour; seeds 4 to 10, polyembryonic.

Coorg mandarin: It is a commercial variety of Coorg and Chickmagalur districts of Karnataka and Wayanad region of Kerala. Fruits are oblate to globose, golden yellow coloured, with depressed base; rind medium to thin having slight adherence; segments 8 to 10; juice abundant with deep chrome colour and attractive flavour, seeds many.

Kinnow mandarin: Crossing between King (*C. nobilis*) and Willow-leaf mandarin (*C. deliciosa*) evolved this cultivar. It is a popular cultivar of Punjab. Fruits are medium-sized, globose to oblate, orange colour with flat base; rind medium to thin, leathery soft, with slight adherence but peelable; segments 9 to 10 and separable fairly easily; juice abundant, bright golden yellow in colour, rich flavour, acidic to sweet; seeds numerous (15 to 25), polyembryonic.

Khasi mandarin: This cultivar is commercially grown in north-eastern region of India. Fruits are depressed, globose to oblate, bright orange yellow in colour, surface smooth, glossy, stalk end even or obtuse, occasionally short necked; rind thick to medium, adherence slight, segments moderate in number, melting; juice abundant, orange coloured with good sour-sweet blend; seeds 12 to 16, polyembryonic; keeping quality good.

Darjeeling or Sikkim mandarin: Popularly grown in Darjeeling district of West Bengal and Sikkim. Fruits are somewhat globose to oblate shape with orange yellow colour; juice moderate, orange in colour.

Ponkan: It is a popular cultivar of South China and Formosa. Trees show alternate bearing tendency. Fruits resemble Nagpur Santra and Coorg mandarins. Fruits mature mid-season.

Clementine: This cultivar has been originated from Algeria. It is an early maturing monoembryonic cultivar with good quality fruits.

Beauty: It is highly popular in Australia. Fruits are of moderate to good in quality, seeds polyembryonic, trees show alternate bearing tendency.

Dancy: This mid-season cultivar is commercially grown in the USA. Trees bear moderate to heavily with alternate bearing tendency. Fruit quality is excellent.

Steve Albert described the following mandarin varieties (https://harvesttotable.com/mandarin_oranges_and_tangerine/)

Dancy or Dancy Tangerine: Rich, spicy flavor. Dark orange-red, smooth, thin rind; peels easily; flesh is deep orange. Smaller than other mandarins and seedy. The traditional Christmas "tangerine." Ripens fall into winter: December and January. Best when grown in desert regions. First planted in Florida in 1867.

Encore: Delicious sweet-tart taste. Juicy, sweet, rich and spicy flavor. Pulp is deep orange and seedy. Yellow-orange rind mottled with dark spots. Ripens spring into summer. The last mandarin to ripen: March, April and May.

Fremont: Rich, sweet fruit and seedy. Medium-size with bright orange rind. Ripens early, fall into winter.

Honey or Murcott: Wonderfully sweet tangerine flavor and juicy. Very small and seedy. Yellow-orange rind, peels poorly. Midseason maturing winter into spring.

Kara: Sweet-tart, sprightly, aromatic flavor with varying seediness. Large fruits of 2½ inches (6.5 cm) in diameter. Ripens winter into spring: March and April. It will be tart if grown near the coast.

Mediterranean or Willow Leaf: Sweet, flavorful, aromatic and juicy. Yellow-orange rind, smooth and glossy; few seeded. Midseason, ripens in spring.

Pixie: Mild, sweet and seedless fruit. Medium-orange flesh with yellow-orange rind. Easy-to-peel. Ripens late. Grows well in intermediate and coastal areas.

Wilking: Rich flavor and juicy. Deep yellow-orange rind with rich yellow-orange flesh. Medium sized. Ripens in midseason: January-April. Grow well in cooler areas.

4.2.2.2 Sweet Orange

Sweet oranges are commonly known as tight skinned orange. Some popular cultivars are described below:

Mosambi: Fruits are medium to large, slightly oblate to globose, colour light yellow to pale orange; rind medium, tight, difficult to peel; flesh firm, juicy, with straw-yellow colour, melting, sweet; seeds 15 to 25 in number.

Sathgudi: Medium to large sized fruits are spherical to globose, with attractive orange yellow colour; rind smooth, finely pitted, leathery; segments 10 to 12; pulp orange coloured, juicy, sweet with acid tinge, and moderate flavour; seeds 10 to 20.

Pineapple: Fruits are medium to large, subglobose, orange coloured at maturity; rind medium thick, bright, glossy, tight; flesh primerose yellow, juicy, sweet with flavour resembling pineapple, good for processing.

Washington Navel: This is a popular cultivar of California. Fruits are large, spherical to ellipsoid, orange yellow in colour, with characteristic navel like nipple protuberance at the stigmatic end; rind thick, easy to peel off, segments separable; flesh soft and melting, deep orange yellow in colour, juice abundant, excellent quality but not suitable for processing, seeds many.

Hamlin: Fruits of this cultivar are medium-sized, oval with smooth surface, orange red in colour, apex rounded; rind thin, smooth; flesh orange colour with a good blend of sour-sweet taste; seeds few (2-5) in number. This early cultivar is grown in Punjab, Uttar Pradesh and Haryana in India.

Malta Blood Red: Fruits are medium to large, ellipsoid to spherical, cadmium yellow in colour with scarlet blush, apex broad; rind thin to medium thick, tight; flesh blood red when fully ripe, juicy, good quality, seeds many.

Shamouti: Fruits are large, oval to ellipsoid with rounded apex, deep orange in colour; rind separates easily; pulp juicy, firm, light orange is colour, sweet; nearly seedless.

Valencia: Fruits are medium to large, oblong to spherical; rind medium thick, deep golden orange in colour, leathery, surface smooth; pulp golden yellow, juicy, slightly acidic with good flavour; seeds few; good for processing.

4.2.2.3 Lemon

Eureka: This virtually thornless cultivar is of Italian origin. Fruits are medium, elliptical to oblong, lemon yellow colour at maturity, rind medium, tight with rugose or pitted surface; flesh tender, pale greenish yellow in colour, juicy, highly acidic with excellent flavour; seeds few to none.

Lisbon: It is of Portuguese origin. Medium-sized fruits are oblong to ellipsoid with prominent nipple, yellow at maturity; rind medium thick, tight; pulp pale greenish yellow in colour, juicy, acidic, excellent flavoured; seeds few.

Galgal: Medium to large fruits are of variable shape with sunflower yellow colour at maturity; rind medium thick, tight, leathery, surface rugose or wrinkled; flesh juicy and acidic; seeds few to many.

Assam Lemon: This is a popular cultivar of north-eastern region of India. Fruits are medium to large, long, elliptic to oblong-obovate; nipple broad, colour greenish- yellow; rind medium, tight, smooth and glossy; pulp white, juicy, acidic with strong aroma; normally seedless but occasionally with few seeds.

Pant Lemon-1: It is a selection from Kagzi Kalan at Pantnagar, U.P., India. Medium-sized fruits are round, smooth, juicy with thin rind.

Baramasi: Fruits of this cultivar are medium, ovoid-globose in shape, light yellow in colour; rind medium to thick; flesh firm, juicy, acidic with good flavour; nearly seedless, occasionally few seeds.

4.2.2.4 Acid Lime

West Indian lime: This cultivar is grown in large scale in Mexico, Florida, West Indies, India and Brazil. It is popularly known as Kagzi lime in India (Bhattacharya and Baruah, 1998). Fruits are small, round obovate or short-elliptical, round base, sometimes with short neck, apex round; rind thin, tight, greenish yellow at maturity; flesh colour greenish yellow, juicy, highly acidic; seeds moderate in number, small, polyembryonic.

Pramalini and Vikram: These cultivars have been developed through clonal selection at Marathwada Agricultural University, Parbhani, India. Fruits are borne in clusters.

PKM-1: This cultivar has been evolved through intensive screening of acid lime clones at Periyakulam, India. Fruits are large and juicy. Trees bear heavily.

Seedless: This late maturing cultivar is a new selection made in Himachal Pradesh in India. Fruits are oblong; rind thin, tight, colour primerose yellow at maturity.

Tahiti lime (*C. latifolia*): This is believed to have been introduced from Tahiti to California, Florida and in other citrus growing regions. In India, it is grown mainly in southern region. Fruits are medium, obovate-oblong, base and apex roundish; rind thin, tight, smooth, colour pale lemon-yellow at maturity; pulp pale greenish-yellow, juicy, acidic; normally seedless (Bhattacharya and Baruah, 1998).

4.2.2.5 Grapefruit

Duncan: Trees are vigorous; fruits are large, oblate to globose, base and apex round; rind medium thick, smooth, light yellow coloured, moderate adherence; flesh tender, off-white coloured, juicy, highly flavoured; seeds many.

Marsh Seedless: Trees are vigorous and productive; fruits are medium to large, round or oblate; rind medium thick, smooth, light yellow in colour; pulp off-white to grayish green, juice abundant, highly flavoured; normally seedless.

Foster: This cultivar has been originated as a pink fleshed budsport of Walters cultivar (Bhattacharya and Baruah, 1998). Fruits are medium to large, oblate to spherical; rind thick, light yellow flushed with pink colour, smooth, leathery; flesh pink coloured, fine, tender, juicy, flavour good; seeds many.

Thompson: It is a pink-fleshed budsport of the cultivar Marsh. Trees are vigorous and productive. Fruits are medium, oblate to spherical, rind medium thick, smooth, light yellow in colour; pulp pink coloured, juicy, flavour good; seedless.

Triumph: Trees are medium but productive. Fruits are medium to small, oblate, globose or ellipsoid; rind medium thick, smooth, pale yellow in colour; pulp juicy, excellent quality; seedy.

Saharanpur Special: Fruits are round to oblate; rind medium thick, surface smooth, yellow; flesh soft, light yellow in colour, juicy, sweetish sour, moderately flavoured; seeds many.

Intergeneric Hybrids of Citrus

Citrus latipes

Citrus macroptera

Citrus ichangensis

Citrus assamensis

Poncirus-Trifoliate orange

Citrus fortunella

Citrange

Citrangedin

Intergeneric Hybrids of Citrus

Citrangor

Citrandarin

Citremon

Intergeneric Hybrids of Citrus

Citradia

Procimequat

Citrus liquat

Orangequat

Tangor

Tangelo

Limonime

Lemandarin

Cultivars of Citrus

Coorg Mandarin

Kinnow Mandarin

Khasi Mandarin

Darjeeling Mandarin

Cultivars of Citrus

Satsuma Group

King

Willow Leaf

Nagpur Mandarin

Cultivars of Citrus

Ponkan

Clementine

Dancy

Mosambi

Sathgudi

Pineapple

Cultivars of Citrus

Washington Navel

Hamlin

Melta Blood Red

Valencia

Shamouti

Lemon

Cultivars of Citrus

Eureka

Lisbon

Galgal

Assam Lemon

Pant Lemon

PKM-1

Seedless Lime

Duncan

Marsh Seedless

Foster

Thompson

Saharanpur Special

The National Citrus Germplasm Repository in China collected accessions belonging to 24 species, 14 botanical varieties and 9 genera. Several promising cultivars and lines, such as oranges Zhongyu 7 and Wanmi 1, were bred and released by using the gene bank accessions (Chen, 1999).

5.0 Breeding and Improvement

A detailed account of breeding of citrus is available in Parthasarathy *et al.* (2019).

Considerable work has been carried out in citrus breeding and quite a large number of hybrids within and between six genera, *i.e., Citrus, Poncirus, Fortunella, Microcitrus, Eremocitrus* and *Clymenia* have been successful (Swingle and Reece, 1967). Since in citrus most of the characters are inherited qualitatively, breeding is much complicated by heterozygosity (Spiegel-Roy and Vardi, 1982) and by self and cross-incompatibility which occur in a number of citrus cultivars. Presence of series of incompatibility alleles in pummelos and their hybrids, in grapefruit hybrids like Mineola, Orlando and Clementine mandarin have been reported. Even with such limitation citrus breeding remains to be one of the major areas of improvement primarily due to the availability of rich germplasm, their wide range of variability and high degree of success of interspecific and even intergeneric crosses.

Selection of superior seedlings is of immense importance. Several good cultivars like 'Kiyomi' tangor and the triploid grapefruit 'Oroblanco' have been evolved (Soost and Cameroon, 1981). In general, triploids tend to low fruit yield (Cameroon and Frost, 1968). Tetraploids may arise spontaneously as nucellar seedlings but are of no economic interest, because they are often very slow growing in habit (Barrett and Hutchison, 1978).

Breeding programme with specific objectives like producing early maturing citrus fruits with high yield and fruit quality, developing rootstocks having disease resistance, wider adaptability, suitability for commercial cultivation of lemon, *etc.* are of considerable importance.

In rootstock breeding, the main emphasis has been given on development of root- stock resistant to tristeza, *Phytophthora*, nematodes, *etc.* Most of the breeding programmes make use of *Poncirus* which is a carrier of resistance to tristeza, *Phytophthora*, nematodes and coldhardiness (Spiegel-Roy and Vardi, 1982), salt-tolerant rootstock has also been found possible in some progenies involving Cleopatra and Sunki mandarins and Rangpur lime (Furr *et al.*, 1963). The genera *Citropsis, Hesperethusa, Microcitrus, Clausena, Clymenia, Atalantia* and *Pleiospermium* are used for evolving new citrus rootstocks (Bitters *et al.*, 1964, 1969b). Barrett (1978) opined that bigeneric and trigeneric crosses of *Citrus, Eremocitrus, Microcitrus* and *Poncirus* were possible. In April, 1974, the US Department of Agriculture released to citrus nurserymen and growers the Swingle Citrumelo citrus rootstock. This rootstock selection was evolved by Walter Swingle in Florida, in 1907. Its parents are Duncan grapefruit (*Citrus paradisi*) x *Poncitrus trifoliata*. It is more salt tolerant than other trifoliate hybrids (Broembsen, 1984). Citrus tristeza virus (CTV), the most severe viral disease of citrus in the world, causes rapid decline and death of trees grafted on sour orange (*Citrus aurantium*) rootstocks, and stem pitting of grapefruit (*C. paradisi*) and sweet orange (*C. sinensis*) trees regardless of rootsock (Garnsey *et al.*,

1987). Natural resistance to CTV occurs in some citrus relatives, including *Atlantia ceylanica, Fortunella crassifolia, Poncirus trifoliata* and *Severinia buxifolia* (Garnsey *et al.*, 1987; Mestre *et al.*, 1997; Yoshida, 1985). Among these species, the CTV resistance in *P. trifoliata* (the trifoliate orange) has attracted the most attention. Genetic analysis showed that the CTV resistance in *P. trifoliata* was controlled by a single dominant gene called *Ctv* (Fang *et al.*, 1998; Gmitter *et al.*, 1996; Yoshida, 1985). Although many CTV-resistant hybrids of citrus and trifoliate orange have been widely used as rootstocks for citrus trees, breeding a marketable CTV-resistant scion cultivar by cossing trifoliate orange with citrus has not been possible because undesirable traits of trifoliate orange remain after several generations of backcrossing with citrus. To overcome this problem, research towards isolating this resistance gene and then transforming it into CTV-susceptible citrus varieties is in progress in Florida. No genetic analysis has been reported for the CTV resistance in *Atlantia ceylanica, Fortunella crassifolia*, or *Severinia buxifolia*. Garnsey *et al.* (1997) reported that some pummelo *Citrus maxima* cultivars were resistant to certain CTV isolates. This finding opens a possible way to breed CTV-resistant citrus scion cultivars by sexual hybrization.

Hybridization work was also undertaken to evolve hybrids of Kagzi lime resistant to canker along with other desirable characteristics by using seedless lemon, citron, kumquat, mandarin and Rangpur lime; on screening, some of the F_1 hybrids were reported to be resistant or highly tolerant to canker disease (Chakrawar, 1981). Another breeding effort in Florida emphasized crosses between sweet orange and *Poncirus* to obtain cultivars with orange type fruit which were also found to be resistant to tristeza (Barrett, 1981).

Spiegel-Roy and Vardi (1982) evolved a selection from Wilking x Valencia, which was found to be of precocious bearing habit with total soluble solids, colour and other characters were considered desirable for fruit processing. The second off-spring of a pummelo (Orblanco) – grapefruit (Marsh) cross Melogold has been released from California (Soost and Cameroon, 1986). Melogold resembles the present white-fleshed grapefruit cultivars but is more pummelo like than Oroblanco. Melogold has been grown only on Troyer citrange and Rough lemon rootstock.

Fruits of triploids with a few seeds are considered to be desirable type for table purpose. It has been reported that occasionally the chromosome of diploid parents became double and some of the resulting tetraploid embryos survived. Such tetraploid plants had been used as seed parents for producing triploids. The triploid embryos can be germinated satisfactorily by using *in vitro* technique. A new triploid mandarin hybrid 'Tacle' has been evolved by crossing monoembryonic, zygotic, diploid Clementine as female with tetraploid orange Tarocco. The trees are vigorous and thorny; the fruit is easy to peel and resembles a large Clementine with smooth, dark orange skin; the pulp is seedless (Starrantino, 1999). 'Shuyou', a selection of the leading local pummelo variety Anjiang Xianyou, is high yielding, producing very large, elongated-ovate, smooth-skinned fruits with fine tender juicy flesh (Wang, 1998).

Matsumoto *et al.* (1999a,b,c,d) described several new cultivars of citrus produced in Japan. Amakusa was produced from a cross between (Kiyomi x Okitsuwase 14), an early ripening mandarin type strain and orange cv. Page. The cultivar shows strong parthenocarpic habit. The fruit acid content is about 1.0 per cent and soluble solids are fairly high (11-12 per cent) in concentration (Matsumoto *et al.*, 1999a). Hareyaka was produced from a cross between Encore (King mandarin x Willow leaf mandarin and Nakano 3 Ponkan (*C. reticulata*). It was registered as Mikan Norin 9. Titratable acidity is mild (1.0-1.2 per cent) and concentration of soluble solids is high (12-13 per cent) (Matsumoto *et al.*, 1999b). Miho-core was produced from a cross between Mihowase (*C. unshiu*) and Eucore (King mandarin x Willow leaf mandarin. It was registered as Mikan Norin 10. The Miho-core was produced from a cross between Mihowase (*C. unshiu*) and Encore (King mandarin x Willow leaf mandarin). It was registered as Mikan Norin 10. The cultivar produces nearly seedless fruits. Titratable acidity is lower (0.8-1.0 per cent) and the concentration of soluble solids is medium (11-12 per cent). It is fairly resistant to citrus canker (Matsumoto *et al.*, 1999c). Youkou was produced from a cross between Kiyomi tango and Nakano 3 Ponkan (*C. reticulata*) and it was registered as Mikan Norin 11, as a superior mandarin cultivar. The cultivar is resistant to both citrus scab and citrus canker (Matsumoto *et al.*, 1999d).

Somatic mutations are common in citrus and through selection of the natural mutants quite a few number of desirable clones have been obtained. Mutations when occurring in single somatic cell often perpetuate in only a part of the tissue and such partial mutants or chimeras have also been reported to occur in citrus. The frequent occurrence of chimeras may lead to clonal impurity and thus bud selection work in propagation becomes important for ensuring clonal purity. Some of the most important citrus cultivars thus arose by somatic mutation. The homohiston variants like Redblush grapefruit were reported by Cameroon and Frost (1968) and certain heterohistons like Shamouti orange by Spiegel-Roy (1979). Selections of natural mutants (Iwamasa and Nishiura, 1981; Hensz, 1981) have been successfully employed on seedlessness (Iyo tangor), season of ripening (Satsuma, Navel), improvement of colour (Ray Ruby grapefruit) *etc.* The variety 'Rhode Summer Navel' has been developed from a spontaneous mutation of *C. sinensis* via selection. Fruits are large in size and seedless, producing a moderate amount of juice with high TSS: acid ratio (Anon, 1998).

In Japan, a few closely related clones of Satsuma mandarin with varied fruit colour and fruit ripening times were obtained through mutation. In USA also mutations had produced Satsuma seedling lines differing in productivity, fruit shape and the ripening time.

The grapefruit clones like Thompson and Foster Pink arose as limb sports on white grapefruit. Gamma irradiation of seed and bud wood performed in Orlando, Florida, resulted in seedless fruit on certain trees of seeded cultivars like Pineapple orange as well as Duncan and Foster grapefruit. Vardi and Spiegel-Roy (1978) developed Shamouti trees of compact habit and early fruiting, and seedlessness in Eureka lemon by irradiation of bud wood with gamma-ray.

Improvement of citrus *in vitro* has also been reported under biotechnology.

6.0 Propagation

Various types of citrus of commercial importance are propagated differently. While budding is almost universally practiced, in case of sweet oranges all over the country and to a certain extent in north-eastern hill regions and Coorg of India, in case of lemons and sweet lime, air-layering and even cutting are being employed for raising plants. Since seeds of most of the citrus species and cultivars are polyembryonic, the nucellar seedlings obtained through seed propagation carry the true characters of the mother plants thus enabling to retain the clonal characters as in vegetative propagation. The phenomenon of polyembryony also helps in raising uniform rootstock seedlings.

6.1.0 Vegetative Propagation

6.1.1 Budding

6.1.1.1. Selection of Rootstocks

Castle (2010) wrote a very interesting review on Citrus rootstocks. According to Castle (2010) The existence of citrus rootstock choices has historically meant that: 1) sour orange could be replaced, thus averting future disasters from large-scale tree loss resulting from citrus tristeza virus infection; 2) citrus orchards could be grown on marginal soils and locations because rootstocks suitable for high pH soils, salinity problems, or colder areas were available; 3) particular objectives could be met by selecting not the best overall choice, but by selecting a rootstock for one specific purpose like large fruit size, for example, the use of rough lemon rootstock for navel oranges or Volkamer lemon for 'Murcott'; 4) rootstocks with complementary tolerances to pests and diseases could be selected to reduce the risk of tree loss; and 5) the harvest season could be extended because of the effect of rootstock on juice Brix/acid ratio, distinctiveness of citrus rootstocks versus other tree fruit rootstocks. The long list of citrus tree performance factors attributable to the rootstock (Table 3) is difficult to match by any other tree fruit.

Table 3: Rootstock-Related Citrus Tree Attributes*

Attribute	*Remarks on Attributes as a Rootstock Selection Criterion*
Yield	A major consideration
Precocity	A special factor as interest in higher-density plantings increases
Yield efficiency	Can be important, but not usually a selection criterion
Fruit quality	A major factor for fresh fruit growers
Fruit size	Important enough to be considered separately from quality
Juice quality	Brix and acid are affected; a major factor for both fresh and processing
Tree growth	Usually considered in terms of vigor and eventual tree size, a criterion that is increasing in importance
Compatibility	Scion-rootstock vegetative compatibility is often important
Ease of propagation	Largely a matter of seed production and degree of nucellar embryony
Shoot flushing	A new criterion related to Huanglongbing and spread by psyllids
Mineral nutrition	Not usually a selection criterion, but there are rootstock effects

Attribute	Remarks on Attributes as a Rootstock Selection Criterion
Salinity	In some instances, an important selection criterion
Clay soil	High content or sol horizons can affect a rootstock decision
High soil pH	A very important factor with trifoliate orange-based rootstocks
Wet soil (flooding)	Not usually a selection factor but can be important
DroughtFactor	Modern irrigation methods usually preclude this as a selection
Cold (freezes)	Often considered in regions threatened by chronic cold
Citrus blight	At one time, a major consideration, but less so now.
Phytophthora rots	Still an important factor in rootstock decisions
P. palmivora/root weevil complex species	A problem specific to Florida involving a particular of Phytophthora fungus and Diaprepes root weevil.
Root weevils	A troublesome problem at times in some areas
Burrowing nematode	A problem in Florida for which specific rootstocks are used
Citrus nematode	A more universal problem with specific rootstock options
Tristeza virus	A serious threat with some rootstock options
Exocortis and xyloporosis viroids	Not generally a threat today with clean bud wood

The list is not intended to be comprehensive but to show the most important attributes representative of the world's citrus industries

* Castle (2010).

6.1.1.2 Selection of Bud

Along with standardization of rootstocks, proper selection of bud wood and use of certified disease-free bud wood materials are of great importance. In the process of bud selection, it is required to ensure that clonal purity (freedom from sports and chimeras), good physiological vigour and high yielding capacity are maintained and the bud material should be free from transmissible diseases. More than fifteen viruses are known to infect citrus trees in various parts of the world and at least four of these, such as tristeza including seedling yellow strain, greening (a mycoplasma disease), exocortis and a strain of psorosis (Rao *et al.*, 1975) have been reported to occur in India. Since the viruses can easily be transmitted through the bud wood, it is necessary to index the mother plant of bud wood sources for freedom of viruses and it can be done through the help of indicator plants, identified for expressing definite and clear symptoms for different virus-mycoplasma disease. The standard indicator plants employed for the indexing are Kagzi lime, West Indian lime for tristeza virus; Etrog citron and Rangpur lime for exocortis virus; sweet orange for psorosis virus and sweet lime and Orlando tangelo for xyloporosis virus. The virus-free bud wood obtained through indexing should be propagated and multiplied for further distribution. The mother plants raised from certified bud wood materials may act as future source of healthy bud wood if they are grown in isolated areas free from virus infection, protected from further infection through control of insect vectors of various viruses. In case of mechanically transmitted viruses, where no known insect vector exists, frequent disinfection of propagation tools such as grafting/budding knife, pruning shears, *etc.* are essential for avoiding transmission of the disease. Through the use of nucellar clones of standard citrus cultivars, the requirements of bud selections can be met to a great extent.

6.1.1.3 Raising of Seedling for Rootstock

It is widely accepted that good citrus production depends mainly on the availability of suit-able and compatible rootstocks. Thus rootstocks provide orchards with useful eco friendly tool to manipulate the vigour and performance of trees. Rootstocks directly affect the ability of plants to uptake the water and nutrients from the soil. The long-term performance of rootstocks and their significant effects on leaf nutrient levels have been studied for different climatic conditions across the world (Srivastav *et al.*, 2005; Toplu *et al.*, 2012). The present study aimed to evaluate the horticultural performance of lemon cv. 'Kagzi Kalan' on eight rootstocks under typical subtropical conditions for maximizing the production and fruit quality. The experiment was conducted on five years old Kagzi Kalan lemon plants budded on eight rootstocks. The variables collected over two successive years included vegetative growth, fruiting density, fruit quality, and leaf nutrient concentrations. Results revealed a significant effect of rootstocks on all the variables studied. Tree height and canopy volume were found higher on rough lemon and RLC-4 rootstocks, while trunk cross sectional area was higher on Billikichlli and RLC-4 rootstocks than other rootstocks. RLC-4 and Attani-2 rootstocks had higher fruiting density, while it was lowest on Troyer citrange. RLC-4 produced heaviest fruits, while trees on rough lemon and RLC-4 had significantly higher juice recovery. RLC-4 and Karna Khatta improved TSS and acid contents in the fruit juice. RLC-4 and Troyer citrange proved their superiority in terms of higher ascorbic acid content. Foliar N concentration in scion cultivars was found higher on RLC-4, rough lemon and Attani-2 rootstocks, while the foliar K content was higher on RLC-4 and rough lemon. Jatti Khatti was more efficient in accumulating Mg in leaf tissues, whereas Troyer citrange restricted accumulation of Na in leaf tissues. Sour orange was more capable of accumulating Cu and Zn in leaf tissues, while other rootstocks had similar uptake of Cu. Karna Khatta, rough lemon and RLC-4 absorbed more Fe, while the foliar Mn content was highest on rough lemon rootstock. Troyer citrange appeared poor performer in terms of fruiting density and nutrient concentration. A positive significant correlation was found between leaf Ca concentration and canopy volume. Hence, considering growth performance, fruit quality and nutrient concentration, RLC-4, rough lemon and Karna Khatta rootstock could be suggested for lemon under alike ecological conditions (Dubey and Sharma, 2016).

It is desirable to raise citrus nursery in light, well drained but fertile soil, free from soil borne pathogens and nematodes. Bold seeds collected from desired trees may be sown in primary nursery beds properly prepared with enough organic manure. It is desirable that the upper top portion of the primary nursery beds are covered with sand (about a 3-5 cm layer) and the seeds before sowing are treated with fungicides to avoid seed-borne infections. Usually 3-4 weeks time is required for germination in secondary nursery beds. In the secondary nursery, proper spacing of about 20-30 cm is maintained between plants and rows, and the 'off' type seedlings of sexual or hybrid origin are eliminated. The elimination process ensures greatly towards getting true-to-type nucellar seedlings which are more or less uniform, upright and vigorous. The optimum temperature for germination ranged from 25°C for *P. trifoliata* to 32.5°C, for Rangpur lime x Troyer citrange. The overall average optimum germination temperature was reported as 29.7°C (Rouse, 1997).

Bending was compared to cutting off for effectiveness in forcing growth of sweet orange (cv. Hamlin) and mandarin (cv. Ninkat) scions budded on Carrizo citrange, Swingle citrumelo and 17 new hybrid citrumelo rootstocks (Bowman, 1999). For both scion types, more than twice as many plants from the bending treatment than the cut treatment had growing scion buds at 12 weeks. The survival and yield of usable plants at 35 weeks of age were also significantly better for the bending treatment than for the cut treatment. There was an overwhelming advantage to using the bending treatment instead of cutting off in forcing scion bud growth for propagating citrus on citrumelo rootstocks.

6.1.1.4 Budding and Grafting

It is the most common method of vegetative propagation in citrus, particularly, in sweet orange, mandarins, grapefruits, *etc.* Different types of budding like patch or forkert or T-budding may be followed with high rate of success. The time of budding varies from place to place and the right stage of budding is when the plant has good sap flow and the cambium tissue is highly active. Well-swollen unsprouted buds, from about one-year old non-angular branches are preferred, and the presence or absence of small portion of wood along with the individual bud does not really matter for the success in bud union. Budding should preferably be done at a height of about 20-25 cm from the ground level which usually becomes ready for orchard planting in 6-9 months after budding. Labanauskas *et al.* (1976) found that Valencia orange trees budded on Cleopatra mandarin and Troyer citrange rootstocks at a height of 15 cm gave the highest yield of fruits and the yield decreased with the increasing budding height. Two-year old seedlings of Cleopatra mandarin and Troyer citrange were budded with Cambell Valencia at the height of 5, 15, 30, 45, 60 or 90 cm and planted in May. The higher the trees were budded, in general, the lower were their yields. The critical budding height was 30 cm (Bitters *et al.*, 1972). The grafting success of Orlando tangelos was found highest on Cleopatra mandarin rootstocks (100 per cent) followed by sour orange x grapefruit (90 per cent), Rangpur lime (88 per cent) and Troyer citrange (79 per cent). Scion height after 90 days was also highest on Cleopatra mandarin (Gonzalez and Figueroa, 1996). Pummelo cv. Wubuyou when grafted on *Poncirus trifoliata* at 10 or 20 cm height, showed vigourous growth in China compared with that grafted at 30 to 70 cm. However, cold hardiness was found to increase by high grafting (Wan *et al.*, 1994).

Budwood of Satsuma mandarin cv. Gongchuan was used to topwork 6-year-old trees of orange cv. Hamlin, which suffered with canker disease. In the third year, topworked trees started to bear good crops (12.8-25.4 kg/tree compared with 14.5-16.3 kg/tree in the control). Topworked trees exhibited no signs of disease, whereas control trees had 34.5-51.2 per cent diseased leaves. Fruit quality of Gongchuan was quite good (Jing, 1998).

6.1.2 Cutting and Layering

In case of lemons, sweet lime, citron, *etc.* cutting and layering (air layering) are commonly employed and plants thus raised develop shallow but good root system. In lemon, cuttings from the middle portion of one-year shoots rooted better (Gabricidze, 1970). Semi-hardwood cuttings of Washington Navel orange rooted profusely when treated with IBA (Shafrir, 1970) while softwood cuttings of sweet lime rooted better than hardwood ones in clay and loam rooting media (Grewal and Singh, 1975). According to Prati *et al.* (1999), the propagation of Tahiti lime through stem cuttings was an efficient and fast method. The highest rooting percentage for sour orange rootstock was obtained when cuttings were prepared in May and treated with IBA at 2000 ppm, but the highest survival was recorded in cuttings taken in June and treated with IBA at 4000 ppm. Cleopatra mandarin rooted better when cuttings were prepared in May and treated with IBA at 1000 ppm. The highest rooting percentage for *C. limonia* was obtained from cuttings taken in April and treated with IBA at 1000 ppm, while cuttings taken in June and treated with IBA at 1000 ppm showed the highest rate of survival. In general, rooting abilities of the rootstocks were in the order of *C. limonia* > Sour oranges > Cleopatra mandarin (Abou Rawash *et al.*, 1998).

The air layering operations are usually done during early monsoon and the propagating materials become ready in about 3 months. After removal from the mother plants, it is preferable to keep the layers in the nursery bed for about a month before final transplanting at the orchard site. For certain specific advantages, however, budding on desired rootstocks may also be followed for these species as well.

6.1.3 Propagation through Tissue Culture

Studies on citrus tissue and organ culture *in vitro* were initiated about three decades back. Gurgel and Sabrino (1951) successfully grew citrus embryos in Sach's medium. The technique of growing isolated citrus embryos in artificial media was described by Hurosvilli (1957). *In vitro* culturing of isolated vesicle stalks from *C. limon* was carried out successfully by Kardon (1963). Murashige *et al.* (1968) established callus culture from albedo and juice vesicles of *C. limon* fruit. Although callus was formed, it failed to differentiate. Nucellus isolated from each monoembryonic cultivar like Chandler pummelo and Temple orange gave rise to 10-15 embryos *in vitro* which developed into complete seedlings (Bitters *et al.*, 1969). The plants arising from embryogenesis of the nucellus *in vitro* were found to be free from most of the pathogenic viruses (Bitters *et al.*, 1972). Pseudobulbils, embryoids and other proliferations of varying sizes and forms were obtained by culturing nucelli of various Clementine clones and lemon, bergamot and citron cultivars on a media containing Murashige and Skoog's (MS) nutrient, supplemented with adenine or methyl aminopurine. Many plantlets were obtained from Clementine and lemons but bergamot produced a few and citron none (Starrantino *et al.*, 1977). Differentiation of callus tissue obtained from stem explants of *C. madurensis* into shoots and roots was recorded by Grinbalt (1972). Chaturvedi and Mitra (1974) obtained rooted plants from isolated stem and leaf tissue of *C. grandis* and *C.*

Propagation

Patch Cutting

Patch Budding

T-budding

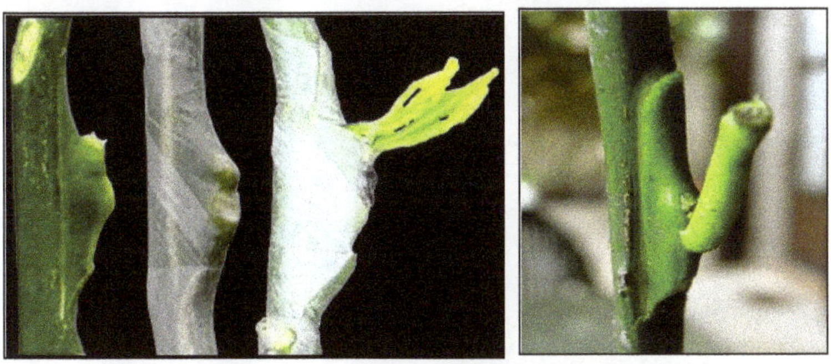

Chip Budding

Air Layering

Propagation by Tissue Culture

Protoplast Fusion

Somatic Embryogenesis

sinensis. Stem segments of different citrus species were cultured in different media supplemented with micronutrients and growth regulators and significant success in obtaining rooted plants was recorded by Bhansali and Arya (1980) and Kitto and Young (1981).

Kochba *et al.* (1972) obtained whole plants by culturing 1 to 8 week old ovules and nucelli of Shamouti and Valencia orange *in vitro* in Murashige and Tucker's nutrients supplemented with kinetin, IAA, coconut milk or GA_3; a higher kinetin/IAA ratio or addition of coconut milk favoured stem elongation more than root formation whereas lower kinetin/IAA ratio favoured root formation and inhibited stem elongation. Embryos, pseudobulbils and embryogenic callus were obtained in *in vitro* culture of undeveloped ovules excised from ripe fruits of the Navel orange group and lemon (Starrantino and Russo, 1980). Large number of seedlings was produced by subculturing the embryos, pseudobulbils and embryogenic callus on fresh MS media with added 6-methylaminopurine. Ben-Hayyim and Neumann (1983) observed that addition of glycerol (140 to 420 mM) to Murashige and Tucker's medium increased the growth of nucellar calli from Shamouti orange, sour orange, Duncan grapefruit and Villafranca lemon ovules.

Parthasarathy and his colleagues carried out extensive studies on micropropagation of various Citrus species including micrografting, vitrification and in vitro acclimatization. They have standardized the protocol for a range of *Citrus* species. In one of their studies undertaken to find out optimum concentration of BAP using quadratic model, shoots (0.5-1.0 cm long) from in-vitro proliferated plants of 5 Citrus species, *viz., C. madurensis, C volkameriima, C. sinensis, C. limon and Poncirus trifohata x C. sinensis* (Troyer citrange) were cultured in MS supplemented with varying concentrations of BAP and kinetin. Morphological observations recorded 4 weeks after culture initiation indicated that there was a significant interaction between species and cytokinins, and species and concentration. Among species *C. volkameriana* recorded highest plant weight, while *C. limon* recorded lowest plant weight. BAP was a better cytokinin for shoot proliferation, whereas kinetin for producing a fewer but healthier plantlets. The optimum concentration of BAP for producing maximum shoots was worked out using quadratic response equation. Troyer citrange responded to a high level of BAP, while *C. sinensis* (Malta) responded to a relatively low level of BAP (Parthasarathy *et al.*, 2001).

A micro-grafting technique was described by Tusa and co-workers (1979 and 1980) with a success ranged between 25 per cent and 100 per cent depending on the stock/scion combination. In India the first successful standardization of micrografting in Citrus was reported by Parthasarathy *et al.* (1997). In vitro micrografting of 0.2 to 0.3 mm long shoot tips of Khasi mandarin orange on 2 weeks old etiolated seedlings of various citrus species achieves 67 percent of successful grafts. Pretreatment of scions with 10 ug/l of 2,4-D greatly increased the success rate. Histological studies using electron microscope show graft union and vascular connection between root stock and scion. Grafts acclimatized in culture jars were transferred to soil and successfully established.

The graft combination and compatibilities have great relevance in the graft success. Grewal *et al.* (1994) found orange cv. Mosambi on *C. depressa* was the most successful combination. Ioannou *et al.* (1991) found that among rootstocks, citrumelo gave the highest grafting success (41 per cent) followed by sour orange (37 per cent) and rough lemon (29 per cent) than those from the greenhouse (32 per cent). Dass *et al.* (1997) studied the effect of different rootstocks, *viz.*, rough lemon, Troyer citrange and Carrizo citrange of different ages was grafted with shoot-tip scion of Nagpur mandarin. The success of *in vitro* micro-grafting was dependent on the type of rootstocks used. Overall success of *in vitro* shoot-tip grafting was more in Troyer citrange followed by Carrizo and rough lemon.

6.1.4 Rootstock

The rootstocks have most prominent effect on growth, productivity, fruit quality and longevity of the tree on one hand and influence the susceptibility of the trees to various insects and diseases on the other hand (Chadha, 1970). Srivastava *et al.* (1994a) observed root CEC as a marker for vigour of citrus rootstocks which also governed the chloride exclusion or accumulation behaviour (Srivastava *et al.*, 1998). The studies on influence of different rootstocks (32 nos.) on the incidence of the die back disease showed that some of the rootstocks such as Rough lemon (*Citrus jambhiri*), Belladakithuli (*Citrus madaraspatna*) and Baduvapuli (*Citrus pennivesiculata*) showed less intensity of symptoms as compared to mandarin seedlings (Aiyappa *et al.*, 1962). The citrus decline is caused due to rootstock scion incompatibility which may be either viral or genetic. Tristeza virus was reported to be present in Bombay and Andhra Pradesh by Vasudeva and Capoor (1958) and Reddy and Papa Rao (1960). Aiyappa and Subramaniam (1959) indicated the presence of some virus infection in the mainland area of Mysore state. Greening disease had also been suspected for causing decline (Fraser, 1967). Nauriyal and Chadha (1966) also reported the presence of inverse pitting in a large number of sweet orange trees examined by them and they suspected these symptoms to be caused by a virus. In India, citrus plantations established using rough lemon rootstock was not only susceptible to salinity and root and foot rot but also became vulnerable to nematode infestation which could be the major reasons for citrus decline. But despite so many rootstock trials no rootstock could provide a permanent alternative to rough lemon rootstock which could have better resistance.

Some of the commonly used rootstocks are sour orange, rough lemon, trifoliate orange, citranges, Rangpur lime, mandarins, *etc.*, brief characteristics of which are as below.

6.1.4.1 Sour Orange (C. aurantium)

Sour orange had been extensively used in the past as rootstock of sweet orange, grape fruit, mandarin orange and lemon (Batchelor and Rounds, 1948). Trees on sour orange rendered good yield (Wutscher, 1978) with better fruit quality (Blondel, 1977). However, it is susceptible to tristeza virus, citrus nematode and burrowing nematode (Baines *et al.*, 1960). Trees on sour orange were usually smaller than those on rough lemon and were moderately cold hardy (Gardner and Horanic, 1958).

6.1.4.2 Rough Lemon (C. jambhiri)

Rough lemon is probably of hybrid origin (Chapot, 1975) and is highly polyembryonic. This is the best known of all rootstocks in South Africa (Broembsen,

1984) and largely used in different countries of the world. Trees on rough lemon stocks turned out high yield with good fruit size but fruits were of poor quality – rough, thick and poorly-coloured skin (Ziegler and Wolfe, 1961). It is tolerant to tristeza and relatively tolerant to saline and calcareous soils but susceptible to foot rot (Grimm and Hutchison, 1973) and blight (Smith, 1975).

6.1.4.3 *Trifoliate Orange (Poncirus trifoliata)*

A deciduous relative of *Citrus*. The trifoliate orange is an excellent rootstock for Satsuma in Japan but is susceptible to exocortis disease and blight (Reitz, 1984) but tolerant to most other major virus diseases, *Phytophthora* foot rot and citrus nematode (Zeigler and Wolfe, 1961). Trees on trifoliate gave high yield with excellent fruit quality (Vuillaume *et al.*, 1981) which were also resistant to frost (Korsantiya *et al.*, 1985).

6.1.4.4 *Mandarin (C. reticulata)*

The most common mandarin rootstock is Cleopatra. It is the most salt tolerant of the common citrus rootstock cultivars with the ability to exclude sodium and chloride taken up by the foot system (Broembsen, 1984) but to accumulate boron (Cooper, 1961). Cleopatra is tolerant to tristeza, exocortis and xyloporosis and fairly tolerant to *Phytophthora* root rot. However, the yield is comparatively low on this rootstock. Casamayor *et al.* (1982) obtained cumulative production of 60.2 tonnes per hectare in the first five crops of Orlando tangelo on sour orange compared with 58.5 tonnes per hectare on Cleopatra mandarin.

Sunki mandarin is used as a rootstock for orange in Brazil (Sobrinho *et al.*, 1978). It is susceptible to foot rot (Bitters, 1974), sensitive to exocortis (Olson *et al.*, 1962) but tolerant to tristeza (Salibe, 1947b).

6.1.4.5 *Sweet Orange (C. sinensis)*

A moderately cold hardy rootstock having wide range of soil adaptation. Trees on sweet orange are little affected by blight (Lawrence and Bridges, 1973) and are resistant to tristeza, exocortis, xyloporosis, sour orange scab and mal secco (Klotz, 1973). Trees on sweet orange rootstock had higher N, P and Cu content in leaf (Wutscher, 1974) and the quality of fruit was also good on sweet orange rootstock (Bitters, 1974).

6.1.4.6 *Rangpur Lime (C. limonia)*

A vigorous hardy rootstock with good adaptability to wide range of soil, particularly for heavy soil. It is tolerant to tristeza and also to salt (Cooper, 1961). However, trees on Rangpur lime are susceptible to foot rot (Carpenter and Furr, 1962), exocortis and xyloporosis. Rangpur lime rootstock is used for orange, grapefruit and tangelo scions. The quality of fruit on Rangpur rootstock is fairly good (Salibe, 1947a).

6.1.4.7 *Troyer and Carizzo Citrange*

Citranges (sweet orange x trifoliate orange hybrid) are now used widely as rootstock to replace sour orange. Trees on citrange are salt sensitive (Cooper, 1961).

Of the citranges, Troyer is mostly used in California which is tolerant to tristeza and inter-mediate resistant to foot rot (Klotz *et al.*, 1972). Orange and grapefruit trees on Troyer are very productive and fruit quality is good (Hutchison, 1978). The name Carizzo was given in 1938 to a citrange tree produced from seedlings of the original hybrids raised near a town called Carrizo Springs in Texas (Savage and Gardner, 1965). In Florida, Carrizo is a popular rootstock and is considered as burrowing nematode resistant (O'Bannon and Hutchison, 1974).

6.1.5 Stock-scion Relation

Considerable information on rootstock-scion relationship among different citrus species and cultivars are available. In a study conducted in Australia, Stafford (1971) attributed the poor performance of Carrizo citrange as a rootstock for Lisbon and Eureka lemon to virus infection and incompatibility. Tanaka *et al.* (1982) observed that Satsuma and Sanbokan were good in interstock behaviour while Natsudaidai scion shoots often died. Valencia orange trees on *Swinglea glutinosa* and *Citropsis gillentina* showed incompatibility (Valle *et al.*, 1982). The trifoliate orange was found to be most dwarfing rootstock for grapefruit, Washington Navel orange and Valencia. The most promising rootstock for close density planting of nucellar Marsh grapefruit and Valencia were Rubidoux trifoliate orange, Rusk citrange, sweet orange x Rubidoux and Rangpur x Troyer citrange (Castle, 1980). Citrumelo C 4475, a hybrid between grapefruit and trifoliate orange, was found useful as a dwarfing rootstock for Valencia orange in Florida (Gardner and Horanic, 1967-68). They also stated that Citrandarin C 40197 and Willets citrange rootstock produced dwarf Valencia orange trees which were very productive for their size. Nucellar Minneola tangelo (*C. paradisi* x *C. reticulata*) budded on Alemon (*C. macrophylla*) produced 40 per cent higher yield than on sour orange rootstock. Alemon was recommended as a rootstock for Minneola and similar high-acid citrus cultivars of Israel, in situations where the threat of tristeza virus was slight (Levy and Lifshitz, 1995). In Turkey, *C. volkameriana* proved to be a good rootstock for Clementine, Satsuma and Fremont mandarin. Trees on *C. volkameriana* showed high yield but trees on *P. trifoliata* gave the highest quality fruits with the brightest rind (Akgul and Tuzcu, 1993). In Cyprus, Frost Valencia orange produced significantly higher yield on *C. volkameriana* compared with other 15 rootstocks studied (Gregoriou and Economides, 1994). In case of Marsh nucellar grapefruit, trifoliate orange rootstock excelled over rough lemon and Cleopatra mandarin in terms of yield and fruit quality (Foguet *et al.*, 1972). The TSS of Ellendale tangor was the highest on trifoliata (Devington and Duncan, 1978). Orange and grapefruit on Rangpur showed high productivity, better quality and good survival (Cohen, 1970). *Citrus volkameriana* was reported to be a better rootstock than sour orange for lemon and mandarin orange, because it resulted in increased fruit size and weight, improved fruit quality and reduced peel thickness (Squillaci, 1964) and also a good rootstock for Temple orange in arid regions (Hankey, 1975). The best rootstock for Kinnow mandarin in the arid region of India had been found to be Jatti Khatti followed by Karna Khatta (Chohan *et al.*, 1978). The performance of Ambersweet (1/2 orange, 3/8 mandarin and 1/8 grapefruit), a new citrus hybrid developed in Florida on rootstocks Cleopatra mandarin and citrumelo was studied by Zekri (1997). It was observed that citrumelo rootstock promoted higher yield, earlier fruit maturity and better fruit and juice quality than Cleopatra. In Chile, Newhall orange grafted on *Poncirus trifoliata* cv. Rubidoux produced highest number of fruits/tree (59.4). The lemon cv. Limoneira 8A grafted on *P. trifoliata* cv. Flying Dragon had the least vegetative development and the greatest number of fruits (189.5) while growth and numbers of Clementine fruits were greatest (75.6) on citrumelo rootstock (Joublan *et al.*, 1997).

In a rootstock trial at Kerikeri, New Zealand, begun in 1993 using Miyagawa Satsuma mandarin as scion, Machin *et al.* (1999) compared ten rootstocks (citranges, citrumelos, *Poncirus trifoliata*, mandarins and *Citrus limonia* x citrange). Fruit yields ranged from 50.7 kg/tree on *Poncirus trifoliata* cv. Rich 16-6 to 102.6 kg/tree on Cleopatra mandarin. Average fruit weight ranged from 73 g on C-35 citrange to 88 g on Troyer citrange. Brix:acid ratio ranged from 10.7 on Cleopatra mandarin to 12.7 on Rusk citrange.

Lemon cv. Limoneira 8A grafted on *Poncirus trifoliata* cv. Flying Dragon showed a crown volume of 30 per cent of that produced on traditional sour orange after 12 years and the tree reached 3.5 m in height (Foguet *et al.*, 1996). In Valencia orange, the highest yield was recorded on Troyer (Cox, 1974) and Mortan citranges (Wutscher and Shull, 1973). Vuillaume *et al.* (1981) found no rootstock to be universally good for all scion species. Thus, Clementine did best on *C. taiwanica*, Tahiti lime on *P. trifoliata*, Orlando tangelo on Cleopatra mandarin and Pineapple orange on Carrizo citrange. Fruits on *P. trifoliata* showed good organoleptic qualities but yields of orange, tangelo and Clementine on this rootstock were poor. The habit of growth of the scion and intake of mineral elements from the soil also depend much on the rootstock used. A good rootstock for citrus should ensure an economic orchard life of about 30-50 years along with its tolerance or resistance to soil and virus diseases. In addition to it, it must be polyembryonic, easily propagatable with adequate and cheap sources of seeds.

The reactions of major citrus rootstocks to important viruses, soil-borne diseases, nematodes, *etc.*, had been worked out by various workers (Bitters, 1960; Ford and Fedar, 1969; Koltz *et al.*, 1965; Capoor and Rao, 1967). On the basis of their resistance against tristeza, psorosis, exocortis, *Phytophthora* and greening, *C. jambhiri*, *C. limon* and Cleopatra mandarin were found suitable as rootstock for sweet orange (Chowdhury *et al.*, 1980). Moreira and Salibe (1968-69) found the following rootstocks tolerant against tristeza: Sunki, Kinnow and Willow-leaf mandarin, Citrumelo 4475, Moreton and Troyer citrange, Seminole and Orlando tangelo, *C. volkameriana*, *C. karna* and some trifoliate hybrids. On old citrus soils infected with *Phytophthora parasitica*, Lisbon lemon performed best on *C. volkameriana* (Carpenter *et al.*, 1981). Acid lime budded on trifoliate orange was found to be more tolerant to canker than acid lime on Karna Khatta (Kale *et al.*, 1996). Troyer and Carrizo citranges were considered to be tolerant rootstocks to tristeza virus (Nava Ayala *et al.*, 1996). In China, Eureka lemon showed high resistance to citrus exocortis viroid and citrus tristeza when grown on Hongju (*Citrus reticulata*) rootstock. It performed very well on red soil (Yang and Jiang, 1993). Chadha (1970) in India and Broembsen (1984) from South Africa made an attempt to summarize the chief characteristics of major citrus rootstocks, so far as their susceptibility and resistance to common citrus diseases, nematode and salts were concerned (Table 4).

Table 4: Response of different Rootstocks to Diseases, Nematodes and Salts*

Rootstock	*Comparative tolerance to*						
	Tristeza	*Xylopo-rosis*	*Exo-cortis*	*Gum-mosis*	*Citrus Nematode*	*Burrowing Nematode*	*Stalts*
Rough lemon	T	T	T	S	S	S	Moderate
Trifoliate orange	T	T	S	T	T	S	Poor
Sweet orange	T	T	T	S	S	S	Moderate
Sour orange	S	T	T	T	S	S	Moderate
Rangpur lime	T	S	S	T	S	S	Moderate
Sweet lime	S	S	—	S	S	S	Moderate to poor
Cleopatra mandarin	T	T	T	T	S	S	Good to poor
Grapefruit	S	T	T	S	S	S	Moderate
Troyer citrange	T	T	S	T	—	S	—

T = Tolerant; S = Susceptible.

* Chadha (1970)

Salt damage usually manifested in citrus as leaf burn and defoliation was associated with accumulation of toxic levels of Na^+ and/or Cl^- in leaf cells (Storey and Walker, 1999). The greater salt resistance of rootstocks, such as Rangpur lime (*Citrus limonia*) and Cleopatra mandarin, is associated with their capacity to limit the accumulation of Cl^- in leaves. Other rootstocks, like trifoliate orange (*Poncirus trifoliata*), demonstrate a capacity to restrict Na^+ transport to the shoot at low salinity. The traits which restrict root to shoot transport of Cl^- and Na^+ in citrus have been shown to be heritable. Sodium and Cl^- concentrations in leaves are modulated by both the scion and rootstock although the rootstock has the greatest influence.

Since most of the citrus species are polyembryonic, they produce both gametic and nucellar seedlings. The gametic seedlings are highly variable and possess low vigour, while the nucellar seedlings provide uniform materials with comparatively more vigour. It is now established that most of the known citrus viruses are eliminated in the process of nucellar embryony and the nucellar seedlings are initially virus free. This phenomenon is of great horticultural importance, as it provides an opportunity of escaping viruses in the nursery stock. When the tedious process of bud wood certification for virus-free scion wood cannot be followed, it may be advisable even to use the nucellar seedlings of certain commercial citrus types for direct orchard planting. Py (1951) reported variation in polyembryony among different citrus species, sometimes even among the cultivars. Thus, in mandarin (*C. reticulata*), the cv. Clementine was found to be primarily monoembryonic, while cultivars like Dai Dai and Willow-leaf were polyembryonic with an average number of about 4.5 embryos per seed. Shaddocks appeared to be strictly monoembryonic, while hybrids of this species and grapefruit were slightly polyembryonic with an average of 2.5 embryos per seed. The true lemons (*C. limon*) were weakly polyembryonic, while citron (*C. medica*) showed a range from 1 to 3.91 embryos. The trifoliate orange contained an average of 2.46 embryos per seed.

The rootstocks *viz.* Orlando tangelo (*C. reticulata* x *C. paradisi*), Morton citrange (*P. trifoliata* x *C. sinensis*) and *C. volkameriana* were recommended for improvement of fruit yields and economic returns in lime cv. Tahiti (Graca *et al.*, 1997). The new hybrid citrus rootstock – CNR P1 (*C. depressa* x *P. trifoliata*) was reported to be resistant to *Phytophthora*, and tolerant to poor soil conditions (Geraci, 1997), while Borneo Red Lime (*C. limettioides*), *C. canaliculata* and calamondin were susceptible to *Phytophthora* and should not be used on heavy soils (Geraci and Giuffrida, 1997). Barros *et al.* (1997) recommended to use *P. trifoliata* cv. Seleta in Rio de Janeiro for high density planting in smaller orchards. Fruit quality was better, and symptoms of chilling injury and percentage decay following cold storage in Tarocco dalmuso orange were least in fruits from trees on sour orange rootstocks (Lovicu *et al.*, 1999). For Shamouti ornage, Carrizo citrange was the best rootstock for fruit yield and quality as well (Tuzcu *et al.*, 1999). Sour orange is highly susceptible to citrus tristeza virus, while Volkamer lemon, Morton citrange, *C. amblycarpa*, Swingle citrumelo and Carrizo citrange were reported to be tolerant to tristeza (Georgiou and Gregoriou, 1999).

Ziyang Xiangcheng (*C. junos*) could thrive well on soils having pH of 4.0-8.5 without leaf yellowing. The scion cultivars Satsuma mandarin, Jincheng orange, Navel orange and pummelo grafted on Ziyang Xiangcheng performed very well (Ma, 1998). Rootstock also exerts some influences on nutrient element composition of leaves. It has been observed that the variation in micronutrients is greater than those in macronutrients. Leaf nutrient concentrations of Xianfeng orange trees were highest on Ichang papeda (*C. ichangensis*), *P. trifoliata* and mandarin rootstocks (Zhou *et al.*, 1998).

Variability of sweet orange grafted onto *Feronia limonia* (*Limonia acidissima*) rootstock, with special emphasis on fruit quality characteristics was studied by Sirisena (1998). The rootstock seedlings were sorted into 3 groups based on length of spines. Growth rate and intensity of seedling pigmentation were higher in the longer spine rootstock group. Magnitude of growth difference at the bud-union, canopy size, fruit weight and cumulative fruit yield/year of the grafted plants were higher when long spine rootstocks were used. However, the same group of rootstocks produced fruits with very poor skin quality, juice with low citric acid and low Brix content. Improved citric acid, Brix contents and fruit skin quality were observed in the orange plants grafted onto rootstock seedlings with short spines.

Twenty-one selections consisting of 13 numbered hybrids and 7 named cultivars were tested as rootstocks for Valencia orange on sandy soil of the Florida peninsula, USA. Trees on Vangasay lemon, HRS 812 (*Citrus sunki* x Benecke trifoliate orange) and HRS 942 (*C. sunki* x Flying Dragon trifoliate orange) produced the highest cumulative yields in 1995-1998 (317.3, 305.7 and 290.9 kg/tree respectively). Trees on 10 rootstocks, including the widely used commercial rootstocks, Swingle citrumelo and Carrizo citrange, were intermediate in cumulative fruit production. Trees on

5 rootstocks, including Sun Chu Sha, Gou Tou1 and Tachibana had low yields and trees on HRS 939 (Flying Dragon trifoliate orange x Nakorn pummelo) and Sour orange 2 were extremely dwarfed and were minimally productive because of tristeza virus disease. Fruits from trees on HRS 954 and HRS 952 (both of which are Pearl tangelo x Flying Dragon trifoliate orange) had the highest total soluble solids (TSS) concentration and fruits from trees on Vangasay and Gou Tou 1 had the lowest TSS concentration (Wutscher and Bowman, 1999).

7.0 Area and Production

Citrus is the world's leading tree-fruit crop. World production in 2016 was to a tune of 124 million metric tonnes (https://en.wikipedia.org/wiki/Citrus_production). Almost four-fold increase in production was found in USA, Brazil, Japan and Spain. The United States are the world's largest producer (with the three major producing areas: Florida, Southwestern US and Texas). Brazil also increased production by 7.4 per cent annually (Passos and Salibe, 1978). Substantial increase in Satsuma cultivation was also observed since 1950. In 2018, citrus fruit production for World was 152 million tonnes. Citrus fruit production of World increased from 38.5 million tonnes in 1969 to 152 million tonnes in 2018 growing at an average annual rate of 2.91 per cent. https://knoema.com/atlas/World/topics/Agriculture/Crops-Production-Quantity-tonnes/Citrus-fruit-production).

In India, citrus is grown in 3.89 lakhs ha with a production of 44.3 lakhs tonnes (Anon, 1996). The most important commercial citrus types in India are the mandarin orange (*Citrus reticulata*), followed by sweet orange (*Citrus sinensis*) and acid lime (*Citrus aurantifolia*) with a total production of 13.4, 8.9 and 7.5 lakh tonnes respectively. Commonly grown commercial cultivars of mandarin are Coorg, Nagpur, Khasi and Darjeeling which have ecological difference but are akin to each other (Chadha and Singh, 1989). The states of Andhra Pradesh, Maharashtra, Karnataka and Punjab only contribute more than 65 per cent towards total production (Anon, 1994a).

The area, production and productivity trend under citrus in different citrus growing states (Table 5) provides an idea about the extent to which the cultivation is successful and sustainable. In terms of area, Maharashtra tops the list (73,115 ha) followed by Andhra Pradesh (56,166 ha), Punjab (37,464 ha), Karnataka (35,406 ha), Uttar Pradesh (22,320 ha) and Bihar (17,317 ha). But the perusal on productivity trend gives an entirely different scenario. Higher productivity is obtained in Tamil Nadu (17.6 tons/ha) followed by Madhya Pradesh and Gujarat (16.0 tons/ha), Andhra Pradesh (15.0 tons/ha) and Bihar (10.0 tons/ha). During the last 35 years, the Indian citrus industry attained a quantum expansion in area from 0.91 lakhs ha in 1961 to 3.89 lakhs ha in 1996, while production increased from 8.24 to 44.3 lakhs tonnes. The indigenous species are commonly grown in homesteads and introduced species/cultivars, which have adopted in different geographical regions are the basis of commercial plantations (Chadha and Singh, 1989). The soil and climate are the two important determinants deciding the success or failure of any citrus cultivar in a given location. There are distinct citrus belts having agro-ecological preferences. The differentiating reasons for such preferences are not known and warrant thorough investigation.

The agro-ecological regions of Deccan (Telengana) plateau and eastern Ghats in hot semi-arid climate with red and black soils having growing period of 90 to 150 days were observed as the most productive belt covering an area of 56,166 ha with a total citrus production of 842.5 thousand tonnes. Whereas, the eco-regions of northern and eastern plains under hot sub-humid (dry and moist) alluvial-derived soils with a growing period of 150 to 210 days spreading to an area of 17,317 ha producing 173.2 thousand tonnes were noted as least productive belt. The cultivar wise potentially identified suitable areas for future commercial cultivation suggested that sweet orange offers a maximum scope of area expansion followed by Kinnow mandarin, Nagpur mandarin and Khasi mandarin in the order (Srivastava and Kohli, 1997a).

Table 5: Total World Area and Production during 2018*

Country	*Element*	*Unit*	*Value*
Afghanistan	Area harvested	Ha	249
Afghanistan	Production	tonnes	2802
Algeria	Area harvested	Ha	476
Algeria	Production	tonnes	627
American Samoa	Production	tonnes	32
Angola	Area harvested	ha	30351
Angola	Production	tonnes	311450
Australia	Area harvested	ha	152
Australia	Production	tonnes	1538
Azerbaijan	Area harvested	ha	919
Azerbaijan	Production	tonnes	4418
Bhutan	Area harvested	ha	2036
Bhutan	Production	tonnes	14498
Cameroon	Production	tonnes	78
Cayman Islands	Production	tonnes	59
China, mainland	Area harvested	ha	179653
China, mainland	Production	tonnes	5793396
China, Taiwan Province of	Area harvested	ha	4261
China, Taiwan Province of	Production	tonnes	71495
Colombia	Area harvested	ha	36309
Colombia	Production	tonnes	602871
CÃ´te d'Ivoire	Area harvested	ha	3458
CÃ´te d'Ivoire	Production	tonnes	43930
Cuba	Area harvested	ha	364
Cuba	Production	tonnes	1891
Cyprus	Area harvested	ha	31

Country	Element	Unit	Value
Cyprus	Production	tonnes	323
Ecuador	Area harvested	ha	6931
Ecuador	Production	tonnes	16678
Egypt	Area harvested	ha	432
Egypt	Production	tonnes	5884
Eswatini	Area harvested	ha	43
Eswatini	Production	tonnes	232
Ethiopia	Area harvested	ha	377
Ethiopia	Production	tonnes	2825
French Polynesia	Area harvested	ha	29
French Polynesia	Production	tonnes	574
Gabon	Area harvested	ha	242
Gabon	Production	tonnes	1429
Greece	Area harvested	ha	151
Greece	Production	tonnes	845
Grenada	Area harvested	ha	175
Grenada	Production	tonnes	1252
Guam	Production	tonnes	113
Guatemala	Area harvested	ha	13930
Guatemala	Production	tonnes	75006
Guinea	Area harvested	ha	45118
Guinea	Production	tonnes	242449
Guyana	Area harvested	ha	322
Guyana	Production	tonnes	778
Honduras	Area harvested	ha	356
Honduras	Production	tonnes	887
India	Area harvested	ha	61025
India	Production	tonnes	773250
Iran (Islamic Republic of)	Area harvested	ha	27700
Iran (Islamic Republic of)	Production	tonnes	616864
Israel	Area harvested	ha	3489
Israel	Production	tonnes	22870
Italy	Area harvested	ha	1469
Italy	Production	tonnes	28641
Japan	Area harvested	ha	3806
Japan	Production	tonnes	69030
Jordan	Area harvested	ha	114
Jordan	Production	tonnes	722
Kazakhstan	Area harvested	ha	51
Kazakhstan	Production	tonnes	806

Country	Element	Unit	Value
Kenya	Area harvested	ha	12363
Kenya	Production	tonnes	84533
Kyrgyzstan	Area harvested	ha	4
Kyrgyzstan	Production	tonnes	89
Malawi	Area harvested	ha	433
Malawi	Production	tonnes	2662
Malaysia	Area harvested	ha	734
Malaysia	Production	tonnes	3432
Malta	Area harvested	ha	54
Malta	Production	tonnes	417
Martinique	Area harvested	ha	25
Martinique	Production	tonnes	236
Mauritius	Area harvested	ha	224
Mauritius	Production	tonnes	670
Mexico	Area harvested	ha	39888
Mexico	Production	tonnes	184277
Micronesia (Federated States of)	Area harvested	ha	
Micronesia (Federated States of)	Production	tonnes	
Morocco	Area harvested	ha	2059
Morocco	Production	tonnes	18241
Nepal	Area harvested	ha	671
Nepal	Production	tonnes	5752
New Zealand	Area harvested	ha	980
New Zealand	Production	tonnes	8476
Nicaragua	Area harvested	ha	191
Nicaragua	Production	tonnes	576
Nigeria	Area harvested	ha	818805
Nigeria	Production	tonnes	4071176
Palestine	Area harvested	ha	194
Palestine	Production	tonnes	1338
Peru	Area harvested	ha	3840
Peru	Production	tonnes	69030
Philippines	Area harvested	ha	19778
Philippines	Production	tonnes	113552
Qatar	Area harvested	ha	37
Qatar	Production	tonnes	166
Republic of Korea	Area harvested	ha	187
Republic of Korea	Production	tonnes	1100
Samoa	Area harvested	ha	148
Samoa	Production	tonnes	711

Country	*Element*	*Unit*	*Value*
Saudi Arabia	Area harvested	ha	4487
Saudi Arabia	Production	tonnes	40428
Sierra Leone	Area harvested	ha	19618
Sierra Leone	Production	tonnes	111613
South Africa	Area harvested	ha	281
South Africa	Production	tonnes	11662
Spain	Area harvested	ha	1127
Spain	Production	tonnes	1709
Suriname	Area harvested	ha	288
Suriname	Production	tonnes	3808
Syrian Arab Republic	Area harvested	ha	7121
Syrian Arab Republic	Production	tonnes	195876
Thailand	Area harvested	ha	2998
Thailand	Production	tonnes	14088
Trinidad and Tobago	Area harvested	ha	621
Trinidad and Tobago	Production	tonnes	6899
Tunisia	Area harvested	ha	13461
Tunisia	Production	tonnes	143098
Turkey	Area harvested	ha	185
Turkey	Production	tonnes	2052
United Arab Emirates	Area harvested	ha	140
United Arab Emirates	Production	tonnes	2806
United Republic of Tanzania	Area harvested	ha	16842
United Republic of Tanzania	Production	tonnes	46122
United States of America	Area harvested	ha	1312
United States of America	Production	tonnes	28514
Uzbekistan	Area harvested	ha	720
Uzbekistan	Production	tonnes	11863
Wallis and Futuna Islands	Production	tonnes	48
Yemen	Area harvested	ha	35
Yemen	Production	tonnes	275
Zimbabwe	Area harvested	ha	122
Zimbabwe	Production	tonnes	564

* FAOSTAT.

An agro-ecological approach which combines the effect of physiography, soil, climate and growing period (Khanna, 1989; Ghosh, 1991; Sehgal *et al.*, 1990) will be very helpful in identifying the agro-ecological preferences for different citrus cultivars grown in the country. The suitability of soil and climate may be concurrently earmarked for exploring the newer potential areas so that the acreage under citrus cultivation can be scientifically expanded without facing early citrus decline problem due to unsuitable soil and climate. To attain this objective, the perspective land use planning is highly imperative to be prepared for major citrus cultivars being grown in the country. The data on area and production of citrus compiled in 1994 are given in Table 6.

Table 6: Area, Production and Productivity along with per cent Share Towards Total Citrus Production in different Citrus Growing States in India*

State	Area (000'ha)	Production (000'tonnes)	Productivity (tonnes/ha)	Per cent Share
North-west India				
Gujarat	8.0	128.0	16.0	4.30
Haryana	3.4	27.4	8.1	0.92
Himachal Pradesh	11.5	9.3	0.8	0.31
Punjab	37.5	306.1	8.2	10.27
Rajasthan	9.1	36.1	4.0	1.22
North-east India				
Arunachal Pradesh	5.5	9.0	1.6	0.30
Assam	12.0	85.0	4.5	2.86
Manipur	2.2	9.0	4.0	0.30
Meghalaya	7.1	48.2	6.8	1.62
Mizoram	5.8	24.1	4.11	0.81
Nagaland	1.4	1.8	1.2	0.06
Sikkim	6.3	16.6	2.6	0.56
East India				
Tripura	12.7	44.5	3.5	1.49
Bihar	17.3	173.1	10.0	5.81
Orissa	12.8	60.0	4.6	2.01
North India				
Uttar Pradesh	22.3	60.1	4.8	2.02
Central India				
Madhya Pradesh	9.54	152.7	16.0	5.13
Maharashtra	73.1	418.7	5.7	14.06
South India				
Andhra Pradesh	56.2	842.5	15.0	28.28
Karnataka	35.4	318.6	9.0	10.70
Tamil Nadu	9.6	17.04	17.6	5.72
Total	**311.1**	**2941.6**	**148.0**	

* Information Bulletin (1994), National Horticulture Board, Govt. of India.

The mandarins (Santras) occupy the largest area, followed by sweet oranges, limes and lemons. The major mandarin growing areas in India are Coorg of

Karnataka State; Nagpur belt of Maharashtra; Darjeeling district of West Bengal; East and West districts of Sikkim; Kamrup, Dibrugarh and Lakhimpur districts of Assam; Khasi, Jaintia and Garo hills of Meghalaya; Jampui hills of Tripura; Tamenlong, Tipaimukh and Jiribam of Manipur; Mokokchung and Tuensang districts of Nagaland and certain parts of Mizoram and the foothills of Himachal Pradesh. A strain commonly known as Desi Santra is also grown to a certain extent in this area. In the Coorg and Wayanad tracts of Karnataka, the cultivar is known as Coorg orange, in the Nagpur belt it is known as Nagpur Santra, in Darjeeling it is called Darjeeling orange; and in Sikkim the cultivar in vogue is known as Sikkim orange. In Assam, the mandarin cultivar grown is known as Sumthira, while in Meghalaya it is known as Khasi mandarin or Soh Niamtra.

8.0 Soil and Climate

8.1.0 Soil

Citrus can grow well in a wide range of soils, but unsuitability of the soil could also lead to failure of the crop. Soil properties like soil reaction, soil fertility, drainage, free lime and salt concentrations, *etc.*, are some of the important factors that determine the success of citrus cultivation. It thrives well in deep, loose, well-aerated soils devoid of any hard pan layers of calcium carbonate in the root zone. As citrus roots are lacking in fine root hairs, humus should be added for soil improvement (Neumann, 1965). The productivity of orange and grapefruit on sour orange rootstocks in Israel, were found to increase with up to 9-10 per cent air space at field capacity in the 25-75 cm soil layer (Platt, 1966). Ford (1963) reported that in Florida, trees planted on 17 inches deep organic soil grew better than on 8 inches deep soil.

The citrus soils differ from other cultivated soils in many respects. The cultivated soils remain fallow for 3–6 months every year, which results in depletion of soil organic matter as very little carbon (C) is added during the fallow phase, while biological oxidation of existing C continues at the same rate as in citrus soils (Sharma and Singh, 2001). Taxonomically, highest quantum of citrus production from citrus is harnessed from the soils represented by soil orders Entisols, Alfisols, Ultisols, and Oxisols (Srivastava and Singh, 2009).

Ideal pH for citrus is considered to be between 5.5 and 7.5 but evidences show that with suitable management it can be grown with success in highly acidic (pH 4.5) soil and those containing free lime (pH 8.5). In California, a pH value is maintained between 5.5 and 7.0. Le Roux (1965) suggested the optimal pH value of 6.5 to 7.0 for South Africa. Citrus is sensitive to soil factors such as high pH, presence of excessive salts in the root zone, inefficient drainage, presence of hard pan in sub-soil (calcium carbonate concentrations), free lime, deteriorated soil structure *etc.* (Kanwar and Randhawa, 1960) and an integrated effect of these soil factors lead to chlorosis induced decline in productivity (Dhingra *et al.*, 1965). Citrus trees are susceptible to salt injury and they cannot thrive well in saline-alkaline soil. Lime induces chlorosis and presence of excessive free lime renders phosphorus, iron, manganese and zinc less available. In Kinnow growing areas, where the soil pH is beyond 8.5, either regular application of gypsum or heavy use of organic manure or both is used as soil amendment on regular basis. The presence of hard pan in substrata of Kinnow growing soils is a common phenomenon. In such soils provision of drainage has to be kept mandatory. Hamlin orange on *Carrizo citrange* rootstock showed good growth and higher yield in sandy Alfisol in a pH range of 4.6-8.0. Below pH 4.6, tree size and yield depressed (Obreza, 1993).

The deteriorating condition of citrus orchards in India had been attributed mainly to poor physical condition of soil and unbalanced nutrition (Bandale *et al.*, 1951; Dhingra and Kanwar, 1963). A survey of soils in the citrus belt of Punjab by Kanwar *et al.* (1965) showed that out of 9317 acres of planted area, 50 per cent was suitable, 12 per cent marginally suitable and another 37 per cent entirely unsuitable for citrus plantation. Therefore, before demarcating an area for citrus plantation, it is important to establish parameters to evaluate the possible constraints and its suitability.

Citrus growing soils vary widely in their texture, structure, available water capacity, sub-soil properties and soil fertility status. Kakde (1956) observed that performance of orange orchards was directly related to physical characteristics and chemical composition of soils. Studies on soil physicochemical analysis of citrus belt of Punjab revealed that soils from all the series were light textured with loam having low water retention capacity and required frequent irrigation. The pH ranged from 8.1 to 8.5 and EC in the root zone of soil supporting healthy citrus plants was 0.4 ds/m and any value above it might lead to chlorosis (Dhingra *et al.*, 1965). On the basis of survey conducted by Kanwar and Randhawa (1960), it was suggested that soil pH should be below 8.5 for successful cultivation of citrus. Nilangekar and Patil (1982) observed poor growth of orchard due to presence of high clay, silt, $CaCO_3$ in sub-soil, low organic carbon, available nitrogen, phosphorus, low exchangeable Mg, high $CaCO_3$, high salinity and ESP in the surface. Chapman (1961) reported ideal soil for citrus as deep, uniform, loamy, well drained, moderately level soil, free of lime and slightly acidic, but such ideal soils are not frequently encountered. In a review of works done on the soil suitability criteria for citrus, Srivastava and Kohli (1997b) revealed that a minimum of 2 feet soil depth, coarse texture with heavier sub-surface, well drained, water table below 6 feet throughout the growing period and free lime not more than 10 per cent were found as suitable parameters for any of the commercial cultivars grown in India. Singh *et al.* (1967) considered that ideal soil for citrus would be medium light or light loam with slightly heavier sub-soil. Chakraborty and Barua (1983) observed that the soils of citrus growing belts of hill districts of Assam were very deep, well structured and the texture varied from sandy clay loam to silt clay in surface horizon. While studying the soils of healthy and declining trees in sweet orange orchards of Punjab, Mann *et al.* (1978) reported that there was not much difference between physico-chemical characteristics under both chlorotic and healthy orchards. However, none of the characteristics except EC and $CaCO_3$ had bearing on growth of Kinnow trees (Brar *et al.*, 1986). Kanwar and Randhawa (1960) observed soil EC below 0.5 ds/m as safe limit for citrus. Nunez Moreno and Veldez-Gascon (1994) observed a wide difference in soil properties of low and high yielding orchards of north-western Mexico. The average value of soil properties in high yielding orchards were: EC 3.8 ds/m, Na 12 me/l, Mg 3.8

me/l and ESP 7.2 while in high yielding orchards EC, Na, Ca, Mg and ESP were observed as 1.1 ds/m, 6.0, 4.0, 1.2 me/l and 4.6 respectively. The soils having clay content more than 60 per cent beyond 30 cm soil depth were observed unsuitable for cultivation of Nagpur mandarin under hot sub-humid tropical climate of Central India (Dass *et al.*, 1998).

High concentrations of salts (*i.e.* sodium and/or chloride) in the root zone appear to be harmful. Experimental reports indicated a considerable reduction of stomatal conductance and photosynthetic rate and assimilation in citrus trees with high foliar concentrations of sodium and chloride (Lloyd *et al.*, 1987; Nastou *et al.*, 1999). El-Desouky and Atawia (1998) observed that seed germination percentages decreased and the number of days required for germination increased with the increasing level of salinity (from 1000 to 6000 ppm) in different citrus species. The maximum tolerable salinity level was determined to be 4000 ppm. The dry matter distribution in different plant organs and the synthesis of photosynthetic pigments were altered by salinity. Nastou *et al.* (1999) recorded the foliar symptoms of salt toxicity as marginal necrosis with subsequent necrosis of older leaves. Salinity induced thickening of the leaf blade was due to an increase in the number and volume of mesophyll cells, while the volume of intercellular spaces decreased. The entire gamut of citrus nutrition remains an unfinished exercise unless dealt with the issues like assessing soil salinity and aluminium toxicity on soluble salt rich (high pH) and divalent bases deprived (low pH) soils, respectively, considering the extreme sensitivity of citrus under both the soil conditions (Srivastava, 2014).

Citrus orchards in Egypt have been laid out on desert sand (Brown, 1953). Darcel (1952) in characterizing citrus soils in Florida suggested not to grow citrus in heavy, clayey, extremely dry and moist soils with impermeable underground strata retaining stagnant water. In India, while the major mandarin orange growing belt of north-eastern hills comes under acidic soils of sandy to clay loam in nature, the other very famous mandarin growing belt of Nagpur region comes under non-acidic and heavy black soils. The citrus soil of Punjab is light with a pH range of 6.4 to 9.0. It appears that loamy soil with comparatively heavier subsoil or even heavy soil with good drainage arrangement can be ideal for citrus.

Citrus trees do not like water-logged condition. Low oxygen tensions, as a result of flooding, caused marked reduction of stomatal conductance and growth of sour orange seedlings after only 4 days, whereas rough lemon seedlings did not respond until after 8 days of flowering (Syvertsen *et al.*, 1983). Physiological responses to flooding might be operative through reduced hydraulic conductivity of flooded roots (Syvertsen and Lloyd, 1994).

8.2.0 Climate

Citrus belongs to the tender evergreen subtropical group and thrives well in frost-free subtropical to semi-tropical climate. However, occasional light frost can be tolerated by most of the citrus species. Being evergreen, citrus has no specific requirement for winter chilling, but cessation of growth activity during winter months helps in flower bud induction, resulting in spring flowering. Different citrus species exhibit a range of tolerance to temperature fluctuations, and the total heat during the growing season is extremely important for normal growth and productivity of various citrus species and cultivars. Climate effect on citrus Fruit quality is basically product of rainfall and accumulation of heat (AH) from flowering to harvest (Aular *et al.*, 2017). Citrus can be grown in different climatic zones ranging from tropical, subtropical, arid and semi-arid areas Abobatta (2019) wrote a detailed review on potential impacts of global climate change on citrus cultivation. According to him, citrus grows properly in the range from 12.8 to 37ÚC, which considered the optimum temperature for citrus growth and fruiting, while, sever high temperature (above 44-45ÚC) stopped citrus growth completely. Low temperature considered the limiting factor for the geographical distribution of citrus, low temperature could stoop metabolism activity, whereas, chilling and frost cause severe injuries and demolish the whole tree. Temperature fluctuation affects negatively growth, decreases total yield, and reduces fruit quality particularly when occurred during the maturity stage, also, there are negative effects of heat waves on the production of different citrus varieties particularly seedless varieties like Navel orange, and some Mandarin and lemon cultivars. On another side, rising carbon dioxide has positive effects on the growth of citrus seedlings and trees productivity.

Studies on the effect of AH on 'Valência' and 'Natal' orange trees FQ indicated that the highest AH were associated with the lowest TA, SS and SS/TA ratio values (Volpe *et al.*, 2002); while evaluating FQ of 'Shogun' tangerine observed that higher mean temperatures, lower precipitation and low soil moisture where negatively affected to fruit development and quality (Chelong and Sdoodee, 2013). The climatic factors like temperature, moisture (rainfall and atmospheric humidity), wind and light intensity are of principal importance for citrus, of which temperature plays a key role. Usually, a low temperature of 20° - 24°F (-6.6°C – 4.4°C) is considered to be injurious to youg trees, while mature old trees are killed at a temperature of about 12° – 16°F (-11.1° to -8.8°C). While the kumquats, mandarins, *C. latipes*, *C. ichangensis* can tolerate low temperature, most of the citrons, lemons and limes are least resistant to cold injury. Grapefruits are most resistant to high temperature and can thrive well in even up to 120°F (48.8°C), citrons and lemons start showing injury at 95°-100°F (35 to 37.7°C). Wutscher (1973) found that for sour orange seedlings vigorous root and shoot growth occurred at soil temperature between 23.9° and 26.7°C, whereas no growth occurred at soil temperature of 10°C. Cary (1970) has shown that 25°C is a near-optimum soil temperature for root growth. Hot winds and excessive heat during flowering and fruit set period are highly detrimental for good bearing and causes fruit drop and sun-burn of fruit. In Satsuma, high temperatures (a mean daily temperature of >25°C, and a maximum daily temperature of >28°C) at the time of flower-fall is considered to be the limiting factors associated with severe fruitlet drop (Huang *et al.*, 1993). Atmospheric humidity has bearing mainly on the physical characters of fruit whereas temperature imparts effects on fruit quality. Low humidity usually favours better colour development of fruits, while in more humid conditions fruits are more juicy with thin rind (Cooper *et al.*, 1963). Citrus is usually considered to be a light-loving plant and the trees are sensitive to shading. Trees exposed to high light intensity usually exhibit paler foliage colour, while those under shade develop deep green foliage. Trees under shade are usually poor fruit yielder but partial shading often results in higher fruit quality. Under higher mean annual temperature condition citrus fruits mature early, fruit size is bigger and the acidity development in the fruit juice remains lower. Fluctuation between day and night temperatures intensifies colour development and accentuates sugar accumulation and acid formation; whereas in the humid tropics even ripe fruit remains green (Anon, 1995). Different climatic factors influence the vegetative growth of citrus plants, their productivity and physico-chemical characteristics of the fruits. Best quality fruit grows in semi-arid, subtropical regions with less than 20 inches of rain, when orchards are irrigated. The period from flowering to ripening also varies considerably on the basis of climatic conditions of different locations. In Japan, the temperature is high in summer and rather low in winter. Thus citrus growing is limited by coldness in winter and citrus zone is restricted to places warmed in winter by the currents. Most citrus orchards are found at lower altitudes: in California at elevations below 700 m, and in Spain up to 250 m whereas in Southeast Asia, trees flourish at altitudes up to 200 m.

Although the temperature responses as observed in various experiments were somewhat variable, a critical analysis of the results suggest a slight, but significant, response of citrus stomata to temperature in the range 20° to 35°C with maximum conductance around 30°C (Syvertsen and Lloyd, 1994).

It has been observed that 90 per cent of the direct shortwave radiation on a clear day and 20 to 50 per cent of diffuse radiation on a cloudy day are absorbed in the outer 1 m of canopy depth of a mature orange tree (Green and Gerber, 1967). The morphology of citrus leaves is influenced by radiation load and leaves exposed to full sunlight are sometimes folded along the mid vein (Syvertsen and Smith, 1984) resulting in half of the lamina shading itself for long periods throughout the day (Syvertsen and Lloyd, 1994). Drought caused up to 20 per cent reduction in leaf expansion, reduced photosynthesis, and moderately reduced tree growth (Mataa *et al.*, 1998). However, the flowering response increased (Pire and Rojas, 1999) and fruit set improved (Mataa *et al.*, 1998) in different species of citrus trees with the level of drought stress. Floral shoots and, especially, the number of flowers per branch exhibited a strong relationship to drought stress (Pire and Rojas, 1999).

9.0 Cultivation

9.1.0 Planting

Usually, citrus is planted in pits of 50 x 50 x 50 cm or 75 x 75 x 75 cm sizes in square system with spacing of 5 to 8 m depending upon the species and rootstocks. For a dwarfing rootstock like trifoliate orange or a species like lime (*Citrus aurantifolia*) a spacing of 5 x 5 m is considered to be adequate. For pummelos and grapefruits and for vigorous rootstocks like rough lemons or karna khatta wider spacing is usually recommended. In the north-eastern region of India, seedling-origin Khasi mandarin orange orchards are usually very closely spaced and in the old orchards, trees are found to grow even 2-3 metres apart in an irregular way. Mandarin orange being more or less upright in growth habit, particularly, when raised from seeds can be planted at a distance of 5 to 6 metres. Naik (1949) suggested a spacing of 7 to 8 metres for budded plants of sweet orange cv. Sathgudi, lemon and pummelo. It was observed that 6 x 4.5 m spacing was optimum for Pearl

Tangelo (Arora and Yamdagni, 1985) and 6 x 6 m for Jaffa (Sarowa *et al.*, 1985) for economic crop production with respect to vegetative growth, yield and quality of fruit. In Haryana conditions, a spacing of 5 x 4.5 m was found to be optimum for Kinnow mandarin budded on Jatti Khatti (*C. jambhiri*) rootstock (Chundawat and Arora, 1981), while a closer spacing of 3 x 3 m was found to be unsuitable for Nagpur mandarin budded on rough lemons (Deshmukh *et al.*, 1981). In Australia, an average annual yield in excess of 60 tonnes/ha could be obtained from 6-year old Valencia trees at a density of 2500-5000 trees/ha. These yields were more than double than those obtained at the more conventional spacing of 667 trees/ha (Cary, 1982). In a comparison between double spacing (7 x 5 x 1 m, 7 x 4.5 x 1.5 m and 7 x 4 x 2 m) and traditional spacing (7 x 6 m), Donadio *et al.* (1999) observed that the cumulative yields were greater in all the double spacing treatments than with single spacing, while the fruit quality was not affected.

The effect of different spacings on the yield of pummelo cultivars Wanbaiyou, Liangpinyou and Shatianyou, grafted on *Poncirus trifoliata* and planted in 1972 in China), was investigated by Wang *et al.* (1999). At closer spacings, production and returns were higher. Returns from the 2 x 1.5 m plot were double that of the 2 x 3 m plot and 3-times that of 3 x 4 m plot. To obtain high yields, it was recommended that trees are originally planted at a spacing of 2 x 1.5 m then thinned by the removal of a row to 2 x 3 m after 5-6 years.

Though the planting is usually done during the monsoon season, it is better not to plant at the time of heavy rains to avoid any water logging near the planting pits. Weather should not be too wet or too dry at the time of planting. In case of lanky plants, staking should be done and at post-monsoon period, the plant bases may be covered with mulches so as to conserve moisture in the root zone. Supplementary irrigation in the early years of planting is essential during the dry months.

9.1.1 High Density Planting

High Density Planting in *Citrus* has been reviewed in detail by Chhetri and Kandel (2019) According to them, tree size and vigor are important characters from plant density view. Commercial farmer interest is increasing towards smaller, less vigorous trees which are efficient in economics and more suitable for high density planting. The factors *viz.* rootstock, spacing, tree height, pruning system are integrated by light interception. Higher the light interception by plant higher will be the yield. Closer planting, taller trees and closer row spacing increased light interception. Good light distribution also promotes the higher yield per unit area. For successful farming there should be good light interception and good light distribution. Trees planted in a north and south direction ensures better light distribution. It should be clear that not all locations and tree combinations are suitable for high density plantings. Annual pruning in recommend technique, use of dwarf rootstocks *viz.* trifoliate, Troyer Citrange, Assam lemon *etc.*, proper spacing, use of plant growth regulators (Paclobutrazol and GA_3) and nutrient management are the basic components for successful high density planting. It has been proved traditional density planting on a 20′ x 20′ spacing would take 12 years to pay back all initial development costs (considering average returns). The same variety planted on a 10′ x 20′ would pay back in eight years. Because these numbers have been proven valid, many growers today face the challenge of how to manage the high density grove they have planted. High density planting technology is gaining popularity because of earlier production and net returns, increasing land values and higher taxes of land, efficient use of nutrient and water due to greater root densities efficient pesticide application and easier weed control. According to Goswami *et al.* (2014), it is the novel technique to obtain high productivity both in short and long duration crops and the important five components of high density planting are dwarfing rootstocks and inter stocks, training and pruning, dwarf scion varieties, use of chemicals/plant growth regulators and suitable crop management practices (Goswami *et al.*, 2014).

Generally, *Citrus* trees are planted in rectangular and square methods with a dimension of 3 m x 4 m and 4 m x 4 m respectively. Later, dimension increases to 6 m x 8 m and 8 m x 8 m after 15 –20 years, respectively by thinning out of trees (Sheheilah, 2013). Light penetration and light interception are most considering factors under high density planting. High yields with better fruit quality of mandarin orange can be obtained from high density planting if orchards have proper light distribution in bearing canopy (Goswami *et al.*, 2014). Rootstock affects the tree size, quality and yield of the fruits (Tucker and Muraro, 1990). Pruning operations depends on factors like variety, age of plant, vigour, fruiting habits and growing condition). Heading back and thinning out are two types of practices in pruning operation of mandarin orange (Tucker *et al.*, 1994). Pruning adjust tree shape, balance ratio between frame work and bearing canopy, alter top/root ratio and also alter carbohydrate status of plant (Tucker *et al.*, 1994). Ghosh *et al.* (2016) reported that 25 cm pruning from terminal portion of lemon cv. Assam lemon shoot (light pruning) gives the better yield than less centimeter pruning and no pruning. If suitable variety of rootstocks is used 800 plants /acre (4046.856 m^2) can be successfully grown. In normal condition 140-300 trees are planted in one acre (Tucker and Muraro, 1990). Singh and Singh (2018) mentioned that planting of *Citrus* (kinnow) at spacing 1.8 m x1.8 m (3000 plant/hectare) gives 20 ton/hectare which is 200 per cent more yield than traditional yield. Cary (1977) reported that when 701 tree/ha was planted as high density planting in *Citrus*, it gives yield more than 60 ton/ha. A few Forner-Alcaide hybrid selections exist which also confer a dwarfing response on scions, in particular FA 517 and FA 418, whose agronomical behaviour has been tested under field conditions. Both rootstocks show lower canopy volumes, but higher yield efficiency when compared with Carrizo Citrange, the most extended *Citrus* rootstock in Spain. Moreover, they produce good fruit quality and optimal response when cultured in alkaline soils, one of the main factors that limit crops in Spanish soils (Leuga *et al.*, 2013). *Feronia limonia* was found to be a highly dwarfing and precocious *Citrus* root stock and suitable for high density planting besides Troyer Citrange as dwarf and Karna Khatta as semi dwarf rootstock for *Citrus* spp (Goswami *et al.*, 2014).

Acid lime cv. Pramalini plantation was evaluated under ultra-high density planting (UHD - 1600 trees ha^{-1} at 2.5 x 2.5 m) and high density planting (HDP - 800 trees ha^{-1} at 5 x 2.5 m) systems over conventional or control (CON - 400 trees ha^{-1} at 5 x 5 m) planting spacing (Ladaniya *et al.*, 2020). The climate of the study area was tropical, dry, sub-humid having yearly average maximum and minimum temperature 34.2°C to 20.5°C and precipitation of 1128 mm. Field soil having montmorillonitic mineralogy was 40 cm deep. Data on plant growth, soil properties, insect pest (leaf miner), disease (bacterial canker), physico-chemical parameters of fruits and fruit yield were recorded three years after planting of seedlings in the experimental field. It was found that under both UHD and HDP, EC gradually decreased over the period of five years. The highest organic carbon (1.13-1.42 per cent) and major nutrients were observed under UHD and leaf N and K contents in HDP. Dense and vigorous plant canopy with maximum (1.49) leaf area index was found in HDP and lowest was under UHD (1.18). Although insect – pest and disease incidence was higher in close spacing plantation than CON, physico-chemical fruit parameters were not affected by spacing. The fruit yield on per plant basis was always significantly lowest under UHD. However, on area basis, under UHD plantation fruit yield (8.24–35.36 t ha^{-1}) was more than two folds as compared to control (3.08-11.64 t ha^{-1}) for different years. Although during last year of experiment the yield under ultra-high density planting (35.36 t ha^{-1}) was significantly higher than high density planting (25.09 t ha^{-1}), the B:C ratio was nearly similar in both the cases. Therefore, it may be concluded that the high density planting system may be preferred during early years of production.

9.2.0 Irrigation

Irrigation is of vital importance of citrus orchards and it is considered to be one of the most critical cultural operations. Water scarcity is one of the major causes of low productivity and decline of citrus orchards. Deficit irrigation (DI) is a recently proposed water saving technique in irrigated agriculture. The impact DI versus full irrigation (FI: 100 per cent crop water requirement) was evaluated in citrus orchards under a hot sub-humid climate of central India (Panigrahi and Srivastava, 2016). Two DI strategies applied to citrus trees were DI1: 20 per cent FI during initial fruit growth period (IFGP) + 40 per cent FI during final fruit growth period (FFGP) + FI during rest of the period, and DI2: 70 per cent FI during entire irrigation season. Fully irrigated trees had the highest vegetative growth. However, DI1 produced 18 per cent higher fruit yield with superior quality fruits, resulting 30 per cent improvement in water productivity under this treatment compared to FI. Fruit yield prediction based on vegetative growth and leaf physiological parameters of the trees using principal component regression analysis technique was found reasonable accurate. These results suggest for adoption of DI1 in citrus orchards of central India and elsewhere having similar agro-climate of this study region (Panigrahi and Srivastava, 2016). Since soil acts as the reservoir of water and since the water holding capacity differs considerably among different soil types, the irrigation requirement also varies greatly from place to place. The coarse-textured sand or sandy loam soils store less water and hence in such soils more frequent irrigation is required. In fine-textured clayey or silt soils, on the other hand, more water is required to wet but the frequency of irrigation should be less. In addition to soil type, irrigation requirement will depend on the species, rootstock, age, bearing capacity, topography, water table position, rainfall, atmospheric humidity, temperature *etc.* Shamouti oranges were found to produce a yield of 47-66 tonnes/ha with a minimum of 500-600 mm irrigation on the coastal plain and 78 tonnes/ha in Nagev with 800-860 mm (Shalhevel *et al.*, 1976).

9.2.1 Irrigation Management

Faulty irrigation management is one of the important components for low citrus productivity. The large scale drying of citrus is mainly due to scarce water resources, frequent drought and lowered water table in mandarin growing areas of Vidarbha and Central India (Dass, 1989). Water management using trickle irrigation system, proper scheduling coupled with moisture conservation technique impart greater dry matter production per unit volume of water used. The irrigation requirement of Shamouti orange on sour orange rootstock ranged between 600 and 660 mm in 7-10 irrigations applied at intervals of 21 to 30 days and on sweet orange rootstock as 500-550 mm water annually at intervals of 20 to 26 days for optimum growth with 50 t/ha production (Shalhevel *et al.*, 1976). Sharma *et al.* (1978) reported the possibility of using free proline and relative water content as irrigation indices in citrus. The water requirement of Nagpur mandarin was determined as 651.9, 849.0 and 997.3 mm/year for young, middle age and matures trees, respectively, using the predicted potential evapotranspiration based on agro-meteorological data (Ghadekar *et al.*, 1989). Water requirement of 1 to 3 and 5 to 8 years' old Nagpur mandarin over nine months (October-June) was observed to increase according to age and canopy size. The maximum water use efficiency was observed with irrigation at 100 to 130 cumulative pan evaporation in Kinnow mandarin (Sharma *et al.*, 1987). Makhija *et al.* (1986) observed that the water need for 6-year-old Kinnow mandarin varied from 539 to 1276 mm depending upon the level of irrigation with average consumptive use of water in two years as 615 mm. A number of comparative studies were taken up in the past to find out the most effective index for scheduling irrigation. The irrigation at 20 kPa tensiometer reading was best as compared to calendar schedule (Stolzy *et al.*, 1963). Ray and Sharma (1990) found that irrigation at -0.05 kPa produced maximum biomass. Klein (1983) concluded that 12 to 23 per cent water could be observed by using soil water potentials by comparing irrigation scheduling through drip irrigation system. The growth of Valencia oranges slowed down at 32 cm and 50-60 cm soil suctions at 30 cm depth in light and medium textured soil, respectively. Smajstrla *et al.* (1986) found that young Valencia orange tree growth was greatest when irrigations were scheduled at 20 cb for no grass and 40 cb for the grass treatments. Ghosh *et al.* (1985) recommended four irrigations at the rate of 60 l/tree at fortnightly intervals during the months of April and May as optimum for economic yield of sweet lime raised in alluvial wasteland of Dehradun Valley. With trickle irrigation system, the applied water is transferred directly to root zone of plants and 20 to 50 per cent water saving has been recorded depending on soil and climatic conditions under drip irrigation (Fritz, 1970). Faton (1970) obtained better plant growth, greater yield, reduced weed growth, less evaporation and less leaching with 16 gallon water applied through drip irrigation to each 4-year-old lime tree at two weeks intervals compared to 320 gallon water under flood irrigation. Better tree growth and less water consumption were recorded in Valencia orange under drip irrigation in sandy soil (Rodney *et al.*,

1977), whereas increased tree growth and yield were recorded in young Valencia orange under drip irrigation method with emitter placed at a distance of 1 meter from the trunk (Azzena *et al.*, 1988).

Smajstrla *et al.* (1984) obtained the highest yields using spray-jet trickle irrigation systems which irrigated the largest fraction of the tree root zone. Yield increase ranged from 39 per cent for drip irrigation to 64 per cent for 2-spray jets per tree. Tashbekov *et al.* (1986) observed that drip irrigation and under-tree sprinkling produced the highest yields with the smallest water requirement. The application rate for drip irrigation of 4-year-old lemon trees was 7400 cubic m/ha annually. Cevik (1987) compared the amount of water applied per tree as over sprinkling and drip irrigation at 22.01 and 10.3 cubic m/season and found the best effect of drip irrigation on growth. Internal fruit quality of Mihowase Satsuma (*Citrus unshiu*) was marginally acceptable when irrigation was not carefully controlled. The deficit irrigation, to a soil water tension of -0 kPa at 60 cm depth, initiated at (i) 0-2 weeks after physiological fruit drop (APFD) in a sandy loam soil or (ii) 0-4 weeks APFD in clay loam soil, significantly increased total soluble solids (TSS) compared with normal irrigation level, when soil water tension reached –30 kPa at 60 cm depth (Peng and Rabe, 1998). Trickle irrigation with two laterals (supplying 3.6 l/h) produced greater tree growth and yields of Mineola tangelo than trickle irrigation with one laterals (supplying 2.6 l/h) or microsprinkler irrigation supplying 70 l/h (Dasberg and Erner, 1997). Trickle and spray irrigation systems were compared by Intrigliolo *et al.* (2000). Trees were sprinkler irrigated at intervals of 7 and 21 days, or irrigated from drip lines with 3 and 5 drippers per tree or from ring drip systems with 3 and 5 drippers per tree. Irrigation volumes were calculated by class A pan evaporation and crop coefficients. Results showed that the least water was consumed and yields were highest with the ring system with 5 drippers per tree. Water consumption was highest with sprinkler irrigation every 21 days. Yields were lowest with line irrigation with 3 drippers per tree. Trickle systems saved 17 per cent of water compared with sprinkler systems.

Huang *et al.* (1997) suggested the irrigation schedule for growing mandarin hybrid Tangkan and *C. reticulata* cv. Ponkan in Taiwan. They observed that during March-June, soil moisture promoted spring flush and increased canopy size. During the fruit development stage, soil moisture was required to increase fruit size, but during the later stages of fruit maturation too much soil moisture reduced the accumulation of sugars and increased juice acidity. It was suggested that irrigation was required to maintain soil moisture tension below 30 cbar at 30 cm soil depth in March-June, if rainfall was lacking; irrigation was required to keep soil moisture tension between 40 and 60 cbar at 30 cm soil depth during June-October (fruit development); irrigation should be controlled to keep soil moisture tension above 80 cbar after October, and irrigation should be stopped during the last month before harvest to avoid a decline in sugar content and an increase in juice acidity.

Drip irrigation with or without Tyvek mulching decreased fruit's total acidity and thereby tended to increase the TSS-acid ratio of Mihowase Satsuma fruits (Peng and Rabe, 1998a). Experimental results indicated that 780 mm of water was sufficient in a normal year for optimum yield and tree growth in Washington Navel orange which was corresponding to 56 per cent of class A pan evaporation. Neither tree growth nor yield was affected by the number of drippers/tree (Eliades, 1998). In Egypt, a photovoltaic driven low pressure drip irrigation system has been designed to combat the water and energy crisis. A 530 W solar generator was installed to operate the drip system. The efficiency of conversion from solar to hydraulic energy was 3 per cent. By matching the solar generator to the load profile of the pump and by regular cleaning of the modules from airborne dust the efficiency of the system could be increased considerably (Muller *et al.*, 1998).

In deciding irrigation application in citrus, the following points need to be considered:

(i) About 60-80 per cent of root activity in citrus is confined to first 60 cm top soil and citrus roots spread far beyond the vegetative spread of the trees. The zone of water application is important and required to be standardized for irrigation economy. Experimental evidence showed that there was no significant yield difference in treatments between wetting of 80 and 40 per cent rooting zones. Neither too much wetting of total root zone, nor wetting of a small percentage of root zone is desirable.

(ii) During the periods of rapid growth, flowering and fruit set, the citrus trees are sensitive to moisture stress. The fruit size usually gets reduced if moisture deficiency occurs frequently when developing fruits are on the tree. During spring and summer months, soils may be allowed to dry out somewhat but wilting should be avoided during bloom and fruit setting period. Stress during the early fruit drop period (April-June in the northern and October-December in the southern hemisphere) causes excessive fruit abscission (Mantell, 1977; Bielorai, 1978). Goell (1986) reported that irrigation treatment influenced both early vegetative growth and yield of grapefruit and trees receiving irrigation at 20 days interval had a higher fruit density (numbers/unit canopy volume).

(iii) Citrus trees are highly sensitive to excess moisture and water logging conditions. They require good soil aeration, and excessive irrigation may result in poor soil aeration leading to reduced growth and yield, predisposing the trees to root-rot and other diseases. Excess irrigation also results in the leaching of essential nutrients from soil as well as in accumulation of certain undesirable salts if the irrigation water is not of good quality. Citrus trees are sensitive to salinity and the total soluble salts in the irrigation water should not exceed 1000 ppm. Over-irrigation of citrus causes decline of orchards in many cases.

(iv) While irrigating, care should be taken so that the irrigation water does not touch the tree trunk directly and a dry area needs to be maintained near the tree trunk. Saturation of soil beneath the canopy and direct contact of irrigation water with the trunk may predispose the tree to root-rot (*Phytophthora* sp.) and certain other diseases. Water logging, particularly, near the tree trunk can be avoided by providing mounds of earth around the tree trunk well below the bud union.

9.2.2 Time and Quantity of Irrigation

Time of irrigation can be determined by various methods. Usually through feeling soils at rooting zone and by examining the general appearance of leaves for signs of wilting, irrigation is applied. Irrigation should not be held back till usual wilting symptoms are apparent. Determination of soil moisture content of the areas occupied by roots gives a fairly accurate indication of irrigation needed. Through determinations of moisture content of field samples and after comparing them with field capacity and wilting point values, time of irrigation can be regulated properly. The so-called moisture meters or tensiometers can also be employed for the purpose. Systematic trials conducted in the USA and certain other countries indicate that maximum fruit yield can be obtained by minimizing or preventing moisture stress, particularly during the first part of growing season including fruit setting period. Certain amount of moisture stress during latter part of growing season, however, increased the soluble solids and vitamin C contents of fruit juice. Studies conducted in India (Chinappa *et al.*, 1977) indicated that water stress of 30 to 40 days' duration in December-January followed by scheduling of irrigation after flowering at 50 to 75 per cent depletion of available soil moisture in clay loam soil resulted in optimum yield and growth of Nagpur mandarin orange under hot, dry climatic conditions. From another study conducted in the same region of Maharashtra (India) with 8-year-old Nagpur mandarin showed that moderate quantity of irrigation (1500 l/tree) when applied at 120 cm from tree trunk resulted in better sized fruit and higher yield (Deole *et al.*, 1977). Singh and Singh (1981) claimed that irrigation at 30 days' intervals reduced the incidence of granulation in sweet orange cv. Valencia Late by 20 per cent.

About the method of irrigation, Singh *et al.* (1961) suggested that young trees up to 8 years might be profitably irrigated by basin system. Other irrigation methods applied are flood, furrow and sprinkler methods. The application of irrigation in right time and in right quantity is more important than the method of irrigation. Usually, under North Indian conditions weekly irrigation during March to June and fortnightly irrigation during November to February are practised. Furthermore, flood irrigation does not saturate the soil fully due to the sudden application of a large amount of water. The irrigated water will partly be lost to the groundwater due to preferential flow in macro pores making it unavailable to crops.

The agricultural management of citrus orchards is changing from flood irrigated managed orchards to drip irrigated organic managed orchards (Hondebrink *et al.*, 2017). Eastern Spain is the oldest and largest European producer of citrus, and is representative of the environmental changes triggered by innovations in orchard management. In order to determine the impact of land management on different soil quality parameters, twelve citrus orchards sites were selected with different land and irrigation management techniques. Soil samples were taken at two depths, 0-2 cm and 5-10 cm for studying soil quality parameters under the different treatments. Half of the studied orchards were organically managed and the other six were conventionally managed, and for each of these six study sites three fields were flood irrigated plots and the other three drip irrigated systems. The outcome of the studied parameters was that soil organic matter (SOM) and aggregate stability were higher for organic farms. Bulk density and pH were only significantly different for organic farms when drip irrigation was applied in comparison with flooded plots. C/N ratio did not vary significantly for the four treatments. Although there are some points of discussion, this research shows that a combination of different management decisions leads to improvement of a couple of soil quality parameters. Organic management practices were found to be beneficial for soil quality, compared to conventional management for soils with comparable textures and applied irrigation water (Hondebrink *et al.*, 2017).

9.2.3 Irrigation and Fertigation

The effect of irrigation (7 rates) and N fertilization (urea, applied at 0-537.5 g/tree in 43 applications (weekly) on orange cv. Pera yield was investigated in Brazil (Bertonha *et al.*, 1999). Irrigation was performed through a microsprinkler and controlled according to the evapotranspiration of the crops, estimated by a class A pan. Water was applied whenever the evapotranspiration reached 7.3 mm. Crop yield as affected by N and irrigation presented a quadratic relationship, and was highest (3.2 boxes or 40.8 kg/tree) following the application of 5095.6 litre of water and 345.9 g N per tree.

Scarcity of water and nutrients in orchards are the major causes of low productivity and decline of citrus in tropics. Holzapfel *et al.* (2001) investigated the effects of water and fertilizer application through DI in Thompson Navel orange in a sandy soil and concluded that drip-fertigation increased yield and quality of fruits compared to conventional irrigation and fertilizer application in the crop. Duenhas *et al.* (2005) reported that the fertigation with reduced fertilizer rates did not result in loss of productivity and fruit quality compared with that in conventional fertilization in sandy soil in Valencia orange. Moreover, they observed that fertilizer application methods had no effect on the nutritional status of the citrus tree. With a hypothesis that the optimal amount of water and nutrients application through drip irrigation (DI) could save substantial irrigation and fertilizers comparative to that under traditional basin irrigation (BI) with band placement of fertilizer (BPF), Panigrahi and Srivastava, (2017) conducted studies in citrus in vertisol of central India. Three DI regimes: I1, I2 and I3 scheduled at 50 per cent class-A pan evaporation rate (Ep), 75 per cent Ep, and 100 per cent Ep along with three fertilizer doses: F_1, F_2 and F_3 at 50 per cent recommended dose of fertilizer (RDF, N: P_2O_5: K_2O =600 g: 200 g: 100 g), 75 per cent of RDF and 100 per cent of RDF were applied to mandarin plants respectively. BI with band placement of fertilizer (BPF) at 100 per cent RDF was taken for comparison. All the drip-fertigation treatments (except I1F_1) produced higher plant growth and fruit yield, with better quality fruits compared with BI+BPF. The highest fruit yield (16.39 t ha^{-1}), water use efficiency (3.9 kg m^{-3}) and fertilizer use efficiency (87.3 kg kg^{-1}) were observed in I2F2. The concentrations of N, K and Fe in leaves were significantly higher in I2F2 compared to other treatments. The leaf photosynthesis, stomatal conductance (Gs) and transpiration were the highest in I3F3. However, leaf water use efficiency (LWUE, photosynthesis/transpiration) was the highest in I2F2. The overall results of this study demonstrated that the application of optimum quantity of water and fertilizers (I2F2) through DI could be

Cultivation Techniques

Planting of Citrus and Citrus Orchards

Irrigation Methods in Citrus

Training of Citrus

Structural Pruning of Young Trees

Growth of Citrus in Pruned Trees

Fertilizer Application and Nutrient Deficiency

Integrated Nutrient Management in Mandarin at Akola, Maharashtra

Tree View and Fruit Bearing in Treatment with 100 per cent RDF + VAM 500g + PSB 100g + *Azospirillum* 100g + *T.harzianum* 100g/plant

Rejuvenated Orchard of Mandarin at Tinsukia, Assam

Nitrogen Deficiency

Potash Deficiency

Sulphur Deficiency

a productive and water efficient option in citrus production in vertisols (Panigrahi and Srivastava, 2017).

9.2.4 Tolerance to Saline Water

Damage to citrus crop due to excessive accumulation of salts is well recognized in many parts of the world growing citrus, especially in semi-arid and arid regions where saline water is largely used for irrigation. Citrus trees are quite sensitive to excess salts (Srivastava and Ram, 2000), and tolerance to soil salinity is correlated with its ability to restrict the entry of toxic ions [Sodium (Na), chlorine (Cl), and boron (B)] into roots and onward transport to shoots. Visible symptoms of salinity include: leaf bronzing, defoliation, leaf chlorosis similar to iron induced chlorosis, small leaves, small fruits, and die-back of young twigs (Cole, 1985). The effect of irrigation with saline water on 6 citrus genotypes was evaluated by Levy *et al.* (1999) in a short-term field experiment at the Ramat ha Negev Desert Agro-Research Centre. Salinity levels ranged from 2.0 to 6.4 dS m^{-1}. Salt-tolerant *Citrus* species and *Citrus x Poncirus* hybrids were tested for their possible use as rootstocks for commercial citrus cultivars irrigated with brackish water. All tested genotypes survived the highest salinities. At all salinity levels, the best chloride excluder was Cleopatra mandarin (*Citrus reshni*), and the worst was sour orange (*C. aurantium*). Gou Tou Cheng (*C. aurantium* hybrid) and Rangpur (*C. limonia*) x Troyer citrange (*C. sinensis* x *Poncirus trifoliata*) RT803 were promising genotypes.

9.3.0 Manuring and Fertilization

Citrus is a nutrient-loving plant and about 15 elements have been known to have important roles to play for proper growth and development of citrus. In addition to the major nutrients like N, P, K, Ca, Mg and S, citrus requires micronutrients Zn, Cu, Mn, Fe, B, Mo, *etc.* Inadequate plant nutrition causes serious disorder in citrus and may eventually lead to decline of the orchards.

Extensive work has been done on citrus nutrition and the mineral nutrition of citrus has been thoroughly reviewed by Smith (1966) and Chapman (1968). Chapman *et al.* (1945) reported that about 18 tonnes of citrus fruits removed about 21 kg nitrogen, 5 kg phosphorus, 41 kg potassium, 19 kg calcium, 3.6 kg magnesium, 2.3 kg sulphur, 40 g boron, 9 g copper, 50 g iron, 13 g manganese and 13 g zinc. The figures amply demonstrate the essentiality of maintaining proper nutrient balance in the citrus orchards, so as to maintain desirable tree health and productivity. Since the nutrient availability and requirement depend on many factors like the species and cultivar, soil fertility, soil type, extent of leaching of nutrients and soil erosion, and on different climatic parameters like rainfall, temperature, *etc.* proper maintenance of fertility of orchard soils is somewhat a complex problem. Judicious fertilizer for different locations can thus be worked out only through location-based nutritional trials coupled with continuous determination of soil and leaf nutrient status.

A review of literature would indicate in Citrus that potassium has been the most evaluated macronutrient, whereas calcium and magnesium were the least. Potassium was the element with the largest influence on fruit characteristics, followed by nitrogen and phosphorus. The most consistent result related to potassium was the positive effect on fruit fresh mass. Zinc was the most evaluated micronutrient, followed by boron, which was the most influential micronutrient on quality, especially due to its positive effect on fruit fresh mass and juice percentage (Aular *et al.*, 2017).

9.3.1 Role of Major and Micronutrients

9.3.1.1 Nitrogen

Nitrogen plays a key role in growth and production of citrus. Due to its deficiency the young leaves become undersized and develop light green to yellowish green colour, general growth gets retarded and foliage becoming sparse. Trees under severe deficiency for a prolonged period fail to produce commercial crop. The most prominent symptoms of inadequate N in Valencia orange are thinness of foliage and premature leaf abscission (Smith, 1970). Adequate amount of nitrogen is required for flowering and fruit setting (Jones and Smith, 1964). Usually, soil containing less than 110 kg/acre of easily oxidizable nitrogen (N) is rated low in available nitrogen and is expected to respond to application of nitrogenous fertilizers. In Florida, annual dressing of 150 to 200 lb N/acre was found to be sufficient for maximum production of high quality orange fruits (Smith, 1959). The optimum nitrogen rate for mature Navel orange in Queensland was 900 g per tree, provided it was applied in winter (Mungomery *et al.*, 1980). Nitrogen from nitrates is more quickly available to the plants and continued application of ammonium sulphate may increase sub-soil acidity. Studies in California on heavier soils showed that ammonical N determined the growth under highly acidic conditions but not when used with lime (Parker and Jones, 1951). High N from ammonium nitrate was found to reduce the root concentration in young orchards on virgin land where the subsoil pH ranged from 4.3 to 4.6 (Ford *et al.*, 1957). It is advisable to apply soluble forms of nitrogen in split doses, particularly in the high rainfall areas, and it is commonly recommended that the half of the required nitrogen should be applied in the form of organic matter. Foliar spray of urea of low biuret content may be beneficial to a limited extent.

Collado Fernandez (2000) grew Navelina orange cuttings in sand and irrigated with a nitrate solution (100 per cent N as nitrate) or an ammonium nitrate solution (40 per cent N as ammonium, 60 per cent as nitrate)) at rates of 5 or 11 meq N/l. The effect of various duration of application of the ammonium nitrate solution (5 meq/l) was also assessed (irrigation from May to the end of the trial in January, from July to January, or from September to January, with N applied as nitrate solution until the start of ammonium nitrate application. The results suggested that application of N as ammonium nitrate could be used commercially to produce smaller fruits of a more desirable size for the market with thinner rind.

In grafted Citrus species, the rootstock nutrient use efficiency is very important. Nitrogen use efficiency (NUE) definitions, commonly used in literature, are evaluated in response to nitrate availability in four citrus rootstocks, Rough Lemon (*Citrus jambhiri* Lush) (RL), Sweet Orange (*Citrus sinensis* L.) Osbeck) (SwO), Cleopatra Mandarin (*Citrus reshni* Hort. ex Tan.) (CM) and Sour Orange (*Citrus aurantium* L.) (SO) (Sorgona *et al.*, 2006). The application of diverse definitions determines different characterizations in N-efficiency among rootstocks. Nitrogen utilization efficiency (NUtE) and nitrogen efficiency ratio (NER) determine equal level of nitrogen efficiency among all rootstocks. Total nitrogen accumulation (TNA), nitrogen uptake efficiency (NUpE), shoot dry weight (SDW) and total leaf area (TLA) response curves produce the same NUE characterization: SO and SwO were nitrate use efficient and inefficient rootstocks, respectively, while the RL and CM exhibit superior and inferior genetic potential, respectively.

9.3.1.2 Phosphorus

Citrus appears to receive phosphorus more readily from the soil than many other plant species and phosphorus usually does not become a limiting factor for citrus growing. Phosphorus deficiency leads to reduction of foliage density and causes deterioration in the physico-chemical characters of the fruits. Fruits usually become large with coarse and thick rind (Embleton *et al.*, 1956). Fruits from deficient trees are more acidic and drop prematurely.

Response of phosphorus fertilization is clearly detectable only in the severely deficient areas, and in highly acidic soils phenomenon of phosphorus fixation results in the non-availability of applied fertilizers to a great extent. In the acidic soils, rock phosphate containing free calcium carbonate may be used as phosphate fertilizer, rather than the usual superphosphates. For correcting the deficiency, only the required dose should be applied, since excess of phosphorus may interfere in the absorption of zinc, iron and copper. It is advisable to apply phosphorus only after soil and leaf analysis.

9.3.1.3 Potassium

The deficiency of potassium causes chlorosis of mature leaves through fading of chlorophyll in blotches near the leaf apex or the margins, gumming of twigs, premature fruit shedding and reduction in fruit size. Koo and Reese (1972) obtained the smallest tree size with high ratio of premature fruit drop and small and poor quality fruits of Pineapple orange when K was omitted from the fertilizer.

Potassium deficiency can be corrected through soil application of fertilizer like potassium chloride, potassium sulphate, *etc.* Excess potash increases acidity in the fruit juice and should be avoided. Reese and Koo (1975) noticed that the K requirements of Hamlin, Pineapple and Valencia oranges were not the same. Thus, Hamlin orange showed maximum yield response to K at 167 kg/ha/year, whereas Pineapple orange yielded the maximum at 112 kg/ha/year.

9.3.1.4 Magnesium

Magnesium deficiency is chiefly associated with acidic and highly leached soils and symptoms are commonly clear in fruit bearing limbs, particularly in the fruit maturing season. Yellow bands are observed on either side of the midrib. In acute deficiency, the entire tree looks yellow and leaf shedding takes place. Seeded cultivars are reported to be more prone to magnesium deficiency and deficient trees are known to be more cold susceptible. Magnesium usually moves from the older leaves to new growth and developing fruits especially of seedy cultivars.

In non-acidic soil, magnesium deficiency can be corrected by application of magnesium sulphate or magnesium carbonate in the soil, while foliar spray of magnesium nitrate is useful for quick action. The Mg deficiency in Valencia orange and grapefruit trees on calcareous soil can be corrected by annual application of magnesium nitrate as foliar spray at the rate of 10 or 15 lb per 100 gallons of water. If the soil is acidic, application of suitable quantity of dolomite lime may help in improving the magnesium status.

9.3.1.5 Calcium

Calcium deficiency symptoms are not very distinct and the symptoms are rather rare under field conditions. Leaves on calcium deficient trees are usually smaller in size and somewhat thickened. Premature drop of leaves and die back of twigs are associated with the deficiency of calcium. The deficiency can be corrected by the application of domestic lime or gypsum. Spencer and Koo (1962) were able to prevent calcium deficiency by application of limestone or triple superphosphate which contains calcium.

9.3.1.6 Sulphur

No clear-cut and distinct symptoms could be identified for sulphur deficiency and the requirement appears to be met indirectly through N, P and K fertilizers carrying S.

Micronutrients

Studies conducted on Khasi Mandarin (*C.reticulata*) in North East India showed that during the foliar application of Zn+Cu+B increased plant height, stem girth, maximum plant canopy spread, highest fruit set per cent, yield, highest number of fruits per plant, fruit physical and biochemical parameters like largest fruit length, fruit diameter, fruit weight, minimum fruit acidity and highest TSS/Acid ratio, while highest TSS was observed with treatment which received micronutrient mixture of Zn+Mn+Cu+B against control

9.3.1.7 Zinc

Zinc deficiency is considered to be the most widespread and damaging micronutrient deficiency of citrus. Deficiency symptoms develop only in the new growth. The leaves are small-sized, chlorotic, rather crowded on short stems giving a bunched appearance. The chlorosis pattern is quite distinct, where the areas between main lateral veins become whitish-yellow. Mild early stage of zinc deficiency resembles those of manganese deficiency, however, when becomes more severe, the mottling becomes more pronounced, and twig die back symptom starts appearing. In acute cases, there is considerable die back of smaller twigs and the tree starts declining. High soil pH can also reduce the availability of the micronutrients manganese (Mn) and zinc (Zn) in soil.

A study was made on Satsuma Mandrin to investigate citrus tissue's response to varying applications of Fe-EDDHA, zinc (Zn) and manganese (Mn) combinations and the resulting effects on citrus fruit yield and quality in calcareous soil in China (Zhang *et al.*, 2014). A field experiment was conducted for three years with Satsuma mandarin trees planted in calcareous soil. Four treatments were conducted: (i) control; (ii) soil Fe application; (iii) soil Fe + Zn application; and (iv) soil Fe + Zn + Mn application. Soil available Fe and Zn concentrations were raised through the application of Fe-EDDHA and Zn, whereas the soil available Mn concentration was not increased through the application of Mn. Applying Fe-EDDHA raised the leaf active Fe content significantly without influencing the leaf and fruit Fe concentrations. The Zn concentrations in leaf and fruit decreased when Fe-EDDHA was applied exclusively, but increased when Fe + Zn and Fe + Zn + Mn were applied in the third year. There was no significant difference in leaf and fruit Mn concentrations among all treatments between 2010 and 2012. Soil Fe, Fe + Zn and Fe + Zn + Mn application increased fruit number by 23.5 per cent, 26.7 per cent and 24.9 per cent, respectively, and soil Fe + Zn application increased the yield in 2012. Fruit TSS, TA and Vc content were higher through the application of Fe + Zn and Fe + Zn + Mn in 2012. The results revealed that Fe and Zn play different roles that affect fruit yield and quality in calcareous soil. The citrus yield was mainly affected by Fe, whereas fruit quality was improved by Zn. The application of Fe-EDDHA alone could induce Zn deficiency, so the co-application of Fe-EDDHA and Zn fertilizer was useful for obtaining higher yield and quality in Satsuma mandarin in calcareous soil.

9.3.1.8 Manganese

Manganese deficiency is indicated when the immature leaves are mottled with light green or yellowish green areas between the major veins. The veins and small adjacent areas of the veins remain green and unlike zinc deficiency, the leaf size remains normal (Camp and Fudge, 1939). The deficiency symptoms are more pronounced on the shady side of the tree and shaded inside leaves show clear symptoms. Only in case of severe deficiency chlorosis pattern persists in the older mature leaves. Since manganese deficiency frequently occurs in combination with deficiencies of zinc or iron, its symptoms are often over-shadowed. Defoliation, loss of vigour and lower yield are the results of severe manganese deficiency.

Manganese deficiency occurs commonly in acidic coastal soil, where manganese content is low and in alkaline soil, where manganese gets fixed in an insoluble form unavailable to the plant.

Manganese deficiency can be effectively corrected through foliar sprays of 0.4 to 0.6 per cent manganese sulphate in spring or summer, when new growth occurs. Bar Akiva (1965) found that manganese oxide at 0.12 per cent was less effective (35 per cent improvement) than manganese sulphate at 0.50 per cent which gave 93 per cent improvement, when applied for correction of Mn deficiency in Shamouti and Valencia oranges.

9.3.1.9 Copper

Copper deficiency is not associated with leaf chlorosis. In severely deficient trees droplets of gum appear on succulent shoots, and pockets of dried gum are found in the twigs. In the fruits, gum pockets are present at the inner corners of the fruit segments next to the central pith. Oranges usually show scabby lesions on the rind, while lemons lack juiciness. Twig drying and die back often occur. Bar Akiva and Lavon (1967) noticed a high ratio of chlorophyll a:b in Cu-deficient Eureka lemon leaves. Where copper injury is not important, regular Bordeaux sprays are effective and deficiencies may be corrected through foliar spray of mixture of copper sulphate (0.4 to 0.5 per cent), hydrated lime and water. Lee (1964) suggested spraying the first flush of orange and lemons after the symptoms appear with a 3:1:4 Zn-Cu-lime mixture to control Cu deficiency.

9.3.1.10 Boron

Foliar symptoms of boron deficiency are non-specific and definite diagnosis can be made only through leaf analysis. Usually, in boron deficiency, tree remains less vigorous with sparse vegetative growth and with abundant blooms. Fruits usually become hard, and brownish gum deposits are observed in the fleshy coat of the fruit. The central core or pith of the fruits may be gum soaked and seeds may abort and turn dark (Sato *et al.*, 1962). Since the range of deficiency and toxicity is very narrow in case of boron, only soil and plant analysis can be used as a correct guide for boron application. Deficiencies can be corrected by foliar application of 0.1 per cent borax or boric acid. Foliar spray of 0.2 per cent borax improved general tree condition, reduced die back and decreased leaf chlorosis in Kinnow mandarin trees (Singh and Misra, 1980).

9.3.1.11 Iron

Iron deficiency, which is often called as 'iron chlorosis' or 'lime induced chlorosis' causes a pattern of fine network of green veins against a cream coloured or nearly white background of leaf lamina. In severe cases, the leaves become small, and twig and branch die back occurs. Chapman and Kelly (1943) considered that splitting and corking of midrib and veins of leaf, downward curling of leaves at right angles to midrib, lacklustre or slightly bronzed appearance, early dropping of leaves were some of the symptoms of iron deficiency. Usually chlorosis becomes more pronounced during autumn and winter and in many cases only a single branch or a section of tree may be affected, while the rest remains normal in appearance. Normally, in calcareous soils with high pH iron deficiency occurs.

Iron deficiency has proved to be one of the most difficult to correct. The foliar application of iron is not useful for correcting the deficiency and application of costly input like iron chelate may help in controlling the deficiency. Different citrus rootstocks vary in their ability to take up iron. The trifoliate orange (*Poncirus trifoliata*) is known to absorb less amount of iron, while sour orange has better ability to do so. Cleopatra mandarin was found to utilize iron in alkaline soil, where drainage is not a problem.

9.3.1.12 Molybdenum

Molybdenum deficiency symptoms become conspicuous only in fully enlarged 3 to 4-month-old leaves, where light green spots appear (Jones and Smith, 1964). The spots are fairly small on mandarins, while in grapefruit the spots are larger in size. The leaves that show symptoms usually drop and as a result deficient trees assume thin and unthrifty look. Trees grafted on grapefruit are more prone to molybdenum

Nutrient Deficiency

Magnesium Deficiency

Calcium Deficiency

Iron Deficiency

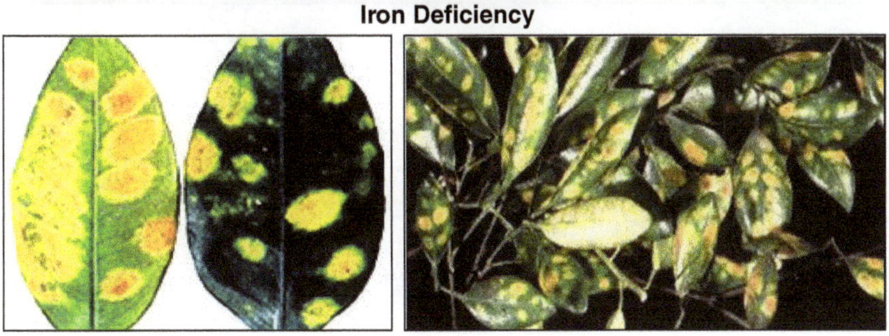

Molybdenum Deficiency

Nutrient Deficiency and Mulching

Boron Deficiency

Copper Deficiency

Zinc Deficiency

Mulching in Citrus

deficiency. The deficiency can be corrected through foliar application of sodium or ammonium molybdate. Soil application has not been effective, especially when the pH is low (Leonard, 1952).

9.3.2 Tissue Analyses and Nutrient Requirement

Like many other crops, leaf has been found to be the most satisfactory diagnostic tool to analyse the nutrient status of citrus trees. Citrus leaf analysis technique has been well recognized for diagnosis to deficiencies, for determination of optimum fertilizer rates and for assisting in the interpretation of fertilizer trial results. Since many factors such as leaf age, season, growth cycle, cultivar and rootstock may influence the levels of nutrients in the leaf, for correct interpretation of leaf analysis the factors should be well considered. Smith and Reuther (1950), Chapman and Brown (1950), Cradock and Weir (1969) and others found that leaves of same age from different flushes did not differ greatly in mineral content and usually spring flush leaves of about 4 to 7-month-old were preferred for analysis. Some workers specify selection of leaves from fruit bearing terminals but investigations have shown that there are significant differences in nitrogen, phosphorus, potassium and calcium contents between leaves taken from behind fruit and leaves from non-fruiting shoots (Cradock and Weir, 1969). Usually leaves from non-fruiting terminals are used for determination of status of mineral elements. Chapman (1960), De Villiers and Beyers (1961), Reuther and Smith (1954) and others suggested standards for classification of nutrient status in citrus and after adopting and revising the values available from different sources. Bhargava and Raghupathi (1993) standardized the foliar nutrient standards for orange and lemon (Table 7).

Table 7: Foliar Nutrient Standard for Orange and Lemon*

Nutrient Element		*Orange*			*Lemon*		
		Low	*Sufficient*	*High*	*Low*	*Sufficient*	*High*
N		2.0-2.19	2.20-3.5	>3.5	1.9-2.19	2.2-2.7	>2.7
P		0.10-0.11	0.12-0.5	>0.5	0.08-0.09	0.1-0.3	>0.3
K	per cent	0.90-1.19	1.20-3.0	>3.0	0.70-0.99	1.0-2.0	>2.0
Ca		0.90-1.09	1.10-4.0	>4.0	1.00-1.49	1.5-4.0	>4.0
Mg		0.20-0.29	0.30-0.50	>0.5	0.15-0.19	0.2-0.5	>0.5
Fe		40.59	60-150	>150	50-59	60-100	>100
Mn		22-24	25-200	>200	15-19	20-200	>200
Zn	ppm	22-24	25-200	>200	15-19	20-50	>50
Cu		4-5	6-100	>100	3-4	5-100	>100
B		20-24	25-100	>100	18-19	20-200	>200
Mo		—	—	—	0.03-0.29	0.3-3.0	>3.0

* Bhargava and Raghupathi (1993).

Good yield in mandarins was obtained when leaf N,P, K, Ca and Mg contents ranged between 2.5 and 2.8, 0.19 and 0.20, 1.5 and 1.7, 2.5 and 3.0, 0.30 and 0.35 per cent respectively (Takdze, 1981). The relation between nutrient uptake and yield of fruits showed that 12 to 13 year old healthy mandarin orange plants produced 65.3 kg fruit per plant when leaf N, P and K contents varied between 1.95 and 2.10, 0.09 and 0.12, 0.90 and 1.03 per cent, respectively (Hussain *et al.*, 1985). Nutrient (P, K, Ca, Mg, Zn, Cu, Mn and Fe) analysis of leaves from fruiting and non-fruiting mandarin twigs (5- to 7-month-old), growing in India, indicated that there was not much difference between both the twigs except for Ca, Zn and Cu contents. Leaves of bearing twigs contained higher concentrations of Zn and Cu and lower concentration of Ca than those of non-bearing shoots (Singh *et al.*, 1997). Data on leaf nutrient status of Coorg mandarin, Kinnow and Nagpur mandarin are presented in Table 8. Santos *et al.* (1999) compared chelated foliar fertilizers with conventional application of salts in the absorption of boron, manganese and zinc in orange 'Pera' trees (*Citrus sinensis*). The micronutrient foliar application increased absorption and foliar contents of Mn and Zn, but not B. The presence of chloride increased the concentration of Zn in leaves in relation to sulphate or sulphate plus potassium chloride. Results also showed that Yogen and MS-2 were not efficient as a source of Mn in this study.

The nutrient element composition of leaves varies with various factors like species, growth flushes, month and age of tree (Lyngdoh *et al.*, 1994, 1996; Sanyal *et al.*, 1994; Sema *et al.*, 1998, 1999). The changes in nutrient element composition of Khasi mandarin leaves due to various factors are given in Table 8. It was concluded that the best time for taking leaf samples for lemon cv. Assam lemon was 2- to 4-month-old for macronutrients (Sema *et al.*, 1998), whereas 5-6, 2-3 and 7-8-month-old leaves were ideal for estimation of Fe, Zn and Mn concentrations, respectively (Sema *et al.*, 1999). In general, spring flush exhibited fewer fluctuations than rainy and winter seasons (Sema *et al.*, 1998, 1999). Leaf N content and leaf colour index increased, but leaf P and K contents decreased with increased N application.

Table 8: Leaf Nutrient Guide for different Citrus Cultivars Grown in Varying Agro-climates of India

Nutrient Element	*Coorg Mandarin Terminals**		*Kinnow mandarin***		*Nagpur Mandarin****
	Fruit Bearing	*Vegetative*	*Optimum Value*	*Deficient Value*	*Optimum Value*
N (per cent)	1.85-2.48	2.08-2.79	2.80	2.30	2.24-2.40
P (per cent)	0.09-0.15	0.09-0.18	0.15	0.09	0.07-0.10
K (per cent)	0.76-1.89	0.78-2.08	1.57	0.72	1.18-1.56
Ca (per cent)	3.18-5.25	2.88-4.60	-	-	1.32-1.55
Mg (per cent)	0.28-0.51	0.25-0.47	-	-	0.48-0.67
Fe (ppm)	77.0-165.8	103.0-160.2	103.00	61.00	110.24-132.12
Mn (ppm)	49.6-123.6	54.0-220.8	38.00	20.00	29.36-42.41
Zn (ppm)	31.4-84.2	31.4-82.6	62.00	18.00	18.60-29-80
Cu (ppm)	13.6-140.4	15.2-177.2	9.80	4.00	8.26-14.09
B (ppm)	—	—	109.00	21.00	—

* Anon (1991), ** Chahil *et al.* (1991), *** Srivastava and Singh (1998) and Srivastava and Kohli (1997b).

TSS, TSS: acid ratio, the leaf a and b colour indices, percentage of mesocarp and specific gravity of fruits were negatively correlated with leaf N content, whereas titratable acidity content was positively correlated with leaf N content (Tachibana and Yahata, 1998). Increases in leaf N contents in 'Monte-negrina' tangerine were positively related to increases in production and to decreases in average fruit weight (Panzenhagen *et al.*, 1999).

Table 9: Changes in Nutrient Element Composition of Khasi Mandarin Leaves Due to different Factors*

	Nutrient Contents									
	Per cent						*ppm*			
	N	*P*	*K*	*Ca*	*Mg*	*Na*	*Fe*	*Cu*	*Zn*	*Mn*
Age of trees (year)										
25-35	2.72	0.079	1.47	0.67	0.12	0.025	84.36	12.89	171.21	50.86
35-45	2.62	0.074	0.65	0.75	0.16	0.020	92.88	14.76	156.18	57.39
45-55	2.50	0.079	1.43	0.65	0.13	0.021	87.71	15.75	160.21	54.17
C.D. at 5 per cent	N.S	0.004	0.54	0.076	0.025	N.S	N.S	1.72	N.S	N.S
Growth flush										
Spring	3.01	0.088	1.32	0.62	0.12	0.022	83.43	15.07	166.79	59.49
Winter	2.37	0.067	1.05	0.75	0.16	0.023	92.54	13.86	158.28	54.06
C.D. at 5 per cent	0.224	0.005	0.09	0.024	0.011	N.S.	4.65	N.S.	N.S	N.S.
January	2.56	0.063	1.31	0.65	0.08	0.006	58.11	10.11	240.44	45.67
February	2.50	0.040	1.11	0.76	0.10	0.006	65.78	14.50	156.22	46.89
March	2.17	0.064	1.51	0.76	0.09	0.008	72.78	12.89	159.78	50.11
April	2.80	0.130	1.59	0.66	0.35	0.011	77.58	12.33	130.72	38.28
May	2.51	0.083	1.37	0.88	0.28	0.015	81.11	13.67	160.55	39.00
June	2.17	0.066	1.06	0.56	0.18	0.027	74.33	15.00	178.56	42.78
July	2.55	0.079	1.02	0.58	0.10	0.022	90.80	12.61	133.17	59.89
August	2.79	0.109	1.08	0.62	0.12	0.046	180.61	14.45	132.95	96.05
September	2.90	0.123	0.91	0.79	0.11	0.061	168.98	26.83	167.28	91.34
October	2.58	0.083	1.04	0.72	0.11	0.034	40.78	15.89	193.89	49.22
November	2.62	0.045	1.33	0.69	0.10	0.024	72.17	14.67	147.89	46.78
December	3.09	0.043	1.04	0.57	0.08	0.009	72.78	10.67	167.94	43.67
C.D. at 5 per cent	0.339	0.012	0.20	0.129	0.046	0.019	20.76	2.99	30.27	10.18
Interactions										
Age x Flush	N.S	N.S.	N.S.	0.041	N.S.	N.S.	N.S.	N.S.	N.S.	N.S.
Age x Month	N.S.	0.20	0.35	N.S.	N.S.	N.S.	N.S.	5.17	N.S.	N.S.
Flush x Month	0.57	0.016	0.28	0.182	0.065	N.S.	29.36	4.23	42.81	14.40
Age x Flush x Month	N.S.	N.S.	N.S.	N.S.	N.S.	N.S.	N.S.	N.S.	N.S.	N.S.

* Lyngdoh *et al.* (1994, 1996); Sanyal *et al.* (1994).

9.3.3 Fertilizer Recommendations in Various Citrus Growing Countries

In Philippines application of nitrophoska, a complete fertilizer containing 15 per cent N, 1 per cent water soluble acid phosphate and 26.5 per cent soluble potash, at the rate of 2 or 3 kg/tree to 25-year-old poor growing Batangs mandarin resulted in appreciable growth promotion and sustainable yield increase within 2 or 3 years (Torres, 1936). On the basis of results obtained from a long-term experiment laid out in Nelspruit, South Africa, Oberholzer (1946) recommended that as a rule, a citrus tree in South Africa should be given 0.5 lb nitrogen a year if under 5 years and 1 lb at the age of 5-10 years. Older trees showed increased yield following application of 2 lb N. For citrus in New South Wales (Australia), 2 or 3 lb N/tree should be applied in two split doses – two-third at the spring ploughing and one-third in summer. Superphosphate at the rate of 2 cwt/acre was also recommended (Benton, 1943). For 15- to 20- year-old orange trees with an annual production of 150 to 200 kg fruit/tree growing in Tunisia, Maillard and Hibbon (1957) recommended application of 4 kg of 12-10-15 fertilizer mixture in early spring, 750 g of urea (45 per cent) in May and the same amount of urea in July. In Japan, the current recommendations for Satsuma are application of N at the rate of 180 to 250 kg/ha for early cultivars and 200 to 300 kg/ha for the late cultivars. Autumn application should be done at least one month before the soil temperature falls below 10°C (Iwakiri and Nakahara, 1981). From the result of an 8-year trial on sandy soil of Argentina, Schatz and Drescher (1971) recommended a N: P_2O_5: K_2O: MgO formula at 2: 1: 1: 0.5 ratio, with 40 g N/tree for each year of Valencia tree life. After studying the results of a 20-year fertilizer experiment in California, Platt (1960) recommended application of 1 to 1.5 lb N/tree/year. Half of the total quantity of N should be applied in the form of organic manures as it improves soil tilth. However, soil condition in California sometimes makes it difficult for citrus to absorb the required amount of nitrogen, and the rates are higher than in Florida (Chapman, 1951).

In China, the traditional time to apply manure is the early winter, with supplementary fertilizers applied during flowering, fruit swelling and after harvest. Studies have shown that this is not the best method for the production of high yields of good quality fruits. Applying basic manures in the summer, just after the secondary physiological stage fertilizers should be applied according to tree growth and fruit load (Ruifang, 1997). Treatment with 21 kg of NPK (16 per cent N, 6 per cent P and 8 per cent K) fertilizer + 10 kg manure was found to be the best for increasing yields and improving fruit quality of 7-year-old mandarin cv. Dayeweizhang (Quing *et al.*, 1997). Suitable fertilizer rates for mandarin in China were reported to be 675-750 kg N, 375-450 kg P_2O_5 and 375-450 kg K_2O per hectare (Chen *et al.*, 1998). The effect of time of application of fertilizers on the yield of 12-year-old Satsuma mandarin cv. Weizhang trees was investigated in China (Ouyang, 1998). Good yields were obtained following application of fertilizers in July (25-50 kg composted rape cakes or faeces) + 1-2 kg urea + 2-4 kg Ca_2PO_4 [$Ca_3(PO_4)_2$ or $CaHPO_4$] + 1-2 kg K_2SO_4 / tree, foliar application of 0.3 per cent urea + 0.2 per cent KH_2PO_4 once a month in March-April and twice in May-June, foliar application of 0.1 per cent borax during flowering, application of 2,4-D after harvest, and application of fertilizers (25-50 kg composted pig or cow manure) + 0.5-1 kg urea/tree before October.

In a 3-year study in Florida, the orange cultivars Valencia, Parson Brown, Hamlin and Sunburst were treated with 168, 224 or 280 kg N $ha^{-1}yr^{-1}$ as a broadcast application of N:P:K dry soluble granular fertilizer (Paramasivam *et al.*, 2000). These rates did not significantly affect the fruit nutrient status, juice quality, or nutrient concentrations in the spring flush. The concentrations of N, P, K, Ca and Mg in the fruits decreased as they enlarged from June to November. However, micronutrient concentrations in the fruits increased from June to August or September and then decreased. Fruit dry weight was greatest in Valencia followed by Hamlin, Parson Brown and Sunburst. Soluble solids concentrations were higher in Valencia and Sunburst than in the other 2 cultivars. Other juice quality parameters did not significantly differ among the cultivars.

In Algeria, Clementine clone 2749 on Troyer citrange and *Poncirus trifoliata* showed the highest yield of 122.0 and 108.0 kg/tree, respectively by application of NPK rates at 600: 150: 600 g/tree/year (Dris, 1997). Lemon cv. Femminello siracusano on sour orange rootstock in Italy showed highest yield by application of N at 80 to 160 g N/tree. A leaf N content in the range of 2.1-2.4 per cent was considered to be optimum (Intrigliolo and Intelisano, 1997). A field trial was carried out on Shamouti orange at Carmiya, southern Isreal, to determine the effect of KH_2PO_4 (Peak) and KNO_3 + Magdilon (standard treatment) treatments on the fruit size of Shamouti oranges (Mussak *et al.*, 2000). Foliar application of Peak + Alvamid (low biuret urea) significantly increased fruit size compared with the standard treatment. Foliar application of Peak increased P and K concentrations in leaves (dry weight basis) and increased the number of large-sized fruits/tree.

9.3.3.1 Fertigation

In trickle irrigation system fertigation is of potential advantage in citrus orchard (Koo and Smastrala, 1984). Bielorai *et al.* (1984) observed higher fertilizer use efficiency. Louse (1990) observed significant impact on yield of Navel orange and Valencia fruit and quality with N fertigation. In a 6-year trial with 13-year-old Navel orange tree the effects were compared to fertigation with Triosol, comprising water soluble NPK fertilizer (3:1:5) at 2 kg/tree applied once and the highest yield was recorded with a single fertigation with Trisol (Fouche and Bester, 1987).

9.3.4 Manuring and Fertilizer Schedules in India

For sustained production and for maintenance of proper orchard health it is essential to apply manures and fertilizers to citrus orchards regularly. For non-bearing trees, fertilizer application may be done in an area more than drip circle, leaving 15-30 cm radius around the tree trunk. It is advisable to apply the fertilizers in split doses and application should be made when there is sufficient moisture in the soil. In the old bearing orchards, since the entire area becomes covered with citrus roots, fertilizers may be applied by broadcasting from trunk to trunk leaving about 30 cm around the tree trunk.

In Punjab, nitrogen is found to be in deficient levels in all the citrus growing areas and based on the work done in the State, the following fertilizer recommendations have been made by the Punjab Agricultural University (Randhawa, 1970).

Age of Tree	*Dose per Tree*		*Fertilizer Basis (g)*
	FYM (kg)	*Nutrient Basis (g) N*	*Ammonium Sulphate (20 per cent N)*
1-3 years	5-20	50-150	250-750
4-6 years	25-50	200-250	1000-1250
7-9 years	60-90	300-400	1500-2000
10 years and above	100	400-800	2000-4000

From the long term systematic nutritional trials conducted at the Citrus Experiment Station of the Indian Institute of Horticultural Research located in Coorg region of Karnataka, the manurial schedule recommended for Coorg mandarin orange and for other parts of India has been given in Tables 10 and 11.

Table 10: Manure and Fertilizer Application Schedule for Coorg Mandarin (kg per tree)

Manure/Fertilizer	*Time of Application*	*1st* Year*	*2nd Year*	*3rd Year*	*4th Year*	*5th Year*	*6th Year*
Cattle manure or farmyard manure (FYM)	February-March	—	5.0	10.0	15.0 to 20.0	20.0 to 25.0	25.0 to 30.0
Calcium ammonium nitrate/ Ammonium sulphate Mixture	March-April	0.225	0.100	0.225	0.70	0.70	0.90
Single superphosphate	—	—	0.225	0.450	0.55	0.55	0.80
Muriate of potash	—	—	0.065	0.140	0.20	0.20	0.20
Calcium ammonium nitrate/Ammonium sulphate Mixture	June-Pre-monsoon	0.225	0.350	0.700	0.80	1.00	1.00
Single superphosphate	—	—	0.450	0.800	0.85	0.85	1.00
Muriate of potash	—	—	0.065	0.150	0.30	0.25	0.30
Calcium ammonium nitrate/Ammonium sulphate Mixture	September-October	0.225	0.225	0.450	0.60	0.80	0.80
Single superphosphate	—	—	0.225	0.450	0.60	0.60	0.70
Muriate of potash	—	—	2.065	0.150	0.20	0.25	0.35
Dolomite/agricultural lime	—	—	1-2	2-3	3	3	3
Actual nutrient per year							
N	—	—	0.160	0.325	0.50	0.60	0.70
P_2O_5	—	—	0.165	0.310	0.38	0.40	0.50
K_2O	—	—	0.140	0.225	0.50	0.52	0.60

* 1st Year: A mixture of 1:1:0.5 prepared by mixing calcium ammonium nitrate/ammonium sulphate, single superphosphate and muriate of potash

Table 11: Fertilizer recommendation for optimum productivity of citrus cultivars grown in some states of India*

Scion	*Rootstock*	*Recommended Fertilizer (g/tree)*			*Yield Obtained (Fruits/tree)*
		N	*P_2O_5*	*K_2O*	
Punjab: Bhatinda					
Kinnow mandarin	Jatti Khatti	400	200	400	713
Jaffa orange	Jatti Khatti	400	200	200	292
Assam:					
Khasi mandarin	Sohmyndong	300	250	300	—
Maharastra: Akola					
Nagpur mandarin	Jambhiri	600	200	400	684
N.R.C.Citrus, Nagpur					
Nagpur mandarin	Jambhiri	600	200	100	659
Acid lime (seedling)	—	800	200	100	67.5 (g/tree)
Karnataka:					
Gonicoppal					
Coorg mandarin	Rangpur lime	400	200	200	500
Andhra Pradesh:					
Tirupati					
Sathgudi	Rangpur lime	400	200	200	500
Tamil Nadu:					
Periyakulam					
Acid lime (seedling)		600	200	300	115.7 (g/tree)

*All India Coordinated Research Project on Fruits (1982-93).

Bajwa and Kaura (1952) reported that for 7- to 8-year-old sweet orange trees application of about 600 g of nitrogen per tree in the form of 50 per cent N as FYM plus 5 per cent as ammonium sulphate resulted in maximum fruit yield, and the yield increased with such treatment was about 60 per cent more in comparison with unfertilized control trees. They advocated application of farm yard manure in January and ammonium sulphate in two split doses in mid February and mid April.

From the fertilizer trials conducted in a 15-year-old Khasi mandarin orchard of seedling origin situated in acid soils of northern slope of Khasi hills, Meghalaya in India, Ghosh *et al.* (1981) reported that,

(i) A fertilizer dose of 300 g N, 250 g P_2O_5 and 300 g K_2O per tree per year was found to be optimum for best economic yield;

(ii) Foliar application of Mg (2 per cent magnesium sulphate) and B (0.1 per cent boric acid) during the end of April before full expansion of spring flush leaf significantly increased the fruit yield; addition of different nutrients like N, P, K, Mg, Ca, Zn, B, *etc.* did not affect the fruit quality significantly.

(iii) Phosphorus nutrition appeared to play an important role for higher productivity and a positive correlation of leaf phosphorus status with fruit yield was observed.

Ammonium sulphate applications acidified the soil, requiring a greater quantity of liming (Panzenhagen *et al.*, 1999). Various N rates and sources did not significantly affect the fruit size, nor juice quality parameters. The improved management of irrigation and fertilizer placement and timing of application contributed to minimize leaching of NO_3^-N below the root zone (Alva and Paramasivam, 1998).

Micronutrients are usually supplied through foliar sprays. Better absorption and effects of foliar sprays are observed when the new spring flush leaves are half expanded. Under New Delhi conditions, Raturi and Mukherjee (1981) observed that three sprays of Zn (0.5 per cent zinc sulphate neutralized with lime) were necessary for maintaining satisfactory level of Zn in all the growth flushes of 15-year-old sweet orange cv. Hamlin trees budded on Jatti Khatti rootstock. Zinc sprays controlled effectively chlorosis and die back symptoms, and through three foliar sprays of zinc sulphate plus 1 per cent urea a marked increase in fruit yield could be achieved over control in the second and third years. The chlorotic conditions and die back in citrus could effectively be corrected by spraying with micronutrients, such as zinc and copper (Dikshit, 1958); zinc, iron and boron (Chowdhury, 1954; Aiyappa *et al.*, 1961); molybdenum and boron (Aiyappa *et al.*, 1959); zinc, copper and iron (Kumar and Sharma, 1960) and zinc plus boron (Singh and Mishra, 1980). Mann and Takkar (1981) found that foliar application of zinc sulphate significantly increased fruit size, fruit weight, total soluble solids and vitamin C content of juice of sweet orange and claimed that unneutralized spray solutions of 0.15 and 0.30 per cent zinc sulphate were equal or more efficient than 0.45 and 0.60 per cent neutralized (with hydrated lime) zinc sulphate in increasing zinc concentration in leaf.

The yield and quality of Kinnow mandarin improved with soil and foliar application of zinc (Table 12). However, foliar application (0.75 per cent) was found to be more effective (Pardeep Wali and Sharma, 1997). It has been suggested that the microelement application should be done in the order of Zn > Mn > Fe > Cu (Wu *et al.*, 1998).

Table 12: Effect of different Methods of Zinc Application on Yield and Quality in Kinnow*

Treatment	*Yield (No. of Fruits)*	*Weight of Fruit (g)*	*Juice (Per cent)*	*Acidity (Per cent)*	*TSS (°Brix)*
Soil application in bands	188	105	37.1	1.5	20.0
Soil application in holes	209	106	41.3	1.4	20.0
$ZnSO_4$ 0.4 per cent spray	281	118	47.5	1.3	21.0
$ZnSO_4$ 0.75 per cent spray	285	123	48.3	1.1	21.0
Unsprayed control	143	98	36.1	1.5	20.0
C.D. at 5 per cent	12.21	3.26	0.59	0.08	0.96

* Pardeep Wali and Sharma (1997).

9.3.5 Organic Citrus Production

Another fast change in the citrus agriculture in the world is the development of organic farming strategies to supply a high quality product to the markets of developed countries. Organic farming is more than a fashion and is well established. Right now, 3 per cent of the Valencia citrus production is under organic farming rules, and no-tillage, reduction of pesticide application, use of machinery to weed, and mulching with chipped pruned branches is widespread among non-organic farmers to avoid expenses, labour and to increase subsidies. Research has been conducted to compare organically managed farms and conventional farms (Marriott and Wander, 2006; Keesstra *et al.*, 2016a, 2016b) as well as for different irrigation systems (drip versus flood irrigation) (Swietlik, 1992; Nelson *et al.*, 2011). However, the combined effect of organic or conventional farming in combination with different irrigation types is not studied yet. Hondebrink *et al.* (2017) found that organically (OR) managed orchards had a higher aggregate stability than conventionally (CO) managed orchards. The C/N ratio is an indicator of the decomposition processes of plant residues in the soil. It was observe values of C/N ratios of around 10, which indicate fast decomposition rates. Drip irrigated farms, either with CO or OR, had the highest C/N ratios, thus the slowest decomposition rate. This could be explained by the fact that more recalcitrant organic matter is present (Hondebrink *et al.*, 2017).

9.4.0 Pruning

The citrus growers of China and Japan followed the large-branch pruning techniques. The procedure is used to correct poorly trained trees, and trees having many badly oriented branches. The most popular system in Japan, the 3-6-9 (vase) system has 3 limbs at an angle of 45°-50° located on the trunk, 2 sub-limbs on each limb, and 2 large branches on each sub-limb. Fruiting branches and twigs are distributed on the large branches and limbs. All the competing, large branches, branches growing vertically and densely are removed. This system maintains high yields and good quality (Shaonan, 1997). In Mexico, Tahiti lime trees were pruned 4-6 times between February and July by removal of 10 cm shoot tips to stimulate vegetative shoot production. Evaluation of shoot production between March and November showed that pruning in May resulted in a 44 per cent increase in shoot production in autumn and increased production of winter fruits (Curti-Diaz, 1996). Mechanical pruning and skirting are widely used to control tree size and improve fruit quality, but their effects on yields have been inconsistent. Hedging and topping 'Ruby Red' grapefruit (Fucik, 1977) and hedging 'Valencia' orange trees (Bevington and Bacon, 1976) reduced yields in the first year after pruning, but increased yields in subsequent seasons. Annual topping of oranges and Satsuma mandarins reduced yields in the two subsequent seasons after pruning (Whitney *et al.*, 1982). In contrast, hedging one side of Satsuma mandarins (Iwagaki, 1981) and two sides of several citrus cultivars (Raciti *et al.*, 1981) and Valencia orange trees (Fucik, 1982) did not affect yields. Similarly, skirting Navel orange (El-Zeftawi, 1976) and lemon trees (Phillips *et al.*, 1990) did not affect yields.

Pruning also has had variable effects on internal and external fruit quality. Internal quality of Satsuma mandarin was not affected by hedging and topping every 3 years (Iwagaki and Ideta, 1977). Similarly, hedging did not affect internal quality of several citrus cultivars in Italy (Raciti *et al.*, 1981). Topping height had only minor effects on fruit size, colour, and juice content of oranges grown on 'Rusk' citrange (*P. trifoliata* x *C. sinensis*) rootstock; however, when trees were grown on a more vigorous root- stock 'Milam' (*C. jambhiri*) topping at the lower height produced poor quality fruit (Wheaton *et al.*, 1995). Hedged and topped Valencia orange trees (Leyva *et al.*, 1987; Raciti *et al.*, 1981) had fewer incidences of fruit with wind scar than did non-pruned trees.

The effects of pruning on fruit size are variable. Topping and light pruning of mandarins increased fruit size (Gilfillan, 1995; Zaragoza and Alonso, 1981; Zaragoza *et al.*, 1992). Similarly, Hamlin orange trees hedged at a 30° angle had larger fruits than trees hedged at 10° (Phillips, 1972). In contrast, neither hand nor mechanical pruning increased fruit diameter of Washington Navel orange (Zaragoza and Alonso, 1981). Topping and hedging reduced fruit size of lemons (Francis *et al.*, 1975) and mandarins because more fruits were set on the non-pruned sides (Raciti *et al.*, 1981). No difference in fruit size was found in orange trees hedged on both sides (Wheaton *et al.*, 1984).

Skirting also has variable effects on fruit quality. Skirting improved orange fruit quality by reducing fruit contact with the soil, thus decreasing damage from soil- borne organisms (El-Zeftawi, 1975). Skirted lemon trees also had less incidence of fruit infected with brown rot (*Phytophthora citrophthora*) than did fruit from non-skirted trees; however, packout was similar (Phillips *et al.*, 1990). Skirting did not affect fruit diameter, yield, juice content, TSS and TA of Navel oranges in Australia (El-Zeftawi, 1976).

Morales *et al.* (2000) stated that pruning generally improved fruit size and decreased yields of Orlando tangelo trees in Florida. Skirting had no effect on canopy temperature and RH, generally decreased yields, and did not improve fruit internal or external quality. The combination of pruning and skirting also decreased yields and improved fruit size but generally did not affect fruit quality.

Jiang (2000) found that using spring shoots for bearing was much better than using autumn shoots for mid-season Satsuma mandarin cultivars; production could be increased by 15-20 per cent. The techniques for regulating fruiting on the spring shoots are by pruning. In the off-year the branches without flowers should be headed back and 2-3 spring shoots left which have no more than 5 leaves. The branches with flower buds on the spring shoots should be thinned out before flowering. In the on-year the autumn shoots of the previous year in the flower part of the canopy should all be removed, while in the upper part of the canopy all the autumn shoots with more than 5 leaves from the previous year should be thinned out.

9.5.0 Weed Control

Effective weed control is perhaps the single largest factor which is likely to determine the future prospects of citrus industry. In some of the developed countries non- tillage practice of weed control has already given encouraging results. Where soil erosion is not a problem, it was found that a zero tillage herbicide weed control soil management treatment could be used for up to 28 years without adversely affecting tree growth and yield (Cary, 1968; Cary and Weerts, 1977). Among several chemicals used, the chemicals like simazine, diuron, atrazine, bromacil and terbacil have been reported to be effective for controlling weeds in citrus plantation (Dhillon *et al.*, 1976; Jawanda *et al.*, 1977; Bhullar and Ghanbari, 1985). It was found that spraying of simazine (8.0 kg/ha) followed by gramoxone (3.0 l/ha) was the best treatment for effectively checking the weed population for a longer period (Goswami *et al.*, 1985). Josan and Chohan (1985) in an experiment with sweet orange found that application of diuron at 2 kg/acre was most effective, which killed 67.58 per cent monocot weeds and 81.78 per cent dicot weeds by pre-emergence application of diuron at 2.5 kg/ha once every year during July under the trees followed by 3 to 5 spraying of paraquat at 2.5 l/ha as post-emergence at 10-15 cm height of weed stage was found effective against 10 to 11 hand weedings for both annual and perennial weeds. It was also found that chemical weed control with diuron had reduced the cost of weeding by 72 per cent compared with hand weeding at the end of five years (Pathak and Srivastava, 1985). In Kinnow mandarin, bromacil and diuron at 6.0 kg a.i./ha, terbacil at 4.5 kg a.i./ha and atrazine and simazine at 6.0 kg a.i./ha were found to be most effective in controlling the weed population (Bhullar and Ghanbari, 1985).

Application of trifluralin at 12.5 lb, oryzalin at 16 lb and norflurazon at 8 lb/acre proved effective in controlling weeds of oranges cv. Hamlin and citrange cv. Carrizo, without apparent phytotoxicity to the trees (Singh and Tucker, 1984). Ivakh and Raskin (1984) obtained best results with bromacil or terbacil at 2 kg/ha, applied twice, or a mixture of 4 kg terbacil + 4 kg simazine/ha. Several dicotyledonous weeds which are generally difficult to control can be effectively controlled by use of satarne (fluroxypyr). Application of satarne in citrus orchards of Spain was found effective to control of *Amaranthus retroflexus*, *Portulaca oleracea*, *Malva* spp., *Convolvulus arvensis*, and *Parietaria officinalis* (Velilla, 1997). Experience of raising citrus trees for over 30 years in China revealed that the best period for spraying herbicides was just after cultivation and application of fertilizers. Spraying of 100 ml akex 820 (butralin) + 50 l water killed weeds in a 666.7 m^2 area. If the inter-row spaces were covered with dried grass or straw, the effect lasted for half a year (Yue Tong, 1997). In orange (cv. Valencia) orchards of Brazil, 2 application of sulfosate or glyphosate at 0.86 kg a.i./ha was found effective to control *Panium maximum* when it was present in a low stand (16 to 30 days after cutting). Application of paraquat or diuron were, however, not very effective (Foloni and Silva, 1997). Spraying with 260X glyphosate solution effectively controlled weeds with more than 12 leaves in Satsuma orchard, but not weeds that were flowering or seed bearing. No phytotoxicity was detected following the herbicide application (Zhang, 1998).

9.6.0 Mulching

Mulching (75 per cent cover) with UV-absorbent polyethylene sheets increased soil moisture content and soil temperature (5-30 cm depth) resulting in an increase in yield and percentage of fruit set of Washington Navel orange. It also decreased both percentage of June and total fruit drop, increased fruit size, fruit juice, volume and decreased fruit rind thickness and total acidity (Ghali and Nakhlla, 1996). The maintenance of the optimum soil water potential (SWP) for Satsuma using polyethylene (PE) film mulching and/or root pruning was investigated by Hyun *et al.* (1993). They observed that at harvest, the soil under the PE film was 28 per cent saturated, which was 9.5-20.2 per cent lower than the other treatments and the fruits produced with PE mulch contained more soluble solids (11.6 °Brix compared with 9.3-9.7 °Brix for other treatments). Mulching had a clear effect on flower bud formation in Satsuma trees, in general, and in the lower part of the crown in particular. Starch content of summer shoots was increased by mulching (Takase and Mizuno, 1997).

9.7.0 Intercropping

In the early stages of establishment of a citrus orchard till bearing, the interspaces can be economically utilized with short duration crops, preferably of leguminous types. Selection of intercrops is of utmost importance. Exhaustive intercrops which deprive the citrus trees of essential nutrients and moisture and the intercrops having intercultural requirements that are antagonistic to the requirements of citrus trees should be avoided. Undesirable intercrops can adversely affect the citrus plants through exhaustion of nutrients, deterioration of physical properties of soil, incompatible irrigation, root injury due to deep soil working, shading and by acting as the alternate hosts for diseases and pests. As far as practicable, the intercrops should be erosion resistant, profit yielding and non-soil depleting types. Bajwa and Ali (1945) observed that growing of crops like wheat, maize, jowar, bajra and sugarcane as intercrops had harmful effects on citrus in North India, while Ghosh *et al.* (1981) recorded that intercropping with maize, ragi, buckwheat and turmeric in the north-eastern hill region had undesirable effects on mandarin orange trees. Srivastava *et al.* (1997) suggested soybean as most suitable intercrop for Nagpur mandarin grown in black clay soils of central India.

Guara/wheat was found as useful intercrops for Jaffa, which increased tree growth, yield and fruits of better quality (Sarowa *et al.*, 1985). Ghosh *et al.* (1985) suggested planting of leguminous vegetables like pea that increased the leaf N content of mandarin orange and thereby tree vigour and yield. While cultivation of radish, cabbage, maize, ginger and potato as intercrops was found detrimental (Table 13). Chadha (1970) opined that excessive and indiscriminate intercropping was one of the causes of citrus decline in Punjab.

Leguminous intercrops are usually considered safe for citrus orchards, since they are less exhaustive and help in adding nutrient through nitrogen fixation. Crops like pea, cowpea, gram, beans are considered safe intercrops. Usually the citrus growers who happen to be near the cities prefer to grow vegetable and fodder crops due to ready market and better profit and in such cases, up to prebearing age, pea, beans, tomato, *etc.* may be grown during *rabi* season and crops like cowpea, gourds (bottle and bitter), lady's finger, *etc.* can be cultivated as *kharif* crops. Fodder berseem as intercrop in citrus orchard has, however, been found to be unsuitable mostly due to its frequent irrigation requirement, whicht imparts adverse effect on citrus.

Table 13: Leaf N (Per cent dry weight) and Yield of Fruits Due to different Intercropping* (Age of the plant 12-13 years)

Crops	*Nitrogen Content of Leaf at Harvesting*	*Yield of Fruits/ Plant (kg)*
Orange alone	2.85	32.5
Radish + orange	2.27	17.8
Pea + orange	2.95	37.0
Cabbage + orange	2.54	21.9
Potato + orange	2.75	26.5
Maize + orange	2.20	13.1
Ginger + orange	2.25	25.5
Soybean + orange	2.70	32.0

* Ghosh *et al.* (1985).

10.0 Flowering, Pollination, Fruit Growth and Development

10.1.0 Flowering

In citrus any bud is a potential fruit bud and flowering takes place mostly in spring and in growing season when soil moisture and temperature both remain favourable. Although the acid group, limes, lemons and citrons, flower almost throughout the year, two major flushes of flowers usually occur in other species of citrus. The intensity of flowering is influenced greatly by the period of growth cessation, and the amount of the preceding bloom or crop. In the North Indian conditions, where the temperature goes down substantially during winter months, major bloom of almost all the species occurs during early spring (February-March) when the atmospheric temperature starts rising after the cold winter and soil moisture condition also improves (Hayes, 1970). In South India, where there is no well-defined winter with very low atmospheric temperature, the flowering season is longer and not very distinct. It is very common to get two crops, occasionally three also, in many citrus types grown in South India (Naik, 1949). The flowering can, however, be regulated by withholding soil moisture, or through fruit thinning by chemicals and adjustment of fruit harvesting. The practice of withholding water from citrus trees as an aid to promote flowering may cause severe injury to the trees, and hence, use of chemicals is most desirable. In Israel, Eureka lemons sprayed five times with CCC at a concentration of 25 ppm considerably increased flowering and production of lemons (Monselise and Halevy, 1964; Monselise *et al.*, 1966). BOA was found to be relatively more effective on branches that were older than six months, while CCC and SADH were found to be more effective on branches that were six months old or younger. Goren and Monselise (1969) found chloramphenicol succinate, 5-fluorodeoxyuridine and 5-bromo-3-butyl-6-methyluracil to be effective in inducing flowering in orange. The increased bloom was accompanied by the shortening of internodes which is the characteristic of all flowering branches. Paclobutrazole at 2.5 to 10 g/tree when applied in November to Frost Dancy trees on sour orange rootstock, reduced shoot length and increased flowering and yield from 14.6 kg in the control to 20.4-21.8 kg (Delgado *et al.*, 1986a) while in Valencia orange there was a reduction in vegetative vigour with shortening of shoots and internodes. In spite of a small fruit set, yield was significantly increased due to greater blooming (Delgado *et al.*, 1986b). Pruning seems to have little effect on time and amount of bloom.

In the absence of sufficient low temperature in central India, soil-water stress is usually practised for induction of flowering in Nagpur mandarin (*Citrus reticulata*) by imposition of withholding irrigation for a certain period and then resuming irrigation at near field capacity moisture level. Under such conditions, the success of flowering induction is governed by the right quantum of water stress as a function of combined effect of soil physico-chemical properties (Srivastava *et al.*, 1997). The reasons for the irregular flowering behaviour are not yet fully understood especially from soil point of view. Dass *et al.* (1998) observed the high correlation (r=0.644) between quantum of soil moisture depletion up to 60 cm soil depth and lowering intensity in regular flowering orchards which was non-significant in irregular flowering orchards of central India. Experimental evidences suggested that flower bud formation in Satsuma mandarin was inhibited by foliar $GA_{1/3}$ level in October and that high IAA and ABA contents in December also promoted the percentage of leafless inflorescences and the number of flower buds per node (Koshita *et al.*, 1999).

The time of blossom bud differentiation has been found to differ from year to year with climatic variations and also from cultivar to cultivar within a species. A brief outline of the time of blossom bud differentiation in sweet orange cultivars at different places is presented in Table 14.

Table 14: Time of Blossom Bud Differentiations in Sweet Orange Cultivars at different locations

Cultivars	*Locality*	*Time of Differentiation*	*Reference*
Pineapple	Florida, USA	January, 29	
Pineapple	Florida, USA	January, 18	Abbott (1935)
Pineapple	Florida, USA	January, 12	
Washington Navel and Valencia	Griffith, NSW Australia	Early spring	West and Bernard (1935)
Valencia Late	Lyallpur (Pakistan)	January, 31	Randhawa and Dhinsa (1947)
Sweet orange	China	3rd week of November	Lin and Wu (1957)
Washington Navel	Japan	Middle of December to early January	Fujita and Yagi (1956)
Valencia	Japan	Late January	
Fukuhara	Japan	Early February	
New Summer	Japan	Middle of March	
Navel	Japan	January	Ito *et al.* (1958)
Blood Red	Pantnagar, UP, India	January, 9	Babu and Kaul (1972)
Jaffa	India	January, 24	

10.2.0 Pollination

In general, pollen development is a normal phenomenon in citrus, excepting in a few important citrus cultivars like Washington Navel, Satsuma mandarin, Bears lime, *etc.* where no viable pollen is produced. Variation in the longevity of pollen grains of different citrus species has been observed. Nair and Singh (1969) observed that the pollen grains from mature buds of *C. medica* had higher germination potentiality than the dehisced ones. In cultivars where pollen is produced in abundance, self-pollination is a general rule. However, in orchard with mixed planting of different cultivars, cross

pollination is not uncommon. The stigma remains receptive for 6-8 days. Honey bees are known to act as the pollinating agent in citrus. Cross pollination is reported to increase seediness in certain citrus cultivars like Shamouti orange, Clementine mandarin and Mineola tangelo and it has been reported that in the later two cultivars cross-pollination has resulted in increased fruit yield. In Washington Navel and Marsh Seedless grapefruit hand pollination with pollen from seedy cultivars had been reported to cause better fruit set. Both male (*e.g.* Washington Navel, Satsuma) and female (*e.g.* Marsh grapefruit, Eureka and Lisbon lemons) sterility, partial self-incompatibility and normal male and female fertility conditions prevailing in the genus *Citrus* which have resulted in several species, such as Siamese pummelo (Aala, 1953), Orlando tangelo (Krezdorn and Robinson, 1958), Italior, Nepali Oblong and Lucknow Seedless lemon (Naidu, 1953), pummelo (Nauriyal, 1952), Clementine mandarin (Soost, 1956) and Sweet lime (Singh and Dhuria, 1960), but Nagpur Santra was found to be self-compatible as well as cross compatible except with grapefruit (Kedar and Gopal Krishna, 1977). The breeding system of citrus is as diverse as the species themselves; self-incompatibility, self-sterility and parthenocarpy are common. As a result, solid planting tend to result in poor fruit set. In most commercial plantings, pollinator trees are provided to increase fruit set (Hearn *et al.*, 1968).

Sweet lime yielded only 8 per cent under selfing as compared with 40-41 per cent fruit set in cross pollination using Duncan grapefruit and Pineapple sweet orange pollen (Kumar *et al.*, 1976). DeLange and Vincent (1979) recommended cross pollination as the best method for overcoming poor yields in Clementine mandarin. A similar effect due to cross-pollination in Star Ruby grapefruit (Burger, 1985), local pummelo 'Goliath' (Brosh and Israeli, 1986) and tangerine cultivar Khieo Waan (Ketsa, 1988) has been reported. Venkateshwarlu and Lavania (1985) reported that cross pollination of Pant Lemon-I with different pollen parents resulted in very high fruit set and fruit retention as compared with that obtained through hand-selfing, natural selfing and open pollination. Of the parents, Italian and Nepali Oblong lemons were found to be the best pollinizers for Pant Lemon-1 (Table 15).

It was observed that under free pollination, cultivars with stigmas above anthers and with longer pistils set more fruits. The results indicated that the predominance of allogamy in these varieties was an important factor for fruit setting (Domingues and Tulmann Neto, 1999). The results obtained by Wallace and Lee (1999) demonstrated the need for careful consideration of interplanting of cultivars and management of pollinators such as the honey bee to maximize fruit production and quality in mandarin.

Table 15: Effect of Different Pollen Parents, Hand-selfing, Natural Selfing and Open Pollination on Fruit Set and Final Fruit Retention of Pant Lemon-1*

Pollen Source	*Initial Fruit Set (Per cent)*	*Final Fruit Retention (Per cent)*
Citrus paradisi		
New Zealand orange	50.93	23.99
Citrus limon		
Bhadri	74.77	45.31
Eureka	50.71	29.60
Florida Rough	49.50	29.02
Italian	82.74	51.28
Kagzi Kalan	5.44	9.21
Nepali Oblong	85.00	51.15
Pant lemon-1		
a. Hand selfing	4.55	1.70
b. Natural selfing	4.10	1.60
c. Open-pollination	7.40	4.20

* Venkateshwarlu and Lavania (1985).

10.2.1 Use of Growth Regulators

Not much work has been done with reference to use of growth regulators for limiting the crop load. A reasonable amount of thinning using NAA and ethrel has been noticed (Asi and Ali, 1970). Application of NAA (350 ppm) 35 days after full bloom induced reasonable amount of thinning followed by hand thinning and ethrel (200 ppm) and regulated the crop in the subsequent year. Application of 2,4-D (10 ppm), cycocel (100 ppm) and 2,4,5-T (20 ppm) was not very effective to induce thinning and regulate crop production of Kinnow (Sharma and Awasthi, 1990). The growth regulators such as mepiquat chloride (1000 ppm, one spray) and cycocel (2000 ppm, 2 sprays) were found highly effective in improving the flowering intensity of irregular flowering of orchards of Nagpur mandarin (Singh *et al.*, 1997). In Balady, mandarin application of paclobutrazol at 2000 ppm at the beginning of flower bud induction reduced the intensity of alternate bearing and regulated cropping in the following year (Malaka, 1996). Flower and fruit thinning of off-season tangor (cv. Murcott) trees with 150 ppm ethephon + 0.6 per cent urea twice in the month of January significantly decreased yields in the current year (314 to 46 fruits/tree) but had no significant effect in the following years. Flower and fruitlet thinning of 'on' season trees decreased yield in the current year (from 1.609 to 49 fruits/tree) but increased yields from 2 to 229 fruits/tree in the following year (Marinho and Souza, 1997). Application of 2,5,6-TPA (trichloro-2-pyridyloxylacetic acid) at 10 mg/l after June drop gave the best results in terms of fruit size, yield and fruit quality characteristics in clementine mandarin (Georgiou, 1998); whereas 3,5,6-

TPA caused thinning which was the main reason for increasing fruit size (Gravina *et al.*, 1998). GA_3 reduced the alternate bearing tendency of Kinnow mandarin trees particularly when applied at 20 ppm (Farmahan, 1997).

The effect of PBO (new plant growth regulator produced in China; 12 g/tree applied 15 days before flowering) on fruit quality of 3-year-old Blood oranges on *Poncirus trifoliata* rootstock was investigated by Zhang *et al.* (1998). In 1997, 20-year-old trees of Jincheng orange on *P. trifoliata* rootstocks were sprayed with PBO at 7 days after the end of flowering. Yield of treated Jincheng trees was increased by 104.1 per cent and yield of treated Blood oranges was increased by 32.4 per cent. Treatment of Jincheng with PBO improved fruit quality (soluble solids, sugar content and skin colour).

10.3.0 Fruit Set

In most citrus species, poor fruit set and high fruit drop result in poor yield, especially under unfavourable environmental conditions, such as late frosts, drought or excessive rain, *etc.* During summer months there is considerable loss of soil moisture resulting in severe fruit drop. Mulching with leaves (Krishnamurthy, 1959) or black polythene (Sinha *et al.*, 1978) has been found to reduce fruit drop. Nei Lei *et al.* (2000) reported that girdling or ringing increased the fruit set percentage in Shatianyou pummelo by 164.3 per cent and the yield was increased by 21.8 per cent; but there was no appreciable effect on the fruit weight, soluble solids content, acidity and ascorbic acid content.

Many growth regulators have been tried in citrus with varying degree of success. Stewart and Klotz (1947) and Erickson (1951) found no increase in fruit set and fruit size in Valencia orange by aqueous sprays of 2,4-D while 2,4,5-T had been found to increase fruit set markedly and improved the size and quality of fruits in mandarin (Singh and Randhawa, 1961). Titratable acidity, ascorbic acid and sugar content of the fruit increased as a result of treatment with 2,4-D, 2,4,5-T or NAA (Singh and Randhawa, 1960, Sinha *et al.*, 1977). 2,4,5-T at 10 and 15 ppm increased fruit set over control in Pineapple and Valencia Late orange (Sharma and Chopra, 1978).

Experiments conducted in California showed that an increase in fruit set resulted when GA_3 was applied to flowers or to individual fruits of Bearss lime, Eureka lemon and Washington Navel orange (Hield *et al.*, 1958). In other investigations, gibberellin sprays had been found beneficial in Clementine mandarin (Soost and Burnett, 1961), Valencia orange (Coggins *et al.*, 1960), grapefruit (Coggins *et al.*, 1962; Chundawat and Randhawa, 1972), sweet lime (Randhawa *et al.*, 1959), Jaffa and Pineapple sweet oranges (Randhawa and Sharma, 1962). Although increased fruit set has been achieved as a result of treatment with GA_3, some undesirable effects of gibberellin treatment on fruit quality was also recorded. Thus, gibberellin-treated fruits had rough and thick skin and were less juicy as compared with untreated fruits.

10.4.0 Fruit Drop

High rate of fruit drop is a serious problem of citrus in India (Randhawa and Dhillon, 1965). Many chemicals have been tried to reduce fruit drop. While NAA was found ineffective in reducing fruit drop in Washington Navel orange (Jones, 1951), 2,4-D at a concentration of 8 ppm, reduced the drop of mature fruits (Stewart *et al.*, 1952). Similarly, sprays of 2,4-D at 60 ppm applied in June or July reduced fruit drop during summer or early fall (Hield *et al.*, 1965). In Satsuma mandarin, AVG (200 ppm) sprayed at petal fall significantly inhibited fruit drop. The application of AVG inhibited the release of ethylene from young fruits for 2 weeks after treatment and decreased ACC content in fruits for 3 weeks (Ogata *et al.*, 1997). Application of GA_3 (200 ppm), at 1 month after anthesis increased fruit set by 65 per cent in Ponkan mandarin. Abscission of fruits was delayed by one month when sprayed at 1 month after anthesis. Fruits with and without pedicels abscised from trees sprayed at anthesis but spraying at 1 month after anthesis caused fruits without pedicels to abscise. Fruits on sprayed trees were smaller than the controls and contained fewer seeds (Tominaga, 1998). Investigations carried out at IARI revealed that 2,4,5-T (30 ppm) was effective in Jaffa and Mosambi and in Pineapple 2,4-D (15 ppm) was most effective (Randhawa *et al.*, 1961) when spraying was done in October. Similarly, GA_3 at 50 ppm and 75 ppm were effective in Jaffa and Pineapple respectively, in reducing June drop (Randhawa and Sharma, 1962). Randhawa *et al.* (1959) found gibberellic acid, 2,4-D and 2,4,5-T being effective in reducing fruit drop in sweet lime, while GA_3 and 2,4-D proved effective in reducing fruit drop in Darjeeling mandarin (Kar *et al.*, 1985). The other recommendations are given in Table 16. Spraying with 50ppm 2,4-dichloropheno-xypropionic acid [2,4 DP (dichlorprop)] preparations (Power and Magdilon) in May (26 mm diameter fruits), increased the fruit weight of Satsuma Myovaze mandarins by up to 15 per cent over the controls without affecting total yield (Mossak *et al.*, 1999).

Table 16: Suggested Growth Regulators for the Control of Fruit Drop in some Sweet Orange Cultivars

Cultivars	*Treatment*	*Reference*
Navel	2,4-D at 5 to 25 ppm	Stewart *et al.* (1947)
	2,4-D at 20 ppm	Jones (1951)
Valencia	2,4-D at 4 to 40 ppm	Stewart *et al.* (1952)
Pineapple	2,4, 5-T at 10 to 23 ppm	Sites (1954)
Blood Red	2,4, 5-T at 25 ppm	Bhullar (1981)
Mosambi	2,4, 5-T at 30 ppm	Randhawa *et al.* (1961)
Jaffa	2,4-D at 20 ppm	Jawanda *et al.* (1972)
Sathgudi	2,4-D at 5 to 10 ppm	Rao (1961)

The hormonal signals controlling fruitlet abscission induced by sugar shortage in citrus were identified in Satsuma mandarin (*Citrus unshiu*) cultivars Clausellina and Okitsu (Gomez-Cadenas *et al.*, 1999). In developing fruitlets, defoliations reduced soluble sugars (by up to 98 per cent), but did not induce nitrogen and water deficiencies. Defoliation-induced abscission was preceded by rises (up to 20-fold) in the levels of abscisic acid (ABA) and 1-amino-cyclopropane-1-carboxylic acid (ACC) in fruitlets. The data indicated that fruitlet abscission induced by carbon shortage in citrus is regulated by ABA and ACC originating in the fruits, while gibberellins are apparently implicated in the maintenance of growth. In this system, ABA may act as a sensor of the intensity of the nutrient shortage that modulates the levels of ACC and ethylene, the activator of abscission.

10.5.0 Fruit Growth and Development

Satsuma mandarin cv. Miho grown under plastic tunnel (where temperature is relatively higher compared with outdoor) continued to grow at an accelerated rate and they were 48 per cent larger than those grown outdoors. From 2 weeks after anthesis, fruits from tunnel houses were significantly more spherical than those grown under ambient conditions and this difference persisted until harvest. Higher temperatures also increased the accumulation of sugars in tunnel house fruits during stage 1 growth. Total sugar level in tunnel house fruits remained significantly higher than those in open field throughout development. Difference in total sugar levels were found largely due to sucrose, with minor contributions from fructose and glucose. Higher sucrose synthase activity in the vascular tissue of tunnel house fruits during stage 1 was associated with higher sugar accumulation in these fruits. Between 8 and 14 weeks after anthesis, fruits in tunnel houses had higher citrate concentrations in the juice sacs. However, the concentration of citrate in fruits from tunnel houses decreased more rapidly during fruit maturation, and was lower at harvest than that of fruits grown under ambient conditions (Richardson *et al.*, 1997).

Concentrations of the macronutrients (K, Ca, Mg, Na, P and S) and micronutrients (Fe, B, Zn, Mn and Cu) in whole fruit, pulp, rind and albedo of fruit from young and mature Bellamy Navel orange trees were measured at fortnightly intervals during fruit development (Storey and Treeby, 2000). The general order of abundance of the macronutrients was K>Ca>Mg=P>S>Na and the abundance of the micronutrients followed the order B>Fe>Mn=Zn>Cu in whole fruit.

Fruit softening is attributed to the dissolution of cell wall components (Huber, 1983). Sinclair and Joliffe (1961) reported a reduction in pectic substances in citrus rind during fruit development and maturation. Eaks and Sinclair (1980) reported that cellulose and hemicellulose levels were higher in immature than in mature orange peel. Monselise *et al.* (1976) and Jona *et al.* (1989) demonstrated an effect of GA on cell wall components of creasing peel. Naohara and Manabe (1994) investigated the changes of pectin during storage of Satsuma mandarin, separately examining fruit rind, membrane and fruit flesh. In addition, cell wall composition of citrus vesicles and peel was also measured (Braddock, 1983; Yoshida and Ueda, 1984).

Changes in rind firmness and cell wall polysaccharide composition were measured in fruit with i) a soft rind, (Satsuma mandarin, *Citrus unshiu* cv. Aoshima), and ii) a firm rind (hassaku, *C. hassaku*), from August to January of the following year by Muramatsu *et al.* (1999). Rind firmness was similar in both species in August, but hassaku had significantly firmer rind than did mandarin from September to January. Both flavedo and albedo tissues were extracted, and the extracts were hydrolyzed and fractionated to yield four fractions: (hot water, EDTA, hemicellulose, and cellulose). In flavedo tissue, sugar concentration was highest in the cellulose fraction, and lowest in the hemicellulose fraction. The concentration in all fractions decreased as the fruit developed and matured. Although the sugar concentration in the cellulose and EDTA fractions of both species was similar in August, it was significantly higher in both fractions in hassaku than in mandarin in January. The sugar concentration of each fraction from albedo tissue was in the order: cellulose > hemicellulose > hot water > EDTA. The range of variation in cell wall sugars in albedo tissue was smaller than that in flavedo tissue.

11.0 Pests, Diseases and Disorders

Citrus Decline

Citrus decline as popularly known in citrus belts of Indian subcontinent has been the subject of considerable research world over. Citrus decline is also known by various other names such as blight in Florida, declinio in Brazil, declinamiento in Argentina, and marchitamiento repentino in Uruguay, depending upon causal factor. It is referred to as frenching decay, chlorosis, neglectosis, and amachamients in Mexico (Srivastava and Singh, 2009). The problem of citrus decline is often confused with die-back and most of the time, citrus decline is considered similar to citrus die-back, considering the fact that no sharp distinction exists between them even today. However, for all practical or operational purposes, the term die-back denotes a lethal condition leading to rapid death of the plant from top to downward due to one or more pathogenic causal factors, whereas the decline refers to a gradual reduction in orchard's productivity which may occur due to nonpathogenic factor (Ghosh,1985). The visual symptoms are variable, often nonspecific, and unreliable to determine the cause and effect relationship. Among the factors responsible for citrus decline, malnutrition is frequently ascribed to chlorotic condition of trees. This condition usually develops after initial few years of excellent 'on' years. Various symptoms comprise of: light green interveinal areas with the midrib, causing yellowish condition of leaf at advanced stages. In this way, growth of the plant is partially retarded, and the plant bears short twigs with narrow leaves. The shoots have a tendency to die-back and even cause the whole of the tree to dry completely in the subsequent years. Of late, multi-micronutrient deficiency in Marathwada region of Maharashtra, India was established to be the causal factor for sweet orange decline (Srivastava and Singh, 2009).

In another review detailing biotic factors, Meena *et al.* (2018) describes biotic causes, namely, rootstock, phanerogamic parasites, intercrops, insects-pest, nematodes and diseases (fungal, bacterial and viral). Major causes of citrus decline in Northeastern India: 1. Citrus trunk borer; 2. Occurrence of greening disease exists in north eastern citrus growing region (Ghosh *et al.*, 1982). Greening and tristeza are present in Khasi mandarin, tristeza is also present in Kagzi lime and Assam lemon in the NE of India by Bhagabati *et al.* (1989). Das *et al.* (2007) also reported greening disease in NE states of India through PCR technique. This is the main cause of citrus decline in NE states; 3. High Incidence of CTV (Singh *et al.*, 2017) in NE states of India; 4. Phytophthora rot and 5. Overall neglect of grove.

Flowering and Pollination in Citrus

Citrus sinensis

Citrus reticulata

Citrus grandis

Citrus jambhiri

Citrus aurantium

Citrus aurantifolia

Pollination by Bees

Pests of Citrus

***Citrus psylla* Infestation**

Leaf Miner Infestation

11.1.0 Pests

Citrus is affected by a large number of insect pests which contribute towards citrus decline (Chadha *et al.*, 1970). They are: citrus *psylla*, citrus blackfly, whitefly, citrus leaf miner, bark eating caterpillar, trunk borer, citrus aphid, mealy bugs, citrus thrips, scales, and mites infesting all stages of growth right from seedlings in nurseries (Aiyappa *et al.*, 1962).

More than 120 species could be recorded in India. In Punjab about 22 species could be recorded (Bindra, 1966; Singh and Sohi, 1957), out of which 14 species were found to be of importance. In Karnataka State more than 50 species were found to damage citrus, while in the north-eastern region 51 species of insect pests were found to be associated with citrus, mandarin orange in particular. The following are some important citrus pests with suggested control measures.

11.1.1 Citrus Psylla (*Diaphorina citri*)

Citrus psylla is a serious pest on young flushes that causes heavy flush drop affecting the fruit set extensively. In severe infestation drying of twigs occurs. It also secretes white crystalline honey dew on which sooty mould grows. More importantly, the insect is an active vector of phytoplasma like organism that causes the dreaded greening disease (Capoor *et al.*, 1967) and thus contributes to the decline of citrus orchards through direct and indirect spread of greening (Ghosh, 1985).

It is a small (3-4 mm long) flying insect of grey colour, having broken brown band in the wings. A single female insect can lay about 800 eggs after the pre oviposition period of about 24 hours. The nymphs hatch out after about 4-6 days in summer and 22 days in winter. The nymphs reach adult stage after 11-25 days and there is no pupal stage. Citrus psylla attacks all species and cultivars of citrus and the nymphs and adults cause damage by sucking the plant sap. The nymphs suck the cell sap from tender shoots, leaves and flowers causing curling of leaves, defoliation and drying of twigs. The nymphs also secrete whitish crystalline honey dew which helps in the growth of fungi. The adult insects may live up to 6 months and there are about 10 overlapping generations in a year. As a result, population build up takes place within a short period of time, making the control of the pest quite difficult. However, some natural enemies like five species of lady bird beetles, Syrphid fly, Chrysopa, certain spider mite were found to cause harm to the psylla. For controlling the insect through chemicals, spraying with malathion (0.05 per cent), monocrotophos (0.025 per cent), carbaryl (0.1 per cent), phosphamidon (0.025 per cent) and parathion (0.025 per cent) have been recommended by various workers. Azadirachtin (0.03 per cent) and neem (*Azadirachta indica*) oil gave high level (90 per cent) of control of citrus psylla (Chakravarthi *et al.*, 1998).

In India (Punjab), *Coccinella septempunctata rependa*, *Menochilus sexmaculatus*, *Chilocorus nigritus*, *Brumus suturalis*, *Chrysopa*, Syrphid fly have been reported predating on psylla (Singh, 1993 and Tandon, 1993). Among the parasitoids the eulophid *Tetrastichus radiatus* has been found effective in regulating the pest population. Besides, *T. phyllocnistoides* and *Diaphorina aligharensis* have been reported from Vidarbha (Tandon, 1993). As high as 95 per cent parasitism of citrus psylla by *T. radiata* has been reported (Singh, 1993).

11.1.2 Citrus Leaf Miner (*Phyllocnistis citrella*)

It is a serious pest in the nurseries as well as in young and old orchards and has been found to occur throughout the country. It damages all citrus species and cultivars. The adults are minute silver white moth with black eyes. The larvae usually mine the underside of leaves making serpentine mines which are silver coloured because of entrapped air. The mining effect badly distorts the leaves resulting in their curling and defoliation during severe attack. The adults lay about 62-127 eggs during 3 days of their life and the entire life cycle is completed in 14-21 days in May-August and 38-65 days during December-March (Bindra, 1970). There are about 16 overlapping generations per year and during cool winter months populations are much reduced. Weekly sprays of monocrotophos (0.035 per cent) during flushing period or spray of phosphamidon (0.03 per cent EC) have been found to be effective in controlling the insect. In Spain, tubefenozide showed good control of leaf miner. Best results were obtained at the rates of 14.4 and 18 g/hectoliter mixed with summer mineral oil. Tests against mites and other beneficial insects indicated an excellent selectivity of tebufenozide (Molla and Jousseaume, 1997). Spraying with abamectin (1.8 per cent, 20 cm^3/100 l water) or imidacloprid (35 per cent, 25 cm^3/100 l water) gave 90 per cent control of *P. citrella* (Salas *et al.*, 1997).

Citrus leaf miner, although basically the seedling pest at times invades the orchards in new flush epidemically causing enormous loss to the vegetative growth of the plant. More than 80 per cent nurseries are infested by the pest in central India (Kalidas and Srivankar, 1995). Epidemic of the pest was also reported from Warud area of Vidarbha (Ghuguskar *et al.*, 1981) in 1981 summer during which more than 87 per cent leaf damage was recorded and also more recently in the *Ambia* flush of 1998. In India, a rich fauna of bioagents of the pest is found which can prove to be the greatest asset in controlling the pest menace. Parasitoids, like *Tetrastichus phyllocnistoides, Citrospillus quadristriatus, Elasmus clarripennis, Bracon phyllocnistoides* and *Simipieses purpurea* have been recorded causing 30-40 per cent parasitism. Black chalchid larvae *Ageniaspis* sp. was found causing 80 per cent parasitism of the moth (Anon, 1991).

Application of insecticides as stem paints or drenches were considered to be most suitable for use in an integrated pest management programme of citrus leaf miner (Iordanou and Charalambous, 1998). Biological control by the inundative release of parasitoid *Ageniaspis citricola* has been reported (Pomerinke and Stansly, 1998; Willink *et al.*, 1998).

Two laboratory bioassays were developed for testing insect pathogens to control *Phyllocnistis citrella* larvae in isolated citrus (cv. Marsh and Inman) leaves. Silwet L-77 and 3 commercial preparations of *Bacillus thuringiensis* (Bt) (Condor, Biobit and Florbac) were tested. Commercial Bt preparations were active when applied topically at high concentrations or when injected at lower concentrations. When Silwet L-77 (0.01 per cent LC_{10}) was added to Bt in the topical bioassay, the activity of Bt was significantly greater than without Silwet L-77. Upto 90 per cent mortality occurred at 0.01 per cent Silwet L-77 + 20 per cent Bt. Mortality is correlated positively with Bt dose when topically applied with 0.01 per cent Silwet L-77 (Shapiro *et al.*, 1998).

Limited effectiveness of chemical control against the citrus leafminer *Phyllocnistis citrella* directed the research work to biological control by protecting local beneficials and introducing exotic parasitoids (Abassi *et al.*, 1999). The studies were carried out in grapefruits in Morocco. The local parasitoids include two species: *Cirrospilus pictus* and *Pnigalio* sp., the introduced ones were *Ageniaspis citricola* and *Semielacher petiolatus*. All these entomophagous insects contributed to reduce the citrus leaf- miner populations but the main impact was due to the imported beneficials specially *S. petiolatus*.

11.1.3 Aphids

Two major species of citrus aphid attacking citrus are brown citrus aphid (*Toxoptera citricida*) and black citrus aphid (*T. aurantii*). The real importance of aphids is due to their relation with the tristeza virus. Brown citrus aphid is the efficient vector of tristeza that is 'quick decline', which played havoc with the citrus industry of Argentina, Brazil and California and has been reported to be present in Maharashtra (Nagpal, 1959), Andhra Pradesh (Reddy and Papa Rao, 1960), Delhi (Nariyani *et al.*, 1967) and also in Punjab (Anon. 1965). Four species of aphids, namely, brown citrus aphid – *Toxoptera citricida* (*Aphis citricidus*), *Toxoptera aurantii* (black citrus aphid), *Aphis pomi* (green apple aphid) and *Myzus oersucae* (green peach aphid) have been reported to occur in citrus in India (Pruthi and Mani, 1945). However, the brown and black citrus aphids are more prevalent and cause major damage to citrus.

The aphids infest the lower surface of the leaves, tender shoots and young fruits. Curling of infested twigs and leaves are usually common. The black citrus aphid produces about five new ones daily for a period of one to three weeks. For the control of aphids, insecticides like oxydemeton methyl (0.025 per cent), monocrotophos (0.025 per cent), parathion (0.03 per cent), methyl demeton (0.03 per cent) and phosphamidon (0.035 per cent) have been found to be effective.

Chemical control of aphid as a vector is the most acceptable approach to greening control. Insecticides like azodrin 40 (monocrotophos), dimethoate, aldicarb, quinalphos, fenvalerate, fenthion, phosalone, fenitrothion and endosulfan were found quite effective (Khanna and Mathur, 1978; Tandon, 1992). Extracts/oils of seed kernels of *Azadirachta indica, Pongamia glabra, Madhuca longifera, Ricinus communis, Cymbopogan nordus* leaves, *Alpinia galanga* bulbs and several others alone or in combination have been tried successfully against citrus insect pests (Borle, 1974; Pande, 1982; Anon, 1989; Chin *et al.*, 1990; Katole *et al.*, 1993; Tandon and Jothi, 1993; Savitri *et al.*, 1997). However, exhaustive studies on this aspect are required to be undertaken.

11.1.4 Scales

Citrus scales, the sap sucking insects, are immobile in the adult stage and are of two distinct types – armoured scales, which have hard cover separate from the body providing protection to the body underneath, and soft scales, which have no separate cover but sometimes a hard skin or a protective waxy secretion. Pruthi and Mani (1945) reported 49 species of scales on citrus in India. The scale insects can be classified as:

11.1.4.1 Armoured Scale

(a) Red scale (*Aonidiella aurantii*): The cover of adult females is circular, flattended, orange to red in colour, and the body under scale is somewhat conical or pear shaped.

(b) Florida red scale (*Chrysomphalus aonidum*): Cover of adult females is circular and female body under the scale is yellowish in colour.

(c) Yellow scale (*Aonidiella citrina*): Very similar to red scale but less damaging, shade-loving species. Usually found on undersides of leaves on lower and inner branches; Yellowish brown in colour.

(d) Oriential yellow scale (*A. orientalis*): Female usually circular with about 2 mm diameter, yellowish brown in colour.

(e) Glover's scale (*Lepidosaphes gloverii*): Scale cover of 2.5-3 mm long, thin and parallel sided. Female deposits eggs beneath the scale cover in two rows.

11.1.4.2 Soft Scales

(i) Soft brown scale (*Coccus hesperidus*): Mature scale is flattened, ovoid, mottle brown to pale yellow in colour, having a waxy parchment-like upper skin. The mature female deposits a few larvae beneath her body each day for one to two months and several generations of soft brown scale develop during the year. It infests foliage and young twigs and produces honey dew copiously, resulting in sooty mould association.

(ii) Soft green scale (*Coccus viridis*): Adult female is ovate, about 3 mm in length and pale green in colour. Each adult female gives birth to 300-500 young ones (crawlers), which settle in young leaves, twigs and fruits. Adult females live for 2-3 months and there are four overlapping generations. Sooty mould association is very common with the insect.

It is difficult to control hard armoured scale insects with insecticide sprays. Soft scales, however, can effectively be controlled by spraying parathion (0.03 per cent) emulsion, dimethoate plus kerosene oil @ 150 ml plus 250 ml respectively, in 100 litres of water, carbaryl 0.05 per cent plus 1 per cent oil, malathion (0.1 per cent), *etc.* Single application of systemic insecticide like monocrotophos (0.1 per cent) can also control soft green scale. Spray of organophosphorus insecticide (like rogor) plus oil sprays can keep the red, yellow scales under control.

11.1.5 Mealy Bugs (*Pseudococcus filamentous, Planococcus citri*)

Mealy bugs are closely related to scales and have been reported from various parts of the country. The body of the insect is segmented and is covered with white mealy wax. Mealy bugs suck plant sap and may cause fruit drops. It also secretes honey dew on which sooty mould develops. Mealy bugs can considerably be controlled through similar sprays as recommended for citrus scales. The citrus mealy bug, *Planococcus citri* is a major pest of citrus orchards in Crete, Greece. The release of predator *Cryptolaemus montrouzieri* effectively reduced populations of *P. citri* and a predator: prey ratio of 1:15 was found effecive (Hamid and Michelakis, 1997).

Pests of Citrus

Aphid infestation

Fruit Sucking Moth Infestation

Trunk Borer Infestation

Lemon Butterfly Infestation

Pests of Citrus

Whitefly Infestation

Fruit Fy Infestation

Mite Infestation

11.1.6 Citrus Bark Borer (*Inderbela tetraonis, Inderbela quadrinotata*)

The bark and shoot borers are serious pests in citrus, especially in neglected and old orchards. The caterpillars feed on bark and thus destroy the translocation tissues of the bark. Since they feed at night, they generally escape notice but their presence on infested trees is indicated by the presence of hanging loose mass of the fine pieces of woods and pellets of excreta, mixed with silky adhesive materials on the branches and stem of the tree. Two species of bark eating caterpillar (*Inderbela quadrinotata* and *I. tetraonis*) are quite serious in neglected orchards. Of late, the infestation is spreading even in the young orchards of Nagpur mandarin alarmingly. Twigs and branches get partly girdled, such branches tend to flower and fruit profusely more than normal and break or decline fast without carrying the crop to maturity (Rajput and Haribabu, 1993). As high as 40-60 per cent orchards suffer from this pest in Vidarbha (Thakre *et al.*, 1984; Shivankar, 1997). The yield and fruit quality are affected (Bindra, 1969). A single borer can kill one tree.

The caterpillars can be controlled by inserting kerosene, petrol, carbon disulphide or chloroform plus creosote (2:10) into their holes and plugging the holes with mud. Removal of webbing, followed by painting or spraying of bark all around the damaged portion with Paris green has also been found to be effective in controlling the insect.

The bark borers can be controlled effectively with 1-2 sprays of systemic or contact insecticides available. Two natural enemies of the pest namely *Syrphus* sp. and *Coccinella septempunctata* were found predating upon the pest. Of these *Syrphus* sp. was predominant (12.72 per cent in relation to aphid population) throughout the pest incidence. Removal of frass and spraying with monocrotophos (0.05 per cent) is an effective control measure. The best time for control is September-October and January-February. The alternate hosts in the vicinity like jamun, guava, litchi, pomegranate, and jujube may also be treated in case of severe infestation (Bindra, 1970).

11.1.7 Citrus Trunk Borer (*Monohomus versteegi*)

Citrus trunk borer (*Monohomus versteegi*) is a serious insect pest of mandarin orange in north eastern region and Sikkim. In Assam 15 to 60 per cent orchards are under the spell of this pest (Chaudhary, 1954). It is equally serious and responsible for citrus decline in West Bengal (Nayer *et al.*, 1976) whereas about 20 per cent mortality of mandarin plantations has been reported from Assam (Dutta, 1959).

The grub bores into the trunk near the tree base and makes tunnels near the pith. The tree dies eventually due to infestation. The insect can be controlled by injecting petrol or carbon disulphide into the bored holes and plugging them with wet soil. Pruning and burning of severely infected branches along with the grub will help in checking the populations build up of the insect.

Similar other borers, such as lime tree borer (*Chelidonium cinctum*), orange shoot borer (*Obera lateapicalis*), orange stem borer (*Chelidonium argentatum*), orange beetle (*Stromatium barbatum*), *etc.* have also been reported from various parts of the country. Multiplication of borers can be checked by keeping the orchard clean, treating the soil around the tree with chlorpyriphos and by spraying with chemicals like methyl parathion (0.05 per cent) or endosulfan (0.5 per cent), carbaryl (0.2 per cent) *etc.*, during oviposition periods.

Machado and Raga (1999) evaluated the efficacy of chemical and biological insecticides against the citrus trunk borer *Macropophora accentifer* in commercial orchards of sweet orange in Brazil. The following insecticides and doses were tested on infested branches or trunks: *Metarhizium anisopliae* (1 x 10^8 conidia/ml), triazophos 400 BR (0.04 g) and imidacloprid 100 AL (0.5 and 1.0 g). Survival analyses were carried out 15 days after application. Imidacloprid (0.5 and 1.0 g) and *Metarhizium anisopliae* reduced the number of larvae by 94, 84 and 88 per cent, respectively. Application of triazophos reduced larval numbers by only 51 per cent.

11.1.8 Citrus or Lemon Butterely (*Papilio* sp.)

Several species of the genus *Papilio* have been reported to occur in different parts of India, of which *P. demoleus* is the most common and serious species. The caterpillars are brownish or black in colour with irregular whitish stripes at young stage. The fully grown caterpillars feed on the leaves and defoliate the branches. Ordinary nursery stocks and young trees are badly damaged by the insect which often feed on fruit stalk causing fruit drop. Hand picking of larvae, spraying with carbaryl (0.1 per cent) WP may be helpful in controlling the insect.

11.1.9 Citrus White and Black Flies

About 30 species of citrus white flies have been reported throughout the world, out of which a few species like *Aleurocanthus husaini* (Hussain's white fly), *A. spiniferus* (orange spiny white fly), *A. citrifolii, Dialeurodes citri, D. elongata* are important in India. The adults of white flies are minute creatures with mealy wings held over the body in a roof-like manner. The adults and the nymphs suck the juice of the leaves and as a result the leaves turn pale. Due to secretion of honey dew, sooty mould also develops on the leaves. The intensity of attack is more on the shady sides of the tree and dry hot summer adversely affects the nymphs of the insect.

Several species of parasites (*Eretmocerus* spp., *Prospaltella* sp.) and predators (*Brumus suturalis, Scymmus* sp.), *etc.* are natural enemies of citrus white flies as the larvae and adults of these parasites and predators feed on white fly eggs and nymphs. Pruthi and Mani (1945) recommended biological control and advocated even the sprays of suspension of entomophagus fungi for the control of white fly. According to Bindra (1970) effective control could be achieved by spraying a 0.03 per cent emulsion of isobenzan and parathion.

Epidemic of citrus blackfly (*Aleurocanthus woglumi*) on Nagpur mandarin during the seventies, eighties and early nineties had put the citrus industry of central India in doldrums. Widespread growth of black sooty mould (*Capnodium* sp.) on the voluminous honey dew secreted by the numerous citrus blackfly nymphs and adults covering the entire trees of orange or lemon, locally known as Kolshi, not only inhibits photosynthetic activity but also affects the appearance and quality of the fruit (Thakre *et al.*, 1985). First and second nymphal instars of the pest are more vulnerable to control, hence the spray schedule should be undertaken at

50 per cent egg hatching stage that ensures maximum coverage of the younger stages of the pest in two sprays to be effected at 15 days intervals with any of the following insecticides, like monocrotophos 1 ml, acephate 0.8 ml, phosphamidon 0.7 ml, phosalone 1.5 ml and dimethoate 2 ml/l of water till drenching that ensures thorough coverage of the underside of leaves (Anon, 1994b).

Time of control: Period of 5 per cent egg hatching

Ambia: First fortnight of April

Mrig: Second fortnight of July

Hasta: First fortnight of December

Use of predators like *Mallada boninensis* and *Serangium parcesetosum* and a parasitoids *Encarsia* sp. and *Eretmocerus gunturensis* hold promise in the control of the pest biologically. The pest activity is maximum during June-July, September-October and March-April and the spray at such time with 0.3 ml phosphamidon or quinalphos 1 ml or thiometan 1 ml/l of water gives effective control of the pest. The host specific eulophis parasitoid *Taxarixia radiata* holds promise for the control of the pest because as high as 40-50 per cent and even above 90 per cent parasitism was recorded at certain locations in Vidarbha (Shivankar and Kohli, 1996).

11.1.10 Fruit Flies

Fruit fly infestations have been recorded from limited areas of comparative cooler and wet zones. In Coorg region of Karnataka, the incidence of oriental fruit fly (*Dacus dorsalis*) had been noticed in citrus orchards intercropped with coffee. By making punctures in the rind of semi-ripe fruits, the female inserts 2-15 eggs at a time and about 200 eggs are laid in a period of a month. The maggots develop in the fruit pulp, and such fruits usually drop prematurely. Secondary infestation of fungus and bacteria in the punctured fruits results in fruit rotting as well. In Sikkim and certain other north-eastern states in India, fruit flies sometimes cause very serious damage of mandarin orange through premature fruit drop.

The fallen infested fruits are to be destroyed for reducing the insect population. Use of poison baits in the orchards starting about two months before harvesting of fruit helps in controlling the insect, and the control of oriental fruit fly could be achieved by applying bait spray containing malathion or trichlorofon spray or fenthion (0.05 per cent) plus crude sugar (1 per cent).

The phosphine cylinder gas formulation ECO_2FUME (phosphine with CO_2 as a carrier gas) was used by Williams *et al.* (2000) as a basis for experimental fumigations of oranges infested with larvae of Queensland fruit fly (*Bactrocera tryoni*) in a 900-litre chamber. Uninfested oranges were included in all fumigations to check for possible adverse effects. A 16-h fumigation at 20°C with an initial phosphine concentration of 0.98 g m-3 resulted in 96.4 per cent mortality of fly larvae.

11.1.11 Mites

Three species of mites have been considered as major pests in citrus and according to Ghai (1964), 13 species of mites have been found to be associated with citrus in India. The insects feed on the leaves, produce multiple grey spots, weaken the leaves and cause defoliation. In Punjab, the common citrus mite, namely *Eutetranychus banksi* and oriental red mite (*E. orientalis*) have been reported to occur. The species like *Tetranychus sexmaculatus* and *T. telarius* were found to be prevalent in Assam. The species *E. banksi* can be controlled by spraying 0.02 per cent emulsion of malathion, parathion or cabophenothion (Sethi, 1967).

For control of red spider mite and oriental red mite, insecticides like monocrotophos (0.5 per cent) and dicofol (0.05 per cent) have been found to be effective. In the orange orchard of Brazil, good control of mites (*Brevipalpus phoenicis*) was observed by application of halfenprox (0.8 and 1.6 g a.i./100 l) which showed a reduction in population of 95 per cent after 127 days of application (Raga *et al.*, 1997). Three spray applications of sulphur colloid suspension could control the citrus red spider mite (*Panonychus citri*). The first spraying should be made after harvest, second at the bud break stage and third in early May. Such a spraying regime was found more effective than applying other acaricides and it was also observed that sulphur colloid suspension had the added effect of improving fruit colour (Youzhu, 1977).

11.1.12 Fruit Sucking Moth

Fruit sucking moths, belonging to different genera (*Otheris, Achoea, Calpe, Anua, Ophideres, etc.*) cause some damage by sucking juice of citrus. Ripe fruits of sweet orange, mandarin and sweet limes are commonly attacked. The insects pierce the fruit rind and expose it to secondary infections of diseases and infestation of flies. The fruit usually drops within a few days and thus becomes unmarketable. It is claimed that through elimination of alternate host plant from the vicinity of citrus orchards the insect infestation could be considerably reduced. Disposal of fallen fruits, which attract the moths, may also help in reducing the insect population. Baiting with malathion (0.05 per cent) and crude sugar and fruit juice also helps in controlling the citrus fruit sucking moth.

The other citrus insect pests of lesser importance in India are citrus stinkbug (*Rhinchocoris humeralis*), white ants (*Odontotermes obesus*), red-tree ant (*Oecophylla smaragdina*), thrips (*Thrips hawaiiensis*), *etc.* The importance of citrus psylla and aphids which have been reported as vectors of greening mycoplasma and tristeza virus respectively, has attained significance in the recent past and the combating of these insect vectors are very essential to check further spread of these deadly mycoplasma-viral diseases of citrus.

Fruit sucking moth, *Otheris* sp. is the regular and serious pest of ripening citrus fruits causing 25-40 per cent fruit drop, particularly during September-October (Shivankar and Kohli, 1996). Citrus mite *Eutetranychus orientalis* is another pest which not only damages the leaves but also the fruits causing the unpleasant blemishes on the peel, deteriorating the quality. Some predatory mites have also been recorded (Shivankar and Kohli, 1996). Lemon butterfly, *Papilio demoleus* feeds voraciously on the vegetative growth throughout the year. The outbreaks of the pest were reported in 1940, 1969, 1982 and 1983 and more recently in July, August 1996 from Vidrabha (Thakre *et al.*, 1984-85 and Shivankar and Kohli, 1996). Bark eating caterpillar *Inderbela quadrinotata* which was a minor pest a few years ago has now assumed the serious dimensions by infesting around 40-65 per cent both old and young orchards necessitating an immediate attention of the growers and also the scientists. Pest incidences have been invariably noticed in the declining orchards in central India. The larval parasitization by *Apanteles* sp. was recorded in case of both lemon butterfly and bark eating caterpillar (Shivankar and Shyam Singh, 1997).

11.1.13 Nematodes

In China, seven species of nematode including *Tylenchulus semipenetrans* were identified in diseased pumello (*C. maxima*) orchards. The growth of infected trees growth was severely depressed, the leaves were yellowing, non-glossy and falling and the apical twig production was slow or nil. Application of 3 per cent furadan (carbofuran) at 150 kg/ha to the soil and irrigation in March-April and August-September helped to control the disease (Ye and Jian, 1997). In Cyprus, grapefruit infected with *T. semipenetrans* were effectively controlled by application of rugby (cadusafos) at 4.9 kg a.i./ha which showed a yield increase of 8.5 per cent. Soil solarization using 80 or 160 m transparent or black polyethylene sheet with 50, 75 and 100 per cent soil cover, during June to August showed significant reduction in pupulation of *T. semipenetrans* in Washington Navel orange orchards of Egypt. Soil solarization, either with transparent or black sheet, substantially increased soil temperature, soil moisture, heat stored and significantly reduced nematode infestation (Ismail *et al.*, 1997). Application of sawdust at 20 t/ha reduced population of *Tylenchulus semipenetrans* compared to control (Al-Hinai and Mani, 1998).

Seven selections of cross Cleopatra mandarin (*Citrus reshni*) x *Poncirus trifoliata* and one of *C. volkamariana* x *P. trifoliata* were considered as highly resistant to citrus nematode (Verdejo-Lucus *et al.*, 2000).

11.2.0 Diseases

Large number of diseases has been reported to occur in citrus. Fawcett (1938), Knorr *et al.* (1957), Wallace (1959) dealt with the disease problems of citrus elaborately, while citrus diseases of India and their association with citrus decline or die back complex have been reported by Reddy (1968), Raychaudhari *et al.* (1969), Fraser (1966), Chohan and Knorr (1970), Nariani and Singh (1971), Capoor (1975) and others. The diseases can be caused by viruses, bacteria and fungi and the important diseases with their suggested control measures are discussed below:

Fungal Diseases

11.2.1 Phytophthora Rot (Foot Rot, Collar rot, Crown rot, Root rot, Brown Rot, *Phytophthora gummosis*)

Almost all the citrus tracts face the problem of *Phytophthora* spp. causing varying degree of root rot, crown rot, gummosis and brown rot of citrus in India (Bajwa, 1941; Chowdhury, 1951; Naqvi, 1988). Besides *P. parasitica* and *P. citrophthora* (Naqvi, 1988), *P. palmivora*, *P. arecae*, *P. nicotinae* are also reported to cause enormous damage to citrus plantations. Predisposing conditions for *Phytophthora* diseases like poor drainage, excessive irrigation, water logging and prolonged contact of water with tree trunk and use of susceptible rootstocks should be avoided (Naqvi, 1988). The fungi are soil inhabitants and flourish well under high moisture conditions. Citrus trees exposed to wet soil for an extended period are more prone to phytophthora infection.

Phytophthora disease produces symptoms of decline in citrus trees through rotting of rootlets, girdling of the tree trunk and defoliation. The first symptoms are dark staining of the bark and the invaded bark is at first light olive brown in colour, firm, with a water-soaked appearance. The fungus affects the bark and can penetrate in a limited way into the wood. In advanced stages, the bark cracks, shreds in length-wise strips as it dries and gum starts exuding, particularly in the late spring. Yellowing and die back of limbs above the injury occur after the disease is well advanced. Lemon, sweet orange, grapefruit and mandarin are highly susceptible, while the rootstocks like sour orange, Cleopatra mandarin, trifoliate orange, Rangpur lime, *etc.* have been found to be tolerant (Klotz, 1950).

Raising of nursery stock in phytophthora-free conditions, avoiding stagnation of water around tree trunk and good drainage provision, painting of Bordeaux paste on the trunk up to about 50-60 cm, use of tolerant rootstocks and high budding at about 30 cm above ground level are some of the preventive measures to be adopted for escaping the disease. In case of infection, if the disease can be detected before the bark covering more than half of the circumference of a stem is damaged, scraping of the affected portions with a little extra healthy tissue without injuring the wood and application of Bordeaux paste (1:2:16-20 water) or Bordeaux paint (1:2:3 linseed oil) on the scraped as well as healthy portion around the infected zone, may help in combating the disease and recovery of the plant.

Spray and drench of metalaxyl or fosetylal provide effective control of phytophthora diseases (Naqvi and Sawant, 1994). Use of *Trichoderma* culture also controls *Phytophthora* sp. at the disease development stage (Sawant *et al.*, 1995). The role of the association of *Trichoderma* spp. isolates and 7 rhizobacteria isolates on *P. parasitica* (*P. nicotianae* var. *parasitica*) and *P. citrophthora* on *Citrus* was tested (Amorim *et al.*, 1999). Under greenhouse conditions, seedlings of *Citrus* sp. were treated with *Trichoderma harzianum* and *T. koningii* isolates (TA) suspensions of rhizobacteria – *Pseudomonas putida*, *P. fluorescens*, *Bacillus subtilis* and *Pseudomonas fluorescentes* (RS) and the association of rhizobacteria with *Trichoderma* isolates (RTA). The seedlings were transplanted to pre-infested substrates with *P. parasitica* and *P. citrophthora* isolates. Treatment RTA controlled the 2 pathogen species but RS only controlled *P. citrophthora*.

11.2.2 Powdery Mildew (*Oidium tingtaninum*)

Powdery mildew is a common disease of citrus in India. On infection whitish powdery growth develops on the young leaves (both sides) and twigs. The affected leaves get distorted in shape and reduced in size. In case of severe infection, leaf falling, premature fruit dropping and die back symptoms become apparent. Warm and humid weather favour the growth of the fungus and usually during spring and autumn the disease symptoms are more severe.

Dusting of sulphur powder (200-250 mesh) or spraying of wettable sulphur (1 kg in 200 litres of water) can control the disease, if applied at the initial stage of

Pests and Diseases of Citrus

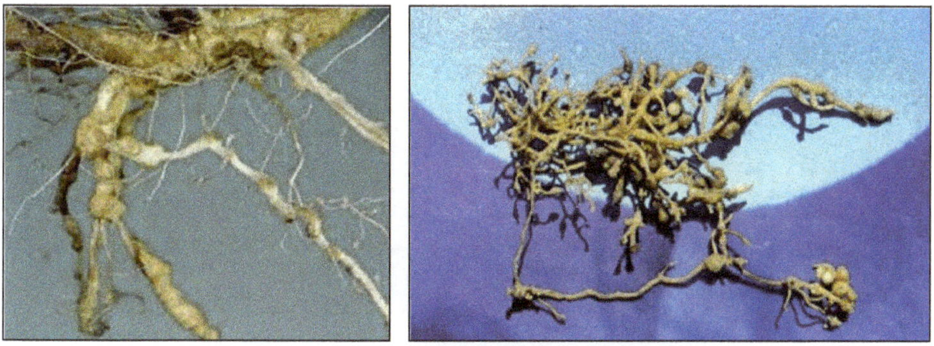

Nematode Infestation

***Phytophthora* Rot Infection**

Powdery Mildew Infection

Pink Disease Infection

Diseases of Citrus

Scab Infection

Anthracnose Infection

Greening Infection

Canker Infection

the disease. In case of severe infection, repeated application of the fungicides like bavistin or calyxin may be essential. Powdery mildew could effectively be controlled by spray application of tridemorph, triadimefon, inocap and benzimidazole fungicides at 10 days interval after first sign of attack (Narasimhan *et al.*, 1984).

11.2.3 Pink Disease (*Corticium salmonicolor* or *Pellicularia salmonicolor*)

Pink disease has been reported from different parts of India, and besides citrus, the desease infects other species like rubber, coffee, cocoa, tea, cinchona, mango, cashew, jackfruit, *etc.*

In the early stage of infection, branches and leaves start drying and the affected branches get covered with fine silvery white mycelium which later turns pink in colour. Brownish discoloration of the bark and wood and shredding of bark into longitudinal strips are noticed in the infected trees. Desiccation, chlorosis, dropping of leaves and final drying of twigs are commonly observed, during August to October, when the disease usually appears.

Pruning and burning of diseased branches and spraying of Bordeaux mixture (1 per cent) or other copper fungicides like blitox (0.2 per cent) also control the disease.

11.2.4 Scab (*Elsinoe fawcettii*)

Scab disease, which is often confused with bacterial canker, has been reported to cause limited damage to citrus, through development of corky lesions on fruits, leaves and young twigs. The leaves with severe infection become distorted, wrinkled and infected fruits become hard and often drop prematurely. This is one of the major diseases in the hilly tracts where low temperature and high humidity prevail. In the north-eastern hilly region and parts of West Bengal the disease is commonly observed in the mandarins, citrons, lemons, *etc.*

Removal of infected leaves, twigs, fruits and spraying of Bordeaux mixture or 0.3 per cent blitox have been found to be effective in controlling the disease.

11.2.5 Felt Disease (*Septobasidium pseudopedicellatum*)

It is a common disease of Assam and other north eastern states of India and of limited occurrence in other parts of the country. In case of infection, the twigs and branches are encircled by soft felt like leathery fungal growth, which may extend up to the petioles and leaf bases and fruit stalks. The causal fungus remains on the surface and below the felt like growth colonies of scale insects thrive.

Usually, older and weak trees in neglected orchards are infected and in north-eastern region of India mandarin (Khasi orange), pummelo (Robabtenga), Adajamirs (*C. assamensis*), Assam lemon *etc.*, are commonly attacked by this fungus. Spraying with Bordeaux mixture controls the disease.

11.2.6 Fruit Stem Anthracnose (*Colletotrichum acutatum*)

The stem of the infected fruit first becomes light-yellow in colour, then brown, wither and drop down. Spraying with chagufen + water (boiled and filtered) then adding water + Jinggangmeisu + salt gave up to 96 per cent control (Liu, 1998).

In addition to the above diseases, certain other fungal diseases like damping off (*Pythium, Rhizoctonia* spp.), wilts (*Fusarium, Verticillium* spp.) are common in the nurseries, sooty mould (*Opodium citris*), melanose (*Diaporthe citri*), ganoderma root rot (*Ganoderma lucidum*), mushroom root rot (*Armillaria* sp.), dry root rot (*Macrophomina, Fusarium, Diplodia* spp.) *etc.*, have been reported to occur in certain areas.

Proper orchard management, clean cultivation and spraying of certain general fungicides like Bordeaux mixture may keep most of the diseases under control.

Bacterial Disease

11.2.7 Greening

Greening, a comparatively new malady, is widespread in many species of citrus and is considered by many to be the second most serious disease. The disease is caused by the new group of organism known as mycoplasma, bacteria-like-organisms (BLO) and not by a virus as believed earlier.

The infected plant shows chlorosis of leaves resembling zinc deficiency (sometimes dotted with green islands), thickening of leaf blade, shortening of twig internodes, off-season blooming, leaf shedding and die back. The chlorosis usually is bound on one side of the midrib and on another side by a lateral vein. The chlorosis diffuses towards the leaf margin. Fruits of the infected trees are usually small-sized, lopsided, remain green for longer period, contain curved columellas, aborted seeds and fall prematurely.

The disease can spread through infected bud wood and the insect vector, citrus psylla (*Diaphorina citri*). The presence of the disease could be confirmed through indexing on indicator plants like sweet orange cultivars Valencia, Pineapple, Navel orange, *etc.*, or through chromatographic techniques developed by Schwarz (1968).

The infected trees are required to be removed for preventing further spread of the disease. Control of the insect vector through pesticides and use of certified disease-free planting materials are the measures to be adopted to keep the disease under check. It has been claimed to achieve temporary recovery from greening disease through injection of certain chemicals like tetracycline. The presence of the disease has been confirmed in almost all the major citrus growing belts of India (Fraser, 1966; Nariani *et al.*, 1967; Ghosh *et al.*, 1980).

11.2.8 Citrus Canker (*Xanthomonas citri*)

Citrus canker is widespread in India and is a very serious disease in acid lime. The disease is caused by a gram negative bacterium which affects all above ground plant parts like leaves, twigs, fruits and thorns. The canker lesions appear as minute water-soaked roundish spots which enlarge slightly and turn brownish and corky. A yellow-brown halo surrounds the lesion, particularly in the leaves. The size of the lesion varies from host to host; on acid lime and sweet orange they are about 2-3 mm in diameter whereas on grapefruit leaves the diameter may even reach 15 mm. Several lesions may coalesce to form a patch. The disease can spread by splashing of bacteria-laden drops of water falling on leaves, twigs and fruits during rains and even through winds. Citrus leafminer insect also indirectly helps in the spread of the disease. The infected fruit gets less market value and in severe cases becomes even unmarketable. Mandarins were supposed to be tolerant to citrus canker but now certain strains are observed pathogenic on Kinnow plantation at Akola (Maharashtra). Pruning of infected twigs before monsoon and spray of 1 per cent Bordeaux mixture was found effective. Streptocycline in combination with Bordeaux mixture or copper oxychloride also give effective control (Kale *et al.*, 1994). Improved selected clones of acid lime like Saisarbati and Pramalini are tolerant to canker.

Pruning and burning of diseased twigs, specially before monsoon, protective spraying of 1 per cent Bordeaux mixture to emerging flushes or application of streptomycin sulphate (500 ppm) are recommended for control of the disease. Control of citrus leaf miner is also essential to reduce the incidence of the disease.

Total eradication programme of citrus canker has successfully been carried out in Australia, South Africa and the US coastal areas around the Gulf of Mexico. Quarantine on infected plant materials has kept Florida free of citrus canker since the end of 1929 by a 16-year eradication programme (Anon, 1983).

Viral Diseases

More than 15 viral diseases of citrus have been reported from India. Among them tristeza, psorosis, crinky leaf and infectious variegation are wide spread (Raychaudhary and Ahlawat, 1984; Reddy and Papa Rao, 1960). *Xyloporosis impietratua*, leathery leaf, woody gall, tumour, multiple sprouting, rubbery wood, *etc.* are limited to certain parts only (Ahlawat and Raychaudhari, 1988). Planting of virus free planting material is the basic need of the industry for virus elimination. It may be achieved either through shoot tip grafting or by nucellar plants (Kapur and Sohi, 1988). Budwood thermotherapy also eliminates the viruses. Citrus aphid (*Toxoptera citricida*) the vector for tristeza should be controlled to check the spread of these viruses (Randhawa and Srivastava, 1986).

In India, CTV cross protection was initiated in early 1970s. Based on lime reaction alone Balaraman and Ramakrishnan (1980) selected mild CTV strain and used them for cross protection demonstration trial at several locations of Bangalore, Tirupati and Periyakulam. Though initial five to six years it was giving promising results in terms of growth of acid lime plants; afterwards it failed, and no significant yield differences were recorded. Recently mild strain cross protection for CTV has been initiated at NRC for Citrus, Nagpur and IIHR, Bangalore. Similarly, CTV cross-protection is successful in Australia (Broadbent *et al.*, 1991), Japan, Florida (Lee *et al.*, 1987), Spain, *etc.* (Koizumi *et al.*, 1991).

11.2.9 Tristeza (Quick Decline)

Tristeza (a word that describes the sad appearance of a tree) virus caused disastrous effects to the citrus industry in Argentina, Brazil, USA (California), *etc.*, and between 1930 and 1950, more than 20 million citrus trees were lost due to infection of this disease. The lime disease in Gold Coast, bud union decline in Australia, stem pitting of grapefruit in South Africa, tristeza in Brazil, quick decline in California and podredumber de las raicills in Argentina, all were found to be caused by tristeza virus and its related strains. Fawcett and Wallace (1946) reported that 'quick decline', a disease of sweet orange on sour orange rootstock was caused by a virus transmitted by tissue grafts, while Meneghini (1946) showed that tristeza was caused by a virus transmitted by oriental citrus aphid, *Toxoptera citricida*. In India, the experimental evidence of the virus was suspected much earlier. Also, six species of aphids were found to act as vectors of the virus in India (Capoor, 1975).

In the infested trees, leaves develop various deficiency symptoms, and leaf falling, root decay, twig die back and ultimate death of the tree are also conspicuous. Affected trees have a tendency to blossom heavily in the off-season. Specific symptoms of tristeza infection are honeycombing, pitting of inner face of the bark of the trunk, severe pitting develops in the species like lime (Mexican Key lime, Indian Kagzi lime), citron, *Citrus pennivesciculata, etc.* while in sweet orange, mandarin orange, rough lemon, sweet lime, *etc.*, stem pitting is usually absent. The definite test for the presence of tristeza requires greenhouse assay through indexing on indicator plants like Indian Kagzi lime, Mexican lime, *etc.*, where definite symptoms of vein clearing and stem pitting confirm the presence of the virus.

The disease can be kept under control by use of tolerant rootstocks, such as rough lemon, sweet orange, Cleopatra mandarin, Rangpur lime, trifoliate orange, citranges, *etc.* and by avoiding the uses of susceptible rootstocks like sour orange, sweet lime, *etc.* The bud wood to be used for propagation should be free from the viruses and through insecticidal spray the aphid (vector) population can be reduced considerably. In the recent past, certain milder strains of the virus could be isolated and attempts have been made to cross-protect the seedlings against the severe strains through inoculation of milder strains, particularly in case of acid lime. Standardization of such pre-immunization techniques for other commercial citrus species and cultivars may be of much practical significance.

11.2.10 Xyloporosis (Cachexia)

Citrus types that are susceptible to xyloporosis virus are tangelos, sweet limes, kumquats and mandarins. The virus causes the development of peg-like outgrowths from the inner face of the bark that fits into corresponding pit-like depressions in the wood. Gum pockets develop in the bark. The virus is bud transmissible and no insect vector is recognized. Through use of seeds or raising budded plants from disease-free certified bud sticks on tolerant rootstocks, the disease can be kept under control.

11.2.11 Exocortis (Scaly butt)

The virus in bud is mechanically transmissible, but no insect vector is known. It is carried symptomless in sweet orange, mandarin orange, grapefruit and rough lemon, but very distinct bark-scaling symptoms appear in the trifoliate orange and in some of its hybrids like citranges. Rangpur lime and certain citrons have also been reported to be susceptible to the virus. So far the disease is not that serious for citrus industry in India, but with more use of rootstocks like trifoliate orange, citrange, Rangpur lime, the disease can pose threat, if proper precaution is not taken in time. Infected trees on susceptible rootsocks are stunted and show bark scaling, mostly in the rootstock portion.

Since the virus is transmitted through infected bud wood or mechanically through contaminated budding knives and pruning shears, use of disease-free scion material, avoiding susceptible rootstocks and disinfecting the budding knives, pruning shears, *etc.* are some of the useful recommendations for control of the disease.

11.2.12 Psorosis

Psorosis is not a serious disease in India and unlike the other viruses that cause bud union related troubles, psorosis affects cultivars irrespective of the rootstock used. The infected trees show leaf flecking in young leaves, and certain strains cause scaling of bark in the branches and trunk regions. For confirming the presence of the virus through indexing, sour orange, sweet orange, mandarin, Mexican lime can be used as indicator plants. No insect vector could be found so far. It is essential to destroy the infected plants and use certified disease-free bud wood materials for propagation.

11.2.13 Other Viral Diseases

In addition to the above mentioned diseases, there are quite a good number of other viral diseases which cause considerable damage to citrus. The diseases like stubborn (acorn) occurring in the USA and Mediterranean countries, bud-union crease reported from the USA, South Africa, Venezuela, Brazil, Egypt, *etc.* woody gall (vein enation), tatler leaf, Satusma dwarf, yellow shoot (Li-ku-bin), petrifaction *etc.* are of some importance in different countries of the world but of little or no concern presently in India.

11.3.0 Physiological Disorders

11.3.1 Granulation

Granulation is a physiological disorder of the juice sacs of citrus fruits wherein they become comparatively hard, assume a grayish colour and become somewhat enlarged (Bartholomew *et al.*, 1934), with an increase in pectin, lignin and other polysaccharides resulting in considerable decrease in juice, TSS and acid contents and increase in rag and peel proportions (Singh and Chohan, 1982) and water content of both pectic and fruit sacs and an increase in respiratory activity (Zong *et al.*, 1974). Several factors, such as high relative humidity (Lloyd, 1961) high temperature during spring (Matsumato, 1964), time of picking (Sinclair and Jolliffe, 1961a), tree vigour (Matsumato, 1964), *etc.*, influence granulation in citrus. The nutritional composition of leaves and fruits has been found to be related to the degree of granulation (Munshi *et al.*, 1978). Seed content of fruit has definite relation with granulation. Highly granulated fruits have less number of viable seeds and vice versa (Singh *et al.*, 1985). This disorder was first reported from California. Later, it was reported from almost all the citrus growing countries of the world like Brazil, South Africa, Australia, Japan, Israel, Egypt, Vietnam, West Indies, India, Cuba, *etc.* Some success has been reported in reducing the incidence of granulation by lime spray at 35 to 40 lb per 100 gallon water (Bartholomew *et al.*, 1941), less frequent irrigation (Lloyd, 1961), delay in the onset of granulation by spraying of 2,4-D at 16 ppm (Sinclair and Jolliffee, 1961b) and treatment with Zn+Cu+K (0.25 per cent each) three times between August and October (Singh and Chohan, 1982). The different cultivars also showed varying degree of granulation. Dancy was found highly susceptible to granulation, followed by Hamlin sweet orange, and minimum susceptibility was obtained in Kinnow mandarin (Singh *et al.*, 1985).

11.3.2 Rind Pitting

Superficial rind pitting (SRP) of Shamouti orange causes serious damage to fruits and tremendous losses to growers. While some SRP developed on the tree and could be identified in the packing house, the majority of SRP symptoms developed 3-5 weeks after harvest, during shipment and marketing. Ethylene increased the incidence of SRP. Fruits with SRP had lower rind K contents than healthy fruits. Spraying trees with 9 per cent Bonus 13-2-44, a K fertilizer increased leaf K concentration and reduced the incidence of SRP significantly. Storage at 5°C also restricted the development of SRP. A combination of preharvest Bonus spray with post-harvest low temperature storage gave good results (Tamin *et al.*, 2000).

12.0 Harvesting and Yield

12.1.0 Stage of Maturity

Lado *et al.* (2018) wrote an exhaustive review on key determinants of citrus fruit quality: metabolites and main changes during maturation. According to Lado *et al.* (2018), because of the genetic diversity among the genus *Citrus*, the the definition of standard maturity indexes of fruit quality is difficult and changes in peel and pulp, make. Commercial maturity indexes in the citrus industry are usually based on peel coloration, percentage of juice, soluble solids/acidity ratio but their relevance may differ among varieties and the specific requirements of the markets. There is also a marked influence of environmental and agronomic conditions such as light and temperature, rootstock selection and plant nutrition, among others. Besides commercial requirements, a more comprehensive definition of fruit quality should also consider organoleptic and nutritional properties that are determined by a complex interaction among a number of bioactive components. Citrus fruit are an excellent source of many phytochemical, including ascorbic acid, carotenoids (antioxidant and pro-vitamin A), polyphenols, flavonoids, limonoids, terpenoids, *etc.*, which greatly contribute to the health-related benefits of these fruits. Criteria and definition of the main maturity indexes for citrus fruit worldwide are described, as well as changes during fruit maturation in key components affecting organoleptic and nutritional properties. Moreover, the involvement of hormonal and nutritional signals and their interaction in the regulation of external and internal maturation of citrus fruit, as well as the influence of environmental and agronomic factors are also important (Lado *et al.*, 2018). Determination of the criteria for citrus fruit maturation is complex as it involves two quite different tissues and systems: internal changes occurring in the fruit flesh and external color modifications taking place in the fruit peel. Therefore, ordinary citrus fruits such

Diseases and Disorder of Citrus

Control plant (CTV-wt) | Alemnow (*C.macrophylla*) | Sour orange (*C.aurantium*) | Duncan grape-fruit (*C. paradisi*)

Tristeza Infection

Exocortis Infection

Psorosis Infection

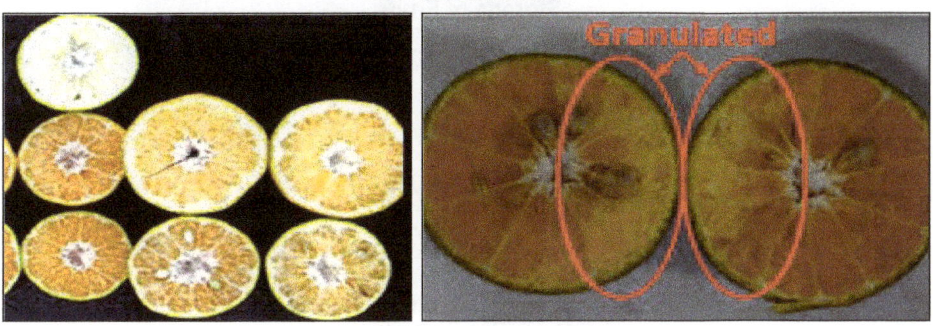

Granulation Problem in Citrus

Disorder and Harvesting of Fruits

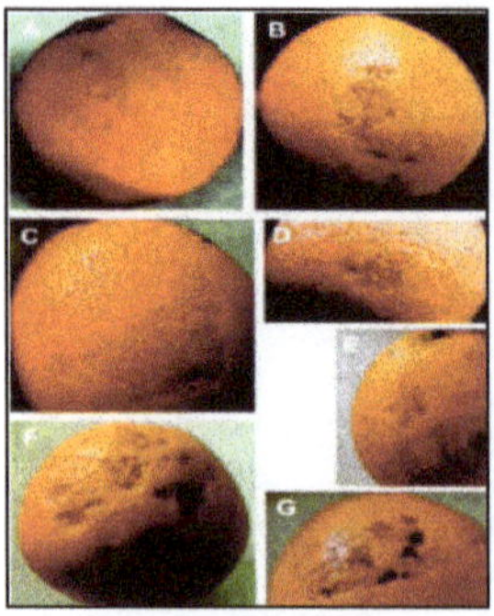

Rind Pitting

Harvesting of Fruits

Harvesting of Fruits

Storage of Fruits

Marketing of Fruits

as sweet oranges, mandarins, lemons and grapefruits are considered to be mature when their external coloration, juice content and soluble solids:acidity ratio and other internal constituents have reached a minimum level of visual acceptance or palatability (Lado, 2014). Moreover, citrus fruit show a non-climacteric ripening behavior and should be harvested when the internal maturity has been achieved since no further relevant changes occur in fruit flesh composition after harvest. In the EU, the main maturity indices considered as minimum standards for citrus quality are juice content (per cent), total soluble solids (TSS; °Brix), TSS:acid ratio and proportion of the fruit surface that present typical coloration (Table 17).

Table 17: Variations in the Main Maturity Indices Required in EU Countries and the USA for Fruits of the more Important Commercial Citrus Varieties*

Citrus Species	*TSS (°brix)*	*Minimum Acidity (Per cent)*	*TSS : Acid Ratio*	*Juice (Per cent)*	*Surface Coloration*
ORANGES	8.0	0.4 – 0.7	8-10	>33	>4/5
Navel oranges			8-8.5	>33	
Blood oranges			7	>30	
Others					>3
Satsumas			6.5	>33	
Clementines			7.0	>40	
Hybrids/others			7.5		
LEMONS				>20	
GRAPE FRUITS	6-7		5.5 – 7.0	>35	>2/3

* Lado *et al.* (2014).

The various physico-chemical parameters of maturity were found to be influenced by agroclimatic conditions and the cultural practices, and the maturity standards vary from region to region (Devkota *et al.*, 1982) in the different citrus species. A remarkable feature of Kinnow fruit growth was characterized by its sigmoid growth pattern (Pal *et al.*, 1977), while the growth of sweet orange was reported to be linear (Jawanda and Bajwa, 1961). Kinnow showed a continuous increase in fruit diameter, reaching its maximum by the end of December, while orange fruit cased growth beyond November 30 in case of Mosambi and December 15 in Pineapple. Valencia on the other hand, after a steady increase in growth up to December 15, showed further increase during February-March (Jawanda and Bajwa, 1961). In Kinnow, the colour break occurred much before the maturity was reached (Devkota *et al.*, 1982). The period from last week of January to mid February was considered optimum for picking Kinnow fruits under Hoshiarpur (Punjab) conditions. In the hills of Darjeeling district, the mandarins are harvested from mid-November to late-December depending on the altitude. The fruits ripen earlier at lower altitude. The fruits may be retained for sometime after ripening for supplying late fruit. For early market the orange at lower altitude is often harvested with the appearance of colour on the skin. Lemon and limes are harvested when almost fully developed but the skin is green. Studies on maturity standards in Kinnow at Palampur (Joolka and Awasthi, 1980) revealed that fruit size remained constant after January 5, while the percentage of juice and TSS became constant after January 25, whereas acidity decreased along with the maturity in most of the citrus fruits (Soule *et al.*, 1967). Jawanda *et al.* (1973) reported this ratio to be 12:1 and 14:1, respectively for inner and outer fruits of Kinnow mandarin. Bhullar (1982) concluded that as a maturity standard for Kinnow, Srinagar and Emperor mandarins, a TSS/acid ratio of 13.0, 12.7 and 10.3, respectively, was suggested when the external rind colour was nasturtium-orange in Kinnow and Srinagar and carrot-red in case of Emperor mandarin.

The fruit of Satsuma mandarin, when mature, are light to deep orange in colour and the surface becomes pebbled and bumpy (Srivastava *et al.*, 1972). Shanker (1972) observed that the fruits of New Zealand grapefruit matured late and became ready for harvesting by the last week of November. The fruits can be kept on the tree till first week of March without any granulation which fetch better price. The fruits of tangelos ripen by the middle of December and the fruits can be retained on the tree without any granulation till the last week of January (Shanker, 1973). Gupta *et al.* (1974) reported that rind of grapefruit changes from green to pale yellow as the fruit advances towards ripening.

Machado and Mendez (1999) carried out experiment to determine the harvesting time of Valencia orange. To evaluate the maturation process, 2 fruit quality parameters were investigated: juice yield and TSS/total acidity. Rootstock did not influence fruit quality, but did influence ripening. For Valencia grafted on Caipira orange, harvesting could be started in July (1368 degree days). For Valencia on *P. trifoliata*, harvesting could be started at the end of June (953 degree days), but could be continued until the first week of January.

Fruits of 11 citrus cultivars were evaluated by Kender *et al.* (1999) for their response to the experimental abscission material, metsulfuron-methyl (2 mg/l), as an aid to mechanical or hand harvest. Cultivars evaluated included Ambersweet, Glen Navel, Hamlin and Valencia oranges, Robinson tangerine, Sunburst tangerine, Murcott and Temple tangors (*C. reticulata* x *C. sinensis*), Orlando tangelo (*C. reticulata* x *C. paradisi*), Ray Ruby and Marsh grapefruits. Fruits of 6 of the 11 cultivars were effectively loosened by metsulfuron-methyl (Hamlin, Valencia, Orlando, Murcott, Temple and Ray Ruby). Addition of an adjuvant was necessary for abscission activity in fruit and leaves. Trees sprayed with metsulfuron-methyl in combination with an adjuvant (Kinetic) had higher per cent cumulative fruit drop, higher internal ethylene, and lower fruit detachment forces (FDF) than trees sprayed with metsulfuron-methyl alone. Sunburst tangerine responded poorly to the abscission material in the presence or absence of Kinetic. Leaf loss was greatest in trees sprayed with metsulfuron-methyl and adjuvant, intermediate in trees sprayed with metsulfuron-methyl alone, and least in control trees.

12.2.0 Method of Harvesting

In most of the citrus species, the fruits remain fresh for several weeks without any deterioration after attaining full maturity (Hayes, 1970), which facilitates harvesting according to one's choice and demand in the market. In India, the fruit picker collects the fruits manually by climbing on a ladder when the trees are tall, with a collecting bag on his shoulder. In Japan, fruits are collected with scissors and in order to convey the harvested fruits from orchard to road, cableways are used (Singh, 1966). Small motor-handcarts and motor-rearcars are also used, and recently conveying machines, such as monorail conveyors and roller conveyors are being used. In the USA, mechanical harvesting systems consisting of abscission chemicals, shakers, windrover and pick-up machines are used, particularly for processing industry. Large air shakers require 250-350 kW of power and weigh of about 10 mt. Under the best conditions they remove about 200-400 kg fruit per litre fuel consumed and can remove fruit at the rate of 20-40 mt per hour, but the fruit should then be picked up from the ground (Reitz, 1984). A mechanical device to de-stemming of orange has been developed in California, USA (Chen, 1994). The de-stemmer comprises of a roller-cutter assembly and an oscillating conveyor. Oranges are fed through a singulating hopper onto the leading edge of the cutter assembly. Fruits are then spun, oscillated and conveyed forward along the cutting cylinders. Any stem that enters the space between the rotating discs is cut by high-speed blades to a length of 3 mm. The machine works best within a disc speed range of 60-75 rpm. and a conveyor frequency range of 1.8-2.0 Hz. Within these ranges the machine successfully de-stems about 80-90 per cent oranges, with the calyx remaining intact.

Experiments on mechanical harvesting of Bergamot orange were carried out in Italy (Giametta and Zimbalatti, 1994). Two shaking machines, with vibrating heads of different sizes (about 700 and 160 kg), and a mechanized frame for intercepting fruits were utilized. The results showed that mechanical harvesting caused an increase in work productivity and in the percentage of orange detached (about 60-70 per cent). Moreover, chemical analysis of mechanically harvested fruits proved that mechanical intervention did not harm the quality of the final product.

A potential, direct-drive, double-spiked-drum canopy shaker has been developed by Peterson (1998) to harvest oranges from high density groves. Each drum is 2.44 m in diameter and has 6 horizontal whorls spaced 300 mm apart on a vertical shaft. Each whorl has 16 nylon rods, 32 mm in diameter, spaced at equal angles. Maximum rod penetration is 1 m, shaking frequency was 4-5 Hz, and maximum horizontal displacement of the rod tip is of 250 mm. The shaker is towed by a tractor along a tree row at a travel speed of 1.4-3.2 Km/h. In the canopy space penetrated by the shaking rods, mature fruit removal averaged 71-91 per cent. The size of the drum can be enlarged as the height is increased and fruit catching and conveying components be added under this system for commercial use.

12.3.0 Yield

The yield of citrus varies not only in the different types and cultivars but it also varies in a particular cultivar with age, care and management, location and the rootstock used. The fruit yield (number of fruits per plant) in several groups and cultivars of citrus in Punjab recorded by Chohan *et al.* (1966) are given below:

Tangelo	*cvs*	*No. of Fruits/ Plant*	*Tangerine*	*cvs*	*No. of Fruits/ Plant*
	Orlando	212		Dancy	540
	Pearl	173		Honey	302
	Minneola	127	Hybrid		
	–	–	–	Kinnow	131
	–	–	–	Wilking	282
Sweet orange	–	–	–	–	–
	Cambell Valencia	128	–	–	–

The average yield of grapefruit in New Zealand was 400 fruits per tree (Shankar, 1972) while it was 600-700 for tangelo (Shankar, 1973) under Allahabad conditions. A healthy plant of mandarin orange, 15- to 29-year-old produces 350-500 fruits; higher yield is not uncommon. Neglected orchards yield much less and the fruit size is also reduced (Hussain *et al.*, 1985). Sulladmath and Iyer (1981) recorded a yield of 10-23 kg fruit per plant in Seville lemon, while Bhan (1972) reported a variation in yield between 600 and 1500 fruits per plant of Kagzi lime.

13.0 Post-harvest Management

After the fruit is taken into the packing shed, it is carefully washed as soon as possible. In California, the growers use a fungicidal cleaning solution of 0.5 to 1.0 per cent soap and 2 per cent soda or 0.5 to 1 per cent sodium-0-phenylphenate, and dip fruits for 4 to 5 minutes. After light brushing and washing fruits are dried, rewaxed and polished with soft brushes in order to substitute the natural wax coat lost in washing.

Even though fungicides are commonly used as a post-harvest treatment to control fruit decay, nowadays, environment friendly technologies, such as heat treatments, are used viable replacements. Queb-Gonzalez *et al.* (2020) evaluated the effects of post-harvest heat treatments (traditional and microwave-assisted) on mandarins intentionally inoculated with *Penicillium digitatum*. For the studied heat treatments, the target temperature was 50 °C, which was held for 2.5 min. After heating, mandarins were cooled and stored at 25 °C for 13 days. MW treatments effectively prevented mold growth during storage, while HW only delayed it. Control mandarins (without treatment) showed the highest significant weight loss. Neither thermal treatment nor storage affected fruit juice pH ($p > 0.05$). Treated mandarins had significantly lower vitamin C content than control fruits throughout storage, and all mandarins lost firmness by the 13th day ($p < 0.05$). Control and MW-treated mandarins had lower citric acid content; however, they retained color, total soluble solids (TSS) and had a higher maturity index. While HW mandarins did not have changes in citric acid content, they had higher TSS, and lower maturity index. MW-assisted treatments were effective at inactivating molds and helped retain some nutritional and physical-chemical characteristics of mandarins. However, juice of MW-treated mandarins was not preferred by judges in the sensory tests, the juice was rated lower than that obtained from the other treatment. Post-harvest heat treatments may constitute a helpful application to control 'mandarin' fungal decay.

Porat *et al.* (2000) reported that a minimum exposure period of 20 s at 56°C isrequired to inhibit *Penicillium digitatum* spore germination *in vitro*. *In vivo* studies, carried out by rinsing and brushing citrus fruits 24 h after artificial inoculation with a *P. digitatum* spore suspension, indicated that hot water brushing (HWB) at 56, 59 and 62°C for 20 s, reduced decay development in the infected wounds to only 20, 5 and less than 1 per cent, respectively, of that in untreated control fruits or fruits treated with tap water. Storage experiments using various organically grown citrus fruits (Minneola tangerines, Shamouti oranges and Star Ruby grapefruits) showed that HWB at 56°C for 20 s reduced decay development by 45-55 per cent. The HWB treatment at 56°C did not cause surface damage, and did not influence fruit weight loss or internal quality parameters.

Fresh acid lime (*Citrus aurantifolia* Swingle) fruits have a constant year round demand nationally and internationally, as its multi-disciplinary uses in domestic culinary, food processing industry, indigenous medicine, cosmetics and health care products. Globally, limes are largely consumed in fresh form which roughly occupies 85.7 per cent of the world lemon and lime (Samaradiwakara *et al.*, 2018). The post-harvest qualities of lime fruits deteriorate quickly after harvest limiting long distance marketing and hence, it is worthy to extend the post-harvest life through preservation of their natural quality. Application of post-harvest treatments has been reported to delay senescence, reduce physiological disorders and decay of lime during storage (Kaewsuksaeng *et al.*, 2015; Win *et al.*, 2006). Samaradiwakara *et al.* (2018) studied the Post-harvest storage of lime fruits, harvested at 119, 133, 147 and 161 days after fruit set (DAFS). They were evaluated based on changes in physiological weight loss (PWL), peel lightness (L*), hue (h°), Chroma (C*) values and overall visual quality rating (VQR) along with juice pH, titratable acidity (TA) and total soluble solids (TSS) under ambient condition (30-34 °C, 70-75 per cent RH). The results showed that stage of maturity had significant effect on changes in PWL, VQR, L* and h° values with no significant effect on juice quality parameters during the storage. The maximum storage life of nine days was observed in fruits harvested at 133 DAFS with significantly higher VQR (3.71±0.18) and h° (116.44±2.18). Both colour and PWL per cent influenced the visual quality of fruits. Thus, lime fruits harvested at 133 DAFS had the longest shelf life of 9 days while stages of 119, 147 and 161 DAFS had shelf life of 6, 7 and 3 days, respectively under ambient storage. The most suitable time for harvesting lime fruits is 133 days after fruit set. Moreover, both immature (119 DAFS) and over mature (161 DAFS) fruits deteriorate rapidly after harvest.

13.1.0 Ripening and Storage

13.1.1 Ripening

Delay in ripening is caused by gibberellin or auxin treatment, and ethephon enhances the process. Monselise (1975) observed that application of GA_3 delayed colour and softness of peel but did not influence internal maturity due to the peculiar structure of the citrus fruits. Treatments with GA_3 prolonged the picking season of lemons (Coggins and Heild, 1968) and prevented early senescence of peel in Shamouti orange (Monselise and Sasson, 1977), also reduced peel disorders of different cultivars of orange and mandarin. Pre-harvest treatments with ethephon on promoting colour and ripening are not used in citrus, because, while promoting colour ethephon does not influence internal composition of the fruit and causes excessive leaf abscission (Monselise, 1983). Citrus differs from typical climacteric fruits and an increase in respiration does not change fruit colour and quality after harvest (Ables, 1973). However, ethylene increased the respiration rate in citrus at various stages of maturity, and an increase in both respiration and ethylene production were noted when young oranges were removed from the tree (Eaks, 1970). The peel colour of citrus fruits is dependent upon the chlorophyll and various carotenoid pigments present within the flavedo. The chlorophyll can partly or entirely obscure the carotenoid pigments, where chlorophyll is not present the carotenoids determine the external colour (Proctor and Caygill, 1985). Due to warm climatic conditions early season citrus cultivars in South Africa are with inadequate rind coloration. This entails the conversion of chloroplasts to chromoplasts and an increase in rind sugars. Ethylene-releasing chemicals (*e.g.* ethephon or ethrel) are used commercially to enhance rind coloration. Spraying of 250 ppm ethrel at 2 weeks before harvest significantly improved the rind colour. This was already evident 1 week after the application and was maximal at the time of harvest (Roux *et al.*, 1997). Ethylene can act in two ways, firstly through its effect on biosynthesis of the carotenoid, β-citraurin (Stewart and Wheaton, 1972) and secondly through causing the degradation of chlorophyll. The formation of β-citraurin is sensitive to temperature and is inhibited above 30°C. At decreasing temperatures of 25, 20 and 15°C a decreasing amount of ethylene is required to produce maximum colour.

Citrus fruits often attain internal ripeness while the peel is still green. The conventional degreening procedure is to use ethylene gas (Weaver, 1972). Ethrel has successfully been used for degreening of citrus fruits at post-harvest stage (Arora *et al.*, 1973; Chauhan and Rana, 1978; Rana and Chauhan, 1976). Lemons dipped in ethephon at a concentration of 1000 ppm, attained a marketable yellow colour within 7 days after treatment, while 50 ppm ethephon was found to be optimum for ripening of grapefruits, tangerins and oranges. Josan *et al.* (1981) reported that treatment with ethrel at 2500 ppm after three days of storage was effective for degreening of Italian Round lemon. Grapefruit cv. Marsh, orange cv. Hamlin and tangerine cv. Robinson were degreened by post-harvest exposure to ethylene gas (Ahrens and Barmore, 1986).

13.1.2 Storage

Citrus fruits can be stored well for a few days at room temperature, and through cold storage preservation techniques most of the citrus fruits can be stored for several months without any appreciable deterioration in marketable quality. Since the peel of fruits plays an important role in regulating physiological loss, the extent of moisture loss during storage is not same for citrus fruits of different species. The optimum storage temperature that ensures the longest storage life with minimum spoilage of fruits and their quality also varies considerably with different types of citrus.

Green and fully ripe yellow fruits of mandarin orange could be stored successfully at 8-10°C under 85-90 per cent relative humidity without impairing the fruit quality (Srivas *et al.*, 1963). Eaks (1976) found that orange could be stored for 12 weeks at 5°C, but storage at 0°C caused chilling injury. Tahiti limes could be stored well up to 2 months at 6°C, but at over 7°C they lost their green colour within 4 weeks and for 6 weeks at 10 or 15.6°C and 98-100 per cent relative humidity without yellowing (Spaldingand Reeder, 1976). In case of Kinnow mandarin, 2.2-3.9°C temperature was found best for storing with polythene wrapper for 60 days (Singh, 1981). The most suitable temperature for storage of Kagzi lime was 18°C (Bose *et al.*, 1965). Pekmezii (1983) reported that the 10°C temperature and 85-90 per cent relative humidity was found to be the best storage conditions of Kutdiken lemon. Under such conditions, it was possible to store the fruit without losing much quality for a period of 9 months from its harvest. Cold storage with intermittent warming promoted long-term storage of lemons for 6 months or even more (Cohen, 1983). Long-term storage of lemons at low temperature (>13°C) causes chilling injury (CI), a physiological disorder. Internal browning develops in the membranes of the segments and core (known as 'membranosis'), and deep brown spots, which develop on the fruit surface, may coalesce and cause pitting. Chemical changes occurred in the injured fruits caused reduction in juice and acid content and development of decay (Cohen, 1986). Liu *et al.* (1998) reported that the optimum temperature for long term storage of Ponkan was 12.5-15°C. Ponkan fruits harvested at the optimum maturity and stored at the optimum temperature had a reasonable storage life span of 3 months. If percentage decay can be kept low, the storage life might be extended to 4 months.

Responses to cold storage and to marketing conditions of Salustiana orange were investigated. Late harvested fruits were stored at 2 or 8°C and 90-95 per cent relative humidity (RH) for 5, 10 and 15 weeks. After each cold storage period, an additional week of shelf-life conditions (SL) at 20°C and 75 per cent RH was simulated. Determination of weight loss, decayed fruits, evolution of CO_2 and C_2H_4, main chemical parameters of the juice and assessment of overall visual appearance and chilling injury incidence were carried out at fixed intervals. Fruits stored at 2°C appeared to be resistant to chilling injury and stored better than those held at 8°C, as the former treatment resulted in lower weight loss and better appearance during storage and SL period. On the other hand, a more pronounced loss of produce due to rots was registered for 2°C-stored fruits compared with 8°C stored fruits. The taste of fruits was always judged acceptable, with slight differences between the 2 treatments (Piga *et al.*, 2000).

The effects of temperature (3, 4 or 5°C) and RH (97 or 90 per cent) on the maintenance of quality and incidence of rot in Montenegrina tangerines (mandarins) during cold storage for 4 weeks were investigated by Brackmann *et al.* (1999). After storage, fruits were held for 5 days at 20°C. After cold storage, fruits stored at 4 and 5°C and 97 per cent RH had higher tritratable acidities, but this behaviour was not observed after 5 days at 20°C. No differences between treatments were observed for TSS and rot incidence.

Treatment with different chemicals has also proved effective in extending the storage life of citrus. Mann and Randhawa (1978) observed that wax coated Kinnow mandarins remained fresh without much loss in weight. In orange, Lodh *et al.* (1963) observed delayed maturity by treatment with 2,4-D. Application of ethrel at 1000 ppm and storage at 24°-28°C in 68 per cent relative humidity enhanced the colour development with lower chlorophyll content and reduced *Penicillium* infection in mandarin (Rahana *et al.*, 1973). The sweet orange fruits treated with aureofungin (100 ppm) and bavistin (0.2 per cent) could be stored for 68 days without any spoilage (Ghosh, 1980). The period of storage of Eureka lemon increased by treatment with 2,4-D and 2,4,5-T in wax emulsion (Stewart *et al.*, 1952), while in Persian lime 2,4,5-T was very effective (Halton, 1953). Treatment with MH markedly prolonged the storage period in Kagzi lime (Bose *et al.*, 1965). The shelf life of Baramasi lemon was enhanced by 35 days when treated with waxol (12 per cent) + captan (0.1 per cent) by reducing physiological loss in weight and decay loss (Singhrot *et al.*, 1987a), while in Kinnow fruits, minimum physiological loss in weight was observed in waxol-coated fruits and packed with diphenyl paper as paper lining (Singhrot *et al.*, 1987b). Josan *et al.* (1983) stated that Kinnow fruits could be kept in good condition for 90 days by wax coating (6 per cent) and kept in wooden boxes lined with perforated polythene sheets. Film-wrapped (polyolefin film or polyvinylchloride) Dancy mandarins could be kept in cold storage (at 5°C, 85-90 per cent RH) for up to 8 weeks (Saucedo-Veloz *et al.*, 1997). Storage experiments indicated that fruits could be stored at low O_2 (3-6 per cent) and CO_2 (2.5-4 per cent) concentrations for >5 months (Sun, 1998). The ideal storage conditions of different citrus fruits are given in Table 18.

Table 18: Storage Life of Citrus Fruits

Types	*Temperature (°C)*	*Relative Humidity (Per cent)*	*Duration*	*References*
Lime	9-10	85-90	6-8 wks	Hardenburg *et al.* (1990)
	9-10	90	4-8 wks	
Lemon (green)	10-14	90	8-20 wks	Snowdon (1990)
Lemon (yellow)	11	90	12-24 wks	
Sweet orange	2-4	90	8-16 wks	

Excessive bitter taste in citrus juice especially after the juice extraction and and dusring storage of such juice is a major problem in citrus industry worldwide because it reduces the quality and commercial value of the product. Bitterness due to flavonoids and limonoids poses a major problem for the citrus industry. Many debittering technologies have been developed based on chemical, physical and microbial processes *e.g.* raising pH of the juice, addition of β-cyclodextrin monomer, use of adsorbent XAD-16, suppression of bitterness by addition of sweetening agents and conversion of bitter principles to non-bitter components by the action of immobilized bacteria (Kore *et al.*, 2018). Most of them are either costly, ineffectine or inconvenient and hardly being followed by the processor. The consumer acceptability of pummelo juice and it is affected badly due to the presence of bitter principles (neohesperidin and naringin) may be prevented by the adoption of low cost techniques like the diffusion of fresh juice up to 70° Brix so that ultimate TSS get extended and remains acceptable during storage (Kore and Chakraborty, 2015).

14.0 Biotechnology

Among the fruit crops, the investigations on biotechnology of citrus have been systematic and its application for the improvement of this important fruit holds promise. Successful application of protoplast culture to produce somatic hybrids of interspecific and intergeneric crosses has led to the development of hybrids for use as rootstocks with resistance to biotic and abiotic factors. The development of promising triploid scions will eventually revolutionize the citrus industry. Understanding the mechanism of genetic transformation and the use of coat protein gene would lead to the development of cultivars which will resist the viral diseases causing citrus decline the world-over. Development of molecular markers to identify species and hybrids would be of immense importance in citriculture.

Biotechnology on various aspects of citrus has been thoroughly reviewed by Parthasarathy *et al.* (2001). Since the application of biotechnology has proved useful in the improvement of citrus, relevant research results are summarized hereunder.

14.1.0 Tissue Culture

Tissue culture research in *Citrus* has been mainly directed in three directions, *viz.*, micropropagation, somatic hybridization using protoplast fusion and genetic transformation.

14.1.1 Micropropagation

Important cultivars of citrus have been propagated using modern genetic approaches. Superior rootstocks for citrus are being developed by interspecific and intergeneric protoplast fusion. The recovery of haploid plants from anther culture and of triploid plants from citrus should aid crop improvement (Litz, 1990). Micropropagation of citrus has also been described under propagation.

The commonly used media in citrus tissue culture are Murashige and Skoog (MS), Murashige and Tucker (MT) and Gamborg (B5). Parthasarathy and Nagaraju (1996) cultured the shoot explants obtained from *in vitro* germinated embryos of seven citrus species, *viz.*, *C. reticulata*, *C. madurensis*, *C. reshni*, *C. volkameriana*, *C. taiwanica* and *C. sinensis* cvs. Malta and Mosambi and *C. limon* on MS and B5 media supplemented with BAP. They found that MS was the best medium for all the species except *C. reshni* for the proliferation of the shoots. Sim *et al.* (1989) cultured cotyledon, epicotyl, shoot-tips, nodal segments and leaves from 2- to 4-month-old seedlings or from mature 15-year-old grafted cv. Blanco plants and grown at pH 5.7 on full strength MS medium with 2 per cent sucrose and BA, NAA, 2,4-D, IBA or a mixture. Regenerated shoots were readily produced from seedling epicotyl, leaf, shoot-tip and nodal stem segments and also from shoot tips and nodal stem segments of mature plant. However, no organogenesis occurred with mature leaf explants. For rooting of microshoots, different auxins such as NAA, IBA and IAA were used. Even transferring the microshoots to auxin and cytokinin-free medium was reported to induce rooting. Barua *et al.* (1996) were successful in *ex vitro* rooting and simultaneous acclimatization of microshoots of *C. assamensis, C. latipes,* and *C. indica* proliferated *in vitro* in soilrite. The rooted plants were transferred to greenhouse in polythene bags in soil. Hazarika *et al.* (1995) developed an efficient acclimatization protocol. The protocol involves direct culturing of 4-week-old *in vitro*-grown proliferated shoots in soilrite topped over sterile FYM. The shoots rooted in 7-15 days depending upon the species. The plantlets were rooted in the above carriers in sterile tissue culture bottles. After 4-6 weeks caps were loosened and taken to mist house for transfer.

14.1.2 Anther Culture

Hidaka *et al.* (1979) studied *in vitro* differentiation of haploid plants using anthers of *Poncirus trifoliata* on MS medium supplemented with 0.2 and/or 2.0 mg/l of 2,4-D, IAA, NAA and kinetin. After 3 weeks of culture, heart-shaped or cotyledonary embryoids appeared from the anthers. Embryoids were formed most efficiently from the anthers of early uninucleate stage, and on the media containing 0.2 mg/l IAA. The callus formation occurred on all media and increased by the addition of 2,4-D. Roots were induced by the transfer of the embryoids from the induction media containing growth regulators to the medium lacking growth regulators. Hidaka *et al.* (1981) studied plantlet formation from anthers collected from immature flower buds of *C. aurantium* on MS with or without IAA, kinetin and yeast extract at 28°C in darkness. Embryoid formation was generally evident 14 weeks after inoculation of anthers of sour orange and 'Choshu-to'. Anthers containing pollen grains at the later uninu-cleate stage formed embryoids effectively on media with 0.02 mg/l kinetin. Addition of yeast extract was not effective. Nuclear divisions of pollen grains and multinucleate pollen grains were observed in the sour orange anthers, while embryoids from pollen grains have not yet been observed microscopically. After incubation at 16-h light, some embryoids differentiated shoots, developing into plantlets, while some of the other embryoids produced more embryoids from the hypocotyl region. Root and shoot induction was stimulated by transfer from the induction media containing growth regulators to a medium devoid of growth regulators. He (1984) also induced plantlets from anthers of Trovita orange (*C. sinensis*). Anthers with microspores at uninucleate stage were cultured on MS with IAA and kinetin. Embryoids appeared directly from within anthers after about 10 weeks of culture. The media containing kinetin and IAA at 0.002 or 0.02 mg/l were effective in embryoid formation.

14.1.3 Protoplast Culture

Wang *et al.* (1986) were successful in retaining its capacity of embryogenesis and plant regeneration using an embryogenic callus line from mature seed of *C. sinensis* (orange) which were subcultured for 10 years. An embryogenic cell suspension from the callus line was established and maintained for 2 years. A high frequency of cell division was induced from protoplasts isolated from the cell suspension. Somatic embryos and plantlets were regenerated from the protoplast culture.

Hassaballa *et al.* (1988) isolated protoplasts enzymatically from etiolated cotyledons of dark-grown 2- to 3-week-old seedlings of Volkamer lime. The protoplasts were cultured in liquid over solid or bead agarose culture media.

Optimal planting density was in the range of 5000 cells/ml. Colonies with 36.3 per cent plating efficiency were transferred to MS medium containing either 1 mg/l zeatin or 10 mg/l BA to induce shoot formation. Cultures were maintained in the growth chamber for 10 weeks under a 16-h photoperiod at 25°C. Shoots were transferred to MS culture medium supplemented with 10 mg/l gibberellin and kept in the light at 23°C for 3 weeks. Roots emerged when shoots were transferred to pots of compost in a greenhouse. Recovered plants were morphologically uniform and normal and had chromosome number of 2n=2x=18 in their root-tips.

Jumin and Nito (1996) isolated protoplasts from embryogenic callus of many rutaceous plants, *viz.*, *Fortunella polyandra*, *Atalantia bilocularis*, *Hesperethusa crenulata*, *Glycosmis pentaphylla*, *Triphasia trifolia* and *Murraya koenigii* cultured on MT basal medium containing 5 per cent sucrose supplemented with different concentrations of BA and 0.6 M sorbitol. The highest planting efficiencies for all species were obtained on MT basal medium supplemented with 0.001 mg/l BA. The highest percentage (51 per cent) of globular somatic embryos was obtained for *F. polyandra* on MT medium supplemented with 0.001 mg/l BA. MT basal medium containing 0.001 mg/l BA was the most suitable for the development of globular somatic embryos into heart-shaped somatic embryos with cotyledon-like structures. The highest percentages of shoot formation for all 6 species were obtained using 0.1 mg/l gibberellic acid. A complete protoplast-to-plant system was developed for *F. polyandra*, *A. bilocularis* and *T. trifolia* which could facilitate the transfer of nuclear and cytoplasmic genes from these species into cultivated citrus through protoplast fusion.

14.1.3.1 Protoplast Fusion and Somatic Hybrids

A review about symmetric and asymmetric somatic hybridization and cybridization in citrus is presented by Bona *et al.* (2011). Somatic hybridization by nuclear or cytoplasmic fusions leads to three distinct possible outcomes — [a] "symmetric" hybrid: fusion of whole nuclei, [b] "asymmetric" hybrid: fusion of parental nuclei, one or both deficient for one or more chromosome(s), and [c] "cybrid": fusion without a nuclear contribution by one parent, *i.e.*, an extremely asymmetric hybrid. Somatic hybridization in citrus has been a powerful tool for the production of novel allotetraploid somatic genotypes that combine desirable characteristics of two parents. These hybrids open new opportunities for sexual hybridization at the tetraploid level, and for interploid crosses for the production of potentially seedless triploids, among other uses. Somatic symmetric hybrids have great potential for rootstock improvement and are excellent as breeding parents, but they may have limited direct application as scion cultivars because they typically contain the entire genomes of both fusion parents, and are likely to exhibit the desirable and undesirable traits of both parents. Asymmetric hybridization is very promising as it allows partial genome transfer and may be better tolerated than a whole-genome transfer. Somatic hybridization seems to be a great tool for citrus breeding. However, the release of new improved cultivars produced by this technology is what is going to prove its effectiveness (Bona *et al.*, 2011).

Grosser *et al.* (1988a,b) reported that from *Severinia disticha* seedling epicotyls were grown in 1/10 strength MT medium with 0.55 mg/l 2,4-D, 0.22 mg/l kinetin,

20 ml/l coconut water and 50 g/l sucrose; protoplasts were prepared by treating 6-month-old callus with enzymes in BH3 protoplast culture medium at 28°C for 2-4 hours on a shaker. Protoplasts were prepared from a habituated orange (*C. sinensis* cv. Hamlin) suspension cell culture, derived from ovule culture. Equal volumes of the 2 protoplast solutions were mixed and fused using polyethylene glycol (PEG). Fused protoplasts were cultured in BH3 medium in sealed containers at 28°C in the dark. The fusion cultures were supplemented with a 1:2 (v/v) mixture of BH3 medium and liquid MT medium containing 125 g/l sucrose and 0.5 g/l malt extract 6 weeks after fusion and transferred to low light. The sucrose level was decreased gradually. Somatic embryos were enlarged on MT medium with 1,500 mg/l malt extract and 25 g/l sucrose and germinated on MT medium with 1 mg/l GA_3 with or without coumarin. Ungerminated embryos were sliced, cultured on MT medium with 5 mg/l BA, and multiple shoots were rooted and transferred to soil.

Mourao *et al.* (1996) induced fusion after protoplast fusion with PEG and regenera-ted somatic hybrid plants from sweet orange (*C. sinensis*) cv. Succari with tangerine (*C. reticulata*) cv. Dancy, tangelo (*C. paradisi* x *C. reticulata*) cv. Minneola, tangor (purported *C. sinensis* x *C. reticulata*) cv. Murcot, tangelo cv. Page, tangelo cv. Minneola x Clementine (*C. reticulata*), and mandarin (*C. reticulata*) cv. Ponkan. Succari protoplasts were isolated from ovule-derived embryogenic cell suspension cultures, and protoplasts of the mandarin-type parents were isolated from seedling leaves. Somatic hybrid plants were identified on the basis of leaf morphology, root-tip chromosome number and isoenzyme and RAPD analyses. These tetraploid plants may have direct cultivar potential; however, their greatest value will be as pollen parents to be crossed with selected monoembryonic diploids to produce seedless triploids. Mesophyll protoplasts of Changshou kumquat (*Fortunella obovata*) were fused electrically with embryogenic protoplasts of Dancy tangerine (*Citrus reticulata*). Intergeneric somatic hybrid plants were obtained after culture of the fusion-treated protoplasts. Cytological examinations showed that the plants were tetraploids with 36 chromosomes (Liu and Deng, 2000).

Grosser *et al.* (1994) regenerated tetraploid somatic hybrid plants from the following attempted parental combinations: Cleopatra mandarin (*C. reticulata*) + Argentine trifoliate orange (*P. trifoliata*); sweet orange cv. Succari (*C. sinensis*) + Argentine trifoliate orange; sour orange (*C. aurantium*) + Flying Dragon trifoliate orange; sour orange + Rangpur lime (*C. limonia*); and Milam lemon (purported sexual hybrid of *C. jambhiri* x *C. sinensis*) + Sum Chu Sha mandarin (*C. reticulata*) following polyethylene glycol-induced fusion of protoplasts. Regenerated plants were classified according to leaf morphology, chromosome number, and peroxidase, phosphoglucomutase, and phosphoglucose isomerase (glucose-6-phosphate isomerase) leaf isozyme profiles. As expected, all of the somatic hybrid plants were tetraploid (2n=4x=36), and all five selections were propagated and entered into commercial citrus rootstock trials.

Four allotetraploid somatic hybrids of citrus, with potential for rootstock improvement, have been evaluated for their response to *Citrus tristeza virus* (CTV) infection (Abbate *et al.*, 2019). CTV is the most important viral pathogen affecting citrus production worldwide. Somatic combinations of 'Milam' lemon (*Citrus jambhiri* Lush.) + Sour orange (*C. aurantium* L Osb.), Calamondin (*C. madurensis* Lour.) + 'Keen' sour orange (*C. aurantium L.*), Calamondin + 'Femminello' lemon (*C. limon* L. Burm. F.) and Cleopatra mandarin (*C. reshni* Hort. ex Tan.) + 'Femminello' lemon, were studied. Plants were grafted with CTV-infected "Valencia" sweet orange budwood. Two different CTV strains collected in Sicily, considered as "mild" and "severe", were used to inoculate candidate rootstocks. The goal of this work is to select a rootstock alternative to CTV susceptible Sour orange, still the prevalent rootstock in the Mediterranean basin. DAS-ELISA and real-time PCR assays confirmed a decreased level of viral replication in tested somatic hybrids, as compared to the susceptible genotypes sour orange and Citrus alemow. The Calamondin+'Keen' sour orange genotype did not support any replication in either CTV strain. Somatic hybridization is confirmed to be an effective tool to obtain functionally new rootstocks. Our results can be considered a starting point to open new approaches for the Mediterranean citrus industry. Indeed, the four somatic hybrids have been propagated for agronomical multisite evaluation trials to further assess if their horticultural performance, fruit holding capacity, and soil adaptation are adequate to replace sour orange.

14.1.4 Somatic Embryogenesis

In citrus embryos arise from nucellus or integument adventitively which is taken advantage for producing true-to-type plants. Pollination and fertilization are essential for the induction of nucellar embryogenesis even though there are reports of success in induction of embryogenesis using unfertilized ovules (Mitra and Chaturvedi, 1972). Gmitter and Moore (1986) regenerated plants via embryogenesis using immature ovules of citrus cultivars; sweet orange cvs. Hamlin, Pell Navel, Pineapple, grapefruit cv. Marsh, sour mandarin cv. Owari, lime cv. Key and lemon cv. Bearss in MS with 500 mg/l malt extract (ME) and 0.01 mg/l 2,4-D alone or with 0.1 mg/l BA or daminozide. Embryos were germinated in MS with 1 mg/l gibberellin and transferred to soil. Plant regeneration via secondary or long-term callus-derived embryos was as efficient as from primary embryos. Nagasawa *et al.* (1999) found 20 mg/l adenine, 500 mg ME and 2 g gellan gum in basal MT medium induced embryogenesis in immature ovules of 'Ohta' ponkan (*C. reticulata*). Parthasarathy and Nagaraju (1992) using excised ovules (8-10 weeks old) of Khasi mandarin and Cleopatra mandarin revealed that MS medium alone or supplemented with malt extract or IBA induced cotyledonary embryoids while NAA alone delayed further development of torpedo embryoids. Cotyledonary embryoids obtained from such cultures after 45-60 days and subcultured in MS medium supplemented with BAP at 0.75 mg/l produced maximum number of embryoids as well as shoots (Parthasarathy and Nagaraju, 1993) and kinetin was better for germination of embryoids (Rahman *et al.*, 1996).

14.2.0 Biochemical and Molecular Markers

Isozyme analysis has been found to be a useful tool for determining the genetic variability, identification of nucellar and zygotic seedlings and identification of hybrids. More than 20 isozyme loci have been genetically characterized in citrus. Extensive studies carried out by many workers have shown that cultivars within the same species could have arisen through mutation and given species status, and isozyme profiles have shown that sweet orange and grapefruit have similarities (Hirai and Kajiura, 1987; Roose, 1988). Isozyme analysis is increasingly being used to distinguish nucellar and zygotic seedlings. The identification of intergeneric and interspecific somatic hybrids obtained by protoplast fusion has also been done using isozyme analysis (Grosser and Gmitter, 1990; Tusa *et al.*, 1990).

DNA markers based on polymerase chain reaction (PCR) are becoming increasingly popular for various uses for which isozymes are used besides the construction of linkage maps and the marker-assisted selection. RAPD has been successfully used to measure the genetic diversity (Sawazaki *et al.*, 1997; Coletta *et al.*, 1998) and to understand the genetic origin of cultivars (Russo *et al.*, 1998). Motomura *et al.* (1996) studied somatic hybrids produced by protoplast fusion between embryogenic callus cells of Hazzara (Abohar) and mesophyll cells of *Microcitrus australis*. These plants proved to be somatic hybrids by nuclear rDNA analysis. Mitochondrial DNA analysis showed that all 10 hybrids contained the Hazzara (Abohar) plastome. Analysis of chloroplast DNA showed that most of them carried specific fragments derived from Hazzara. Genome analysis for economically important fruit quality components has also been reported. Fang *et al.* (1997) found that acidless trait in pummelo 2240 is controlled by a single recessive gene called *acitric*.

14.3.0 Genetic Transformation

To date there have been several reports of citrus transformation. Two different groups have made use of protoplast systems to induce genes in citrus. Kobayashi and Uchimiya (1989) obtained low numbers of kanamycin resistant calli by treating protoplasts from a suspension cell line of orange with DNA and polyethylene glycol (PEG). The DNA contained a gene for neomycin phosphotransferase II (*npt II*), which confers resistance to kanamycin in plant cells. The calli stably expressed NPT II enzyme activity, and were shown by Southern hybridization to contain an intact kanamycin resistance gene. Genetic transformation of Washington Navel and Trovita sweet orange calli has been achieved by co-cultivation with *Agrobacterium tumefaciens* (Hidaka *et al.*, 1992). Transformation was also achieved by cocultivation with *Agrobacterium* of epicotyl explants from Carrizo citrange (Pena *et al.*, 1995, 1998); *Poncirus trifoliata* (Kobayashi *et al.*, 1997) seedlings; internodal stem segments of greenhouse-grown seedlings from Pineapple sweet orange (Pena *et al.*, 1998) and by particle bombardment of tangelo (*C. reticulata* x *C. paradisi*) embryogenic cells (Yao *et al.*, 1996). Citrus-tristeza virus (CTV) is the most devastating citrus virus. Studies have shown that protection against a virus may be achieved by the introduction of the cloned viral coat protein (*CP*) gene into transgenic plants (Gutierrez *et al.*, 1992). The *CTV CP* gene was isolated, sequenced and cloned into a plant expression vector. Transgenic plants of Carrizo citrange (*Citrus sinensis* x *Poncirus trifoliata*) and sour orange (*Citrus aurantium*) expressing the *CTV CP* gene were identified based on β-glucuronidase (EC-3.2, 1.31) reporter gene expression, polymerase chain reaction, Southern hybridization and immunoblotting using a CTV CP-specific antibody (Moore *et al.*, 1993).

Soil salinity significantly limits citrus production in many areas worldwide and Carrizo citrange, considered an excellent citrus rootstock is very sensitive to salt stress, which restricts its use in salty soils. Carrizo citrange was successfully transformed by Cervera *et al.* (2000a) using *Agrobacterium tumefaciens* with the halotolerance gene *HAL2*, originally isolated from yeast and implicated in salt tolerance mechanism. Final selection of *HAL2* transgenic plants was made on the basis of PCR data for the detection of *HAL2* gene, instead of GUS expression assays, since integration of truncated portions of pBII21. HAL2, T-DNA fragment, where the left border (*uid*A cassette region) was lost, was probed to be very frequent. Ten regenerated plants, positive for *HAL2* by PCR, were selected. The transgenic nature of these plants was confirmed by Southern and Northern analyses. This is the first time that a gene from yeast has been stably integrated and expressed in citrus plants.

A collection of 70 transgenic citrus plants for the *uid*A (GUS) and *nptII* (kanamycin kinase) genes have been maintained under screenhouse conditions over a period of 4-5 years (Cervera *et al.*, 2000b). A detailed scanning of the plants allowed to detect four phenotypic off-type plants and a large variation of transgene integration and expression patterns among the population. Off-type plants were analysed and characterized as nucellar tetraploids, probably originating from tetraploid strarting tissues rather than from somaclonal variation events.

14.4.0 Genome Editing

Gmitter *et al.* (2018) presented an overview on genome editing in Citrus. According to them the most widely used genome editing system is CRISPR-Cas9, which stands for clustered regularly interspaced short palindromic repeatsCRISPR associated protein 9. This is a short piece of bacterial DNA that is part of the defense mechanism used to fight invading viruses, by recognizing and "cutting" their DNA or RNA. By clever molecular biology techniques, this bacterial defensive machinery has been used now to make small changes in the DNA sequences of genomes of many different organisms, including fungi, plants, animals and even human. The concept has been evolving since the late 1980s and first demonstrated to be effective in citrus in 2014. Since then, many research projects have been initiated to use CRISPR in hope of making HLB-resistant plants. Genome editing differs from standard GMO plants in a very fundamental way. GMOs involve the incorporation of foreign DNA from other organisms such as bacteria, viruses, other plants, even insects and animals. By contrast, genome editing can be accomplished in such a way that there is absolutely nothing foreign that remains in the plant, only its own natural DNA sequences with a small change in the "spelling."

14.5.0 *In vitro* Conservation

Survival of citrus tissues subjected to freezing-thawing treatments has been reported by many workers. Cryopreservation of embryogenic cultures is one of the most valuable tools not only for the conservation of genetic resources of citrus which has a great economic importance but to conserve plant materials in a form ready for further genetic manipulations. Chen and Wang (1995) stored germplasm using seedlings of *C. madurensis* and cultures were maintained without subculture for five years. In similar studies, seedlings of the hybrid *C. sinensis* x *C. reticulata* x

C. tangerina were stored in 2.5 cm x 20 cm tubes for 13 years without subculture. Further they found that citrus germplasm can be stored for over 10 years without sub-culture and 15 m^2 culture-room could store about 1000 accessions.

15.0 References

Aala, F.T. (1953) *Philipp. J. Agric.*, **18**: 101-103.

Abbassi, M., Harchaoui, L., Rizqi, A., Nadori, E.B. and Nia, M. (1999) In: *Proc. fifth International Conf. on Pests in Agric.*, Parl 3, Montpellier, France, pp. 609-615.

Abbate, L., Panno, S., Mercati, F., Davino, S. and Bosco, S. F. D. (2019) *Eur J Plant Pathol.*, **153**: 837–847. https: //doi.org/10.1007/s10658-018-1599-0

Abbott, C.E. (1935) *Amer. J. Bot.*, **22**: 464-485.

Ables, F.B. (1973) In: *Ethylene in Plant Biology*, Academic Press, New York, pp. 169-70.

Abobatta W.F. (2019) *MOJ Eco Environ Sci.*, **4**(6): 308 312. DOI: 10.15406/mojes.2019.04.00168

Abou Rawash, M., El-Wakeel, H.F., Kassem, N. and Mohaned, E.A. (1998) *Ann. Agric. Sci.* (Cairo), **43**: 523-537.

Ahlawat, Y.S. and Raychaudhuri, S.P. (1988) *Proc. 6th.Int. Citrus Cong.*, Tel Aviv, Israel, pp. 871-879.

Ahmed, S. (1960) *Punjab Fruit J.*, **23**: 4-9.

Ahrens, M.J. and Barmore, C.R. (1986) *Proc. 22nd Int. Hort.* Cong., Davis, California, Abst. No. 370.

Aiyappa, K.M. and Srivastava, K.C. (1965) *Oranges, Lemons and Limes*, Farm Inf. Bull., Govt. of India, New Delhi.

Aiyappa, K.M. and Subramaniam, C.K. (1959) *Mysore Agric. J.*, **34**: 159-162.

Aiyappa, K.M., Dikshit, N.N. and Bojappa, K.M. (1959) *Mysore Agric. J.*, **34**: 70-76.

Aiyappa, K. M., Srivastava, K. C. and Bojappa, K. M. (1962) *Indian Hort.*, **19**: 70-79.

Aiyappa, K.M., Srivastava, K.C. and Bojappa, K.M. (1961) *4th Hort. Res. Workers' Conf.*, Poona.

Akgul, F. and Tuzcu, O. (1993) *Doga Turk Tarim ve Ormancilik Dergisi.*, **17**: 359-371.

Al-Hinai, M.S. and Mani, A. (1998) *Nematologia Mediterranea.*, **26**: 179-184.

Alva, A.K. and Paramasivam, S. (1998) *Soil Sci. Soc. Amer. J.*, **62**: 1335-1342.

Amorim, E.P.da R., Itamar, S. and De, M. (1999) *Phytopathologica.*, **25**: 335-338.

Anonymous (1965) *Facts About Pest Control*, USDA.

Anonymous (1977) *Rep. Res. Workshop on Citrus*, Bangalore, pp. 48-52.

Anonymous (1983) *Plant Industry News.*, **24**: 3.

Anonymous (1989) *Botanical Pesticides, The Sustainable Agricultural Newsletter.*, **1**: 4.

Anonymous (1991) *Res. Rep. on AICRP of Tropical Fruits*. IIHR, Bangalore.

Anonymous (1994a) *Citrus: Special Horticultural Information Service*. National Horticulture Board, Ministry of Agriculture and Cooperation, Gurgaon, pp. 96-118.

Anonymous (1994b) *Final Repot NRCC Project*, No. 3.1.

Anonymous (1995) *Annual Rep. on Entomology Research Work*, Regional Fruit Research Station, Katol. Submitted to Plant Prot. Res. Rev. Committee, pp. 1-67.

Anonymous (1996) *Annual Statistics*, CCP: CI/ST96: FAO, p. 24.

Anonymous (1998) *Plant Var. J.*, **11**: 45-46.

Arora, J.S., Pal, R. N. and Jawanda, J. S. (1973) *Punjab Hort. J.*, **13**: 13-17.

Arora, R. K. and Yamdagni, R. (1985) *Proc. 3rd Nat. Citrus Sem.* Calcutta, p. 5.

Asi, A.A. and Ali, N. (1970) *J. Agric. Res.*, Pakistan, **8**: 281-290.

Aular, J. M. Casares, M. and W. Natale (2017) *Científica.*, Jaboticabal, **44**(4): 64-72.

Azzena, M., Deidda, P. and Dettori, S. (1988) *Proc. 6th Int. Citrus Cong.*, Tel Aviv, Israel, 2: 747-751.

Babu, H. and Kaul, G. L. (1972) *Punjab Hort. J.*, **12**: 93-96.

Baines, R.C., Bitters, W.P. and Clarke, O.F. (1960) *Plant Dis. Reptr.*, **4**: 281-285.

Bajwa, B.S. (1941) *Punjab Fruit J.*, **5**: 889-891.

Bajwa, B.S. and Ali, Niaz. (1945) *Punjab Fruit J.*, **9**.

Bajwa, B.S. and Kaura, N.R. (1952) *Indian J. Hort.*, **9**: 29-32.

Bandale, J.R., Narayana, N. and Kibe, M.M. (1951) *Poona Agril. Coll. Magz.*, **42**: 3-10.

Bar-Akiva, A. (1965) *Phyton.*, 22: 131-173.

Bar-Akiva, A. and Lavon, R. (1967) *Israel J. Agric. Res.*, **17**: 16.

Barmore, C.R. and Grierson, W. (1982) *Proc. 21st Int. Hort. Cong*, Hamburg, Vol. 1, pp. 396-406.

Barrett, H.C. (1978) *Proc. Int. Soc. Citriculture*, **2**: 1-8.

Barrett, H.C. (1981) *Proc. Int. Soc. Citriculture*, **2**: 24-9.

Barrett, H.C. and Hutchison, D.J. (1978) *Econ Bot.*, **32**: 27-45.

Barros, J.C., Da S.M. de, Graca, J., Celestino, R.C.A. and Castro, N.G. (1997) *Comunicado Tecnico – Empresa de Pesquisa Agropecuaria do Estado Rio de Janeiro*, No. 233, p. 4.

Bartholomew, E.T., Sinclair, W.B. and Turrell, F N. (1941) *Calif. Agric. Exp. Sta. Bull.*, No. 647.

Bartholomew, E.T., Sinclair, W.B. and Raby, E.C. (1934) *Calif. Citrogr.*, **19**: 88-89.

Barua, A., Nagaraju, V. and Parthasarathy, V.A. (1996) *Ann. Pl. Physiol.*, **16**: 124-128.

Batchelor, L.D. and Rounds, M. B. (1948) In: *The Citrus Industry*, Vol. II, Univ. of Calif. Press, Berkley.

Ben-Hayyim, G. and Neuman, H. (1983) *Zeitschrift fur Pflazenphysiol.*, **110**: 331-337.

Benton, R.J. (1943) *Fruit World.*, Melbourne, **44**: 4,10.

Ben-Yehoshima, S. (1978) *Proc. Int. Soc. Citriculture*, **1**: 110-115.

Bertonha, A., Frizzone, J.A. and Martins, E.N. (1999) *Acta Scientiarum.*, **21**: 537-542.

Bevington, K.B. and Bacon, P. (1978) *Australian J. Exp. Agric. & Anim. Husb.*, **18**: 591-596.

Bevington, K.B. and Bacon, P.E. (1976) *Australian Citrus News.*, **50**: 2.

Bhagabati, K.N., Ahlawat, Y.S., Chakraborty, N.K., Borthakur, B.C. (1989) *Indian Phytopathology.*, **42**: 552-555.

Bhan, K.C. (1972) *Proc.1st All India Sem. on Citriculture*, Nagpur, p.24.

Bhansali, R.R. and Arya, H.C. (1980) *Proc. Soc. Citriculture*, Griffith, NSW, Australia, pp. 135-140.

Bhargava, B.S. and Raghupathi, H.P. (1993) In: *Methods of Analysis of Soils, Plants, Waters and Fertilizers* (Ed. H.L.S. Tandon) FDCO, New Delhi, pp. 49-82.

Bhattacharya, R.K. and Baruah, B.P. (1998) In: *A Text Book on Pomology*, Vol. III (Ed. T.K. Chattopadhyay), Kalyani Pub., Ludhiana, India, pp. 1-71.

Bhattacharya, S.C. and Dutta, S. (1956) *ICAR Monograph No. 20.*

Bhullar, J.S. (1981) *Progressive Hort.*, **13**: 137-138.

Bhullar, J.S. (1982) *Punjab Hort. J.*, **22:** 35-40.

Bhullar, J.S. and Ghanbari, A. (1985) *Punjab Hort. J.*, **25:** 31-35.

Bhutani, D.K. (1979) *Insects and Fruits*, International Book Distributors, Dehradun, pp. 136.

Bielorai, H., Deshberg, E. and Bruin, M. (1984) *Proc. Int. Soc. Citriculture.*, **1**: 118-120.

Bindra, O.S. (1966) *Punjab Hort. J.*, **6**: 108-116.

Bindra, O.S. (1969) *Res. J. Punjab Agric. Univ.*, pp. 64-78.

Bindra, O.S. (1970) *Jt Publication of PAU & Ohio USAID*, pp. 64-78.

Bitters, W.P. (1960) *Calif. Citrogr.*, **45**: 349, 363-364.

Bitters, W.P. (1974) *Proc.1st Int. Citrus Short Course*, Sept. 24-29, 1973. Univ. of Forida.

Bitters, W.P., Brusca, J.A. and Cole, D.A. (1964) *Calif. Citrogr.*, **49**: 443-448.

Bitters, W.P., Brusca, J.A. and Cole, D.A. (1969b) *Proc Int. Soc. Citriculture*, Vol. I. pp. 16-26.

Bitters, W.P., Cole, D.A. and McCartg, C.D. (1981) *Proc. Int. Soc. Citriculture*, Shimizu, Japan, Vol. I. pp. 109-110.

Bitters, W.P., Rangan, T.S. and Murashige, T. (1969a) *Calif. Citrogr.*, **54**: 139-141.

Bitters, W.P., Murashige, T., Rangan, T.S. and Nauer, E. (1972) *Proc. 5th Conf. Int. Org. Citrus*, Virul Univ. Press, Gainesville, pp. 267-271.

Blondel, L. (1977) *Proc. Int. Citrus Cong.*, **2:** 185-190.

Bonavia, F. (1890) *The Cultivated Oranges and Lemons, etc. of India and Ceylon*, W.H. Alleh & Co., London.

Borle, M. N. (1974) *Indian J. Ent.*, **40**: 344-350.

Bose, T.K., Roy, B.N., Sengupta, M.S. and Bhoumik, B. (1965) *Hort. Adv.*, **7**: 126-180.

Bowman, K.D. (1999) *HortScience.*, **34**: 142-143.

Brackmann, A., Lunardi, R. and Waclawovsky, A. (1999) *Revista Cientifica Rural.*, **4**: 38-42.

Braddock, R.J. (1983) *Food Technol.*, **37:** 85-87.

Brar, S.P.S., Kumar, D., Singh, B. and Chohan, G.S. (1986) *Indian J. Hort.*, **43**: 210-215.

Bretschneider, E. (1898) *History of European Botanical Studies in China*, Imperial Russian Academy of Science Press.

Broadbent, P., Bevington, K.B. and Coote, B.G. (1991) *Proc. 11th Conf. Int. Organization of Citrus Virologists*, Riverside, California, pp. 64-70.

Broembsen, L.V. (1984) *Citrus and Subtrop. Fruit J.*, **610**: 5-12.

Brosh, P. and Israeli, E. (1986) *Alon Hanotea.*, **40**: 1079-1089.

Brown, C.H.P. (1953) *World Crops.*, **5**: 441-444.

Burger, D.W. (1985) *Scientia Hort.*, **25**: 71-76.

Calvert, D.V. and Reitz, H.J. (1966) *Proc. Fla. Sta. Hort. Soc.*, **79**: 1-6.

Cameroon, J.W. and Frost, H.B. (1968) In: *The Citrus Industry* (Eds. W. Reuther, L.D. Batchelor and H.J. Webber), Univ. of California, Vol.II, pp. 325-381.

Camp, A.F. and Fudge, B.R. (1939) *Fla. Agr. Exp. Sta. Bull.*, No. 335, pp. 1-55.

Capoor, S.P. (1975) *Indian J. Hort.*, **32**: 1-6.

Capoor, S.P. and Rao, D.G. (1967) *Proc. Int. Symp. Subtrop. and Trop. Hort.*, Bangalore, p. 9.

Capoor, S.P., Rao, D.G. and Shivankar, S.M. (1967) *Indian J. Agric. Sci.*, 37: 572-576.

Carpenter, J.B. and Furr, J.R. (1962) *Phytopath.*, **52**: 1277-1285.

Carpenter, J.B., Burns, R.M. and Sedlack, R.F. (1981) *Citrograph.*, **66**: 287-288.

Cary, P.R. (1977) *Proceeding of International of Society of citriculture.*, **1**: 162-165.

Cary, R.P. (1968) *J. Hort. Sci.*, **43**: 299-315.

Cary, R.P. (1970) *Hort. Res.*, **10:** 20-23.

Cary, R.P. (1982) *Proc. 21st Int. Hort. Cong.*, Hamburg, Vol. I, pp. 437-447.

Cary, R.P. and Weerts, P.G.J. (1977) *Proc. Int. Soc. Citriculture*, Vol. I, pp. 39-43.

Casamayor, R., Bello, L., Ramirez, P. and Garcia, R. (1982) *Proc. Int. Soc. Citriculture*, Shimizu, Japan, pp. 28-31.

Castle, W. (2010) *HortScience.*, **45**(1): 11-15.

Castle, W.S. (1980) *Proc. Fla. Sta. Hort. Soc.*, **93**: 24-27.

Cervera, M., Ortega, C., Navarro, A., Navarro, L. and Pena, L. (2000a) *J. Hort. Sci. Biotech.*, **75**: 26-30.

Cervera, M., Pina, J.A., Juarez, J., Navarro, L. and Pena, L. (2000b) *Theor. Appl. Genet.*, **100**: 670-677.

Cevik, B. (1987) *Doga Tarum ve Ormaniciuk.*, **11**: 42-43.

Chadha, K.L. (1970) *J. Res. Punjab Agric. Univ.*, Ludhiana, India, pp. 9-25.

Chadha, K.L. and Singh, H.P. (1989) *Proc. Citriculture in North-West India*, pp. 21-65.

Chakraborty, D.N. and Barua, J.P. (1983) *J. Indian Soc. Soil Sci.*, **31**: 287-295.

Chakravarthi, V.P., Savithri, P., Prasad, P.R. and Naidu, V.G. (1998) In: *Advances in IPM for Horticultural Crops* (Eds. P.P. Reddy, N.K.K. Kumar and A. Verghese), Bangalore, India, pp. 32-33.

Chakrawar, V.R. (1981) *Abst. Nat. Symp. Trop. and Subtrop. Fruit Crops*, Bangalore, p. 11.

Chapman, H.D. (1951) *Calif. Citrogr.*, **36**: 223-253.

Chapman, H.D. (1960) *Calif. Agric. Expt. Stat. Manual*, No. 53.

Chapman, H.D. (1961) *Indian J. Hort.*, **18**: 251-276.

Chapman, H.D. (1968) In: *The Citrus Industry* (Eds. W. Reuther, L.D. Batchelor and H.J. Webber), Univ. of California, Vol. II, pp. 127-289.

Chapman, H.D. and Brown, S.M. (1950) Hilgardia, 19.

Chapman, H.D., Brown, S.M. and Bayner, D.S. (1945) *Calif. Citrogr.*, **30**: 162-163.

Chapot, H. (1975) *The Citrus Plant*, Tech. Monograph No. 4, pp. 6-13.

Chaturvedi, H.C. and Mitra, G.C. (1974) *HortScience.*, **9**: 118-120.

Chauhan, K.S. and Rana, R.S. (1978) *Indian J. Hort.*, **31**: 154-156.

Chelong, I. and Sdoodee, S. (2013) *Kasetsart Journal: Natural Science.*, **47**: 333-341.

Chen, D.Q., Chen, M.H., Ban, J.Q., Xu, X.J., Zhang, Q.X. and Hu, S.L. (1998) *South China Fruits.*, **27**: 10-11.

Chen, P. (1994) *Applied Engineering in Agriculture.*, **10**: 765-769.

Chen, Z. and Wang, J. (1995) *J. Fujian Agric. Univ.*, **24**: 376.

Chen, Z.S. (1999) *South China Fruits.*, **29**: 14-15.

Chhabra, H.K., Bindra, O.S. and Singh, I. (1978) *Haryana J. Hort. Sci.*, **7**: 136-138.

Chhetri, L.B. and Kandel, B.P. (2019) *J. Agric. Studies.*, **7**(2): 63-74.

Chin, C.C., Chiu, S.C. and Ko, S.C. (1990) *J. Agric. Res.*, China, **37**: 430-439.

Chinappa, K.S., Gopalkrishna, N., Deshmukh, P.P. and Joshi, A.T. (1977) *Abst. Int. Symp. Citriculture*, Bangalore, p. 10.

Chohan, G.S., Nauriyal, J.P. and Bakshi, J.C. (1966) *Punjab Hort. J.*, **6**: 46-61.

Chohan, G.S., Sharma, J.N. and Thatai, S.K. (1978) *Indian J. Hort.*, **35**: 212-215.

Chohan, H.S. and Knorr, L.C. (1970) *Res. J. Punjab Agric. Univ.*, Ludhiana.

Chowdhury, K.G., Mali, V.R. and Sawant, D.M. (1980) *J. Maharashtra Agric. Univ.*, **3**: 40-44.

Chowdhury, S. (1951) *Indian J. Agric. Sci.*, **16**: 570-571.

Chowdhury, S. (1954) *FAO Plant Protection Bull.*, **2**: 151-153.

Chundawat, B.S. and Arora, R.K. (1981) *Abst. Nat. Symp. Trop. and Subtrop Fruit Crops*, Bangalore, p. 27.

Chundawat, B.S. and Randhawa, G.S. (1972) *Indian J. Hort.*, **29**: 277-282.

Coggins, C.W.Jr. and Heild, H.Z. (1968) In: *The Citrus Industry* (Eds. W. Reuther, L.D. Batchelor and H.J. Webber), Univ. of California, Vol.II.

Coggins, C.W.Jr., Heild, H.Z. and Burns, R.M. (1962) *Proc. Amer. Soc. Hort. Sci.*, **81:** 223-226.

Coggins, C.W.Jr., Heild, H.Z. and Garber, M.J. (1960) *Proc. Amer. Soc. Hort. Sci.*, **76**: 193-198.

Cohen, E. (1982) *Abst. 21st Int. Hort. Cong.*, Hamburg, No. 1340.

Cohen, E. (1986) *Abst. 22nd Int. Hort. Cong.*, Davis, California, No. 367.

Cohen, M. (1970) *Proc. Fla. Sta. Hort. Soc.*, **83:** 78-84.

Cole, P.J. (1985) *Irrigation Science.*, **6**: 63–71.

Collado Fernandez, M. (2000) *Alimentaria.*, **37**: 91-95.

Colletta, F.H.D., Machado, M.A., Targon, M.C.P.N., Moreira, M.C.P.G. and Pompeu, J.J. (1998) *Euphytica.*, 102: 133-139.

Cooper W.C., Ascension, P., Furr, J.R., Hilgeman, R.H., Cahon, G.A. and Boswell, S.B. (1963) *Proc. Amer. Soc. Hort. Sci.*, **82**: 180-182.

Cooper, W.C. (1961) *Proc. Fla. Sta. Hort. Soc.*, **74**: 95-104.

Cox, J.E. (1974) *Agric. Gaz. NSW Australia.*, **85**: 48.

Cradock, F.W. and Weir, R.G. (1969) *NSWDA Bull.* S-40.

Curti-Diaz, S.A. (1996) *Agrociencia.*, **30**: 405-409.

Darcel, F.C. (1952) *Trop. Agric.*, **29**: 4-6, 77-87.

Das, A.K., Rao, C.N. and Singh, S. (2007) *Curr. Sci.*, **92**: 1759–1763.

Das, T., Mitra, G.C. and Chatterjee, A. (1995) *Phytomorphology.*, **45**: 57-64.

Dasberg, S. and Erner, Y. (1997) *Acta Hortic.*, **449**: 125-131.

Dass, H.C., Srivastava, A.K., Lallan Ram and Shyam Singh. (1998) *Indian J. Agri., Sci.*, **68**: 692-694.

Dass, H.C. (1989) *Paper presented at the Citrus shows cum seminar R.F. Res. Station*, Abhor, Jan., 6-7.

De Candolle, A. (1886) *Origin of Cultivated Plants*, Kagan Paul, London.

De Lange, J.H. and Vincent, A.P. (1979) *Citrus and Subtrop. Fruit J.*, **547**: 9-14.

De Villiers, J.I. and Beyers, C.J. (1961) *Amer. Inst. Biol. Sci.*, Washington, DC, **3**: 107-119.

Delgado, R., Casamayor, R., Rodriguez, J.L., Cruz, P. and Fajaro, R. (1986a) *Acta Hortic.*, **179**: 537-544.

Delgado, R., Casamayor, R., Rodriguez, J.L., Cruz, P. and Fajaro, R. (1986b) *Acta Hortic.*, **179**: 545-548.

Deole, D.G., Joshi, A.T. and Deshmukh, P.P. (1977) *Abst. Int. Symp. Citriculture*, Bangalore, p. 10.

Deshmukh, P.P., Shelke, B.D., Tijare, R.R. and Badge, T.R. (1981) *Abst. Nat. Symp. Trop. and Subtrop. Fruit Crops*, Bangalore, p. 27.

Devkota, R.P., Grewal, S.S. and Dhatt, A.S. (1982) *Punjab Hort. J.*, **22**: 131-135.

Dhillon, H.S., Bhutani, V.P. and Gautam, D.R. (1976) *Indian J. Weed Sci.*, **2**: 134-139.

Dhingra, D.R. and Kanwar, J.S. (1963) *Punjab Hort. J.*, **3**: 54-59.

Dhingra, D.R., Kanwar, J.S., Bhumbla, D.R. and Sehgal, K.L. (1965) *PAU J. Res.*, **2**: 143-160.

Dikshit, N.N. (1958) *Sci. and Cult.*, **24**: 91-94.

Domingues, E.T. and Tulmann Neto, A. (1999) *Scientia Agricola.*, **56**: 163-170.

Donadio, L.C., Cabrita, J.R.M. and Sempionato, O.R. (1999) *Laranja.*, **20**: 107-177.

Dris, R. (1997) *Acta Hortic.*, **448**: 375-381.

Dubey, A.K. and Sharma, R.M. (2016) *Sci. Hort.*, **200**: 131–136

Duenhas L H, R.L. Villas Bôas, M. Souza Cláudio, V.A.M.Oliveira, and A.B. Dalri (2005) *Journal of Brazilian Association of Agricultural Engineering.*, **25**: 154–160.

Dutta, S. (1958) *Indian J. Hort.*, **15**: 146-153.

Dutta, S. (1959) *Indian Hort.*, **4**: 28-32.

Eaks, I.L. (1970) *Plant Physiol.*, **45**: 334.

Eaks, I.L. (1976) *Calif. Agric.*, **30**: 18.

Eaks, I.L. and Sinclair, W.B. (1980) *J. Food Sci.*, **45**: 985-988.

El-Desouky, S.A. and Atawia, A.A.R. (1998) *Alexandria J. Agric. Res.*, **43**: 231-254.

Eliades, G. (1998) *Tech. Bull. – Cyprus Agric. Res. Inst. No. 193*, pp. 8.

El-Zeftawi, B.M. (1975) *Australian Citrus News.*, **50**: 4.

El-Zeftawi, B.M. (1976) *Sci. Hort.*, **5**: 59-63.

Embleton, T.W., Jones, W.W. and Kirkpatrick, J.D. (1956) *Proc. Amer. Soc. Hort. Sci.*, **67**: 191-201.

Erickson, L.C. (1951) *Proc. Amer. Soc. Hort. Sci.*, **58**: 46-52.

Fang, D.Q., Federici, C.T. and Roose, M.L. (1997) *Genome*, **40**: 841-840.

Fang, D.Q., Federici, C.T. and Roose, M.L. (1998) *Genetics*, **150**: 883-890.

FAO (1997) *Citrus fruit, fresh and processed*, Annual Statistics, pp. 1-33.

Farmahan, H.L. (1997) *South Indian Hort.*, **45**: 30-33.

Faton (1970) *Citrograph.*, **55**: 173-175.

Fawcett, H.S. (1938) *Citrus Diseases and their Control*, McGraw-Hill Co., New York, p. 656.

Fawcett, H.S. and Wallace, J.M. (1964) *Calif. Citrogr.*, **32:** 50, 88-89.

Foguet, J.L., Campo, S.A., Oste, C.A., Alvarez, S. and Toll Jubes, J. (1972) *Revista Indust. Agricola de Tucuman.*, **49**: 35-41.

Foguet, J.L., Gonzalez, J.L., Vinciguerra, H.F. and Blanco, A.S. (1996) *Adv. Agroindustrial.*, **17:** 10-12.

Foloni, L.L. and Silva, N.G. (1997) *Proc. Brighton Crop Protection Conf. Weeds*, Brigton, UK, pp. 712-722.

Ford, H.W. (1963) *Proc. Amer. Soc. Hort. Sci.*, **82**: 177-179.

Ford, H.W. and Feder, W.A. (1969) *Proc. Int. Citrus Symp.*, **2**: 941-948.

Ford, H.W., Reuther, W. and Smith, P.F. (1957) *Proc. Amer. Soc. Hort. Sci.*, **70**: 237-244.

Fouche, P.S. and Bester, D.H. (1987) *Citrus and Sub-tropical Fruit J.*, **62**: 8-12.

Francis, H.L., Miller, M., Boswell, S. and Colladay, C. (1975) *Citrograph.*, **61**: 24-26.

Fraser, L.R. (1967) *Report to the Dept. of External Affairs*, Canberra, Australia.

Fraser, L.R. (1966) *Citrus Dieback in India Report.*, Dept. Ext. Affairs, Canberra, Australia, p. 95.

Fritz, W.D. (1970) *RUHR, Stick Scoff Ag. Bouchum.*

Fucik, J.E. (1977) *Proc. Int. Soc. Citriculture.*, **1**: 172-176.

Fucik, J.E. (1982) *J. Rio Grande Valley Hort. Soc.*, **35**: 139-147.

Fujita, K. and Yagi, T. (1956) *Bull Kanagawa Agric. Expt. Sta. Hort. Sect.*, **4:** 23-28.

Furr, J.R., Carpenter, J.B. and Hewitt, A.R. (1963) *J. Rio Grande Valley Hort. Soc.*, **17**: 90-107.

Gabricidze, Z.S. (1970) *Subtrop. Kul'tury*, No.2, pp. 109-115.

Gardner, F.E. and Horanic, G.E. (1958) *Proc. Fla. Sta. Hort. Soc.*, **71:** 81-86.

Gardner, F.E. and Horanic, G.E. (1968) *Proc. Fla. Sta. Hort. Soc.*, **80:** 85-88.

Garnsey, S.M., Barrett, H.C. and Hutchison, D.J. (1987) *Phytophylactica.*, **19**: 187-191.

Garnsey, S.M., Su, H.J. and Tsai, M.C. (1997) *Proc. 13th Conf. Int. Organ. Citrus Virologist*, pp. 138-146.

Georgiou, A. (1998) *Agricoltura Mediterranea.*, **128**: 307-312.

Georgiou, A. and Gregoriou, C. (1999) *Sci. Hort.*, **80**: 113-121.

Geraci, G. (1997) *Rivista di Fruticoltura e di Ortofloricoltura.*, **59:** 27-29.

Ghadekar, S.R., Dixit, S.V. and Patil, V.P. (1989) *PKV Research J.*, **13**: 143-148.

Ghai, S. (1964) *Entomology in India*, Ent. Soc. India Publ., pp. 529-600.

Ghali, M.H. and Nakhlla, P.G. (1996) *Ann. Agric. Sci.*, Moshtohor, **34**: 1099-1116.

Ghosh, A., Dey, K., Bhowmick, N., Medda, P.S., and Ghosh, S.K. (2016) *Vegetos.*, **29**(1). https: / /doi.org/10.5958/2229-4473.2016.00007.0

Ghosh, D., Mitra, S.K. and Bose, T.K. (1985) *Proc. 3rd Nat. Citrus Sem.*, Calcutta, pp. 1-2.

Ghosh, S.P. (1985) *Horticulture in Northeast India*. New Delhi: Associate Publishing Company.

Ghosh, S.P., Verma, A.N. and Sheo Govind (1980) *Punjab Hort. J.*, **20**: 106-111.

Ghosh, S.P., Verma, A.N. and Sheo Govind (1981) *Abst. Nat. Symp. Trop. and Subtrop. Fruit Crops*, Bangalore, p. 49.

Ghosh, S.P. (1991) *Agro-climatic zone specific research*, ICAR, Krishi Anusandhan Bhawan, New Delhi, p. 5.

Ghuguskar, H.T., Radke, S.G. and Borle, M.N. (1981) *PKV. Res. J.*, **5**: 143-147.

Giametta, G. and Zimbalatti, G. (1994) *Proc. Int. Agricultural Engineering Conf.*, Thailand, pp. 82-89.

Gilfillan, I.M. (1995) *Citrus J.*, **5**: 27-28.

Gmitter Jr., F.G. and Moore, G.A. (1986) *Plant Cell Tissue Organ Cult.*, **6**: 139-147.

Gmitter, F.G., Xiao, S.Y., Huang, S., Hu, X.L., Garnsey, S.M. and Denz, Z. (1996) *Theor. Appl. Genet.*, **92**: 688-695.

Gmitter, F., Zhang, Y. and Grosser, J. (2018) *Citrus Industry*, pp. 12-15.

Goell, A. (1986) *Abst. 22nd Int. Hort. Cong.*, Davis, California, No. 513.

Gomez-Cadenas, A., Mehouachi, J., Tadeo, F.R., Primo-Millo, E. and Talon, M. (2000) *Planta.*, **210:** 636-643.

Gonzalez, A. and Figueroa, L.A. (1996) *J. Agric. Univ. Puerto Rico.*, **80**: 203-205.

Goren, R. and Monselise, S.P. (1969) *Planta.*, **88**: 364-368.

Goswami, A.M., Saxena, K. and Kurein, S. (1993) *High density planting in citrus. in. Advances in Horticulture*, K. L. Chadha and O. P. Pareek (eds.). Malhotra Publishing House, New Delhi,India.

Goswami, A.M., Saxena, S.K. and Sharma, B.B. (1985) *Abst. 3rd Nat. Citrus Sem.*, Calcutta, p. 32.

Gountur, D.L. and Brown, M. (1983) *Econ. Res. Div.*, Fla. Dept. Citrus Working Paper Series, Oct. 25.

Graca, J., Barros, J.C. Da S.M.De, Celestino, R.C.A. and Castro, N.G. (1997) *Comunicado Tecnico-Empresa de Pesquisa Agropecuaria do Estado do Rio de Janeiro*, No. 234, p. 4.

Gravina, A., Arbiza, H., Bozzo, A., Laserre, E., Arias, M. and Ronca, F. (1998) *Acta Hortic.*, **463**: 413-418.

Green, B.A. and Gerber, J.F. (1967) *J. Amer. Soc. Hort. Sci.*, **90:** 77.

Gregoriou, C. and Economides, C.W. (1994) *Acta Hortic.*, **365**: 57-68.

Grewal, H.S., Dhatt, A.S. and Gosal, S.S. (1994) *Ann. Biol.*, **10**: 1-6.

Grewal, S.S. and Singh, G. (1975) *J. Hort. Sci.*, **4**: 141-146.

Grierson, W. (1973) *Int. Inst. Refrig. Comm.*, **2**: 51-65.

Grierson, W. and Hatton, T.T. (1977) *Proc. Int. Soc. Citriculture*, Vol. I., p. 227.

Grimm, G.R. and Hutchison, D.J. (1973) *Plant Dis. Reptr.*, **47**: 669-672.

Grosser, J.W. and Gmitter, G.G.Jr. (1990) *Plant Breed Rev.*, **8**: 339-397.

Grosser, J.W., Gmitter Jr., F.G. and Chandler, J.L. (1988a) *Plant Cell Rep.*, **7**: 5-8.

Grosser, J.W., Gmitter Jr., F.G. and Chandler, J.L. (1988b) *Theor. Appl. Genet.*, **75:** 397-401.

Grosser, J.W., Louzada, E.S., Gmitter Jr., F.G. and Chandler, J.L. (1994) *HortScience.*, **29**: 812-813.

Gupta, O.P., Singh, J.P. and Arora, R.K. (1974) *Indian Hort.*, **19**: 7-8.

Gurgel, J.T.A. and Sabrino, J.S. (1951) *Ana Esc. Super Agr. Luis de Queiroz Univ.*, Sao Paulo, **8:** 727-746.

Gutierrez, E.A., Moore, G.A., Jacno, C., McCaffery, M. and Cline, K. (1992) *Phytopathology.*, **82**: 1147.

Halton, T.T. (1953) *Proc. Fla. Sta. Hort. Soc.*, **6**: 200-202.

Hamid, H.A. and Michelakis, S.E. (1997) *Bull. OILB/SROP*, **20:** 7-12.

Hankey, D.D. (1975) *Citrograph.*, **60**: 427-442.

Hardenburg, R.E., Watada, A.E. and Wang, C.Y. (1990) *The Commercial Storage of Fruits, Vegetables and Florist and Nursery Stocks*, USDA, ARS, Agricultural Handbook 66, 130 pp.

Hassaballa, I.A., El-Gindy, F.A. and Power, J.B. (1988) *HortScience.*, **23**: 757.

Hayes, W.B. (1970) *Fruit Growing in India*, Kitabistan, Allahabad.

Hazarika, B.N., Nagaraju, V. and Parthasarathy, V.A. (1995) *Abst. Nat. Symp.*, Role of Plant Biotech. Improving Agric. Challenges and Opp., 23-25 March, Jaipur, p. 55.

He, H., Pan, R.C., Han, M.L. and Li, G.G. (1998) *Acta Bot. Yunnanica.*, **20**: 459-463.

Hearn, C.J., Reece, P.C. and Fenton, R. (1968) *Proc. Fla. Sta. Hort. Soc.*, **81**; 94-98.

Hensz, R.A. (1981) *Proc. Int. Soc. Citriculture*, Vol. 1, pp. 257-266.

Hernandez, M., Harnandez, I., Mesa, A. and Forteza, I. (1987) *Agrotecnia de Cuba.*, **19:** 75-79.

Hidaka, T. (1984) *J. Jap. Soc. Hort. Sci.*, **53:** 1-5.

Hidaka, T., Takayanagi, R., Shinozaki, S., Fujita, K. and Omura, M. (1992) *Proc. 7th Int. Citrus Congr.*, Acireale, Italy, 8-13 March, pp. 323-328.

Hidaka, T., Tamada, T. and Shichijo, T. (1981) *Proc. Soc. Citriculture.*, **1**: 153-155.

Hidaka, T., Yamada, Y. and Shichijo. T. (1979) *Jap. J. Breed.*, **29:** 248-254.

Hield, H.Z., Coggins, C.W.Jr. and Garber, M.J. (1958) *Calif. Agric.*, **12**: 9-11.

Hield, H.Z., Coggins, C.W.Jr. and Garber, M.J. (1965) *Hilgardia.*, **36:** 297-311.

Hilgeman, R.H., Sharples, G.C. and Dunlap, J.A. (1975) *Congresso Mundian de Citricultura*, Muracia, Spain, pp. 33-37.

Hilgeman, R.H. and Hewland, S. (1955) *Proc. Agric. Arizona.*, **7:** 10-11.

Hirai, M. and Kajiura, I. (1987) *Jap. J. Breed.*, **37**: 337-388.

Hodgson, R.W. (1967) In: *Citrus Industry*, Vol. I (Eds. W. Reuther, H.J. Webber, and L.D. Batchelor), Univ. California, Berkeley.

Holzapfel, E.A., Lopez, C., Joublan, J.P. and Matta, R. (2001) *Chilean Journal of Agricultural Research.*, **61**: 51-60.

Hondebrink, M.A., Cammeraat, L.H. and Cerda, A. (2017) *Science of the Total Environment.*, **581–582**: 153–160.

Hooker, J.D. (1872) *Flora of British India*, Reeve & Co., London.

Huang, E., Chang, S. and Wu, W. (1997) *Proc. Symp. on Enhancing Competitiveness of Fruit Industry*, Taiwan, No. 38, pp. 51-68.

Huang, S.P., Wu, G.L. and Li, S.Y. (1993) *China Citrus.*, **22**: 3-5.

Huber, D.J. (1983) *Hort. Rev.*, **5**: 169-219.

Hume, H.H. (1959) *Citrus Fruits*, New York.

Hurosvilli, K.G. (1957) *Subtrop. Kul'tury*, No.1, pp. 78-97.

Hussain, T., Ghosh, D., Mitra, S.K. and Chattopadhyaya, P.K. (1985) *Abst. 3rd Nat. Citrus Sem.*, Calcutta, p. 13.

Hutchison, D.J. (1978) *Proc. Int. Soc. Citriculture*, **2**: 523-525.

Hyun, H.N., Lim, H.C., Han, H.R. and Moon, D.G. (1993) *J. Korean Soc. Hort. Sci.*, **34**: 368-377.

Intrigliolo, F. and Intelisano, S. (1997) *Acta Hortic.*, **448**: 499-507.

Intrigliolo, F., Giuffrida, A. and Germana, C. (2000) *Rivista di Frutticoltura e di Ortofloricoltura.*, **62:** 43-46.

Ioannou, M., Kyriakou, A. and Ioannou, N. (1991) *Tech. Bull. Cyprus Agric. Res. Inst.*, **139:** 7.

Iordanou, N. and Charalambous, P. (1998) *Tech. Bul.-Cyprus Agric. Res. Inst.*, No. 197, p.16.

Ismail, A.E., Ghali, M.H., Nakhlla, F.G. and Aboul-Eid, H.Z. (1997) *Pakistan J. Nematology.*, **15:** 71-87.

Ito, H., Fujita, K. and Ogaki, T. (1958) *J. Hort. Assoc.*, Japan, **27**: 94-106.

Ivakh, M.I. and Raskin, M.S. (1984) *Subtrop. Kul'tury*, No.6, pp. 116-119.

Iwagaki, I. (1981) *Proc. Int. Soc. Citricult.*, Vol. I, pp. 169-171.

Iwagaki, I. and Ideta, M. (1977) *Bul. Shikoku Agr. Expt. Sta.*, **30:** 1-15.

Iwakiri, T. and Nakahara, M. (1981) *Proc. Int. Soc. Citricult.*, Shimizu, Japan, **2**: 537.

Iwamasa, M. and Nishiura, M. (1981) *Proc. Int. Soc. Citriculture*, Shimizu, Japan, **1**: 141-153.

January Lado, J., Gambetta, G. and Zacarias, L. (2018) *Sci. Hortic.*, **233**: 238-248.

Jawanda, J.S. and Bajawa, B.S. (1961) *Punjab Hort. J.*, **1**: 29-33.

Jawanda, J.S., Arora, J.S. and Sharma, J.N. (1973) *Punjab Hort. J.*, **13:** 3-12.

Jawanda, J.S., Sinha, M.K. and Uppal, D.K. (1972) *Indian J. Hort.*, **29**: 269-276.

Jawanda, J.S., Vij, V.K. and Sharma, J.N. (1977) *Indian J. Agric. Sci.*, **45**: 282-285.

Jenny Zoremtluangi1, J., Saipari, E. and Mandal, D. (2019) *Res. on Crops.*, **20**(2): 322-327.

Jiang, R.S. (2000) *South China Fruits.*, **29:** 6.

Jing, X.D. (1998) *South China Fruits.*, **27**: 24-25.

Jona, R., Goren, R. and Marmora, M. (1989) *Sci. Hortic.*, **39**: 105-115.

Jones, L.T. (1951) *J. Agric. West Aust.*, **28:** 72-76.

Jones, W.W. and Smith, P.F. (1964) In: *Hunger Signs in Crops* (Ed. H.B. Sprague), David Mckay Co. Inc. Washington.

Josan, J.S. and Chohan, G.S. (1985) *Proc. 3rd Nat. Citrus Sem.*, Calcutta, pp. 33-34.

Josan, J.S., Sharma, J.N. and Chohan, G.S. (1981) *Indian J. Hort.*, **38**: 178-183.

Josan, J.S., Sharma, J.N. and Chohan, G.S. (1983) *Indian J. Hort.*, **40**: 183-187.

Joublan, M.J.P., Venegas, V.A., Wilckens, E.R. and Guerrero, C. (1997) *Agro-Ciencia.*, **13**: 149-157.

Jumin, H.B. and Nito, N. (1996) *Plant Cell Rep.*, 15: 332-336.

Kaewsuksaeng, S., Tatmala, N., Srilaong, V. and Pongprasert, N. (2015) *Post-harvest Biology and Technology.*, **100**: 1-7.

Kakde, J.R. (1956) *Agric. Soil Mag.*, **30**: 19-24.

Kale, K.B., Kolte, S.O. and Peshney, N.L. (1994) *Indian Phytopath.*, **47**: 253-255.

Kale, K.B., Raut, J.G. and Dudhe, Y.H. (1996) *PKV Res. J.*, **20**: 105-106.

Kalidas, P. and Shivankar, V.J. (1995) *Final Report Nat. Res. Centre for Citrus*, Amravati Road, Nagpur-440 010.

Kanwar, J.S. and Randhawa, N.S. (1960) *Hort. Adv.*, **4**: 61-67.

Kanwar, J.S., Sehgal, J L. and Dhingra, D.R. (1965) *Punjab Agric. Univ. Bull.*, **56**: 27.

Kapur, S.P. and Sohi, H.S. (1988) *Tree Protection*, pp. 49-61.

Kar, P., Hossain, T., Sen, S.K., Bose, T.K. and Mitra, S.K. (1985) *Abst. 3rd Nat. Citrus Sem.*, Calcutta, pp. 35-38.

Katole, S.R., Thakre, H.S. and Mahajan, R.K. (1993) *J. Maharashtra Agric. Univ.*, **18:** 68-71.

Kedar, V.P. and Gopal Krishna, N. (1977) *Indian J. Hort.*, **34**: 385-386.

Keesstra, S., Pereira, P., Novara, A., Brevik, E.C., Azorin-Molina, C., Parras-Alcantara, L., Jordan, A. and Cerda, A. (2016b) *Science of the Total Environment.*, **551–552:** 357–366.

Keesstra, S.D., Bouma, J., Wallinga, J., Tittonell, P., Smith, P., Cerda, A., Montanarella, L.,Quinton, J., Pachepsky, Y., van der Putten, W.H., Bardgett, R.D., Moolenaar, S., Mol, G. and Fresco, L.O. (2016a) *The significance of soils and soil science towards realization of the UN sustainable development goals* (*SDGs*). SOIL Discussions. (cf. Hondebrink *et al.*, 2017).

Kelin, I. (1983) *HortScience.*, **18**: 942-944.

Kender, W.J., Hartmond, U. and Burns, J.K. (1999) *Hort. Tech.*, **9**: 412-416.

Ketsa, S. (1988) *Natural Sci.*, **22**: 225-227.

Khanna (1989) *Survey of Indian Agriculture*, The Hindu, Madras, pp. 28-53.

Kitto, S.L. and Young, M.J. (1981) *HortScience.*, **16**: 305-306.

Klotz, L.J. (1950) *Calif. Univ. Circ.*, No. 396.

Klotz, L.J. (1973) *Colour Handbook of Citrus Diseases*, Univ. of California, Riverside.

Klotz, L.J., Bitters, W.P. and Wolfe, T.A. (1965) *Calif. Agric.*, **19**: 10-12.

Klotz, L.J., Calavan, E.C. and Weathers, L.G. (1972) *Calif. Agr. Expt. Sta. Circ.*, p. 559.

Knorr, L. C., Sint, R. F. and Ducharme, E. P. (1957) Florida Agr. Expt. Sta. Bull., 587: 157.

Kobayashi, S. and Uchimiya, H. (1989) *Japanese J. Genet.*, **64**: 91-97.

Kobayashi, S., Ohgawara, T., Saito, W., Nakamura, Y. and Omura, M. (1997) *J. Japanese Soc. Hort. Sci.*, **66:** 453-458.

Kochba, J., Speigel-Roy, R. and Safran, H. (1972) *Planta.*, **106:** 237-245.

Kohli, R.R. and Srivastava, A.K. (1997) *Citrus nutrition - A Retrospection Agri. Review.*, **18**: 128-138.

Kohli, R.R., Srivastava, A.K. and Huchche, A.D. (1997) *Indian Farming.*, **47**: 25-27.

Koizumi, M., Kuhara, S., Leki, H., Kano, T., Tranaka, A. and Iwanami, T. (1991) *Proc. 11th Conference International Organization of Citrus Virologist*, Reverside, California, pp. 125-127.

Koo, R.C.J. and Reese, R.L. (1972) *Proc. Fla. Sta. Hort. Soc.*, **84:** 11-16.

Koo, R.C.J. and Smajstrala, A.G. (1984) *Proc. Fla. Sta. Hort. Soc.*, **97**: 33-36.

Kordan, H.A. (1963) *Nature.*, **198**: 867-869.

Kore V.T. and Chakraborty I. (2015) *J. Food Sci. Technol.*, **52**(9): 6073-6077.

Kore, Vijaykumar T., Tawade, Sima S., Devi, Lembisana H., Chakraborty, I. (2018) *Amer. Res. J. Food Nutri.*, **1**(1): 11-21.

Korsantiya, N.G., Sanikidze, A.B. and Pirtskhalaishvili, S.K.U. (1985) *Subtrop. Kul'tury*, No.4, pp. 80-85.

Koshita, Y., Takahara, T., Ogata, T. and Goto, A. (1999) *Sci. Hortic.*, **79**: 185-194.

Krezdorn, A.H. and Robinson, F.A. (1958) *Proc. Fla. Sta. Hort. Soc.*, **71**: 86-91.

Krishnamurthy, S. (1959) *Indian J. Hort.*, **16**: 221-227.

Kumar, R., Singh, J.O. and Gupta, O.P. (1976) *Haryana J. Hort. Sci.*, **5:** 25-30.

Kumar, S.A. and Sharma, O.P. (1960) *Indian J. Hort.*, **17**: 243-245.

Labanauskas, C.K., Bitters, W.P. and McCarty, C.D. (1976) *HortScience.*, **11**: 117-118.

Ladaniya, M.S., Marathe, R.A., Das, A.K., Rao, C.N., Huchche, A.D., Shirgure, P.S. and Murkute, A.A. (2020) *Sci. Hortic.*, **261**: 108935. https: //www.sciencedirect. com/science/article/abs/pii/S0304423819308210?via per cent 3 ihub

Lado, J., Rodrigo, M.J. and Zacarias, L. (2014) *Stewart Post-harvest Review.*, **2**: 2-6.

Lawrence, F.P. and Bridges, G.D. (1973) *Univ. Fla. Coop. Ext. Ser. Cir.*, p. 394.

Le Roux, F.H. (1965) *Fmg. South Afr.*, **41**: 65,66,69,70.

Lee, B.W. (1964) *Calif. Citrogr.*, **49**: 326-328.

Lee, R.F., Brlansky, R.H., Garnsey, S.M. and Yokomi, R.K. (1987) *Phytophylactica.*, **19:** 215-218.

Legua, P., Forner, J.B., Hernandez, F. and Forner-Giner, M.A. (2014) *Sci. Hortic.*, **174**: 60–64.

Leonard, C.D. (1952) *Citrus Fruit, Farm Chemicals*, p. 130.

Levy, Y. and Lifshitz, J. (1995) *Sci. Hortic.*, **61**: 131-137.

Levy, Y., Lifshitz, J., Malach, Y. de and David, Y. (1999) *HortScience.*, **34**: 878-881.

Leyva, D., Alvarez, E. and Sanchez, P. (1987) *Citricos y Otros Frutales.*, **10:** 53-62.

Lin, H.C. and Wu, U.W. (1957) *Acta Bot. Sin.*, **6:** 134-40.

Litz, R. (1990) *In vitro.*, **26**: 2, 25A.

Liu, B. (1998) *South China Fruits.*, **27**: 23.

Liu, F.W., Pan, C.H., Hsueh, S.M. and Hing, T.H. (1998) *J. Chinese Soc. Hort. Sci.*, **44**: 239-252.

Liu, J.H. and Deng, X.X. (2000) *Scientia Agricultura Sinica.*, **33**: 98-100

Lloyd, J., Kriedemann, P.W. and Aspinall, D. (1989) *Plant Cell Environ.*, **12**: 529.

Lloyd, J., Kriedemann, P.W. and Syvertsen, J.P. (1987) *Aust. J. Plant Physiol.*, **14**: 387.

Lloyd, R.H. (1961) *Citrus News.*, **37**: 21-25.

Lodh, S.S., Dey, S. and Bose, A.N. (1963) *Sci. and Cult.*, **29**: 48-49.

Louse, Fergusum (1990) *Rept. Citrus Research Centre and Agricultural Station*, Univ. of California, Riverside, pp. 20-22.

Lovicu, G., D'Hallewin, G., Pala, M., Farci, M., Mura, D., Pinna, R., Sedda, M. and Manueddu, E. (1999) *Informatore Agrario.*, **55**: 75-77.

Lyngdoh, G.B., Kar, P.L. and Sanyal, D. (1994) *Crop Res.*, **7**: 49-53.

Lyngdoh, G.B., Kar, P.L. and Sanyal, D. (1996) *Haryana J. Hort. Sci.*, **25:** 13-15.

Ma, J.Q. (1998) *South China Fruits.*, **27**: 11.

Mabberlley, D.J. (2004) *Blumea journal of plant taxonomy and plant geography.*, **49**(2-3): 481-498.

Machado, L.A. and Raga, A. (1999) *Revista de Agricultura.*, Piracicaba, **74**: 229-234.

Machado, S.O. and Mendez, M.E.G. (1999) *Revista Cientifica Rural.*, **4**: 91-95.

Machin, T., Sutton, P., Anderson, P., Hackett, B. and Mooney, P. (1999) *Orchardist.*, **72**: 64-66.

Mageed, K.J.A., Sharma, B.B. and Sinha, A.K. (1988) *Indian J. Agric. Sci.*, **58**: 284-286.

Maillard, S. and Hibbon, J.S. (1957) *Fruits et Prim.*, **27:** 247-252.

Makhija, M., Sharma, B.B. and Sinha, A.K. (1986) *South Indian Hort.*, **34**: 129.

Malaka, A.S. (1996) *Ann. Agric. Sci.*, Cairo, **41**: 951-960.

Mann, M.S. and Takkar, P.N. (1981) *Abst. Nat. Symp.Trop. and Subtrop. Fruit Crops*, p. 83.

Mann, Mjk. S., Munshi, S.K., Bajwa, M.S. and Takkar, P.N. (1978) *Punjab Hort. J.*, **18**: 41-47.

Mann, S.S. and Randhawa, J.S. (1976) *Progressive Hort.*, **10**: 35-41.

Mantel, A. (1977) *Proc. Int. Soc. Citricult.*, Vol. I, pp. 74-79.

Marinho, C.S. and Souza, M.De. (1997) *Pesquisa Agropecuaria Brasileira.*, **32**: 293-296.

Marriott, E.E. and Wander, M.M. (2006) *Soil Science Society of America Journal.*, **70**(3): 950-959.

Mataa, M., Tominaga, S. and Kozaki, I. (1998) *J. Japanese Soc. Hort. Sci.*, **67**: 28-34.

Matsumato, K. (1964) *Mem. Ehmi Univ.*, Sect. VI, 10: 535-555.

Matsumoto, R., Okudai, N., Yamamoto, M., Yamada, Y., Asada, K., Oiyama, I., Ikemiya, H., Murata, H., Yoshinaga, K., Koizumi, M. and Iwanami, T. (1999a) *Bull. National Institure of Fruit Tree Science.*, **33**: 37-46.

Matsumoto, R., Okudai, N., Yamamoto, M., Yamada, Y., Takahara, T., Oiyama, I., Ishiuchi, D., Asada, K., Ikemiya, H., Murata, H., Yoshinaga, K., Uchihara, S., Ieki, H. and Shichijo, T. (1999b) *Bull. National Institute of Fruit Tree Science.*, **33**: 47-56.

Matsumoto, R., Okudai, N., Yamamoto, M., Yamada, Y., Takahara, T., Oiyama, I., Ishiuchi, D., Asada, K., Ikemiya, H., Murata, H., Yoshinaga, K., Uchihara, S. and Ieki, H. (1999c) *Bull. National Institute of Fruit Tree Science.*, **33**: 57-66.

Matsumoto, R., Okudai, N., Yamamoto, M., Takahara, T., Yamada, Y., Oiyama, I., Ishiuchi, D., Asada, K., Ikemiya, H., Murata, H., Kunga, T., Shichuo, T.,

Yoshinaga, K., Uchihara, S., Koizumi, M. and Iwanami, T. (1999d) *Bull. National Institute of Fruit Tree Science.*, **33**: 67-76.

McDonald, R.E. and Camp, T.H. (1973) *US Dept. Agri.* ARS-3-25, p. 6.

Meena, A.K., Dutta, F., Mingnam Ch. Marak and Meena, R.K. (2018) *Int. J. Curr. Microbiol. App. Sci.*, **7**: 2807-2815.

Meneghini, M. (1946) *Biologoco.*, **12**: 285-287.

Mestre, P.F., Asins, M.J., Pina, J.A. and Navarro, L. (1997) *Theor. Appl. Genet.*, **95**: 1282-1288.

Mitra, G.C. and Chaturvedi, H.C. (1972) *Bull. Torrey. Bot. Club.*, **99**: 184-189.

Molla, P. and Jousseaume, C. (1997) *Proc. Quadtrieme Conf. Int. Sur. Les Ravageurs en Agric.*, France, pp. 339-346.

Monselise, S.P. (1975) *Colloque CNRS* (Ed. R. Ulrich, R), 238, Paris, pp. 97-104.

Monselise, S P. and Halevy, A. H. (1964) *Proc. Amer. Soc. Hort. Sci.*, **84**: 141-146.

Monselise, S.P., Goren, R. and Halevy, A.H. (1966) *Proc. Amer. Soc. Hort. Sci.*, 85: 195-200.

Monselise, S.P. (1983) *Proc.21st Int. Hort. Cong.* Hamburg, pp. 468-478.

Monselise, S.P. and Sasson, A. (1977) *Proc. Int. Soc. Citriculture*, Vol. 1, pp. 232-237.

Monselise, S.P., Weiser, M., Shafir, N., Goren, R. and Goldschmidt, E.E. (1976) *J. Hort. Sci.*, **5**: 341-351.

Moore, G.A., Gutierrez, E.A., Jacono, C., McCaffery, M. and Cline, K. (1993) *HortScience.*, **28**: 512.

Morales, P., Davies, F.S. and Littell, R.C. (2000) *HortScience.*, **35**: 30-35.

Moreira, S. and Salib, A.A. (1968-69) *Proc. Int. Citrus Symp.*, Riverside, **1**: 351-357.

Motomura, T., Moriguchi, T., Akihama, T., Hidaka, T. and Omura, M. (1996) *J. Japanese Soc. Hort. Sci.*, **65**: 214.

Mourao, F.A.A., Gmitter Jr., F.G. and Grosser, J.W. (1996) *Fruit Var. J.*, **50**: 76-80.

Muller, J., Koller, K., Algohary, S. and Hegazi, A. (1998) *Landtechnik.*, **53**: 138-139.

Mungomery, W.V., Jorgensen, K.R. and Barnes, J.A. (1980) *Proc. Int. Soc. Citriculture*, Griffith, NSW, Australia, pp. 285-288.

Munshi, S.K., Singh, R., Vij, V.K. and Jawanda, J.S. (1978) *Sci. Hortic.*, **9**: 357-367.

Muramatsu, N., Takahara, T., Ogata, T. and Kojima, K. (1999) *HortScience.*, **34**: 79-81.

Murashige, T., Nakanon, R. and Tucker, D.P.H. (1968) *Phytomorphology.*, **17:** 1-4.

Mussak, I., Greenberg, J., Yacov, B. and Cadmon, E. (1999) *Alon Hanotea.*, **53**: 506-508.

Mussak, J., Osorovits, J., Mezuman, U., Geffen, A. and Frid, A. (2000) *Alon Hanotea.*, **54:** 38-40.

Myers, L. H. and Grierson, W. (1978) *Proc. Int. Soc. Citriculture*, p.1.

Nagasawa, K., Tomomatsu, K., Kunitake, H. and Kimatsu, H. (1999) *Proc. School Agric. Kyushu Tokai Univ.*, **18**: 11-18.

Nagpal, R.L. (1959) *Citrus Industry.*, **40**: 14-15.

Naidu, K.C. (1953) *Andhra J. Agri.*, **2**: 236-240.

Naik, K.C. (1949) *South Indian Fruits and their Culture*, P. Varadachary and Co., Madras, pp. 477.

Nair, P.K.K. and Singh, B. (1969) *J. Indian Bot. Soc.*, **48**: 221-228.

Naohara, J. and Manabe, M. (1994) *J. Food Sci.*, **59**: 578-580.

Naqvi, S.A.M.H. and Sawant, S.D. (1994) *Nat. Meet of Phytophthora Worker*, Calicut.

Naqvi, S.A.M.H. (1988) *Annual Report*, 1988-89, National Research Centre for Citrus, Amravati Road, Nagpur-440 010.

Naqvi, S.A.M.H. (1993) *Proc. Nat. Sem. Export Oriented Hort. Production*, PKV, Akola, pp. 182-185.

Narasimhan, V., Subramanian, K. S., Shanmugam, N. and Jeyarajan, H. (1984) *Pesticides.*, **18**: 61-62.

Nariani, T.K. and Singh, G.R. (1971) *Proc. Indian Nat. Sci. Acad.*, **37**: 365-371.

Nariani, T.K., Raychaudhuri, S.P. and Bhatta, R.B. (1967) *Indian Phytopath.*, **20**: 146-150.

Nastou, A., Chartzonlakis, K., Therios, I. and Bosabalidis, A. (1999) *Adv. Hort. Sci.*, **13**: 61-67.

Nauriyal, J.P. (1952) *Curr. Sci.*, **21:** 347.

Nauriyal, R.P. and Chadha, K.L. (1966) *Indian J. Hort.*, **24**: 60-62.

Nava Ayala, J., Viloegas Monter, A., Barrientos, P.F. and Martinez Garza, A. (1996) *Agrociencia.*, **30:** 573-576.

Nayer, K.K., Ananthakrishnan, T.N. and David, T.N. (1976) *General and Applied Entomology*, Tata McGraw Hill Publishing Co. Ltd., New Delhi, p. 350.

Nelson, S.D., Young, M., Enciso, J.M., Klose, S.L. and Setamou, M. (2011) *Subtropical Plant Science.*, **63**: 14-22.

Neumann, K.H. (1965) *Citrus*, Springer Verlag, Vienna.

Nie, L., Liu, H.X. and Chen, L.G. (2000) *South China Fruits.*, **29**: 7-8.

Nilangekar, R.G. and Patil, V.K. (1982) *J. Indian Soc. Soil Sci.*, **30**: 213-215.

Nunez-Moreno, J. H. and Valdez-Gascon, B. (1994) *Commu. Soil Sci. & Plan. Anal.*, **25**: 1747-1757.

O'Bannon, J.H. and Hutchison, D.J. (1974) *Proc. Int. Citrus Short. Course*, Univ. of Florida, pp. 22-29.

Oberholzer, P.C.J. (1946) *Bull. Dept. Agric. South Afr.*, No. 271.

Obreza, T.A. (1993) *J. Production Agric.*, **6**: 546-552.

Ogata, T., Fujita, H., Shiozaki, S., Horiuchi, S., Kawase, K. and Kato, A. (1997) *J. Japanese Soc. Hort. Sci.*, **66**: 229-234.

Olson, E. O., Cooper, W. C., Maxwell, N. and Shull, A. V. (1962) *J. Rio. Grande Valley Hort. Soc.*, **16**: 44-51.

Oppenheimer, J.D. (1932) *Citrusfruchte*, Berlin-Leipzig.

Ouyang Ch°Chun (1998) *South China Fruits.*, **27**: 21.

Pal, R.N., Singh, R., Vij, V.K. and Munshi, S. K. (1977) *Indian J. Hort.*, **34**: 4-7.

Pande, N.D. (1982) *Indian J. Ento.*, **48**: 228-232.

Panigrahi, P. and Srivastava, A.K. (2017) *Journal of Integrative Agriculture.*, **16**(5): 1184-1194.

Panigrahi, P. and Srivastava, A.K. (2017) *Sci. Hortic.*, **210**: 6-13.

Panzenhagen, N.V., Koller, O.C., Sartori, I.A. and Portelinha, N.V. (1999) *Pesquisa Agropecuaria Brasileira.*, **34**: 527-533.

Paramasivam, S., Alva, A.K., Hostler, K.H., Easterwood, G.W. and Southwell, J.S. (2000) *J. Pl. Nutrition.*, **23**: 313-327.

Pardeep Wali and Sharma, O.N. (1997) *Haryana J. Hort. Sci.*, **26**: 213-215.

Parker, E.R. and Jones, W.W. (1951) *Calif. Agr. Exp. Sta. Bull.*, p. 722.

Parthasarathy, V.A. and Nagaraju, V. (1992) *J. Tree Sci.*, **11:** 146-148.

Parthasarathy, V.A. and Nagaraju, V. (1993) *Ann. Plant Physiol.*, **7**: 169-174.

Parthasarathy, V.A. and Nagaraju, V. (1996) *Indian J. Hort.*, **53:** 171-174.

Parthasarathy, V.A., Nagaraju, V. and Parthasarathy, U. (2000) In: *Biotechnology of Horticultural Crops*, Vol. I. (Eds. V.A. Parthasarathy, T.K. Bose, P. Das, S.K. Mitra and S. Mohandas), Naya Prokash, Calcutta, pp. 119-160.

Parthasarathy, V.A., Nagaraju, V. and Rahman, S.A.S. (1997) *Folia Hort.*, 9: 87-89.

Parthasarathy, V.A., Barua, A., Nagaraju, V. and Parthasarathy, U. (2001) *Indian J. Hort.*, **58**: 336-341.

Parthasarathy, V.A., Singh, S.K., Vinoth, S. and Aswath, C. (2019) *Breeding of Horticultural Crops*: Vol 2: Tropical Fruits. Today & Tomorrows Printers and Publishers. p. 149-213.

Passos, O.S. and Salibe, A.A. (1978) *Proc. Int. Soc. Citricult.*, p. 6.

Pathak, A.C. and Srivastava, K.C. (1985) *Abst. 3rd Nat. Citrus Sem.*, Calcutta, pp. 34-35.

Pekmezii, M. (1982) *Abst. 21st Int. Hort. Cong.*, Hamburg, No. 138.

Pena, L., Cervera, M., Juarez, J., Navarro, A., Ortega, C., Pina, J.A., Duran-Vila, N. and Navarro, L. (1998) *Acta Hortic.*, **463**: 61-68.

Pena, L., Cervera, M., Juarez, J., Ortega, C., Pina, J.A., Duran-Vila, N. and Navarro, L. (1995) *Plant Sci.*, **104**: 183-191.

Peng, Y.H. and Rabe, E. (1998a) *J. Hort. Sci. Biotech.*, **73:** 229-234.

Peng, Y.H. and Rabe, E. (1998b) *J. Fruit Sci.*, **15**: 128-132.

Peterson, D.L. (1998) *Applied Engineering in Agriculturae.*, 14: 455-458.

Phillips, P.A., O'Connell, N.V. and Menge, J.A. (1990) *California Agr.*, **44**: 6-7.

Phillips, R.L. (1972) *Proc. Fla. Sta. Hort. Soc.*, **85:** 48-50.

Piga, A., D'Aquino, S. and Agabbio, N. (2000) *Fruits.*, **55**: 37-44.

Pire, R. and Rojas, E. (1999) *Fruits.*, **54**: 177-182.

Platt, J. (1966) *Soil Sci.*, **102**: 82-84.

Platt, R.G. (1960) *Calif. Citrogr.*, **45**: 246-251.

Pomerinke, M.A. and Stansly, P.A. (1998) *Florida Entomologist.*, **81**: 361-372.

Porat, R., Daus, A., Weiss, B., Cohen, L., Fallik, E. and Droby, S. (2000) *Post-harvest Biology and Tech.*, **18**: 151-157.

Prati, P., Mourao Filho, F. De A.A., Dias, C.T. dos S. and Scarpare Filho, J.A. (1999) *Scientia Agricola.*, **56**: 185-190.

Proctor, F.J. and Caygill, J.C. (1985) *Proc. 39th Easter School*, Univ. Nottingham, UK, pp. 317-332.

Pruthi, H.S. and Mani, M.S. (1945) *Imp. Coun. Agr. Res. Sci.*, Mongraph, 16.

Py, C. (1951) *Fruits d'outre Mer.*, **6**: 321-326.

Queb-Gonzalez, D.B., Malo, A.L., Sosa-Morales, M. E. and Rojas, R.V. (2020) *Heliyon.*, **6** (1).

Quingguo, Z., Jin, Z. and Jisbeng, Z. (1997) *China Fruits*, No.4, p. 52.

Raciti, G., Spina, P., Scuderi, A. and Intrigliolo, F. (1981) *Proc. Int. Soc. Citricult.*, Vol. I, pp. 175-180.

Raga, A., Sato, M.E., Ceravolo, L.C. and Rossi, A.C. (1997) *Revista de Agricultura.*, **72**: 363-373.

Rahaman, A.A.S., Nagaraju, V. and Parthasarathy, V.A. (1996) *Ann. Plant Physiol.*, **10**: 45-49.

Rahana, K.V.R., Sateoja, S., Setty, G.R., Mnjundasswamy, A.M. and Moorthy, K.V.N. (1973) *Indian Fd. Pack.*, **27:** 5-9.

Rajput, C.B.S. and Haribabu, R.S. (1993) *Citriculture*, Kalyani Publishers, Ludhiana, pp. 326.

Rana, R.S. and Chauhan, K.S. (1976) *Punjab Hort. J.*, **16**: 30-32.

Randhawa, G.S. and Dhillon, B.S. (1965) *Indian J. Hort.*, **22**: 33-35.

Randhawa, G.S. and Dhinsa, H.S. (1947) *Proc. Amer. Soc. Hort. Sci.*, **50**: 165-171.

Randhawa, G.S. and Sharma, B.B. (1962) *Indian J. Hort.*, **19:** 83-91.

Randhawa, G.S., Jain, N.L. and Sharma, B.B. (1961) *Indian J. Hort.*, **18**: 277-284.

Randhawa, G.S., Singh, J.P. and Dhuria, H.S. (1959) *Indian J.Hort.*, **16**: 206-209.

Randhawa, G.S. and Srivastava, K.C. (1986) *Citriculture in India*, Hindustan Publishing Corporation (India), pp. 23.

Randhawa, N.S. (1970) *Res. J. Punjab Agric. Univ.*, Ludhiana, pp. 34-43.

Rao, D.G., Pandey, P.K., Chakarborty, N.K. and Capoor, S.P. (1975) *Indian J. Hort.*, **32**: 117-123.

Rao, S.N. (1961) *4th Hort. Res. Conf.*, May, 1961.

Rao, X.X., Jian, Z.H., Xiao, S.H. and Yang, L. (1998) *South China Fruits.*, **27**: 24.

Ratori, G.B. and Mukherjee, S.K. (1981) *Proc. Nat. Symp. Trop. and Subtrop. Fruit Crops*, Bangalore, p. 48.

Ray, P.K. and Sharma, B.B. (1990) *Indian J. Hort.*, **38:** 123-128.

Raychaudhari, S.P. and Ahlawat, Y.S. (1984) *Proc. Int. Soc. Citriculture*, pp. 442-445.

Raychaudhuri, S.P., Naraini, T.K. and Lele, V.C. (1969) *Proc. Ist. Int. Citrus Symp.*, **3**: 1433-1437.

Reddy, G.S. (1968) *ICAR Tech. Bull. (Agric.)*, ICAR, New Delhi.

Reddy, G.S. and Papa Rao, A. (1960) *Andhra Agric. J.*, **7**: 5.

Reese, R.L. and Koo, R.C.J. (1975) *J. Amer. Soc. Hort. Sci.*, **10:** 195-198.

Reltz, H.J. (1984) *Outlook on Agriculture.*, **13**: 140-146.

Reuther, W. and Smith, P. (1954) In: *Fruit Nutrition* (Ed. N. F. Childers, N), Somerset Press, Somerville, New Jersey, p. 907.

Richardson, A.C., Marsh, K.B. and Macrae, E.A. (1997) *J. Hort. Sci.*, **72**: 919-929.

Rodney, D.R., Roth, R.L. and Gardner, B.R. (1977) *Proc. Int. Soc. Citricult.*, Vol. I, pp. 106-110.

Roose, M.L. (1988) *Proc. VI Int. Citrus Symp.*, **1**: 155-165.

Rouse, R.E. (1997) *Proc. Interamerican Soc. Trop. Hort.*, **41**: 136-139.

Roux, C.Le., Huysamer, M. and Rabe, E. (1997) *J. South African Soc. Hort. Sci.*, **7**: 51-54.

Ruifang, H. (1997) *South China Fruits.*, **26**: 19.

Russo, M.P., Rosa, M.L., Astuto, A. and Reupero, G.R. (1998) *Adv. Hort. Sci.*, **12**: 85-88.

Salas, H., Figueroa, D. and Wallink, E. (1997) *Adv. Agroindustrial.*, **17**: 22-24.

Salibe, A.A. (1947a) *Proc. Inst. Int. Citrus Short Course*, Univ. of California, pp. 44-50.

Salibe, A.A. (1947b) *Proc. Inst. Int. Citrus Short Course*, Univ. of California, pp. 68-77.

Samaradiwakara, S.D., Champa, W.A.H. and Eeswara, J.P. (2018) *Tropical Agricultural Research.*, **30**(4): 125-131.

Santos, C.H., Duarte Filho, J., Modesto, J.C., Grassi Filho, H. and Ferreira, G. (1999) *Scientia Agricola.*, **56**: 999-1004.

Sanyal, D., Kar, P.L. and Lyngdoh, G.B. (1994) *Crop Res.*, **8:** 287-290.

Sarowa, P.S., Mann, S.S. and Brar, E.S. (1985) *Proc. 3rd Nat. Citrus Seminar*, Calcutta, p. 4.

Sato, K. and Others (1962) *Bull. Hort. Res. Sta.*, Hiratsuka, Ser., No.1, pp. 37-64.

Saucedo-Veloz, C., Arana-Errasquin, R., Chavez-Franco, S., Oerez, A. and Reyes, M.I. (1997) *7th Int. Controlled Atmosphere Res. Conf.*, California, USA, No.17, pp. 238-243.

Savage, E.M. and Gardner, F.E. (1965) *Citrus Industry.*, **46**: 5-7, 26.

Savitri, P., Chakravarthi, V.P. and Rajendra Prasad, P. (1997) *The Hindu*, May 22, 1997, p. 23.

Sawazaki, H.E., Muller, G.W. and Sodek, L. (1997) *Revista Brasileira de Biologia.*, **57**: 337-342.

Schatz, A.S. and Drescher, R.W. (1971) *Seric. Tecnica Estacion Experimental Agropecuaria Concordia*, No. 40, p. 14.

Schwarz, R.E. (1968) *Proc. 4th Conf. Internat. Organ. Citrus Virol.*, 1966, pp. 118-127.

Sehgal, J.L., Mondal, D.K., Mondal, C. and Vadivelu, S. (1990) *Tech. Bull. NBSS & LUP*, Pub.24, pp. 24.

Sema, A., Kar, P.L. and Sanyal, D. (1998) *The Hort. J.*, **11:** 17-23.

Sema, A., Kar, P.L. and Sanyal, D. (1999) *The Hort. J.*, **12:** 13-19.

Sethi, S.L. (1967) *J. Can. Ent.*, **60:** 180-181.

Shafrir, M. (1970) *Narel. Hassadeh.*, **50**: 1209-1212.

Shalhevel, J., Mantell, A., Bielorai, H. and Shimshi, D. (1976) *Int. Irrig. Inf. Center*, Pub. No.1 Bet Dagan, Israel, pp. 63-78.

Shanker, G. (1972) *Indian Hort.*, **17:** 13-14.

Shanker, G. (1973) *Indian Hort.*, **18**: 7-8.

Shaonan, C. (1997) *South China Fruits.*, **26:** 11-12.

Shapiro, J.P., Schroeder, W.J. and Stansly, P.A. (1998) *Florida Entomologist.*, **81**: 201-210.

Sharma, B. K. and Singh, N. (2001) *Indian Society of Soil Science.*, **49**: 373–377.

Sharma, R.K. and Awasthi, R.P. (1990) *Indian J. Hort.*, **47**: 162-166.

Sharma, S.D. and Chopra, S.K. (1978) *Indian J. Hort.*, **35**: 195-201.

Sharma, V.K. and Mahajan, K.K. (1997) *Nat. Sem. on Orchard Management for Sustainable Production of Tropical Fruits*, held on 10-11 March, 1997, Patna, Bihar.

Sheheilah, F. (2013) *Identification of export opportunities for Syrian Citrus growers to EU: A supply chain perspective*. Dissertion at department of Agricultural economics and Rural development, Georg-August University of Gottingen (cf. Chhetri and Kandel, 2019).

Shivankar, V.J. (1997) *Abst. Nat. Symp. on Citriculture held at National Research Centre for Citrus*, Amravati Road, Nagpur, Nov. 17-19, 1997, p. 78.

Shivankar, V.J. and Shyam Singh (1997) *Abst. Nat Symp. on Citriculture held at National Research Centre for Citrus*, Amravati Road, Nagpur, Nov. 17-19, 1997, p. 61.

Sim, G.E., Goh, C.J. and Loh, C.S. (1989) *Plant Sci.*, **59:** 203-210.

Sinclair, W.B. and Jolliffe, V.A. (1961b) *J. Food. Sci.*, **26**: 276-282.

Sinclair, W.B. and Jolliffe, V.A. (1961a) *J. Food Sci.*, **26**: 125-130.

Singh, A.K., Meetei, N.T. and Singh, B.K. (2017) *Virus Dis.*, **28**(4): 401-407.

Singh, J.P. and Dhuria, H.S. (1960) *Indian J. Hort.*, **17**: 9-20.

Singh, J.P. and Randhawa, S.S. (1961) *Indian J. Hort.*, **18**: 285-293.

Singh, K. (1966) *Punjab Hort. J.*, **6**: 117-122.

Singh, K.K., Grewal, K.S., Kumar, R. and Singh, R.R. (1985) *Progressive Hort.*, **17**: 282-284.

Singh, K.K., Grewal, K.S. and Jawanda, J.S. (1961) *Punjab Hort. J.*, **1**: 135-140.

Singh, M. and Tucker, D.P.H. (1984) *Proc. Fla. Sta. Hort. Soc.*, **97**: 57-53.

Singh, M.I. and Chohan G.S. (1982) *Punjab Hort. J.*, **22:** 52-59.

Singh, M.P., Dhillon, R.S., Sharma, K.K. and Singh, S.N. (1980) *J. Res. Punjab Agric. Univ.*, Ludhiana, 17: 127-134.

Singh, R. and Mishra, K.K. (1980) *Punjab Hort. J.*, **20:** 143-148.

Singh, R. and Nath, N. (1969) *Proc. 1st Int. Citrus Symp.*, pp. 435-440.

Singh, R. and Singh, S. (1981) *Proc. Nat. Symp. Trop. and Subtrop. Fruit Crops*, p. 43.

Singh, S. and Sohi, G.S. (1957) *Govt. Agric. College Mag.*, Ludhiana, 4: 19-34.

Singh, S.N. (1981) *Punjab Hort. J.*, **21:** 23-26.

Singh, S., Krishnamurthi, S. and Katyal, S.L. (1967) *Fruit Culture in India*, ICAR New Delhi, pp.162.

Singh, S.C., Patiram, R.M., Prasad, R.N. and Chandra, A. (1997) *J. Hill Res.*, **10:** 158-160.

Singh, Shyam, Lallan Ram, Srivastava, A.K., Huchche, A.D. and Dass, H.C. (1997) *Abst. Nat. Symp. on Citriculture*, National Research Centre for Citrus, Nagpur, November 17-19, p. 23.

Singhrot, R.S., Sharma, R.K., Sandooja, J.K. and Singh, J.P. (1987a) *Haryana J. Hort. Sci.*, **16**: 25-30.

Singhrot R.S., Singh, J.P., Sharma, R.K. and Sandooja, J.K (1987b) *Haryana J. Hort. Sci.*, **16**: 31-39.

Sinha, R.B., Singh, R.K. and Pandey, S. (1978) *Indian J. Hort.*, **35**: 188-194.

Sinha, R.B., Singh, R.K. and Singh, J.P. (1977) *Indian J. Hort.*, **34**: 319-325.

Sites, J.W. (1954) *Proc. Fla. Sta. Hort. Soc.*, pp. 56-59.

Slack, J. Turpin, J.W., Duncan, J.H. and McKay, O.L. (1978) *Proc. Int. Soc. Citriculture*, 1978.

Smajstrla, A.G., Koo, R.C.J. and Weldron, J.H. (1984) *Amer. Soc. Agril. Engrs.*, **84**: 2107.

Smajstrla, A.G., Parsons, L.R., Aribi, K. and Yelledis, C. (1986) *Proc. Fla. Sta. Hort. Sci.*, **98**: 25-32.

Smith, P.F. (1966) In: Fruit Nutrition (Ed. N.F. Childers) Somerset Press, Somerville, New Jersy.

Smith, P.F. (1970) *J. Amer. Soc. Hort. Sci.*, **95**: 15-17.

Smith, P.F. (1975) *J. Amer Soc. Hort. Sci.*, **100**: 368-369.

Smith, P.F. and Reuther, W. (1950) *Proc. Amer. Soc. Hort. Sci.*, **55**: 61-72.

Smith, P.F. (1959) *Citrus Mag.*, 22: 16-17, 23.

Snowdon, A.L. (1990) *A Colour Atlas of Post-harvest Diseases and Disorders of Fruits and Vegetables.* Vol.I. General Introduction and Fruits, Wolfe Scientific Ltd., London, p. 302.

Sobrinho, J.T., Pompeu, J.Jr., Rodriguez, O. and Bardin, D. (1978) *Proc. Int. Soc. Citriculture*, Vol. 3, pp. 74-93.

Soost, R.K. (1956) *Proc. Amer. Soc. Hort. Sci.*, **67**: 171-175.

Soost, R.K. and Burnett, R.G. (1961) *Proc. Amer. Soc. Hort. Sci.*, **77**: 194-201.

Soost, R.K. and Cameroon, J.W. (1981) *Proc. Int. Soc. Citricult.*, **1**: 59-60.

Soost, R.K. and Cameroon, J.W. (1986) *Calif. Agric.*, **40**: 30-31.

Sorgona, A., Abenavoli, M.R., Gringeri, P.G. and Cacco, G. (2006) *Sci. Hortic.*, **109**: 389-393.

Soule, J., Grierson, W. and Blair, J. G. (1967) *Agri. Ext. Service*, Inst. Food and Agril, Univ. of Florida, Gainseville, USA, Cir., No. 315.

Spalding D.H. and Reeder, W.F. (1976) *J. Amer. Soc. Hort. Sci.*, **101**: 367-370.

Spencer, W.F. and Koo, R.C.G. (1962) *Proc. Amer. Soc. Hort. Sci.*, **81**: 202-208.

Spiegel-Roy, P. (1979) *Euphytica.*, **28**: 361-336.

Spiegel-Roy, P. and Vardi, A. (1982) *Abst. 21st Int. Hort. Cong*, Hamburg, No. 407.

Squillaci, A. (1964) *Tech. Agric.*, **16**: 72-93.

Srivas, S.R., Purthi, J.S. and Siddappa, G.S. (1963) *Food Sci.*, **12:** 340-343.

Srivastav, M., Dubey, A.K. and Sharma, R.R. (2005) *Indian J. Agric. Sci.*, **75**: 333–335.

Srivastava, A K., Kohli, R. R., Ram, Lallan, Huchche, A.D. and Dass, H.C. (1994) *Indian J. Agric. Sci.*, **64**: 324-325.

Srivastava, A.K. and Shyam Singh (1998) *Fertilizer Use Efficiency in Citrus*, Tech. Bull. No. I. National Research Centre for Citrus, Nagpur, pp. 1-66.

Srivastava, A.K. and Ram, L. (2000) *Indian J. Agric. Sci.*, **70**: 679-681.

Srivastava, A.K., Kohli, R.R., Ram, Lallan and Dass, H.C. (1998) *Indian J. Agric. Sci.*, **68**: 39-41.

Srivastava, A.K., Ram, Lallan, Huchche, A.D., Kohli, R.R. and Dass, H.C. (1994b) *Indian J. Hort.*, **51**: 32-36.

Srivastava, A.K., Shyam Singh and Srivankar, V.J. (1997) *Proc. Nat. Symp. on Citriculture*, held at National Research Centre for Citrus, Nagpur, November 17-19, (Abst. p. 34).

Srivastava, A.K., Shyam Singh, Ram, Lallan, Kohli, R. R., Huchche, A. D. and Dass, H. C. (1997) *Proc. Nat. Sym. on Citriculture*, National Research Centre for Citrus, Nagpur, Nov. 17-19, 1997 (Abst. p. 52).

Srivastava, A.K. (2014) *Current Horticulture.*, **2**(1): 03-04,

Srivastava, A.K. and Kohli, R.R. (1997a) *Nat. Sem. on Orchard Management for Sustainable Production of Tropical Fruits*, March 10-11, 1997, Patna, Bihar (Abst. pp. 19).

Srivastava, A.K. and Kohli, R.R. (1997b) *Soil Suitability Criteria for Citrus – An Appraisal, Agric. Reviews*, **18**: 134-146.

Srivastava, A.K. and Shyam Singh. (2009) *Journal of Plant Nutrition.*, **32**: 197-245.

Srivastava, K.C., Bopaiah, M.G. and Kushalappa, K.G. (1972) *Indian Hort.*, **17**: 11.

Stafford, L.M. (1971) *Aust. J. Exp. Agric. Anim. Husb.*, **11**: 118-122.

Starrantino, A. (1999) *Rivista di Frutticoltura e di Ortofloricoltura.*, **61**: 45-47.

Starrantino, A. and Russo, F. (1980) *HortScience.*, **15:** 296-297.

Starrantino, A., Spina, P. and Russo, F. (1977) *Giornale Botanico*, Italiano, **111**: 358.

Stewart, I. and Wheaton, T.A. (1972) *J. Agric. Fd. Chem.*, **20**: 448-449.

Stewart, W.S. and Klotz, L.J. (1947) *Bot. Gaz.*, **109**: 150-162.

Stewart, W.S., Klotz, L.J. and Heild, H.Z. (1947) *Calif. Citrog.*, **32**: 314.

Stewart, W.S., Palmer, J.E. and Hield, H.Z. (1952) *Proc. Amer. Soc. Hort. Sci.*, **59**: 372-375.

Stolzy, L.H., Taylor, O.C., Gavber, M.J. and Lambard, P.B. (1963) *Proc. Amer. Soc. Hort. Sci.*, **82:** 119-123.

Storey, R. and Treeby, M.T. (2000) *Sci. Hortic.*, **84**: 67-82.

Storey, R. and Walker, R.R. (1999) *Acta Hortic.*, **478**: 39-81.

Sulladmath, U.V. and Iyer, C.P.A. (1981) *Abst. Nat. Symp. Trop. and Subtrop. Fruit Crops*, Bangalore, p. 9.

Sun, Y.T. (1998) *South China Fruits.*, **27**: 16-17.

Swietlik, D. (1992) *J. Amer. Soc. Hort. Sci.*, **117**: 22-27.

Swingle, W.T. (1948) *Citrus Industry, Univ. of California*, pp. 129-474.

Swingle, W.T. and Reece, P.C. (1967) *The Citrus Industry* (Eds. W. Reuther, L.D. Batchelor, and H. J. Webber), Univ. of California, Vol. I, pp. 190-430.

Syvertsen, J.P. and Lloyd, J.J. (1994) In: *Handbook of Environmental Physiology of Fruit Crops*, Vol. II. Sub-tropical and Tropical Crops (Eds. B. Schaffer and P.C. Anderson), CRC Press, Boca Raton, Florida, pp. 65-99.

Syvertsen, J.P. and Smith, M.L. (1984) *J. Amer. Soc. Hort. Sci.*, **109**: 8-7.

Tachibana, S. and Yahata, S. (1998) *J. Japanese Soc. Hort. Sci*, **67**: 671-676.

Takase, S. and Mizuno, K. (1997) *Res. Bul. Aichi-Ken Agric. Res. Centre.*, **29**: 199-205.

Takdze, R.M. (1981) *Subtrop. Kul'tury*, No. 1, pp. 61-68.

Tamim, M., Goldschmidt, E.E., Goren, R. and Shachnai, A. (2000) *Alon Hanotea.*, **54**: 152-157.

Tanaka, M., Morimoto, J., Harano, H. and Maesaka, K. (1982) *Proc. Int. Soc. Citriculture*, Shimizu, Japan, Vol. I, pp. 127-130.

Tanaka, T. (1954) *The Japanese Soc. For Promotion of Science*, Tokyo, p. 152.

Tandon, P.L. (1992) *Fruit Crops* (Eds. K.L. Chadha and O.P. Pareek), Vol. 3, pp. 1527-1555.

Tandon, P.L. and Jothi, B.D. (1993) *Indian J. Hort.*, **50:** 337-340.

Tashbekov, Kh. K., Sai dov, I.I. and Kreidik, B.M. (1986) *Technologiya Grosheniya*, Programmirovanie Urochaya, Moscow, USSR, 1986, pp. 147-150.

Thakare, K.R., Radke, S.G., Borle, M.N. and Ghuguskar, H.T. (1984-85) In: *Two decades of Research on Citrus Pest Management in Maharashtra*, pp.5.

Tolkowski, S. (1938) *Hesperides: A History of the Culture and Use of Citrus Fruits*, John Bale Sons & Curnow, London.

Tominaga, S. (1998) *Acta Hortic.*, **463**: 301-305.

Toplu, C., Uygur, V., Kaplankýran, M., Demirkeser, T.H. and Yýldýz, E. (2012) *J. Plant Nutr.*, **35:** 1329-1340.

Torres, J.P. (1936) *Phillip. J. Agric.*, **7**: 193-204.

Tucker, D.P.H. and Muraro, R.P. (1990) *A High Density Planting, Proceeding of Florida State Horticulture Society*,103,55-59.

Tucker, D.P.H., Wheaton, T.A. and Muraro, R.P. (1994) *Citrus Tree Pruning and Practices*, University of Florida, Fact Sheet HS-144 June 1994. P.9

Tusa, N., Grosser, J.W. and Gmitter, Jr., G.G. (1990) *J. Amer. Soc. Hort. Sci.*, **115**: 1043-1046.

Tusa, N., Pasqual, F. De and Radogna, L. (1979) *Tech. Agric.*, **31**: 285-292.

Tusa, N., Pasqual, F. De and Rodogna, L. (1980) *Proc. Int. Soc. Citricult.*, Griffith, NSW, Australia, pp. 143-145.

Tuzcu, O., Yieldirim, B., Duzenoglu, S. and Bahceci, I. (1999) *Turkish J. Agric. & Forestry.*, **23**: 125-135.

Valle, N. Del, Herrera, O. and Rios, A. (1982) *Proc. Int. Soc. Citricult.*, Shimizu, Japan, Vol. 1, pp. 143-147.

Vardi, A. and Spiegel-Roy, P. (1978) *Proc. Int. Soc. Citriculture*, **2**: 51-57.

Vasudeva, R.S. and Capoor, S.P. (1958) *FAO Plant Protect. Bull.*, **6**: 91.

Velilla, J. (1997) *Proc. Congr. Spanish Weed Sci. Soc.*, Spain, pp. 389-392.

Venkateshwarlu, G. and Lavania, M.L. (1985) *Haryana J. Hort. Sci.*, **14:** 182-186.

Verdijo-Lucus, S., Sorribas, F.J., Forner, J.B. and Alcaida, A. (2000) *HortScience.*, **35**: 269-273.

Verma, A.N. and Ghosh, S.P. (1979) *Indian Hort.*, **23**: 2-4.

Volpe, C., Schoffel, E. and Barbosa, J. (2002) *Revista Brasileira de Fruticultura.*, **24**(2): 436-44.

Vuillaume, C., Morean, B. and Aubert, B. (1981) *Fruits.*, **36**: 2-4.

Wallace, H.M. and Lee, L.S. (1999) *J. Hort. Sci. Biotech.*, **74**: 82-86.

Wallace, J.M. (1959) *Citrus Virus Diseases*, Univ. of California, Berkeley.

Wan, L.C., Chen, Z.S., Wu, T. and Nie, H.T. (1994) *China Citrus.*, **23**: 18-19.

Wang, C.T. (1998) *China Fruits.*, **2**: 55.

Wang, D., Xiao, C.M., Fan, Z. and Li, X. (1999) *South China Fruits.*, **28**: 10-11.

Weaver, R.J. (1972) *Plant Growth Subtances in Agriculture*, W. H. Freeman and Co., USA, pp. 202-211.

West, E.S. and Bernard, C. (1935) *J. Comm. Sci. Ind. Res.*, Washington, 8: 93-100.

Wheaton, T.A., Whitney, J.D., Castle, W.S., Muraro, R.P., Browning, H.W. and Tucker, D.P.H. (1995) *J. Amer. Soc. Hort. Sci.*, **120**: 861-870.

Whitney, J.D., Wheaton, T.A., Castle, W.S. and Tucker, D.P.H. (1983) *Trans. Amer. Soc. Agr. Eng.*, **26**: 704-709.

Williams, P., Hepworth, G., Goubran, F., Muhunthan, M. and Dunn, K. (2000) *Post-harvest Biol. Technol.*, **19:** 193-199.

Willink, E., Figueroa, D., Zamudio, P. and Salas, H. (1998) *Avance Agroindustrial*, **18**: 14-18.

Win, T.O., Srilaog, V., Heyes, Kyu, K.L., J. and Kalyanarat, S. (2006) *Post-harvest Biol. Technol.*, **42:** 23–30.

Wolff, J. (1977) *Proc. Int. Soc. Citricult.*, p. 328.

Wu, W., Li, Y.G., Liu, W.F. and Chen, J.Q. (1998) *J. Southwest Agric. Univ.*, **20**: 198-202.

Wutscher, H.K. (1973) *Rio Grande Valley Hort. Soc.*, 27: 34-39.

Wutscher, H.K. (1974) *Proc. Inst. Int. Citrus Short Course*, pp. 97-113.

Wutscher, H.K. (1978) *Proc. Int. Soc. Citricult.*, **2:** 526-529.

Wutscher, H.K. and Shull, A.B. (1973) *Trop. Reg. Amer. Soc. Hort. Sci.*, **17**: 66-73.

Wutscher, H.K.and Bowman, K.D. (1999) *HortScience.*, **34**: 622-624.

Yang, G.L. and Jiang, X.M. (1993) *China Citrus.*, **22**: 39.

Yao, J.L., Wu, J.H., Geave, A.P. and Morris, B.A.M. (1996) *Plant Science Lim.*, **113:** 175-183.

Ye Jian Win and Jian Cheng, C. (1997) *South China Fruits*, **26**: 16.

Yoshida, T. (1985) *Bull. Fruit Tree Res. Sta.* (Okitsu) B. No. **12:** 17-26.

Yoshida, Y. and Ueda, M. (1984) *J. Japanes Soc. Hort. Sci.,* **53**: 354-361.

Youzhu, Y. (1997) *South China Fruits.,* **26:** 26.

Zaragoza, S. and Alonso, E. (1981) *Proc. Int. Soc. Citricult.,* **1**: 172-175.

Zaragoza, S., Trenor, I., Alfonso, E., Primo-Millo, E. and Augusti, F.S. (1992) *Proc. Int. Soc. Citricult.,* **2:** 725-728.

Zekri, M. (1997) *Fruits.,* **52**: 141-148.

Zhang, C.Y. (1998) *South China Fruits.,* **27**: 28.

Zhang, GeCheng, Li, JiXiang, and Luo, F. Yung (1998) *South China Fruits.,* **27**: 26.

Zhang,Y., C.X. Hu, Q.L.Tan, C.S. Zheng, H.P. Gui, W.N. Zeng, X.C. Sun, X.H. Zhao. (2014) *Sci. Hortic.,* **174**: 46–53.

Zhou, X.W., Li, Z.Y., Chen, X.N., Lu, B. and Fu, K.Y. (1998) *J. Fruit Sci.,* **15**: 34-38.

Ziegler, L.W. and Wolfe, H.S. (1961) *Citrus Growing in Florida,* Univ. of Florida Press, pp. 64-83.

Zong, R.J., Shao, P.F. and Dai, L.Y. (1974) *Scientia Agric. Sin.,* **3**: 60-64.

2

POMEGRANATE

M.Sankaran, K.Dhinesh Babu,
Jai Prakash and A.R. Karale

1.0 Introduction

Pomegranate (*Punica granatum* L., Punicaceae) is an ancient, beloved plant and fruit. The name "pomegranate" follows the Latin name of the fruit Malum granatum, which means "grainy apple." The generic name *Punica* refers to Pheonicia (Carthage) as a result of mistaken assumption regarding its origin. Pomegranate is an important fruit crop of arid and semiarid regions of the world. It is commercially grown for its fruits which are sweet-acidic in taste and used for dessert purpose. The fresh fruit is of exquisite quality while its processed products, such as bottled juice, syrup and jelly are highly appreciated. *Anar-rub* is a product locally prepared from the juice by adding sugar and heating to a thick viscous consistency. It keeps well and is used like tomato sauce or ketchup. The pomegranate and its usage are deeply embedded in human history, and utilization is found in many ancient human cultures as food and as a medical remedy. Despite this fact, pomegranate culture has always been restricted and generally considered as a minor crop.Recent scientific findings corroborate traditional usage of the pomegranate as a medical remedy and indicate that pomegranate tissues of the fruit, flowers, bark, and leaves contain bioactive phytochemicals that are antimicrobial, reduceblood pressure, and act against serious diseases such as diabetes and cancer (Holland *et al.*, 2009)

2.0 Composition and Uses

2.1.0 Composition

Pomegranate is a rich source of a variety of phytochemicals, which are responsible for its strong antioxidative and anti-inflammatory potential. El-Nemr *et al.* (1990) analyzed the chemical composition of pomegranate fruits and recorded that the edible parts represented 52 per cent of the total fruit weight, comprising 78 per cent juice and 22 per cent seeds. The fresh juice contained 85.4 per cent moisture, 10.67 per cent total sugars, 1.4 per cent pectin, 0.1 g/100 ml total acidity (as citric acid), 19.6 mg/100 ml free amino nitrogen and 0.05 g/100 ml ash. The seeds were a rich source of total lipids, protein, crude fibre and ash, representing 27.2, 13.2, 35.3 and 2.0 per cent, respectively, and also contained 6.0 per cent pectin and 4.7 per cent total sugar.

Leena Borah (1990) reported the following juice composition of pomegranate cv. Ganesh:

- ✰ Moisture (per cent): 87.20
- ✰ TSS (°Brix): 12.00
- ✰ Acidity (per cent): 0.31
- ✰ pH: 2.95
- ✰ Reducing sugar (per cent): 9.93
- ✰ Total sugar (per cent): 10.66
- ✰ Ascorbic acid (mg/100 g): 22.00
- ✰ Tannins (per cent): 0.092

- Ash (per cent): 0.52
- Calcium (mg/100 g): 12.07
- Phosphorus (mg/100 g): 4.91
- Iron (mg/100 g): 0.34

El-Shaaraway and Nahapetian (1983) reported that pomegranate seed contains about 15 per cent oil with a high refractive index, iodine value and very low melting point. The oil has a potential for industrial use. The seeds also contain 1.09 mg oestrone and 0.036 mg coumestrol (a non-steroidal oestrogen)/100 g of seeds (Moneam *et al.*, 1988).

Melgarejo *et al.* (2000) analyzed individual organic acid and sugar compositions of 40 cultivars of pomegranate. On an average, for both seasons, citric acid was predominant with a range of 0.142-2.317 g/100 g (average content for all cultivars, 0.282 g/100 g). Malic acid was the second most abundant with a range of 0.135-0.176 g/100 g (average of 0.139 g/100 g). Total sugars ranged between 11.43 g/100 g and 13.5 g/100 g. Glucose, fructose, sucrose and maltose were detected, although maltose (first reported here) was not quantifiable. Fructose ranged between 5.96 g/100 g and 7.04 g/100 g, with an average of 6.58 g/100 g, quite similar to that of glucose (6.14 g/100 g). Sour cultivars showed the lowest fructose and glucose contents. The average sucrose content was 0.01 g/100 g.

Normakhmatov and Rakhmanberdyeva (1999) isolated polysaccharides from pomegranate (cultivars Shirobal, Dashnobal and Fergana) fruit peel and seeds. Fruit peel and seeds were rich in water-soluble polysaccharides and pectin substances.

Nasr *et al.* (1996) determined the polyphenolic content of pomegranate peel and isolated proanthocyanidins from developing fruits of 8 Tunisian cultivars. Proanthocyanidins were an important component of total phenols. Total phenols, ellagic tannins and gallic and ellagic acids were determined in the peel of Chelfi fruits and shaded fruits contained a higher content of total phenols and proanthocyanidins and a lower content of anthocyanins than peel from fruits exposed to the sun.

Six anthocyanin pigments were found to be responsible for the red colour of pomegranate cv. Mollar juice. These were analyzed quantitatively and qualitatively by HPLC and identified as delphinidin 3-glucoside, delphinidin 3, 5-diglucoside, cyanidin 3-glucoside, cyanidin 3,5-diglucoside, pelargonidin 3-glucoside and pelargonidin 3,5-diglucoside. The fruit skin contained only the cyanidin and pelargonidin derivatives (Gil *et al.*, 1995).

2.2.0 Uses

The edible part of pomegranate is known as 'arils'. Aril is nothing but the juicy sac encasing a soft seed of pomegranate. Ripe pomegranate fruits are consumed fresh. Juice extracted from fruits makes an excellent drink. Ready-to-serve (RTS) beverage obtained by diluting the juice and addition of standard preservatives has great demand especially during hot summer season. Apart from these, an attractive jelly, *anar-rub* (pomegranate jam) and syrup can also be manufactured from ripe fruits.

Uses of Pomegranate

Juice (Left) and Jam (Right) from Pomegranate

Jelly (Left) and Syrup (Right) from Pomegranate

Wine from Pomegranate

Species of Pomegranate

Punica granatum

Punica protopunica

The fruit juice easily ferments with addition of yeast and is used for the production of wine. The juice of wild pomegranates in Azerbaijan is used in the manufacture of citric acid and sodium citrate for medicinal purpose.

Arils of sour type pomegranate are dried for making *'anardana'*, which is used as acidifying agent for cooking. It is an important ingredient of Muglai cuisine in North India. It gives a good mouthfeel upon consumption.

The bark of the stem and root contain a number of alkaloids belonging to the pyridine group. It is also employed in the therapeutics in dysentery and diarrhoea.

Wealth from waste: The waste materials of pomegranate industry *viz.*, seeds, peel (rind), pruned shoots, *etc.* are not allowed to go waste and converted useful in making wealth from waste.

The seeds obtained after juice extraction are used for production of virgin pomegranate seed oil through cold press technique. Seed oil is useful for making edible soft gel capsule through microencapsulation. Seed oil is useful in arresting hair fall and preventing baldness.

Pomegranate peel is useful in preparation of mouth-wash that helps to keep oral healthcare due to its antimicrobial properties. Peel powder is useful in improving skin glossiness and making facial preparations.

The shoots obtained after pruning of mother plants are used as hardwood cuttings for raising pomegranate saplings in the nursery.

3.0 Origin and Distribution

Pomegranate is one of the favorite table fruits of tropical and subtropical regions. Cultivation of this fruit crop by mankind dates back to antiquity. The usage of pomegranate is deeply embedded in human history with references in many ancient cultures of its use in food and medicine.

The crop is native of Iran (Primary Centre of Origin), while Afghanistan, Pakistan and Himalayan ranges of India are considered to be the Secondary Centres of Origin. It performs well under Mediterranean countries. It is extensively cultivated in India, Iran, Turkey, Egypt, Tunisia, Morocco, Spain, Afghanistan and Baluchistan province of Pakistan. Besides, it is cultivated to some extent in Myanmar, China, Japan, USA (California), central Asia, and India. It is also found growing in Bulgaria (Kondav, 1969), and in southern Italy (Slykov, 1968).

4.0 Species and Cultivars

4.1.0 Species

Pomegranate belongs to the family Punicaceae. The genus has two species: (i) *Punica protopunica* and (ii) *Punica granatum*. *Punica protopunica* is found wild in the Socotra island and *Punica granatum* is cultivated in tropical and subtropical regions.

Punica granatum has been classified into two subspecies, *chlorocarpa* and *porphyrocarpa*. *Chlorocarpa* is found in Transcaucasus region and *porphyrocarpa* in central Asia.

4.2.0 Cultivars

Several types of pomegranates cultivated in India are distinguished by the shape of the fruit, colour and thickness of rind, colour of the seeds (aril) and the taste.

Bhagawa is the predominant cultivar occupying largest area under cultivation in India. On the basis of yield and physico-chemical characters, various cultivars were reported from different states in India, *viz.*, Bhagawa, Phule Bhagawa Super, Solapur Lal, Ganesh, Arakta, Mridula in Maharashtra; Dholka in Gujarat; Jalore Seedless, Jodhpur Red, Jodhpur White and Seedless (Bedana) in Rajasthan; Jyoti, Bassein Seedless and Madhugiri in Karnataka; Chawla, Nabha, Country Large Red in Haryana; Yercaud, CO-1, Vellodu in Tamil Nadu.

At present, the commercial cultivars cultivated on extensive scale includes, Bhagawa, Phule Bhagawa Super, Ganesh, Arakta, Mridula, Dhoka, Jalore Seedless, Jyoti, Kandhari, *etc.*

The different cultivars of pomegranate grown in India, may be categorized as Sweet type cultivars or Sour type cultivars. Besides, Pomegranate is also grown for ornamental purpose.

Sweet Type Varieties

Alandi or Vadki

Fruit medium in size, fleshy testa, deep pink arils with sweet, slightly acidic juice, seeds are hard.

Bhagawa

It is a selection from the F_2 progenies of the cross, Ganesh X Gulesha Red. It has bold red arils, fruit surface is deep rose pink colour, rind is thick (3.4mm). It is the most popular cultivar, occupying largest area under cultivation (Jadhav and Sharma, 2007). This cultivar has been gradually replacing Ganesh and other varieties due to its high demand in the international market (Chandra and Meshram, 2010).

Bassein Seedless

Fruit surface smooth, skin colour yellow with red tinge, shape round, taste sweet (TSS 16-17oBrix), juicy, aril pink in colour, average fruit weight 250-300 g, moderately susceptible to leaf spot.

Dholka

Fruit large in size, fleshy testa, pinkish white or whitish with sweet juice, seeds soft, juice acidic. Suitable for processing. Popular variety of Gujarat.

Ganesh

Prolific yield, medium-large sized fruits, yellowish with red tinge on surface, soft seeds, pinkish arils and sweet juice, agreeable taste. Selection from Alandi by Dr. G.S. Cheema. This soft seeded variety of pomegranate has revolutionized the Indian pomegranate industry. The most popular cultivar of Maharashtra till release of Bhagawa in 2003.

G-137

Fruit is large sized, fruit surface yellowish pink, less acidic and more sweeter than Ganesh. It is a clonal selelction from Ganesh (Keskar, 1990).

Jalore Seedless

The tree has spreading habit. Round fruit, yellow with red tinge in colour, average fruit weight 295-430 g, taste sweet, (TSS 15-16 oBrix), aril light pink to pink in colour, juicy, susceptible to fruit borer. Early maturing, popular variety of Rajasthan.

Jyoti

Semi-spreading habit. Medium sized fruits, Fruit smooth, round, yellowish red in colour, average weight 270-300 g, arils pink- light red, juice is sweet (14-17 oBrix), soft- medium seeds, It has a yield potential of over 18 tonnes fruits per ha (Sulladamath, 1985). Selection from seedling population of Bassein Seedless and Dholka. Popular variety of Karnataka.

Kabul

Fruit large in size, rind deep red mixed with pale yellow, thick, fleshy testa, dark red with slightly bitter juice.

Kandhari

Fruit large in size, rind deep red, fleshy testa, blood-red or deep pink with sweet, slightly acidic juice, seeds hard.

Mridula

It was developed at Rahuri, from the open-pollinated progeny raised from the F_2, derived from Ganesh x Guleshah Red. Dwarf tree, fruits have red surface, semi-smooth surface, round in shape, average weight 210-250g, aril sweet (12-15oBrix), deep red in colour, moderately juicy, Seeds softer than Ganesh.

Muskat Red

Fruit small to medium in size, rind somewhat thick, fleshy testa, with moderately sweet juice, seeds not very hard.

Paper Shell

Fruit medium in size, rind thick, fleshy testa, reddish to pink with sweet juice, seeds soft.

Phule Bhagawa Super

It is a selection from Bhagawa; It matures about 2 weeks earlier than Bhagawa with higher TSS and shelf life. It was released by MPKV, Rahuri in 2013.

Ruby

Ruby is a multiple hybrid developed at the Indian Institute of Horticultural Research, Bangalore (Jalikop and Sampath Kumar, 2001) for aril colour and seed mellowness. The hybrid develops dark red arils in winter and dark pink or red in summer. Ruby derives certain fruit quality attribute from Ganesh while incorporating genes for red colour from a Russian variety Gulesha Rose Pink. Fruits resemble in size and shape to those of Ganesh. However, the skin colour is reddish-brown, Rind is thin (2.4 mm), fruit contains bold arils (37.2 g/100 arils), dark red, sweet (16 oBrix) juice. Average fruit weight 270 g and yield 16-18 tonnes per hectare.

Spanish Ruby

Fruit small to medium in size, rind thin, fleshy testa, rose coloured, seeds soft.

Solapur Lal

A bio-fortified variety of pomegranate developed through hybridization, jointly by NRC on Pomegranate, Solapur & IIHR, Bengaluru. It matures in 160-165days, TSS >17.6°B, uniform red skin, deep red arils, It has more anthocyanin, ascorbic acid., iron and zinc compared to Bhagawa. Parents: Bhagawa x [(Ganesh x Nana) x Daru]. It was released by NRCP, Solapur in 2017 (Anon., 2017).

Yercaud

A superior clone, Acc. No. 455 with pinkish red arils and medium hard seeded type was named as 'Yercaud-1' and released by the Tamil Nadu Agricultural University for commercial cultivation. It has Pinkish red arils, medium hard seeded (Syed *et al.*, 1985).

Among the sweet type varieties, about 25 varieties are commercially cultivated in India nowadays. It includes, Bhagawa, Gnesh, Ruby, Jalore Sedless, G-137, ARakta, Mridula,

Varietal Performance in India

Study of different pomegranate cultivars indicated that those designated as soft-seeded (Bedana, Bassein Seedless, Ganesh and Dholka) meet the following criteria: testa thickness <0.5 mm, whole seed density <0.6 g/ml, testa density <0.4 g/ml and testa weight <50 per cent of whole seed weight (Purohit, 1985). Among the 10 cultivars assessed by Godara *et al.* (1989), in Haryana state of India, Chawla and Nabba were the best dessert cultivars, having large fruit size, soft to medium-hard seeds, high TSS, and average values of percentage juice and acidity.

Jagtap *et al.* (1992) reported on 7 physico-chemical characteristics of fruits in 29 indigenous and exotic cultivars of pomegranate. The best performing cultivars were G137 for fruit weight (214 g), Alandi for aril size and Ganesh for juice content (58.5 per cent). Aril content as a percentage of fruit was highest in Kabul (72 per cent); similar values were observed in Ganesh and Muscat.

Prasad and Bankar (2000) evaluated 9 pomegranate cultivars in the arid condition of Rajasthan. They recorded highest mean fruit yield of 18.7 kg per plant in Jalore Seedless (Table 1) followed by G-137 (17.45 kg). The cv. Jalore Seedless also showed its superiority with respect to fruit size, juice content, softness of seed and quality characters of fruits.

Five hybrid progenies of pomegranate (Ganesh x Jyothi, Ganesh x Kabul, Ganesh x YCD1, Jyothi x Ganesh, Ganesh x Bedana Bosco) were evaluated. Ganesh

Cultivars of Pomegranate

Dholka

Ganesh

Jyoti

Cultivars of Pomegranate

Kabul

Mridula

Kandhari

Paper Shell

x Bedana Bosco gave the highest single fruit weight (160 g) while Ganesh x Jyothi gave the highest 100 aril weight (19 g). No significant differences was observed in TSS content (Nageswari *et al.*, 1999).

Varieties from other Countries

The cultivars, Schahvar which has thin rind and Robab with thick rind are grown in Iran (Firuzch and Ludders, 1979). Of the 46 cultivars evaluated, for the gravelly soils of the Fergana Valley in Uzbekistan the promising ones were Bala-Myursal, Kara-Bala-Myursal, Gyuleisha Azerbaidzanskaya, Bashkalinskii, Al Shirin Nar, Kai-Achik-Anor, Shirin Nor and Kazake Anor (Saidaliev, 1986).

Al-Khatani (1992) evaluated eleven pomegranate cultivars; Taifi-A was given the highest scores for sensory evaluation. The cv. Kab El-Jameel contained significantly more edible portion and more juice and had lower pH and higher acidity than any other cultivar.

In Huizhan County in China, Jiang *et al.* (1995) rated some of the local cultivars as of the highest quality, including Huopao, Huahongpi, Lupi and Ruo. Although Huopao did not produce the largest fruits (432 g on average, compared with 444 g, reaching 875 g in Huahongpi), it produced the largest fleshy seeds (100-seed weight of 67.4 g vs. 49.3 g in Huahongpi). Yixian Dahong is an important cultivar grown mainly in the mountainous area in China. It produces large (250-520 g) fruits with a bright red skin and bright red, juicy, acid-sweet fleshy seeds with a soluble solids content of 15-17 per cent. Trees are productive and tolerant to alkaline soil and drought (An *et al.*, 1994).

Dong and Yang (1994) recorded that fruits of cv. Tiepitian, a popular cultivar weighing 500 g on average, but can reach 1520 g. The fleshy seeds are bright red, large and juicy with a soluble solids content of 14-15.5 per cent. Fruits have a long shelf-life and trees are disease resistant.

Ornamental Varieties

Double Flower

It produces large, attractive flowers, wherein numerous stamens are modified into petals. Some types do not set fruits, highly acidic fruits

Nana

Dwarf, grows to a height of 1-2 ft, leaves and flowers are miniature type, fruits are tiny (20-40g/fruit), highly acidic, 5 per cent acidity, hard seeds. Fully grown plant looks like a 'bonsai tree of pomegranate' (Jalikop, 2007).

Sour Type Varieties for Anardana

Amlidana

It was obtained by crossing Ganesh x Nana. Fruit medium sized, yellowish surface, pink arils, sour due to high acidity content.

Phule Anardana

Released by MPKV, Rahuri for anardana purpose, pinkish arils, sour type

Solapur Anardana

High yielding variety for anardana purpose, Red surface, deep red arils, acidity 4.8 per cent, medium hard seeds, rich in anthocyanin, released from ICAR-NRCP in 2017.

Exotic Varieties

Wonderful

It is the oldest variety, deep uniform red fruit with thick rind, medium hard seeds, 1.5-1.75 per cent acidity, suitable for juice purpose.

Early Wonderful

Large, deep red fruit, medium hard seeds, ripens 2weeks ahead of Wonderful.

Granada

It is bud sport mutant of Wonderful, ripens one month earlier than Wonderful, medium hard seeds.

5.0 Breeding and Improvement

Breeding of pomegranates in India was done for disease resistance (Jalikop Jalikop *et al.*, 2005; 2006), low acidity and high fruit quality under hot arid environment (Samadia and Pareek, 2006); fruit yield (Manivannan and Rengasamy, 1999); juice production (Jalikop and Kumar,1998); aril color (Wavhel and Choudhari, 1985); fruit weight, flesh color, seed size, and juice content (Karale *et al.*, 1979); TSS (Choudhari and Shirsath, 1976); and seed softness (Jalikop and Kumar, 1998). Two of the main commercial export cultivars from India, 'Mridula' and 'Bahgwa', are the result of a selection from progenies of a cross between 'Ganesh' and the red 'Gul Shah Red' pomegranate cultivar from Russia (Mahatma Phule Agricultural University 2007).

Improvement of pomegranate can be done by three methods, *viz.* (i) collection of superior germplasm from indigenous and exotic sources, (ii) improvement by selection and (iii) controlled hybridization.

5.1.0 Collection of Superior Germplasm

In India, the varietal improvement programme of pomegranate was initiated in Maharashtra state as back as in 1944, but collection of exotic cultivars from Iraq, Afghanistan, Iran, Russia and other countries failed to establish.

Purohit (1983) reported that attempts to introduce superior deciduous cultivars in the tropics failed because of development of disorders caused by lack of temperature below 7°C in winter and absence of dormancy. The deciduous cultivars hardly produced 5-10 flowers per plant per year in tropics and out of this nearly 75 per cent flowers were male and intermediate and hence did not set fruits. Therefore, improvement of evergreen pomegranate suitable for tropical regions was initiated in 1980 at Bangalore.

5.2.0 Improvement by Selection

Heterozygosity and cross-pollination are the two important factors which bring about the variation in the seedlings and this naturally-induced variation (natural hybrid) is used for the selection of good types of new strains.

Selection of seedlings from Alandi in Pune district and Dholka in Gujarat was taken up. Dr. Cheema did the pioneering work in making selection from elite plants collected from Alandi which gave rise to GBG-1, now renamed as 'Ganesh' as a chance seedling. The observed superiority in Ganesh (GBG-1) in respect of prolific yield, medium-size fruits with soft seeds, better pulp and sweet juice with an extremely agreeable flavour is not comparable to imported fruits of exotic cultivars particularly in regard to size and quality of fruits. Sawant (1973) made an attempt to identify desirable types by clonal selection from Ganesh pomegranate and reported clones G-137, G-133, G-107 and G-132 as superior to Ganesh. These selections are more promising considering the fruit quality.

The success in any breeding programme depends on the extent of genetic variability present in various biometric characters in the plant population. The assessment of variability is, therefore, the first step in the breeding programme. Since, most of the characters of economic importance are polygenic in nature and are highly influenced by the environmental fluctuations, it is difficult to judge whether observed variability is heritable or due to environment. This suggests the imperative need of partitioning the phenotypic variation into its heritable and non-heritable components. Manohar *et al.* (1981) studied phenotypic and genotypic variabilities in 16 cultivars of pomegranate at Udaipur and concluded that individual plant selection for rind weight, per cent acidity, fruit weight, aril/fruit and fruit yield/tree would be satisfactorily effective as these characters exhibit high heritability as well as high genetic advance.

The Muskat cultivar of pomegranate is not only famous in Kolhar region of Ahmed-nagar district but in the entire Maharashtra due to its large fruits, sweetness and soft seediness. The orchards in this region have been developed from the seedlings and there is considerable heterozygosity amongst the plants. Due to this heterozygosity, there appears to be large variation in fruit characters which offers a good scope for improvement of pomegranate by resorting to seedling selection and subsequently propagating them clonally.

Exploitation of the existing variation among the seedling plants of Muskat resulted in selection of 4 promising seedlings having higher yield potential and excellent fruit qualities. These selections are superior to Ganesh in respect of consumers' preference for fruit weight, grain weight and other quality components, the only defect being light pink colour of the aril (Bhapuar, 1976; Patil, 1976; Karale, 1977 and Kolhe, 1980). Naik (1979) surveyed gardens around Kolhar and has reported five promising seedlings, *viz.*, P-23, P26, SK-1, P-16 and P-13. G137, a clone selected from the variety Ganesh, had heavier fruits (289.9 g) and seeds (219.4 g), more juice (88.7 per cent), less acidity and more reducing sugars (12.38 per cent) than its parent. The fruits of G137 kept well and were yellowish-pink and smooth surfaced (Keskar *et al.*, 1990).

Clonal selections of Ganesh and seedling selections of Muskat do not, however, possess the deep red colour of the aril which is normally evident in exotic cultivars.

Three soft-seeded types from seedling population of cv. Alandi were identified by Ramu *et al.* (1996), one of which was a highly promising selection designated RCRI. Ten-year-old plants of RCRI have given yields of 267 fruits/tree, with an average fruit weight of 220 g giving 58.7 kg fruit/tree. This corresponds to an estimated yield of 28058 kg fruit/ha, and compares well with average yields of 10000-12000 kg/ha for the cultivars.

The cultivars Duo Hong 1, Duo Qing 11 and Duo Bai 2, obtained by single-plant selection for Shandong province, showed wide adaptability, gave high stable yields, and produced high-quality fruit with attractive appearance, large seeds and high sugar content. Duo Hong 1 has red fruit skin, Duo Qing 11 green fruit skin and Duo Bai 2 yellowish fruit skin. Duo Qing 11 has good storage and transport quality. Xian (1999) described pomegranate cv. Daluzi, a sport of cv. Qingpiruanzi in China. Arils are green in colour (light red in Qingpiruanzi), very juicy, with a soluble solids content of 14 per cent and a honey sweet flavour, of very good eating quality.

Of 72 cultivars, 25 were selected as standard table cultivars and 19 for juice proce-ssing, using the weighed selection scale method (Onur, 1983). Suitable cultivars were classified into trial groups which were determined by the combination of maturity dates (early, mid-season, late) and fruit acid contents (sweet, sweet-sour, sour). Considerable variation was noted between the trial groups with regard to fruit and aril sizes and yields, fruit and rind colours, and soluble solids content of the juice. The relationships between rind thickness and aril size, rind thickness and juice yield, and fruit weight and fruit width and length were also noted. Maturity dates were not considered in the classification of juice cultivars.

5.3.0 Hybridization

Some Russian cultivars of pomegranate were collected in 1973 and assessed for their adaptability, yield and fruit quality. None of these cultivars was found adaptable to the environmental conditions in Maharashtra and all were poor yielders. However, these cultivars had deep red colour of aril. For introducing this character into Ganesh, hybridization programme was taken up and Ganesh was crossed with four Russian cultivars, Shirin Anar, Gulesha Rose Pink, Gulesha Red and Guleshah (Anon., 1979). Hybrid seedlings (140) thus obtained was planted in 1980 and their growth, flowering and fruiting behaviour, yield and fruit quality were recorded. Kale (1986) screened these hybrids and seven hybrids were found promising for some desirable characters. However, none can be used directly because every one possesses one or the other defect in it. Further breeding programme may lead to development of superior types which might be comparable with exotic types.

Purohit (1987) studied the effect of pollen parent on seed hardness and crosses were made between the soft-seeded *Punica granatum* cultivars Ganesh and Bassein Seedless and the hard-seeded cultivars Kabul Yellow, Bedana Sedana and

Cultivars of Pomegranate

Spanish Ruby

Ruby Red

Bhagwa

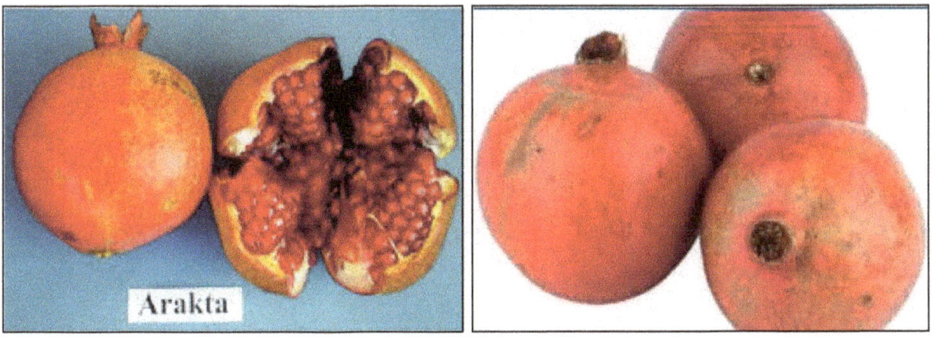

Arakta

Propagation of Pomegranate

Hard Wood Cutting

Cleft Grafting

Air Layering

Lupania. Pollination by a soft-seeded cultivar slightly decreased seed hardness in hard-seeded cultivars in comparison with pollination by a hard-seeded cultivar, while pollination by either hard or soft-seeded cultivar increased seed hardness in soft-seeded cultivars. It is concluded that soft-seeded cultivars should be planted at a distance from other cultivars to maintain fruit quality.

6.0 Propagation

In the middle of 20th century, many pomegranate plantations were established using seedlings as planting material in India. They showed variation in characters *viz.*, low yield, poor quality fruits, *etc.* due to seed propagation. To avoid the variation and to have uniformity in yield, quality, *etc.*, it is essential to establish the orchards with vegetatively propagated/clonal planting materials. Nowadays, air-layers, hardwood cutting raised plants and tissue culture plants are used as planting materials.

6.1.0 Cutting

Pomegranate can be propagated through cuttings round the year, using pruned wood in polyhouses under high humidity. Pruned wood obtained from plants during rest/stress period can be used for preparation of cuttings. Hardwood cutting, semi-hard wood cutting and *in situ* hard wood cutting have been reported in pomegranate. This method is farmers' friendly and less laborious.

The common method of vegetative propagation is by rooting of hardwood cuttings. The cuttings are taken from fully mature wood, about one-year-old and the suckers which arise from the base of the stem are usually utilize ed for this purpose; 20-25 cm long cuttings showed good rooting (Rajanna, 1981). Root formation is markedly improved by using growth substances. Phadnis (1974) reported higher percentage of rooting and larger number of roots by treatment with IAA at 200 ppm. Basal and medium cuttings rooted to the extent of 75 per cent by soaking the base of cuttings in 50 ppm IBA for 18 hours and to 66 per cent by quick dip for 20 seconds in 5000 ppm IBA (Sulladamath, 1985). Bose and Mandal (1972) obtained 85-100 per cent rooting in pomegranate using IBA at 3000 ppm in talc under mist and the rooted cuttings also showed very high percentage of survival. Bicket and Bilderback (1982) reported that softwood cuttings of pomegranate when treated with 0.5 per cent potassium salt of IBA as quick dip and planted in peat-perlite media under intermittent mist gave high percentage of rooting.

In an experiment on propagation, stem cuttings from basal, sub-apical and apical-shoot portions were treated with IBA at 0, 2500, 5000, 7500 and 10000 ppm by the quick-dip method. Basal cuttings with a diameter of 1.0 to 1.25 cm and treated with 5000 ppm IBA gave the highest per cent survival, number and length of roots, and number of shoots. Sub-apical cuttings treated with 5000 ppm IBA were next best. Apical cuttings failed to sprout or root at any IBA concentration (Purohit and Shekharappa, 1985). Hardwood cuttings taken from3-year-oldplantsofthe cultivar Bassein Seedless and treated with IBA at 2500 ppm or paclobutrazol at 1250 or 2500 ppm singly or in combination, showed best rooting (55 per cent) with IBA + paclobutrazol at 2500 ppm (Reddy and Reddy, 1989). Singh (1994) obtained 51.91

per cent rooting in 1-year-old hardwood cutting of cv. Jalore Seedless with IBA at 1000 ppm; Seradix-B was the second best treatment (46.15 per cent rooting). The longest roots were observed in the Seradix-B treatment.

The beneficial effect of IBA at different concentrations and method of treatment on rooting in hardwood cuttings of pomegranate was reported by Panda and Das (1990), Sandhu *et al.* (1991) and Jain and Parmar (1996). Treatment with IBA + NAA proved more effective than the application of any of the growth substances (Reddy and Reddy, 1990). In China, He (1999) reported that cuttings from 1-year-old shoots rooted without any treatment and survival reached 95 per cent. In rooting of pomegranate cutting, river silt was the best rooting medium and hardwood cuttings rooted better than semi-hardwood cuttings (Baghel and Saraswat, 1989). Moisture level in the rooting media improved the rooting percentage to a significant level. Chauhan and Singh (1971) observed that the rooting percentage increased from 39.2 to 90 by increasing moisture level of the medium from 5.80 to 20.14 per cent.

6.2.0 Air-layering

Another method of propagation of pomegranate is air-layering. It is popularly known as 'gootee' or marcottage. Air layering is the most commercial method of vegetative propagation in the Deccan Plateau of India with success rate above 85 per cent during July-September. For air layering, 0.8 -1.5 cm thick 6 -18 months old upright branches are girdled by removing 2-3 cm wide ring of bark followed by application 2000-3000 ppm IBA on upper part of the cut. The girdled stem is wrapped with *Sphagnum* moss and covered with a small polyethylene sheet and tied with jute string. Roots become visible after 4-6 weeks period. Well rooted layers are detached from the mother plants from the lower girdle after 8-10 weeks and planted in polythene bags filled with potting mixture in the nursery.

7.0 Area and Production

As a cultivated crop the pomegranate is grown to a limited extent in selected locations in many states of India. Maximum area (about 1.47 lakh hectares) under this crop is in Maharashtra, particularly in Nashik, Solapur, Ahmednagar and Pune districts and in other areas. The other states, where pomegranate cultivation is important are Gujarat, Karnataka, Andhra Pradesh, Rajasthan, Telangana, Tamil Nadu, Chhattisgarh, Himachal Pradesh, *etc.* The area and production of pomegranate in India during 2017-18 are given in Table 1.

The versatile adaptability, hardy nature, low maintenance cost, steady and high yield, great nutritional and therapeutic values, better keeping quality and inbuilt drought tolerance, *etc.* makes it highly amenable for its rapid spread on wider scale.

8.0 Soil and Climate

8.1.0 Soil

Pomegranate is not very particular about its soil requirement and can be grown on diverse soil types. The trees give very good yield in deep loamy or alluvial soil, although it thrives well on comparatively poor soil where other fruit crops fail to

Table 1: State-wise Area and Production of Pomegranate*

States	2015-16		2016-17		2017-18	
	A (000 Ha)	P ('000 MT)	A (000 Ha)	P ('000 MT)	A (000 Ha)	P ('000 MT)
Andhra Pradesh	8.39	113.28	7.95	119.14	9.47	139.02
Chhatisgarh	0.42	0.97	0.59	3.51	0.76	4.34
Gujarat	18.54	278.10	23.18	350.82	30.51	461.75
Himachal Pradesh	2.48	1.99	2.67	2.74	2.77	3.15
Jammu & Kashmir	0.00	0.01	-	-	-	-
Jharkhand	0.01	0.04	0.01	0.01	0.02	0.04
Karnataka	27.26	319.34	26.58	378.14	25.97	268.23
Kerala	0.01	0.07	0.01	0.09	0.01	3.33
Madhya Pradesh	6.28	54.20	9.35	102.93	9.68	114.27
Maharashtra	128.40	1486.11	140.72	1616.50	147.91	1789.46
Nagaland	0.08	0.46	0.09	0.55	0.09	0.50
Odisha	0.24	0.89	0.24	0.90	0.26	0.92
Rajasthan	2.47	10.05	2.86	10.38	4.44	26.58
Tamil Nadu	0.48	15.92	0.57	15.83	0.56	13.06
Telangana	1.84	25.02	1.21	11.28	1.49	19.87
Others	0.01	0.00	0.01	0.01	0.01	0.01
Total	196.89	2306.44	216.03	2612.82	233.93	2844.52

* Horticulture Statistics at a glance (2018).

flourish. It can tolerate soils which are slightly alkaline, though it can also be grown in medium or light black soils of at least 60 cm depth. It is a salt-hardy fruit tree. In the central Asia, pomegranate is mainly found in alluvial soil with a low salt content (Kulenkamp *et al.*, 1985).

In sandy soil, application of bitumen emulsion (0.5 or 1.0 per cent a.i.) with the top 2 cm of moistened sandy soil around pomegranate cv. Manfalouti trees increased fruit size, total acidity, TSS content and total sugars and in general, the 0.5 per cent a.i. treatment gave the best results (Heikal, 1991)

Field experiments on acid soils in Tamil Nadu, with soft-seeded pomegranate showed that the application of 0.945 kg dolomite/tree twice a year produced the highest fruit yield/tree (13.08 kg) as well as significantly enhanced TSS, reducing and non-reducing sugars, acidity and ascorbic acid content in juice (Muthumanickam and Balakrishnamurthy, 1999).

8.2.0 Climate

Pomegranate has a versatile adaptability to wide range of climatic conditions. It can grow in the plains as well as on the hills up to an elevation of about 1850 metres. The tree grows best in semi-arid climate where cool winter and hot summer prevail. In areas of low temperature, the tree is deciduous but in tropical and subtropical conditions it is evergreen or partially deciduous. The tree requires hot and dry climate during the period of fruit development and ripening. The optimum temperature for fruit development is about 38°C. The tree cannot produce sweet fruits unless the temperature is high for a sufficiently long period. The quality of fruit is adversely affected in humid climate.

Pomegranate is a 'winter-hardy and drought-tolerant plant' and can thrive under arid conditions. Although, it is highly drought resistant, the plant bears well only under irrigated condition. Kucherona (1976) recorded that pomegranate cultivars Ak Dona, Kzyl Dona and Boris were winter-hardy. Two-year-old shoots of hardy cultivars were not damaged even at 2°C. He also observed high resistance in all cultivars to dehydration. Kulenkamp *et al.* (1985) however, found noticeable damage due to frost when temperature dropped below –14° or –15°C. Aridity and frequent anomalies of the climate cause leaf shedding and fruit cracking. It can withstand frost but is injured by temperature below –11° C. The plants of pomegranate also withstand during high temperature (45°C) during summer in arid region (Westwood, 1978).

9.0 Cultivation

9.1.0 Planting and Plant Density

The layout is done following square or hexagonal system. Pits about 90 cm^3 are dug at the required distance and filled with mixture of organic manure and soil. The planting materials are planted in these pits during monsoon. Generally, in the poor soils of Deccan, the plants are given a spacing of 3.5-5 metres; in lighter but deeper soils the distance is increased up to 5 x 5 metres.

Apart from pit system, bedding system-a new method of planting the saplings in bed is also followed (Marathe *et al.*, 2017), mainly in case of soils having medium drainage facility. The beds of convenient length, 4 feet width and 1 foot height is made for planting the saplings.

Planting density is the most important yield contributing factor which can be manipulated to attain the maximum production per unit area. The optimum spacing is important for the maximum utilization of land and good income over a long period. Faiziev (1973) conducted a spacing trial in pomegranate with cultivars Kazakai, Ak Dona and Achik Dona using 5 x 3m, 4 x 2.5m or 4 x 1.5 m spacing which gave 666, 1000 and 1666 plants/ha, respectively. The highest yield was obtained from the closest spacing. Planting density had no adverse effect on fruit weight. Kutuzova (1984) recommended a spacing of 3 x 2.5 or 4 x 2 m (1333 and 1250 plants/ha, respectively) for high and economic yield of good quality fruits.

In an investigation at Mahatma Phule Agricultural University (MPAU), Rahuri, India with a plant population of 1000, 666, 500 and 400 plants/ha, Bagate (1979) observed that as the plant density was increased, yield per hectare also increased without affecting fruit quality, and a density of 1000 plants/ha gave 2.30 times higher yield and 2.44 times more profit compared with normal plant population of 400/ha. It was also recommended that for higher yields for the first four to five years after planting, a distance of 5 x 2 m may be adopted and alternate plants may be removed afterwards maintaining a planting distance of 5 x 4 m.

Most of the farmers in Maharashtra follow 15x10ft spacing while, in Gujarat 12x8 ft spacing is adopted for establishing orchard of pomegranate variety Bhagawa.

9.2.0 Training and Pruning

9.2.1 Training

Pomegranate plants may be trained under multi-stem training or single stem training system. In general, pomegranate plants are allowed to grow as bush with a number of main shoots arising at ground level and growers find it convenient to allow 4-5 such shoots to develop. However, single stem training was also followed by farmers for better grade fruits. To attain this, all the side shoots should be removed at the time of planting, the main stem is pinched at a height of about one metre and up to 25-30 cm below the cut surface, 3-4 branches well-distributed on all sides should be allowed to grow to form main limbs of the tree. The desired shape of the tree is formed within 2 to 3 years of planting.

This system is practiced by relatively less number of farmers in India but widely practiced in countries such as Israel, USA, *etc.* It facilitates easy intercultural operations (Holland *et al.*, 2009).

Pomegranate tree has a tendency to throw out suckers and as such it is rather difficult to train the plant to a single stem. Secondly, single stem training method has its own disadvantage. The crop is highly susceptible to stem borer and as such training on single stem is found to be hazardous. Allowing too many stems also hinders the inter-culture operation and therefore, maintenance of 3-4 stems per plant is recommended.

In pomegranate cv. Daquinpitian which has a narrow branching angle, Zhang *et al.* (1997) found that widening the angles was an effective means of restricting growth, and promoting early bearing and high yields. The best angle was 60-90o.

9.2.2 Pruning

Pomegranate does not usually require pruning except for removal of suckers (water sprouts), dead and diseased branches and developing a sound framework of the tree. It is essential to remove the suckers as soon as they arise and one of the main causes of poor bearing in commercial orchards is the failure to remove these shoots in time.

The fruits are borne on short spurs produced all along the slow growing mature wood. These bear fruits for 3 to 4 years. Only a limited pruning of the bearing tree is, therefore, required (Singh, 1969). Dutta (1965) suggested that annual pruning in winter during dormant period should be confined to shortening of the previous season's growth to encourage fruiting.

For obtaining proper shape of tree and getting a good crop, a set of new shoots should be allowed to develop every year on all sides of the tree and gradual growth of new shoots should be encouraged by restricted cutting back of the bearing shoot.

It should be borne in mind that shortening of the older wood is likely to reduce the yield.

In an experiment on pruning of pomegranate, Pawar *et al.* (1994) reported that shoot length and number of leaves per shoot increased with pruning intensity. Pruning also delayed bud sprouting, flower appearance and harvesting.

The highest yield was obtained from unpruned trees, although the number and percentage of better grade fruits increased with the severity of pruning. Fruit size, juice content and TSS (°Brix) content increased with increasing severity of pruning. Rind and aril colour, percentage aril content and seed hardness were not influenced by pruning treatments. They also found that the percentage of sun-scorched fruits and the percentage of fruits showing internal breakdown decreased as the pruning intensity increased. Pruning did not affect the incidence of cracked fruits.

9.3.0 Irrigation

The plants in newly established orchards require regular irrigation. Once the plants are well established, they can wihstand drought to a considerable extent. Regular irrigation is essential from flowering to ripening of fruits, as irregular moisture condition results in dropping of flowers and small fruits. It may also cause development of cracks on mature fruits which reduces market value and consumer's acceptance.

Bangal and Khaire (1982) studied water saving by use of drip irrigation in comparison with traditional check basin system along with sugarcane trash as a mulch in both the systems. They found that check basin required 108 ha cm of water whereas drip with mulch required 40.31 ha cm of water. Patil *et al.* (1982) observed water saving to the extent of 44 per cent in drip irrigation system compared with check basin and 64 per cent when mulch was used in drip system. They further reported that though there was slight reduction in yield, the water use efficiency was more in drip method which indicated the possibility of doubling the area under the crop with same quantity of water.

Trickle irrigation was applied to pomegranates at rates of 0-100 per cent of water evaporation replenishment (based on cumulative pan evaporation) (Afria *et al.*, 1998). The highest transpiration rates, leaf areas, yield and fruit TSS were recorded in the 60 per cent treatment. Chopade and Gorantiwar (1998) found that the yield of pomegranate differed significantly between irrigation treatments and was highest (65.53 q/ha) from drip irrigation applying water equivalent to 10 per cent allotted area per tree.

In trials at Parbhani, India 14-year-old trees of pomegranate cv. Muskat were irrigated with 75 mm water at IW/CPE ratios of 0.4, 0.6, 0.8 or 1.0 (Lawand and Patil, 1994). Tree volume, shoot length, and leaf area were maximum at an IW/CPE of 1.0, but flower number and fruit yield/tree were highest at an IW/CPE of 0.8. Irrigating at an IW/CPE ratio of 0.8 rather than 1.0 during flowering and fruit development and of 0.6 during *bahar* treatment (November-December) produced the greatest number of flowers and fruit yield/tree.

Planting and Training of Pomegranate

Planting

Single Stem Training

Multistem training

Training

Pruning of Pomegranate

Pomegranate after Pruning

In an experiment in Egypt carried out on 15-year-old pomegranate cv. Manfalouty trees by Abou-Aziz *et al.* (1995a), it was observed that June-December irrigation regime when soil reached 60 or 40 per cent of field capacity resulted in the highest fruit numbers and yields (306.8 and 69.2 kg/tree, respectively averaged over two years), the least fruit splitting (3.7 per cent) and the highest TSS:acid ratio (20.9). Fruit yields and most quality parameters did not differ significantly between the 40 and 60 per cent field capacity treatments. In a similar investigation on irrigation regime Abou-Aziz *et al.* (1995b) also found that overall perfect flower production peaked on 26 April and male flower production and initial fruit set peaked on 6 May. The most dense (least spread out) perfect flower production and fruit set during early spring (April to early May) occurred in the April-October, 60 per cent field capacity (FC) treatment, followed by the May-November, 40 per cent FC treatment. Male flower production during this period was least dense in the June-December, 40 per cent FC treatment.

Results of the trial on irrigation method conducted by Behnia (1999) in Iran showed that drip irrigation reduced water consumption from 50-66 per cent of surface irrigation supply, and increased yield by 21-24.5 per cent, the introduction of drip did not cause any water stress during growth, even with trees 10 to 16 years old.

The effects of watering interval (irrigation to field capacity every 4, 7 or 14 days) and cycocel (chlormequat) application rate (0, 500 or 1000 ppm) on the nutritional status of the transplants of pomegranate cultivar Manfalouty were reported by Ahmed (1994). Leaf mineral concentrations (P, K, Ca, Mg, Fe, Mn and Zn) decreased with increasing interval between irrigation, except for N concentrations, which increased. Total carbohydrate, starch and polysaccharide contents were reduced by increasing water stress while total soluble sugars showed opposite trend.

Meshram *et al.* (2019) reported that among six microirrigation systems evaluated for pomegranate, maximum plant height, stem diameter, LAI, No. of flowers and yield were recorded in subsurface drip irrigation system (SDI system) with 2 inline laterals (30x30cm) followed by (40x40cm).

9.4.0 Manuring and Fertilization

Apart from using organic manure in the pit before planting, about 20 kg farmyard manure and ash should be incorporated around a bearing tree at the onset of monsoon (Cheema *et al.*, 1954).

Phadnis (1974) recommended application of 50 kg FYM and 3-5 kg oil cake or 1 kg of sulphate of ammonia prior to flowering for healthy growth as well as fruiting. Experiments carried out at the Mahatma Phule Krishi Vidyapeeth (MPKV), Rahuri, India over a period of eight years indicated the following schedule to be optimum for good growth and economic production (Anon., 1979).

Quantity of Nutrients/Plant	*Age of Plants*
250 g N + 125 g P_2O_5 + 125 g K_2O	1½ to 2 years
500 g N + 125 g P_2O_5 + 125 g K_2O	2½ to 3 years
500 g N + 125 g P_2O_5 + 250 g K_2O	3½ to 4 years
625 g N + 250 g P_2O_5 + 250 g K_2O	Above 4 years

Lal (1975) observed application of 0.5 kg N and 0.25 kg P_2O_5 per tree to be the optimum under the agroclimatic conditions of Udaipur in Rajasthan. Pathak and Pundir (1981) recommended application of N, P and K at 240, 160 and 60 kg, respectively, per hectare for a 12-year-old orchard. In a 3-year-old plantation, 200 kg N, 120 kg P and 60 kg K per hectare gave the best result (Pareek, 1981).

In 3-year trials on pomegranate (cv. Local Selection) Singh *et al.* (1988) observed that fruit size, TSS (16.46 per cent) and yield (16.4 kg/tree) were highest in trees receiving the highest NPK rates (N 1000 g, P_2O_5 1000 g and K_2O 1500 g/tree). The unfertilized control yielded 13.3 kg/tree.

Prasanna Kumar and Dhandar (1996) reported that 4-year-old plants of cv. Bassein Seedless produced higher yield by the application of N, P and K at 500 g, 500 g and 250 g/plant, respectively singly and the yield was significantly higher in combined treatment at the above rates (Table 2). Treatment with N and K_2O markedly improved the yield and increased the fruit weight in pomegranate cultivars Jyothi and RCRI (Padmavathamma and Hulamani, 1999).

Table 2: Interaction Effects of N, P and K on Fruit Yield (kg/plant) of Pomegranate*

Treatments		*Nitrogen (g/plant)*			
Phosphorus (g/plant)	*Potash (g/plant)*	*0*	*250*	*500*	*750*
0	0	1.35	2.00	4.44	6.00
	125	1.50	2.50	5.40	5.50
	200	0.95	3.05	5.15	5.35
250	0	1.10	3.30	5.45	4.95
	125	1.40	5.20	9.40	9.20
	250	1.85	5.75	8.58	9.04
500	0	1.60	5.54	6.41	8.03
	125	1.90	5.92	7.09	8.94
	250	2.25	6.58	9.37	9.19
C.D. at 5 per cent = 1.05					

* Prasanna Kumar and Dhandar (1996).

Foliar sprays of 0.25 per cent each of $ZnSO_4$, $FeSO_4$ and $MnSO_4$ combined with 0.15 per cent boric acid significantly increased the yield from 18.5 kg/plant in the control to 26.37 kg/plant and juice content from 65.6 per cent to 74.8 per cent. The highest TSS (17.35oBrix) was obtained by applying $ZnSO_4$ (0.4 per cent) combined with boric acid (0.20 per cent) (Balakrishnan *et al.*, 1996).

The manures and fertilizers are applied in the basin one metre away from the trunk at the time of *bahar* treatment, in split doses, in case of bearing orchard.

9.4.1 Tissue Analysis and Nutrient Requirement

In pomegranate most of the nutrients were found to stabilize in the 8th pair of leaf from growing tip and may be taken as standard for nutritional diagnosis

Irrigation and Fertilizer Application

Drip Irrigation

Fertilizer Application

Mulching in Pomegranate Orchard

Polthene Mulching

Straw Mulching

(Bhargava and Dhandar (1988). However, Munde *et al.* (1980) reported the stable age of leaf for Fe, Mn and Zn as 4, 5 and 11-12 months age, respectively. Afria *et al.* (1998) working with cv. Jodhpur Red recorded maximum nutrient accumulation (N, P and K) in young leaves compared with matured and old leaves. They suggested sampling of leaves in February for N, June for P and April for K estimation. Lal and Chauhan (1979) noted the highest fruit yield with leaf N content of 2.35 per cent. Doring and Ludders (1987) reported that plants watered with NaCl and $Na2SO_4$ solutions of different concentrations mainly accumulated Na in roots and Cl and SO_4 in leaves, the concentrations increasing with the amount of salt applied. Na accumulation by leaves was greater when supplied with Cl than SO_4 salt, and at equivalent supply rates, more Cl and SO_4 were absorbed and translocated.

Wang *et al.* (1995) studied the Na content of pomegranate plants which increased slightly up to a soil salt concentration of 0.3 per cent NaCl, but increased more with 0.4 per cent NaCl. NaCl treatments also increased the K content producing higher K:Na ratios in treated plants than in controls. In pomegranate Na accumulated in the root phloem, while K preferentially entered the xylem.

Maity *et al.* (2019) reported that the uptake pattern of macronutrients followed the order of Ca> N > K > Mg > S > P. Most of the uptake of N, K, Ca, Mg and S from the soil occurred between pre-pruning and bloom of the crop whereas the highest P uptake took place from fruit development to fruit maturity. The demand for micronutrients particularly, Fe, Mn and B was highest during fruit enlargement stage while that of Zn was during the fruit development stage.

9.5.0 Mulching

Chattopadhyaya and Patra (1992) found that black polyethylene mulch resulted in the best plant growth, earliest flowering and fruiting and highest yield (164.9 q/ha). Among different organic mulches (Wheat straw, sugarcane baggas and safflower straw) evaluated for pomegranate, Sugarcane baggas was found to be the best mulching material for pomegranate plants in terms of no. of fruits/plant, fruit weight, Leaf area index and water use efficiency (Meshram *et al.*, 2018).

9.6.0 Intercropping

Since pomegranate takes about 4 to 5 years to come to good bearing, intercropping is highly desirable. Low growing vegetables and pulse crops may be grown as intercrops in between young trees till 4 to 5 years of their age.

10.0 Flowering, Pollination, Fruit Growth and Development

10.1.0 Crop Regulation

In order to regulate flowering, irrigation is withheld for about two months prior to the normal flowering season. After two months, pruning followed by foliar spray of ethrel 39.5 per cent SL @ 2 ml/l, manures or fertilizers are applied and light irrigation is given. Three to four days later, heavy irrigations at normal interval are followed. The tree readily responds to this treatment by producing new growth and blooms and bears a good crop.

Normally, the pomegranate starts fruiting in about 4 years after planting. According to seasonal changes, there are three flowering seasons in the Deccan, India *viz.*, June-July (*mrig bahar*) coinciding with the break of monsoon; February-March (*ambia bahar*) and September-October (*hasta bahar*). Although pomegranate may be indu-ced to bear fruit in any of the seasons, ordinarily, only one *bahar* is taken from a tree, and the season of fruiting to be adopted is mainly determined by the demand, price and the availability of water.

In the Deccan, *mrig bahar* is taken where water is scarce during the hot weather and, therefore, the flowering is so forced that the maximum requirement of water falls during the rainy season or immediately after the rains when plenty of water is available. For *mrig bahar*, treatment of withholding irrigation from December to April-May results in sufficient suppression of growth. The land is ploughed up or a shallow digging is given in April up to a depth of 10 cm. The trees usually shed their leaves by March and remain in dormant condition up to May. The manures and ferti-lizers are applied and the first light irrigation is given in the middle of May and there-after one or two light irrigations may be given before the rains set in. The trees start growth by June with the onset of rains and flowers and fruits are produced. The fruits ripen in October and the harvesting continues up to December. In dry regions of western Maharashtra, the fruits get spoiled because of black spots and cracks due to high humidity in the rainy season.

Ambia bahar is also taken in the Deccan, where enough water is available during hot weather. In this case, the fruits are available during June and July and no irrigation is given after the start of the rainy season. The trees shed their leaves by October- November when a shallow digging or ploughing is done. The manures and fertilizers are applied in the month of December-January, the first irrigation is given in January and the flowers appear within a month after irrigation. In dry regions of western Maharashtra *ambia bahar* has been found to be better in yield and quality of fruits compared with *mrig bahar* fruits (Rane *et al.*, 1982a). For *hasta bahar*, the trees have to be made dormant during August-September. This is rather uncertain because of the rains which occur during this period and, therefore, not usually practised. Under Bangalore conditions, *mrig bahar* (fruit maturing in October-November) is found superior to *ambia bahar* in all respects (Anon., 1985).

Sonawane (1986) studied the possibilities of regulating flowering between *hasta bahar* and *ambia bahar*. He found that the fruits from November and December flowering fetched the highest price followed by those from October, February and January flowering and the highest net return was obtained from fruits of October flowering. In trials with the cultivar Ganesh, Sonawane and Desai (1989) induced cropping by irrigating the trees in the first week of October, November, December, January or February after rest induced by withholding irrigation for 50 days, for harvesting in March-April, April-May, May-June, June-August and July-September, respectively. The highest yield of marketable fruits was obtained with January cropping, followed by February and October cropping. Considering the comparable yields, fruit prices and irrigation needs, it is suggested that October cropping could be substituted for January cropping.

Sonawane *et al.* (1994) also observed the condition of the fruits produced after flower induction in October, November, December, January or February by staggering the irrigation schedule. Diseased (rotten) fruits or insect affected fruits were commonest in January flowering crops. Fruit cracking was more of a problem in crops following flowering in January and February than in November and December. Sun scorching of fruits was commonest in crops from flowering in December.

Results of five year study with cv. Ganesh have shown that it was highly profitable to take *hasta bahar* crop under rainfed condition (Raturi and Hiwale, 1991). The fruits harvested from *hasta bahar* crop were found to be of superior in quality (Table 3).

Table 3: Physico-chemical Characters of Fruits in different Cropping Seasons in Pomegranate cv. Ganesh*

Cropping Season (Bahar)	*Fruit Weight (g)*		*Fruit Length (mm)*		*Fruit Diameter (mm)*		*TSS (Per cent)*		*Acidity (Per cent)*	
	88-89	*89-90*	*88-89*	*89-90*	*88-89*	*89-90*	*88-89*	*89-90*	*88-89*	*89-90*
Mrig	141.57	192.85	69.27	73.34	65.99	67.83	15.60	14.97	0.45	0.43
Mrig + Hasta	129.57	192.85	67.53	73.12	63.83	69.00	16.46	15.92	0.38	0.36
Hasta	115.14	182.57	63.88	73.66	59.71	68.75	17.94	17.91	0.27	0.23
C.D. at 5 per cent	N.S.	N.S.	N.S.	N.S.	N.S.	N.S.	0.68	0.86	0.056	0.057

* Raturi and Hiwale (1991).

For standard fruit size, it is advisable to maintain specific number of fruits on each plant since overloading adversely affects the size and fruit quality. The number to be maintained would depend on the age and vigour of the tree. In Muskat cultivar, the crop load of 50 to 60 fruits per tree was found to be optimum (Rane *et al.*, 1982b).

10.2.0 Flowering

In the evergreen pomegranate cultivars, flower buds of the spring flush are borne on mature wood of one-year-old shoot whereas the flowers which appear during July-August are borne on the current year's growth between July and August.

Cai *et al.* (1993) observed seven stages during flower bud differentiation with ovule differentiation being the most critical. Differentiation started in early June andrequired between 2 and 11 months to complete. The optimum temperature for differentiation was a monthly average of 20 +_5°C, but differentiation did not proceed below 10°C.

In Japanese Dwarf cultivar, vegetative growth and flower initiation take place simul-taneously. The flowers are found mostly in clusters, either terminally or in axils of the leaves (Nath and Randhawa, 1959a). While Bavale (1978) and Nalawadi *et al.* (1973) reported that flowers are mostly axillary and usually solitary when terminal. The inflorescence in pomegranate has been reported to be cyme. Rendle

(1938), Pool (1941) and Lawrence (1951) described the inflorescence in pomegranate to be terminal, solitary or cymose (Singh, 1977), while Nath and Randhawa (1959a) observed the inflorescence to be dichasial cyme and due to heavy drop of secondary and tertiary buds they appear to be solitary in clusters.

Nalawadi *et al.* (1973) and Phadnis (1974) reported three flowering seasons in western India. Vachaspati and Chauhan (1973) recorded usually two flowering seasons in North India. The flowering period of different cultivars is also quite variable. Under Delhi conditions, Dholka, Kandhari, Muskat and Patiala flowered only once a year while GBG-1 and Japanese Dwarf flowered twice. Bavale (1978) indicated that Russian cultivars, because of their deciduous nature, behaved differently and flowered sparsely during May-June. Umrao and Singh (1995) recorded longest flowering period in evergreen cultivars Dholka and Ganesh, the latter flowering all the year round and Dholka flowered from February until October. All the deciduous cultivars flowered from March-April until July-August.

In Egypt, El-Sese (1988) investigated flowering of 3 cultivars of pomegranate. In cv. Arabi, flowering could begin as early as late March compared with the first week of April in cultivars Manfalouty and Nab El-Gamal. Approximately 80-85 per cent of the perfect flowers were produced between the onset of flowering and the end of the full bloom period. Fruit set during the early flowering stage was >70-80 per cent; it declined to 40-50 per cent during the full bloom period but then increased to >85 per cent during lateflowering. Fruit retention at harvest increased from 9-33 per cent during early flowering and full bloom to 85 per cent during late flowering.

In a study conducted in 3 successive seasons at the Assiut University, Egypt, on the flowering and fruiting behaviour of 4 pomegranate cultivars (Wardi, Arabi, Nab El-Gamal and Manfalouty), El-Kassas *et al.* (1998) observed that the flower buds of all 4 cultivars were mixed and were borne in the axils of the outermost leaves (termed as incomplete terminally bearing habit). Flowers formed on short shoots or spurs of >1-year-old and these bearing units were always borne on slow growing mature bran-ches not less than a year old. Two-year-old spurs, particularly those 1-3 cm long, carried a higher percentage of perfect flowers and a lower percentage of male flowers than one-year-old spurs. A high proportion of perfect flowers opened (>80 per cent per tree) during the early waves of flowering (3-5 weeks). Fruits set from these early blooming flowers were of better quality than those set later. The ratio between male and female flowers depended on cultivar and season, and under the conditions of this study male flowers were found to be 1.5-4.4 times as abundant as perfect flowers on the same tree.

Chaudhari and Desai (1993) found that treatment with GA_3 one month before expec-ted flowering produced significant increase in male flowers (72.03 per cent) compared with 61.14 per cent in controls and a decrease in hermaphrodite flowers (13.86 per cent compared with 20.90 per cent in control). Ethrel (250 ppm) produced higher percentage of herma-phrodite flowers than under control. In a trial on 7-year-old pomegranate cv. Ganesh trees, flowers were removed manually from day 75 of flowering onwards at 14-day intervals or growth regulators [GA, NAA, maleic hydrazide, ethrel (ethephon) or carbaryl] were applied to check new flower induction or reduce the growth of new shoots on which new flowers might appear

(Ahire *et al.*, 1994). All treatments reduced total flower production compared with the control except for manual thinning alone which increased it. Carbaryl at 0.7 per cent applied on day 75 had the greatest inhibitory effect on flower production. In all cases the majority of flowers dropped were male.

10.3.0 Floral Biology

The time of dehiscence of anther varied in different cultivars and no general sequence was found in the time of anthesis (Nath and Randhawa, 1959b; Josan *et al.*, 1979a). In cv. Patiala, dehiscence started prior to anthesis, whereas in Muskat White it started afterwards. Dehiscence was found to be affected by temperature and atmospheric humidity. Evreinoff (1957) reported that the development of stamens is normal and pollen is fertile in both hermaphrodite and male flowers of pomegranate. The fertility of pollen grains in hermaphrodite flowers was found to be higher than in the male flowers (Nath and Randhawa, 1959b). The stigma attained maturity one day before anthesis and receptivity remained at peak on the same and second day after anthesis and it gradually went down till the third day after which it abruptly became non-receptive. Josan *et al.* (1979b) noted that stigma remained receptive from one day before anthesis to five days after the completion of anthesis and that pollen germination was over 90 per cent. Babu (2010) briefly reviewed the floral biology of pomegranate.

10.4.0 Pollination

Both self- and cross-pollination are recorded in pomegranate. Gammie and Patwardhan (1924), Nalawadi *et al.* (1973) and Bavale (1978) indicated that pomegranate is a cross-pollinated crop. Greater percentage of fruit set was observed by hand pollination and pollination under natural condition, *i.e.*, open pollination (Nath and Randhawa, 1959c; Josan *et al.*, 1979c). Pross (1938) working with four pomegranate cultivars in central Asia considered pomegranate to be a self-pollinated crop, while Singh (1977) stated that pomegranate is often cross-pollinated.

In an investigation on the method of pollination and fruit set in 7 cultivars of pomegranate, flowers were either allowed to open pollinate or self-pollinate with or without emasculation of the anthers. The 2 self-pollination treatments were bagged immediately after pollination. Fruit set was 79 per cent and 43.3 per cent for intact open and self-pollinated flowers or 26.4 per cent and 66.2 per cent for the same after emasculation. Karale *et al.* (1993) suggested that the pomegranate, which is normally thought of as a cross-pollinated crop, is capable of both open and self-pollination. The hermaphrodite flowers have a higher fruit set than the intermediate type flowers.

10.4.1 Flower Thinning

Hussein *et al.* (1994) applied sevin (carbaryl) and NAA for thinning of flowers of pomegranate cv. Manfalouty and observed that in mid-May or on 1 June in Assiut, Egypt, increased fruit weight, TSS and sugar contents and reduced peel percentage (by weight) and total acidity compared with untreated control.

10.5.0 Fruit Growth and Development

Shulman *et al.* (1984) observed a single sigmoid fruit growth pattern of cv. Mule's Head while in cv. Wonderful it was reported to be more or less linear. Prasad *et al.* (1999) studied the fruit growth and development of cv. Jalore Seedless. They observed a linear increase in size, diameter, volume and weight of fruit (Table 4) from fruit set till the harvest of fruits except the specific gravity which decreased gradually indicating sigmoid growth curve pattern of fruit. The colour of the fruit changed from greenish to deep pink with red and yellow patches at maturity. The aril/rind ratio increased from 0.652 to 2.060 from fruit set to the harvesting of fruits. The TSS and TSS/acidity ratio of aril increased with the advancement of maturity while acidity decreased at maturity (Table 5). It is suggested to harvest cv. Jalore Seedless between 120 and 135 days after anthesis for best dessert quality.

Table 4: Physical Changes in the Fruits during Growth and Development*

Date of Observation	*Length of Fruit (cm)*	*Diameter of Fruit (cm)*	*Weight of Fruit (g)*
4.9.1992	4.20	4.10	40.45
19.9.1992	4.75	4.72	56.45
5.10.1992	5.28	5.22	81.82
20.10.1992	5.56	5.60	96.70
5.11.1992	6.20	6.50	175.50
5.12.1992	6.81	7.27	265.58
20.12.1992	7.91	8.30	274.98
5.1.1993	8.01	8.32	279.57
C.D. at 5 per cent	1.51	1.39	8.22

* Prasad *et al.* (1999).

Table 5: Qualitative Changes in the Fruits during Growth and Development*

Date of Observation	*Aril : Rind Ratio*	*Rind Thickness*	*Per cent of Juice*	*Per cent of Seed*	*TSS (°brix)*	*Acidity (Per cent)*
4.9.1992	0.632	0.40	24.83	N.D.	10.05	0.76
19.9.1992	0.834	0.40	37.36	20.65	10.50	0.44
5.10.1992	1.389	0.38	39.74	20.77	11.50	0.40
20.10.1992	1.461	0.39	41.32	19.52	12.87	0.38
5.11.1992	1.522	0.34	45.17	10.02	14.00	0.35
20.11.1992	1.803	0.32	49.26	19.14	16.25	0.33
5.12.1992	2.084	0.32	50.12	18.31	18.12	0.30
20.12.1992	2.064	0.30	51.08	16.39	18.50	0.29
5.1.1993	2.060	0.29	51.03	15.80	18.65	0.30
C.D. at 5 per cent	0.020	0.02	2.61	2.02	1.11	0.03

ND – Not detectable; * Prasad *et al.* (1999).

Flowering and Pollination in Pomegranate

Male (Left) and Hermaphrodite (Right) Flower

Fruit Growth and Bagging in Pomegranate

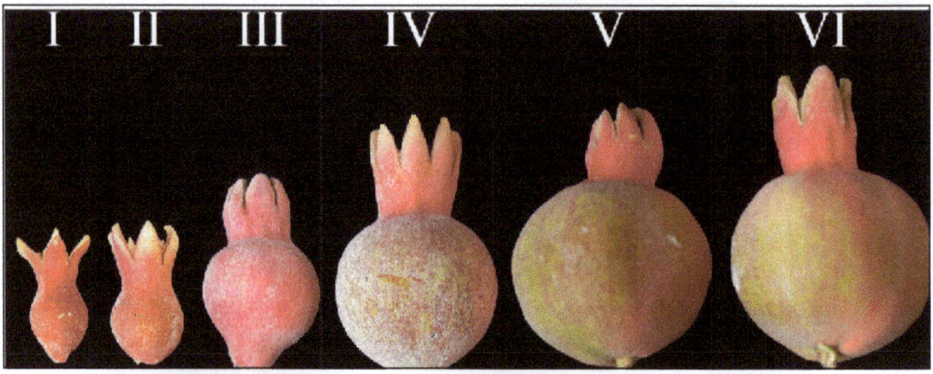

10.5.1 Effect of Growth Regulators

The results on application of ethrel, GA, NAA, MH, carbaryl and manual removal of flowers on fruit drop, fruit number and yield in pomegranate cv. Ganesh were repor-ted by Desai *et al.* (1993). Trees treated with NAA (250 ppm) and carbaryl (0.7 per cent) produced a higher number of large fruits (>200 g), and the highest net return. Manual removal of flowers and treatment with GA (20 ppm) produced the highest number of large fruits and highest mean weight (224.2 g). Spraying of 2,4-D at 20 ppm on cv. Ganesh at flowering and again 45 and 90 days later increased fruit length, girth, weight and volume compared with water-sprayed controls. This treatment also greatly increased aril weight and fruit juice content and resulted in the highest fruit yield (22.52 fruits/tree, compared with 9.21 fruits/tree in control) (Venkatesan and Mohideen, 1994).

Effect of glutathione applied at 25, 50 or 75 ppm to Manfalouty pomegranate trees 3 weeks after anthesis or 4 weeks later were reported by Mostafa (1998). Glutathione application had significant positive effects on fruit weight and induced earlier maturation. Moreover, all treatments significantly increased both total soluble solids (TSS) and TSS:acid ratio and decreased total acidity in fruit juice. The treatments also considerably decreased the percentage of fruit cracking. The best overall treatment was 50 ppm glutathione applied on the earlier date.

10.5.2 Effect of Bagging

Hussein *et al.* (1994) reported that bagging of pomegranate fruits reduced the percen-tage of fruitfly (*Virachola livia*) and significantly increased fruit yield per tree, individual fruit weight and size, TSS content, total sugars content and TSS:acid ratio. Juice acidity and tannin content were reduced by bagging.

Early bagging (one week after fruit set) has better results than late bagging (one month after fruit set).

A bagging experiment was also conducted on cultivars Tiepiqing and Hongpitian with a view to improving the appearance by reducing rust patches. Different kinds of bag were used. Bagging clearly improved rind colour and the brown patch area on the rind of cv. Tiepiqing was reduced by 10-58.5 per cent with different kinds of bag and fruit cracking was reduced by 2.0-3.0 per cent (Zhu *et al.*, 1998).

11.0 Pests, Diseases and Disorders

11.1.0 Pests

Pomegranate is attacked by a large number of pests, fruit being more vulnerable than any other parts and those which cause considerable damage are discussed below.

11.1.1 Pomegranate Butterfly or Fruit Borer (*Virachola isocrates; V. livia*)

It is the most important pest, widely distributed all over India and is found wherever pomegranates are grown. The conspicuous symptoms of damage are offensive smell and excreta of caterpillars coming out of the entry holes. The excreta is found stuck around the holes. The affected fruits ultimately fall down and are of no use, even if the fruits are picked before falling down. The extent of damage may be as much as to 50-90 per cent.

Application of metacid 50 EC 1 ml/l or carbaryl 0.2 per cent (sevin 50 per cent WP 4 g/l) or phos-phamidon 0.03 per cent (dimecron 100 EC, 0.3 ml/l of water) at fortnightly interval is effective. Three alternate sprays of phosphamidon and dimethoate at 0.05 per cent at monthly interval were recommended by Shukla and Prasad (1983). Four synthetic pyrethroids were tested by Kakar *et al.* (1987) for the control of the lycaenids *V. isocrates* and *Deudorix epijarbas* on pomegranates. Foliar applications of cypeme-thrin and permethrin (at 150 g a.i./ha), fenvalerate (at 50 g a.i./ha) and deltamethrin (at 7.5 g a.i./ha) effectively controlled both species. Dubey *et al.* (1993) recommended two sprays of 0.14 per cent endosulfan to control fruit borer.

The effectiveness of *Trichogramma chilonis* in the control of *Virachola livia* on pomegranate was studied by Thakur *et al.* (1993). Following 5 successive releases of the parasitoid in the test sites, infestation by *V. livia* was significantly reduced.

11.1.2 Bark Eating Caterpillar (*Inderbela tetraonis; I. quadrinotata*)

This insect bores the bark and feed inside old trees and those not maintained well are more prone to the attack of this pest. Usually, there is only one caterpillar in each hole but there may be 10-12 holes in a badly infected tree and such trees will bear no fruit.

In field studies, out of 10 insecticides tested (along with petrol and kerosene), fenvalerate at 0.01 per cent and quinalphos at 0.05 per cent injected into larval holes were most effective in controlling the bark eating caterpiller (*Indarbela quadrinotata*) (Mole and Tambe, 1990). Shevale (1991) found that dichlorvos at 0.08 per cent was the most effective, closely followed by monocrotophos at 0.08 per cent, fenvalerate at 0.04 per cent and carbaryl at 1 per cent in controlling bark eating caterpillar.

Keeping the orchard clean and avoiding overcrowding of trees help to minimize the attack by this pest.

11.1.3 Stem Borer (*Olemecampetus bilobus*)

Another common insect which is found in neglected orchards. The caterpillar makes a hole and bores through the branches. Usually, the main trunk or main branches are affected. The presence of the pest is noticed through the hole on the bark and a network of dried excreta is seen outside the hole. It comes out at night and feeds on bark. The measures suggested for controlling the bark eating caterpillars are also effective in controlling the beetles as well.

11.1.4 Leaf Footed Plant Bug (*Leptoglossus membranaceus*)

The adult bugs are observed to be primarily responsible for causing damage to the fruits. With the strong probosis they pierce the rind of the ripening fruits and eventually the infested fruits drop down prematurely (Jadhav *et al.*, 1976). It can becontrolled by spraying 0.1 per cent malathion.

Pests of Pomegranate

Pomegranate Butterfly

Fruit Borer Infestation

Stem Borer Adult (Left) and Infestation (Right)

Pests of Pomegranate

Leaf Footed Plant Bug Infestation

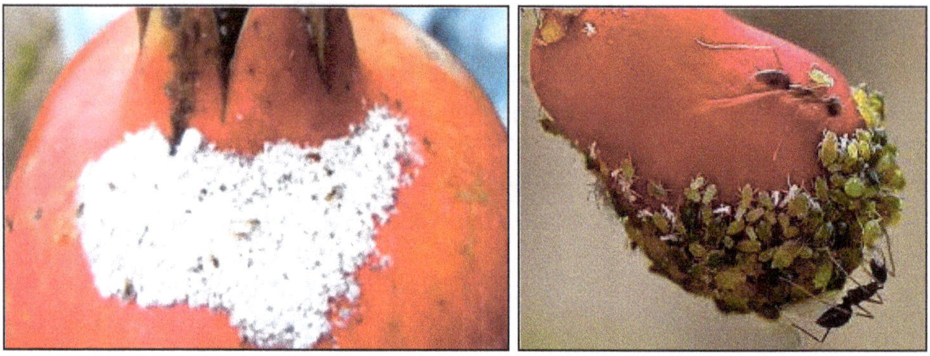

Mealy Bug (Left) and Aphid (Right) Infestation

Thrips Infestation

11.1.5 Sap Sucking Insects

Mealy bugs, scale insects, white flies, thrips, aphids, mites are some of the insects reported from various parts of India on pomegranate leaves. Due to the damage by these insects, the trees are devitalized and shed off buds and flowers and small fruits. To prevent the spread of mealy bugs and scale insects the affected parts should be pruned and destroyed in the initial stage of attack. If the infestation becomes severe, spraying with 0.08 per cent diazinon, parathion or 0.1 per cent malathion has been suggested (Jadhav and Ajri, 1985). White flies and aphids can be controlled by 0.03 per cent dimethoate or phosphamidon. Mites are controlled by 0.05 per cent kelthane, 0.05 per cent aromite or 0.03 per cent dicofol.

Non-toxic dormant oil sprays at 3 and 2 per cent effectively reduced aphid and mite popu-lation and also increased the yield (Markarimi, 1999).

11.1.6 Bag Worm (*Clania cramen*)

The caterpillers construct a case over their body and live within nibbling leaf lamina. To check the damage caused by the caterpillars, they should be collected and destroyed mechanically. In case of severe infestation, 0.05 per cent endosulfan is effective.

11.1.7 Carob Moth (*Ectomyelois ceratoniae*)

The control of carob moth on pomegranates was studied in Turkey by Mart and Kilincer (1994) using *Bacillus thuringiensis* and chemical and mechanical methods. After egg hatching application of *B. thuringiensis* (as Thuricide and Biobit), cyhalothrin and mechanical control resulted in 64.8, 67.2, 69.5 and 67.6 per cent mortality, respectively.

11.1.8 Nematode

Twelve varieties of pomegranate are reported as being susceptible to *Meloidogyne incognita* (Verma *et al.*, 1986).

A survey of 17 pomegranate nurseries by Siddqui and Khan (1986) revealed the presence of 12 genera of plant parasitic nematodes. *Meloidogyne incognita* and *M. javanica* were the most widespread and prevalent. Fenamiphos, oxamyl and aldicarb at 20, 10 or 5 g a.i./plant, were evaluated for control of plant parasitic nematodes. Fenamiphos at 20 g a.i./plant was the best treatment for controlling *Meloidogyne* spp.

11.2.0 Diseases

Pomegranate suffers from the die-back of twigs caused by *Pleuroplaconema punicae*; discolouration of fruits and seeds by *Aspergillus castaneus*, dry rot by *Phomopsis* sp. and leaf and fruit spot by *Cercospora* and *Gloeosporium* spp. Beside these, canker and die-back caused by *Ceuthospora phyllosticta* and leaf spot by *Pestalotia* sp. have also been reported.

Sherkar (1979) undertook investigation on important fungal diseases of pomegranate in Maharashtra and found that pomegranate orchards were severely attacked by large number of fungi causing fruit spot, fruit rot and leaf spot.

11.2.1 Fruit Spot

This is a fungal disease caused by *Drechslera rostrata* (Lande, 1975). The affected fruits develop small irregular spots surrounded by greenish yellow border making the appearance of fruits less attractive. In severe infestation (after formation of blotch) the discolouration of the surface slowly extends to inner tissues and even up to seeds, causing the aril brownish in colour. Such fruits are not edible.

The occurrence of this disease could be prevented by pruning the dead and affected twigs or branches. Spraying with dithane M-45 or captan 500 g in 200 l of water after fruit formation is recommended.

11.2.2 Fruit Rot

It is a fungal disease caused by *Glomerella cingulata* (Singh and Chohan, 1972) and is characterized by discolouration of fruits starting from lower part or sometimes from the sides. After few days, discoloured area becomes brown to black. The infected area enlarges and mostly half or sometimes the whole fruit shows rotting within a week.

Another fruit rot disease caused by *Phomopsis* sp. occurs mostly in the rainy season. When the flowers are affected by this disease, they fail to set fruit, while the young fruits may drop. Its presence is detected by the appearance of yellow to black spots all over the fruit, particularly, concentrated at the point of attachment to the twig.

Fruit rot disease caused by *Aspergillus foetidus* has also been reported by Dutta and Ghosh (1963).

The occurrence of the fruit rot disease could be prevented by pruning of the affected twigs and fruits and burning them. Sherkar (1979) reported that benlate, bavistin (0.5 per cent), dithane M-45 or dithane Z-78 (0.25 per cent) should be sprayed and three or more sprays are required for effective control.

11.2.3 Leaf Spot

The disease is caused either by fungus or by bacteria and is incited by *Xanthomonas punicae* and various fungi like *Pseudocercospora punicae, Curvularia lunata, Colletotrichum gloeosporiodies, etc.* At the early stage, black coloured indefinite spots appear on the upper surface of the leaves which extend rapidly if the weather is favourable. Affected leaves fall off. Blackish or brownish irregular spots appear on the fruit. Spraying with 0.2 per cent dithane M-45, dithane Z-78 or captan 0.1 per cent at aninterval of 15 days has been suggested.

Bacterial leaf and fruit spot caused by *Xanthomonas campestris* cv. *punicae*, was effectively controlled by paushamycin at 500 ppm + copper oxychloride at 0.2 per cent, applied 3 times at fortnightly intervals (Suriachandraselvan *et al.*, 1993).

11.2.4 Post-harvest Diseases

Considerable damage is caused by the rotting of fruits due to infection with various fungi in storage and markets in different parts of India. Kanwar and

Pests of Pomegranate

Whitefly Infestation

Spider Mite Infestation

Carob Moth (Left) and Nematode (Right) Infestation

Diseases of Pomegranate

Leaf (Left), Flower (Middle) and Fruit (Right) Spot Infection

Cercospora leaf spot infection

Fruit Rot Infection

Thakur (1972) reported the following post-harvest fungal diseases and their causal organisms:

Disease	*Casual organism*
1. Aspergillus rot	*Aspergillus nidulans*
2. Internal rot	*Aspergillus clavatus*
3. Soft rot	*Rhizopus arrhizus* and *R. stolonifer*
4. Spicaria rot	*Spicaria* spp.
5. Black mould	*Aspergillus niger*
6. Botryodiplodia rot	*Botryodiplodia theobromae*
7. Penicillium rot	*Penicillium expansum*

All the above fungi proved to be essentially wound parasites except *R. arrhizus, R. stolonifer, P. expansum* and *A. niger* which were able to initiate infection on both injured and uninjured fruits. Application of 2,4-D at the rate of 500 ppm as preinfection dip and spraying of sulphur compounds on packing straw for control of soft rot during storage has been suggested (Kanwar and Thakur, 1973).

11.3.0 Disorders

11.3.1 Fruit Cracking/or Splitting

Cracking of fruits is a serious problem in pomegranate which occurs more frequently in dry atmosphere of the arid regions. The cracked fruits though sweeter, lose keeping quality and are unfit for shipment and liable to rot (Symconides, 1930). The cracking of pomegranate is believed to occur due to sudden change in soil moisture content (Symconides, 1930; Cheema *et al.*, 1954). At the time of fruit ripening, if the soil becomes too dry and then irrigated heavily, cracking may result (Hayes, 1953). According to Pant (1976), increase in air temperature is the cause of fruit cracking.

It has also been assumed that fruit cracking is related to the deficiencies of calcium and boron. This malady is also reported as a varietal character in pomegranate (Palmenoe, 1972) and it was least in cvs. Karkai, Guleshah and Bedana. Vashistha (1982) reported that percentage of unmarketable cracked fruits was more in Jodhpuri compared with Saharanpuri. Godara (1982) reported that out of 41 strains, P.S. 75-K-3 was tolerant to fruit cracking and cv. Appuli was found resistant to cracking (Purohit, 1983). The extent of cracking in different cultivars, however, varies. Rind thickness and texture of fruits in various cultivars seem to influence the intensity of cracking (Pareek, 1981).

Adequate and regular irrigation and interculture throughout the bearing period may reduce cracking. Bacha and Ibrahim (1980) reported that application of 5 per cent pinolene as vapour guard (an antitranspirant) 4-5 weeks before harvest reduced fruit splitting. This chemical had no appreciable effect on yield or chemical composition of fruit. Spraying of GA_3 at 120 ppm was found to reduce the fruit cracking in pomegranate cv. Mahabad (Sepahi, 1986).

Singh *et al.* (1993) reported the effect of different treatments on fruit cracking in pomegranate. With irrigation treatment, minimum fruit cracking (14 per cent) was obtained when irrigated at weekly intervals. Mulching reduced fruit cracking, with dried grass or farmyard manure being the most effective. Dried grass resulted in the highest yield. Of the spraying treatments, 0.005 or 0.002 per cent boric acid, 1 per cent KNO_3 or 1 per cent $MgSO_4$ were the most effective in reducing fruit cracking and increasing yield.

11.3.2 Pomegranate Decline

Pomegranate decline was observed in Iran by Banihashemi (1998) in newly establi-shed 3- to 7-year-old orchards. The first disease symptoms (yellowing, wilting and death of some branches) usually appeared in the early summer. He concluded that a sudden drop in temperature in late autumn and early winter in the milder areas where the trees are not fully dormant, are the main contributing factors to decline.

12.0 Harvesting and Yield

12.1.0 Harvesting

The fruits are ready for harvest 5-7 months after the appearance of blossoms. They are harvested when the skin turns slightly yellow and the fruit gives a metallic sound when tapped. The ratio of soluble solids to total acid was considered by Moeyked (1971) and Tressler and Jaslyn (1971) to be the best chemical criterion of maturity for harvesting. Fruits of cv. Wonderful reached maturity for commercial harvest when the soluble solids content attained a fairly constant level of 15 per cent (Ben-Arie *et al.*, 1984).

Pomegranate cv. Bejestan fruits were harvested 4 times at weekly intervals. TSS content and titratable acidity were measured before and after storage at 1, 3 or 5°C and 85 per cent RH for 5 months (Sherafatian, 1994). Fruit quality was evaluated by a panel. From the results, the optimum harvesting date was determined when TSS content reached 17.5 per cent.

12.2.0 Yield

The tree starts yielding from the 4th year onwards giving 20-25 fruits per tree. Yield increases progressively and in the 10th year a tree may produce 100-150 fruits per year. The average yield in a well managed orchard may be as much as 200-250 fruits per tree. Economic yield is obtained up to an age of 25-30 years.In order to obtain early high yield in 5 cultivars of pomegranate, Dong *et al.* (1999) suggested planting at 4 m x 2.5 m and use of high rates of fertilizer. Spraying with GA_3 at 80 mg/l or 0.3 per cent borax + 0.2 per cent urea and artificial pollination improved fruit set and yields of 13.66 t/ha were obtained after 5 years.

Haggag *et al.* (1999) reported from Alexandria, Egypt that sprays with dormex (hydrogen cyanamide) (0, 0.5, 1, 2 or 4 per cent) in January or February on cv.

Banati reduced the number of days for the onset of bud break and to 50 per cent bud break. It also increased fruit yield and quality, and accelerated fruit ripening. Yields increased as dormex concentraion increased.

Diseases of Pomegranate

Anthracnose Infection

***Aspergillus niger* Infection**

***Botryodiplodia* Rot Infection**

Disorders of Pomegranate

Fruit Cracking Symptom

Decline Symptoms

13.0 Post-harvest Management

13.1.0 Ripening and Storage

13.1.1 Ripening

Pomegranate fruits are considered as non-climacteric based on the pattern of CO_2 and ethylene production (Elyatem and Kader, 1984). Lee *et al.* (1974a) also reported that ripening of pomegranate showed a non-climacteric pattern of respiration, the respiration rate decreasing continuously between 25 and 125 days after flowering. A pseudo-climacteric pattern of respiration was found in very young immature fruit (Ben-Arie *et al.*, 1984).

During ripening of 6 pomegranate cultivars, Hernandez *et al.* (2000) found that six anthocyanin pigments were responsible for the red colour of juice. In the early fruit-ripening stages, delphinidin 3,5-diglucoside was the main pigment, followed by cyanidin 3,5-diglucoside, while in the later stage, the monoglucoside derivatives cyanidin 3-glucoside and delphinidin 3-glucoside increased considerably. The pelargonidin derivatives were always present in small amounts.

13.1.2 Storage and Value Addition

Pomegranate fruits keep well for a long time. The fruits kept well for 2 months at 0°C, for one month at 4.5°C and for 15 days at room temperature (Heikal *et al.*, 1966). Dhumal (1984) reported that treatment with bavistin (0.2 per cent) and storage in polythene bag (100 guage) prolonged the shelf-life upto 30 days at room temperature. Experiments also indicated that fruit stored at 0o and 4.5°C and at 80-85 per cent relative humidity did not undergo any shrinkage or spoilage for seven months. They may be sprayed with 2 per cent lypol solution and aerated properly before using. Generally, treating the harvested fruits with chemicals for protecting against the fungal rot during storage is not in practice. But if they are dipped in bavistin at 0.05 or 0.1 per cent and topsin 0.1 per cent, the post-harvest fungal rot caused by *A. niger* and *Aspergillus* spp. (green pigment) will not appear up to 10 days of storage (Padule and Keskar, 1988). In bulk storage, the fruit may be packed in layers in wooden crates and stored at 0°C and 80 per cent RH using rice straw or paper as packing material (Mukherjee, 1958).

From the results of the experiment on storage of fruits, Askary and Sahedy (1994) concluded that cv. Malas Torsh had the best storage quality and that a temperature of $1^{\times}$ was best for a long storage period. Sherafatian (1994) concluded that the best storage temperature of cv. Bejestan fruit was 3°C.

Effect of controlled atmosphere (CA) storage of pomegranate (cv. Hicaz) was studied by Kupper *et al.* (1995). Minimal losses of quality and weight were observed in fruits stored in 6.0:3.0 per cent (CO_2:O_2) CA regime for 6 months. Fruits stored in air at 6°C were acceptable after 5 months' storage. Fruits stored in air at 8 or 10°C had a storage life of 50 days whereas fruits stored at these temperatures under CA conditions had a storage life of 130 days.

In the experiment on controlled atmosphere storage of fruits of cv. Mollar, Artes *et al.* (1996) observed that CA storage reduced weight loss and reduced the development and degree of severity of husk scald, especially when the fruits were transferred to warm temperatures, without significant differences relating to the atmosphere used. It was concluded that Mollar pomegranates can be adequately stored for 2 months in atmospheres of 5 per cent O_2 and 0 or 5 per cent CO_2 at 5°C and 95 per cent RH without an ethylene scrubber. Holcroft *et al.* (1998) in the investigation on storage of pomegranate (cv. Wonderful) concluded that moderate CO_2 atmospheres (10 kPa) prolonged storage life and maintained the quality of pomegranates, including adequate colour intensity of the arils.

Xing *et al.* (1997) reviewed the factors influencing development of husk scald of pomegranate fruits during storage. They included storage temperature, relative humidity, atmospheric composition, harvesting time, polyphenol oxidase (catechol oxidase) activity and dihydroxyphenol content in the husk.

Al-Khatani (1992) found that drying at 40°C and 25 per cent RH seriously damaged the fruits and those of De Jativa, Molar, Succary and Taifi-R cultivars softened on the fourth day of drying. Fruits dried at 30°C and 33 per cent RH or 20°C and 47 per cent RH remained acceptable at 20 ± 2°C and 47 per cent RH for up to 3 months or more, depending on the cultivar. Soibam *et al.* (2014) developed cane wine blended with pomegranate juice in order to add colour, nutritive constituents and better acceptability.

14.0 Biotechnology

Advances in biotechnological applications, in particular tissue culture of pomegranate were first reviewed by Omura (1991). Since then, a number of in vitro studies have been carried out on pomegranate such as micropropagation through shoot proliferation by means of meristem culture, somatic embryogenesis, synthetic seed production and utilization, plant regeneration via shoot organogenesis either indirectly through callus cultures or directly by adventitious organogenesis in explants, induction of altered ploidy and genetic transformation. Naik and Chand (2011) wrote a comprehensive review on tissue culture of tissue culture.

14.1.1 Problems of Tissue Culture

14.1.1.1 Browning of Culture

High phenolic exudation owing to activation of explant and media browning has made it hard for workers dealing with pomegranate. Mostly when mature plant parts such as nodal explants, shoot tips, leaf segments are used for culture establishment of pomegranate, the browning is encountered and the elimination of browning is necessary for the establishment of cultures. The methods commonly used to overcome the harmful effect of browning in pomegranate included the use of adsorbing agents such as activated charcoal or polyvinylpyrrolidone (PVP), addition of antioxidants to the culture medium or soaking the explants in antioxidant solution, transfer of explants to fresh medium at frequent intervals or a combination of approaches with varying degree of success (Naik *et al.*, 1999; Naik and Chand, 2011). The intensity of browning was increased with increased position and the length of explants. Minimum browning intensity was observed in 1st nodal explants having 1.5 cm length. However, explants of 3rd node with 2.5

Bearing Orchard of Pomegranate

Packaging, Marketing and Storage of Pomegranate

cm length registered higher establishment (68.5 per cent) and growth of explants. Furthermore, the most effective browning control was observed in subculturing of nodal explants twice, at the first day and third day of inoculation, which also found better in establishment of explants followed by activated charcoal 200 mg/L into the medium. Maximum length of shoots (3.9 cm) was recorded in 1st position of node with 2.5 cm length of explants (Singh and Patel, 2016). Different explants, duration of mercuric chloride treatment and antioxidants were tried for culture initiation. Among the various explants, maximum aseptic culture (55.00 per cent) establishment was obtained with double nodal explants containing third + fourth nodes. The effective surface sterilization with lowest contamination (15 per cent) was achieved with $HgCl_2$ (0.10 per cent) for 3 minute. Lesser intensity of browning and higher culture establishment (54 per cent) was observed with 2000 mg/l activated charcoal (Guranna and Hoolageri, 2017).

14.1.1.2 Recalcitrance

Non responsiveness to tissues in culture is called recalcitrance. Naik and Chand 2003) showed that pomegranate is a difficult species, if not a recalcitrant one, at least for in vitro regeneration through organogenesis. Of the several factors responsible for in vitro recalcitrance of a plant species, production of ethylene in in vitro culture is a key factor. Efforts were made to ameliorate this problem in pomegranate using ethylene inhibitors namely silver nitrate or aminoethoxyvinylglycine (AVG) (Naik and Chand 2003; Kanwar *et al.*, 2010).

14.2.0 Micropropagation

Various factors such as genotype, explant type, season, media and growth regulators have been reported to influence the micropropagation of pomegranate. The most commonly used basal medium for omegranate micropropagation while half strength MS (1/2 MS), woody plant medium (WPM; Lloyd and McCown 1980) and B5 (Gamborg *et al.*, 1968) are useful in some cultivars. 6-Benzylaminopurine (BAP) has been the most commonly used plant growth regulator (PGR) either alone or in combination with a low concentration of cytokinin kinetin (Kin) or auxin, NAA (Naik and Chand, 2011).

Mahishi *et al.* (1991) cultured shoot tip explants on supplemented MS medium and then transferred to Lloyd and McCrown woody plant medium for rapid growth and elongation of shoots, which were rooted on modified MS medium before transfer to potting mixture. An 80 per cent success rate was achieved with the establishment of plant-lets in a 1:1:1 v/v peat, perlite and sand mixture.

Drazeta (1997) cultured apical and axillary vegetative buds as primary explants on MS medium containing different combinations of plant growth regulators. Best production of shoots was achieved on medium containing BAP [benzyladenine] (1 mg/l) and NAA (0.1 mg/l), but shoots exhibited vitrification, so subsequent cultures were placed on a medium containing BAP at 0.5 mg/l and NAA at 0.1 mg/l. Best growth and shoot multiplication were observed on media containing relatively low concentrations of NAA (0.1 mg/l) and BAP at <0.5 mg/l. Shoots were rooted on media containing IBA (0, 0.1, 0.5, 1 or and 2 mg/l). Rooted plants were transferred to a soil substrate in the greenhouse. After 4 weeks plants had hardened off. The highest degree of hardening off was exhibited by shoots rooted on a medium containing IBA at 0.1 mg/l.

Yang and Ludders (1993) induced adventitious shoots from callus from leaf and stem explants of *in vitro*-cultured shoots of dwarf pomegranate cv. Nana (grown as an ornamental as well as for its fruits) after 2 months of culture. Leaf segment and stem explants were initially cultured on modified MS basal medium supplemented with the following growth regulators: (i) BA, zeatin, kinetin or 2iP at 0.1-1.5 mg/l; (ii) IAA, IBA or NAA at 0.1-1.0 mg/l plus 0.5 mg BA/l. Adventitious shoot elongation was stimulated most effectively when the initial calluses were transferred from the shoot induction medium to MS basal medium supplemented with 0.5 mg BA/l and 0.1 mg IBA/l. Elongated shoots rooted easily on half-strength MS medium. Rooted shoots were transplanted to soil and kept in a gtreenhouse under mist conditions. Such plants began to flower within 2-3 months, indicating that *in vitro* culture of leaf and stem explants is a potential method for mass propagation of dwarf pomegranate.

Suspension cultures were produced from calluses derived from leaf explants and cultured at a range of growth regulator concentrations. Addition of 5 µM BA and 1µ M NAA stimulated proliferation of suspension culture. After plating on 0.6 per cent agarose medium colonies formed most frequently on medium containing 2 µM BA and 1 µM NAA. Adjustment of medium pH from 5.8 to 4.8 improved cell proliferation and colony formation. No somaclonal variation was detected in 136 regenerated plant-lets (Omura *et al.,* 1990).

Naik *et al.* (1999) reported a rapid and efficient procedure for *in vitro* clonal propagation of an elite cultivar of pomegranate (cv. Ganesh) using nodal stem segments of a mature tree. Bud break and axillary shoot proliferation was induced in the nodal segments on MS medium supplemented with benzyladenine (BA), zeatin riboside (ZR) or thidiazuron (TDZ), with ZR being most effective. The greatest number of shoots was developed on a medium containing 2.0 mg/1 ZR. However, the shoots failed to elongate. This problem was overcome by transferring the shoot clumps to a medium with a lower level of ZR. TDZ was the least effective of all the cytokinins tested. Rooting was induced in 86 per cent of the regenerated shoots in a half-strength MS medium supplemented with 1.0 mg/1 IBA. Upon transfer of the rooted shoots to an auxin-free half-strength MS medium the primary roots elongated further and additionally new roots developed. The plantlets were acclimatized and established in soil.

A complete protocol for *in vitro* regeneration of pomegranate, using cotyledonary nodes derived from axenic seedlings was reported by Naik *et al.* (2000). Shoot development was induced from cotyledonary nodes on MS medium supplemented with 2.3-23.0 µM BA or kinetin. The maximum number of shoots (9.8 shoots/explant) was developed on a medium containing 9.0 µM BA. Shoots formed *in vitro* were rooted on half-strength MS supplemented with 0.054-5.4 µM NAA. Zeatin riboside (ZR) was reported to induce bud break and axillary shoot proliferation in mature nodal segments of an elite cultivar Ganesh (Naik *et al.,*1999).

In-vitro Propagation of Pomegranate

Propagated from nodal segment. a shoot development from nodal segments of pomegranate cultivar 'Malas Saveh' after 10 day of culture on WPM medium containing 4.7 μM Kinetin + 0.54 μM NAA, b development of multiple shoots after 20 day of culture on the same medium, c elongation of shoots after 30 day of culture on the same medium, d root development in shoot cultured on half strength WPM medium containing 5.4 μM NAA after 3 weeks of culture. e a well-rooted plantlet ready for acclimatization. f a healthy and well-hardened plant.

(ValizadehKaji, B., Ershadi, A. and Tohidfar, M. (2013) *Physiology and Molecular Biology of Plants*, DOI:10.1007/s12298-013-0193-3)

In-vitro Propagation of Pomegranate

Propagated by anther culture (Naik, S.K. and Chand, P.K. (2011) *Plant Cell Reports* 30: 701–721.

An optimum culture medium for micropropagation of pomegranate, cvs. 'Malase Yazdi' and 'Shirine Shahvar' was reported and protocols have been developed for in vitro shoot proliferation and rooting of the cultivars (Mulaei *et al.*, 2019). Axillary buds of young shoots were used as explants. The shoots were multiplied in vitro on Murashige and Skoog (MS) supplemented with 0, 1.5, 2, and 2.5 $mg.l^{-1}$ of N6-benzyladenine (BA). The micropropagated/cloned shoots were rooted on MS and half strength (1/2) of MS containing indole-3-butyric acid (IBA) at three concentrations (0, 0.5, and 1 $mg.l^{-1}$) and 0.5 $mg.l^{-1}$ naphthalene acetic acid (NAA). The cultivar 'Malase Yazdi' multiplied better than 'Shirine Shahvar'. In both cultivars, the maximum shoot number with maximum shoot height was produced on medium with 2 $mg.l^{-1}$ BA. Control medium free from plant growth regulator also supported shoot multiplication. The cultured shoots exhibited the greatest number of leaves on a medium containing 2.5 $mg.l^{-1}$ BA. The cloned shoots of 'Malase Yazdi' rooted the best on half-strength MS medium with 0.5 $mg.l^{-1}$ IBA while shoots of 'Shirine Shahvar', rooted on 1 $mg.l^{-1}$ IBA and 0.5 $mg.l^{-1}$ NAA in half MS medium (Mulaei *et al.*, 2019). Protocols have been reported by other workers (Guranna *et al.*, 2017; Bachake *et al.*, 2019; Kumar *et al.*, 2019).

14.3.0 Somatic Embryogenesis

Somatic embryogenesis has enormous potential for clonal multiplication, synthetic seed production and genetic transformation. To date, both direct and indirect somatic embryogenesis and plant regeneration of pomegranate has been reported from five different laboratories using different explants (Naik and Chand, 2011). Somatic embryogenesis has been reported in seed explant and petal tissue (Raj Bhansali, 1990; Nataraja and Neelambica, 1996). Embryogenic cell lines can be selected by using staining techniques. Embryos produced roots after 8 weeks on MS medium containing 5 mg/l IAA, IBA or NAA. However, shoots were developed when embryogenic cultures were placed on medium having 5 mg/l BAP or Kin. Embryos rapidly proliferated into secondary embryos at 5 mg/l and IAA (Raj Bhansali, 1990). Complete plantlets were obtained on half-strength MS medium containing 1 mg/l IAA or IBA. A 6-week dark incubation period was essential for embryogenic callus induction (Kanwar *et al.*, 2010).Nataraja and Neelambika (1996) induced somatic embryos from petal explants on MS medium supplemented with various growth regulators individually or in different combinations. Numerous embryoids formed from friable callus on MS medium with 5 mg IAA, IBA or NAA/l; after 8 weeks they had only developed roots, but those formed on MS medium with 5 mg BAP [benzyladenine] or kinetin/l developed shoots. Explants cultured on MS medium supplemented with 5 mg IAA and 5 mg BAP/l differentiated numerous embryoids but they developed multilobed or fused cotyledons. However, callus induced on MS medium with 1 mg IAA or 1 mg IBA/l on subculturing in the same medium with half-strength salts and sucrose (4 per cent) produced both roots and shoots, resulting in plantlets.

A high frequency of shoot organogenesis was achieved on MS basal medium supplemented with 1 mg/l BA. In nodal cuttings, the best adventitious bud formation was observed on 1/2 MS medium upplemented with 0.5 mg BA + 0.1 mg

NAA/l. This medium also promoted shoot proliferation. NAA was more effective than BA in shoot elongation (Naik and Chand, 2011).

Somatic embryos are considered as the choicest propagules for preparation of synthetic seeds. However, improper maturation and/or germination of somatic embryos have been one of the major bottlenecks for the utilization of synthetic seeds. This has underpinned a new perspective in synthetic seed technology: the use of non embryogenic unipolar meristematic plant propagules, such as shoot apices, axillary buds or nodes (Naik and Chand, 2011). Considering the inadequate information or inordinately low embryo to plant conversion rate in pomegranate, Naik and Chand (2006) used in vitro nodal segment for production of synthetic seeds of pomegranate. They prepared synthetic seeds by encapsulating nodal segments from in vitro shoot cultures derived from mature nodal explants (source A) or axenic cotyledonary nodes (source B) in calcium alginate hydrogel containing MS medium supplemented with 4.4 lM BAP and 0.54 lM NAA. A combination of 3 per cent sodium alginate and 100 mM calcium chloride was most suitable for formation of ideal synthetic seeds. Sprouting (shoot development) of encapsulated nodal segments of both the sources was observed, but one step germination, *i.e.* both shoot and root formation was achieved, only with encapsulated nodal segments of source B.The encapsulated nodal segments could be stored up to 30 days at 4 °C without marked loss of sprouting ability. Since a desirable feature of synthetic seeds is their ability to retain viability in terms of sprouting/germination even after a period of storage, the synthetic seed system developed by Naik and Chand (2006) could facilitate transfer of sterile and elite materials in the form of synthetic seeds to extension centers and laboratories of distance places while maintaining their viability (Naik and Chand, 2011).

Development of an efficient method to regenerate plants adventitiously in a reasonably high frequency directly from explants is a desirable requirement for high-throughput genetic transformation (Naik and Chand 2003). To date, direct shoot organogenesis in pomegranate has been rarely reported (Naik and Chand, 2011). Jaidka and Mehra (1986) were the first to report direct regeneration (without intervention of callus) of shoots and whole plants from cotyledon, leaf and stem explants of pomegranate cv. Kandhari anar. The regeneration of shoots and whole plants was achieved on MS or Nitsch's (1969) medium supplemented with 2.0 mg/l NAA and 2.0 mg/l BAP. Later a reproducible protocol was developed by Naik and Chand (2003) for adventitious shoot organogenesis followed by complete plant regeneration from the hypocotyl and cotyledon explants of pomegranate (cv. Ganesh).

Indirect organogenesis involving induction of callus with exogenous application of plant growth regulators was reported by Guranna *et al.* (2017). They found nodal segment was superior for induction of callus when cultured on MS basal medium consisting of BAP 5 mg/l + NAA 0.4 mg/l. Early shoot initiation, more number of shoots per explant and maximum shoot length was noticed when proliferated calli were cultured on MS basal medium containing BAP 2 mg/l + NAA 0.1 mg/l + GA_3 0.5 mg/l. Early *in vitro* root initiation, highest per cent rooting and maximum number of roots per plantlet were recorded when microshoots were cultured on full strength

MS medium supplemented with IBA 3 mg/l.Successful acclimatization of the in vitroregenerated plantlets has been reported in pomegranate using different potting substrates with varying degree of success (Naik and Chand,2011). Vermiculite was used by Omura *et al.* (1987) for acclimatization of normal endemic and dwarf pomegranate plantlets in the green house. Sharon and Sinha (2000) reported the establishment of plantlets in the pots filled with either vermiculite or a 1:1 mixture of vermiculite and soil in the net house with 100 per cent survivability of plantlets in both the potting medium when 80 per cent RH, 14–18 klux light intensity and temperature below 30°C were constantly maintained inside the net house (Naik and Chand, 2011). A sterile mixture of peat: soilrite (1:1) moistened with 1/4 MS macro and micro-salts without organics facilitated a maximum survival (78.7 per cent) of plantlets (Singh and Khawale, 2006). The survival of plantlets was 94.12 per cent in a potting mix of sterilized loamy soil and sand (1:1) of which 67.85 per cent survived upon subsequent transfer to loamy soil (Kanwar *et al.*, 2010). Vermicompost was used by Naik *et al.* (1999, 2000) and Naik and Chand (2003) for acclimatization of plants (cv. Ganesh) before establishment in soil. 60–68 per cent of plantletssurvived following transfer to vermi-compost and 75–84 per cent of these plants transferred to soil survived. There was 80 per cent success in establishment of the plantlets in perlite under mist in green house (Al-Wasel, 1999).

14.4.0 Genetic Transformation

There are very few reports available on genetic transformation in pomegranate. The effects of different sterilization time and different growth regulators on the in vitro leaf and stem segment regeneration were studied, and the advantages of two regeneration systems were compared (Chen *et al.*, 2012). The results showed that it was easier for stem segments to be sterilized than for the leaf, but the green and compact callus from leaves was more likely to be obtained and then more adventitious buds differentiated. The callus from stem segments was crisp, and the adventitious buds were less. The adventitious buds from the stem segment callus grew more quickly, but the time consuming of leaf regeneration system was shorter. The adventitious buds could be induced from a few emerald green leaves, which could be used for transformation, but the adventitious buds could not be induced directly from stem segments. There was not any significant difference between stem and leaf explants in callus sensitivity to antibiotics.The leaf callus regeneration system might be more appropriate for the genetic transformation (Chen *et al.*, 2012).

The first report on genetic transformation was reported by Terakami *et al.* (2007). The dwarf pomegranate (*Punica granatum* L. var. *nana*) is a dwarf ornamental plant that has the potential to be the model plant of perennial fruit trees because it bears fruits within 1 year of seedling. They established an *Agrobacterium*-mediated transformation system for the dwarf pomegranate. Adventitious shoots regenerated from leaf segments were inoculated with *A. tumefaciens* strain EHA105 harboring the binary vector pBin19-*sgfp*, which contains neomycin phosphotransferase (*npt II*) and green fluorescent protein (*gfp*) gene as a selectable and visual marker, respectively (Terakami *et al.*, 2007). After co-cultivation, the inoculated adventitious shoots were cut into small pieces to induce regeneration, and then selected on MS medium supplemented with 0.5 μM α-naphthaleneacetic acid (NAA), 5 μM N^6-benzyladenine

(BA), 0.3 per cent gellan gum, 50 mg/l kanamycin, and 10 mg/l meropenem. Putative transformed shoots were regenerated after 6–8 months of selection. PCR and PCR-Southern blot analysis revealed the integration of the transgene into the plant genome. Transformants bloomed and bore fruits within 3 months of being potted, and the inheritance of the transgene was confirmed in T_1 generations. The advantage of the transformation of dwarf pomegranate was shown to be the high transformation rate. The establishment of this transformation system is invaluable for investigating fruit-tree-specific phenomena (Terakami *et al.*, 2007).

Kanwar *et al.* (2008) also transformed wild pomegranate using Agrobacterium strain LBA4404 harboring pBI121 carrying β-glucuronidase (gus) and nptII genes. Integration of the transgene (*Npt II*) was confirmed in four out of five putatively transgenic plants by using PCR analysis.

An efficient Agrobacterium-mediated transformation method was developed for pomegranate (*Punica granatum* L.), a difficult-to-transform plant (Kaji *et al.*, 2014). In vitro shoot segments were inoculated with Agrobacterium tumefaciens strain LBA4404 harboring the binary vector pBI121 carrying the neomycin phosphotransferase (nptII) gene as a selectable marker and β- glucuronidase (gus) gene as a reporter. After 28 d in WPM selection medium containing 50 mg L-1 kanamycin, 59 new shoots proliferated. gus analysis was performed on these putative transgenic shoots, of which 32 stained positive. Positive staining shoots were cut and cultured in selection medium for 2 subsequent subcultures until final Gus analysis. After three months of the selection period, 6 putative transgenic shoots were obtained. Presence of the Gus and nptII genes was confirmed by polymerase chain reaction. Southern blot analysis confirmed that TDNA was stably integrated into the genome of three out of six PCR-positive plants. The transgenic plants were rooted and successfully acclimatized (Kaji *et al.*, 2014).

In order to transform the tissues, we first optimized callus induction and shoot regeneration efficiency. Induction of calluses was best achieved from the intermodal stem sections on WPM medium supplemented with 12 µM BA and 8 µM NAA. The highest frequency of shoot regeneration (69.33 per cent) and number of shoots (7.16) per piece of callus were obtained when calluses were incubated on WPM medium supplemented with 12 µM BA and 2 µM NAA. Callus pieces were inoculated with *Agrobacteriu tumefaciens* strain *EHA105* harboring the binary vector *pBin19* carrying the neomycin phosphotransferase (*nptII*) gene as a selectable marker and green fluorescent protein (*GFP*) gene as a reporter. After 4 weeks in WPM selection medium (the best callus induction medium supplemented with 50 mg L^{-1}kanamycin), putative transgenic calluses were obtained. From such calluses transformed shoots began to appear approximately 3 week after the kanamycin resistance transgenic callus pieces had been placed on the best shoot regeneration medium containing 50 mg L^{-1} kanamycin. *GFP* fluorescence was observed in putatively transformed calluses and shoots. Molecular analysis confirmed the integration of the *nptII* transgene in transformed shoots. The transgenic shoots were rooted and successfully acclimatized (Kaji and Abbasifar, 2017).

Helaly *et al.* (2014) developed an in vitro regeneration of plantlets of four commercial pomegranate genotypes (Manfalouty, Tahrir, Badr and Araby) grown in Egypt. Shoot tips were selected and separated from 7-year-old mother plants as explants. The MS media supplemented with different dosage of growth substances were examined under controlled conditions. Gene transformation mediated by Agrobacterium tumefaciens was examined for only two genotypes (Manfalouty and Araby) which recorded highest and lowest regeneration capacity, respectively. The results showed that regeneration of pomegranate was mainly controlled by genotype, whereas the response of the explants was largely affected by the environmental factors. The data also show that the embryonic calli were formed within six weeks in the presence of MS media supplemented with 1 mg L-1 of 2,4-D. Shoots emerged successfully from the embryonic callus in the presence of 6 mg L-1 BA. Addition of 1 mg L-1 IBA to the MS basal medium increased the shoot forming, roots ratio (100 per cent) and plantlet growth in all pomegranate genotypes. The highest recorded regeneration frequency was 79.75 per cent in genotype Manfalouty followed by 70 per cent, 63 per cent and 55 per cent in Tahrir, badr and Araby, respectively. The regenerated putative transgenic plantlet inoculated and co-cultivated with Agrobacterium tumefaciens were able to grow vigorously in the media containing 50 mg l-1 Kanamycin with high degree of gene stability. Fresh weight (FW) of the regenerated plantlets was increased, whereas the concentration of H_2O_2 was decreased due to the infection with A. tumefaciens in both genotypes. Similarly, the infected plants showed higher specific activities of CAT and SOD compared to control. The PCR analysis showed the integration of the transgene into the infected plant genome corresponding to the relevant sequence of both of the NPT-11 gene and the 35 S promoters (Helaly *et al.*, 2014).

Transgenic approach for resistance to bacterial blight disease caused by *Xanthomonas axonopodis* pv. *punicae* appears to be promising to minimize the losses caused by disease. Hosamani *et al.* (2016) attempted to develop an efficient protocols were developed to get healthy and well-formed plants from juvenile and mature-origin explants of the pomegranate cv. 'Bhagwa' and transformants with PFLP gene. One of the strategies to lead plants become resistant against bacterial pathogens is employing a transgene, like plant ferredoxin-like protein (PFLP). Different treatment combinations of hormonal concentrations were taken for leaf, petal, nodes and cotyledonary explants to standardize an efficient *in vitro* regeneration protocol and find out the best treatment for faster regeneration. *Agrobacterium tumefaciens* carrying gene pCAMBIA construct with the constitutive CaMV35S promoter, PFLP gene, terminator and *nptII* selectable marker (Kanamycin resistance), was used for transformation of explants. Putative transformants were identified on selection medium containing kanamycin at different concentration. Integration of transgene and expression at various levels were confirmed using PCR. Out of 4 putative transformants analyzed, three plants showed amplification for PFLP gene specific primer (Hosamani *et al.*, 2016).

14.5.0 Molecular Markers

A large number of studies have been reported on molecular markers (Chauhan and Kanwar (2012).

14.5.1 Genetic Diversity and Identification of Cultivars

The level of variation in various fruit traits was described among 14 Jordanian pomegranate landraces and the genetic relatedness was investigated using RAPD and SSR markers (Owais and Abdel-Ghani, 2015). Euclidean distances among studied landraces ranged from 3.33 to 12.01, with a mean of 7.65. Fruit and aril traits explained the variation in the first component (28.92 per cent), while other traits were present in the second (17.615) and third (12.81 per cent) component, and therefore contributed less to the variability. Genetic distances based on RAPD scores ranged from 0.24 to 0.66, indicating that considerable level of divergence exists among studied pomegranate landraces. The set of SSR markers used in this study was monomorphic, which might be due to the fact that available SSR markers are too few to identify polymorphic SSR markers to differentiate between landraces present within a small geographic area. Multivariate analysis showed that dendrograms constructed based on fruit related traits as compared with that based on RAPD scores were not consistent. Pomegranate landraces displayed high variability in fruit and aril related traits, which could be considered a valuable source of genes for commercial uses. Results revealed the presence of small seeded variety that has large arils (Owais and Abdel-Ghani, 2015).

Similarly, 16 of the 100 primers showed reliable polymorphic patterns in 24 Iranian pomegranate genotypes (Sarkhosh *et al.*, 2006). Interspecific variability of both RAPD marker genotypes and fatty acid composition was detected among pomegranate cultivars from the South Anatolia region of Turkey (Ercisli *et al.*, 2007). Comparisons between the morphological and molecular data suggested that the genetic diversity of these cultivars was accurately reflected by the RAPD markers (Durgac *et al.*, 2008). Sarkhosh *et al.* (2009) studied the associations between 37 fruit morphology traits and DNA markers in 21 soft-seeded accessions from Iran by using 29 random decamer primers. The study suggested that fruit characters and RAPD markers could be used for genetic discrimination of soft seeded accessions of pomegranate. The hybrids were tested for their paternity using 21 RAPD markers of which thirteen primers revealed that 7 hybrids out of 15 hybrids have similar pattern as their parent (Singh *et al.*, 2019).

New microsatellite markers were used to characterize a set of 78 pomegranate accessions from Turkey (Çali°kan *et al.*, 2017). Using six SSR primers, a total of 41 alleles were characterized with an average of 4.6 alleles per locus and mean probability of identity (PI) value of 0.366. These data indicated a high level of polymorphism in pomegranate germplasm. Five synonymous groups could be detected among 30 accessions. This microsatellite-based key is a first step towards a database for marker assisted identification of pomegranate accessions in Turkey. The present study provides essential information to devise a pomegranate core germplasm collection without duplication of plant material, to sustainably manage pomegranate breeding programs, and to establish conservation strategies for preserving local pomegranate genetic resources (Çali°kan *et al.*, 2017). Fluorescent simple sequence repeats (SSR) based on DNA sequencing is a powerful and automated platform for the detection of genetic polymorphism. Luo *et al.* (2018) reported for the first time, 13 SSRs using a fluorescent SSR panel according to the reference genome information, and detected the genetic diversity and population structure of 136 pomegranate varieties. The average number of alleles per locus was 6.31, gene diversity ranged from 0.16 to 0.37 with an average of 0.28, and polymorphic information content (PIC) varied from 0.14 to 0.29 with a mean value of 0.22. The mean genetic distance (GD) between varieties was 0.32. Population structure analysis divided the varieties into three different subpopulations (Q1, Q2, and Q3). The cultivars in Q1 all came from China, and almost all cultivars came from foreign countries combined with Chinese were grouped into Q2. The only accession originated from Myanmar was grouped to Q3. The results indicated that 13 SSR markers were polymorphic. The varieties originating from the same geographical region were more closely related, and extensive gene flow had taken place between the varieties from different regions. The results may be useful for the efficient selection breeding and variety protection (Lou *et al.*, 2018).

Combining different marker system are greatly better and more effective considered an important priority step toward diversity study and characterization (Hamdy and El Zayat, 2017). Based on the study they can point out that, the two-widespread varieties in Egypt Manfalouti, and Nab El-Gamal, formed a high homogeneity in genetic similarity with the Wonderful, as an essential global species for breeding programs, efforts widening the genetic base of pomegranates and the introduction of new genotypes in Egypt (Hamdy and El Zayat, 2017).

14.5.2 Expressed Sequence Tags (EST)

The cultivation of soft seeded pomegranate is an important direction in pomegranate breeding. To comprehensively understand the molecular mechanisms involved in the formation of soft seeded pomegranate (*Punica granatum* var. Hongmanaozi), Zhang *et al.* (2017) established an expressed sequence tag (EST) resource. Two thousand valid sequences were generated, from which 907 unigenes were obtained after initial assembly using the CLUSTALX program. Among these unigenes, 51 showed no similarity to any protein in the public databases, 433 matched with proteins of unknown function, and 423 matched with proteins of known or putative functions. The 423 unigenes were further classified into 13 categories. Among these categories, protein synthesis, cell structure, protein destination and storage, secondary metabolism, signal transduction and transporters accounted for 8 per cent, 8 per cent, 4 per cent, 7 per cent, 6 per cent and 17 per cent, respectively. We also successfully developed 10 highly polymorphic expressed sequence tag simple sequence repeat (EST SSR) markers for pomegranate. The results provide a new tool for future activities in pomegranate breeding (Zhang *et al.*, 2017).

To understand the seed softness character in pomegranate (*Punica granatum* L.) fruit, Zarei *et al.* (2016) have studied gene expression associated to cell wall formation in the seed of soft- and hard-seeded pomegranate genotypes at five different stages from fruit set to ripening. For this purpose, some cell wall formation related genes were isolated and identified from pomegranate and their expression patterns were compared in two soft- and hard-seeded genotypes. They reported that for some of the cell wall biosynthesis genes the expression levels increased at stages where the seed cover was becoming hard, then decreasing gradually during fruit ripening. CCR and CAD, two specific monolignol biosynthesis genes, showed higher expression in the hard-seeded genotype during the critical stages of seed hardening, but COMT, a specific S-unit monolignol, showed a completely different pattern with higher expression in the soft-seeded genotype. Cellulose synthesis -related genes (CelSy and SuSy) showed higher expression in the soft-seeded genotype at critical stages of seed development. However, EGase had higher expression at early stages and rapidly decreased to the undetectable level during fruit development, and CHS, another phenylpropanoid gene, showed no significant variation over time. The differences between seed hardness was detectable at 60 DAF in the soft- and hard-seeded genotypes. The lignin and cellulose content increased in the seeds of both genotypes during fruit development. The soft-seeded pomegranate had lower lignin content, while had relatively higher amount of cellulose compared to the hard-seeded one. These results suggest that lignin and cellulose have complementary roles in the cell wall structure of pomegranate seed. Taken together, the data suggest that the difference between the ratio of S to G monolignols could be the main cause of the seed softness in pomegranate (Zarei *et al.*, 2016).

14.5.3 Genomics

SSR markers have been applied to cultivar identification, genetic diversity assessment, genetic mapping, and molecular evolution in plants. In spite of the recent progress in the development of single nucleotide polymorphisms (SNPs) for pomegranate (Ophir *et al.*, 2014; Qin *et al.*, 2017), SSR markers remain useful for practical breeding applications due to their reliability, high polymorphism, and ease of use (Liu *et al.*, 2020) Up to now, the SSRs of pomegranate were mainly developed from enriched genomic libraries (Curro *et al.*, 2010; Soriano *et al.*, 2011; Parvaresh *et al.*, 2012), a time-consuming and laborious process (Liu *et al.*, 2020).

Chang *et al.* (2019) reported genome editing and gene discovery in pomegranate hairy roots using Clustered Regularly Interspaced Short Palindromic Repeats (CRISPR)/CRISPR-associated protein 9 (Cas9) (CRISPR/Cas9), coupled with transcriptome and biochemical analyses. Single guide RNAs (sgRNAs) were designed to target two UDP-dependent glycosyltransferases (UGTs), *Pg*UGT84A23 and *Pg*UGT84A24, which possess overlapping activities in β-glucogallin (a galloylglucose ester; biosynthetic precursor of HTs) biosynthesis. A unique accumulation of gallic acid 3-*O*- and 4-*O*-glucosides (galloylglucose ethers) was observed in the *PgUGT84A23* and *PgUGT84A24* dual CRISPR/Cas9-edited lines (*i.e.*, *ugt84a23 ugt84a24*) but not the control (empty vector) or *PgUGT84A23*/ *PgUGT84A24* single edited lines (*ugt84a23* or *ugt84a24*). Transcriptome and real-time qPCR analyses identified 11 UGTs with increased expression in the *ugt84a23 ugt84a24* hairy roots compared to the controls. Of the 11 candidate UGTs, only *Pg*UGT72BD1 used gallic acid as substrate and produced a regiospecific product gallic acid 4-*O*-glucoside. This work demonstrates that the CRISPR/Cas9 method can facilitate functional genomics studies in pomegranate and shows promise for capitalizing on the metabolic potential of pomegranate for germplasm improvement.

Liu *et al.* (2020) identified 36,792 simple sequence repeat (SSR) markers in the pomegranate genome. Dinucleotide repeats are the most abundant SSRs, and of the mono-, di-, tri-, and tetranucleotide repeats, A/T, AT/AT, AAT/ATT, and AAAT/ATTTwere the most dominant motif type. There were only 1.80 per cent SSR motifs located in the coding regions, while most (98.20 per cent) SSR motifs were located in noncoding regions. We confirmed the reliability of 50 SSR markers by PCR amplification, from which 11 highly polymorphic SSR markers were selected to study the genetic diversity of 218 pomegranate accessions. A total of 63 alleles was detected for the 11 SSR primers, ranging from 2 (PG152) to 9 (PG093, PG080) across the pomegranate accessions. The polymorphism information content values of each SSR primer pair ranged from 0.22 to 0.70, with a mean value of 0.45. Based on population structure analysis, these accessions were clustered into three main populations and 42 accessions were chosen as the core set (Liu *et al.*, 2020).

The limited genomic resources have restricted further elucidation of genetics and evolution of these interesting traits (Yuan *et al.*, 2018). A 274 Mb high quality draft pomegranate genome sequence, which covers approximately 81.5 per cent of the estimated 336 Mb genome, consists of 2177 scaffolds with an N50 size of 1.7 Mb and contains 30 903 genes was reported by Yuan *et al.* (2018). Phylogenomic analysis supported that pomegranate belongs to the Lythraceae family rather than the monogeneric Punicaceae family, and comparative analyses showed that pomegranate and *Eucalyptus grandis* share the paleotetraploidy event. Integrated genomic and transcriptomic analyses provided insights into the molecular mechanisms underlying the biosynthesis of ellagitannin based compounds, the colour formation in both peels and arils during pomegranate fruit development, and the unique ovule development processes that are characteristic of pomegranate. This genome sequence provides an important resource to expand our understanding of some unique biological processes and to facilitate both comparative biology studies and crop breeding (Yuan *et al.*, 2018).

The availability of the draft genome sequence of pomegranate *cv.* Dabenzi presents unprecedented opportunities for the development of largescale genomic resources, such as DNA markers for genotyping applications. Patil *et al.* (2020b) identified a new set of highly polymorphic simple sequence repeat (SSR) markers by targeting the SSR motif lengths of ≥ 24 bp. A total of 1,73,633 SSRs were identified in the 296-Mb pomegranate genome assembly, reflecting an average density of 527.97 SSRs/Mb. Of these, 43,853 SSRs belong to Class I category (> 20 bp). Concerning the abundance of repeat types in the current dataset, dinucleotide (NN) repeats (31.19 per cent) were the dominant class among all SSRs identified in the genome, followed by tetranucleotide (NNNN: 20.5 per cent) and trinucleotide repeats (NNN: 16.8 per cent). The top two SSR motifs in NN category were AT/AT (64.90 per cent) and AG/CT (28.51 per cent), whereas AAT/ATT (34.66 per cent) and AAG/CTT (28.91 per cent) were the most abundant among NNN repeats. Primer pairs were designed for a total of 2856 Class I SSRs and 110 primers were then assayed initially on eight pomegranate genotypes for polymorphism survey. Polymorphic fragments were obtained for 82 SSRs (77.36 per cent), of which a subset of 13 informative SSRs was further employed to investigate genetic diversity among 46 pomegranate genotypes. Approaches, such as population structure, cluster and PCA elucidated genetic relationships among 46 diverse pomegranate genotypes. To concluded Patil *et al.*

(2020a) developed the first set of genome-wide SSRs in pomegranate that will serve as a powerful genomic tool for future genetic studies. These SSRs have widespread applications in QTL mapping and marker-assisted selection for breeding.

A total of 17,439 mature miRNAs (~ 21 nt) earlier generated through RNA seq in the pomegranate were used for in silico analysis. After complexity reduction, a total of 1922 representative mature miRNAs were selected and used as query sequences against pomegranate genome to retrieve 2540 homologous contigs with flanking regions (~ 800) (Patil *et al.,* 2020b). By using pre-miRNA prediction web server, a total of 1028 true contigs harbouring pri-miRNAs encoding 1162 pre-miRNAs were identified. Survey of these sequences for SSRs yielded a total of 1358 and 238 SSRs specific to pri-miRNA and pre-miRNAs, respectively. Of these, primer pairs were designed for 897 pri-miRNA and 168 pre-miRNA SSRs. In pri-miRNA sequences, hexa-nucleotides repeats were found to be most abundant (44.18 per cent) followed by mono- (18.41 per cent) and di-nucleotide (17.01 per cent), which is also observed in pre-miRNA sequences. Further, a set of 51 randomly selected pre-miRNA-SSRs was examined for marker polymorphism. The experimental validation of these markers on eight pomegranate genotypes demonstrated 92.15 per cent polymorphism. Utility of these functional markers was confirmed via examination of genetic diversity of 18 pomegranate genotypes using 15 miRNA-SSRs. Further, potential application of miRNA-SSRs for discovery of trait specific candidate genes was showed by validating 51 mature miRNA against publically available 2047 EST sequences of pomegranate by target and network analysis. In summary, the current study offers novel functional molecular markers for pomegranate genetic improvement (Patil *et al.,* 2020b).

However, these resources are fragmented, and almost of all of the scaffolds or contigs are arbitrarily ordered and oriented. This greatly impedes map-based cloning studies and hampers the identification of crucial intergenic regions related to valuable traits, including the divergence between soft- and hard-seeded pomegranate varieties (Lou *et al.,* 2020). Complete and highly accurate reference genomes and gene annotations are indispensable for basic biological research and trait improvement of woody tree species. Hence, Lou *et al.* (2020) integrated single-molecule sequencing and high-throughput chromosome conformation capture techniques to produce a high-quality and long-range contiguity chromosome-scale genome assembly of the soft-seeded pomegranate cultivar 'Tunisia'. The genome covers 320.31 Mb (scaffold N50 = 39.96 Mb; contig N50 = 4.49 Mb) and includes 33 594 protein-coding genes. We also resequenced 26 pomegranate varieties that varied regarding seed hardness. Comparative genomic analyses revealed many genetic differences between soft- and hard-seeded pomegranate varieties. A set of selective loci containing SUC8-like, SUC6, FoxO and MAPK were identified by the selective sweep analysis between hard- and soft-seeded populations. An exceptionally large selective region (26.2 Mb) was identified on chromosome 1. The assembled pomegranate genome is more complete than other currently available genome assemblies (Lou *et al.,* 2020). The study indicated that genomic variations and selective genes may have contributed to the genetic divergence between soft- and hard-seeded pomegranate varieties (Lou *et al.,* 2020).

15.0 References

Abou-Aziz, A.B., El-Kassas, S.E., Boutros, B.N. and El-Sese, A.M. (1995a) *Assiut J. Agric. Sci.*, **26**: 116-128.

Abou-Aziz, A.B., El-Kassas, S.E., Boutros, B.N. and El-Sese, A.M. (1995b) *Assiut J. Agric. Sci.*, **26**: 129-147.

Afria, B.S., Pareek, C.S. and Singh, K. (1998) *Indian J. Hort.*, **55**: 139-143.

Ahire, G.Z., Desai, U.T., Chaudhari, S.M., Masalkar, S.D. and Kale, P.N. (1994) *J. Maharashtra Agric. Univ.*, **19**: 153-154.

Ahmed, R.D. (1994) *Annals Agric. Sci. Moshtohor*, **32**: 1651-1663.

Al-Khatani, H.A. (1992) *J. Amer. Soc. Hort. Sci.*, **117**: 100-104.

Al-Wasel ASA (1999) *J King Saud Univ* **11:** 3–14.

An, G.C., Wang, J.J. and Wang, G.Z. (1994) *J. Fruit Sci.*, **11**: 62-63.

Anonymous (1979) *Ann. Res. Report, Dept. of Hort.*, Mahatma Phule Krishi Vidyapeeth, Rahuri.

Anonymous (1985) *Pomegranate Improvement Programme and Progress – A report from Univ. Agril Science*, Bangalore.

Anonymous. (2018) Horticulture Statistics Division, Ministry of Agriculture, Government of India, p. 458.

Artes, F., Gines, Marin, J. and Martinez, J.A. (1996) *Zeitschrift fur Lebensmittel-Untersuchung und Forschung*, **203**: 33-37.

Babu, K.D. 2010 *Fruits, Vegetables, Cereal Sci. Biotechnol,.* **4:** 45-50.

Bacha, M.A. and Ibrahim, I.M. (1980) *Egyptian J. Hort.*, **6**: 135-140.

Bachake SS, V.B. Jadhav,P.P. Deshpande *et al.* (2019) *J. Pharmacog. Phytochem.* **8(3):** 2548-2553.

Bagate, A.H. (1979) *M.Sc. (Ag) thesis* submitted to Mahatma Phule Krishi Vidyapeeth, Rahuri.

Baghel, B.S. and Saraswat, B.K. (1989) *Indian J. Hort.*, **46**: 458-462.

Balakrishnan, K., Venkatesan, K. and Sambhandamurthi, S. (1996) *Orissa J. Hort.*, **24**: 33-35.

Bangal, G.B. and Khaire, P.S. (1982) *Paper submitted at the All India Pomegranate Res. Workers Meetting* held at MPAU, Rahuri, 102 July.

Banihashemi, Z. (1998) *Phytopathologia Mediterranea*, **37**: 127-132.

Bavale, S.V. (1978) *M. Sc. (Ag) thesis* submitted to Mahatma Phule Krishi Vidyapeeth, Rahuri.

Behnia, A. (1999) In: *Irrigation under conditions of water scarcity*, pp. 13-17.

Ben-Aire, R., Segal, N. and Guelfat-Reich, S. (1984) *J. Amer. Soc. Hort. Sci.*, **109**: 898-901.

Bhapuar, S.K. (1976) *M.Sc. (Ag.) thesis* submitted to Mahatma Phule Krishi Vidyapeeth, Rahuri.

Bhargava, B.S. and Dhandhar, D.G. (1988) *Prog. Hort.*, **19**: 196-199.

Bicket, J. and Bilderback, T.E. (1982) *Amer. Nurseryman*, **156**: 85-86.

Bose, T.K. and Mandal, D.P. (1972) *Indian Hort.*, **17**: 25-26, 30.

Cai, Y.L., Lu, X.U., Zhu, L.W., Wang, D.X. and Zhu, X.T. (1993) *Acta Hort.*, **420**: 23-26.

Çali°kan, O., S. Bayazit, M. Öktem And Ali Ergül. (2017) *Turk J Agric* **41:** 142-153

Chandra, R., Pal, R.K., Deshmukh, R., Singh, N.V., and Maity, A. (2015) *Indian J. Arid Hort.*, **9** (1-2): 1-6.

Chang, L., S. Wu and L. Tian (2019) *Hortic Res.*, **6:** 123.

Chattopadhyay, P.K. and Patra, S.C. (1992) *South Indian Hort.*, **40**: 309-312.

Chaudhari, S.M. and Desai, U.T. (1993) *J. Agric. Sci.*, **63**: 34-35.

Chauhan, K.S. and Singh, J. (1971) *Indian J. Agric. Sci.*, **41**: 293-296.

Cheema, G.S., Bhat, S.S. and Naik, K.C. (1954) *Commercial Fruits of India with Special Reference to Western India*, Macmillan & Co.

Chen, Y.H., Tan, B. Li, H.T., Hu, Q.X., Ye, X., Meng, H.J. and Feng., J.C. (2012) *J. Fruit Sci.*, **29:** 598-604.

Chopade, S.O. and Gorantiwar, S.D. (1998) *Annals Plant Physiol.*, **12**: 98-102.

Curro S, Caruso M, Distefano G, Gentile A, La Malfa S (2010) *Am J Bot* **97:** e58–e60.

Damiano, C., Pedro, M.D.A. and Frattarelli, A. (2008) *Prop. Orn. Plts.*, **8** (1): 3-8

Debergh, P.C. and Maene, L. (1981) *Sci. Hort.*, **14**: 335-345.

Desai, U.T., Ahire, G.Z., Masalkar, S.D. and Choudhari, S.M. (1993) *Annals Arid Zone*, **32**: 161-164.

Dhumal, J.D. (1984) *M.Sc. (Ag.) thesis* submitted to Mahatma Phule Krishi Vidyapeeth, Rahuri.

Dong, R.B., Li, Y.L., Zhao, X.C., Wu, K.X. and Sun, Z.C. (1999) *China Fruits*, **1**: 43.

Dong, Y.C. and Yang, L.X. (1994) *J. Fruit Sci.*, **11**: 260.

Doring, J. and Ludders, P. (1987) *Gartenbauwissenschaft*, **52**: 26-31.

Drazeta, L. (1997) *Review of Research Work at the Faculty of Agriculture, Belgrade*, **42**: 49-59.

Dubey, J.K., Amith Nath and Thakur, J.R. (1993) *Indian Forest.*, **119**: 928-931.

Durgac C, M. Ozgen, O. Simsek, Y.A. Kacar,Y. Kýyga, S.Celebi,S.Gunduz Kand Serce (2008) *Afr J Biotechnol.*, **7(9):** 1294–1301

Dutta, B.G. and Ghosh, G.R. (1963) *Indian Phytopath.*, **16**: 223.

Dutta, S.C. (1965) *Horticulture in Eastern region of India*, Farm Information Unit., Dir. of Extn., New Delhi, p. 155.

El-Kassas, S.E., El-Sese, A.M., El-Salhy, A.M. and Abdalla, A.A. (1998) *Assiut J. Agric. Sci.*, **29**: 147-162.

El-Nemr, S.E., Ismail, I.A. and Ragab, M. (1990) *Nahrung*, **34**: 601-606.

El-Sese, A.M. (1988) *Assiut J. Agric. Sci.*, **19**: 320-336.

El-Shaaraway, M.I. and Nahapetian, A. (1983) *Fette Scifen Anstrichitted*, King Faisal Univ., Saudi-Arabia, **85**: 123-126.

Elyatem, S.M. and Kader, A.A. (1984) *Scientia Hort.*, **24**: 287-298.

Ercisli S, Agar G, Orhan E, Yildirim A, Hizarci Y (2007) *Biochem Syst Ecol* **35:** 764-769.

Evreinoff, V.A. (1957) *J. Agric. Trop. Bot. Appl.*, **4**: 124-138.

Faiziev, D. (1973) *Referativnyi Zhurnal 7*, **55**: 748.

Firuzch, P. and Ludders, P. (1979) *Erwerbsobstbau*, **21**: 260-262.

Fougat, R.s., Paldey, S.B., Shmad, T. and Godhani, P.R. (1997) *J. Applied Hort.*, **3:** 23-29.

Gammie, G.A. and Patwardhan, G.B. (1924) *Bombay Dept. Agric. Bull.*, No. 30.

Gil, M.I., Garcia-Viguera, C., Artes, F. and Tomas-Barberan, F.A. (1995) *J. Sci. Food Agric.*, **68**: 77-81.

Godara, N.R. (1982) *Paper presented at the All India Pomegranate Res. Workers Meeting held at MPAU*, Rahuri, 1-2 July.

Godara, R.K., Godara, N.R., Jitender Kumar and Surinder Kumar (1989) *Res. Dev. Rep.*, **6**: 76-80.

Guranna, P. and H. C.Hoolageri (2017) *Annual Research & Review in Biol.*, **21 (5):** 1-7.

Guranna, P., I. Hosamani, R. Sathyanarayana, R. Hegde and K. Hipparagi. (2017) *Biotechnol. J. Int.*, **20**(3): 1-8.

Haggag, M.N., Ezz, T.M. and El-Kobbia, A.M. (1999) *Alexandria J. Agric. Res.*, **44**: 175-184.

Hamdy and M.A.S. El-Zayat (2017) *Plant Omics*, **10:** 237-245.

Hayes, W.B. (1953) *Fruit Growing in India*, Kitabistan Allahabad, p. 344.

He, X.Y. (1999) *J. Zhejiang For. Sci. Technol.*, **19**: 48-50.

Heikal, E.K. (1991) *Annals Agric. Sci. Mostohor*, **29**: 067-979.

Heikal, H.A., Ismail M.M. and El-Mahmoudy, L.T. (1966) *Agric. Res. Rev. Cairo*, **44**: 86-96.

Helaly, M.N., El-Hosieny, H., Tobias, N., Alsudays, I., Omar, S.A., Elsheery, N.I. (2014) *Aust. J. Crop Sci.*, **8**: 307-316.

Hernandez, F., Melgarejo, P., Tomas-Barberan, F.A. and Artes, F. (2000) *Eur. Food Res. Technol.*, **210**: 39-42.

Holcroft, D.M., Gil, M.I. and Kader, A.A. (1998) *J. Amer. Soc. Hort. Sci.*, **123**: 136-140.

Holland, D., Hatib, K. and Bar-Ya'akov, I. (2009) In: Janick, J. (ed.) *Hort. Rev.*, Wiley and Sons, Vol. 35, pp. 127-191.

Hosamani, A., S. Mohandas, V. Manjula, K. Prakash and Geetha M.S. (2016) *Electronic J. Plant Breed.* **7:** 864- 870

Hussein, A.A., Abdel-Rahman, A.G. and Ahmed, R.D. (1994) *Annals Agric. Sci. Moshtohor*, **32**: 949-957.

Hussein, M.A., El-Sese, A.M., El-Mahdy, T.K., Abo-El-Sabour, B. (1994) *Assiut J. Agric. Sci.*, **25**: 41-51.

Jadhav, L.D., Ajri, D.S., Kadam, M.V. and Dorga, S.K. (1976) *Entomologists Newsl.*, **6**: 57.

Jadhav, S.S. and Ajri, D.S. (1985) *Indian J. Entomol.*, **47**: 63-65.

Jagtap, D.B., Desai, U.T. and Kale, P.N. (1992) *J. Maharashtra Agric. Univ.*, **17**: 399-401.

Jain, P.K. and Parmar, K.L. (1996) *JNKVV Research J.*, **27**: 56-58.

Jalikop, S.H. (2007) *Euphytica*, **158**: 201-207.

Jalikop, S.H. (2010) *Fruit, Vegetable, Cereal Sci. Biotechnol.*, **4** (2): 26–34.

Jalikop, S.H. and Kumar, P.S. (2000) *Indian Hort.*, **45**: 19-20.

Jalikop, S.H. and Sampath Kumar, P. (2001) *Indian Hort.*, **45**: 19-20.

Jiang, S.G., Wang, J.L. and Yang, T.L. (1995) *J. Fruit Sci.*, **12**: 126-128.

Josan, J.S., Jawanda, J.S. and Uppal, D.K. (1979a) *Punjab Hort. J.*, **19**: 59-65.

Josan, J.S., Jawanda, J.S. and Uppal, D.K. (1979b) *Punjab Hort. J.*, **19**: 66-70.

Josan, J.S., Jawanda, J.S. and Uppal, D.K. (1979c) *Punjab Hort. J.*, **19**: 134-138.

Kaji,B.V., Ershadi,A. and Tohidfar, M. (2014) *Int. J. Hort. Sci. Technol.* **1**(1): 31-41.

Kaji, B.V. and Abbasifar, A. (2017) *Biocatalysis Agril Biotechnol.*, **10:** 46-52.

Kakar, K.L., Dogra, G.S. and Nath, A. (1987) *Indian J. Agric. Sci.*, **57**: 749-752.

Kale, R.H. (1986) *M.Sc. (Ag.) thesis* submitted to Mahatma Phule Krishi Vidyapeeth, Rahuri.

Kanardov, V.I., Kolesova, N., Khrabrov, G. and Yu, M. (1990) *Sadovodstvo i Vinogradarstvo*, **6**: 37-39.

Kantharajah, A.S., Dewitz, I., and Jabbari, S. (1998) *Erwerbsobstbau*, **40**: 54-58

Kantharajah, A.S., Dewitz, I. and Jabbari, S. 1998) *Erwerbsobstbau*, **40**: 66-68.

Kanwar, K., Deepika, R. and Chauhan, R. (2008) In*: Nat. Sem. Physiol. Biotechnol. Approaches to Improve Plant Productivity,* held on March 15–18, 2008, Haryana CCSHAU, Hisar, India, p. 52.

Kanwar, K., Joseph, J., Deepika, R. (2010) *Plant Cell Tissue Organ Cult* **100:** 199–207.

Kanwar, J.S. and Thakur, D.P. (1972) *Indian J. Mycol. Pl. Path.*, **2**: 189-191.

Kanwar, J.S. and Thakur, D.P. (1973) *Sci. Cult.*, **39**: 274-277.

Karale, A.R. (1977) *M. Sc. (Ag) thesis* submitted to Mahatma Phule Krishi Vidyapeeth, Rahuri.

Karale, A.R., Supe, V.S., Kaulgud, S.N. and Kale, P.N. (1993) *J. Maharashtra Agric. Univ.*, **18**: 364-366.

Keskar, B.G., Karale, A.R., Dhawale, B.C. and Choudhari, K.G. (1990) *J. Maharashtra Agric. Univ.*, **15**: 105-106.

Kolhe, V.H. (1980) *M. Sc. (Ag) thesis* submitted to Mahatma Phule Krishi Vidyapeeth, Rahuri.

Kondav, N.K. (1969) *Izd. Balgar Akad. Naukite*, Sofija, p. 68.

Kucherona, T.P. (1976) *Subtrop. Kultury*, 4: 104-107.

Kulenkamp, A., Lein, G. and Borisenko, V. (1985) *Beitrage zur Tropischen Landwirtschaft and Veterinarmedizin*, **23**: 244-255.

Kumar, K., Arora, P.K., Brar, J.S., Bhatia, D. and *Kumar*, A. (2019) *Indian J. Hort.*, **76(2)**: 273-278.

Kupper, W., Pekmezci, M. and Henze, J. (1995) *Acta Hort.*, **398**: 101-108.

Kutuzova, A.S. (1984) *Subtrop Kul'tury*, **1**: 123-127.

Lal, N. (1975) *Ph. D. thesis* submitted to Univ. Udaipur, Udaipur.

Lal, N. and Chauhan, K.S. (1979) *Abst. Int. Symp. Arid Zone Res. Dev. Jodhpur*, India.

Lande, P.S. (1975) *M. Sc. (Ag) thesis* submitted to Mahatma Phule Krishi Vidyapeeth, Rahuri.

Lawand, B.T. and Patil, V.K. (1994) *J. Maharashtra Agric. Univ.*, **19**: 220-223.

Lawrence, H.M. (1951) *Taxonomy of Vascular Plants*, Macmillian and Co., New York.

Lee, S.W., Kum, K.S. and Kim, S.D. (1974a) *J. Korean Soc. Hort. Sci.*, **15**: 57-63.

Leena Borah (1990) *M.Sc. (Food Technol) thesis*, University of Mysore, India.

Liu, C.,J.Li, and G. Qin. (2020) *Tree Genetics & Genomes* **16:** 36. https: //doi. org/10.1007/s11295-020-1428-4

Luo, X., S.Cao, Z. Hao, L. Hou, D. Cao, J.Zhang, H. Li, J. Niu, H. Xue and L. Chen. (2018)*The J. Hort. Sci. Biotechnol.*, https: //doi.org/10.1080/14620316.2018.14 32994.

Luo, X., Li, H., Wu, Z. Yao, W., Zhao, P., Cao, D., Yu, H. and Li, K., Krishna, P., Zhao, D., Fuhong Z. X., Chen, L., Wang, Q., Jing, D., Cao, S. (2020) *Plant Biotechnol. J.* 18: . 955–968.

Mahishi, D.M., Muralikrishnan, A., Shivashankar, G. and Kulkarni, R.S. (1991) In: *Proc. Intern. Seminar on New Frontiers in Horticulture*, organized by Indo-American Hybrid Seeds, Bangalore, India, November 25-28, 1990, Netherlands, Kluwer Academic Publishers, Dordrecht, pp. 215-217.

Mahishni, D.M., Muralikrishna, A., Shivashankar, G. and Kulkarni, R.S. (1991) *Kluwar Academic Publisher, Dordrecht, Netherlands*, pp. 215-217.

Maity, A., Babu, K.D., Sarkar, A. (2019) *Scientia Hort.*, **252**: 138-148.

Manohar, M.S., Tikka, S.B.S. and Nathulal (1981) *Indian J. Hort.*, **38**: 187-190.

Marathe, R.A., Sharma, J., Babu, K.D, Murkute, A.A. *et al.* (2017) *Natl. Acad. Sci. Lett. DOI 10.1007/s40009-017-0567-0*

Markarimi, A. (1999) *Iranian J. Agric. Sci.*, **30**: 1-9.

Mart, C. and Kilincer, N. (1994) *Ege Universitesi Ziraat Fakultesi, Bitki Koruma Bolumu, Izmir*, Nirai Mucadele Arastirma, pp. 511-518.

Melgarejo, P., Manuel, S. and Artes, F. (2000) *Eur. Food Res. Technol.*, **211**: 185-190.

Meshram, D.T., Gorantiwar, S.D., Sharma, J. and Babu, K.D. (2018) *J. Agromet.*, **20** (3): 196-201

Meshram, D.T., Gorantiwar, S.D., Sharma, J. and Babu, K.D. (2019) *Scientia Hort.*, **246:** 686-692.

Moeyked, K. (1971) *M.S. thesis* submitted to Pahlavi Univ., Shiraz, Iran.

Mole, U.N. and Tambe, A.B. (1990) *Plant Prot. Bull. (Faridabad)*, **42**: 7-8.

Moneam, N.M.A., El-Sharaky, A.S. and Badeldin, M.M. (1988) *J. Chromatogr.*, **438**: 438-441.

Moriguchi, T., Omura, M., Matsuta, M. and Kozaki, I. (1987) *HortScience*, **22**: 947-948.

Mostafa, F.M.A. (1998) *Assiut J. Agric. Sci.*, **29**: 201-214.

Mukherjee, K. (1958) *Sci. Cult.*, **29**: 94.

Mulaei, S., Jafari, A., Shirmardi, M. and Kamali, K. (2019) *Int. J. Fruit Sci.*, DOI: 10.1080/15538362.2019.1680334.

Munde, S.S., Patil, V.K. and Chauhan, S.D. (1980) *J. Maharashtra Agric. Univ.*, **5**: 90-91.

Murkute, A.A., Patil, S., and Mayakumari (2003) *Agri. Sci. Dig.*, **23**: 29-31.

Muthumanickam, D. and Balakrishnamurthy, G. (1999) *Indian J. Hort.*, **56:** 213-218.

Naik, S.K., Pattnaik, S. and Chand, P.K. (1999) *Scientia Hort.*, **79**: 175-83.

Naik S.K., Chand P.K. (2003) *J Plant Physiol* **160**: 423–430.

Naik SK, Chand PK (2006) *Scientia Hort.*, **108:** 247–25.

Naik SK, Pattnaik S, Chand PK (1999) *Scientia Hort.*, **79**: 175-183.

Naik SK, Pattnaik S, Chand PK (2000) *Scientia Hort.*, **85:** 261–270

Naik, S.K. and Chand, P.K. (2006) *Scientia Hort.*, **108**: 247-252.

Naik, S.K. and Chand, P.K. (2011) *Plant Cell Rep.*, **30:** 707-721.

Naik, S.K., Pattnaik, S. and Chand, P.K. (1999) *Scientia Hort.*, **79**: 175-183.

Naik, S.K., Pattnaik, S. and Chand, P.K. (2000) *Scientia Hort.*, **85**: 261-270.

Naik, V.B. (1979) *M.Sc. (Ag) thesis* submitted to Mahatma Phule Krishi Vidyapeeth, Rahuri.

Nalawadi, J.G., Farooqui, A.A., Dasappa Reddy, M.A.N., Gubhaiah Sulkeri, G.S. and Nalini, P.S. (1973) *Mysore J. Agric. Sci.*, **7**: 213-325.

Nasr, C.B., Ayed, N. and Metche, M. (1996) *Zeitschrift fur Lebensmittel-Untersuchung und Forschung*, **203**; 374-378.

Nataraja, K. and Neelambika, G.K. (1996) *Indian J. Exp. Biol.*, **34**: 719-721.

Nath, N. and Randhawa, G.S. (1959a) *Indian J. Hort.*, **16**: 61-68.

Nath, N. and Randhawa, G.S. (1959b) *Indian J. Hort.*, **16**: 121-135.

Nath, N. and Randhawa, G.S. (1959c) *Indian J. Hort.*, **16**: 136-150.

Normakhamatov, R. and Rakhmanberdyeva, R.K. (1999) *Chem. Nat. Compounds*, **35**: 96-97.

Omura, M. (1991) Pomegranate (*Punica granatum* L.). In: Bajaj YPS (ed) *Biotechnology in agriculture and forestry*, vol 16, Trees III.

Omura M, Matsuta T, Moriguchi and T, Kozaki I (1987) *HortScience* **22:** 133–134.

Omura, M., Matsuta, N., Moriguchi, T. and Kozaki, I. (1990) *Bull. Fruit Tree Res. Sta.*, **17**: 19-33.

Onur, C. (1983) *Research and Training Centre of Horticultural Crops*, Ministry of Agriculture, Forestry and Village Affairs, Erdemli, Turkey.

Ophir R, Sherman, A., Rubinstein, M., R.Eshed, Schwager, M.S. Harel-Beja R., Bar-Ya, I., akov, I and Holland, D. (2014) *PLoS One* **9**(2): e88998.

Owais, S.J. and Abdel-Ghani, A. H. (2015) *Int. J. Agric. Biol.*, DOI: 10.17957/IJAB/15.0101

Padmavathamma, A.S. and Hulamani, N.C. (1999) *Karnataka J. Agric. Sci.*, **11**: 1126-1128.

Padule, D.N. and Keskar, B.G. (1988) *Maharashtra J. Hort.*, **4**: 73-76.

Palmenoe, M. (1972) *Prilog Poznaunja i. voduo stirorti sipkana podruciv Bara jugostoveusko – va carsotua*, **5**: 117-118.

Panda, J.M. and Das, R.C. (1990) *Orissa J. Hort.*, **18**: 32-37.

Pant, T. (1976) *Udyanika*, **2**: 29-35.

Pareek, O.P. (1981) *Project Rept. Proc. Ist Nat. Workshop on Arid Zone fruits Res.*, Haryana Agricultural University, Hissar.

Parvaresh M, Talebi M, Sayed-Tabatabaei B-E (2012) *Scientia Hort.* **138:** 244–252

Pathak, S.P. and Pundir, J.P.S. (1981) *Udyanika*, **4**: 7-11.

Patil, B.N. (1976) *M. Sc. (Ag.) thesis* submitted to Mahatma Phule Krishi Vidyapeeth, Rahuri.

Patil, B.R., Mane, T.A., Pampattiwar, P.S. and Bangal, G.B. (1982) *Paper presented at all India pomegranate Res. Workers Meeting*, Mahatma Phule Krishi Vidyapeeth, 1-2 July.

Patil, P.G., Singh, N.V., Parashuram, S. *et al.* (2020a) *Trees* **34,** 987–998

Patil, P.G., Singh, N.V., Parashuram, S. *et al.* (2020b) *Physiol Mol Biol Plants* **26,** 683–696.

Patil, V.M., Dhande, G.A., Thigale, D.M., Rajput, J.C. (2011) *Afr. J. Biotechnol.*, **10**: 18130-18136.

Patil, V.M., Dhande, G.A., Thingale, D.M. and Rajput, J.C. (2011) *Afr. J. Biotechnol.*, **10:** 18130-18136

Pawar, S.K., Desai, U.T. and Chouchari, S.M. (1994) *Annals Arid Zone*, **33**: 45-47.

Phadnis, N.A. (1974) *Indian Hort.*, **19**: 8-13.

Pool, R.J. (1941) *Flowers and Flowering Plants* (2nd Edition), New York.

Prasad, R.N. and Bankar, G.J. (2000) *Indian J. Hort.*, **57**: 305-308.

Prasad, R.N., Bankar, G.J. and Vashishtha, B.B. (1999) *Haryana J. Hort. Sci.*, **28**: 69-70.

Prasanna Kumar, B. and Dhandar, D.G. (1996) *Indian J. Hort.*, **53**: 114-117.

Pross, A.G. (1938) *Sov. Subtropics*, **3**: 27-37.

Purohit, A.G. (1983) *Paper presented at 2nd Workshop on Arid Zone fruit Res.*, Udaipur, 8-10 July.

Purohit, A.G. (1985) *Indian J. Agric. Sci.*, **55**: 367-368.

Purohit, A.G. (1987) *Indian J. Agric. Sci.*, **57**: 753-755.

Purohit, A.G. and Shekharappa, K.E. (1985) *Indian J. Hort.*, **42**: 30-36.

Qin G, C.Xu,R. Ming, H.Tang, R.Guyot,E.M. Kramer, Y.Hu,X. Yi, Y.Qi and X.,Xu (2017) *Plant J.* **91**: 1108–1128

Raj Bhansali, R. (1990) *Ann. Bot.*, **66**: 249-253.

Raj Bhansali, R. (2001) In: *Biotechnology of Horticultural Crops*, Vol. I (Eds. V.A. Parthasarathy, T.K. Bose, P.C. Deka, P. Das, S.K. Mitra and S. Mohandas), Naya Prokash, Calcutta, India, pp. 341-343.

Rajanna, N. (1981) *Thesis Abst.*, **9**: 279.

Ramu, B.S.S., Reddy, P.N., Hussain, S.A. and Patii, P.B. (1996) *Curr. Res. Univ. Agric. Sci.*, **25:** 72-73.

Rane, D.A., Karale, A.P., Wavhal, K.M. and Kamble, K.B. (1982a,b) *Paper presented at All India Pomegranate Res. Workers Meeting* held at Mahatma Phule Krishi Vidyapeeth, Rahuri, 1-2 July.

Raturi, G.B. and Hiwale, S.S. (1991) *Indian J. Hort.*, **48**: 13-15.

Reddy, Y.N. and Reddy, T.T.N. (1990) *J. Maharashtra Agric. Univ.*, **15**: 153-155.

Reddy, Y.T.N. and Reddy, Y.N. (1989) *Progressive Hort.*, **21**: 106-110.

Rendle, A.B. (1938) *Classification of Flowering Plants*, Vol. III, Cambridge Univ. Press.

Saidaliev, U.S. (1986) *Referativnyi Zhurnal*, **3**: 55: 725.

Sandhu, A.S., Minhas, P.P.S., Singh, S.N. and Kambhoj, I.S. (1991) *Indian J. Hort.*, **48**: 302-304.

Sarkhosh A, Zamani Z, Fatahi R, Ebadi A (2006) *Sci Hort.*, **111:** 24–29.

Sawant, R.B. (1973) *M.Sc. (Ag.) thesis* submitted to Mahatma Phule Krishi Vidyapeeth, Rahuri.

Sepahi, A. (1986) *Iran Agric. Res.*, **5**: 93-99.

Sharon M. and S.Sinha (2000) *Ind J Plant Physiol.,* **5:** 344–348.

Sherafatian, D. (1994) *Seed and Plant,* **10**: 25-31.

Sherkar, B.V. (1979) *M. Sc. (Ag.) thesis* submitted to Mahatma Phule Krishi Vidyapeeth, Rahuri.

Shevale, B.S. (1991) *Plant Protection Bull. (Faridabad),* **43**; 31-32.

Shukla, R.P. and Prasad, V.G. (1983) *Entomon.,* **8**: 381-383.

Shulman, Y., Fainberstein, L. and Lavee, S. (1984) *J. Hort. Sci.,* **59**: 265-274.

Siddiqui, Z.A. and Khan, M.W. (1986) *Pakistan J. Nematol.,* **4**: 83-90.

Singh S.K., Khawale, RN. (2006) In: *Plant Biotechnology and its applications in tissue culture* (Eds. A. Kumar and S. Roy), IK International Pvt. Ltd., New Delhi, pp: 107–113.

Singh, J. (1977) *M. Sc. (Ag.) thesis* submitted to Punjab Agricultural University, Ludhiana.

Singh, N.V., Singh, S.K., and Patel, V.B. (2007) *Indian J. Agric. Sci.,* **77**: 509-11

Singh, P. and R.M. Patel. (2016) *African J. Plant Sci.,* **10** (2): 43-49.

Singh, P., Patel, R.M. and Kadam, S. (2013) *Afr. J. Biotechnol.,***12:** 2863-2868.

Singh, P.S. and Chohan, J.S. (1972) *Curr. Sci.,* **41**: 651.

Singh, R. (1969) *Fruits,* National Book Trust., p. 136.

Singh, R.P., Sharma, Y.P. and Awasthi, R.P. (1993) *Progressive Hort.,* **22**: 92-96.

Singh, R.R., Singh, H.K. and Chauhan, K.S. (1988) *Progressive Hort.,* **20**: 77-79.

Singh, R.S. (1994) *Curr. Agric.,* **18**: 87-89.

Singh, T.J., T.Gupta and S.Sharma. (2019) *Sci.Hort.,* **247:** 436-448.

Slykov, G.N. (1968) *Trudy Priklad, Bot Genet. Selek.,* **39**: 322-350.

Soibam, H., Singh, A. V. and Chakraborty, I. (2014) *Life Sci. leafl.,* **52:** 16-23

Sonawane, P.C. (1986) *M. Sc. (Ag.) thesis* submitted to Mahatma Phule Krishi Vidyapeeth, Rahuri.

Sonawane, P.C. and Desai, U.T. (1989) *J. Maharashtra Agric. Univ.,* **14**: 341-342.

Sonawane, P.C., Desai, U.T., Chaudhari, S.M. and Deore, B.P. (1994) *J. Maharashtra Agric. Univ.,* **19**: 161-162.

Soriano J.M, E. Zuriaga E, P. Rubio,G. Llácer, R.Infante and M.L. Badenes (2011) *Mol. Breed* **27:** 119–128.

Sulladamath, U.V. (1985) *Progress Rept. 3rd Nat. Workshop on Aric Zone Fruit Res.,* Mahatma Phule Krishi Vidyapeeth, Rahuri, 5-8 July.

Suriachandraselvan, M., Jayasekhar, M. and Anbu, S. (1993) *South Indian Hort.,* **41**: 228-229.

Suryanarayana, V. and Venkateswara Rao, K. (1982) *Nat. Sem. Pl. Propagation,* Calcutta, pp.: 70-71.

Syed, S., Ramdas, S. and Muthuswami, K. (1985) *South Indian Hort.*, **33**: 67.

Symconides, P. (1930) *The Cyprus Agric. J.*, **25**: 91.

Terakami, S., Matsuta, N., Yamamoto, T. *et al.* (2007) *Plant Cell Rep* **26:** 1243–1251.

Thakur, J.N., Pawar, A.D. and Rawat, U.S. (1993) *Plant Protection Bull.*, **45**: 54-55.

Tressler, D.K. and Jaslyn, M.A. (1971) *Fruit and Vegetable Juice Processing Technology*, AVI. Pub. Co., Conn.

Umrao, V.K. and Singh, G.N. (1995) *Recent Hort.*, **2:** 141-146.

Vachaspati, S. and Chauhan, K.S. (1973) *Farmer and Parliament*, **18**: 12.

ValizadehKaji, B., Ershadi, A. and Tohidfar, M. (2013) *Physiol. Mol. Biol. Plants*, DOI 10.1007/s12298-013-0193-3.

Venkatesan, K. and Mohideen, M.K. (1994) *South Indian Hort.*, **42**: 239-244.

Verma, R.R. (1986) *Indian J. Nematol.*, **15**: 247.

Wang, L.J., Ma, K., Jiang, W.B., Ling, Z.F. and Wang, Y.L. (1995) *Acta Hort.*, **22**: 336-340.

Westwood, M.N. (1978) *Temperate zone pomology*. W.H. Freeman and Company, San Francisco, USA, p. 248.

Xian, K.Z. (1999) *South China Fruits*, **28**: 41.

Xing, H., Hu, Q.X., Kou, L.P. and Xu, H. (1997) *J. North Forestry College*, **12**: 93-96.

Yang, Z.H. and Ludders, P. (1993) *Angewandte Botanik*, **67**: 151-156.

Yuan, Z.,Y. Fang, T. Zhang *et al.* (2018) *Plant biotechnol.*,**16 (7):** 1363-1374. https: // doi.org/10.1111/pbi.12875.

Zarei, A., Z.Zamani, R.Fatahi *et al.* (2016) *Scientia Hort.*, **205:** 7-16.

Zhang, S., L. Chen, R. Huang, A.Isha and L.Dong (2017) *Plant Breed.*, **136:** 994-999.

Zhang, H. (1997) *J. Fruit Sci.*, **14**: 191-192.

Zhu, Q.Z., Yin, X.L., Wang, Y., Sun, J.M. and Shi, Y.H. (1998) *China Fruits*, **4**: 28.

③

LITCHI

Sanjay Kumar Singh, S.C. Maity and S. K. Mitra

1.0 Introduction

The Litchi or Lychee (*Litchi chinensis* Sonn.) is the most important sub-tropical evergreen tree and belongs to the Soapberry family Sapindaceae and sub-family Nephelaceae which has 125 genere and nearly 1000 species. The genus, *Litchi*, has two species, *Litchi philippinensis* and *Litchi Chinensis*, usually known as Pearl of India (Mehta, 2017). Litchi fruit is famous for its excellent quality, characteristic pleasant flavour and for attractive red colour. The litchi in its long history has been awarded many distinctive honours. Chinese have long considered it their most unique gift fruit. The earliest monograph on any horticultural subject deals with the litchi. Martinia (1655), as quoted by Mehta (2017), called the litchi "The King of Fruits". Litchi reached India through Myanmar and North East region during the 18th Century. India is the second largest producer of Litchi which accounts for about one fifth of the global production and has a good export potential. Bihar is the leading state in litchi production in the country whereas Assam, Orissa, Jharkhand, Punjab, Himanchal Pradesh, Tripura, Uttar Pradesh, Uttarakhand and West Bengal are the other main Litchi growing states. Bihar accounts for about three – fourth of the litchi production of the country. Litchi has a vast untapped potential in the domestic as well as the global market. The fruit crop pays substantially to the employment and economy of millions of people South East Asia (Mehta, 2017).

2.0 Composition and Uses

2.1.0 Composition

The food value of litchi mainly lies in its sugar and acid contents which again vary due to cultivars and climate. The moisture content in litchi fruit varies from 77 to 83 per cent (Kadam and Despande, 1995). The sugar content in different cultivars ranges from 6.74 to 18.86 per cent in India (Singh and Singh, 1964); 12 to 15 per cent in Florida and 11.8 to 20.6 per cent in Hawaii (Miller and Bazore, 1945). Besides sugar, litchi contains 1.1 per cent protein, 0.1 per cent fat, considerable amount of calcium and phosphorus and vitamin C, B_1, and B_2 (Table 1). The provitamin A content is virtually nil, but some cultivars contain a reasonable content of niacin (Galan Sauco, 1989).

The acidity in litchi fruit varies from 0.20 to 0.64 per cent in 12 Indian cultivars (Mathew and Pushpa, 1964). Acidity decreases during the process of ripening and storage. Malic acid is the predominant nonvolatile acid in litchi (Table 2). The ascorbic acid content in litchi varies widely such as 40.2 mg/100 g in Kwai Mi and 80.8 mg/100 g in Brewster (Wenkam and Miller, 1965), 44 mg/100g in Calcutta Late (Chadha and Rajpoot, 1969), 90 mg/100 g in unspecified variety (Thompson, 1955).

2.2.0 Uses

The litchi makes an excellent canned fruit. China, Taiwan and Thailand have substantial canning industries, and about 20 per cent of the total Chinese production is canned (Menzel, 1991b). The preservation of fruits in syrup is a traditional practice, the skins are peeled off, the seeds are removed and fruits, either whole or halved,

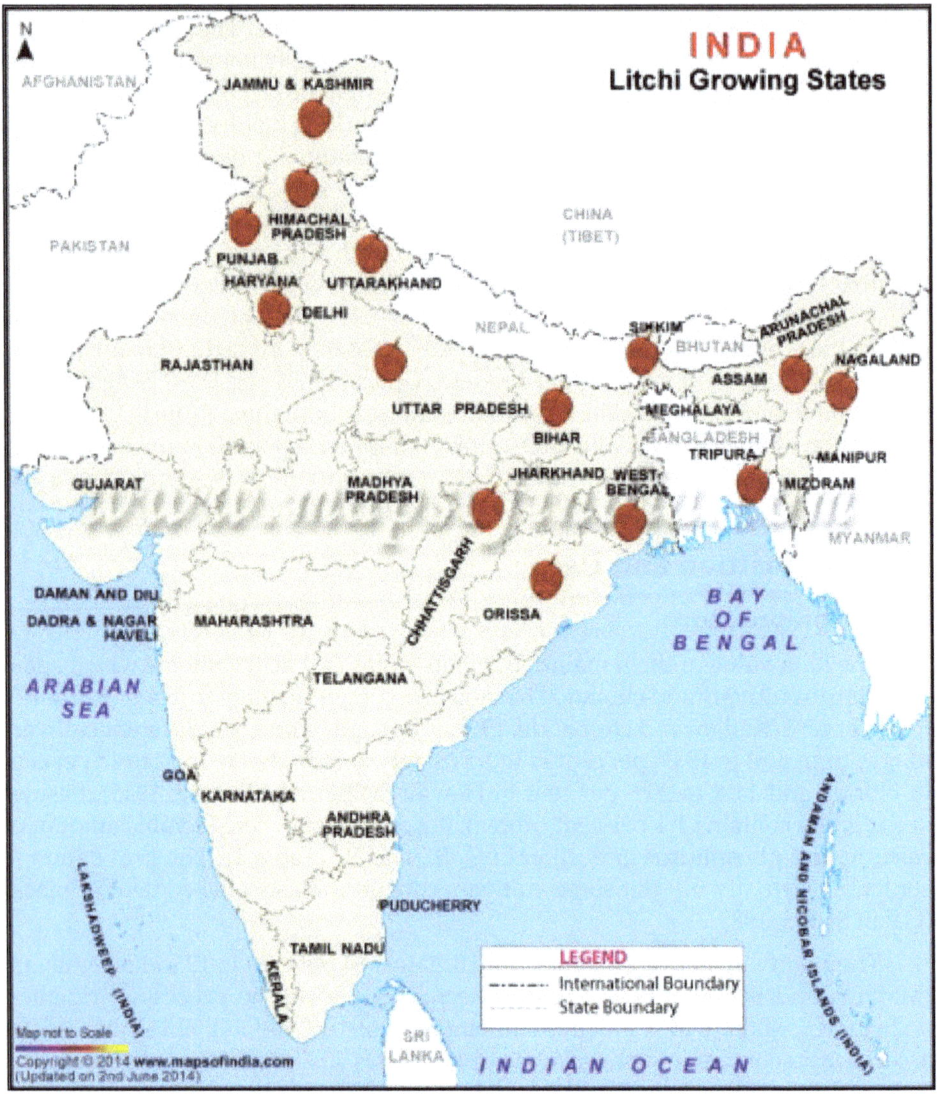

Litchi Producing States in India

are immersed in syrup. Preservation in honey is also possible and is practiced in China (Galan Sauco, 1989). A highly flavoured squash is also prepared from the inferior fruits which are liked by many during the summer months. Various other products, such as pickles, preserves, sherbet, ice cream and wine are also made from litchi in China. Dried litchi, commonly known as 'litchi nut', is very popular among the Chinese.

Table 1: Nutritive Value of Litchi Fruits (per 100 g edible pulp)

Constituents	*Fresh Pulp*	*Dried Pulp*	*Reference*	*Fresh Pulp*	*Reference*
Energy (Calory)	63-00&64-00	277-00	Nath *et al.* (2017)	-	-
Moisture (per cent)	81-90&84-83	17-90 & 22-30		80.60	Wills *et al.* (1986)
Protein (g)	0-68&1-00	2-90 & 3-80		1.10	
Fat (g)	0-30&0-58	0-20 &1-20		0.10	
Carbohydrate (g)	13-31&16-40	70-70&77-50		17.50	
Fibre (g)	0-23&0-40	1-40		0.10	Menzel and Simpson (1993)
Ash (g)	0-37&0-50	1-50&2-00		-	
Calcium (mg)	8-00&10-00	33-00		7.00	
Phosphorus (g)	30-00&42-00	-		41.00	
Iron (mg)	0-40 1-70	-		1.30	
Sodium (mg)	3-00 3-00	-		-	Wills *et al.* (1986)
Potassium (mg)	170-00	1]100		-	
Thiamine (mg)	28-00	-		0.05 mg	
Riboflavin (mg)	0-05 0-05	-		0.07 mg	
Nicotinic acid (mg)	0-40	-		-	
Niacin	-	-		0.5 mg	
Ascorbic acid (mg)	24-00&60-00	42-00		49.00	

Litchi is a source of vitamin C (72 g/100 g). It is believed that the daily need of vitamin C in an adult can be met by consuming just nine litchis. Other constituents of litchis are minerals: 14 per cent daily value (DV) of Cu, 9 per cent DV of P and 6 per cent DV of K. Similar to most fruits, lychees have little non saturated fat and sodium and are also cholesterol free. The level of polyphenols in lychees, as observed in a French study, was reported to be greater when compared to numerous other fruits. In terms of phenols, monomers and dimers of flavan-3-ol were the main components, constituting 87 per cent of the phenols. The levels of phenols decrease with storage. The major anthocyanins present are malvidin-3-acetylglucoside, cyanidin-3-rutinoside and cyanidin-3-glucoside. Friedelin, stigmasterol and stigmasteryl acetate are supported in the plant (Rastogi and Mehrotra, 1999) and are reduced with storage or browning. Traces of malvidin-3-glucoside are also present. Overconsumption of lychee may result in fainting spells or skin rashes. The Chinese used the seeds found in the Malay Peninsula as an anodyne, and these are strongly recommended to treat neurological diseases and orchitis. The lychee also has anticancer activity (Wang *et al.*, 2011) and strong free radical scavenging activity (Takuya *et al.*, 2008).

Table 2: Organic Acids in Litchi Fruits[1]*

Organic Acid (meq/100 g)	*Based on Methyl Ester Data (meq/100 g)*	*Based on TMS Data[2]*
Malic	4.16	3.57
Citric	0.52	0.04
Succinic	0.04	0.25
Levulinic	0.01	trace
Phosphoric	-	0.2
Glutaric	-	0.04
Malonic	0.02	-
Lactic	0.02	-
Ascorbic[3]	0.28	0.28
Volatile acids[4]	0.13	0.13
Total acid	5.18	4.51
Titrable acids[4]	4.60	4.60

[1]Each value is the mean of 7 or more replicates, [2]Separation of organic acids as derivatives of trimethylsylil, [3]Colorimetric value, [4]Values obtained titrimetrically.

* Chan and Kwok (1974).

2.3.0 Myth on Toxicity of Litchi Fruit

The litchi fruit is being consumed from as early as 2000 BC in China and approximately from the last 400 years in India. AES (Acute Encephalitis Syndrome) is being reported from last two decades. The numerous health benefits and its use in various traditional Chinese medicine systems along with potential hypoglycaemic effects of litchi seed are documented. However, hypoglycaemia alone may not be able to explain encephalopathy, which usually persists in spite of infusion of glucose. As per reports, only about 60 per cent of the affected children show symptoms of hypoglycaemia. In contrast even, hyperglycaemia has been reported from many AES affected children in Muzaffarpur. AES cases were reported in 2013 and 2014 but despite a very good crop of litchi in 2016 season, very few AES cases (about 15) were reported in 2016. Further, many AES cases are reported from as early as March last week and first week of April and as late as July when litchis have either not even set or barely set (less than 1 g in size) or when their season is already over. This supports and substantiates the view that AES is not related to litchi consumption. It is shocking that why children are being affected by eating litchi only in Muzaffarpur. Why children at other places are not getting affected?

Litchi Uses

Litchi Canning

Litchi Juice

Litchi in Syrup

Honey from Litchi Flowers

Litchi Uses

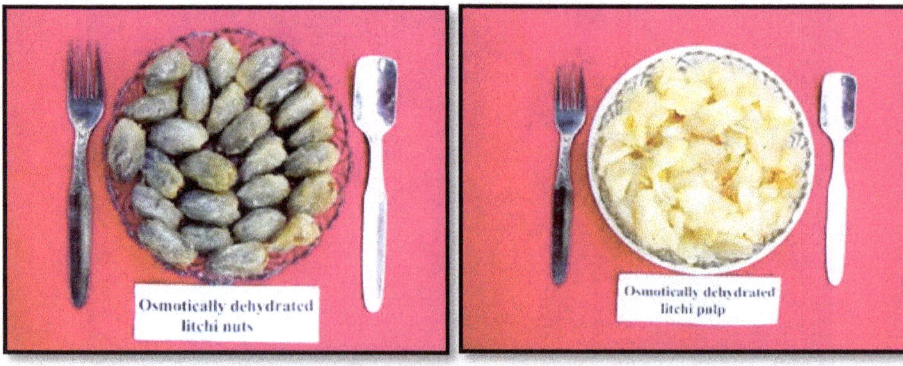

Osmotically Dehydrated Litchi Nuts and Pulp

Sherbet from Litchi

Icecream from Litchi (left) and Wine from Litchi (right)

Litchi is grown worldwide in several countries of the world, many states in India and moreover many districts in Bihar (Nath *et al.*, 2016).

As we know China is the leading producer of litchi. However, the AES reports have come only from Muzaffarpur, India, and from Vietnam and Bangladesh. It is highly surprising that why the cases of AES have erupted only from Bihar and one district of West Bengal, *i.e.* Malda if it is indeed occurring only due to litchi consumption. Even if we talk of only Bihar, districts Bhagalpur, Samastipur, Darbhanga, Begusarai, Katihar and Purnea remained unaffected from AES despite having good crop of litchi. Moreover, the AES cases have been reported in past from Gaya district as well where litchi is not cultivated. In the past, All India Institute of Medical Sciences, Patna and United Nations Children Fund have stressed that the AES is occurring due to enterovirus while investigating the AES cases. They observed that the enterovirus is spreading due to faecal and oral contamination. This situation warrants a serious thought as if AES is occurring due to litchi then why all the regions with concentrated litchi pockets are not affected.

Only 62 per cent of the total admitted children had low blood glucose levels." Previous reports have showed that high blood sugar levels have also been recorded in AES affected children in Muzaffarpur in 2015. If AES is indeed occurring due to hypoglycaemia caused by litchis, no explanation is conceivable for the abnormally high blood sugar levels in the patients observed in 2015. The MCPG (methylene cyclopropyl glycine) present in litchi might have caused hypoglycaemia in the affected children, but hypoglycaemia alone may not be held responsible for AES occurrence. Hypoglycaemia alone may not be able to explain encephalopathy, which usually persists in spite of infusion of glucose.

Bandyopadhyay *et al.* (2014) had concluded that the evidence gathered by them points towards a viral etiology although the causative virus remained unidentified. They recommended further studies using high throughput sequencing technology and re-sequencing micro-arrays for identification and characterization of the causal. They reported that the hypoglycaemia caused by eating litchis might have aggravated the AES rather than actually causing it.

The AES occurrence coincides with the litchi harvest season. This period is also marked byvery high temperatures and hot winds prevalent during this time of the year. Heat stroke may also be a contributory factor for AES. It is reported that each affected child seemed to be an isolated case in a village having approximate population of about 2500. This again brings questions as to only one child ate too many litchis than others and missed the evening meal as well. If it is indeed caused by only by litchi consumption, it must result in AES in children of similar nutritional status and belonging to the same age group on consuming litchi fruit. It is worth mentioning that in some of the studies, among the affected AES children somewere in the age group of below 1 year who cannot eat litchi fruits (Dinesh *et al.*, 2013). It is indeed mind-boggling that if AES is occurring due to consumption of litchi, then how it is affecting children below the age of 1 year for whom eating litchi is unthinkable. This cannot be overlooked and plausible reason need to be found out by the researchers for this phenomenon.

We are of the view that the association between litchi and AES occurrence is far from being established irrefutably. Further studies and critical examinations are required to understand the causal agents and their mechanism of action for the occurrence of AES before it can be unequivocally related only to litchi consumption. In our view, an unidentified virus, the hot climatic condition prevalent at that time of the year along with rampant malnutrition and lack of hygiene might be the causes of occurrence of AES. However, as a precautionary measure, it is advised that attention should be given on the nutrition of children, intake of proper evening meal and consumption of fully mature fresh litchi fruits by children in rural areas of vulnerable districts, particularly in malnourished children (Nath *et al.*, 2016).

3.0 Origin and Distribution

Litchi originated in the region between latitudes 23° and 27° N in the southern China, northern Vietnam and Malaysia (Menzel and Simpson, 1990), where it has been cultivated for over 3000 years. It was introduced to Myanmar and India by the end of 17th century (Goto, 1960; Liang, 1981) and to the West Indies by the 18th century. It was introduced into Australia, South Africa and Hawaii by the end of the 19th century (Menzel and Simpson, 1986b). An attractive and rare sub tropical fruit tree that bears a luscious red fruit, the Lychee originated over 2000 years ago first in the north tropical rainforests and mountain forests of Southern China, where it is a dominant tree species.The second place wild lychee trees began to emerge was in the southern sub tropical low elevations of Southern China in the provinces of Guangdong/Kwangtung and Fujian/Fukien primarily along rivers and near the seacoast. These lowland lychees are most likely a result of fruit/seeds that were washed down rivers from the mountains or were transported by birds and then were deposited naturally in great conditions for lychee growth.There are still villages in Southern China today with lychee trees that are over 1000 years old! There is some evidence to suggest that wild lychees may have also originated in northern mountain regions of Vietnam (http://www.lycheesonline.com/origin.cfm).

It grows wild in elevated and lowland rainforest, especially in Guangdong Province and Hanian Province and was distributed to the tropical and subtropical world from 1600 AD (Menzel and Simpson, 1994). It is grown as a major commercial crop in China, Taiwan, Vietnam, Thailand, India, Madagascar, South Africa and Reunion Islands and to a limited scale in Australia, New Zealand, Indonesia, Mauritius, Israel Spain, the U.S. and Mexico (Menzel and Simpson, 1994).

4.0 Botany

Litchi is an evergreen tree, generally with round and compact canopy. Generally, litchi trees have short, thick, dark greyish brown trunk. However, several litchi trees can grow very tall if not trained and pruned at right growth stages. About 3-4 vegetative flushes are produced each year having several leaves and internodes. The floral flush is initiated after a brief period of chilling requirement (about 200 hours <7°C) is fulfilled, generally in the months of Dec-Jan. Floral bud initiation starts during spring months of Feb-March in India. Floral flush is a terminal inflorescence compound dichasium. In between the vegetative flushes, dormant periods reign. The leaves grow during this period becoming dark green, thickening and extending the tree canopy. Generally, trees show luxuriant vegetative growth during summer months. New flushes may follow as soon as in one and a half month at optimum summer temperatures after the leaves of previous flush mature. It has been observed that low temperatures reduce vegetative growth while pruning promotes it. Different cultivars show different colours of new flush ranging from light green to bright red or reddish brown. However, the leaves turn dark green when mature. In India, China cultivar shows bright red to brown coloured new flushes while the Shahi cultivar shows light green colour flush. At maturity, leaves are thick glossy dark green in colour on the upper surface and greyish green colour on lower surface. Leaves are pinnately compound with 4-7 leaflets arranged opposite to each other, attached with short petioles to the rachis.

Litchi inflorescence is panicle, determinate and composed of several panicles produced on current season shoots. The length of panicle varies from 9-45 cm and flower buds range from 400-3500 per panicle. The flower buds are inconspicuous whitish green in colour. The male flower is first to open followed by female flowers and then hermaphrodite flowers. Panicles are mixed in nature with the lower buds emerging into leaves and the terminal buds producing floral branches.

In India, the peak flowering season is during Feb-March. Flowers are apetalous, have a cup shaped calyx with 4-5 short sepals and 6-7 stamens about 0.3-0.6 cm in width at full bloom. The flowers are borne on the same panicle in succession which is almost synchronous on a tree and for the same variety. There are three types of flowers based on the development and function of pistil and stamens. The type I flower buds are functionally male, generally showing 6-7 stamens and do not have functional ovules. The type II flowers are functionally female, with a pistil having two carpel and two lobed stigmas while the stamens are indehiscent stamens. The type III flowers are functionally male having 6-8 stamens but dysfunctional rudimentary pistil with absent style and stigma. Cultivar and environment influence litchi flowering significantly. Generally, a trend has been seen of higher male flowers with increase in temperature. The floral succession occurs at an interval of about 7-10 days.

Fruits take about 60-80 days to mature. Litchi fruit is a round or heart shaped drupe, with weight ranging between 18 to 30 g in India. The pericarp is thin with flattened tubercles at maturity. The pericarp is leathery in beginning and becomes hard, brittle and dry with passage of time. The fast moisture loss results in quick drying leading to colour change of pericarp from red to brown. Litchi is an arillate fruit. The white juicy aril consists 60-70 per cent of the fruit. Aril is an extension of cells in the outer seed coat. Fruits have one brown coloured seed at maturity, 1.5-4 g in weight depending on cultivar. Fruits with abortive or shriveled seeds are preferred due to high pulp recovery. These fruits are also known as 'chicken tongue'.

Litchi seed is recalcitrant in nature and sensitive to moisture stress. Within the fruit, seed viability is maintained but is rapidly lost within a day or two after separation from the whole fruit. Seeds sown soon after separation from the whole fruit, germinate readily and the optimum conditions required for germination are sand bed under shade with regular irrigation. In a litchi cultivar, 'Early Bedana' with a chicken tongued seed, traditionally considered non-viable (Pandey and Sharma, 1989), germination recorded was as high as 60 per cent when sown in sand bed under shade.

Phenology is "*the science dealing with the influence of climate on the recurrence of such annual phenomena of animal and plant life as bird migrations and budding*". The growth of the lychee tree is rhythmic, meaning that "the shoots have a marked endogenous periodicity of extension." The litchi being a terminal growing crop, girdling helps in limiting the carbohydrate source of the growing fruits within the region between the girdled portion and the panicle (Singh *et al.*, 2017b).

One of three primary types of shoots (*i.e.* actively growing branch tips or laterals regardless of type of growth), is typically induced to develop from initiated buds: vegetative (leaves only), generative (determinate panicles or indeterminate inflorescences), or mixed (composed of both leaves and lateral inflorescences inserted at each node or, in general, a mixture of flower and leaves) (Davenport, 2003). Vegetative flushes (*i.e.* growth occurring in numerous shoots, usually in sections of tree canopy or throughout the entire tree) of growth typically occur one to several times per year on individual stems (*i.e.* branch tips that are in rest). The frequencies of flushes that occur annually depend upon cultivar, size of the tree, and growing conditions (especially related to nitrogen and water availability). Reproductive flushes generally occurs after extended periods of stem rest in the low-latitude tropics or immediately following periods of cool night temperatures in the higher latitude in tropics and subtropics (Davenport, 2000).The light compensation point (LCP), light saturation point (LSP) and apparent quantum yields (AQY) of litchi indicated that it was a heliophilous plant (Zhang *et al.*, 2014).

5.0 Species and Cultivars

5.1.0 Species

Litchi is also known as linchi (Thailand), letsias (the Philippines), litsi (Indonesia), lychee (Australia), laici (Malaysia), laici (southern China), lici (Italy), litchi (Portugal) and Litchi de Chine (France). The original spelling (litchi) appears to have originated in northern China (Tindall, 1994).

The litchi belongs to the family Sapindaceae and subfamily Nepheleae which has about 150 genera and 2000 species (Chapman, 1984b; Menzel, 1991b; Yeap, 1987). Only few members of this family are of horticultural importance. Related members of horticultural importance of the family from South East Asia include the longan (*Dimocarpus longan*), rambutan (*Nephelium lappaceum*), pulsan (*N. mutabile*) and Spanish lime.

The genus *Litchi* has three species *i.e.* subsp. *chinensis*, the commercial litchi, subsp. *philippinensis*, which developed in the Philippines and Papua New Guinea at high altitude and subsp. *javenensis* from the Malay Peninsula and Indonesia (Menzel, 1991b). The subsp. *philippinensis* has long oval-shaped fruit with inedible flesh and long thorn-like protuberances which can be used as rootstock. The subsp. *javenensis* bears fruit somewhat similar to litchi but the aril is thinner (Menzel and Simpson, 1994).

Menzel (1991b) has described the characteristics of litchi tree. The trees are long-lived, medium to large, much branched, round-topped, handsome evergreen, reaching up to 10 metres or more in height with a short stocky trunk. Leaves are compound, alternate, consisting of 4-7 oblong leaflets, each about 7-10 cm in length, glossy dark green above and greyish green under surface. Bark is greyish brown and rough. Vegetative growth is rhythmic and occurs in 3 to 4 recurrent flushes alternating with periods of rest (Menzel, 1991b; Menzel *et al.*, 1988a, b) during which new leaf primordia are initiated from the apical dome (Robbertse *et al.*, 1995a). Occurrence of vegetative flush 1 to 3 months prior to flowering eliminates flowering (Menzel and Simpson, 1994). The inflorescence is a compound raceme developing both from terminal and axillary buds. Flowers are small, apetalous, male, pseudohermaphrodite (functional male) and hermaphrodite. Flowers of different sexes on the panicles do not open simultaneously. They are found to open in different flushes.

The mature litchi fruits, which are one-seeded nuts, usually develop in bunches and vary in shape and size according to cultivar. The fruits tend to be 2.5 to 3.5 cm in diameter and usually oval in shape. The pericarp is papillate like strawberry and turns pinkish red when fruit is ripe. The aril, an edible portion in litchi, is an outgrowth of the outer cell layers of the seed coat (outer integument), which separates easily, lies under the pericarp and completely surrounds the seed. It is fleshy, succulent, transluscent, almost white and soft in texture. The seed is dark brown in colour and 1.0 to 2.2 cm in length and 0.6 to 1.2 cm in width.

5.2.0 Cultivars

The Chinese consider that the litchi has more cultivars than any other fruits. Although over 100 cultivars are grown in China, only about 15 are commercially important. The Chinese recognize extreme difficulty in maintaining the yield and quality of highly prized litchi cultivars under a particular climatic, soil or cultural conditions other than those from the original area of selection. Each district, therefore, has its own major cultivar. Wai Chee cultivar dominates in Guangdong province while Souey Tung and Haak Yip dominate in Fujian province. Other important cultivars are Tai So, Chen Zi (Brewster), Sum Yee Hong, Kwai May and No Mai Chee (Menzel and Simpson, 1986a). In Thailand, litchi is commonly grown in the northern provinces of Chiang Mai and Chiang Rai. Lowland litchi, also grown in the central province of Samut Songkhram are Kom, Kara-Tone, Sa-Lack, Kam-Pun, while the commercial cultivars in the northern provinces are Hong Huay, O-Hia and Kim Cheng (Subhadrabandhu, 1992). The varietal situation in Vietnam is very confusing. One cultivar may be known under several different names depending on the district. The Tu hu, Vai Trung, Tau Lai, Thieu, Kha Xuan, Hoai Chi are some of the cultivars growing in Vietnam (Menzel, 1991a). The major litchi cultivars grown around the world are given in Table 3.

About 50 cultivars of litchi are grown in different states in India. In Bihar, which has the largest area under litchi in India, the important cultivars are Shahi, Kasba, China, Deshi, Purbi and Muzaffarpur. The important cultivar grown in Uttar Pradesh is Dehra Dun. The cultivars recommended for growing in Punjab and Haryana are Saharanpur, Dehra Dun, Muzaffarpur, Rose Scented, Calcutta and Seedless Late. In West Bengal, Bombai, Elaichi, China and Bedana are considered important both for quality and yield.

Table 3: Major Litchi Cultivars Around the World*

Country	*Main Growing Areas*	*Major Cultivars*
China	Guangdong	Wai Chee, Haak Yip, Sum Yee Hong, Kwai May and No Mai Chee
Fujian	Souey Tung, Haak Yip, Tai So and Brewster	
Taiwan	Tai Chung	Haak Yip and Sah Keng
Thailand	Chiang Mai, Lamphun and Fang, Tai So, Wai Chee and Baidum	
India	Bihar state	Shahi, Rose Scented and China
Madagascar	Eastern coastal belt	Tai So
South Africa	Transvaal-Lowveld	Tai So, Bengal (Madras) region
Reunion	Wet coastal/sub-coastal areas	Tai So
Mauritius	-	Tai So
Australia	Eastern coastal strip U.S.A., Hawaii	Fay Zee Siu, Tai So, Bengal, Wai Chee, Kwai May Pink and Salathiel Tai so and Kaimana

* Menzel (1995).

Higher individual fruit weight is a desirable characteristic. Normally, each litchi fruit varies between 18-32 g. High average fruit weight increases the pulp recovery. Cultivar with longer keeping quality should be selected. Uniform ripening is required to complete harvesting operation in one go reducing cost. Early or late varieties are preferred to realize maximum profits. Attractive bright red fruit colour is a desired feature. A thicker fruit pericarp slows down the moisture loss rates and the consequent desiccation thus enhancing the shelf life of the fruit and preventing fast pericarp browning development. Fruits showing 'chicken tongue' *i.e.* presence of aborted or shriveled seed are very much liked due to higher pulp recovery. The characteristics of some important litchi cultivars as described by Chakraborty *et al.* (1980); Ray *et al.* (1985); Menzel and Simpson (1986a) and Ghosh *et al.* (1987) are given below.

There are thirty-five cultivars of litchi grown in different parts of India. Most of these cultivars are selected from Chinese cultivars. The majority of litchi production comes from only six cultivars namely Shahi, China, Muzaffarpur, Rose Scented (Dehradun), Seedless and Bombai. As a matter of fact, each district has its own major cultivars. In Bihar, Shahi, Rose Scented and China. are the most commercial varieties in Muzaffarpur while Kasba and Purbi are the choicest cultivars of eastern parts of Bihar. The other cultivars grown on small scale are Early Bedana, Late Bedana, Desi, Mandaraji, Longia and Kaselia. Among eight cultivars being grown in West Bengal, the cultivars Bombai, Ellaichi Early, Ellaichi Late and China are considered to be more promising both from quality and production point of view. Other cultivars of this area are Deshi, Purbi, Muzaffarpur and Kasba. Late Seedless and Culcuttia are partly resistant to hot and desiccating winds and bear good quality fruits profusely. The physico-chemical composition of some litchi cultivars is given in Table 4.

Table 4: Litchi Cultivars and their Distinctive Characteristics*

Name of Variety	*Maturity Time*	*Characteristic Features*
Shahi	Earliest (seond week of May)	This is the most popular cultivar of North Bihar and Jharkhand, Uttaranchal and UP. The fruits have distinct rose aroma and hence called as Rose Scented. It is known as "Shahi" in Uttaranchal and "Muzaffarpur" in Western Uttar Pradesh. This is an early season maturing cultivar ripens during last week of May to first week of June at various locations. It matures 12-15 May in Jhakhand, 25th May in North Bihar and by end of May in tarai region of Uttaranchal. Trees are very vigorous(7.6 m height and 8.2m canopy spread) and high yielder (90-100 kg/tree) but mature fruits are prone to cracking. Fruits are medium to large in size (3.2 cm length and 3.1 cm diameter), medium in weight (20.49g/fruit), globules-heart or obtuse in shape having rose madder and fuchsia purple backgroundwith red tubercles at ripening. Pulp grayish-white, soft moderately juicy (54.8 per cent) and sweet with 20.0 per cent TSS, 12.79 per cent total sugar and 0.33 per cent total acidity. Seeds are small (1.89 cm length,1.32 cm diameter and 2.07 g weight), smooth, shining round-ovate in shape and blackish chocolate in colour. Rind: pulp: seed ratio by weight is 12.22: 75.93:11.85. The fruits are known for excellent aroma and quality.
Sabour Madhu (Purbi x Bedana)	Late	Fruit is medium with 68.6 per cent pulp, 21.0 per cent TSS with light red colour. Has higher no. of fruits (24), ripened 8 days later than 'Kasba'.
Sabour Priya	Late	It is hybrid of Purbi x Bedana. The fruit has higher pulp recovery and TSS content.
Swarna Roopa	Late maturing but early (3rd week of May) in Ranchi conditions	Cracking resistant cultivar selected at CHES, Ranchi, fruits with small seed and high aril (65-70 per cent) content. The cultivar is suitable for extended harvest as it matures after China and is prized for its attractive fruit colour and resistant to cracking. Bears 18-20 attractive fruits in cluster, protuberances completely flattened at ripening. Fruit is round, medium (3.0 x 2.7cm, 18.95g), highly sweet (TSS 19.0 per cent) and contains high pulp content (76.62 per cent) with medium acidity (0.39 per cent). The pulp contains 12.5 per cent total sugar, 8.5 per cent reducing sugar and 38.30mg/100g ascarbic acid. It has small and flat seed (3.1g) with dark brown colour.
Ajhauli	Early	This is another early maturing variety becomes ready for harvest by 15-18 May and yields about 80-100 kg from a sixteen year old tree. The tree is vigorous in growth. It bears red fruits with average weight of 17.3 g and contains 59.3 mg/100 vitamin C. The TSS of the cultivar is 20 per cent with 0.4 per cent acidity and 12.37 g/100 g sugar. The full grown up tree yields 70-95 kg fruits every year.

Name of Variety	*Maturity Time*	*Characteristic Features*
Bombai	Early (first to second week of May)	It is a regular bearing variety, table purpose in West Bengal, average yield is 80-90 kg/tree; fruits are large in size (3.5 cm long and 3.2 cm diameter); obliquely heart shaped and weigh 15-20 g; Fruit colour attractive carmine red with uranium green skin background; pulp is greyish-white, sweet, soft, and juicy; flesh recovery of 63 to 67 per cent. Fruits are suitable for canning and acharacteristic feature of this cultivar is a tiny under developed fruit attached to the fruit stalk of each fully developed fruit
CHES-2	Late	It is a clonal selection from Bombaia at CHES, Ranchi, Jharkhand. It bears fruits both at the outer and inner canopy, which helps in reducing the sunburn as well as fruit cracking. The fruits are deep red, conical shaped and appear in a cluster of about 15-20. The fruit has an average weight of 21.3 g containing 3.8 g seed and 16.1 g pulp. Low sunburn and cracking percentage.
Calcuttia	Late season last week of June	Heavy yield and tolerance to hot dry winds make it attractive to growers and less susceptible to sunburn and cracking Table purpose, fruits are large in size, oblong with tyrian rose colour and dark tubercles at maturity. Pulp is creamy white, soft, juicy, very sweet with 18.2 °Brix (TSS).
China	Medium-late	This in one of the best cultivar of litchi for North India. It is tolerant to hot waves and fluctuations in soil moisture which escape to fruit cracking. It is also known by the names of Calcuttia in West Bengal. This is a medium-late season cultivar and fruits ripen during end of May in West Bengal, first week of June in Jharkhand and during third week of June in North. Trees of cultivar China are dwarf (4.0 m height, and 6.0m spread) and high yielder (80-100 kg/tree) but prone to alternate bearing. It bears fruits in cluster of 12-18. The plants bear less fruit in southern direction. Fruits are large in size (3.86 cm length and 3.26cm diameter), medium in weight (22.0g/fruit), oblong in shape and tyrant rose in colour with dark tubercles at maturity. Aril is creamy-white, soft, Juicy, sweet having 18.17 per cent TSS, 11.0 per cent total sugar and 0.43 per cent titrable acidity. Seeds are glaucous, dark chocolate in colour, oblong to concave or Plano convex in shape, medium in size (2.9 cm length and 1.5 cm diameter average in weight (3.49 g/seed). The ratio of rind: pulp seed by weight is 16.42: 69.22:14.36.
Dehradun	Third week of June	Table variety cultivated under the name of 'Dehra Rose'. Fruits are medium in size, measuring 3.7 cm in length, 3.5 cm in diameter, weighing 15.2 g and having oblique-heart to conical shape with bright rose pink colour. The pulp of this cultivar is greyish-white, soft, and moderately juicy with 18o Brix TSS, 10.4 per cent sugar and 0.44 per cent acidity. This variety is highly susceptible to sunburn and cracking especially under rainfed conditions

Name of Variety	Maturity Time	Characteristic Features
Early Seedless (Syn. Early Bedana):	Early ripening	It is also known as Early Seedless because of its early ripening and small seeds. The cultivar is very much popular in Uttar Pradesh and Punjab. Trees are medium, attaining an average height of 5.0 m and spread of 6.2 m. It is a medium yielder cultivar (50-60 kg/tree) but bears fruits regularly. Fruits are medium in size (3.2x3.0 cm) and weight (16.33g) having oval or heart shape, rough surface with uranium green skin covered with carmine red tubercles at maturity. Aril is creamy white, soft, juicy (69.0 per cent) and sweet containing 19.8 per cent TSS, 13.6 per cent total sugar and 0.54 per cent titrable acidity. Seed is very small, shrunken, glamorous, dirty chocolate in colour with an average weight of 0.82g. The rind: pulp: seed ratio by weight is 13.06: 83.19:3.75. The overall fruit quality of the cultivar is good
Elaichi	Mid season (first week of June)	Fruits are conical, marigold orange red with an average weight of 12-15g Pulp creamy white in colour, sweet, soft, juicy with agreeable flavour. The cultivar has 18 °Brix TSS, 11.50 per cent total sugar 0.45 per cent acidity and 6.91:1 pulp: stone ratio at ripening. Seeds are relatively small, shining, less susceptible to sunburn and cracking.
Gulabi	Late fourth week of June	The pulp is firm, have a unique rose flavour, greyish white, and sweet; seed large in size; pulp recovery 60 to 65 per cent.
Mandraji	Mid season	Fruits have rough thick skin.
Kasba	Mid season (first week of June)	Fruits are conical, marigold-orange red in colour, fruit weighs between 21-25g; fruit pulp creamy white in colour, sweet, soft and juicy with agreeable flavour; Seeds are relatively smaller than Shahi; recovery 57 to 60 per cent. High yielder and less susceptible to sunburn and cracking
Kaselia	Late maturing	Fruit attain a pink red colour. The pulp content is comparatively low and the seed is big; pulp recovery 50 to 54 per cent and poor taste. This cultivar is also known as 'Khatti' or 'Pickling'.
Late Large Red (Syn. Muzaffarpur)	Late maturing	Fruits are large in size, oval or oblong conical with crimson red tubercles. Pulp is greyish white, soft, moderately juicy with a TSS of 20.3 °Brix; pulp recovery 62 to 65 per cent
Late Seedless (Syn. Late Bedana	Late maturing early to mid June	This cultivar is also known as Late Seedless. This is a late maturing cultivar which usually ripens in third week of June in Uttaranchal and end of May in Jharkhand. The trees are vigorous having an average height of 5.5 m and spread of 7.0m.

Name of Variety	Maturity Time	Characteristic Features
		It is a high yielder cultivar giving an annual yield of 80-100 kg/ tree. Although the fruit size is medium (3.2x 2.9 cm) but average weight is 25.0g. The fruits are conical in shape with vermilion to carmine in colour having dark blackish brown tubercles at maturity. Pulp is creamy white, soft, juicy (65.4 per cent), sweet having 20.0 per cent TSS, 13.8 per cent total sugar and 0.44 per cent acidity. Seeds are very small (1.8x0.9 cm in size and 0.85 g in weight), shrunken, glamorous, chocolate in color having bursiform shape similar to canine of dog. The rind:pulp: seed ratio (by weight) of late seedless is 14.76:81.89:3.35. Overall quality of fruits is very good
Longia	Late maturity	Fruits are medium in size and the aril has an excellent aroma. Due to shy bearing habit, there is a declining preference for this cultivar; pulp recovery 50 to 55 per cent.
Purbi	End of May or first week of June	Table purpose and at maturity, red tubercles appear on pinkish brown background with, pulp recovery 57-60 per cent. It is also known as Bengal and quite popular in Australia Fruits are less susceptible to crackin
Bengal	Mid season maturing	It is a seedling selection of Indian cultivar Purbi and sent to Florida in 1929. Trees are vigorous and spreading. Fruit (23-27 g) are formed in large clusters of up to 50 or more, the thick skin is very rough and attractive bright red. The fruit tip is distinctively pointed, large seeded and very few abortive seeds, aril is often undeveloped and may not cover the seed at the pointed end; pulp recovery of 55 per cent or lower. Branches are thin but reasonably resistant to wind damage Large seeded with low pulp recovery and show uneven ripening
Sabour Bedana	Early June	An excellent variety recently released from Sabour Fruits are large in size (24-30 g); soft, and juicy; small, chicken tongue seeds 80 to 90 per cent; and aril recovery of 75 to 80 per cent.
Deshi	Early season	It is an cultivar with regular high yields and less susceptible to sunburn and cracking. Large sized fruits (22-24 g) suitable for canning.
Pyazi	Early season	Pulp has flavor similar to blend boiled onion with poor edible quality fruits

* Singh *et al.* (2017).

The physico-chemical characteristics of a few Indian strains/cultivars are given in Table 5.

Table 5: Physico-chemical Characteristics of Important Litchi Cultivars*

Cultivar	*Fruit Wt. (g)*	*Pulp Wt. (g)*	*Seed Wt. (g)*	*Cracking (Per cent)*	*TSS (°Brix)*	*Acidity (Per cent)*	*Ascorbic Acid (mg/100 g)*	*Total Sugars (g/100 g)*	*Fruit Yield (kg/ tree)*
Ajhauli	17.34	09.77	2.88	51.20	20.00	0.40	53.80	12.87	70-95
Green	18.10	13.53	3.22	49.10	18.90	0.35	50.40	12.56	80-90
Purbi	20.77	13.40	3.96	7.60	20.60	0.31	53.80	11.00	90-100
Late Bedana	19.50	13.19	2.18	24.50	20.70	0.30	59.60	13.00	80-100
Early Bedana	19.32	14.99	1.47	4.50	19.50	0.24	50.40	13.91	50-60
China	20.30	12.30	3.83	0.00	20.80	0.44	45.30	11.85	90-100
Kaselia	11.56	07.42	2.68	24.80	18.70	0.78	51.30	10.90	60-70
Deshi	17.90	11.63	3.07	33.40	21.40	0.28	43.80	13.37	80-90
Dehradun	16.80	11.30	2.46	32.80	17.80	0.72	30.00	11.30	80-90
Kasba	20.30	14.20	2.80	0.00	18.80	0.46	50.60	14.37	60-70
Dehra Rose	17.77	11.78	2.81	29.70	22.10	0.31	45.60	13.59	80-90
Rose Scented	18.44	13.52	3.42	28.70	21.70	0.30	41.40	14.57	80-90
Longia	13.80	09.68	2.54	0.00	19.80	0.41	47.40	12.52	35-40
Shahi	20.98	12.33	3.88	31.50	22.60	0.26	53.60	13.85	80-100
Trikolia	18.25	13.28	2.18	32.30	22.40	0.32	40.40	13.95	80-105
Bombai	18.93	12.97	3.83	0.00	20.50	0.28	49.40	11.68	80-90
Calcuttia	22.00	13.53	3.49	0.50	18.20	0.43	43.60	11.00	80-100
Swarna Roopa	18.95	14.52	3.10	0.00	19.00	0.39	47.00	12.50	70-80
CHES-2	21.33	14.23	3.26	0.00	19.80	0.20	48.20	11.70	80-90

* Singh *et al.* (2012).

Brewster

This cultivar was obtained from Fujian province, China by Rev. W.M. Brewster and propagated in Florida in 1903. It was subsequently recognized as the Chinese cultivar Chen Zi. Litchi orchards in Florida have been almost exclusively confined to this one cultivar. Brewster requires a relatively severe winter (minimum temperature below 7°C) to initiate flowers. Trees are small and upright, with wide, strong crotch angles and dense foliage. It is one of the few cultivars with distinct lenticels (corky outgrowth) on the branches. The medium-sized fruits (20 to 22 g) are heart-shaped, having a bright pinkish red, thick, rough skin and are borne in small loose clusters. The flesh is slightly fragrant, juicy and sweet when fully ripe. Seeds are small to medium in size, with a large percentage being undeveloped. Flesh recovery is about 65 to 75 per cent.

Deshi

It ripens earlier and a good cultivar of West Bengal and Bihar in India. Fruits are medium in size, 3.4 cm long and 3.0 cm in diameter; average weight 17.2 g. The fruits are oval, oblong or oblong-conical in shape. The fruit is light red with deep red tubercles at ripening. Fruits are juicy; bearing is moderate and regular.

Haak Yip

This is the commercial cultivar of Thailand, Taiwan and Fujian province of China. Trees are medium in size, with dense foliage and long, thin, fragile branches. The heart-shaped fruit is medium in size (20 to 22 g) and formed in large compact cluster (15 to 30 fruits). The purplish red skin is thin and soft. The flesh, which separates easily from the seed is sweet, crisp and slightly aromatic and of excellent quality. Seeds are medium in size and fully developed (flesh recovery of 68 to 76 per cent).

Rose Scented

This is an important cultivar of Uttar Pradesh, Uttaranchal and Punjab. Fruits start ripening by third week of June. Medium vigorous trees (5 m height and 7m spread) produce medium to high yield (80-90kg/tree). Fruits are medium to large in size measuring 3.7 cm length, 3.5 cm diameter, 15.2 g weight having oblique-heart to conical shape. Carmine red coloured fruits of Dehra Rose look very attractive at ripening. Pulp of this cultivar is grayish-white, soft moderately juicy at ripening. Pulp of this cultivar is grayish-white, soft, moderately juicy (61.9 per cent), with 17 per cent TSS, 10.4 per cent sugar and 0.44 per cent acidity. Seeds remain small (2.4cm length, 1.4 cm diameter and 2.4 g weight), shrunken, mostly oblong in shape and dark chocolate in colour. The rind: pulp seed ratio is 18.4: 64.4: 17.2. The major defect of the cultivar lies in its susceptibility to sunburn and cracking.

Tai So

Tai So is the most widely grown cultivar in southern Queensland, Australia, and is comercially grown in South Africa in the name of Mauritius. Trees are vigorous and spreading with an open crown. Fruits are large (22 to 26 g) and somewhat egg-shaped with flat shoulder and a round tip. The thin skin is bright red changing to dull red at maturity. Taste is acidic sweet when immature and sweet when fully ripe. Seeds are medium in size giving a fair flesh recovery (65 per cent).

Wai Chee

This is the most common cultivar of China and has also occupied a large area in Australia. Trees are low, dome-shaped with thick branches and have compact foliage. The small (16 to 18 g), rounded fruits are formed in small, loose cluster. Skin colour is deep red. The flesh is soft, very juicy and sweet. Most seeds are fully developed with flesh recovery of 63 to 73 per cent.

Cultivars of Litchi

Bedana

Bombai

China

Deshi

Kasba

Mclean

Early Large Red

Nafarpal

Cultivars of Litchi

Piyaji

Seedless Late

Rose Scented

Purbi

Elaichi

Muzaffarpur

Muzaffarpur Late

Shahi

Cultivars of Litchi

Swarna Roopa

Mauritius

Exotic Cultivars

Shuang Batlia

Bai Taying

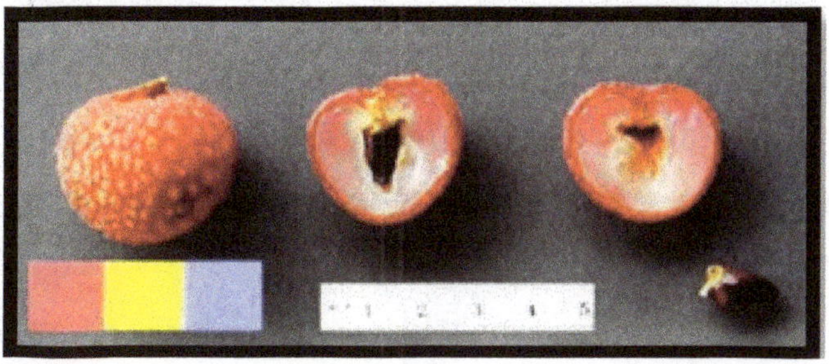

Groff

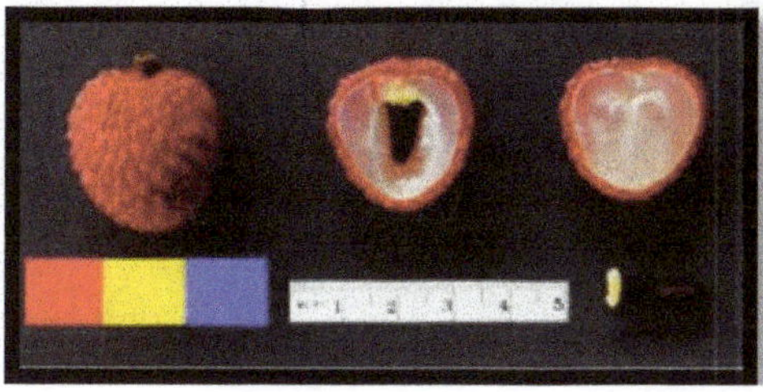

Heung Lai

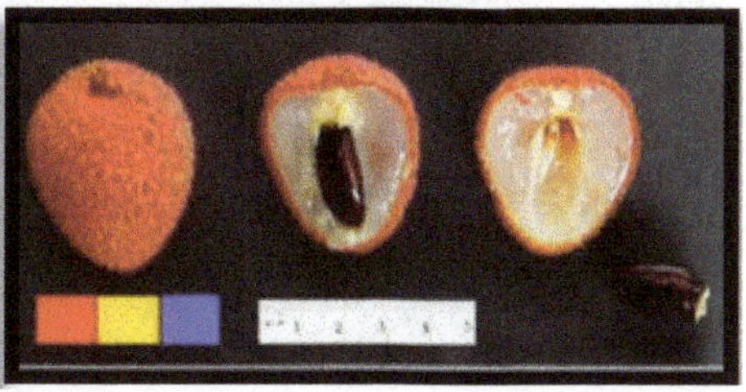

Yu Her Pau

Exotic Cultivars of Litchi

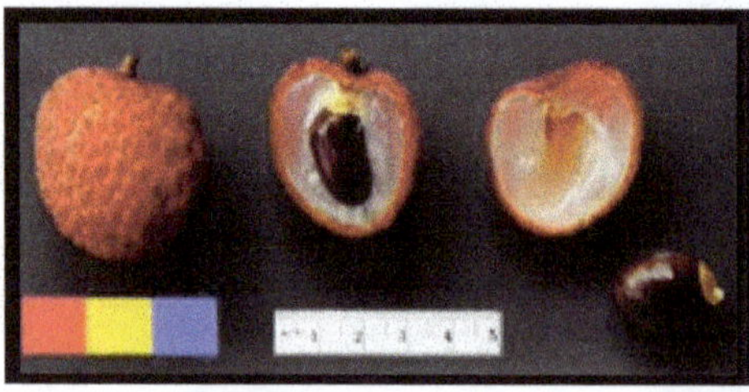

Shan Chee

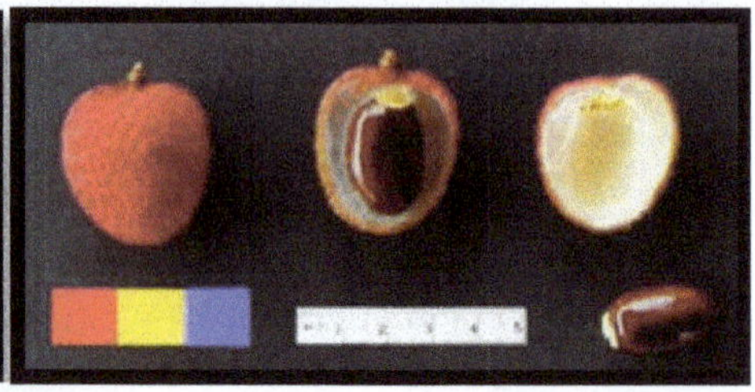

Peerless

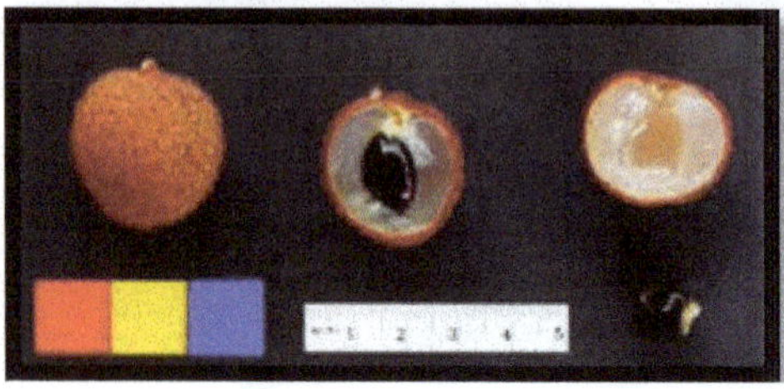

Khom

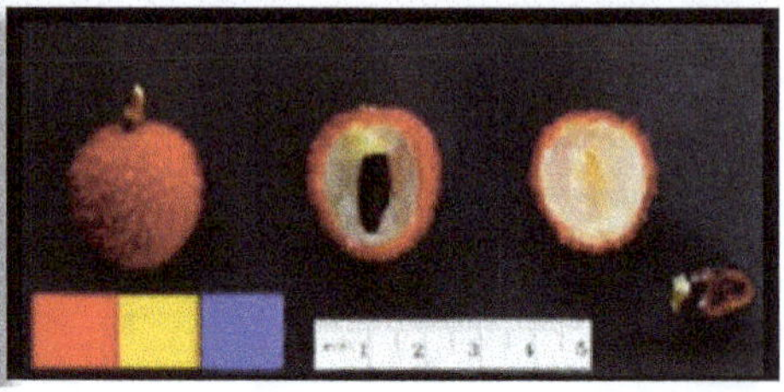

Champoo

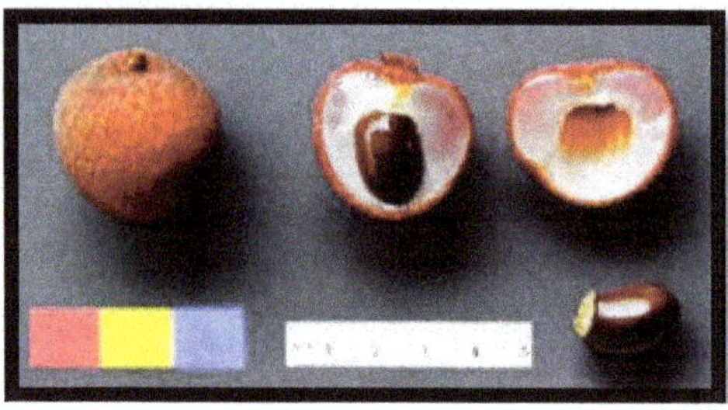

Hak Ip

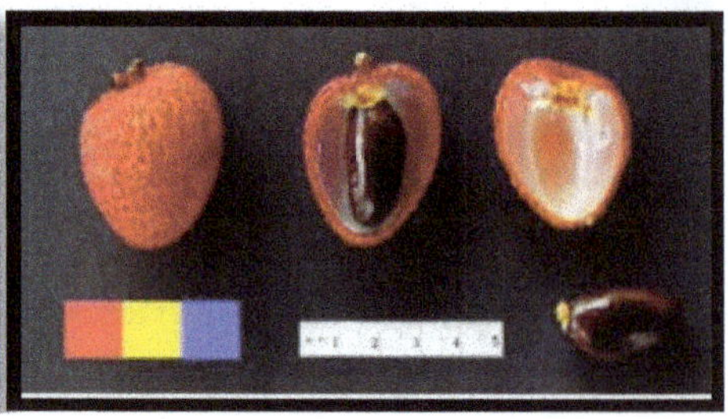

Bengal

Kawai May Pink

Fay See Siu

Baitangying

The tree is dwarf, precocious, early maturing, and high yielder with good quality fruits. It has good tolerance to some adverse environmental factors.

Liquili

It is a high yielding late maturing cultivar from Guangxi province, China. It begins flowering in the middle of April and the fruit ripens in early August. Fruits 15.68-21.39 g contain 15.02 to 18.45°Brix and 37 to 38 mg per 100 g ascorbic acid. It has wide adaptability and resistance to adverse climatic conditions.

The most important varieties grown in eastern China are Nanmuye, Guiwei, Jiangmiaolan and Dahongpao. The fruits of Nanmuye mature in mid-August and are yellowish red in colour, are borne in large, long clusters weighing 400-1100 g; fruit shelling rate is 72.7 per cent and the soluble solids content is 15.49 per cent; trees are productive, producing large fruit clusters weighing 500-1000 g; fruit shelling rate is 81.6 per cent and eating quality is very good. Fruits of Guiwei and Jiangmiaolan are dark red in colour and they mature in early August and late July, respectively (Wong, 1999).

Table 6: Attributes of New Varieties Developed at ICAR-NRC on Litchi, Muzaffarpur, Bihar*

Cultivar/Germplasms	*Special Characteristics*
Gandaki Sampada	☆ A late maturing strain that ripens during mid June. ☆ Fruits are large in size (36.85 g), conical in shape, vermilion to carmine colour, and cracking resistant. ☆ Fruit consists of creamy-white, soft and juicy pulp with 80 to 85 per cent pulp recovery. ☆ It has very high percentage of shrivelled and small seeded fruits. ☆ Yield ranges from 120-140 kg/tree.
Gandaki Yogita	☆ Dwarf plant stature, tolerant to hot waves and fluctuations in soil moisture. ☆ Tolerant to fruit borer and fruit cracking. ☆ Late maturing variety (5-15th June). ☆ Fruits are round in shape, tyrant rose in colour with creamy-white and juicy pulp. ☆ Good yield potential (70-80 kg/tree). ☆ Suitable for high density planting.
Gandaki Lalima	☆ A highly nutrient efficient strain possessing dark green leaves and the capability to withstand climatic aberrations ☆ Late maturing cultivar that ripens during second week of June. ☆ Fruits are conical, bright marigold-orange red in colour. ☆ Fruit weighs between 28-32 g with creamy white pulp. ☆ High yielder with average yield of 130-140 kg/tree.

* Gupta *et al.* (2017).

6.0 Breeding and Improvement

The development of better cultivars is very slow, because it takes several years for most seedlings to begin bearing fruit. When they do fruit, less than 1 percent of the seedlings are worthy of selection. Litchi is believed to originate in China where it was introduced to many countries including Thailand. There are many cultivars known today, these cultivars are mainly derived from the selection of seedling plants (Ray, 2002). Chaudhuri (1940) reported that the haploid chromosome number of litchi varieties was found to be 14. The *Litchi* spp. originated from more than one wild progenitor and the haploid chromosome number may be 14, 15 or 16 or rarely 17. According to Smith (1976) and Lu *et al.* (1987), the chromosome number of *L. chinensis* is 2n = 2x = 30. The somatic chromosome number of rambutan is 2n = 2x = 22 and that of longan 2n = 2x = 20.

The major objectives (Ray, 2002; Singh *et al.*, 2012) for breeding new cultivars of litchi are:

1. High and regular yields and avoidance of biennial bearing.
2. Large fruit size (individual fruit weighing over 20 g) and high number of fruits per panicle.
3. Small seed size, higher aril percentage and or/higher percentage of aborted seeds.
4. Bright red peel colour.
5. Good fruit quality (*i.e.* high sugar content and good flavour).
6. High keeping quality.
7. Resistance to physiological disorders (less prone to fruit cracking at maturity)/or pests like fruit borer.
8. Wider adaptability to ecological conditions.
9. Desirable tree characteristics, such as dwarfness, sterio fruiting

One of the most important factors limiting expansion of litchi production in several countries is the lack of suitable cultivars. New cultivars can only be developed from the selection of seedlings with improved characteristics. Most plant improvement programmes in litchi have been limited to the screening of small populations (<500 plants) of open pollinated seedlings (C. Menzel, personal communication, 1989). However, litchi breeding programme including germplasm collection and selection of seedlings from open and controlled pollination was started in Taiwan in 1982 (Yen *et al.*, 1984). The objectives of the breeding work were as follows according to priority (in parentheses were the cultivars that might provide the gene sources):

1. Various mature seasons to extend the supply of litchi in Taiwan (very early - Sun Yueh Hong, very late - Hawai Li)
2. High yielding and wide adaptability (Hak lp)
3. High quality and sweet fragrance (Yu Her Pau and No Mai Tsz)
4. Large fruit size (Yu Her Pau, Sun Yueh Hong and Sah Keng)

5. High percentage of small-seeded fruits (No Mai Tsz, Yu Her Pau and Sah Keng)
6. Fruit with bright and colourful red skin and smooth protuberances (No Mai Tsz and Kwai Mi)
7. Vigorous (Yu Her Pau),
8. Processing for dried litchi nut (No Mai Tsz) Long storage life,
9. Resistant to pests and diseases (downy blight, fruit borer and longicorn beetle) as well as low percentage of skin-cracking fruits.

In another words, the ideotype required for litchi are dwarf, regular bearing, precocious, wider adaptability with early and late maturity with high yield. There should be dark red skin surface and colour with fragrant aril and seed should be chicken tongued and the variety should be resistance to cracking and sunburn.

Table 7: Special Characteristics of some of the Litchi Cultivars*

Special Traits	*Source of Traits*
Small seed	Bedana, Nuomici, Lingshan Xiangli, Hainan Xiaodingxiang, Guangxi Zhangluoli
Crisp and sweet flavour with low tannin	Lingshan Xiangxi
Early maturity	Shahi, Shanyuedong Red, Early Bedana, Dehra Rose
Large fruit	Edanli (60–70 g), Vnsdasu (40–45 g)
High yields	Heiye (Black Leaf), Baitang-ying (White Super Poppy)
High temperature tolerance	Shanyehong: better flower bud differentiation at 20 C as compared to 12 C to common varieties
Good for canning purpose	Heiye, Xuangxi
Drought tolerance	Tianyan (Sweet Stone)
Good on-tree storage	Huaizhi: fully ripe fruits can be left on trees for fresh picking
Late maturity	Xuehuazi: ripening during early/mid-July, with high-yielding characteristics; Fijian Xiafanli (Fujian Xiafan litchi): maturing during late July or early August, enabling extended supply period in combination with early and intermediate maturity types; Longia, Kaselia, Yogda

* Gupta *et al.* (2017).

6.1.0 Breeding Constraints

The development of improved cultivars in litchi is a lengthy and cumbersome process, and trees when fruited only produce less than 1 per cent of the seedlings worthy of selection. Breeding for a new hybrid cultivar through conventional methods would require about 40 years in fruit trees such as litchi (Zheng *et al.*, 2001). In addition to poor seed viability and wide variability in seedling progenies, erratic flowering poses a hindrance for pollen collection during the breeding process. Although crosses between litchi and longan are successful, the use of longan as parents in a hybridization programme is limited by its biennial bearing tendency.

In crosses involving plants that have a tendency to produce chicken-tongued seeded fruits as the female parent, many of the most valuable progeny are lost before the harvest (Puchooa, 2004). Only a few cultivars have been bred through conventional methods because of the long juvenile period, the apparent lack of genetic variability in the existing germplasm, and the great expenditure that is required in terms of land, time and money (Litz *et al.*, 2005). Thus, the improvement of litchi appears to be confined mainly to selections of improved chance seedlings or genotypes. Future efforts in plant breeding need to emphasize identification of potential parents followed by their reciprocal crossing.

6.2.0 Gene Pool and Heritability

Most of the commercial cultivars originated from Chinese sources and the genetic base of commercial cultivars is relatively narrow (Ray, 1999). The potentialities of the three known litchi subspecies are yet to be exploited.

Fruit and seed weight had strong positive correlations with total sugar ascorbic acid, protein and tryptophan contents and significant negative correlation with acidity and phenol content (Singh *et al.*, 1987). Therefore, selection for fruit and seed characters can produce nutritionally superior genotype. The genetic variability and heritability with respect to fruit characters in nine litchi cultivars were studied by Sarkar *et al.* (1991). Significant genetic divergence for different fruit characters could be observed. The coefficient of variablity was the highest for dry seed weight (44.6-49.8) and broad-sense heritability and genetic advance were 89.1 and 82.3 per cent respectively. High heritability and high genetic advance were recorded for fresh seed weight, fruit weight, fruit volume and fresh and dry pulp weight. High heritability (42.7-79.5 per cent) in other fruit characters led to an inference that these characters could be improved by selection.

The natural extent of out crossing in litchi varied from 65 to 87 per cent depending on the nearness to the pollen source. Inbreeding depression with respect to fruit and seed weight was observed if selfing takes place (Stern *et al.*, 1993). Dwivedi and Mitra (1995) assessed the genetic diversity in a collection of 11 litchi cultivars on the basis of 8 fruit quality traits. The study revealed enormous diversity in the material as indicated by the wide range D2 value from 85.06 (between Bombai and Elaichi) to 26088.03 (between China and Elaichi). The cultivars fell into two clusters when grouped using the Tocher's method. Cluster I was the bigger one, accommodating nine cultivars and two cultivars fell in Cluster II. Crossing between cultivars belonging to Cluster I with the cultivars of Cluster II might lead to heterosis in fruit quality traits.

According to fruit characters, the collections of litchi were divided into three distinct groups:

Group I: (or Hon Li group)

Usually early and matures during early June in Chai Yi area. Fruit oblong to ovoid, bright red, with distinct protuberances and generally large-seeded. Fruit weighs between 20 and 25 g. This group includes Kau Shiung Early, Yu Her Pau Hon Pi, Foo chow, Wan Nien Hong and Chung yuan Hong.

Group II: (or Hak Ip group)

Matures during middle and late June in Chia Yi. Fruit cordate and dark red: protuberances usually turn smooth towards maturity. Fruit weighs about 20 g. This group includes Hak Ip, Kwang Tung, Kang Wei Late, Sah Keng and Sun Yueh Hong.

Group III: (or Kwai Mi group)

Generally, late and matures in early July in Chia Yi. Nearly globular and colourfully red: rough protuberances usually turn smooth towards maturity. Fruit weighs between 17 and 22 g. A specific fragrance is present in aril in most cultivars of this group and its content varies with different cultivars. The acid content of fruit decreases rapidly and is eatable when the skin just turns light red. This group includes Kwai Mi, No Mai Tsz, Hwai Li, Hwai Chih, Kwai Chih, Shang Shu Hwai, Tang Po and Hsin Hsin.

Guinuo, a new litchi cultivar selected from offspring plants of mutated individual seedlings in Guangxi, China. The cultivar was characterized with its big fruit size, excellent eating quality and high yield. Its pulp was low in juicy and crisp. The average weight of fruit, edible portion and content of soluble solid were 37.02 g, 73.36 per cent and 18.55 per cent respectively. (Zhu *et al.,* 2010).

6.3.0 Hybridization

Development of new varieties through intervarietal hybridization is limited in litchi due to slow growth of the hybrid seedlings, long gestation period to produce flowers and less than 1 per cent is worthy selection (Hamilton and Yee, 1970; Menzel, 1983). Thus, the improvement of litchi appears to be confined mainly to selection of improved chance seedlings or genotypes (Ray, 1999). The litchi cultivars developed through selection or hybridization is depicted in Table 19. Recently four new litchi selections were made at China and were recommended for cultivation in Hainan province (Maio Ping Sheng *et al.,* 1998). The selections were named as Aili, Ziniangxi, Mianbaoli and Fengil. Aili is a dwarf selection, producing fruits weighing 24.8 g on an average. Ziniangxi has large fruits (39.1-59.5 g) of very good quality, while fruits of Mianbaoli have a soluble solids content of 17.5 per cent. Four-year-old trees of Fengil produced an average of 11.6 kg of fruits with very high soluble solids content (18 per cent).

Intergeneric hybridization (litchi and longan) failed to evolve any improved hybrid and the resultant seedling progeny was quite variable with small fruit size which appeared to be a dominant character. Diallel crosses between *Nephelium lappaceum, N. rambutan* - akee, *Dimocarpus longan* and *Litchi chinensis* failed to produce any promising results. The intergeneric crosses between longan and litchi were successful but with variable progeny (Ray, 1999, 2004). Reciprocal crosses between commercial cultivars of litchi (Bengal and Kwai May Pink) and longan (MaCleans Ridges and Duan Yu) revealed that hybrid progeny could be produced only when litchi was used as the female parent. Morphologically the hybrid plants were similar to litchi (the maternal parent) but leaves were smaller. Three types of seeds developed in litchi following pollination with longan pollen: (1) Normal seeds with a developed testa and embryo; (2) seeds with aborted embryo but normal testa development, and (3) seedless, where the ovule remained the same size as at anthesis without further development of embryo or testa (Mc Conchie *et al.*, 1994).

Wild species like *N. philippinensis* (syn. *N. intermedium*) can provide an important gene for breeding work. Wild seedless varieties can be introduced in breeding programme which may have a great bearing in fruit production for canning industry. The possibility of the potential of hybrids of *N. mutabile* and other wild species with rambutan may be explored for use as rootstock which could be more resistant to root diseases than the rambutan (Ray, 1999).

7.0 Propagation

7.1.0 Seed

Propagation by seed is not usually recommended since litchi seedlings are generally diverse and most are characterized by a long juvenile period. Nevertheless, new cultivars can only be developed from the selection of seedlings with improved characteristics. Several popular cultivars are seedling selections, *e.g.* Peerless from Brewster (or Chen Zi) in Florida (Knight, 1963), Bengal from Purbi in Florida (Batten *et al.*, 1982) and Groff from Haak Yip in Hawaii (Hamilton and Yee, 1970). Litchi seeds will keep for 4 weeks in the fruit after harvest. They begin to lose viability within a day if removed and none will germinate after 4-14 days, depending upon the relative humidity (Storey, 1973). The seeds showed 100 per cent germination when they were sown fresh or a day after being taken out of the fruits and kept in water. Water-stored seeds, however, remained viable for an appreciably longer period and showed 75 per cent germination on the 7th day and 30 per cent germination on the 15th day of storage (Ray and Sharma, 1985). Soaking of seeds in growth regulator solutions (GA_3, IBA and ethephon) improved the germination rate (Prasad *et al.*, 1996). During seed germination the various starch degrading enzymes showed periodicity in peak activity. α-amylase was significantly active at the beginning of germination and was closely followed by β-amylase. Starch phosphorylase was most active only when the activities of both α and β-amylases were low (Prasad *et al.*, 1994).

7.2.0 Vegetative Propagation

7.2.1 Stem Cuttings

Cuttings most favoured were from 15 to 20 cm long with stem 8-15 mm in diameter. Litchi produced roots successfully from softwood (Ahmed, 1961), semi-hard wood (Paxton *et al.*, 1978) and hardwood cuttings (Sen *et al.*, 1969; Paxton *et al.*, 1978). Paxton *et al.* (1978) obtained 60-80 per cent rooting from semi-hardwood and hardwood cuttings of cv. Bengal, collected in May in southern Queensland, Australia prior to flowering. Applied auxins have been found to increase percentage and uniformity of rooting, hasten root initiation and increase number and length of roots in litchi cuttings (Ahmed, 1961; Sen *et al.*, 1969). Bose *et al.* (1985) noted significant promotion in rooting in semi-hardwood cuttings of cv. Bombai by ringing large shoots 15 days prior to taking the cuttings and treating them at the time of planting with IBA at 10,000 ppm in combination with ferulic acid at 200 ppm. Litchi cuttings can appa-rently root successfully in a variety of media, such as sand, peat, vermiculite, styrofoam and organic matter, provided moisture, drainage, aeration, support, sanitation and pH are adequate. Girdling of 20 cm long terminal and subterminal cuttings 30 days before severing, followed by treatment with IBA (5000 and 10,000 mg/kg talc) and inserted in 1:1 peat + sand and 1:1 polystyrene (granules) + perlite under mist at 29?C and/or with bottom heat (20± 2 °C) markedly enhanced rooting. Single node cuttings (2 cm long) rooted best when treated with 10,000 mg IBA/kg talc (Oosthuizen, 1994, 1995).

7.2.2 Layering

Air-layering or marcottage is, and will probably continue to be the most widely used commercial method of propagation of litchi (Menzel, 1985). Upright branches (1-1.5 cm stem diameter and 30-60 cm long) from well developed trees, free from pests and diseases are selected for marcottage (Loebel, 1976; Datta *et al.*, 1995). Roots developed faster on branches with mature vegetative growth compared with recently flushed wood (Campbell, 1981). Litchis can be marcotted all the year round in Hawaii, although they are usually marcotted in spring (on non-flowering branches) before active growth commences (Yee, 1972). In Florida, marcots were taken in June during conditions of warm temperatures, high humidity and daily rain (Storey, 1973). The best rooting and survival were obtained in spring and early summer compared with autumn and winter in India (Chadha, 1968), South Africa (Joubert, 1970) and Queensland (Campbell, 1981). Several workers have demonstrated that auxin application improved rooting and survival of litchi marcots (Chadha, 1968; Sadhu *et al.*, 1972). Sadhu *et al.* (1972) recorded that phenols (*p*-hydroxybenzoic and ferulic acids) showed synergism with IBA in the rooting. Pretreatment of stock plants with cycocel [chlormequat] (100 ppm) followed by treatment with IBA at 5000 ppm and *p*-hydroxybenzoic acid at 200 ppm during the preparation of air layers resulted in 100 per cent rooting after 60 days and 80 per cent survival of rooted layers after 6 months of planting. Similar rooting and survival percentages were observed following pretreatment with ethrel [ethephon] (250 ppm) and treatment of air layers with IBA at 3000 ppm (Dutta *et al.*, 1995).

The young plants must be planted in individual bags (15x15x25cm) filled with good mixture of soil, vermicompost, sand and Vermiculite. Two or three vegetative flushes must appear before the plants hardened off in the open air prior to final planting. Planting of detached air layers in the root trainer of 350 ml cup size facilitate the better root development in the litchi air layers.

7.2.3 Stooling

Stooling involves cutting a plant to the stump and mounding soil around the base of the newly developed shoots to encourage adventitious rooting. Litchi is particularly suited to this method of propagation, since they have stiff branches which do not bend easily and are capable of producing several shoots each year (Menzel, 1985). Ram and Majumder (1981) compared stooling and marcottage of litchi. Three-year- old trees at close spacing (6944/ha) were cut back to a 75 cm stump in February, earthened up and allowed to root. They were detached in September and planted in nursery. Stooling resulted in greater percentage of rooting and establishment than marcottage.

7.2.4 Grafting

The *Litchi chinensis* is a species of great genetic diversity, manifesting severe graft incompatibilities and that these incompatibilities are also indicative of graft-transmissible pathogens. It is, therefore, proposed that care should be exercised when propagating this species (Murray and Murray, 1994).

7.2.4.1 Apical Grafting

Litchi has been successfully propagated by apical grafting on seedlings and mature trees (top working) in China, Israel, India and South Africa. Bolt and Joubert (1968) obtained good success after splice-grafting onto Tai So seedlings (3-10 mm stem diameter) in South Africa. They used scion wood of 10 cm length with at least 2 slightly swollen buds. The best result (70 per cent success) was obtained with non-terminal wood. Younger terminal wood was too thin, grew slower and dehydrated readily (35 per cent success). Grafts grew faster if leaves were retained on the stock (side shoots were usually removed). It was suggested that scion wood should be cinctured 3 weeks before grafting.

7.2.4.2 Approach Grafting

Propagation of litchi by approach grafting or inarching has also been tried in China. A slice of bark and wood is cut from both stems and the cut surfaces secured together for close cambial contact. After the union occurred (1-10 months) the top of the stock is removed above the graft and the base of the scion is removed below the graft. Approach grafting was also successfully employed in Florida to propagate Chinese, Indian and African cultivars onto Brewster marcots (Cobin, 1954; Chadha, 1968).

7.2.4.3 Wedge Grafting

This is a detached method of grafting in which grafting operation is performed on about one year old seedling. The 3-6 month old scion stick is taken from the shoots. The best time for grafting is rainy season and spring season. For rainy season grafts, short (3-4 buds), round scions of green but completely mature wood (just before the opening of new shoots), are preferred while for spring graft, hard and semi-hardwood is selected. It is advisable to take healthy and disease free scion sticks. However, for better success treatment of scion sticks with 0.1 per cent Benomyl is recommended.

Under favourable temperature conditions, sprouting of takes place within 2-3 weeks of grafting. The new plants having 50-100 cm height may be planted after sufficient hardening within a year of grafting. For better success the rootstock should be raised in polythene bags of 4-5 kg capacity and the successful grafts should be transplanted in to bigger sized bags.

Propagation of Litchi

Propagation by Seed

Air Layering

Cleft Grafting

Whip and Tongue Grafting

Micro-propagation in Litchi

7.2.5 Budding

Nelson (1954) and Lynch and Nelson (1956) obtained success in chip and shield budding in Florida. However, shield budding onto Tai So seedlings was found successful in South Africa (Bolt and Joubert, 1968).

7.2.6 Rootstock

The compatibility of 5 rootstocks for litchi cv. Baitangying was investigated in China using 16-month-old seedling rootstocks and 4- to 6-year-old young trees which were top worked. Hexiachuan, Luyuexue and Baila were found as good rootstocks for Baitangying which showed 78-84 per cent survival of grafted plants (Lu and Li, 1998).

7.2.7 Year Round Production of Litchi Plants under Polyhouse

The nurserymen can adopt following steps for the year round production of litchi under playhouse:

- ✰ Establishment of mother block under controlled conditions at 2x2 m spacing accommodating 125 plants in 500 m^2 polyhouse.
- ✰ Layering operations should start at least two years after establishment of old mother plants. This operation will increase the no. of shoots for further layering and after each operation, number of shoots will be available in multiples.
- ✰ Air layers should be detached from mother plants after proper root development.
- ✰ Removal of polythene strip and shifting of air-layers to 300/500 cc root trainers in shade net or polyhouse. A good substrate should be used for filling of root trainers.
- ✰ Transfer these plants after 2 months in 1 kg capacity polybag and kept in polyhouse.
- ✰ Air-layered plants will be ready in 4-6 months for planting under polyhouse conditions to harden, these newly developed plants put in net house for 15 days.

For year round propagation of litchi plants, the above steps are employed two times in a year. In this way under polyhouse (top poly sheets and side nets) higher number of plants can be produced regularly throughout the year in comparison to open field. A tentative economics of litchi plants production in polyhouse is given below; however, the level of profit depends on the seriousness of the nursery men and marketing of the plants.

To have quick establishment of litchi orchard, now-a-days well grown up preformed plants with well-established root system and canopy architecture are preferred. Such plants are 3-4 years old and developed in bigger pots or containers. The plants are shifted in well-developed pits in a favourable season with least disturbance to root mass soil. Preformed plants may start bearing in 2-3 years of shifting and establishment.

7.2.8 Shifting Techniques of Grown up Litchi Plants

It is also now possible to shift bearing litchi plant of 10-12 years of age. The operation is quite risky hence performed very carefully. The best time of shifting litchi plant is autumn-winter season, when the plants are entering in dormant phase. Before final shifting of such plants, a deep trench is dug out in circular faishon at least 1-1.5 m away from the tree trunk. The trench should be made in 2or,3 stocks to exert stress on the plant. This process may take3-5 week time. Dining the process, some roots are cut which takes it own time to heal and making callus. The foliage of plant is reduced upto secondary or even primary branch level in between leaving one nurse shoot growing apically. The plants are then lifted with huge moss of soil carefully and planted in pits of appropriate size developed well in advance. The help of plant lifting machines like JCB, can also be taken to avoid risk factor while lifting the plants. This method is now becoming popular in the areas, where new plan of construction and development of roads are unavoidable.

8.0 Area and Production

The most important litchi-producing areas are in China, Taiwan, Thailand, India, South Africa, Madagascar and Australia. The area and production in major litchi producing-countries are given in Table 8.

Table 8: Area and Production of Litchi in some Countries

Country	*Area* (hectares)*	*Production** (tonnes)*
China	39,700	200000
Thailand	13,550	10000
India	11,410	90000
South Africa	180	8000
Madagascar	800	50000
Australia	1,000	2000
Mauritius	200	1000
Reunion	200	5000
USA	100	500

* Menzel and Simpson (1986b), * *Menzel (1995), Mitra (2002).

India has second position in litchi production after China, with an area and production of approx. 1,00,000 ha and 7,00,000 MT, respectively during 2018-19. States comprising Bihar, Jharkhand and West Bengal accounts for 85 per cent of the total litchi production in the country whereas in north western part of India (Uttarakhand, Punjab, Himachal Pradesh and Jammu & Kashmir) are the other potential litchi producing states (Table 9). There has been a substantial increase in the area and production of litchi in India during the past 50 years. The area has increased from 9400 ha (1949-1950) to 100,000 ha (2018-2019) (Table 10). The contribution of litchi to the total area under fruit cultivation has increased from 0.75 per cent to 1.70 per cent. Increase in area during 1991–1992 to 2014–2015 (22 years) has been

Table 9: State-wise Area (000 ha), Production (000 MT) and Productivity of Litchi in India

State	*2011–2012*		*2012–2013*		*2013–2014*		*2014-15*		*2016-17*		*2017-18*		*2018-19*	
	A	*P*	*A*	*P*	*A*	*P*	*A*	*P*	*A*	*P*	*A*	*P*	*A*	*P*
Bihar	31.1	236.4	31.28	256.43	31.48	234.20	32.0	198.0	33.0	275.8	33.27	282.27	34.1	290.4
West Bengal	8.9	85.3	9.19	90.00	9.30	93.90	9.4	76.8	9.6	95.4	9.88	79.00	11.3	100.3
Jharkhand	4.8	57.5	5.27	58.24	5.27	58.24	5.4	58.8	5.8	65.1	4.13	64.99	6.6	65.0
Assam	5.3	41.5	5.63	49.64	5.38	48.08	5.5	48.7	5.6	54.1	5.37	50.24	6.4	60.0
Chhattisgarh	4.5	27.1	4.99	30.89	5.36	37.63	5.6	39.2	5.1	32.5	7.35	52.17	5.2	33.2
Uttrakhand	9.5	19.0	9.49	19.16	9.44	30.71	9.7	19.5	11.0	24.9	10.50	24.27	11.1	25.5
Punjab	1.7	24.5	1.75	26.52	1.85	28.00	2.0	32.1	2.5	33.4	2.71	43.96	2.5	30.4
Odisha	4.5	20.1	4.46	20.26	4.47	20.32	4.5	20.3	4.7	25.7	4.47	20.48	5.5	25.1
Tripura	3.2	16.6	3.46	17.97	3.88	20.18	3.9	20.6	4.3	21.6	1.81	6.28	5.6	22.0
Others	3.2	16.6	7.23	10.98	7.74	14.04	3.0	10.6	14.4	33.7	12.85	62.79	17.4	64.9
Total	80.40	538.10	82.70	580.10	84.17	585.30	84.9	523.3	96.00	662.20	92.34	686.45	105.7	716.8

A: Area in 000 ha; P: Production in 000 MT.

* Indian Horticulture Database, National Horticulture Board, (Anonymous, 2019), Govt. of India.

more than 80 per cent, whereas production increase during the same period is to the tune of more than 150 per cent (Anonymous, 2018).

Table 10: Area, Production and Productivity of Litchi in India*

Year	Area (000 ha)	Production (000 MT)	Productivity (MT/ha)
1991-92	49.30	243.80	4.90
2001-02	58.10	355.90	6.10
2002-03	54.10	476.40	8.80
2003-04	53.70	478.50	8.90
2004-05	60.20	448.60	6.10
2005-06	63.20	391.10	6.20
2006-07	64.90	402.60	6.20
2007-08	69.20	418.40	6.00
2008-09	72.00	423.00	5.90
2009-10	74.40	483.00	6.50
2010-11	77.60	497.30	6.40
2011-12	80.40	491.30	7.00
2012-13	83.34	493.52	7.00
2013-14	84.17	585.30	7.00
2014-15	84.10	528.30	6.22
2015-16	90.0	559.00	6.21
2016-17	93.00	568.00	6.10
2017-18	92.34	686.45	7.43
2018-19	105.7	716.8	6.78

*Anonymous (2019).

The concentrated pockets of litchi production in different states have been listed in Table 11.

Table 11: Major Litchi Growing Areas in India

State	Districts
Tripura	West Tripura, North Tripura, South Tripura, Dhalai Tripura
Assam	Bongoigaon, Kamrup, Goalpara, Nalbari, Barpeta, Sonitpur, aogaon,Lakhimpur, Golaghat, Jorhat,Cachar
West Bengal	Murshidabad, 24- Parganas, Nadia, South 24-Parganas, Malda, UttarDinajpur, DakshinDinajpur, Hubli
Bihar	Muzaffarpur, Vaishali, Samastipur, Sitamarhi, East Champaran, West Champaran, Gopalganj, Sheohar, Darbhanga, Madhubani, Purnia, Katihar, Begusarai, Saharsa, Bhagalpur, Araria, Kishanganj, Khagaria, Madhepura, Munger *etc.*
Odisha	Sundergarh, Sambalpur, Angul, Deogarh
Uttarakhand	Udham Singh Nagar, Champawat, Nainital, Dehradun, TehriGarhwal,PauriGarhwal, Haridwar, *etc.*
Punjab	Gurdaspur, Hoshiarpur

9.0 Soil and Climate

9.1.0 Soil

The litchi grows under a wide variety of soils including alluvial soils, loams, heavy clays, organic soils, calcareous soils (with 30 per cent lime) and rock piles (Singh and Jawanda, 1962; Nijjar, 1972; Cull and Paxton, 1983; Chapman, 1984b). In China, the best litchi trees are found in Guangdong province, close to the rivers, on alluvial sands with good drainage and access to the water table (Chapman, 1984a). In southern Queensland, soils of low water-holding capacity and low fertility have made it easier to stress the trees for water and nutrients, and hence reduce tree vigour for cropping (Cull and Paxton, 1983; Chapman, 1984b). The pH of litchi soils in China is usually about 5.5 (Groff, 1921). The pH range should normally be 5.5-6.5. In alkaline soils, application of micronutrients is essential (Tindall, 1994). Litchi can withstand water for a consi-derable period, provided the water does not become stagnant, but will die after prolonged immersion. Many authors have suggested that litchi requires mycorrhiza to grow satisfactorily (Katyal and Chugh, 1961; Anon., 1978; Galan Sauco, 1989), although healthy plants have been examined which was completely devoid of mycorrhizal tubercles (Cobin, 1954). In India, it is suggested that new plants should be grown in soil taken from the vicinity of old trees to introduce the mycorrhiza (Chadha, 1968).

9.2.0 Climate

The litchi is adapted to the warm subtropics, cropping best in regions with brief, cool, dry, frost-free winter and long, hot summer with high rainfall and humidity. Poor production is experienced in many countries because winters are not cool or dries enough to induce a growth-check prior to flowering (Table 12). In some parts of southern Queensland, winter frost was found to damage the panicle and developing flowers and limit fruit production in that season (Menzel and Simpson, 1986b). The litchi usually likes low elevations but can be grown up to an altitude of 800 metres with varying degrees of success. The ideal climatic-conditions for litchi in relation to annual growth cycle in Australia are as follows (Menzel *et al.*, 1986a).

Growth Stage	*Ideal Climate for Growing Litchi*
Flushing	28-30 °C, high relative humidity, heavy rainfall
Dormancy	Minimum 15°C, 50 mm rainfall per month
Flowering	16-20 °C, light rainfall
Fruit set	18-24°C, moderate relative humidity
Harvest	24-28°C, even rainfall, high sunlight, high relative humidity

Pollination is optimum between 19-22°C but may still be satisfactory for fruit set at lower temperature of 15 °C or higher temperature of 27°C (Menzel and Simpson, 1994). A certain degree of water stress is needed for flower initiation. It seems that low humidity is detrimental to litchi fruit set (Batten, 1986a). High temperature (above 38 °C) and relative humidity below 60 per cent resulted in severe skin cracking of fruit (Sharma and Ray, 1987b).

Table 12: Minimum Temperature and Rainfall in Various Litchi Producing Areas Around the World*

Around the World	*Temperature (°C)*	*Rainfall (mm)*	*Factor Limiting Winter Dormancy*
Fuzhou, China	7.8	7.8	-
Guang Zhou, China	9.2	2.4	T
Nambour, Queensland	6.9	91	PT
West Bengal, India**	12.0	9.9	T
Fang, Thailand	11.4	2.8	T
Haifa, Isreal	9.4	175	PT
Homestead, Florida	12.0	47	PT
Honolulu, Hawaii	19.2	107	PT
Nelspruit, S. Africa	6.2	7	-

T: Temperature, PT: Precipitation.

* Menzel *et al.* (1986a), ** Ghosh (1998).

The production of crowded orchard declines when sections of the canopy remain shaded for most of the day. Heavy shade (10 per cent of full sun) during the first three weeks after full bloom has been reported to cause poor harvest (Yuan and Huang, 1988) suggesting that extended overcast weather after fruit set could be detrimental for production under some circumstances (Menzel and Simpson, 1994). From a long duration trial, Zhuang *et al.* (1983) concluded that early flowering was not conducive for satisfactory fruit set because of low temperature in association with overcast weather. The lack of adequate assimilate supply during cloudy conditions may result in poor fruit set, but it is more likely to be related to a direct effect of rain on the anthers or stigmas (Menzel and Simpson, 1994).

In litchi cv. Nuomizi, at light intensities > 60 klx, photosynthesis began to be inhibited. Leaf photosynthate accumulation was in the order of autumn > summer > winter (Deng and Ni, 1994). Changes in leaf and soil temperature and photosynthetic photon flux (PPF) have profound influence on litchi stomata. Optimum stomatal conductance (gs) was recorded at leaf temperature above 35 °C, PPF of about 1200 μmol quanta $m^{-2}s^{-1}$ (about half of non-irradiance in summer in subtropical Australia), high humidity, and predawn leaf water potential (Ψ) above -1.0 MPa (Batten *et al.*, 1992; Menzel and Simpson, 1994). Stomatal conductance and net CO_2 assimilation (A) in cv. Tai So (Mauritius) increased in the morning attaining peak about 0700 to 0800 h and declined thereafter during the middle of the day, sometimes with a recovery late in the afternoon (Menzel *et al.*, 1993). Net CO_2 assimilation is also optimum with Ψ above -1.0 MPa (Menzel and Simpson, 1994). Litchi leaves have low average gs and A as compared with the maximum value recorded for many fruit crops (Menzel and Simpson, 1994).

9.3.0 Climate Resilient Litchi Production

Fluctuating weather patterns have started impacting the famous litchi orchards, resulting in poor harvests and decreased sweetness of the fruit. The quality of litchi was hit hard by unexpected variations of temperatures in April and May, according to farmers and scientists. In the past 3-4 years, instead of the usual high temperature and dry weather, there was a lot of moisture in the air in both the months. Litchi's size has been reduced and its sweetness has also decreased due to variations in temperatures. Litchi fruits require low humidity for healthy growth. "Litchi could be one of the fruits (affected by it) but there is a lack of scientific data to prove how much climate change has affected it however the productivity of litchi is better in India in comparison with its native country China, but still a wide gap exists between present productivity level (7-8 tonnes per ha) and realizable potential productivity (14-15 tonnes per ha).

In Bihar, in the past couple of years, there has been a trend of increasing rainfall and thunderstorm in the month of May. It affects the growth of litchi. Earlier, the months of April and May used to be very dry because of very low rainfall. Prevalence of dry westerly winds would mean low humidity during daytime," Low temperatures have prevailed for longer periods that delay the maturity period by 7-8 days. Fruit sizes have also reduced due to low temperature and its abnormal increase in the last week of May and the first week of June. It exacerbated the ripening process that reduces the size. As much as 25 per cent of the harvest damaged by infestation of red-blended caterpillar and fruit flies compared due to low temperature in April and early May. The repeated thunderstorms results in unripe litchi and mango falling from the trees. Sometimes unprecedented heat played spoilsport by burning the skin of the fruit, which then spoiled due to insect attacks. The extreme weather resulted in small fruits. They should weigh at least 35gm to 40gm instead of the 20gm. The fruit's skin and flesh too remained fragile, as it split because of excessive heat, thereby getting infected by insects."

The higher temperature, good sunshine and adequate soil moisture in the litchi rhizosphere result in improved fruit size and quality. The prolonged cloudy weather and rains during the full bloom hamper normal cross pollination and fruit set and, moist weather leads to severe attacks of mite and other insects as well as incidence of lichen growth on the trunk and branches causing bark splitting, leading to other fungus infestation in the litchi crop. Preharvest low intensity sunshine due to cloudy weather reduces the content of ascorbic acid and sugar in the fruit (Kumar, 2014).

The resilient adaptation strategies for successful litchi cultivation inter alia include use of (1) windbreakers for avoiding damage by heavy windstorm, (2) better root stocks, (3) canopy management, (4) girdling, (5) rejuvenation of old unproductive orchards, (6) mulching and (7) honeybees as potential pollinators. The skilful use of adaptation strategies at appropriate time is the key for successfully overcoming the ill effects of climatic changes (Kumar and Nath, 2013).

10.0 Cultivation

10.1.0 Wind-break

Litchi orchards need to be protected from strong winds which can cause complete flower or fruit drop or in severe cases stunted growth, limb splitting and tree death. Strong persistent wind can reduce growth by up to 50 per cent because of the loss of new growth (Menzel and Simpson, 1986c). So, for proper establishment of a litchi orchard a suitable wind-break should be planted around the orchard boundary and it should be at right angle to the direction of prevailing wind. A row of tall growing trees, such as seedling mango, jamun, eucalyptus, *etc.* should be established between blocks and long ridgelines preferably 12 months before establishment of the orchard. Wind break trees must be planted at least 10 m from the litchi trees to minimise shading and root competition.

10.2.0 Land Preparation and Planting

Land preparation should commence 3-6 months before planting. First, the land should be cleared of all the bushes and other wild vegetations. Depression and other uneven spots should be levelled. After this, the land should be ploughed properly. Correction of soil acidity should be done prior to planting with lime or dolomite, which should be incorporated as deep as possible to obtain maximum effect in the root zone.

Before planting, pits of dimension 1 m^3 should be dug at the desired points, a few weeks before the actual planting. These are then allowed to remain open for a few days and then filled with top soil mixed with manures and fertilizers at the rate of 20 to 25 kg farmyard manure, 2 kg of bone meal and 300 g of muriate of potash. If available, a basket of soil from a litchi orchard which contains mycorrhizal fungus should also be added. The pits are then watered so that the soils settle down. At the time of planting, a small hole is made at the centre of the refilled pit and the desired clone is planted with the help of planting board. Container-grown plants have a very brittle root system and should be handled carefully to minimise the damage of roots. Planting materials should be treated for erinose mite control before planting. It is suggested to use dimethoate at 1 ml/l plus a wetting agent at the same rate. Trees should be immersed in the solutions for 10 seconds before planting. Water should be applied immediately after planting.

Planting of litchi should be done when the weather is neither too wet nor too dry. The best planting time is August-September in West Bengal, Bihar, Assam and Tripura in India. In Australia, planting is usually done in late summer to early autumn or in spring in cold or frost-prone areas. The objective is to ensure quick and even establishment of young trees.

10.3.0 Planting System and Distance

The planting materials may be planted in square or hexagonal system. The former method is commonly followed as cultural operations can be performed more conveniently. Though the hexagonal system accommodates about 15 per cent more trees than the square system, it is not usually followed as the plants become crowded after a few years and cultural operations become difficult.

Under good growing conditions, litchi can spread up to 12 m in diameter in 25 years. In India, litchis are usually planted 10 m apart both ways, *i.e.* in rows and between plants. The distance may be reduced to 8 m when the climate is comparatively dry and the soil is not so fertile. In Australia, the recommended planting distance for Tai So, Bengal and Haak Yip is 11m x 6m, for Kwai May Pink, Salathiel and Wai Chee it is 10m x 5m (Menzel and Simpson, 1986c).

Square system of planting provided to be the most effective for increasing tree volume (121.23 and 130.12 m^3), canopy spread (6.40 and 6.42 m), net assimilation rate (0.313 and 0.369 mg/cm^2) and fruit set (3.64 and 3.91 per cent). The maximum tree volume, canopy spread, net assimilation rate (NAR) were recorded in square system of planting and significantly higher yield (82.30 and 85.52 kg/tree) was obtained in square system, whereas, the maximum yield per hectare (156.73 and 158.75 q/ha) was obtained in double hedge row system of planting (Kumar and Lal, 2007).

According to Pandey *et al.* (2015), among various plant densities, moderate plant density 6x6m accommodating 256 plants/ha performed better over other densities. The highest vegetative growth in terms of canopy area (17.95 m^2), plant volume (53.49 m^3) and fruit production with a tune of 18.18 kg/plant fruit yield, productivity efficiency (0.138) and fruit weight (21.51 g) was obtained in 6x6m spacing. TSS content of the fruit increased with deceasing the plant density, whereas, acidity content showed reverse trend. However, the closest spacing 2x2 m showed poor plant vigor, yield/plant and fruit quality except plant height. 6x6 m planting density showed encouraging result at the early year of fruiting and thus can be concluded that the plant population of 256 per hectare with 6x6 m spacing may be adopted for getting higher yield per unit area.

10.3.1 Planting Techniques and Pit Preparation

Litchi is an evergreen, spreading plant which attains the height of about 10-12 m at its full grown stage. The light penetration on its canopy is also desirable for proper fruiting, hence planting in square system at the distance of 9-12 m. within and between the rows has been advised. This will also help in cultural operations to be performed conveniently. However, planting of litchi in double hedge row system at a distance of 4.5x4.5x9 m accommodating 329 plants/ha and medium density at 6x6 m accommodating about 278 plants/ha have been suggested. Before planting, pit of 90x90x90xcm dimension to be dug at marked place during summer season in the month of April-May and soil should be spread on the ground. During this process the harmful insect and pest will be minimized/controlled. Further the pit should be filled with top soil mixed with about 40kg decomposed FYM/compost, 2 kg neem/karanj cake, 1 kg bone meal/single super phosphate, 200-300 g muriate of potash, 20 g Carbofuron 3G just before the onset of monsoon. About 2 baskets of soil from the root zone of old litchi tree also added which will be helpful for the mycorrhiza growth. Then the soil may be allowed to settle properly during first few rains and get leveled properly. During planting time a hole of about the size of earth ball to be made in the centre of the pit at marked point and planting is done. After proper fixing of the plants, preparation of ring and watering is done immediately for proper establishment of the plant. Further plant is regularly irrigated till it is properly established.

Dipping the root ball of air layers in dip solution formulation containing Rhizobacteria @ 10 ml/litre water for 3-5 minutes help improving the root mass and root characteristics and their by enhances the nursery survival up to 90 per cent. Drenching of plant pot/poly bag with 1 per cent Rhizobacterium solution after 30 days of planting gave good result. The use of right and proper potting mixtures containing various materials like [River bed soil (RBS): Vermi-compost (1: 2)] + vermiculite (50g/kg) and cocopith (50g/kg) + 5g DAP has been found highly effective in production of vigorous planting materials (Purbey and Nath, 2013) up to 82 per cent nursery survival and vigorous plant growth (secondary, tertiary root with plant height and number of leaflets). The nursery men are advised to detach the rooted air layers in 1st week of October, defoliate 80-85 per cent leaves, remove the polyethylene cover from the detached air-layers and safely dip in 0.3 per cent Carbendazim solution to check fungal infection and better establishment.

10.3.2 Role of Mycorrhiza

The initial establishment of litchi orchard requires much care and effort. Litchi plants are slow growing and need protection from water stagnation and frost. The mycorrhizal symbiotic association with litchi plants promotes vegetative and root system growth and aids uptake of water and minerals. Studies have also indicated that litchi orchard soils rich in mycorrhiza help in enhancing the resistance of plants to various biotic as well as abiotic stresses. The mycorrhiza penetrates the root epidermal cells in the cortex while the root hairs, endodermis and vascular tissue do not show its presence. It has been observed that litchi plants in symbiotic association with mycorrhiza have well developed root system and enhance fruit quality and yield. The uptake of phosphorus and potash increases due to VAM.

Altogether, thirteen taxa of AMF have been isolated and identified. Among these, 8 species belonged to the genus *Glomus*, 2 species to *Acaulospora* and 1 species each to *Rhizophagus, Entrophosphora* and *Scutellospora*. *Glomus* was observed to be predominant followed by *Acaulospora* in the rhizosphere soils of litchi. Increased fine root density, canopy spread, chlorophyll content of leaves and leaf area index have been observed at NRCL with application of AM consortia (*G. mossae* + *G. fasciculatum*+ *G. intraradices*). Further, leaflets of inoculated plants has been reported higher concentrations of P, K, Cu, Zn and lower concentrations of Ca, Mg and Mn as compared to control plants. It had been a common practice to incorporate soli (5-7 kg) from the vicinity of old trees to introduce the mycorrhiza or a culture of the fungus in the pit of newly-planted plants which helped in the faster growth of

newly-planted plants. Besides this there are some helper bacteria like *Azotobacter* and other microorganism (*Trichoderma*) that works synergistically with AM fungi to promote plant growth. For organic litchi production, this will be critical inputs. Therefore, it is suggested to apply arbuscular mycorrhiza to improve nutrient acquisition, growth performance and quality production of litchi fruits (Kumar and Sharma, 2016).

10.4.0 Care of Young Plants

Establishment of young plants in the field is one of the major problems in litchi cultivation. Because they are very exacting in their climatic requirements, especially at the early stage, the plants will suffer if adequate attention is not paid. It is, therefore, essential that the young plants should be well protected against severe climatic conditions both in winter and summer. Staking of trees is needed for the first 12 months only, until trees are well established. Rapid vegetative growth during tree establishment can be achieved by frequent application of nutrients. The fertilizer programme for young trees starts when the plants begin to flush (Menzel *et al.*, 1986b).

10.5.0 Irrigation

Adequate soil moisture is considered essential for successful cultivation of litchi. Where annual rainfall is more than 125 cm and is well distributed throughout the year, litchi can be grown without irrigation. The young plants neither establish nor do they make optimum growth if there is shortage of water. Provided nutrition and temperature are optimum, rapid growth during the first four years can be achieved by high rates of irrigation. The timing and rate of irrigation depend on the normal rainfall pattern which varies with orchard and year. The amount of water required is to be estimated by multiplying the evaporation rate (mm per week) by the ground coverage of the tree (m^2). Water should be applied 2-3 times a week in sandy soil and 1-2 times in sandy loam or clay soil for young plants. Trees should be watered every two to three weeks in sandy loam soil and less frequently in a clay soil with greater water holding capacity to maintain productivity (Oosthuizen *et al.*, 1994). A summary of watering as proposed by Menzel and Simpson (1986c) is shown in Table 13.

Table 13: Summary of Watering Young Litchi Trees*

Year	*Canopy Diameter (m)*	*Tree Ground Coverage (sq.m)*	*Evaporation Rate/Week (mm)*	*Total Irrigation Required/Week (l/tree)***
1	0.5	0.2	15 (winter)	3
2	1.0	0.8	15	12
4	2.0	4.0	15	60
1	0.5	0.2	40 (summer)	8
2	1.0	0.8	40	32
4	2.0	4.0	40	160

* Menzel and Simpson (1986c).

** Divide between numbers of irrigation period per week.

The bearing litchi trees require a water stress condition for floral induction. Nakata and Suehisa (1969) concluded that flowering was favoured by dry period before and during floral initiation. However, stressing the trees after panicle emergence did not appear to improve flowering as compared with a stress only up to the first visible sign of floral buds, rather slightly reduced the yield. Stern and Gazit (1993) irrigated the control trees to replace 50-60 per cent of evaporation from a class A pan which ranged from about 6-12 mm day^{-1} in September to 5-6 mm day^{-1} in late November, while in dry treatment, irrigation was withheld from mid-September for 35 days, when some of the leaf margins began to turn brown. The trees were then given 1 mm water day^{-1} until the end of November. The dry

Caring, Irrigation and Fertilizer Application in Litchi Orchard

Care of Young Litchi Plant

Litchi Bearing Orchard

Ring Method of Irrigation

Fertilizer Application

treatment caused significant increase in flowering and fruit production (Table 14). Application of water from the panicle emergence resulted in better fruit set by reducing the premature abscission of flowers and the number of male flowers (Menzel and Simpson, 1991).

Table 14: Effect of Water Stress from Mid September until Late November on Flowering and Yield of Litchi in Israel*

Treatment	*Flowering (0-3)*	*Yield (kg/tree)*
cv. Mauritius		
Control	1.8	23.0
Water-stressed	2.9	50.4
cv. Floridian		
Control	1.6	13.7
Water-stressed	2.8	25.2

Note: Control trees watered daily to replace 50-60 per cent class A pan evaporation. Levels of flowering: 0 = no flowering; 3= heavy flowering *Stern and Gazit (1993)

Water stress reduced the number of fruits per tree, average fruit weight, flesh recovery and yield (Table 15) as compared to the trees irrigated weekly to replace evapotranspiration (Menzel *et al.*, 1993). Water deficits reduced initial fruit set by 30 per cent and final fruit set by 70 per cent and increased fruit splitting by 30-40 per cent (Menzel *et al.*, 1995a).

Litchi in China and Vietnam are generally not irrigated. Rainfall during summer is usually sufficient for successful flowering and fruit growth. Trees are irrigated in Taiwan, Thailand and India during the dry spring-summer period. The highest yield was obtained in Nainital, India, following 7 irrigation applied at 15-day intervals from 20 February (Chandel and Kumar, 1995). In South Africa and Australia, trees are irrigated from flowering (panicle emergence) onwards until the completion of post-harvest flush. Irrigation of young trees should be done by the basin system. As the trees grow, the basin should be gradually enlarged and thus in older plantations irrigation should be done by flooding or furrow method.

Table 15: Effect of Irrigation Strategies on Productivity of Tai So Litchi in South Africa*

Treatment	*Yield (kg/tree)*	*No. of Fruits/ Tree*	*Average Fruit Weight (g)*	*Sugar/ Acid Ratio*	*Aril Recovery (Per cent)*	*Percentage of Fruits Split One Week Before Harvest*
Wet	51.4	2335	23.0	64.9	72.0	10.0
Dry	7.4	457	16.2	64.0	66.9	38.6

Note: Data are the means of eight trees per treatment * Menzel *et al.* (1993)

Chen and Liu (2001) found that drought-resistant 'Dongliu No.1' had thicker leaves and higher specific leaf weight compared with drought-sensitive 'Chenzi'.

Under drought, relative water content (RWC) decreased and electrolyte and inorganic phosphate (Pi) leakages increased, especially in 'Chenzi'. Drought also reduced chlorophyll and carotenoid.

10.6.0 Manuring and Fertilization

Application of nutrients to the soil is essential for satisfactory growth and cropping. Nitrogen is the major nutrient and occupies an important position in the nutrition programme of litchi. The other major nutrients are phosphorus, potassium, calcium and magnesium. Micronutrients (iron, boron, copper, zinc and manganese) are also required in very small amounts to maintain tree performance.

10.6.1 Deficiency Symptoms

The deficiency of different mineral elements showed the following symptoms in litchi (Menzel and Simpson, 1986d). Apart from fruits, a considerable amount of macro and micro nutrient is removed from soils thought annual pruning of shoots. Hence, the application of manure and fertilizer in the required doses is therefore, of paramount importance for sustaining the regular production of the crop. The nutritional requirement in litchi plants depends upon the soil fertility condition. For example the soil of Doon Valley is deficient in N and P where as litchi orchard in Punjab is deficient in N, K, Mg, Zn and Mn. Therefore in these regions the optimum amount of nutrients should be applied after proper soil testing.

The major symptoms of nutrient deficiencies have been presented in Table 16 (Singh *et al.,* 2012).

Table 16: Symptoms of Nutrients Deficiency in Litchi*

Nutrients	*Symptoms on Leaf*	*Additional Symptoms*	*Remedies*
N	Chlorosis of old leaves; curling of leaf margins; increased leaf inclination and texture; decreased leaflet size; chloroplast disorganization; failure of leaf development and defoliation	Un-uniform panicles; general stunting of growth; poor flowering; fruit set and fruit retention; lower yield with smaller size fruits or production of undeveloped fruits	Soil application of nitrogenous fertilizers in time
P	Tip and marginal necrosis of mature leaves proceeding towards the midrib; leaf curling premature desiccation and fall	Fewer vegetative and floral bud development; delayed fruit set	Soil application of phosphatic fertilizers in time
K	Leaf yellowing; necrosis are as first in leaf apices and gradually progress along the base; premature abscission of mature leaves	Stunted plant growth; poor fruit set	Soil application of potassic fertilizers in time
Zn	Bronzing of leaflets; smaller leaflets; yellow inter-veinal are as and green veins (mottling). Premature flower dehiscence and ovule abortion	Long internodes; smaller fruits with reduced aril recovery and sugar content; flower drop and ovule abortion	Soil application of 25g $ZnSO_4$, per plant during September or foliar application of 0.4 per cent $ZnSO_4$, 2-3 times during the period of fruit growth and development

Nutrients	*Symptoms on Leaf*	*Additional Symptoms*	*Remedies*
Fe	General chlorosis of young leaves; spreading of other leaves	Branch die back	Foliar application of 0.4 per cent ferrous sulphate, 2-3 times during flush emergence.
B	Leaves are smaller and withered; thickened petioles; curling may occur; exhibits blossom, wilting and necrosis, reduced pollen production and viability; poor fruit set	Small fruits with reduced pulp percentage; flower buds may fail to develop cracking of fruits	Foliar application of 0.4 per cent borax, 2-3 times during fruit development
Cu	Small leaves; interveinal chlorosis or mottling; small fruit with reduced pulp percentage; induces pollen sterility, thus limits fruit set in crops	Terminal die-back of young shoots	Foliar application of 0.2 per cent copper sulphate 2-3 times during flush emergence.

* Singh *et al.* (2012).

10.6.2 Fertilization Practices in Litchi Orchards

10.6.2.1 Young Trees

Litchi is slow to establish and make early growth. Following establishment, using a complete fertilizer and taking into consideration micronutrient deficiencies in the nutrition programme coupled with optimum water supply will stimulate 4-6 vegetative flushes in a year (Cull, 1986). The rate of each nutrient is increased every year. The fertilizer recommendations for young litchi plants suggested for different countries are presented in Table 17. Koen and Smart (1983) suggested application of fertilizer beyond 20 cm of the trunk, as the litchi is sensitive to fertilizer burn.

Table 17: Fertilizer Recommendations for Young Litchi Trees

Country	*Tree Age (Years)*	*Nutrients (g/tree/year)*			*References*
		N	*P*	*K*	
Australia	1	190	10	25	Menzel *et al.* (1986b)
	2	350	40	100	
	3	530	70	150	
Hawaii	1	34	34	34	Yee (1972)
	2	68	68	68	
	3	136	136	136	
	4	272-410	272-410	272-410	
India	1	75	25	75	Ghosh (1988)
	2	100	25	100	
	3	150	50	150	
	4	250	75	250	
South Africa	1	56	25	25	Koen and Smart (1983)
	2-3	140	25	50	

10.6.2.2 Bearing Trees

Litchi is very responsive to environmental conditions which influence the vegetative, flowering and fruiting cycle of the plant. Consequently, trees from four years onward require careful management to induce regular flowering and fruit set. The fertilizer rate depends on tree size, vigour, crop load, cultivar, spacing, climate and soil type. In West Bengal, India, trees above 6 years were suggested to be fertilized with 600 g each of N and K_2O and 200 g of P_2O_5 per year (Ghosh *et al.*, 1986a). Paxton and Chapman (1980) recommended that trees in subtropical Queensland should not be fertilized during spring, as it promotes growth flushes and prevents premature flower and fruit abscission. Fertilizer rates for well-grown, high-yielding trees in Australia as suggested by Menzel and Simpson (1986d) are given in Table 18.

Table 18: N, P and K Doses for Well-grown Litchi Trees*

Tree Age (Years)	*Canopy Diameter (m)*	*N (g/tree)*	*P_2O_5 (g/tree)*	*K_2O (g/tree)*
4-5	1.0-1.5	200	80	300
6-7	2.0-2.5	300	100	450
8-9	3.0-3.5	400	130	550
10-11	4.0-4.5	500	170	700
12-13	5.0-5.0	600	200	800
14-15	6.0-6.5	800	250	1200
15 +	> 6.5	1000	300	1400

* Menzel and Simpson (1986d).

Increased K levels in the leaves increased photosynthetic rate, particularly at low light or low CO_2 levels and decreased respiration rate, and ascorbic acid oxidase (ascorbate oxidase) and polyphenol oxidase (catechol oxidase) activities (Deng *et al.*, 1994). Annual applications of P and K may not be required in litchi orchards with large amounts of soil extractable P and K (Menzel *et al.*, 1995b). Soil application of potassium (600 g K/tree) resulted in the greatest fruit weight and improved the pulp quality (Lal *et al.*, 1996). Foliar application with zinc sulphate (1.0 per cent) or potassium chloride (2.0 per cent) in cv. Purbi just after fruit set and again after a fortnight produced the best quality fruits in terms of TSS content, sugar-acid ratio and ascorbic acid content (Kumar *et al.*, 1995).

In the initial stage of establishment, however, application of 30 kg well decomposed FYM, 2 kg karanj cake, 150 g urea, 150 g SSP and 150 g mutriate of potash per plant has been found to be optimum. As the plants advances in age, an additional dose of manure @ 5 kg FYM, 150 g karanj cake, 150 g urea, 100 g SSP and 50 g muriate of potash per plant should be added each year with the dose of previous year. Thus, sixteen years old fully grown up Litchi tree should be supplied with 80-100kg FYM, 4 kg karanj cake, 2 kg urea, 1.5 kg SSP and 0.8 kg muriate of potash per plant per year. Application of organic manue in litchi improves the yield and quality of fruits. Application of lime in acidic soils @10-15 kg/tree at an interval of 3 years has been found to increase the yield.

10.6.3 Regulation of Litchi Bearing

Flower induction takes place in the dormant bud, whereas flower initiation takes place in the early part of the growing bud stage (Huang and Chen, 2003). For good flowering, there must be an effective cold temperature (<15 C) for a prolonged period during flower induction. If the temperature is high (>20 C), then leaf primordia grow but the floral organs atrophy. Excessively high temperature (>25 C) can even cause all floral primordial to atrophy, leaving behind only leaves (Cronje, 2010). After flower initiation, flower panicle and flower develop continuously uninterrupted and lead to anthesis, which lasts about 4–6 weeks depending on the temperature and the cultivar (Menzel, 2001). Litchi flowering follows the "male-female-male" pattern. The ratio of male to female flowers depends on the environment and the various cultivars. If temperature is high during flower initiation, then the proportion of female flowers is reduced (Cronje, 2010). In southern China, flowering and fruit set coincide with the rainy season. Wet conditions promote fruit setting but sometimes may affect pollinator activity, pollen tube growth and stigma receptivity. In contrast, in South Africa, the favourable period of fruit set is characterized by low humidity and low rainfall. Under such circumstances, irrigation can help to ensure good flowering and fruit set (Cronje, 2010)

Singh *et al.* (2017a) indicated that the production of an adequate number of flushes by a bearing shoot, with a sufficient area of mature leaves and carbohydrate reserves, is one of the critical factors determining high fruit retention in litchi. Although low temperature appears a pre-requisite for floral induction in litchi but there is strong relationship between flowering and timing of flush emergence and its inhibition during winter. The slight fluctuation in temperature and rainfall pattern [during winter] disturb phasic changes, and harvesting of fruits with few leaves, post-harvest cincturing (Girdling), use of paclobutrazol (PBZ) or KNO_3 can effectively employed in the sub-tropics to induce synchronous flowering.

Further experiments on whole tree basis will be required to identify the potential status of remote carbohydrate resources, and the relative role of carbohydrates reserves and leaf photosynthesis compensation caused by application of paclobutrazol/or KNO_3 or practicing girdling on fruit set and fruit growth in litchi.

Irregular bearing is a major problem in litchi and it can be regularised through girdling (2-3 mm size on 75 per cent primary branches) during end September which improves flowering, fruiting, fruit weight, yield and quality. Application of paclobutrazol (3ml *a.i.*/m^2 canopy surface area) advanced the flower emergence by six to seven days. Ethrel (2ml/l) proved to be the most effective for flowering and fruit induction in litchi cultivar 'Bombai' (Nath *et al.*, 2017).

10.6.4 Tissue Analyses and Nutrient Requirement

Litchi nutrition management has been dependent on leaf standards which are based on limited work for most nutrients. Leaf age and fruiting have dominant effects on litchi nutrient composition (Menzel *et al.*, 1987, 1988a). The concentrations of mobile nutrients (*e.g.* N, P and K) were generally higher in young compared with old leaves and higher before fruiting commenced, while the reverse was true for many of the immobile nutrients. The concentration of the mobile nutrients (N and P) generally fell slightly after harvest during initiation of the new vegetative flush and then increased to reach maximum levels between fruit set and harvest. Menzel *et al.* (1988a) showed that the litchi fruit was a strong sink for N and P and these nutrients were readily mobilized from adjacent leaves. As leaves act as reserves for N, P and K, application of fertilizer before fruit set may build up tree reserves and improve fruit set and development. Poor fruit set of litchi in China, associated with low levels of N and P prior to and after fruit set, was reduced by the application of foliar sprays of urea and potassium phosphate (Menzel and Simpson, 1987a). The concentration of leaf K fell slightly after harvest to reach maximum levels towards the maturation of the post-harvest vegetative growth flush. Levels normally declined substantially during flower and early fruit development in cropping trees, but were fairly stable in trees that remained vegetative during spring or early summer (Menzel *et al.*, 1992a). Tentative leaf standards for litchi have been developed in South Africa and Australia, but no consensus has been reached on a standard sampling time and tissue type. In South Africa, leaves are sampled behind the fruiting cluster about 8-10 weeks after fruit set (Koen and Smart, 1983), while in India (Ghosh *et al.*, 1986a) and Florida (Young and Koo, 1964) they are taken from dormant branches before flower initiation. In Australia, it is recommended that the most recently matured leaf behind the fruiting cluster be sampled 2-6 weeks after fruit set (Menzel *et al.*, 1988b). However, Menzel *et al.* (1992a) further reviewed the leaf nutrition standard of litchi and suggested that the leaves be sampled for diagnostic purpose from flowering branches 1-2 weeks after panicle emergence.

Huang *et al.* (1998) selected 30 litchi orchards to survey changes in leaf nutrient contents and yields of cultivars Haak Yip and No May Tsz and of soil fertility, during the growing season. Some leaf nutrient contents were significantly related to each other. The fruit yield of cv. Haak Yip was described well by leaf N content in a quadratic relation, with the highest yield at 1.71 per cent N. According to the relationship, fruit yield should be guaranteed to be >90 per cent of maximum as leaf N content remained in the range 1.5-1.9 per cent at anthesis. Fruit yield was highly significantly related to contents of P, Mg, Ca and B (first order relationship), with correlation coefficients (r) of 0.498, 0.561, 0.444 and 0.543 respectively, for 52 observations. The anthesis stage is the best period to sample mature leaves. The suggested criteria for leaf nutrients at anthesis for cv. Haak Yip are 1.60-1.90 per cent N; 0.12-0.27 per cent P; 0.60-1.00 per cent Ca; 0.30-0.50 per cent Mg; 50-100 mg Fe/kg; 100-250 mg Mn/kg; 10-25 mg Cu/kg; 15-30 mg Zn/kg and 25-60 mg B/kg.

Field experiments were also conducted in China by Chen *et al.* (1998) with 21-year-old litchi trees to study the effect of N and K on yield. Soil was a clay loam, pH 4.4-4.5, with contents of organic matter of 1.35-1.78 per cent, total N of 0.067-0.086 per cent, total K of 1.03-1.34 per cent, rapidly available N of 76-80 mg/kg and rapidly available K of 36-45 mg/kg in the 0-40 cm layer. Application of N, P and K at rates of 0.84, 0.5 and 1.2 kg/year respectively, increased yield. The optimum contents of N, P and K in old leaves in the autumn were 1.6-1.75, 0.25-0.32 and

0.65-0.95 per cent respectively. The optimum leaf nutrient standards developed in different countries (Table 19) are given below:

Table 19: Litchi Leaf Nutrient Standards

Nutrient	*South African Standard (Cull, 1977)*	*Israel Standard (Galan Sauco, 1989)*	*Australian Standard (Menzel et al., 1992a)*
N (per cent)	1.30-1.40	1.50-1.70	1.50-1.80
P (per cent)	0.08-0.10	0.15-0.30	0.14-0.22
K (per cent)	1.00	0.70-0.80	0.70-1.10
Ca (per cent)	1.50-2.50	2.00-3.00	0.60-1.00
Mg (per cent)	0.40-0.70	0.35-0.45	0.30-0.50
Fe ($\mu g\ g^{-1}$)	50-200	40-70	50-100
Mn ($\mu g\ g^{-1}$)	50-200	40-80	100-250
Zn ($\mu g\ g^{-1}$)	15	12-16	15-30
Cu ($\mu g\ g^{-1}$)	10	-	10-25
B ($\mu g\ g^{-1}$)	25-75	45-75	25-60
Na ($\mu g\ g^{-1}$)	-	300-500	<500
Cl (per cent)	-	0.30	<0.25

According to Sivakumar and Corsten (2007), the relationship between the leaf nutrients and fruit quality attributes showed positive relationships between leaf potassium (K) and anthocyanin content and titratable acidity (TA); leaf phosphorus (P) and pericarp h°, leaf nitrogen (N) and fruit weight; leaf calcium (Ca) and fruit firmness.

Kumari (2009), while studying leaf analysis in Litchi cvs. Purbi, Bedana, Kasba and Shahi, found that the concentration of secondary nutrients varied from 0.099 to 0.225, 0.050 to 0.114 and 0.070 to 0.076 percent for Ca, Mg and S, respectively. However, mean leaf nutrient concentration was 0.162, 0.082 and 0.073 percent for Ca, Mg and S, respectively. Highest concentration of Ca, Mg and S was observed for the 4th pair of leaves. Concentration of Cu, Fe, Mn and Zn in leaf varied from 4 to 21, 192 to 614, 75 to 175 and 15 to 35 ppm, respectively. However, mean leaf nutrient concentration was 12.5, 403, 125 and 25 ppm for Cu, Fe, Mn and Zn, respectively. Highest concentration of Cu and Mn was observed for the 1st pair of leaves whereas Fe for 2nd pair of leaves. Highest Zn concentration was observed for both 2nd and 4th pair of leaves.

10.7.0 Pruning and Training

Litchi trees do not require major pruning or training but for high productivity it is very essential orchard operation. Since commercial harvest in litchi tree is obtained from two third of the lower canopy from the outer periphery, boosting more fruiting branches in this region is the main concern of training and pruning operation. Pruning of young trees is a standard practice in most litchi growing areas of the world. The practice is extended also to mature trees in a few countries, such

Pruning, Rejuvenation and Mulching of Litchi Orchard

Pruning of Litchi

Framework Develpoment

Rejuvenation of Litchi Orchard

Black Polythene Mulch

Organic Mulch

as China, Taiwan. The trees in southern China are generally more open than those in Australia. Pruning and tree training start in the nursery and continue during tree establishment. The main reasons for pruning litchi trees as suggested by Menzel and Simpson (1987a) are:

1. Improves tree structure against wind damage.
2. Increases bearing surface during early years.
3. Reduces pest and disease incidence.
4. Improves the penetration of sprays into the canopy.
5. Facilitates access during routine cultural operations, such as fertilization and weeding.

Branches developing from the main trunk at a sharp angle (crotch angle) may split later and should, therefore, be removed. In case of cultivars which produce only a few but excessively long branches (*e.g.* Tai So and Kwai May Pink), the growing points should be pruned to encourage branching. In China, it is recommended that virtually no wood be taken with the panicle at harvest. Cutting back more than this removes vegetative buds and delays autumn flushing and reduces the next crop (Menzel and Simpson, 1986b). On the contrary, Hasan and Chattopadhyay (1995) opined that litchi fruits might be harvested by removing the panicles along with 7.62 or 15.24 cm of the shoot in order to improve fruit quality in the following season.

The main objective of annual pruning is to control growth of the tree as in case of high density planting and promoting the flowering. The recent findings advocates pruning of litchi is facilitate healthy growth and better yield. However, in the past litchi trees were never pruned. Only at the time of fruit harvesting, a small portion of branches were used to be lopped off along with fruit bunch. This practice was considered to be helpful in improving subsequent fruiting by encouraging terminals shoot production. Sometimes root pruning and exposure also improves the flowering in adult trees. Time of annual pruning is very much important for the next year cropping. It should be done just after fruit harvesting. As soon as the fruit is harvested, the plants put forth new growth, therefore pruning must be completed before the initiation of new growth, therefore pruning must be completed before the initiation of new growth. Pruning after the initiation of new flush results the production of unhealthy shoots which do not mature for flower bud differentiation for the next fruiting season. During pruning tree vigour should be taken into consideration. Winter pruning should be strictly avoided. However the unproductive shoots arising from the main branches under the canopy should be removed as and when it is appearing. Rational supplementary pruning (light pruning) can be done after flower bud formation in spring depending upon the requirement. Reiterative pruning is done to one year old bearing mother branches by removing 1/3 or 4/5 of the main branches. The damaged shoots due to severe cold, also needs to be removed just before the new growth in initiation. After pruning, the leaves should be allowed to decompose in the basin of each tree. This can be done by removing the woody branches after drying of leaves followed by mixing of leaves in the soil before application of regular fertilizer dose.

Pruning treatments in litchi cv. Haak Yip increased the number of inflorescences and advanced flower opening (measured in mid-February) to the extent of 54, 43 and 18 inflorescences and 71, 76 and 8 per cent floral opening for the August, September and November/control treatments, respectively (Campbell, 1994). Pruning of bearing branches of litchi cv. Mauritius trees grown in South Africa was recommended as a method of controlling tree size because it did not affect long term yields and was also environmentally friendly (Oosthuizen *et al.*, 1995). The commercial benefits of pruning on litchi production are yet to be confirmed. Overall, flowering and yield were as good after winter (May/June) pruning as with summer (January/February) pruning in cultivars Bengal, Kwai May Pink and Wai Chee grown at 4 sites in subtropical Queensland, Australia (27°S), but both were less than on un-pruned trees (Menzel *et al.*, 1996).

At Brooklet in northern New South Wales, at the southern end, the optimum time of pruning appeared to be in early- to mid-February, but this was before the previous crop was harvested. Trees hedged in early March had virtually no crop since most of the branches did not complete a flush of growth before the normal time of floral initiation in winter. The results of this study confirm the strong relationship between flowering and shoot development in litchi. Suggestions are provided on options for pruning litchi trees growing in different environments, including warm tropical areas, warm subtropical areas, and cool subtropical areas (Olsen *et al.*, 2013).

Litchi and mango growers can obtain adequate leaf-fruit-ratios with a reliable canopy and plant management system (like practicing girdling, manual removal of winter flushes, and application of paclobutrazol during September-October). Decreasing leaf-fruit ratios increase the photosynthetic efficiency of the leaves but flower induction for the next growth cycle is lower. In contrast, a high leaf-fruit ratio assures a sufficient storage supply, mainly starch, for the following crop, which will inhibit biennial bearing (Singh *et al.*, 2017b).

10.8.0 Rejuvenation of Unproductive Litchi Orchards

Rejuvenation of old, senile and unproductive litchi trees for better productivity and quality without disturbing the original plantation has been standardized through heavy reiterative pruning followed by proper fertilization and plant protection measures. Heading back of unproductive litchi tree is to be done at 2-2.5 m height in the month of August-September followed by judicious thinning and pruning of newly emerged shoots after six months resulted in enhanced production of quality fruits after 3 years (Singh *et al.*, 2012). Space available between the two rows of litchi orchards can be successfully utilized through growing of intercrops like mustard, pea, faba bean *etc.* during the initial period of canopy development in rejuvenated orchard.Steps of Rejuvenation are as follows:

Step-I Selection of unproductive tree

Step-II Reiterative pruning at about 2-2.5m height above the ground during August-September

Step-III Retaining 3-4 outward growing limbs

Step-IV Cut end pasted with Bordeaux paste/Copper fungicide

Step-V Profuse emergence of new shoots on pruned branches after 2 months

Step-VI Regular thinning of emerging shoots, retaining well spaced 8-10 healthy shoots/branch

Step-VII Shoot management to build open umbrella shaped canopy

Step-VIII Fruiting starts 3rd year onwards

After the reiterative pruning, the tree should be fed with half of the optimum fertilizer doses after at least 25-30 days of the operation, followed by irrigation near the root zone, just like the commercial bearing orchards. The manures and fertilizers should be applied through ring method. The dose per tree as an adult bearing stage *i.e.* 75-80 kg well rotten FYM, 2 kg Neem/Castor cake, 1.00kg Urea, 1.5kg Single Super Phosphate and 500g Muriate of Potash should be applied preferably in two split doses one during August-September and another in February-March. The micronutrients like Zinc ($ZnSO_4$ @ 2-3g/l) and Boron (Borax @ 2g/l) may be applied through foliar application from one year after the reiterative pruning.

10.9.0 Canopy Management

Management of optimum stature of litchi tree with compact and stereo bearing canopy is an important aspect of orchard management. Generally, litchi tree grows continuously which results in larger canopy. In view of the present shortage of space and requirement of high density orcharding, promotion of large bearing area, accommodation of more number of plants/unit areas, management of tree height, canopy concentration to provide more fruiting branches and penetration of more light to inner portion is more relevant. Hence, proper shaping of tree from initial stage and pruning of branches after fruit harvest in young plants, shaping is quite convenient at this stage. However, removal of 20 cm shoots after harvesting should be done every year to promote healthy vigorous shoots for the next year crop. Reiterative pruning is practiced to rejuvenate the old orchards whereas corrective pruning by removing undesired, unfruitful and centre leader branches to ensure light and aeration inside the canopy should be adopted to enhance the stereo fruiting and thereby the productivity of the orchard (Nath *et al.*, 2012).

Training is an important operation and should be initiated at the nursery stage itself. Single stem air layered plants should be raised in bags and allowed to grow up to 40-50 cm. The layered plants has strong tendency to produce branches at the ground level which should be prinched or pruned. Further, strong, well spaced, out facing with wider crotch shoot should be allowed to form the main branches. It is necessary to continue shaping by removing all the branches forming crotches with main branches as and when grow (Nath *et al.*, 2016).

The main aim of frame working is to develop sufficient fruiting terminals. Compact and stereo-bearing canopy can only be achieved by well thought out pruning at early stage and shaping of young tree. The canopy should have well illuminated and sufficient number of strong and healthy branches for stereo bearing. To develop good and compact canopy, 25-30 cm fruit bearing shoots at the time of harvesting should be removed. In this way, 2-3 new terminals will develop which will consequently develop into fruiting branches next season. As principle, young trees are lightly pruned to provide larger area for photosynthesis. Severe pruning at young stage should however be avoided as it hinders the tree development and increases the period of juvenility. Besides, removal of dead and diseased shoots, small internal branches which prevent the sun rays to penetrate inside the canopy should be removed. Although, removal of branches along with the bunch had shown its promise, but culturing compact canopy seems to be more useful in other litchi growing countries. The techniques includes, more pruning at top (open window) and outer parts of the tree canopy and less at bottom and inner side of the canopy which promotes stereo fruiting. In Israel however, dwarf stature litchi tree has been maintained by topping and hedging (Singh *et al.*, 2012).

ICAR-National Research Centre on Litchi, Muzaffarpur, Bihar has come out with protocol to double the productivity of litchi by employing knowledge of plant geometry, row orientation, canopy architecture, good agriculture practices and infusion of mechanization, thereby harnessing the abundance of soil radiation in litchi production zone. The centre has developed hedgerow system of planting of litchi at spacing of 8m x 4m covering an area of 32 m^2 by individual plant in place of 64 m^2under conventional system. With the orientation of row from East to West and developing stereo fruiting-enabling canopy architecture through open center to harness photo-synthetically active radiation (PAR) between 08:00 AM to 11:00 AM in morning hours and 03:00 PM to 05:00 PM evening hours on both sides of the hedgerow, respectively. This has enabled production of experimental yield of 19.84 MT/ha in comparison to 8-9 MT/ha under conventional system (Nath, 2019).

Nath *et al.* (2014) studied three types of first order branches (FOB) *viz.*, 2, 3, 4 and second order branches (SOB) *viz.*, 2, 3, 4 and a control (total of ten treatments) to develop canopy architecture for HDP in litchi and found that the FOB will have a crotch angle of 90° on the main trunk at a height of 70 cm, whereas SOB will have a crotch angle of 60°. Both types of branches will be maintained at a length of up to 60 cm. The plant to plant distance is 6x6 m. Final height of plants at full grown stage will be 4.5 m with a canopy spread of 2.25 m on each side. The main framework (FOB and SOB) will cover 1 m distance from trunk from each side. The remaining 1.25 m will be available for growth of third order branches and fruiting terminals from each side.

10.10.0 Weed Control

Weeds in litchi orchards reduce plant growth and vigour by competing with the trees for light, moisture and nutrients. Application of paraquat controlled broad-leaved annual weeds and grasses in the early stage of growth, but it will only scorch mature perennial weeds. Systemic herbicides, such as glyphosate are translocated to the roots of actively growing weeds ensuring complete death. Most of the persistent weeds in litchi orchards (nut grass, couch grass, paspalum) could be controlled by glyphosate (Menzel *et al.*, 1987), and *Cyperus rotundus, Oxalis martiana, Eragrostis curvula* were also controlled effectively by application of paraquat (Ghosh, 1988).

10.11.0 Mulching

As soil moisture is one of the limiting factors for successful cultivation of litchi, use of mulch is very beneficial. It reduces the loss of moisture from the soil, enhances the rate of penetration of rain water or irrigation into the soil and controls the growth of weed, thus eliminating the competition between the weeds and the fruit trees. The use of mulch encourages the development of better root system of young litchi plants at a much faster rate than those growing in a bare soil. Suitable fibrous mulches are wheat, or barley straw, sorghum stubble, peanut shells and bagasse (Menzel *et al.*, 1987). Mulching the tree basins at IIHR, Bangalore, India, resulted in better fruit size with decrease in fruit cracking (Chadha and Pareek, 1991).

10.12.0 Intercropping

Litchi is a slow-growing tree, takes at least 5 to 6 years to come to flowering and fruiting and the commercial fruit bearing starts from about 9-10 years. So the vacant space in the litchi orchard may be utilised effectively by growing intercrops during the early years. This will not only give a good annual income to the growers during the period till the litchi trees start bearing but will also protect the young litchi plants, enrich the soil, improve its physical condition and keep the weeds under control. The choice of intercrops depends upon the climate and soil and also on the marketing facilities. But care should be taken to ensure that these intercrops are not grown at the cost of litchi plants. These should be fertilized separately and protected from the attack of pests and diseases.

Quickly growing crops like vegetables, pulses and berseem can be planted to make use of these initial years when the litchi plants are small. Phalsa and papaya fruits can also be used as intercrop during initial years of orchard establishment. Vegetables like onion, chilli, spinach, tomato, beans, cowpea, radish and cole crops can be easily grown for higher incomes per unit area. Even in mature litchi orchards some shade loving intercrops like turmeric, elephant foot yalm, colocasia can be grown with adequate fertilization. Sunhemp and Sesbania may be planted for enriching the soil fertility and should be ploughed back into soil before flowering.

Inter space utilization protocol for young non-bearing litchi orchards in form of different models including litchi + banana, litchi + okra and litchi + gladiolus have been developed for the benefit of rural mass and litchi farmers. Depending upon the vicinity of the city either one of the model may be adopted. Litchi with banana model has been most remunerative and suitable for consecutive three years with total fresh yield of 36, 28, and 18.2 t/ha of banana as main crop, 1st and 2nd ratoon crop, respectively (Singh *et al.*, 2012). Similarly, litchi + okra-gladiolus crop rotation found best in terms of net return and plant growth followed by cowpea-potato-onion and cowpea-French bean-okra with a net profit of Rs. 1,42,694/-, Rs. 93,180/- and Rs. 43,668/- /ha, respectively up to 3rd year. These models can be adopted at different locations as per their suitability to land and market and need of the farmers.

10.13.0 Litchi Cultivation in Non-traditional Areas

Owing to specificity in climatic requirements suffice by cool conditions for 3-4 months for flower induction; litchi cultivation is localized to only few pockets in India. Recent efforts by ICAR-National Research Centre on Litchi, Muzaffarpur in identifying patches suitable for litchi cultivation shows that, litchi can be grown well in non-traditional areas of South India, either as sole or mixed plantations with coffee in parts of Coorg in Karnataka, Wynad in Kerala and Lower Puleny hills, Kallar and Burliar of Nilgiri hills, besides some parts of Kanyakumari district of Tamil Nadu. In these regions, litchi trees flowers in the month of August - September and fruits mature in the month of December and January in contrary to April and May in Northern India. Shahi, Rose Scented, Early Seedless, Dehradun, Maclean, Green and Calcuttia are the varieties grown with good fruit quality. Climatic survey in high altitudes of Southern India (Karnataka, Kerela and Tamilnadu) revealed the potentiality of these regions in producing quality fruits that was at par with fruits produced in Northern India. By breaking the seasonlity barriers, it was observed that a difference of 6.50 °C in temperature and 981.99 mm rainfall between the traditional and non-traditional litchi growing areas is the prime factors triggering flowering in litchi. The places with more than 900 m MSL like Coorg, Kelpata, Iduki, Ooty are good for off season (November-January) litchi production in India. The potential area of litchi cultivation in South India includes Coorg, parts of Chikmagulur and Waynad. Some farmers grow litchi in rural area of Bangalore, which fruit in the months of May –June. The exact area of litchi in these regions is not well known. In Coorg and Wynad areas it is estimated that more than 200 growers are growing litchi on a smaller scale with 1-10 trees in their gardens/plantation/back yard. But there are few farmers who have planted 1 acre to 10 acres of litchi as intercrop with coffee. In Wynad district of Kerala, off season cultivation is going on. Not many farmers in the south are aware of litchi and the benefits of its plantation. They grow coffee, cardamom and pepper, apart from litchi, mangosteen and rumbutan (Nath, 2019).

As such, there was a growing demand of litchi planting materials among growers in this region. Since availability of genuine and quality planting materials is a problem, looking at the hurdles faced during rail and road transportation, a group of growers from Karnataka air-lifted freshly detached litchi air layers from ICAR – NRCL, Muzaffarpur within 24 hours, with a success rate in survival of air layers to the tune of 80 per cent. With the combined efforts and farmers interest, litchi cultivation is gaining popularity in this area and has potential for providing litchi during off season which certainly helps in crop diversification and income enhancement of farmers of these areas (Nath, 2019).

11.0 Flowering, Fruit Growth and Development

11.1.0 Flowering

11.1.1 Pattern of Floral Initiation

Inflorescences are terminal and the branches on which they are produced are hapaxanthic. The apical bud of a modular branch can only produce one inflorescence, but one or more of the lateral buds, directly below the apical bud, may also be transformed into inflorescences. After the fruits are harvested one or more of the buds remaining below the zone of inflorescence production will flush to produce new modules (branches). The inflorescence is a heterocladic pleiothyrsoid where the dichasia are the flower bearing nuts (Robbertse *et al.*, 1995b).

Results of experiments on 9-year-old trees of cv. Nuomizi showed the last autumn shoots which sprouted in early october had sturdy flower clusters of moderate length and good fruit set, while those which sprouted in late September or in late October, always had short or long, thin flower clusters, with low fruit set. Promoting the last autumn shoots which sprout in early October is the key factor in obtaining good yields from this cultivar (Li and Chen, 1998).

The development of different floral parts takes place in acropetal succession, without a dormant period between floral initiation and anthesis (Shukla and Bajpai, 1974). The time of floral initiation varies with genotype and environmental conditions. In the northern hemisphere it usually occurs between November and February, while in southern hemisphere it usually occurs between June and September.

11.1.2 Tree Age

Litchi trees from air layers may come to flower at the age of 3-5 years (Higgins, 1971) while seedlings take 8-12 years to flower (Loebel, 1976). The grafted trees were, however, found to flower earlier than air layered trees (Bolt and Joubert, 1980).

11.1.3 Vegetative Growth and Dormancy

It is generally believed that litchi trees need a period of vegetative dormancy to flower successfully in spring (Menzel, 1983). This dormancy is normally induced by moisture stress or cold, although it can be artificially induced by manipulative techniques, such as withholding irrigation and fertilizer, the application of growth regulators, root pruning and cincturing (Menzel and Paxton, 1986a). If new flushes appear in autumn, the new shoots usually fail to bear flowers in the next spring.

11.1.4 Temperature and Photoperiod

In litchi, it has been reported that between photoperiod and temperature, the latter has a direct influence on floral initiation. None of the litchi plants exposed to short (8-hour-light) day and long (16-hour-light) day flowered under a night temperature of 75 °F (23.9°C) and flowering occurred only when the night temperature was maintained at 60 °F (15.5 °C) or lower for a period of about two months. Flowers were initiated in the plants kept in the greenhouse (16-hour-light) and outdoors (12.1- 12.8-hour day) after the minimum temperature dropped to and below 65 °F (18.3 °C) for a period of 30 days (Nakata and Watanabe, 1966). Records for 8 years from a Brewster litchi orchard suggest that a soil moisture deficit and recurring temperature between 45 °F (7.2°C) and 32 °F (0 °C) in autumn and winter before flowering promote flowering and fruiting in Florida. The total duration of relatively low temperatures, rather than the frequency of time of occurrence during the critical period, seems to be the determining factor (Young, 1970). Batten and Lahav (1981) stated that vegetative growth of plants was best at 30 °C which ceased below 15 °C, but they flowered only at 10 °C. Flowering in litchi was best after 10 weeks at 15 °C, while periods of 8 h or more per day above 20 °C were detrimental (Menzel and Simpson, 1995). Menzel and Simpson (1991) in their experiment with five litchi cultivars noted high day/night temperatures (20°/15°, 25°/20° and 30°/25°C vs 15°/10 °C) after panicle emergence decreased the proportion of leafless panicles, increased the proportion of leafy panicles and reverted some reproductive

Intercropping and Flowering of Litchi

Intercroping with Onion, Amorphophallus and Brinjal

Initiation of Flowering

Fruit Seting Stage

Marble stage

Hermaphrodite Flowers in Litchi

Three Phases of Flower Production in Litchi

buds to vegetative growth. The reversion to vegetative characteristics at high temperature was accompanied by a reduction in the proportion of female flowers.

11.1.5 Nutrition

The deficiency of N, P and K depressed plant growth and flowering (Goldweber, 1959; Mallik and Singh, 1965), while applications of N, P and K were found to promote growth, flowering and increased yield (Koen *et al.*, 1981; Ghosh *et al.*, 1986a). The N, P and K contents of leaf and shoot in litchi cv. Bombai, declined in the month of November-December *i.e.* perior to panicle initiation and the concentration of various elements was higher in heavy bearing year than less bearing year (Sanyal and Mitra, 1993). It was suggested that all nutrients should be applied well before flowering to build up these reserves which are utilized for cropping (Menzel *et al.*, 1992b).

Dai *et al.* (1998) studied the effect of N on amino acid content and sex-expression of flowers on 22-year-old litchi cv. Lanzhu plants which were supplied with 0.25, 0.5, 0.8 or 1.2 kg N/plant (treatments N1-N4 respectively). The results indicated that with increasing N application the total content of amino acids increased in treatments N1, N2 and N3 but decreased in treatment N4. The effects of N application rate on the contents of different amino acids varied. The contents of valine, phenylalanine, serine, tyrosine and histidine were greatest in treatment N2, while the other amino acids peaked in treatment N3. Valine, phenylalanine and arginine contents were significantly positively correlated to total flower numbers. N application rate was significantly positively correlated with total and male flower numbers but significantly negatively correlated with the ratio of female: total flowers.

In an investigation by Xiong *et al.* (2001), the contents of the macro-elements in autumn shoots of litchi decreased in the order of N, K and P, which were not related to environment and variety, but notably to season. The contents of P and K in leaves of autumn shoots decreased rapidly, but the content of N did not change at the flowering and fruit bearing stages. The autumn shoots developed into fruit branches, and Litchi yield was significantly correlated to contents of N, P and K in leaves of autumn shoots; Litchi yield was significantly correlated to the content of N and P at the flowing stage, it was significantly correlated to the content of K at the maturity stage. Reasonable ratio among N, P and K was one of important measures to achieve high and stable yield in Litchi.

11.1.6 Cincturing

Nakata (1953; 1956) was the first to demonstrate the response of litchi trees to cincturing (girdling) and reported flowering and increased yield by up to 15 times, with cv. Brewster in Hawaii. Cincturing is performed on the lower part of the major branches of the trees. It involves removing a ring of bark and cambium. The ring is cut with a hacksaw, 3 mm wide and 3 mm deep, and the cut-surface is covered with a saturated paste of copper oxychloride to prevent disease infection (Menzel and Simpson, 1987b). The effects of branch cincturing at different stages of vegetative flush maturity on the growth and flowering of two litchi cultivars in subtropical Queensland were investigated. In cv. Bengal there was a marked increase in flowering when a branch with dormant vegetative growth was cinctured and a smaller increase with a branch in early or late flush. Flowering was not improved with a branch in mid-flush. In cv. Tai So, there was a reduction in flowering when a branch in early flush was cinctured (Menzel and Paxton, 1986a). The cincturing has possible beneficial effects on trees capable of less than 70 per cent bloom, but has detrimental effects in those capable of 70 to 100 per cent bloom if the cincture-cut does not remain open for 2-3 months (Menzel and Paxton, 1986b). Cincturing is sometimes practiced in China if trees are about to flush after heavy rainfall in autumn (Menzel and Simpson, 1986b). In cv. Bombai there was a marked increase in flowering by cincturing in autumn (Ghosh, 1988; Sanyal *et al.*, 1996) and cincturing the north and west side shoots showed more promising results in inducing flowering and increasing the hermaphrodite flower production than cincturing the east and south side shoots (Sanyal *et al.*, 1996). However, the cincturing treatments reduced the size of panicle. Spiral girdling of the trunk, performed in early December, improved flowering, fruitset, yield and fruit-quality in young trees of cvs. Nuomici and Guiwei, which were shy or non-bearing. The difference in healing patterns resulting from spiral and closed girdling indicated that the former technique was easier to use and would give more stable results (Zhou *et al.*, 1996). Chen and Luo (1998) reported that spiral girdling or ringing was an effective method for increasing pistillate flower production; ethylene also had some effect for increasing the proportion of pistillate flowers. Wu *et al.* (2000) found that shortening by removing the primary panicles, thinning and short-pruning the secondary panicles reduced the number of male flowers, increased the female/male ratio and proportion of fruiting panicles.

Chang and Lin (2008) found that fruit retention in litchi was sensitive to girdling applied during 3rd week. Most fruit dropped on branches with ≤ two flushes, whereas fruit continued to develop on branches with three flushes and on the controls. There was a gradual loss of fruit when the girdling was applied during 8th week. The number of fruit retained on branches with two and three flushes was similar to the controls. At harvest, regardless of the time of branch girdling, fruit yield and quality increased with increasing number of flushes; shoots with three flushes were similar to un-girdled controls. Leaf efficiency on branches girdled during 3rd week was inferior to that girdled during 8th week. All treatments had similar leaf efficiency when branches were girdled during 8th week. On the other hand, girdling treatment applied during week 3 resulted in variable leaf efficiency among treatment, indicating that fruit were utilizing reserves in bearing shoots. He concluded that bearing shoots of '73-S-20' litchi trees require a minimum number of three flushes for adequate fruit production.

Girdling is, however, ineffective if the trees are in poor health and lack of major soil nutrients may decrease new flush/tree growth, 6 months prior to bud initiation. It should not be done in windy areas as the tree trunks may break which eventually results in the death of the plant. In dry areas it should be done in alternate years or only half of the trunk should be girdled in 1 year. Before girdling, the trees should be given a complete fertilizer in July after harvesting, and for next 3 or 4 weeks, they should be given sufficient irrigation in order to induce vegetative flush (Rajwanshi *et al.*, 2017).

Girdling (Syn. Cincturing) (a ring of bark and cambium was removed from the branch base) plus defoliation (100 per cent leaf removal) in litchi reduces endogenous IAA concentration and increased the transcript level of two IAAresponsive genes (LcAUX/IAA1 and LcSAUR1), one cell wall degrading enzyme gene (LcPG1) and one ethylene biosynthetic gene (Lc-ACO1), in contrast to the decreasing accumulation of auxin response factor ($LcARF_1$) mRNA, with the concomitant increase in fruit drop (Kuang *et al.*, 2012).

11.1.7 Growth Regulators

In Hawaii, the litchi trees were sprayed with NAA in autumn to promote flowering (Shigeura, 1948; Bonner and Liverman, 1953). It was suggested that the treatment caused faster maturation of shoots. Chapman *et al.* (1980) tried with growth retardants to control vegetative growth. Spraying with SADH at 1000 mg/1 though inhibited vegetative growth, but the effect was only temporary and the trees flushed again after 6 weeks. Ghosh *et al.* (1986b) reported that ethrel (1.25 and 2.50 ml/l) advanced the flowering by 7 to 13 days and increased the pistillate flower production. In a comparative study, Sanyal *et al.* (1996) reported that in cv. Bombai, treatments with ethrel (1 ml/l), KNO_3 (104 mg/1) caused 70 per cent of the shoots to flower, however, the effect was not very promising in heavy fruiting year. The flowering was considerably enhanced in the less fruiting year (though not to the extent of heavy fruiting year) by KNO_3. All the chemical treatments *viz.* ethrel, KNO_3 and TIBA (103 mg/1) markedly increased the hermaphrodite flower production. Treatment with TIBA resulted in higher percentage of fertile pollen.

Application of GA_3 (50ppm) and MH (25ppm) during October showed early initiation of flowering panicle and conversion of some vegetative flushes into mixed flowering panicle in litchi which ultimately will improve the yield of plant.

Package of practices for higher production of quality litchi fruits by foliar application of Planofix @2.5ml/10 litre of water or NAA 20 mg/litre of water, one week after fruit set significantly checked the fruit drop. Spray of 4.0 per cent KNO_3 delayed colour break by 5 days in litchi cv. Shahi. Three sprays of Borax @ 4.0 g/litre of water during fruit development stage at 15 day interval maintaining moisture level in the soil was standardized for checking the fruit cracking and quality improvement. Application of KNO_3 (2.0 and 4.0 per cent) and $CaNO_3$ (1.0 per cent) delayed the colour break stage by 4 days and increased the fruit retention by 18.0 and 26.0 per cent, respectively. Covering the canopy with 30 per cent green and 50 per cent white shade net has extends the harvesting period by almost 16 days.

11.2.0 Physiology of Flowering

Flowering in litchi depends mainly on cultivar and climate. The flower bud differentiation in litchi occurs in December and is finished by January. The emergence of inflorescence starts in January–February. The vegetative buds develop first followed by reproductive bud. The litchi flower bud is a mixed bud, initially giving vegetative buds followed by reproductive buds. The inflorescence is a racemose panicle. The litchi orchard is buzz with activity during its peak flowering season in February-March.

There is chronological relationship between flowering period and fruit maturity (delayed flowering leads to a delayed harvest time) and first vegetative activity, flowering and harvest (early vegetative shoot leads to early flowering leading itself to early harvest time) (Jannoyer and Lauri, 2009). The flower buds are borne on terminal or axillary panicles. The inflorescence is borne on current season branches. However, the current season branches should have attained a certain physiological maturity. It has been observed that 12-15 weeks old branches bear flowers. Litchi needs chilling temperatures for flower initiation (Malhotra *et al.*, 2018). The time of cold weather and its intensity, moisture stress, physiological age of branches are known to affect the litchi flowering. The flower panicles are determinate in nature and each panicle consists of several branches. Panicles are mostly borne terminally. However, axillary panicles also develop.

According to Malhotra *et al.* (2018), the reproductive phase in litchi is associated with the ability to alternate between the production of vegetative and reproductive buds. The stress factors like water stress, cold winter are the triggering factors to induce dormancy which promotes flowering. The trees must experience one flush of vegetative growth during summer immediately after harvest and require a minimum number of three flushes for adequate fruit production. The flushes maturing earliest (before the winter period) produce floral shoots, while trees flushes maturing quite late produce vegetative shoots. The stem must take rest for 4 to 6 weeks prior to winter-spring flush for panicle formation in same flush. High leaf N is conducive for frequent flushes of vegetative growth and concentration less than 1.70 percent discourages initiation of vegetative flushes in the fall period. The vegetative flushing just prior to floral induction results in poor or no initiation and often vegetative shoots. The transition from vegetative to floral development is the consequence of changes in the physical and genetical expression in the shoot apex (with high cytokinin concentration). One month before flower bud formation, abscisic acid increases markedly and total cytokinin content increased in the xylem sap reaching maximum during flower bud formation and full bloom. The chlorophyll concentration dependent photosynthesis rate of lychee leaves on younger flushes and adjacent to fruit, is greater than that on shaded older flushes. The available total non-structural carbohydrates or starch is found to be accumulated before flower initiation and leaf flushing. Girdling (generally done when post-harvest flushes are matured) and spray of paclobutrazol (PBZ) and/or KNO_3 induced flowering in China litchi.

The bearing shoot of litchi comprised of three to four flushes and its carbohydrate reserves are essential for adequate cluster production. The age of the vegetative flush, or the degree of dormancy (during cool winter), appears to have a strong influence on the subsequent shoot development. Flush growth should be restricted to 0.6 – 2.0 cm in October-November and that leaves should be removed from flush growth if > 10 cm. These practices prevent alternate fruit set and stabilise yield (Singh *et al.*, 2017b)

The contribution of fruit photosynthesis to the total carbon requirement of a fruit was highest during early development and declined to zero at harvest. The fruit contributed 3 per cent to its own carbon requirement over the season. The leaves reached full size after 25 days after they emerged, but maximum net photosynthetic rate (A) occur until after 50 days. Photosynthesis then remained high for another 150 days. The increase in A as the leaves developed was strongly related to increases in stomatal conductance and chlorophyll concentration, which were highest in dark green leaves. Thus, gas exchange followed the course of chlorophyll synthesis, which lagged behind leaf expansion (Hieke *et al.*, 2002).

According to Kumari, S. (2017), the C:N ratio was found to be significantly higher at flowering stage. Physiological parameters like photosynthesis, stomatal conductance, internal CO_2, chlorophyll, carotene and xanthophyll content were found to be higher at flowering stage than pre-flowering stage.

A CO_2 at leaf temperature at < 24° C is only half of the maximum values, indicating a potential shortage of carbohydrate supply during blooming season and early fruit set in the spring (Chang and Lin, 2007). Assessing changes in biochemical constituents in shoots, *viz.* carbohydrates, phenols, soluble protein and amino acids in relation to flushing and panicle initiation helps in developing a proper understanding of the panicle initiation process in litchi (Das *et al.*, 2004). Higher level of RNA, phenols, and reduced level of amino acids, N, P, K, content favoured flowering. Non-reducing sugars, total soluble sugars, alcohol insoluble N, total N and nitrate reductase activity recorded higher in floral shoots in litchi. Respiration rate, carotenoids content, xanthophylls is found higher at floral bud differentiation (FBD) stages, while total free amino acids reduced during FBD. Leaf nitrogen (N) can also play a critical role in regulating flushing just prior to panicle emergence. It has direct effect on flowering, high N reduced flowering while low N promotes flowering (Thunyarpar, 1998). By restricting leaf N level to below 1.75 to 1.85 per cent prior to panicle emergence, flowering can be promoted in subtropical environment.

The flushes maturing earliest (before the winter period) produce floral shoots, while flushes maturing quite late (flushing just prior to floral induction) produce vegetative shoots. Significant reduction in the winter flushing (November and December) was recorded in paclobutrazol and KNO_3 treated trees as compared to the control. 'China' litchi has lowered photosynthetic rate (*A*), transpiration rate (*E*), internal CO_2 concentration (*Ci*), stomatal conductance (*gs*) and Leaf temperature (*Tl*) than 'Shahi' litchi. Floral shoots had more *A*, *E* but less *Ci*, *gs* over non-floral shoots. Leaves of floral and non-floral branches had reflected significant variation in biochemical status. Chlorophyll a, b and total chlorophyll contents were higher during flowering stage over floral bud differentiation stage irrespective of treatments. 'China' litchi had lesser content of total carbohydrate (total CHO) and reducing sugar (RS) content than 'Shahi' litchi. The total CHO and RS content increased from FBD stage to flowering stage but in contrast the proline and phenol content reduced during the same period. Although low temperature appears a pre-requisite for floral induction in lychee but there is strong relationship between flowering and flush emergence just before FBD. Harvesting of fruits with few leaves, post-harvest cincturing (Girdling), use of paclobutrazol (PBZ) or KNO_3 during September-October may be used effectively in the sub-tropics to inhibit winter flushing and induce synchronous flowering (Singh *et al.*, 2017a).

11.3.0 Floral Biology

The litchi inflorescences are many branched panicles 5-30 cm long, many–flowered; flowers small, yellow-white, functionally staminate or pistillate, calyx four–numerous, corolla absent, stamens usualy six, rarely ten, filaments long in staminate flowers and short in pistillate flowers (Tindal, 1994). Litchi requires a period of cool weather (15–20°C) for successful flower initiation. The flowering period in litchi varies with the genotype and environmental condition (Singh *et al.*, 2012). Generally, the flower bud differentiation in litchi starts in December and is completed by the end of January (Singh *et al.*, 2012). The flower panicle emerges from January to the end of February. Before floral bud initiation, the twig produces vegetative growth that subsequently throws a reproductive bud in the form of a racemose panicle.

There are three types of flower buds in litchi which come in a uniform succession on a tree. Flowering time, number and proportion depend on several factors. Among them, most important are cultivar and climatic conditions. The flowers are small and whitish green or whitish yellow in colour. The length of inflorescence ranges from 9 to 45 cm in length and 8 to 30 cm in width. The flower buds differ on the basis of degree of sexual development *i.e.* functionality of stamens and pistil.

The anthers of male or type I flowers are small and longitudinally dehiscent. The anthers are initially pale white in colour and become brownish at maturity. These flowers show very small non-functional ovules. The type II flowers are hermaphrodite with indehiscent anthers containing non-viable pollen. These flowers are functionally female and they remain receptive to pollen for about 2-3 days. The white coloured sticky stigma glistens during its receptive period. After 2-3 days, the surface dries up, looses its glossiness or shine and the colour changes from bright white to greyish and then brown. The nectar disc found at the ovary base produces secretion for attracting insects. The total number of female flowers varies between panicles. The percentage of female flowers in a panicle is significantly influenced by the cultivar as well as environmental factors. It is also understood that higher the number of female flowers in a panicle, greater is the potential yield of such cultivar. It is also influence d Type III is a functional male flower with closed stigma impermeable to pollen. These flowers remain functional for 7-10 days.

The peak opening time of flowers is early in the morning. The duration of flowering (anthesis to pollination) usually ranges from 26 to 35 days (Chadha and Rajpoot, 1969). The flowers are apetalous and greenish white to greyish in colour. Three types of flowers are commonly found in an inflorescence: male, hermaphrodite and pseudohermaphrodite (functional male). According to Robbertse *et al.* (1995b) male and female flowers are borne in variable positions on the dichasia. The first flowers to open are the no.1 flowers on proximally situated dichasia and usually male phase (M1). The no. 2 flowers (usually female) follow after a few days and no. 3 flowers (usually second male phase M2) a few days later. The no. 4 flowers usually remain in the undeveloped bud stage. Variation of the gender of the flowers in the different positions on the inflorescence can occur. It appears that, to a great extent the sex ratio of the flowers on a given inflorescence is determined by the sex of the no. 1 flowers on the dichasia.

Anthesis and dehiscence continued throughout the day and night, with peak opening early in the morning (Chadha and Rajput, 1969). The female flowers of cv. Mauritius tend to reach full-maturity only 3-5 days after anthesis (Stern *et al.*, 1997). As soon as the stigmas start dividing into lobes, they become receptive and remain so up to 3 days after anthesis (Chaturvedi and Saxena, 1965). The ovary consisted of 2-4 carpels, each with its own stigmatic lobe (Scholefield, 1982).

11.3.1 Pollination

Litchi is a highly cross pollinated crop. The flowers are self-sterile, requiring insects for pollination (Scholefield, 1982). On the other hand, Tott and Swart (1995) reported that litchi has a limited ability for self-pollination, despite being self-fertile.

Entomophilic cross pollination plays a major role in litchi fruit set. During full bloom, the orchard is filled with a characteristic scent and buzz with activity of hundreds of honeybees (*Apis dorsata:* Rock bee; *Apis mellifera:* European bee; *Apis cerana*: Indian honeybee and *Apis florae:* Small honeybee) along with some other small insects who are the major pollinators of litchi. Honeybee boxes in litchi orchards may increase the yield of litchi orchards by 30-40 per cent. Besides this, they help in monetary profits due to the honey, beeswax and other products. The major time of bee foraging is early in the morning till about 11 a.m. However, the bee activity is recorded even later in the day. It is imperative to not spray any harmful chemical which may affect bee foraging activity during flowering time. If at all any safe sprays need to be done they should be carried out during evening. High temperature, intense sunlight, rains, strong winds, or even cloudy day reduce honeybee foraging drastically. These further can also cause significant increase in rates of flower drying as well as flower drop. An intense rain or very strong winds can lower potential fruit yield by causing excessive flower drop. This is especially significant if these occur during the opening of female flower because the female wave comes once only coupled with the fact that flowers remain functional and receptive for a short window of 2-3 days. Moreover, the proportion of female flowers varies with cultivar and season. Wind also affects some pollination in litchi.

Pollination rate in cv. Mauritius followed a consistent pattern; it was low during the first male flowering (M1) and attained a high value only when the pseudo-hermaphrodite flowering (M2) started (Stern and Gazit, 1996). In all cvs, HLH Mauritius, Floridian and Bengal (=Madras) in South Africa, M2 pollen gave relatively high germination (average 40 per cent). There were significant germination differences between the M1 and M2 pollen of HLH Mauritius and Floridian, but none was observed with Madras (Fivaz and Robbertse, 1995). Pollen germination percentage varied greatly between cultivars Nuomici, Huaizhi and Shui-dong and was also significantly affected by weather conditions before staminate flowers opened, especially by mean temperature during the other development stage. Pollen of cv. March Red retained 26.3 per cent germination after desiccated storage for one year at –30 °C (Xian *et al.*, 1994).

The litchi honey is produced on commercial scale, sold at prices at par with some of the premium honey in the world market, because of its uniqueness. The yield of honey with the Italian honeybee species is the highest in Bihar compared to other states, with a production rate of 40 and 60 kgs honey/hive/year, for stationary and migratory beekeeping respectively

11.3.2 Fruit Set

Though the litchi trees flower profusely, only a small percentage of the flowers develop into fruits (Kanwar and Nijjar, 1975; Sharma and Ray, 1987a). Shahi cultivar is regular bearing in nature as opposed to China litchi cultivar which shows irregular bearing. The main reason of irregular bearing is failure of plants to bloom and/or excessive flower and fruit drop. Thus, fruit set in litchi is significantly influenced by cultivar and climate, particularly temperature and intensity during winter months and during flowering. Fruit set is also influenced by irregular climatic vagaries which hinder proper floral development and further pollinators foraging activity in case of normal flowering. There is a definite correlation between fruit set and sex-ratio (Chadha and Rajpoot, 1969). Several factors, such as moisture stress induced by low soil moisture, high or low temperature and strong winds enhance flower drop and reduce fruit set (Paxton and Chapman, 1980). Fruit set in 12-year-old cv. Mauritius was virtually eliminated in branches without leaves suggesting that stored reserve alone cannot support a crop (Roe *et al.*, 1995). CO_2 assimilation and fruit retention responded very quickly to changes in leaf: fruit ratio after fruit set. Virtually no fruit survived when all the leaves were removed, suggesting that stored reserves in isolated branches are inadequate for successful fruit set in litchi, which is mainly dependent on current CO_2 assimilation. It was observed that there were significantly more fruits when 30 leaves were retained (Table 20). Relative fruit retention was below 5 per cent in branches with no leaves and 50-60 per cent in branches with six leaves per fruit (Roe *et al.*, 1997). Ovule abnormality is one of the main factors responsible for poor litchi production (Stern *et al.*, 1996). Hand pollination of the 5-day-old flowers significantly improved fruit set which might be attributed to caging the tree that significantly lowered the temperature and delayed the senescence (Stern *et al.*, 1997). Spraying of putrescine dichloride (5×10^{-5} M and 5×10^{-4} M) in litchi cv. Bombai before anthesis had positive effect on fruit set (Mitra and Sanyal, 1990).

Any obstruction in pollinator activity has a direct bearing on fruit set of litchi. Cultivars can be selected based on their performance and inherent nature of bearing potential to avoid and/or reduce this major bottleneck. Proper pollination and consequent fertilization result in fruit set. Initial fruit set per panicle is very high depending on cultivar and varying from season to season. Nevertheless, fruit drop occurs in 2-3 waves lowering the final number of mature fruits per panicle to 5-30. This may be a mechanism of natural fruit thinning because the trees might not be able to bear a high load such as fruit set of even 150-180 fruits per panicle in the beginning in some cultivars, particularly commercially bearing trees of 15-20 years of Shahi cultivar. Yield from a commercially bearing tree is considered acceptable on producing about 15-20 fruits per panicles and about 60-90 kg fruit/tree.

Table 20: Effect of Leaf and Fruit Removal on the Number of Fruits per Panicle, Relative Fruit Retention, Total Fruit Weight, Average Seasonal Net CO_2 Assimilation and Concentration of Starch in the Leaves and Stems of 'Tai So' litchi (Data are the means of five trees per treatment)*

Treatment	*No. of Fruits per Panicle*	*Relative Fruit Retention (Per cent)*	*Total Fruit Weight (g/panicle)*	*Net CO_2 Assimilation (μmol m^{-2}/s^{-1})*	*Concentration of Starch (Per cent)*	
					Stem	*Leaf*
Leaf removal						
30 leaves	8.2	27.2	111	9.9	4.8	1.8
20 leaves	4.8	16.0	72	12.1	1.8	0.6
10 leaves	4.8	16.0	80	13.1	2.7	2.3
5 leaves	1.0	3.3	16	13.8	1.7	0.9
0 leaves	0.8	2.7	12	-	0.5	-
LSD (P=0.05)	2.2	7.4	35	1.8	N.S.	N.S.
Fruit removal						
30 fruit	8.7	29.3	119	9.9	4.8	3.2
20 fruit	7.0	35.0	96	7.6	4.0	2.9
10 fruit	6.8	68.0	101	6.2	9.8	6.5
5 fruit	3.0	60.0	49	3.1	8.8	2.3
0 fruit	(0)	(0)	(0)	3.7	11.3	1.7
LSD (P=0.05)	2.1	22.2	28	1.8	N.S.	N.S.

* Roe *et al.* (1997).

11.4.0 Fruit Growth and Development

11.4.1 Fruit Growth

In fruit growth of litchi, seeds develop initially at a high rate, followed by membranous mesocarp and aril or arillus, which grow very fast towards the later stage of fruit development. Khan (1929) divided fruit growth into 3 stages according to the development of aril; no aril development; aril development incomplete, with a portion of the testa uncovered; and the aril fully developed, with the seed-coat fully covered. Kanwar and Nijjar (1976) stated that the first phase of fruit growth was manifested by seed development and in the second phase flesh growth occurred in 3 stages lasting 38, 25 and 25 days, and characterized by formation of embryo, rapid growth of pericarp and the rapid growth of flesh respectively.

Fruit growth in litchi depends on current CO_2 assimilation rather than on stored carbohydrates (Roe *et al.*, 1995, 1997). In cvs Muzaffarpur and Calcutta, the fruit weight, diameter, TSS content and ascorbic acid, anthocyanin, reducing non-reducing and total sugars, total pectin and protein concentrations increased from fruit set to ripening. Fruit growth showed a sigmodial pattern. Moisture, acidity and phenol concentrations decreased with ripening (Singh and Singh, 1995).

Changes in endogenous phytohormone concentrations during fruit growth of litchi cultivars Wuheli (seedless fruits) and Huaizhi (seeded fruits) were studied by Zhou *et al.* (1998). Unlike with Huaizhi, the increase in fruit weight of Wuheli occurred mainly during the last stage of fruit development (60-80 days after anthesis). Changes in the cytokinin and gibberellin (GA1+3) contents of fruit peel of Wuheli were similar to those of Huaizhi; abscisic acid (ABA) content was high and GA1+3 content was low during the initial stages of fruit development. The IAA content was higher, but the ABA content was lower, in fruits of Wuheli than in those of Huaizhi after the midway stage in fruit development. It is suggested that the low fruit abscission rate of Wuheli during the intermediate and latter stages of fruit development might be related to the low ABA and high IAA contents.

11.4.2 Fruit Drop

The initial fruit set in litchi is very high but a very small proportion finally matures. The premature fruit drop commences soon after fruit set and continues till fruit maturity, with most fruit abscising in the first 2-4 weeks. The drop of fruit is thought to be due to failure of fertilisation, embryo abortion, internal nutrition and hormonal imbalance and external factors like high temperature, low humidity and strong winds (Menzel *et al.*, 1986d). Sharma and Ray (1987a) in their studies with five litchi cultivars recorded maximum fruit drop during the first fortnight after fruit set and at harvest the fruit retention varied between 3 to 39.6 per cent among the cultivars (Table 21). It is postulated that the high incidence of fruit abscission in litchi could be a physiological rather than a genetic problem (Fivaz and Robbertse, 1995). Treatment with growth regulators like NAA at 20-30 ppm, GA_3 at 20-50 ppm, 2,4-D at 10 to 20 ppm was found effective in minimising fruit drop when sprayed on panicles, before the flowers opened (Shukla *et al.*, 1978; Barua and Mohan, 1984; Singh and Phogat, 1984; Ghosh *et al.*, 1986b,c). Foliar application of NAA at 40 or 50 ppm at pea stage and again 21 days later were recommended to promote fruit set and reduce fruit drop (Brahmachari *et al.*, 1996; Bhat *et al.*, 1997).

Table 21: Fruit Set and its Retention in some Litchi Cultivars*

Date of Observation	*Percentage of Fruit Retention*				
	Shahi	*Rose Scented*	*Purbi*	*China*	*Bedana*
March 27	100	100	100	–	–
April 4	14.27	15.25	12.95	100	100
11	8.91	10.82	7.33	54.6	50.23
18	6.87	8.75	6.29	53.0	40.35
25	5.66	7.88	5.58	46.8	35.25
May 2	5.21	7.49	5.19	45.9	33.21
9	5.06	7.07	4.76	43.8	31.96
16	4.22	5.77	3.26	41.9	28.92
23	4.00	5.54	3.00	40.3	24.70
30	Harvested	Harvested	Harvested	39.6	23.57

* Sharma and Ray (1987a).

The effects of the synthetic auxin 3,5,6-trichloro-2-pyridyl-oxyacetic acid (3,5,6-TPA) on fruit retention and yield of litchi cv. Kaimana (a cultivar characterized by heavy initial set followed by heavy fruit drop) were investigated by tern and Gazit (1999) in Israel. Application of 3,5,6-TPA decreased fruit drop and increased yield by 130-170 per cent compared with untreated control trees. The best response occurred with application of 25 or 50 mg/l 3 weeks after full female bloom (FFB) (when fruits weighed ~ ~0.5 g), as opposed to 2, 4 or 5 weeks after FFB.

In an experiment on 30-year-old trees of Huaizhi litchis, the flower clusters were dipped in a solution of 50 x 10-6 M GA, 100 x 10-6 M 2,4-D or 20 x 10-6 M NAA 20 days before pistillate flower flowering. Some trees were sprayed with different concentrations of MH, GA or GA + 2,4-D, or GA + NAA 35 days after pistillate flower opening. The findings indicated that spraying once with 1 mM MH resulted in 76.7 per cent mini-kernelled fruits, but spraying twice produced 100 per cent mini-kernelled/aborted kernel fruits (Diao and Guan, 1998). Zhang *et al.* (1999) reported that spraying 14-year-old litchi trees of cv. Dahongpao with a solution of 0.15 per cent brassinolide on 5 May and 5 June increased yields from 16.0 kg/tree (controls) to 22.4 kg. Foliar application of micronutrients, such as boron and zinc also reduced the rate of this abscission (Pujari and Syamal, 1977; Verma *et al.*, 1980).

12.0 Pests, Diseases and Disorders

12.1.0 Pests

Like most fruit tree crops, it is usually attacked by two or three key pests, several secondary pests and a large number of occasional pests in localized areas where it is grown. Among biotic stress, insect pests are the major constraints in healthy litchi production (Srivastava *et al.*, 2016).

12.1.1 Litchi Fruit and Shoot Borer (*Conopomorpha sinensis* Bradley)

It is a major pest of litchi in Bihar. The larvae mine the leaves in August to February and migrate to *Eugenia jambolana* or *Cassia tora* during March to April. They come back to litchi in May and bore into the fruit and feed below the petiole (Lal and Sharma, 1978). The damaged shoot and leaf eventually die and drop off.

It is one of the major threats to litchi growers, as causes severe losses to fruit and shoot, to the tune of 24-48 per cent and 7-48 per cent, respectively. The insects (larvae) damage the newly emerged shoot during the September - October resulting in failure of shoot to bloom. Further, it punctures the peduncle of fruits (both developing as well as maturing) during April-May cause severe loss through early fruit drop and appearance of excreta/larvae, when fruit is cut/opened after ripening.

Apart, from *A. cramerella, Carpophilus obsoletus, C. mutilatus, Dacus dorsalis* and *Ephestia cautella* have been reported to bore into the fruit in Bihar, India (Kumar and Lal, 1984). *Myllocerus delicatulus* (Singh, 1976), *Cryptocephalus insubidus* (Singh, 1974) have been reported to feed on leaves in Saharanaur and *Dudua aprobola*, to web young shoots in Punjab (Mann and Singh, 1984). These pests cause maximum damage with the emergence of new vegetative flushes, in general, and during rainy season in particular.

Pruning of trees after harvest, field sanitation and removal of young fallen fruits has been found effective in reducing pest population. Fruit borer can be excluded by enclosing the fruit panicles in nylon mesh bags. Bagging also improves colour and quality of litchi fruits. Manuring the litchi trees with 4 kg of castor and 1 kg of neem cake per tree in the root zone after the first shower of monsoon reduces the insect pests' infestation.

Two spray of systemic insecticide *viz.*, thiacloprid 21.7 SC or imidacloprid 17.8 SL @ 0.5-0.7 ml/L during September at 15 days interval for management of shoot borer infestation on emerging shoots is essential. Neem act as deterrent, hence spraying of neem oil (4ml/L) before flowering may be done to avoid egg laying by fruit borer. Further, spraying of deltamethrin 2.8EC (1ml/L) after fruit attains the clove size (at 12-15 days after fruit set) should be done for fruit borer management. Spraying of novaluron 10 EC 1.5 ml/L or emamectin benzoate 5 per cent SG (0.4 g/L) during aril (pulp) development stage which comes around 30 days after fruit set and last spray of novaluron 10 EC 1.5 ml/l or lambda cyhalothrin 5 EC @ 0.5 ml/l at about 10 days before harvesting.

12.1.2 Bark Eating Caterpillar (*Indarbela tetraonis; Indarbela quadrinotata*)

This is an important insect pest of litchi in India. There are two species of bark eating caterpillars, (*Indarbela tetraonis* Moore and *Indarbela quadrinotata* Walker) and both are polyphagous in nature with wide range of host plants including, longan, rambutan, mango, mulberry aonla, ber, citrus, phalsa guava, jack-fruit, jamun, loquat, pomegranate and rose. Damage is caused by caterpillar, which bore into trunk, main stems and thick branches of litchi tree and destroying xylum tissues resulting into poor growth and fruiting of the tree. Caterpillars remain within the tunnel inside the stem during day, come out in night and feed upon the bark. Older and uncared trees are more affected by the pest. Infestation may be noticed by the presence of ribbons of wood chips, frass and silken thread over the bark surface.The adult moth lays eggs on old branches during May and June in groups of 15-25 which hatch after 8-11 days. The larvae feed on the surface of the bark till September. Afterwards they bore into bark and underlying tissues in the branches. These become full grown in December and pupate in late April.

Plugging the holes with fumigants, such as carbon bisulphide, petroleum or formalin and then plastering with mud kills the catterpillar. As a preventive measure, spraying on trunk and branches with 2ml/l chlorpyriphos may be done for bark eating caterpillar. The caterpillars can be killed by inserting an iron spoke into the tunnels. Remove the webs from tree trunks and put emulsion of DDVP (0.05 per cent) in each hole and plug them with mud. Mix chlorpyriphos 20 EC @ 2 ml /l of water and apply in the bark eating caterpillar infested area with a brush at 15 days interval.

12.1.3 Leaf Roller or Folder [*Statherotis (Argyroploce) leucaspis; Platypeplus aprobola* Meyrick]

Leaf roller/leaf folder is a very serious litchi pest in all litchi growing areas. The incidence of leaf roller can be observed on litchi trees throughout the year, however more population has been observed from August-February. The caterpillars roll and web the tender leaves, feed inside and subsequently pupate. Occasionally they damage the flower buds. Heavy infestations cause extensive leaf damage, especially to developing flushes. The egg, larval and pupal periods last for 2-8, 12-53 and 7-24 days respectively (Mishra and Pandey, 1965). Leaf injured by the leaf roller varied between 16.70 to 71.60 per cent while trees infestation varied 12.88 to 53.54 per cent during August to February.

Because of their sheltered habit it is difficult to kill the insects with a contact insecticide. Combination of a contact insecticide with one possessing a fumigant action may control the insect (Waite, 1986). Cypermethrin (0.006 per cent) is most effective in controlling leaf roller in UP, India (Singh and Kumar, 1994).

12.1.4 Leaf Cutting Weevils

Grey weevil (*Myllocerus* Sp,) and red weevil (*Apoderus blandus*) are polyphagous pest therefore, besides litchi its damage other important crops. Adult has long snout with grey colour, though poor flier but very active feeder on the leaves of litchi. It attacks leaves, shoot and flower.

Adult weevils congregate on the tender leaves and nibble irregular holes on the leaves and sometimes consume the entire leaf leaving the midrib only. The damage of red weevil is more severe at the time of shoot emergence as it prefers newly leaves therefore; newly established orchard/nurseries are more vulnerable for pest attack.

12.1.5 Litchi Looper (*Perixera illepidaria*)

Looper attacks tender leaves in mass and defoliate the new shoots. In severe attack, it completely defoliates the newly emerged flush. Besides litchi so many host plants *viz.* longan, rambutan, mango, and castor have been reported for this pest. Incidence of this pest has been observed from July to December however, highest population recorded during September-October.

12.1.6 Litchi Bug (*Tessaratoma* spp.)

Stink bugs are regular pests in all the litchi growing areas in India and elsewhere in the world. *Tessaratoma* species complex, *viz., Tessaratoma javanica* (Thunberg), *T. papillosa* (Dury), *T. quadrata* Distant, *T. nigripes* Dallas and *T. malaya* Stål (Hemiptera: Heteroptera: Tessaratomidae) are causing huge economic losses in quality and quantity of the litchi, and *T. javanica* is the most destructive among all in India.

12.1.7 Erinose Mite (*Aceria litchii*)

It is one of the most important pests of litchi. The characteristic symptom of this mite is chocolate brown velvety growth on the ventral surface of the leaves (with deep lesion resulting in reduced photosynthesis), in which the mites take shelter, feed and breed. The adults lay eggs singly on the ventral surface of the young leaves or while they are still in the bud. The nymphs hatch in 2-3 days and commence to feed immediately. Both nymphs and adults damage the leaves, inflorescence and young developing fruits through sucking of sap form newly emerged shoots. Due to continuous sucking of sap by the nymphs and adults, leaf tissues become aggravated and formed erineum. The nymphal period lasts for 8-12 days and the

Pests of Litchi

Bark Eating Caterpillar Infestation in Litchi

Fruit Borer Infestation in Litchi

Litchi Mite Infestation (a) & (b) Initial stage, (c) & (d) Advance stage

Pests and Diseases of Litchi

Semilooper Infestation in Litchi

Red Weevil Infestation in Litchi

Leaf Blight or Twig Blight Infection in Litchi

adult longevity varies from 2 to 3 days (Singh, 1993). The affected leaves become distorted and show leaf curl symptom. Infestation progresses through new flushes from older erinose on mature leaves. If these sequences are not broken, terminal growth stops and tree vigour decreases due to a reduction in photosynthesis (Waite, 1986). Temperature and relative humidity are reported to have significant influences on mite population and the largest number exists in April to May and the lowest in November to December (Sharma *et al.*, 1986).

Litchi mite control measures must be preventive. Once the mite is established, it is almost impossible to eradicate, hence depending upon infestation it is recommended that trees should be monitored specifically for erinose mite incidence on a monthly basis schedule. The mite population could be controlled by spraying 0.05 per cent dicofol before and after flowering followed by 0.05 per cent phosphamidon and dimethoate (Sharma and Rahman, 1982). Sharma (1985) reported that spraying of kelthene (1.5 ml/1) before and after flowering gave the most effective result in reducing leaf curl and increasing fruit yield. Two spray of chlorfenapyr 10 EC (3ml/L) or propargite 57 EC (3ml/L) at 15 days interval during July. If required, prune newly infested twigs during October and spray with chlorfenapyr 10 EC (3ml/L) or propargite 57 EC (3ml/L) (Srivastava *et al.*, 2016).

12.1.8. Plant Health Management through IPM

Some new pests *viz.* red weevil (*Apoderus blandus*), semi looper (*Anisodes illepidania*), bag worm (*Eumata crameri*) and green looper (*Thalasso despilaria*) has also been identified out of whichred weevil and semi looperhave been found to be the major emerging foliar pest of litchi in some years. *Trichogramma chilonis* @ 50,000/ha and application of Nimbicidine @0.5 per cent and Cypermethrin @ 0.005 per cent have been found effective in minimizing the damage of fruit borer. Less damaged litchi fruits at harvest has been found by use of pheromone trap + Tricho. Card @50000/ha and *Kamdhenu Keet Niyantrak* (5 per cent) sprays before panicle immergence and flowering time which opens up scope for safe crop management. Integrated management of litchi mite where mechanical and chemical measures are applied to control the mite. General cleaning and pruning of infested twigs/ leaves in the month of June, September and January followed by spraying of dicofol or proparzite (3 ml/lit.) or dimethoate (0.05 per cent) or profenofos (0.05 per cent) twice at weekly interval has successfully control the mite (Srivastava *et al.*, 2016).

12.2.0 Diseases

There are no serious diseases of litchi. Several disease of litchi are listed in the literature worldwide but only some of them are of high economic importance such as brown blight or blossom blight or litchi downy mildew disease (*Peronophythora litchii*), dieback and root rots. Brown blight is a major disease in China and Thailand. Anthracnose (*Colletotrichum gloeosporioides*) disease attack litchi in China, India and Australia. Some of the economically important diseases of litchi in India are leaf spots (*C. gloeosporioides* and *Botryodiplodia theobromae*), anthracnose (*C. gloeosporioides*) and twig blight (*C. gloeosporioides*, and *Gloeosporium* sp.) at pre-harvest stage (Kumar *et al.*, 2011, 2014), and fruit rots caused by several pathogens including *Alternaria alternata*,

Aspergillus flavus, Cylindrocarpon tonkinense, B. theobromae and *C. gloeosporioides* at post-harvest stage (Kumar, 2019).

12.2.1 Leaf, Panicle and Fruit Blight (*Alternaria alternate*)

Leaf blight is a prominent disease in nursery plants. The symptoms starts from tip of the leaf as light brown to dark brown necrosis that advances towards both the margins of the leaf leading to complete necrosis of the affected leaves that dries up subsequently At later stage the same pathogen also cause panicle blight, fruit blight and post-harvest fruit decay/rots.

Follow orchard sanitation by removing and destroying infected leaves wherever possible. For leaf blight, fungicidal spray of copper oxychloride (0.25 per cent) or carbendazim (0.1 per cent) can be done if disease severity increases. Prophylactic spray of difenoconazole (0.025 per cent) or carbendazim (0.1 per cent) or Thiophanate methyl (0.14 per cent) or azoxystrobin (0.023 per cent) may be done for panicle and fruit blight, one spray just after flowering and before fruit set, and another spray 20 days before harvest of fruits.

12.2.2 Leaf Spots (*Botryodiplodia theobromae, Colletotrichum gloeosporioides, Pestalotia pauciseta, Microdiplodia litchi*)

Usually mixed infection of different pathogens appears. Older leaves show higher infection. The leaf spots caused by *Botryodiploda* and *Colletotrichum* start either from tip or the margin of the lamina. These spots are dark, chocolate in colour in case of former while it is brown irregular in outline in case of latter. In case of *Pestalotia*, spots (0.5-2.0 cm) are light coloured and appear on both sides of the leaves. Leaf spot incited by *Microdiplodia* are yellowish brown to brick red in colour, mostly around the margin.

Apply foliar spray of mancozeb (0.25 per cent), or copper oxychloride (0.25 per cent) or thiophanate methyl (0.14 per cent) severity of the disease increases

12.2.3 Anthracnose (*Colletotrichum gloeosporioides*)

Mainly a fruit disease, may also affect leaves, twigs and flowers of litchi. Infected fruit become unmarketable. The disease first appears as brown pinhead lesions usually on the top of semi mature fruit. Circular dark-brown to black sunken lesions later become easily visible on mature fruits. More spots appear on the top and sides of the fruit that slowly covers the fruit surface.

Remove infected leaves, twigs and fruits and provide need based spray either with copper oxychloride (0.25 per cent) or carbendazim (0.1 per cent) or chlorothalonil (0.15 per cent) or difenoconazole (0.025 per cent) or azoxystrobin (0.023 per cent). Pre-harvest spray of fungicides help in extending post-harvest life but while doing so, care must be taken for residual toxicity of chemical applied

12.2.4 Red Rust (*Cephaleuros virescens*)

Red, circular to semicircular spots (3.0-6.5 mm) appear mostly on the leaves and sometimes also on tender stems. Orange yellow to pink velvety coating is formed on the spots on formation of sporangia of the algae. In older leaves the lesions turns light brown to brick-red.

In the Kangra valley in Himachal Pradesh, India, red rust caused by an algal parasite *Cephaleuros parasiticus* = *C. mycoides* was first noticed in 1981. The disease first appeared on young unfolded tender leaves. On the infected young leaves, small lesions of velvety white growth appear on the lower surface. On the upper surface, just opposite the lesions, chlorotic patches occur. As the leaves unfold and increase in size, the velvety growth becomes more prominent and dense and larger areas of leaves are covered with this growth. Old and thick leaves show various types of malformation (depressions and curling). The velvety growth turns light brown and finally dark brown to brick-red. The affected leaves become leathery and brittle. The disease results in a considerable decline in tree vigour and the fruit yield. Six sprayings of lime-sulphur, 3 during autumn and 3 during the spring season at fortnightly intervals give a highly satisfactory control of the disease, reducing twig infection by 94 per cent (Sood *et al.*, 1987).

Foliar spray of copper oxychloride (0.3 per cent) may be done in July and October if severity increases.

12.2.5 Wilt (*Fusarium solani*)

The sudden death phenomenon observed in Australia also occurs in China but it is not a common malady. A similar wilt disease which occurs in Taiwan is caused by *Fusarium* sp. Tree decline in South Africa has been attributed to infestation with nematodes (Menzel and Simpson, 1986b).

Young trees of litchi, less than five year age often wilt in less than a week time. The first symptoms noticed in the field is yellowing of foliage followed by gradual wilting and drying, leading to complete death of the plant within 4-5 days. Some brown spots appear on the root crown and lateral roots' phloem, spreading later to the xylem.

Apply castor cake or neem cake as manures. Apply *Trichoderma harzianum, T. viride, Pseudomonas fluorescens etc.* as biocontrol agents which also act as plant growth promoters. In absence of bio-control agent drench rhizosphere soil with hexaconazole or bavistin (0.1 per cent).

12.3.0 Disorder

12.3.1 Fruit Cracking

Sun-burning and skin-cracking of developing fruit can be serious problems in litchi which are promoted by high temperature, low humidity and low soil moisture (Kanwar *et al.*, 1972; Kanwar and Nijjar, 1975). Temperature higher than 38°C in combination with relative humidity lower than 60 per cent was very favourable for cracking of litchi fruits (Kanwar and Nijjar, 1975; Teaotia and Singh, 1970). Inadequate moisture during the early period of fruit growth results in the skin becoming hard and inelastic (sun-burnt), and it may crack when subjected to increased internal pressure as a result of rapid aril growth following irrigation (Menzel, 1984). The different litchi cultivars showed variations in sun-burning and skin-cracking. In West Bengal, India, the percentage of fruit cracking in eight varieties of litchi was recorded in the order of: Deshi > Early Large Red >

Muzaffarpur > Nafarpal > Elaichi > Bombai > McLean > Bedana (Sanyal *et al.*, 1990). In general, cultivars, which have relatively thin skin, few tubercles per unit area and rounded to flat in shape are less prone to crack (Kanwar *et al.*, 1972). Calcium and boron contents in the pericarp of uncracked fruits were significantly higher than those of the cracked fruits (Sanyal *et al.*, 1990; Li and Huang, 1995) while the Ca, P and Mg contents in the aril of cracked fruits were significantly higher than those of uncracked fruits (Li and Huang, 1995). Li *et al.* (1999) also found a negative correlation between the calcium content in litchi leaves and the rate of fruit cracking. Spraying with a solution containing calcium compounds significantly increased the calcium levels in the leaves and fruit peel and significantly lowered fruit cracking rate. Water stress reduced the calcium accumulation in fruit peel especially during rapid aril growth and thus increased the fruit cracking rate.

Frequent irrigation during the critical period of aril growth (Kanwar and Nijjar, 1975); spraying of zinc sulphate (1.5 per cent) at weekly intervals, starting from pea stage of fruit growth to harvest (Singh and Singh, 1954) or spraying of GA_3 at 40 ppm (Veera and Das, 1974), ethephon at 1.0 to 10.0 ppm (Shrestha, 1981) and 2,4-D at 20 ppm or NAA at 50 ppm (Bhat *et al.*, 1997) reduced the incidence of fruit cracking.

Huang *et al.* (1999) studied time-course changes of fruit cracking percentage, mechanical properties of the pericarp and contents of cell wall components in litchi pericarp by comparing 2 litchi cultivars differing in cracking susceptibility. Mechanical strength of the pericarp of crack-resistant cv. Huaizhi was greater than that of cracking-susceptible cv. Nuomici. The resistance of Huaizhi was also reflected in higher threshold cracking turgor as well as in higher pericarp tensile strength and extensibility. Higher contents of calcium and structural components (cellulose, hemicelluloses and insoluble pectin) were found in the pericarp cell walls of Huaizhi, which might have materially contributed to its higher pericarp strength and thus cracking resistance.

12.3.2 Sunburn

It is also known as lesion browning or pericarp necrosis. The occurrence of sunburn varies with cultivar and environmental factors. Temperatures higher than 38°C and relative humidities below 60 per cent have been reported to accelerate sunburn. The fruits which are inside canopy and shaded by leaves and branches show less sunburn. It initiates in mesocarp region followed by epicarp and endocarp. Shahi and Early large red cultivars have been found to be more affected by sunburn than Bedana cultivar. The sunburn can be reduced by covering the fruits with paper bags which shaded fruits, increased humidity and thus lowered temperature of fruits. Fruits affected with stunted fruit disorder show significantly smaller size and colour change to greenish yellow in litchi.

12.3.3 Fruit Drop

For management of fruit drop and cracking spraying of $ZnSO_4$ (0.2 per cent) at 30 days before panicle emergence induces healthy inflorescence leading to more fruit set. Foliar application of Boron (0.2 per cent), 2-3 times during the period of fruit growth and development enhance fruit retention, minimize cracking, improve

Physiological Disorder, Harvesting and Sorting of Litchi

Sun Burn and Fruit Cracking in Litchi

Harvesting with Ladder

Sorting of Litchi

Manual Harvesting from Small Trees

Excellent Quality Harvested Fruits

Litchi Harvesting with 15 cm Shoot with the Bunch

fruit colour, sweetness and enhance maturity leading to increased percentage of quality production.

Two foliar applications of planofix @ 4 ml/5 litre water may be done at an interval of 15 days from peanut size fruits to reduce fruit drop. Treatment with growth regulators like NAA (20-30 ppm), GA (20-25 ppm), 2,4-D (10-20 ppm) are effective in minimizing fruit drop when sprayed on panicles, before the flower opening.The fruitlet drop can be reduced by the sprays of $ZnSO_4$ and synthetic auxins like 3,5,6- trichloro-2-pyridyl-oxyacetic acid (3,5,6-TPA) on fruits increasing yield (Singh *et al.*, 2016).

13.0 Harvesting and Yield

13.1.0 Stage of Maturity

The maturity of the fruit is determined by flattening of tubercles, comparative smoothness of epicarp, attainment of red colour, TSS to acidity ratio equal or greater than 40 (TSS:total soluble solids) and 50-60 days to maturity depending on the cultivar. TSS of 18-23°B is considered acceptable for consumers. Fruit size greater than 18 g is also an indicator. Fruits should be harvested as soon as they ripen to enable transportation and marketing in faraway places. It also enhances the post-harvest storage period of the fruits. On the other hand, care should be taken to not let the fruit over mature on trees. In this case, the titratable acidity of fruits drop leading to a rather bland fruit taste.

In cv. Calcutta, the fruit maturity is optimal at 55 days after fruit set, when the fruits are bright pinkish, red, with flattened tubercles (Gaur and Bajpai, 1977). The colour of cv. Tai So becomes bright red, changing to dull red when fully mature while in cv. Haak Yip the skin colour becomes dull purplish red at maturity (Batten, 1986b). In cv. Elaichi, fruits with 18.1 total soluble solids, 0.24 per cent acid and having specific gravity of 1.00 are considered as optimum stage of maturity (Biswas and Roy, 1985). Underhill and Wong (1990) reported that the °Brix: acid ratio of the pulp was well correlated to mean eating quality. This ratio has been widely adopted internationally as the commercial maturity standard for litchi.

Litchi cultivar Shahi attains maturity between 73-78 days after fruit set. At the time of maturity, the fruit weights around 20 grams. Although climatic factors influence the time of flowering they do not necessarily affect fruit development or maturity. Harvesting takes around five to eight weeks, depending on the cultivars, production area and condition. The maturity groups of litchi cultivars based on heat unit summation have been worked out and its possible relation with sugar: acid ration at maturity has been derived. Shahi and Rose Scented are early maturing cultivars with less than 1400°C unit summation for maturity (Table 22).

Table 22: Grouping of Litchi Cultivars Based on Heat Unit Summation for Maturity

Degree Days Range	*Cultivars*
<1400	Ajhuli, Trikolia, Early Bedana, Shahi, Rose Sented
1400-1600	China, Purbi, Late Bedana
>1600	Dehradun, Kasba

Thus, the varieties should be harvested after attending proper stage of maturity. Litchi should be harvested in the morning hours or late afternoon when ambient temperatures are lower. Harvesting during hot periods can speed up the rate of water loss. At the time of harvesting the fruits should have the following characters.

Parameters	*Standards*
Fruit weight	20-25 g
Fruit colour	Red on outer and inner surface of the skin
Total degree days	1300-1400°C (Shahi and Rose Scented) 1400-1600°C (China)
Sugar content	15-17 per cent
TSS: acid Ratio	60.0-70.0
Skin	Smooth textured and easily peeled off

13.2.0 Method of Harvesting

The fruits are harvested in bunches from the tree. Selective harvesting is frequently carried out, since all fruits on a tree do not ripen at one time. The harvesting period, for any one cultivar may last for one to two weeks. The panicles should be cut from the tree with secateurs with as little leaf as possible. Experiments have shown that removal of foliage reduces next season's flowering (Batten and Lahav, 1981).

It is a terminal bearing tree with fruit bunches hanging on the periphery of the tree. Litchi fruit tree looks very attractive at the harvest time, when fruits are ripened due to the contrast bright red coloured fruit against dark green compact foliage. Fruits are harvested during summer months of May to July and the major percentage of the crop is used as fresh fruit followed by canning or as juice.Litchi fruit was introduced to Israel from South Africa, California and India in 1930s. Major litchi cultivars are 'Mauritius and Floridian. The orchards in Israel record productivity of 10 t/ha due to good orchard management practices like close planting, fertigation, drip irrigation, pruning and use of growth regulators. The harvest season in Israel ranges between July-August. Litchi was introduced in Bangladesh from Myanmar. Several cultivars were also introduced from India. Muzaffarpur, Bedana, China and Bombai are the major cultivars in Bangladesh with the harvesting period between May to June. The major varieties of litchi are Mauritius and McLean in South Africa. The main season is from November to January. In Vietnam, the main commercial variety is 'Thieuthanhha'. The harvest period ranges from late May to early June. Fruit are available from mid March to mid June in Thailand. Major cultivars in Australia are Kwai May Pink, Feizixiao and Salathiel and harvest season ranges between October to March. 'Bengal' is the main cultivar in Brazil and the fruits are harvested during December and January.

13.3.0 Yield

With proper care and management and under suitable environmental conditions a litchi tree may continue to bear economic crop for over 100 years of age. Heaviest yields, however, occur in years with cool-dry winter and warm wet humid springs (Menzel *et al.*, 1986c). The yield varies among the cultivars with the age of tree. Chau Kay-Ming (1986) from China stated that a maximum yield up to 500 kg per tree is not uncommon for a 100-year-old tree. Paxton and Chapman (1980) recorded a yield of 120 kg in 10-year-old trees of cv. Bengal in Southern coastal Queensland. In India, a full grown litchi tree, on an average, bears about 80 to 150 kg fruits per year (Nijjar, 1972). Four-year-old trees showed the following yields at Maroochy Horticultural Research Station, Nambour, Australia when the trees received cool, dry winter followed by wet spring, the conditions favourable for good flowering and fruiting (Menzel and Simpson, 1986a).

Cultivar	*Yield (kg/tree)*
Tai So	24.5
Bengal	28.5
Wai Chee	21.3
Gee Kee	6.5

Ray *et al.* (1985) in describing the nine important cultivars commercially grown in Bihar, India stated the following yields.

Cultivar	*Yield (kg/tree)*
China	104
Dehrarose	90
Deshi	95
Early Bedana	84
Kasba	106
Late Bedana	90
Purbi	104
Rose Scented	92
Shahi	97

14.0 Post-harvest Management

14.1.0 Packaging

After harvesting, the fruits should be kept in cool, dry and well-ventilated rooms because the litchi fruits ripen under high atmospheric temperature. Fruits for export are cooled immediately after harvest to 0-2°C usually by means of old water immersion or exposure to cooled air (Moreuil, 1973; Cull and Paxton, 1983). This is considered essential for flavour retention and for reducing the rate of enzymatic reactions. A good packaging, careful handling and an efficient transport system will ensure supply of litchi fruits to the consumers in an acceptable condition at a minimum cost. Skin browning from moisture loss and fruit rots is major problems in retail outlets, especially with carryover consignments at the wholesale market (Ledger, 1986). In Thailand, fresh litchis are packed in plastic boxes for export to Singapore. In India, fruits are packed in shallow baskets or crates. Most litchis of

Australia are currently packed in volume filled, bulk packs. In Table 23 the range of packages used for bulk packing in Australia is listed (Ledger, 1986).

Litchi is a non-climacteric fruit which at 25 C exhibits a moderate rate of respiration (30-40 ml CO_2/kg/h) and a low rate of ethylene production (less than 0.1 µl/kg/h). However, it deteriorates rapidly after harvest (Tian *et al.*, 2005). Shelf life at room temperature (30 C) is less than 72 h.

Litchi fruits are treated with sulphur prior to long distance transport. Fumigation with 1.2 to 2 per cent SO_2 for 10 minutes followed by a 2-minute dip in 1N HCl directly afterwards stabilized the red colour and reduced skin browning (Underhill *et al.*, 1992a,b; La-Ongsri *et al.*, 1993). Immediately after SO_2 treatment the pericarp may appear a uniform pale yellow colour, and then turn red again after 1 or 2 days (Jurd, 1964). The gaseous treatment helps to maintain colour by fixing the anthocyanins and acting as a fungicide and preventing browning.

Table 23: Bulk Packages Used for Litchi in Australia*

Package	*Material*	*External Base Dimension*	*Palletisation*	*Fruit Weight*
** AUF 9	Fibreboard	380 x 285 mm	12/layer	4.5-5 kg
Litre bulk pack	Polystyrene (open top and lidded)	383 x 288 mm		
12 punnet	Fibreboard	470 x 360 mm	Not suitable	5 kg
Standard	Fibreboard	485 x 325 ± 5 mm	8/layer	4.5-5 kg
Tray pack	Polystyrene	487 x 335 mm		
Frangible styrobox	Polystyrene	470 x 335 mm		4 x 1 kg or 4 x 1.5 kg

* Ledger (1986); ** Recommended.

14.2.0 Ripening and Storage

14.2.1 Ripening

Litchi is a non-climacteric fruit and usually do not improve in quality after harvest. So the fruits should be allowed to ripen properly on the tree. The development of red pigmentation of litchi fruits was found to be associated with four anthocyanin pigments, namely cyanidin-3-glucoside, cyanidin-3-galactoside, pelargonidin-3-glucoside and pelargonidin-3, 5-diglucoside, as well as other flavonoids (Prasad and Jha, 1978a). There appears to be no correlation between leaf alkaline composition and that of flowers and fruits during ripening (Prasad and Jha, 1978b). Acid phosphatase activity in fruit rose until the respiratory climacteric, whereas alkaline phosphatase activity declined markedly during the same period (Prasad and Jha, 1979a). Oxygen uptake attained a climacteric peak in partially ripe fruits and subsequently declined. Fructose content rose progressively throughout the maturation period. Pyruvate content attained a peak in fruits that were 75 per cent mature, and then declined, rising again at the climacteric and subsequently remaining unchanged (Prasad and Jha, 1979b).

Bhattacharya and Mohan (1977) reported that dipping the fruits of cv. Khatti in ethrel at 250 ppm solution improved the fruit colour and the peeling of fruit also became easier. On the other hand, spraying with urea at 120 g/1 or GA_3 at 50 mg/1 three weeks before harvest delayed the ripening of litchi cv. China by 6 days (Mishra *et al.*, 1985).

Fruit bearing in litchi coincided with highest temperature and least humidity period during April to June (1st week) in Bihar and adjoining areas which leads to the fruit cracking, sun burning and finally reduced the overall fruit quality and substantial amount goes waste due to cracking and sun burning. Therefore, the technique of bagging litchi bunches with perforated butter paper or brown paper bag during fruit development stage found best for all the quality parameters. Bagging of litchi fruit bunches with about 20 per cent perforated butter paper bag or brown paper bags keeping distal end of the bag opened at 40-45 days after fruit set produced about 34 per cent less cracked and spotted fruits and 30 per cent higher Class I fruits (Singh *et al.*, 2011; Purbey and Nath, 2013).

14.2.2 Storage and Browning of Pericarp

Litchi plants have a shelf life of about 72 h, or less at atmospheric temperature. The post-harvest losses for litchi are approximately 25–30 per cent and possibly as much as 50 per cent (Jiang *et al.*, 2001). The number of techniques are introduced to avoid browning as well as to control post-harvest rot and to enhance the shelf life of litchi fruit, including fumigation with sulphur, dips against fungi, materials promoting growth of plants, microbial agents (antagonist; *viz. Bacillus subtilis*), coats of wax and chitosan, radiation and treatments with heat. Of all these techniques, application of sulphur-mediated disinfection and dips having antifungal activities are commercialised (Jiang *et al.*, 2003). The efficacy of the antagonistic yeast (*Candida tropicalis*YZ27) as a biological control agent against post-harvest natural decay and quality retention of litchi (cv. Bombai) was studied by Zhimo *et al.* (2018). Application of the yeast antagonist YZ27 has great potential for development of commercial biocontrol formulations as an alternative of SO_2 fumigation to control post-harvest decay and quality retention of litchi fruit.

At non-refrigerated temperatures, litchi fruit deteriorates rapidly within hour after harvest (Brown, 1986). Dehydration, brown discoloration of the pericarp and rotting reduce the post-harvest self-life and saleability of the fresh fruit (Chang, 1983). Scott *et al.* (1982) demonstrated microscopically the appearance of fine cracks in the skin of exposed fruit soon after harvest. Loss of moisture through these cracks leads to brown discoloration after 3-5 per cent weight loss. The litchi peel consisted of ectocarp, mesocarp and endocarp. The external wall and cuticle of the ectocarp cells are very thin and projected outwards in a half sphere, thus they could be easily damaged mechanically (Pan Xun Cao and Baogui, 1996). It is believed that pericarp browning is caused by a rapid degradation of phenols by the activities of polyphenol (catechol) oxidase (PPO) and peroxidase (POD) which were found highest in the epicarp, with progressively less activity in the mesocarp and endocarp (Underhill and Critchley, 1996). The browning of litchi fruit pericarp caused by water loss was also investigated by Jiang and Fu (1999), in relation to total phenolic concentration, phenylalanine ammonialyase (PAL) activity, reduced glutathione content, polyphenol oxidase (PPO) [catechol oxidase] and peroxidase (POD) activities, and peroxide content. Development of pericarp browning increased with reduced storage relative humidity (RH from 90 per cent to 60 per cent) and increased storage time (from 1 to 3 days), and there was a strong time x RH interaction. Water loss from the pericarp increased PAL activity in the pericarp of fruits stored at 60 per cent or 70 per cent RH with a maximum after 2 days of storage, while activity of this enzyme during storage at 80, 90 or 95 per cent RH remained at lower levels. Reduced glutathione content as an indicator of antioxidant capacities was lower for low storage RH values than for high ones. Accordingly, the results suggest that water loss-induced browning of litchi pericarp may be due to both a reduction in antioxidant capacities and an increase in oxidation of phenolics catalysed by PPO and POD.

Zhang *et al.* (2000) recorded changes in phenolic compounds and anthocyanins in litchi (cv. Huaizhi) fruits during storage at ambient temperature (20-25°C) for up to 7 days and at 4°C for up to 35 days, and on changes in phenolic compounds and anthocyanins in the peel at various browning stages. Flavon-3-ol monomers and dimers were the major phenolic compounds, representing about 87.0 per cent of the total phenolic compounds detected, and cyanidin-3-glucoside was the major anthocyanin, representing 91.9 per cent of anthocyanins. Small amounts of malvidin-3-glucoside were also detected. These compounds were declined during storage and browning. Standardization of post-harvest handling techniques of litchi was reported by Chakraborty *et al.* (2010). Many other methods have been put forward to slow the rate of litchi browning. These include the use of SO_2 (including sulphites), high humidity, acidification of flesh, low temperature, fungicides, growth regulators or modified atmosphere packaging. The results obtained with respect to colour retention or overall storability varied greatly, depending upon the methods employed (Table 24).

Table 24: Methods to control post-harvest browning of litchi fruits

Control Method	*Treatment details*	*Storage Life (Days)*	*References*
Acidification	Dip in H_2SO_4 at pH 0.2, store at 5.5°C and 80 per cent RH.	60	Kesta and Leelawaana (1992)
	Dip in HCL at pH 0.3-0.5, store at 22°C	60	Zauberman *et al.* (1989)
	Dip in 1N, HCL, store at 23°C	72	Underhill *et al.* (1992)
Sulphuring followed by acidification	Dip in sodium metabisulphite (60 g/l) for 10 min, treat with 4 per cent HCL for 2 min, store at 1°C for28 days then at 20°C for 3 days	30-35	Fuchs *et al.* (1993); Duvenhage (1994)
	Fumigate with SO_2 store at low temperature (2-4°C) and 85 per cent RH	2-8	Fourie (1990); Lonsdale and Kremer-Kohne (1990)
Salting	Sprinkle 2 per cent NaCl on fruits	6-7	Popenoe (1948)
	Dip in 2 per cent NaCl and store in polythene bags at 25°C	6	Gaur *et al.* (1978)

Control Method	*Treatment details*	*Storage Life (Days)*	*References*
Waxing	9 per cent wax emulsion + 0.5 per cent sodium orthophenyl phenate, store at 30-35°C and 59-69 per cent RH	6	Garg and Ram (1972)
PGR treatment	Treat with kinetin 5 ppm	8	Sharma and Ray (1987a)
	Alar + $CaCl_2$	7	Nagar (1994)
	Thiabendazol 1000 ppm	6	Chen *et al.* (1986)
	NAA 40 ppm, store at 2-5°C	12	Ghose (1980)
Fungicide treatment	Bavistin-100 ppm	6-10	Bain *et al.* (1983); Sharma and Ray (1987a); Schutte *et al.* (1990)
	Benomyl (0.05-0.1 per cent) at 52°C, store at 20°C	7-11	Swarts and Anderson (1980); Tongdee *et al.* (1982); Scott *et al.* (1982)
	Benomyl 0.1 per cent, store at 5°C	14	Wong *et al.* (1990; 1991)
	Procloraz (0.05 per cent), store at 5°C	35	Wong *et al.* (1990)
Irradiation	Irradiate with 250-300 Gy, store at 4°C	28-30	Akamine and Goo (1972); Mitchel *et al.* (1992)

Post-harvest temperature management is important in extending the shelf-life of litchi (Brown, 1986). In South Africa, litchis are pre-cooled and stored at 0 to 1 °C for sea shipment (approximately 30 days) to London (Swarts and Anderson, 1980). Tongdee *et al.* (1982), however, stated that chilling injury may occur after 30 days at 7 °C. Chaudhury and Banerjee (1959) reported that at 36 to 38 °F (2.2 to 3.3 °C) and relative humidity of 80 to 85 per cent, fruits could be stored for a month. Lin (1982) in Taiwan, observed that unwrapped litchi turned brown within 4 days at 20 °C, while it took 8 days when stored at 5 °C or 0 °C. Rotting of litchi can be controlled by a rapid unheated dip of prochloraz (Brown *et al.*, 1984).

Peng (1998) studied the effects of post-harvest hot water treatment, sulphur (S) fumigation followed by HCl solution dip, packing methods and storage conditions on the quality, weight loss and respiration rate of cv. Heiye fruits. S fumigation + acid dip, S fumigation alone or hot water treatment + acid dip significantly increased fruit peel acidity (pH 5.12, 5.33 and 4.84, respectively) compared to the control treatment (pH 5.65), without affecting the total soluble solids content or acidity of the pulp. Weight loss of fruits was markedly lower in plastic boxes without holes than when fruits were unpacked or packed in boxes with holes. Heat-treated fruits had similar weight loss patterns to control fruits at 4°C. Fruit respiration rate decreased sharply as weight loss decreased during the early phase of storage, but became stable during later stages. Fruits stored at 4°C had markedly lower respiration rates than those stored at 21o or 35°C. The optimum storage conditions of litchi fruits were at 5 °C for 30 to 35 days (Huang *et al.*, 1990). However, fruits stored at 1 °C suffered from chilling injury. The recommended storage conditions for litchi fruits are given in Table 25.

Table 25: Recommended Storage Conditions for Litchi Fruit

Temperature (°C)	*Relative Humidity (Per cent)*	*Duration (Days)*	*Reference*
5	90	28-42	Mercantilia (1989)
5-10	90-95	28-42	Snowdon (1990)
1.7	90-95	21-35	Anon. (1991)

As part of the effort to reduce the use of potentially hazardous chemicals in agriculture a new procedure that may enable SO_2 to be avoided has been developed (Lichter *et al.*, 2000). Instead of fumigation, litchi fruits were sprayed with hot water while being brushed in a revolving drum, after which the fruits were subjected to hydrochloric acid treatment. Fruits that were processed in this way maintained a uniform red colour for at least 35 days, without apparent deterioration in external or internal quality, or taste. Physiological studies demonstrated that polyphenol oxidase (PPO) [catechol oxidase] activity was reduced by the hot water brushing (HWB) procedure as compared with controls.

Packing of fruits in punnets overwrapped with plastic film with 10 per cent CO_2 or vacuum packed, and stored for 28 days at 1°C followed by 3 days shelf life at 20°C showed less skin browning compared to unwrapped fruit, but their flavour was unacceptable (Ahrens and Milne, 1993). Wrapping of litchi fruits of cv. Bombai in perforated polyethylene bags gave the longest storage life with better results at 4°C than at 0 °C *i.e.* up to 18 and 16 days respectively (Mitra *et al.*, 1996).

In a study on storage of fresh fruits of cultivars Huoli and Heiyeli (Lin, 1999) treated with polysaccharide solutions in combination with various disinfectants and stored (500 g) in plastic bags at room temperature, and 79.2-81.5 per cent RH; controls were untreated. Fruits treated with polysaccharides at 1-2 per cent or 100 mg/kg, DM (homemade disinfectant) at 100 mg/kg and MI (homemade disinfectant) at 200 mg/kg, with or without 0.01 per cent emulsifier, gave the best results with 84.2-89.2 per cent sound fruits, a colour index of 2.8-3.1 and weight loss of 2.3-2.71 per cent after storage for 12 days whereas the controls had only 4 per cent sound fruits, 100 per cent browning and a weight loss of 3.1 per cent after storage for 4 days. Lemmer and Kruger (1999) evaluated the post-harvest quality of each potential litchi cultivar (based on yield and time of harvest) to ensure a high quality product suitable for export. New cultivars Wai Chee, Fay Zee Siu, Kwai May Pink and Haak Yip were evaluated and compared with the two widely grown cultivars HLH Mauritius and McLean's Red, in relation to different post-harvest quality criteria. During the 2 weeks of evaluation after the full 24-day storing period, HLH Mauritius achieved the best results in terms of firmness and percentage mass loss. McLean's Red showed the highest percentage mass loss and lowest firmness. Although Wai Chee and Fay Zee Siu could not compete with HLH Mauritius in terms of percentage mass loss and firmness, both these cultivars performed better than McLean's Red.

14.2.3 Pericarp Browning

Pericarp browning is the first visual sign of fruit decline. Browning that occurs during the first few days after harvest is usually caused by dehydration of the pericarp. This problem has been reported in several litchi producing areas and involves the appearance of necrotic spots which may expand and affect a significant portion of the fruit. Fruits start to brown once they lose a few percent of the harvested pericarp fresh weight. Below 50 percent of its initial fresh weight, the pericarp is entirely brown. Browning of the pericarp occurs at ambient temperatures of 20-30 °C within 24 hours of harvest. Water loss (desiccation) of litchi results in brown spots on the bright-red shell (pericarp). Other factors also cause the fruit to brown, including mechanical stresses of various sorts (tugging the pedicel at harvest, sliding the fruit down a rough picking bag, dropping fruit from short heights); microbial and insect attack; and extremes of temperatures. In short, anything likely to accelerate cell breakdown is likely to increase fruit browning. This occurs as a result of damage (due to causes unknown) to the parenchymatous cells of the thin wall of the mesocarp. This damage speeds up a characteristic hypersensitive reaction, leading to necrosis of these cells, which then spreads to the epicarp and endocarp. Skin-browning is associated with an increase in polyphenol oxydase and perioxydase enzyme activity and with ascorbic acid oxidation. PPO cannot oxidise the degradation of anthocyanins, but the presence of phenolic extract stimulates pigment degradation by PPO, leading to the development of brown pigments. A decrease in the ascorbic acid (anti oxidative compound) content of litchi pericarp is associated with an increase in the peel browning index, and more ascorbic acid exerts a preventive action on the anthocyanin degradation. It has been suggested that anthocyanins, PPO and phenols contribute to litchi pericarp browning. Losses may be considerable due to the browning; although the aril is not affected the fruit is not suitable for marketing. In extreme cases, the pericarp may crack. Through the use of fungicides and refrigeration, litchi fruit have a storage life of about 30 days. However, when they are removed from storage, their shelf life at ambient temperature is very short due to pericarp browning and fruit rotting.

Packing fruit into moisture-proof (plastic) bags and punnets can substantially reduce water loss and slow the rate of browning. Cool temperature storage also slows browning. Low temperatures slow evaporation as well as respiration and probably slow tissue senescence. It has been reported that fruit treated with polyamines, then wrapped and stored at 5°C, had lower membrane permeability and less browning than control. A controlled atmosphere of 3 to 5 per cent O_2 and 3 to 5 percent CO_2 has also been shown to slow water loss and reduce browning. Sulphur dioxide fumigation effectively reduces pericarp browning, but some countries do not permit due to concerns over sulfur residues in fumigated fruit. Thus, sulfur-free post-harvest treatments that maintain fruit skin colour are increasingly important. Alternatives to SO_2 fumigation for control of pericarp browning (Kore and Chakraborty, 2014) and fruit rotting are pre-storage pathogen management. Israel has recently promoted a hot water brush/acid/prochloraz treatment, but acid treatments sometimes give an artificial and persistent red colour to the fruit that masks poor eating quality (Pongener *et al.*, 2019).

14.3.0 Value Addition

Litchi growers can easily prepare different processed products of litchi by adopting simple food safety and processing guidelines using locally available resources. Processed products can be preserved for a longer time; hereby, processing not only extends the period of availability of the fruit but also allows for realising good income commensurate to market demand. Chakraborty *et al.* (2010) developed several processed products of litchi like blended litchi beverage, osmo-air dried litchi nuts, pulp, osmo syrup and evaluated the retention of their quality attributes after preparation and also during storage. Osmotically dehydrated litchi products may be a convenient, low cost and viable option for managing the post-harvest losses of extremely perishable litchi during off-season (Chakraborty *et al.*, 2009)

ICAR-NRC on Litchi, Muzaffarpur has developed protocols for different value-added products of litchi (Table 26). The basic product is the fruit juice or pulp which can be marketed on business to business (B2B) mode and becomes the primary ingredient for further processing. Consumer or retail products such as beverages (squash, RTS, crush, cordial, and wine), litchi nut, canned litchi, dehydrated litchi pulp (litchi raisin) are some popular products that can be marketed directly in retail to consumers (Business to consumers, B2C) or even as B2B.

Table 26: Some of the Processed Product of Litchi

Litchi Pulp	✰ Homogenized juice of litchi fruit pulp ✰ Litchi pulp is used in preparation of beverages like squash, nectar, RTS, cordial *etc.* ✰ Permitted class-II preservative added ✰ Store in air-tight containers of convenient size ✰ Storage life up to one year under refrigerated conditions
Litchi Squash	✰ Contains litchi pulp (25 per cent v/v), TSS (40°B), acidity (0.8 per cent), permitted class-II preservative and stabilizer ✰ Diluted before consumption ✰ Best before 4 months ✰ Keep in refrigerator after breaking seal and consume within 2-3 days
Litchi RTS	✰ Ready-to-Serve fruit beverage ✰ Contains litchi pulp (10 per cent v/v), TSS (10°B), acidity (0.3 per cent), and stabilizer ✰ Store under refrigerated conditions ✰ Storage up to 4 months
Canned Litchi	✰ Litchi aril preserved in sugar syrup inside air-tight glass bottle ✰ No chemical preservative added ✰ Best before 1 year
Dehydrated Litchi Pulp	✰ Prepared through osmo-mechanical dehydration ✰ Packed in air-tight packaging ✰ Stored in cool and dry conditions ✰ Storage life > 1 year under refrigerated conditions

Litchi Nut	☆ Whole litchi fruit is dried ☆ Product is packed in air-tight packaging ☆ Storage life > 1 year
Litchi Wine	☆ Alcoholic product prepared through anaerobic fermentation of litchi fruit pulp ☆ Contains 10-12 per cent alcohol ☆ Product retains flavour of litchi
Litchi Honey	☆ Litchi honey is obtained by processing raw honey collected from litchi orchards ☆ Contains >78°BTSS

Litchi pulp /juice are the primary product obtained by squeezing, crushing, pressing the aril (edible portion) of litchi. Litchi pulp can be preserved and stored in containers of convenient size. Preserved pulp can be marketed as a product in the form of business to business dealings or can be used as ingredient for further processing into different litchi beverages.ICAR-NRCL has developed protocol for preparation of litchi squash with inclusion of litchi pulp: 25 per cent, TSS (sugar): 40 per cent, acidity: 0.8 per cent and addition of preservative (SO_2) @350 ppm. All these products are shelf stable and if provided with the correct storage conditions can be stored for relatively longer duration. This provides an added advantage for marketing as per demand and price dynamics.

14.4.0 Marketing

The Indian litchi has period advantage of fruit availability (May-July) when premium quality litchi is not available elsewhere except Thailand (May-June) and Israel (July) but quantum of produce available from these (Two) Countries is very less or negligible. Despite the period advantage, India has negligible share in the world litchi trade. Threforte, certain requirement of international marketing has to be fulfilled. (1) The fruits should be of extra grade (measuring 33 mm diameter); fruits should be properly packed and stored *i.e.* Sulphited litchi fruits are banned in International market. Therefore, Modified Atmospheric Packaging (MAP) along with Control Atmospheric Storage (CAS) condition is required. In Southern India where the litchi matures by December, export can be explored. This period is advantageous for export to European Union where there is high demand during Christmas period. Most of the litchi produced in India particularly in Assam, Tripura, and Bihar are organic in nature and need to be exported as of organic litchi to foreign market. Even for normal produce "Brand promotion" is needed to catch the assured market.

Most of the litchi fruit grown in South-East Asia and the Pacific is sold within the country or close to the areas in which it is produced (Table 27). The Middle East, which includes Bahrain, Kuwait, UAE, Oman, and Saudi Arabia is a major market for litchi. The major suppliers of litchi fruits are Madagascar, South Africa, Mauritius, Thailand, and Australia. Markets of Europe consist of Netherlands, UK, and Germany where they receives major share from South Africa, Madagascar, Thailand and Israel. Litchi has market for fresh fruits but canned, frozen and dried fruits are also exported to some extent.

Maturity Stages and Marketing of Litchi

Different Maturity Stages of Litchi

Marketing of Litchi

Transport of Litchi

Packaging and Storage of Litchi

Packaging and Storage of Litchi

Litchi Storage with SO_2 Fumigation

Table 27: Major Litchi-Exporting Countries*

Exporting Countries	*Export (per cent)*	*Importing Countries*
China	2 per cent	Hong Kong, Japan, Singapore, USA, Australia, Korea, Europe middle East, North America
Taiwan	5 per cent	Philippines, Japan, Singapore, USA, Canada
Thailand	10 per cent	Malaysia, Singapore, Hong Kong, Europe, Canada
Australia	33 per cent	Hong Kong, Singapore, Europe, Arab States
South Africa	50 per cent	Europe (France, Nether land, Germany, U.K
Israel	95 per cent	Europe

* Singh *et al.* (2017).

In India, the marketing of litchi is unorganized leading to a meager export. However, during 2017-18, 100 tons of litchi were exported to different countries (Table 28). The export of litchi to EU has declined inspite of the fact that the Indian litchi is available 2-3 weeks earlier than Thai litchi and 4-5 weeks earlier than Chinese litchi. The major reasons for this decline are the increasing domestic demand for this juicy fruits and better price margin in local market.

Table 28: Export of Litchi from India - Country-wise (Qty in MT; Value in Rs. Lacs)

Country	*2015-16*		*2016-17*		*2017-18*	
	Qty	*Value*	*Qty*	*Value*	*Qty*	*Value*
Nepal	9.40	4.96	52.61	29.56	100.82	43.07
United Arab Emirates	0.03	0.02	20.37	52.68	5.10	10.78
Saudi Arabia	0.05	0.02	0.00	0.00	1.53	3.31
Kuwait	0.12	0.30	0.33	0.46	0.29	0.58
Norway	0.00	0.00	0.00	0.00	0.20	0.12
Bahrain	0.20	0.16	0.00	0.00	0.25	0.09
Others	0.07	0.03	50.71	22.37	0.05	0.03
Total	9.87	5.49	124.02	105.07	108.24	57.98

Source: Anon. (2018).

Litchi producing areas in Gurdaspur and Hoshiarpur districts of Punjab are near to Amritsar international airport. Export of litchi from this area can be enhanced by setting up of packhouse in the area. Technology for CA and MA storage of litchi needs to be standardized perfected so that shelf life of litchi can be extended and then litchi to nearer areas like Gulf countries can be sent by reefer containers through MA cartons. Residue analysis laboratory need to be set up appropriately to minimize the risk of rejection of consignment to begin with.

15.0 Biotechnology

The progress in biotechnological investigation in litchi is not adequate. In tissue culture, success in embryo, anther and protoplast culture has been reported. Biochemical markers have also been used for identifying genotypes. No report on genetic transformation is available.

15.1.0 Tissue Culture

15.1.1 Micropropagation

In litchi, a suitable technique for propagation using tissue culture techniques has not yet been standardized. One of the problems of micropropagation of litchi is the secretion of polyphenol into the medium by the tissues. These polyphenols undergo oxidation to produce tannin substances which kill the tissue (Dodd and Kantharajah, 1991). Another problem is that field-grown litchi used as a source of known genetic material, is heavily contaminated with organisms which are very difficult to remove prior to culture (Kantharajah *et al.*, 1989). Both of these problems are frequently encountered with hardwood species and the usual approach is to develop the experimental culture techniques using young tissues from seeds germinated *in vitro*. This technique has the advantage of providing very clean material as well as producing juvenile tissue which usually responds well in culture.

Shoot segments (0.8 cm long) of cv. Liudong 1 were inoculated onto half strength MS medium with varying concentrations of BA, GA_3 and IAA and cultured at 25-30 °C and 2000 lux for 10 h daily. Maximum regeneration rate of the segments was noted with 1.0 mg BA, 0.2 mg IAA and 0.6 mg GA_3/l. The regenerated segments were then grafted on stocks, cultured in tubes with a seedling age of 15-20 days, began to produce new shoots after 13 days and developed into perfect plants (Yu, 1991).

15.1.2 Embryo Culture

One of the long term objectives is to breed varieties having superior fruit qualities. Litchi fruits that have aborted seeds are termed 'chicken tongues' and are preferred by consumers, since these fruit have a high flesh to seed ratio. In litchi breeding programmes, crosses involving plants which produce these fruits as the female parent, many of the most valuable progeny are lost prior to harvest as a result of abortion. Production of plants by culturing embryos prior to abortion would be expected to yield progeny with a high proportion showing the 'chicken tongue' character (Kantharajah *et al.*, 1992, 1997).

The tissue culture technique known as embryo rescue provides a technique which will prevent the loss of such valuable progeny in breeding programme. The technique involved sterilizing the surface of the young fruit with a high concentration sodium hypochlorite solution, then dissecting the fruit in the sterile environment of a laminar flow cabinet to remove the embryo. Since the embryo is enclosed in a sterile environment by the testa, this technique allowed the isolation of young embryos as small as 2 mm. One of the problems encountered was the browning of very young tissues due to exudation of polyphenols which could be solved by adding polyvenyl- pyrrolidone at 800 mg l-1 to the medium. Using the technique it was possible to rescue embryos as small as 3 mm (Kantharajah *et al.*, 1997). However, more success was achieved with more mature embryos (>3 weeks of age) and the very young embryos had a poor survival rate.

15.1.3 Anther Culture

The technique of anther culture has a number of applications in plant breeding. Fu and Tang (1983) cultured anthers of two litchi varieties on Murashige and Skoog (1962) medium supplemented with 2,4-D and NAA. After about 90 days, callus was formed from the anthers. When this callus was transferred to a medium containing kinetin and NAA, numerous embryoids were formed. The young embryoids were transferred to rooting medium of very low salt concentration and having very low concentration of kinetin and IAA, where rooting was achieved. The work of Fu and Tang (1983) showed that the plants contained 15 chromosomes confirming that haploid plants had been produced.

15.1.4 Protoplast Culture

Yu *et al.* (1996) initiated friable and granular embryogenic calli from young embryos (cv. Xiaofanzhi) which were then subcultured on STS (silver thiosulphate) containing medium for more than 2 years. Protoplasts were obtained from embryogenic cell suspensions with a mixed enzyme solution containing CPW salts, 11 per cent mannitol, 0.8 per cent cellulase R-10 and 0.4 per cent macerozyme R-10. The yield after purification was 10-15 x 10^6/g FW and viability was 92-97 per cent. Embedding the protoplasts in calcium alginate beads and co-culturing them with nurse cells promoted continuous division, small callus formation and direct formation of somatic embryos. The protoplast-derived callus was cultured on a solid medium and the frequency of differentiation into somatic embryos was 100 per cent. Eventhough, the somatic embryos suffered from hyperhydricity and had difficulty in differentiating cotyledons and buds, some of the embryos, after culture on different media for 2-5 months, initiated roots and buds and eventually plants.

15.2.0 Biochemical Markers

Molecular techniques, *e.g.* isozyme analysis, protein analysis or DNA fingerprint that are more effective to identify cultivar names, should be applied to solve this problem. The most advantage of using molecular technique is due to its quickness and young tree can be used to identify cultivars. While morphological method that used reproductive characters is time-consuming, as it cannot be used before the lychee tree starts to bear fruits in the fifth to seventh years. An international board should be established to collect the correct genetic resources, germplasm and genetic data of lychee cultivars. China, where lychee originated and have a lot of lychee cultivars, seems to be the most suitable place to act as the centre of this board. Longan has been suggested as a source of cold or drought tolerance or resistance to erinose mite. Lychee x longan hybrids have been produced in China and Australia, but no commercial cultivars have been released to industry (Singh *et al.*, 2019). Isoenzyme banding patterns produced by 19 litchi cultivars and 6 vegetative selections revealed sufficient differences to distinguish between most cultivars. A considerable number of polymorphic patterns were found, particularly for phosphoglucose isomerase. All

6 vegetative selections were identical in all respects to their parent HLH Mauritius (Cilliers and Visser, 1995).

While the isozyme pattern was used as reliable marker, rapid developments in technology led to the introduction of DNA based molecular markers which have proved to be very efficient. Among these, RAPD and AFLP (amplified fragrant length polymorphism), markers are quite beneficial for assessing genetic diversity. In recent years RAPD markers have been successfully exploited in litchi (Kumar *et al.*, 2006). Comparison of parents displaying differences in DNA markers may be useful for breeders in selection of parents with different genomes resulting in new combinations of characters in their progeny.

15.3.0 Genetic Transformation

The majority of transgenic studies are based on the use of *Agrobacterium tumefaciens*. The gene transfer to litchi has been achieved till date by *Agrobacterium*-mediated transformation. *Agrobacterium*-mediated transformation has been most widely used and is compatible with both the productions of cell suspension cultures and the regeneration of transgenic plants from a variety of litchi cultivars. *In vitro* grown leaf tissues were used for transformation. After four weeks in culture, expression of GFP was apparent when the regenerated callus and the leaves were observed under fluorescence microscope fitted with a blue exciter filter, a blue dichroic mirror and a barrier filter. Although no transformed litchi plantlets were regenerated, screening for GFP gene expression may prove useful to improve transformation efficiency and to facilitate detection of transformed litchi plants.

Previously bacterial chitinase (ChiB) gene were expressed into litchi cv. 'Bedana' which however showed low level of chitinase activity, and consequently transgenic plants showed partial resistance against *Phomopsis* sp. pathogen (Das and Rahman, 2010).

Biotechnology holds several promises in litchi improvement. Tissue culture techniques like anther and zygotic embryo culture can be exploited for raising homozygous lines. Likewise, genetic transformation to raise stable transformants for different characters is gradually been explored. Genetic markers are of special significance as it can aid in conventional breeding approaches of litchi varieties. Despite the successful regeneration of different genotypes, the conversion rate of somatic embryos into normal plantlets remains low. Future research must be focused on enhancing conversion frequency of somatic embryos into normal plantlets and regeneration of plantlets from shoot/nodal segments (Kumar *et al.*, 2019).

16.0 References

Ahmed, S. (1961) *Agric. Pakistan*, **12:** 769-775.

Ahrens, F.H. and Milne, D.L. (1993) *Yearb. South Afr. Litchi Growers' Assoc.*, **5:** 29-30.

Akamine, E.K. and Goo, T. (1977) *Res. Bull. Hawaii Agric. Exp. Sta.*, **169:** 1-20.

Anonymous (1978) *Rept. South China Agricultural University*, Guangdong, pp. 1-160.

Anonymous (1991) *In: SeaLand Service Inc.* Nerw Jersey, USA.

Anonymous, (2018) Horticultural Statistics at a Glance 2018. Ministry of Agriculture & Farmers Welfare, GOI, New Delhi

Anonymous, (2019) NHB Database. Gurugram, Haryana

Bain, J. M., Chaplin, G. R., Scott, K. J., Brown, B. I. and Willcox, M. E. (1983) *Australian Hort. Res. Newsl.*, **55:** 124.

Bandyopadhyay, B., Chakraborty, D., Ghosh, S., Mishra, R. and Rahman, M. (2015) *Clinical Microbiol.*, **4:** 181

Barua, P.C. and Mohan, K.K. (1984) *South Indian Hort.*, **32**: 92-95

Batten, D.J. (1986a) *Acta Hort.*, **175**: 79.

Batten, D.J. (1986b) *Proc. Ist Nat. Lychee Sem.*, Australia, pp. 73-76.

Batten, D.J. and Lahav, E. (1981) *Res. Rep. Trop. Fruit Res. Stn.*, Alstonville, N.S.W., pp. 37-88.

Batten, D.J., Lloyd, J., and Mc Conchie, C. (1992) *Australian J. Plant Physiol.*, **19:** 317.

Batten, D.J., Watson, B.J. and Chapman, K.R. (1982) *Qd. Fruit Veg. News*, **53:** 890-894.

Bhat, S.K., Raina, B.L., Chogtu, S.K. and Muthoo, A.K. (1997) *Adv. Plant Sci.*, **10:** 83-86.

Bhattacharyya, R.K. and Mohan, N.K. (1977) *South Indian Hort.*, **25:** 91-94.

Biswas, S. and Roy, B.N. (1985) *Haryana J. Hort. Sci.*, **14:** 30-32.

Bolt, L.C. and Joubert, A.J. (1968) *Farming S. Afr.*, **44:** 11-13.

Bolt, L.C. and Joubert, A.J. (1980) *Inf. Bull. Citrus Subtrop. Fruit Res. Inst.*, No. 92, pp. 4-5.

Bonner, J. and Liverman, J. (1953) In: *Growth and Differentiation in Plant* (Ed. W.E. Loomis), Iowa State College Press.

Bose, T.K., Dhua, R.S., Mitra, S.K. and Sen, S.K. (1985) *Indian J. Hort.*, **42**: 153-155.

Brahmachari, V.S., Lalduhawama, K. and Kumar, R. (1996) *Orissa J. Hort.*, **24**: 5-9.

Brown, B.I. (1986) *Proc. Ist Nat. Lychee Sem.*, Australia, pp. 77-78.

Brown, B.I., Scott, K.J. and Mayer, D.G. (1984) *Singapore J. Primary Indust.*, **12:** 40-47.

Campbell, R.J. (1994) *Proc. Fla. Sta. Hort. Soc.*, **107**: 348-350.

Campbell, T.P. (1981) *Qd. Fruit Veg. News*, **52**: 156-157.

Chadha, K.L. (1968) *Indian Hort.*, **12:** 13-36.

Chadha, K.L. and Pareek, O.P. (1991) *In: Technologies for minimizing Risk in Rainfed Agriculture.* Indian Society for Extension Education, International Fund for Agriculture Development and ICAR, N. Delhi, pp. 271-312.

Chadha, K.L. and Rajpoot, M.S. (1969) *Indian J. Hort.*, **26:** 124-129.

Chakraborty, I., Joseph, J. and Chaurasiya, A.K. (2010) *Acta Hort.*, **863:** 519-524.

Chakraborty, I., Chaurasiya, A.K. and Saha, J. (2010) *Acta Hort.*, **863:** 637-644.

Chakraborty, I., Pathak, P.K. and Chaurasiya, A.K (2009) *Indian Hort.* **54**(5): 39-40.

Chakraborty, S., Saha, N.K., Prakash, K. and Bisht, H.C. (1980) *Indian Fd. Pack.*, **34**: 17-22.

Chan, H.T.Jr. and Kwok, S.C.M. (1974) *J. Food Sci.*, **39:** 792-793.

Chandel, S.K. and Kumar, G. (1995) *Adv. Plant Sci.*, **8:** 284-288.

Chang, J. C., and Lin, T. S. (2008) *J. American Soc. Hort. Sci.*, **133**(2): 284-289.

Chang, T.C. (1983) *Res. Rep. Food Indust. Res. Dev. Inst. Hsinchu*, Taiwan, N. E-86, pp. 1-11.

Chapman, K.R. (1984a) *Rept. Qd. Dep. Primary Ind.*, pp. 1-123.

Chapman, K.R. (1984b) In: *Tropical Tree Fruits for Australia* (Ed. P.E. Page), Qd. Govern. Printer, Brisbane, pp. 179-191.

Chapman, K.R., Paxton, B. and Cull, B.W. (1980) *Bienn. Rep. Maroochy Hort. Res. Stn.*, No.2, pp. 32-33.

Chaturvedi, R.B. and Saxena, G.K. (1965) *Allahabad Farmer*, **39:** 10-13.

Chau Kay-Ming (1986) *Proc. Ist Nat. Lychee Sem., Australia*, pp. 52-53.

Chaudhury, J.K. (1940) *Curr. Sci.*, **9:** 416.

Chaudhury, J.K. and Banerjee, H. (1959) *Indian J. Pl. Physiol.*, **2:** 141-144.

Chen, F., Li, Y. and Chen, M. (1986) *Acta Horticulturae Sinica*, **13:** 151-156.

Chen, JieZhong and Luo, XiaoZheng (1998) *South China Fruits*, **27:** 26-27.

Chen, L.S. and Liu, X.H. (2001) Acta Hortic., **558:** 297-299

Chen, XiuDao, Chen, JianSheng, Zhang, FaBao and Yang, Shaottai (1998) *Guangdong Agric. Sci.*, No. 2, pp. 27-29.

Cilliers, B. and Visser, G.J. (1995) *J. South Amer. Soc. Hort. Sci.*, **5**: 11-13.

Cobin, M. (1954) *Bull. Fla. Agric. Exp. Stn.*, No. 546, pp. 1-35.

Cronje RB (2010) *In:* The cultivation of litchi. ARC Institute for Tropical and Subtropical Crops, Nelspruit, pp. 8–22

Cull, B.W. (1977) *Report on Overseas Study Tour*, Queensland Dep. Primary Ind., pp. 1-92.

Cull, B.W. (1986) *Proc. Ist Nat. Lychee Sem.*, Australia, pp. 19-20.

Cull, B.W. and Paxton, B.F. (1983) *Qd. Agric. J.*, **109:** 53-59.

Dai LiangZhao, Lin ChangXian, Liu LiRong, Su MuLan and Fu YanGen (1998) *J. Fujian Agric. Univ.*, **27**: 419-421.

Das, B., Nath, V., Rai, M. and Dey, P. (2004) *Indian J. Hort.*, **61:** 1–5.

Das, D. and Rahman, A. (2012) *Plant Cell Tissue Organ Cult.*, **109**: 315–325

Davenport, T.L. (2000) *Hort. Tech.*, **10:** 733-739.

Davenport, T. L. (2003) *Hort. Sci.*, **38:** 1331-1335.

Deng, Y. and Ni, Y. (1994) *Guangdong Agric Sci.*, **2:** 20-22.

Deng, Y., Ni, Y. and Chen, N. (1994) *J. South China Agric. Univ.*, **15:** 80-84.

Diao, ShenBiao and Guan XueFang (1998) *South China Fruits*, **27:** 25.

Dinesh, D.S., Pandey, K., Das, V.N.R., Topno, R. K., Kesari, S., Kumar,V., Ranjan, A., Sinha, P.K. and Das, P. (2013) *Int. J. Curr. Microbiol. Appl. Sci.* **2**(12): 531-538

Dodd, W.A. and Kantharajah, A.S. (1991) *Australia Lychee Yearb.*, **1:** 56-60.

Dutta, P., Biswas, B. and Mitra, S.K. (1995) *Curr. Agric. Res.*, **8:** 54-57.

Duvenhage, J.A. (1994) *Yearb. South Afr. Litchi Growers' Assoc.*, **6:** 36-38.

Dwivedi, A.K. and Mitra, S.K. (1995) *The Hort. J.*, **8:** 113-118.

Fivaz, J. and Robbertse, P.J. (1995) *Yearb. South Afr. Litchi Growers' Assoc.*, **7**: 26-30.

Fourie, P.C. (1990) *Yearb. South Afr. Litchi Growers' Assoc.*, **3**: 23-25.

Fu, L. and Tang, D. (1983) *Acta Genetica Sinica*, **10:** 369-374.

Fuchs, Y., Zauberman, C., Ronnen, R., Rot, I., Weksler, A. and Akerman, M. (1993) *Acta Horticulturae*, **343**: 29-33.

Galan Sauco, V. (1989) *Lychee Cultivation*. FAO Plant Production and Protection Paper 83, FAO, Rome.

Garg, R.C. and Ram, H.B. (1972) *Progressive Hort.*, **3:** 33-39.

Gaur, G.S. and Bajpai, P.N. (1977) *Progressive Hort.*, **9:** 11-17.

Gaur, G.S., Bajpai, P.N. and Azad, C.S. (1978) *Progressive Hort.*, **10:** 63-67.

Ghosh, B. (1988) *Ph.D. Thesis* submitted to BCKV, India.

Ghosh, B., Bose, T.K. and Mitra, S.K. (1986a) *Abst. 22nd Int. Hort. Congress*, California, Abst. No. 1189.

Ghosh, B., Bose, T.K. and Mitra, S.K. (1986b) *Abst. 22nd Int. Hort. Congress*, California, Abst. No. 426.

Ghosh, B., Bose, T.K. and Mitra, S.K. (1986c) *HortScience* 21:717.

Ghosh, D. (1980) *M. Sc. (Ag.) Thesis*, Calcutta Univ., Calcutta.

Ghosh, B., Biswas, B., Mitra, S.K. and Bose, T.K. (1987) *Indian Fd. Pack.*, **41**: 34-37.

Goldweb, S. (1959) *Proc. Fla. Sta. Hort. Soc.*, **72:** 353-156.

Goto, Y.B. (1960) *Pacific Rim Food Conf.*, **1:** 15-23.

Groff, G.W. (1921) *The Lychee and Longan*, Orange Judd, New York, pp. 1-180.

Gupta, A K., Nath, V., Singh, A., Singh, M., Marboh, E.S., Pandey, S. and Pathak, A. (2017) *In:* Lychee Disease Management, Springer Nature, Singapore.

Hamilton, R.A. and Yee, W. (1970) *Proc. Fla. Hort. Soc.*, **83:** 233-235.

Hasan, M.A. and Chattopadhyay, P.K. (1995) *Environ. Eco.*, **13**: 110-116.

Hieke, S., Menzel, C. M. and Lüdders, P. (2002) *Tree Physiol.*, 22: 955–961

Higgins, J.E. (1971) *Bull. Hawaii Agric. Exp. Stn.*, No.44: 1-21.

Huang, H.B. and Chen, H.B. (2003) *J. Fruit Sci.*, **20**(6): 487–492

Huang, W. T., Huang, T. M., Hsiang, W. M., Chang, M. H., Lin, M., Wang, C. H. and Wu, W. L. (1998) *J. Agric. Res. China*, **47:** 388-407.

Huang, X., Kang, D. and Ji, Z. (1990) *J. South China Agric. Univ.*, **11**: 13-18.

Huang, X.M., Wang, H.C., Gao, F.F. and Huang, H.B. (1999) *J. Hort. Sci. Biotech.*, **74:** 351-354.

Hundal, H.S. and Arora, C.L. (1996) *J. Indian Soc. Soil Sci.*, **44:** 294-298.

Indu Mehta (2017) *IOSR J. Humanities Social Sci.*, **22 (8): 21-25.**

Jannoyer, M. and Lauri, P.E. (2009) *Acta Hortic.*, **820:** 395- 402

Jiang, Y. M. and Fu, J. R. (1999) *J. Hort. Sci. Biotech.*, **74:** 43-46.

Joubert, A.J. (1970) *Bull. South Afr. Dep. Agric. Tech. Serv.*, No. 389, pp. 1-22.

Jurd, J.R. (1964) *J. Food Sci.*, **29**: 16-19.

Kadam, S.S. and Deshpande, S.S. (1995) In: *Handbook of Fruit Science and Technology: Production, Composition, Storage and Processing* (Eds.D.K. Salunkhe and S.S. Kadam) Marcel Dekker, Inc, New York, pp. 435-437.

Kantharajah, A.S., Dodd, W.A. and McConchie, C.A. (1989) *Proc. 2nd Nat. Lychee Sem.*, Cairns, Australia, pp. 59-69.

Kantharajah, A.S., Ebert, G. and Ludders, P. (1997) *Erwebsobstbau*, **39:** 116-118.

Kantharajah, A.S., McConchie, C.A. and Dodd, W.A. (1992) *Ann. Bot.*, **70:** 153-156.

Kanwar, J.S. and Nijjar, G.S. (1975) *Punjab Hort. J.*, **15:** 9-13.

Kanwar, J.S. and Nijjar, G.S. (1976) *Haryana J. Hort. Sci.*, **5:** 164-168.

Kanwar, J.S., Nijjar, G.S. and Rajput, M.S. (1972) *Punjab Hort. J.*, **12:** 146-152.

Katyal, S.L. and Chugh, D.V. (1961) *Hort. Adv.*, **5:** 39-41.

Ketsa, S. and Leelawatana, K. (1992) *Acta Hort.*, **321:** 726-731.

Khan, K.S. (1929) *Agric. J.*, **24**: 183-187.

Knight, R.J. (1963) *Yearb. Proc. Fla. Lychee Growers' Assoc.*, **10:** 47-51.

Koen, T.J. and Smart, G. (1983) *Rept. Citrus Sub-Trop. Fruit Res. Inst.*, p.1.

Koen, T.J., Langenengger, W. and Smart, G. (1981) *Inf. Bull. Citrus Sub-trop. Fruit Res. Inst.* No. 103, 9-12.

Kore, V.T. and Chakraborty, I. (2014) *Int. J. Fruit Sci.*, **14:** 205-224.

Kuang, J.F., Wu, J.Y., Zhong, H.Y., Li, C.Q., Chen, J.Y., Lu, W.J. (2012) *Int. J. Mol. Sci.*, 13: 16084-16103

Kumar R. and Nath V. (2013) *In:* Climate-Resilient Horticulture: Adaptation and Mitigation Strategies, Springer, India.

Kumar, M. and Lal, R. L. (2007) *Pantnagar J. Res.*, **5**(1): 45-49

Kumar, M. and Lal, B.S. (1984) *Bull. Ent.*, **25**: 98-99.

Kumar, M., Gupta, M., Shrivastava, D., Prasad, M., Prasad, U.S. and Sarin, N.B. (2006) *Intl J. Agri. Res.*, **1:** 390-400

Kumar, R. (2014) *Acta Hortic.*, **1029:** 145-154.

Kumar, S., Sharma, R.K. and Choudhary, R. (1995) *The Hort. J.*, **8:** 33-37.

Kumar, V. and Sharma, S. (2016) In: Training Manual, MTC, ICAR-NRCL, Muzaffarpur, Bihar

Kumari, R. (2009) *Ph.D Thesis*, BAU, Ranchi, Jharkhand, 133p.

Lal, B.S. and Sharma, D.D. (1978) *Pesticides*, **12**: 40-42.

Lal, R.L., Tewari, J.P. and Mishra, K.K. (1996) *Recent Hort.*, **3:** 18-20.

La-Ongsri, S., Gomolmanee, S. and Onnop, W.A. (1993) *In: Post-harvest Handling of Tropical Fruit*. Proc. Int. Conf. ACIAR, 50.

Ledger, S.N. (1986) *Proc. 1st. Nat. Lychee Sem.*, Australia, pp. 79-81.

Lemmer, D. and Kruger, F. (1999) *Neltropika Bull.*, **305:** 16.

Li, J. G. and Chen, H. (1998) *South China Fruits*, **27:** 24-25.

Li, J. G. and Huang, H. (1995) *J. South China Agric. Univ.*, **16**: 84-89.

Li, J. G., Gao, F. F., Huang, H. B., Tan, Y. W. and Luo, J. T. (1999) *J. South China Agric.Univ.*, **20**: 45-49.

Liang, J. (1981) *J. Agric. Trad. Bot. Appl.*, **28**: 259-270.

Lichter, A., Dvir, O., Rot, I., Akerman, M., Regev, R., Wiesblum, A., Fallik, E., Zauberman, G. and Fuchs, Y. (2000) *Post-harvest Biol. Technol.*, **18**: 235-244.

Lin Xiong, Zhang Xiaoqi, Li Chongyang, Shen Xiuhua. (2001) *J. South China Normal Univ.* (Natural Science), **4:** 98-100.

Lin, B. F. (1999) *J. Guangxi Agric. Biol.*, **18:** 205-208.

Lin, S.C. (1982) *Proc. 21st Int. Hort. Congress. Hamburg*, Abst. No. 1177.

Litz, R.E., Raherjo, S., Matsumoto, K. and Witjaksono, A. (2005) *In:* Litchi and longan: botany, production and usesM CABI, Cambridge, MA, pp. 49–58

Loebel, R. (1976) *Trop. Fruit Bull., N.S.W. Dep. Agric.*, **6**: 1-12.

Lonsdale, J.H. and Kremer-Konhe, S. (1990) *Yearb. South Afr. Litchi Growers' Assoc.*, **3**: 49-51.

Lu, L.X., Chen, J.L. and Chen, X.J. (1987) *J. Fujian Agril. College*, **16**: 224-228.

Lu, Q. F. and Li, H. P. (1998) *South China Fruits*, **27**: 32.

Lynch, S.J. and Nelson, R.O. (1956) *Ceiba*, **4**: 315-336.

Malhotra, S. K, Singh, Sanjay Kumar and Nath, V. (2018) *Indian J. Agric. Sci.*, 88 (9): 1319–30.

Mallik, P.C. and Singh, D.L. (1965) *Indian Agric.*, **9:** 127-132.

Mann, G.S. and Singh, G. (1984) *Int. Pest Control*, **26:** 71.

Mathew, A.B. and Pushpa, M.C. (1964) *J. Food Sci. Technol.*, **1:** 71.

McConchie, C.A., Vithanage, V. and Batten, D.J. (1994) *Ann. Bot.*, **74:** 111-118.

Menzel, C.M. (2001) *South Afr. Litchi Growers Assoc Year* **12:** 9-14.

Menzel, C. (1991a) *Development of lychee production in Vietnam, Technical Report for UNDP*. VIE/86/005 Project, pp. 33.

Menzel, C.M. (1983) *Sci. Hort.*, **21:** 210-215.

Menzel, C.M. (1984) *Sci. Hort.*, **22:** 333-345.

Menzel, C.M. (1985) *Sci. Hort.*, **25:** 31-48.

Menzel, C.M. (1991b) In: *Plant Resources of South-East Asia*. No.2. *Edible Fruit and Nuts*. (Eds. E.W.M. Verheij and R.E. Coronel), Pudoc, Wageningen, p.191.

Menzel, C.M. (1995) *WANATCA Yearb.*, **19**: 41-48.

Menzel, C.M. and Paxton, B.F. (1986a) *J. Hort. Sci.*, **61:** 135-139.

Menzel, C.M. and Paxton, B.F. (1986b) *Australian J. Agric.*, **26:** 255-259.

Menzel, C.M. and Simpson, D.R (1994) *In: Handbook of Environmental Physiology of Fruit Crops*. Vol.II: *Sub-tropical and Tropical Crops* (Eds. B. Schaffer and P.C. Anderson), CRC Press, Inc. Florida, pp. 123-145.

Menzel, C.M. and Simpson, D.R. (1986a) *Qd. Agric. J.*, May-June, pp. 125-136.

Menzel, C.M. and Simpson, D.R. (1986b) *Proc. Ist Nat. Lychee Sem.*, Australia, pp.55-70.

Menzel, C.M. and Simpson, D.R. (1986c) *Proc. Ist Nt. Lychee Sem.*, Australia, pp.11-17.

Menzel, C.M. and Simpson, D.R. (1986d) *Proc. Ist Nat. Lychee Sem.*, Australia, pp.22-36.

Menzel, C.M. and Simpson, D.R. (1987a) *Qd. Fruit Veg. News*, September, pp. 15-18.

Menzel, C.M. and Simpson, D.R. (1987b) *Qd. Fruit Veg. News*, August, pp.15-18.

Menzel, C.M. and Simpson, D.R. (1990) *Plant Varieties J.*, **44:** 197.

Menzel, C.M. and Simpson, D.R. (1991) *J. Hort. Sci.*, **66:** 335-344.

Menzel, C.M. and Simpson, D.R. (1993) *In: Encyclopaedia of Food Science, Food Technology and Nutrition* (Eds. R. Macf, R.K. Robinson and M.J. Sadler), Academic Press, London, pp. 2108.

Menzel, C.M. and Simpson, D.R. (1995) *J. Hort. Sci.*, **70:** 981-987.

Menzel, C.M., Carseldine, M.L. and Simpson, D.R. (1988a) *Australian J. Exp. Agric.*, **28:** 793.

Menzel, C.M., Carseldine, M.L., Haydon, G.F. and Simpson, D.R. (1992a) *Scientia Hort.*, **49**: 33-53.

Menzel, C.M., Chapman, K.R., Paxton, B.F. and Simpson, D.R. (1986c) *Australian J. Exp. Agric.*, **26**: 261-265.

Menzel, C.M., Haydon, G.F. and Simpson, D.R. (1992b) *J. Hort. Sci.*, **67:** 149-160.

Menzel, C.M., Oosthuizen, J.H. and Roe, D. (1993) Unpublished data – as quoted in Menzel and Simpson (1994)

Menzel, C.M., Oosthuizen, J.H., Roe, D.J. and Doogan, V.J. (1995a) *Tree Physiol.*, **15**: 611-617.

Menzel, C.M., Piccone, M.F. and Simpson, D.R. (1986a) *Qd. Fruit Veg. News*, October, pp. 15-18.

Menzel, C.M., Piccone, M.F. and Simpson, D.R. (1986b) *Qd. Fruit Veg. News*, February, pp. 27-29.

Menzel, C.M., Piccone, M.F. and Simpson, D.R. (1986d) *Qd. Fruit Veg. News*, January, pp. 15-17.

Menzel, C.M., Piccone, M.F. and Simpson, D.R. (1987) *Qd. Fruit Veg. News*, January, pp. 15-18.

Menzel, C.M., Simpson, D.R. and Doogan, V.J. (1996) *J. South Afr. Soc.Hort. Sci.*, **6**: 16-19.

Menzel, C.M., Simpson, D.R., Haydon, G.F. and Doogan, V.J. (1995b) *J. South. Afr. Soc. Hort. Sci.*, **5:** 97-99.

Menzel, C.M., Watson, B.J. and Simpson, D.R. (1988b) *Qd. Agric. J.*, **114:** 19.

Mercantilia (1989) *Guide to Food Transport – Fruit and Vegetables*. Mercantilia Pub.

Miao, P. S., Jian, H., Ye, H., Deng, L. and Fu, W. (1998) *China Fruits*, **2**: 53.

Miller, C.D. and Bazore, K. (1945) *Hawaii Expt. Sta. Bull.*, p.96.

Mishra, K.A., Syamal, M.M. and Ray, P.K. (1985) *Indian Fd. Pack.*, **39:** 56-59.

Mishra, S.C. and Pandey, N.D. (1965) *Labdev. J. Sci. Tech.*, **3:** 126-127.

Mitchell, G.E., McLauchtan, R.L., Issac, A.R., Williams, D.J. and Mottingham, S.M. (1992) *J. Food Comp. Anal.*, **5**: 291-311.

Mitra, S., Sarangi, A.B.S. and Kar, N. (1996) *Environ. Eco.*, **14:** 538-542.

Mitra, S.K. (2002) FAO Publication 200/04, pp.5-13.

Mitra, S.K. and Sanyal, D. (1990) *Gartenbauwissenschaft*, **55:** 83-84.

Moreuil, D. (1973) *Fruits*, **28:** 637-640.

Murashige, T. and Skoog, F. (1962) *Physiologia Plantarum*, **15:** 473-479.

Murray, V.G. and Murray, R.D. (1994) *Proc. Fla. Sta. Hort. Soc.*, **107**: 354-358.

Nagar, P.K. (1994) *Post-harvest Biol. Technol.*, **4:** 225-234.

Nakata, S. (1953) *Hawaii Agric. Exp. Sta. Prog. Note*, No. 95, 1-4.

Nakata, S. and Suehisa, R. (1969) *Amer. J. Bot.*, **56:** 1121-1126

Nakata, S. and Watanabe, Y. (1966) *Bot. Gaz.*, **127:** 146-152.

Nataka, S. (1956) *Hawaii Farm Sci.*, **4:** 4-5.

Nath V., Singh, Sanjay Kumar, Pongener, Alemwati, Gupta, A K and Sharma S. (2017) *Training Manual*. ICAR-NRC on Litchi, Muzaffarpur, India, 161pp.

Nath, V., Kumar, G., Pandey, S.D. and Gupta, A.K. (2016) *Acta Hort.*, **1211**: 153-160.

Nath, V., Sharma, S. and Barman, K. (2016) *Indian J. Agric. Sci.*, **86** (3): 283-90.

Nath, V., Thakre, M., Pandey, S. D., Kumar, A., and Kumar, R. (2014) *Acta Hort.*, **1029:** 287-293.

Nelson, R.O. (1954) *Proc. Fla. Lychee Growers' Assoc.*, **1**: 8-11.

Nijjar, G.S. (1972) *Bull. Punjab Agricultural University*, India, pp. 1-24.

Normand, F. and Bouffin, J. (1995) *Fruits*, **50**: 205-214.

Olesen, T., Menzel, C. M., McConchie, C. A., and Wiltshire, N. (2013) *Scientia Hort.*, **156:** 93-98.

Oosthuizen, J.H., Froneman, I.J. and Roe, D.J. (1995) *J. South Afr. Soc. Hort. Sci.*, **5**: 61-63.

Oosthuizen, J.H., Roe, D.J. and Menzel, C.M. (1994) *SA Lietsjekwekersvereniging Jrb.*, **6**: 24-28.

Oosthuizen, W.G. (1994) *Inligtingsbulletin - Instituut vir Tropiese en Subtropiese Gewasse*, **226:** 11-12.

Oosthuizen, W.G. (1995) *Yearb. South Afr. Litchi Growers' Assoc.*, **7:** 41-43.

Pan, X. C. and Baogui, X. (1996) *Acta Hort. Sinica*, **23**: 227-230.

Paxton, B.F. and Chapman, K.R. (1980) *Bienn. Rept. Maroochy Hort. Res. Stn.*, **2**: 34-37.

Paxton, B.F., Saranah, J. and Chapman, K.R. (1978) *Bienn. Rep. Maroochy Hort. Res. Stn.*, **1:** 34.

Peng, Y. H. (1998) *J. Fruit Sci.*, **15**: 217-222.

Pongener, A., Purbey, S. K. and Sharma, S. (2019) *In:* Compendium on Short Course on Advance Plant Protection Tools and Technique for Safer Fruit Production ICAR-NRCL, Muzaffarpur, Bihar, pp. 28-33.

Popenoe, W. (1948) *Manual of Tropical and Subtropical Fruits*, Hofner Press, New York, pp. 312-333.

Prasad, J.S., Kumar, R., Mishra, M., Singh, A.K. and Prasad, U.S. (1996) *HortScience*, **37:** 1187-1189.

Prasad, J.S., Mishra, M., Kumar, R., Singh, A.K. and Prasad, U.S. (1994) *Seed Res.*, **22:** 89-97.

Prasad, U.S. and Jha, O.P. (1978a) *Pl. Biochem. J.*, **5:** 101-107.

Prasad, U.S. and Jha, O.P. (1978b) *Pl. Biochem. J.*, **5:** 44-49.

Prasad, U.S. and Jha, O.P. (1979a) *Indian J. Pl. Physiol.*, **22:** 84-86.

Prasad, U.S. and Jha, O.P. (1979b) *Indian J. Pl. Physiol.*, **22:** 163-165.

Puchooa, D. (2004) *Afr J Biotechnol.*, **3**(11): 576–584

Pujari, M.M. and Syamal, N.B. (1977) *Proc. Bihar Acad. Agric. Sci.*, **25:** 27-30.

Rajwanshi, R., Kumar, M. and Lal, B. (2017) In: Lychee Disease Management, Springer Nature, Singapore.

Ram, M. and Majumder, P.K. (1981) *Sci. and Cult.*, **47**: 332-333.

Rastogi, R.P. and Mehrotra, B.N. (1999) Compendium of Indian medicinal plants, **2:** 1970–1979.

Ray, P.K. (1999) In: *Tropical Horticulture,* Vol. I. (Eds. T.K. Bose, S.K. Mitra, A.A. Farooqi and M.K. Sadhu), Naya Prakash, Calcutta, India, pp. 102-177.

Ray, P.K. (2002) Breeding Tropical and subtropical Fruits. *Narosa Publishing House,* New Delhi, India. pp.338.

Ray, P.K. (2004) Litchi, Production Technology and Post-harvest handling for Export, RAU, Pusa, Samastipur, Bihar, p. 43.

Ray, P.K. and Sharma, S.B. (1985) *J. Agric. Sci., Camb.,* **104:** 247-248.

Ray, P.K., Sharma, S.B. and Mishra, K.A. (1985) *Indian Hort.,* **30:** 9-13.

Robbertse, H., Fivaz, J. and Menzel, C.M. (1995a) *J. Amer. Soc. Hort. Sci.,* **120**: 914-920.

Robbertse, P.J., Fivaz, J. and Menzel, C.M. (1995b) *Yearb. South Afr. Litchi Growers' Assoc.,* **7:** 21-25.

Roe, D. J., Menzel, C.M., Oosthuizen, J.H. and Doogan, V.J. (1997) *J. Hort. Sci.,* **72:** 397-405.

Roe, D.J., Menzel, C.M. and Oosthuizen, J.H. (1995) *Yearb. South Afr. Litchi Growers' Assoc.,* **7:** 11-16.

Sadhu, M.K., Naskar, B.B. and Basu, R.N. (1972) *Indian Agric.,* **16:** 251-257.

Sahni, R. K., Kumari, Sweeti, Kumar, Manish, Kumar, Manoj and Kumar, Amrendra (2020) *Int. J. Curr. Microbiol. App. Sci.* **9**(04): 1827-1840

Sanyal, D. and Mitra, S.K. (1993) *Crop Res.,* **6:** 294-299.

Sanyal, D., Biswas, B. and Mitra, S.K. (1996) *Haryana J. Hort. Sci.,* **25:** 29-34.

Sanyal, D., Hasan, A., Ghosh, B. and Mitra, S.K. (1990) *Indian Agric.,* **34:** 19-22.

Sarkar, T.K., Bandopadhyay, A. and Gayen, P. (1991) *Haryana J. Hort. Sci.,* **20**: 56-59.

Scholefield, P.B. (1982) *Sci. Hort.,* **16**: 263-272.

Schutte, G.C., Botha, T., Kotze, J.M. (1990) *Yearb South Afr. Litchi Growers' Assoc.,* **3:** 10-14.

Scott, K.J., Brown, B.I., Chaplin, G.R., Wilcox, M.E. and Bain, J.M. (1982) *Scientia Hort.,* **16:** 253-262.

Sen, P.K., Basu, R.N., Bose, T.K. and Roychaudhury, N. (1969) *Plant Sci.,* **1:** 216-219.

Sharma, D.D. (1985) *Indian Fmg.,* **35**: 25-26.

Sharma, D.D. and Rahman, M.F. (1982) *Entomon.,* **7**: 55-56.

Sharma, D.D., Singh, S.P. and Akhori, K.K. (1986) *Indian J. Agric. Sci.,* **56**: 59-63.

Sharma, K.K., Sharma, K.N. and Cheema, H.S. (1995) *Haryana J. Hort. Sci.,* **24**: 1-7.

Sharma, S.B. and Ray, P.K. (1987a) *Haryana J. Hort. Sci.,* **16:** 168-174.

Sharma, S.B. and Ray, P.K. (1987b) *Indian Fd. Packer,* **41:** 17-20.

Shigeura, G. (1948) *Rep. Hawaii Agri. Exp. Sta.,* pp. 138-142.

Shrestha, G.K. (1981) *Horticultural Sci.*, **16:** 498.

Shukla, K.S., Misra, R.L., Kaul, M.K. and Prasad, A. (1978) *Haryana J. Hort. Sci.*, **7**: 162-164.

Shukla, R.K. and Bajpai, P.N. (1974) *Indian J. Hort.*, **31**: 226-228.

Singh, A., Abidi, A.B., Srivastava, S. and Singh, I.S. (1987) *Narendra Dev J. Agric. Res.*, **2**: 160-163.

Singh, A., Pandey, S.D. andNath, V. (2012) *The world litchi cultivars,* NRCL, Muzaffarpur. India. pp. 1-65.

Singh, G. (1993) In: *Advances in Horticulture,* (Eds. K.L. Chadha and O.P. Pareek), Malhotra Pub. House, New Delhi, pp. 1501-1526.

Singh, G., Nath, V., Pandey, S.D., Ray, P.K. and Singh H.S. (2012) The Litchi, FAO, New Delhi, India

Singh, H. (1974) *Indian J. Ent.*, **36**: 168.

Singh, H. (1976) *Indian J. Ent.*, **38**: 238.

Singh, K.K. and Jawanda, J.S. (1962) *Punjab Hort. J.*, **2**: 69-74.

Singh, L.B. and Singh, U.P. (1954) *The Litchi,* Government of Uttar Pradesh, India, pp. 14-22.

Singh, L.B. and Singh, U.P. (1964) *The Litchi,* Supdt. Printing and Stationary, Uttar Pradesh, Allahabad.

Singh, M.P. (1952) *Indian J. Hort.*, **9**: 53-58.

Singh, O.P. and Phogat, K.P.S. (1984) *Punjab Hort. J.*, **24**: 83-88.

Singh, P. and Singh, I.S. (1995) *Mysore J. Agric. Sci.*, **29**: 252-255.

Singh, Sanjay Kumar, Kumar Amrendra, Pandey, S.D. and Nath, Vishal (2017a) *Int. J. Advanced Biol. Res.*, **7**(1): 185-189

Singh, Sanjay Kumar, Kumar, A., Pandey, S.D. and Nath, V. (2017c) In: *Gyan Manthan.* **6**: 100-104. *Westville Publishing House* 47, B-5, Paschim Vihar, New Delhi – 110063

Singh, Sanjay Kumar, Nath, V. and Kumar, A. (2017d) In: Training Manual ICAR-NRC on Litchi, Muzaffarpur, India, pp. 48-51.

Singh, Sanjay Kumar, Pandey S D and Nath Vishal (2019) In: Breeding of Horticultural Crops *Today & Tomorrow's Printers and Publishers,* New Delhi - 110 002, India, pp. 431-458.

Singh, Sanjay Kumar; Nath, Vishal; Marboh, E. S., and Sharma, Swati (2017b) *Int. J. Curr. Microbiol. Applied Sci.*, **6**(3): 500-509

Singh, Y.P. and Kumar, V. (1994) *Recent Hort.*, **1**: 36-39.

Sivakumar D., and Korsten, L. (2007) *J. Plant Nutrition,* **30**(10): 1727-1735.

Smith, P.M. (1976) *In: Evolution of Crop Plants* (Ed. N.W. Simmonds), Longman, London, pp. 301-324.

Snowdon, A.L. (1990) *A colour Atlas of Post-harvest Diseases and Disorders of Fruit and Vegetables*. Vol.I. *General Introduction and Fruits,* Wolfe Scientific Ltd., London, p. 302.

Sood, A.K., Sharma, R.D. and Singh, B.M. (1987) *Indian Hort.,* **32**: 19-20.

Srivastava, K., Patel, R. K. and Kumar, Amrendra (2016) *In:* Training Manual, MTC, ICAR-NRCL, Muzaffarpur, Bihar

Stern, R.A. and Gazit, S. (1993) *Acta Hort.,* **349:** 209.

Stern, R.A. and Gazit, S. (1996) *J. Amer. Soc. Hort. Sci.,* **121:** 152-157.

Stern, R.A. and Gazit, S. (1999) *J. Hort. Sci. Biotech.,* **74:** 203-205.

Stern, R.A., Eisenstein, D., Voet, H. and Gazit, S. (1996) *J. Hort. Sci.,* **71:** 661-671.

Stern, R.A., Eisenstein, D., Voet, H. and Gazit, S. (1997) *J. Hort. Sci.,* **72:** 19-25.

Stern, R.A., Gazit, S., El-Batsari, R. and Degani, C. (1993) *J. Amer. Soc. Hort. Sci.,* **118:** 109-114.

Storey, W.B. (1973) *Yearb. Calif. Avocado Soc.,* **56:** 75-86.

Subhadrabandhu, S. (1992) *Acta Hort.,* **292**: 13-23.

Swarts, D.H. and Anderson, T. (1980) *Inf. Bull. Citrus and Subtrop. Fruit Res. Inst.,* **98:** 13-15.

Takuya S, Hiroshi N, Hajime F, Norihiko N, Takako K, Zsolt R, Tetsuya I, Shukoh H, Hideki O (2008) *Biosci Biotechnol. Biochem.* **72**(2): 463-476

Teaotia, S.S. and Singh, R.D. (1970) *Progressive Hort.,* **2:** 21-32.

Thompson, B.D. (1955) *Proc. Fla. Lychee Growers Assoc.,* **2**: 27.

Thunyapar T. (1998) *J. Japanese Soc. Hort. Sci.* **67:** 1161–1163.

Tian SP, Li BQ, Xu Y (2005) *Food Chem.,* **91:** 659-663

Tindall, H.D. (1994) *Hort. Reviews,* **16**: 143-96.

Tongdee, S.C. Scott, K.J. and McGlasson, W.B. (1982) CSIRO *Food Res. Quarl.,* **42:** 25-28.

Tott, A.P.Du and Swart, D.J. (1995) *Yearb. South Afr. Litchi Growers' Assoc.,* **7:** 31-32.

Underhill, S.J.R. and Critchley, C. (1996) *Proc. Australian Post-harvest Hort. Conf.,* Melbourne, Australia, 18-22 September, 1995, pp. 13-14.

Underhill, S.J.R. and Wong, L. S. (1990) *Acta Hort.,* **316:** 245-251.

Underhill, S.J.R., Bagshaw, J., Zauberman, G., Ronen, R., Prasad, A. and Fuchs, Y. (1992a) *Acta Hort.,* **321**: 732-741.

Underhill, S.J.R., Critchley, C. and Simons, D.H. (1992b) *Acta Hort.,* **321**: 718-725.

Veera, S. and Das, R.C. (1974) *Hort. Adv.,* **9:** 11-13.

Verma, S.K., Jain, B.P. and Das, S.R. (1980) *Haryana J. Hort. Sci.,* **10:** 4-10.

Waite, G.K. (1986) *Proc. Ist. Nat. Lychee Sem.,* Australia, pp. 42-48.

Wang, L.J., Lou, G.D., Ma, Z.J., Liu, X.M. (2011) *Food Chem.,* **126**(3): 1081-1087.

Wenkam, N.S. and Miller, C.D. (1965) *Hawaii Agr. Exp. Sta. Bull.*, 135.

Wills, R.B.H., Lim, J.S.K. and Greenfield, H. (1986) *Food Tech. Aust.*, **38**: 118.

Wong, D. (1999) *China Fruits*, pp. 55.

Wong, S., Jacobi, K.K. and Giles, J.E. (1991) *Scientia Hort.*, **46**: 245-251.

Wong, S., Jacobi, K.K. and Peacock, B.C. (1990) *Singapore J. Primary Indust.*, **18**: 15-17.

Wu, D.Y., Lin, X.D., Ye, Q.H. and Wang, W.H. (2000) *J. South China Agric. Univ.*, **21**: 19-21.

Xian, X.U., Zhang, Z.W., Wang, B.Q., Yuan, P.Y. and Qiu, Y.P. (1994) *Guangdong Agric. Sci.*, **4**: 25-27.

Yeap, C.K. (1987) *Yearb. West Australia Nut. Tree Crops Assoc.*, **12**: 16.

Yee, W. (1972) *Circ. Univ. Hawaii Agric. Coop. Ext. Serv.*, No. 336, pp. 1-24.

Yen, Chung-ruey, Laio, Yu-Wan and Tien, Yung-jou (1984) *J. Chinese Soc. Hort. Sci.*, **30**: 214-222.

Young, T.W. (1970) *Proc. Annu. Meet. Fla. Sta. Hort. Soc.*, **83:** 362-367.

Young, T.W. and Koo, R.C.J. (1964) *Proc. Fla. Sta. Hort. Soc.*, **77**: 406-410.

Yu, C., Chen, Z. and Lu, L. (1996) *J. Fujian Agric. Univ.*, **25**: 386-389.

Yu, Y.B. (1991) *J. Fujian Agric. Sci. & Tech.*, **5**: 17-18.

Yuan, R.C. and Huang, H.B. (1988) *Scientia Hort.*, **36**: 281-292.

Zauberman, G., Ronen, R., Akerman, M. and Fuchs, Y. (1989) *Acta Horticulturae*, **268**: 309-314.

Zhang H N, Huang X M and Chen H B. (2014) *Acta Hortic.* **1029**: 245–51.

Zhang, D. L., Quantick, P. C. and Grigor, J. M. (2000) *Post-harvest Biol. Technol.*, **19**: 165-172.

Zhang, G. C., Qing, Y. W., He, Q. and Tang, C. (1999) *South China Fruits*, **28**: 35.

Zheng Q, Hu G, Chen D (2001) *Acta Hortic.*, 558: 143–148

Zhimo V. Yeka., Saha Jayanta, Singh Balveer and Chakraborty Ivi (2018) *Curr. Sci.*, **114**(5): 1100-1105

Zhodng, Y.W. and Wu, S.X. (1983) *Guangdong Agric. Sci.*, No. 4, pp. 1-5.

Zhou, B. Y., Zuo, L., Ye, Y. C., Zhao, X. D. and Ye, Y. X. (1998) *Acta Hort. Sinica*, **25:** 236-240.

Zhou, X., Wu, D. Huang, H., Lai, J. and Huang, D. (1996) *Acta Hort. Sinica*, **23**: 13-18.

Zhu, J.H., Peng, H.X., Lu, G.F., Xu, N., Pan, J.C. and Liu, B.H. (2010) *Acta Hort.*, **863**: 487-490.

Zhuang, W., Yan, J., Lin, X., Chen, R. and Chen, J. (1983) *J. Fujian Agric. College*, **12**: 297.

④

PERSIMMON

A.P. George, R.J. Nissen and V.A.Parthasarathy

1.0 Introduction

The persimmon (*Diospyros kaki*) is a native of China. *Diospyros* L. belongs to the Ebenaceae, a plant genus that includes over 500 species that are distributed worldwide (Zhu *et al.*, 2019; Duangjai *et al.*, 2006) and is one of the largest angiosperm genera (Frodin, 2004). Some of its species (*e.g.*, *Diospyros kaki* Thunb) produce edible fruit, and others are sources of ebony wood, stock, medicinal materials or ornamental plants (Wang *et al.*, 1997). The most well-known species is *D. kaki*, which originated from China (Kanzaki, 2016). One fossil leaf of wild persimmon (*Diospyros miokaki*) that is 25 million years old was found in the Shandong Province of China and stored in the Shanwang Paleontological Museum (Wu, 2005). The persimmon fruit is unique because of the hyper accumulation of proanthocyanidins during fruit development, causing the mature fruit of most cultivars to have an astringent taste (Zhu *et al.*, 2019).

Most development of the crop has occurred in Japan where it is, regarded as the national fruit; more than 250,000 tonnes are produced annually. Smaller, but expanding persimmon industries are being developed in Italy, Brazil, California and Israel (George and Nissen, 1984). Although the persimmon is widely distributed throughout Asia, the fruit has not gained popular appeal in most of the Western countries. The main reason for this is that in the past only, poor quality astringent cultivars had been grown. The fruit of these astringent cultivars is inedible until completely soft and ripe. Astringency or bitterness is caused by the soluble tannin content of the fruit, which decreases as the fruit ripens. Even a moderately soft fruit may taste astringent. In more recent years, non-astringent cultivars have gained greater popularity. Non-astringent cultivars are completely free of astringency at maturity and the fruit may be eaten while still firm or at the desired stage of softness. Unlike astringent cultivars, most non-astringent cultivars also have excellent shelf-life. In comparison with other subtropical tree crops, the persimmon has few production problems. The future of this fruit depends on the selection of high quality cultivars and specific marketing techniques to ensure a firm non-astringent persimmon reaching the consumer.

2.0 Composition and Uses

2.1.0 Composition

According to Ito (1962) persimmons contain approximately 79 per cent water, 0.4 per cent protein, 0.4 per cent crude fibre and 0.7 per cent pectin. Persimmons are high in vitamins A and C, containing 2710 International units of vitamin A per 100 g of fruit and 11 mg of ascorbic acid per 100 g of fruit (Griffith *et al.*, 1982). The average composition of persimmon fruits is presented in Table 1.

Sugar content of fruits varies significantly with cultivar and maturity stage. Fruits contain mainly fructose, glucose and small amount of sucrose (Senter *et al.*, 1991) and unidentified oligosaccharide (Daood *et al.*, 1992). Malic acid is predominant in all cultivars, followed by citric acid (Senter *et al.*, 1991; Daood *et al.*,

1992). Malic acid content increases with maturity and citric acid content decreases (Senter *et al.*, 1991). Changes in nonvolatile acid contents during fruit development were investigated by Ayaz and Kodhoglu (1998). Malic and citric acids were predominant in all development stages. During harvesting fumaric acid was the most abundant acid in the fruit. Presence of other organic acids such as isocitric, ascorbic, fumaric and gallic acids have been reported. Metabolic processes at full ripe stage led to considerable loss in soluble sugars and organic acids (Daood *et al.*, 1992). The ascorbic acid concentration of cv. Jiro was 121.49 mg/100 g FW, which was 2.45 times that of the astringent cultivar Fangshi. But two pollination-variant, non-astringent cultivars, Zenjinaru and Nishimura Wase, had much lower ascorbic acid concentrations than the others (Fei *et al.*, 1994).

Table 1: Major Constituents of Persimmon Fruits (on fresh weight basis)*

Constituents	*Non-astringent*	*Astringent*
Moisture (per cent)	82.4	79.1
Protein (per cent)	0.58	0.43
Crude fibre (per cent)	0.49	0.37
oBrix	16.2	19.5
Sucrose (per cent)	1.05	11.95a
Glucose (per cent)	6.79	4.65a
Fructose (per cent)	7.29	4.40a
Soluble pectin (per cent)	0.68	0.68
Soluble tannin (per cent)	0.09-0.49b	1.41

a: Content in vacuolar sap of parenchyma cells and expressed on per cent mass/volume (m/v).

b: Basis in mg/100 g FW

* Ito (1962, 1980), Fei *et al.* (1994), Yonemori *et al.* (1997)

According to Fei *et al.* (1994), the non-astringent cultivars had higher contents of vitamin B_1, niacin and β-carotene than those of astringent ones. The carotenoids were identified as *cis*-mutatoxanthin, antheroxanthin, zeaxanthin, neolutein, cryptoxanthin, α-carotene and β-carotene and fatty acid esters of cryptoxanthin and zeaxanthin (Daood *et al.*, 1992). Carotenoid contents and carotenoid profiles varied among cultivars. β-carotene was the predominate provitamin A isomer in 11 of the 15 cultivars with β-cryptoxanthin predominate in the rest (Homnava *et al.*, 1990). The fruits were an excellent source of retinol (Homnava *et al.*, 1990; Daood *et al.*, 1992). The flesh of stored persimmon cv. Fuyu fruits contained saturated and unsaturated fatty acids at 39.7 and 60.4 per cent level respectively (Lee and Kim, 1994). The contents of traceelements like Fe, Zn, Se, were noticeably higher in non-astringent cultivars (Fei *et al.*, 1994).

The characteristic astringency of immature persimmons is due to water soluble tannins (kaki-tannin) present in the flesh of the fruit (Kitagawa and Glucina, 1984). Non-astringent cultivars have a fewer number of smaller sized tannin cells compared with astringent cultivars (Miyabayashi, 1941; Ito, 1980; Yonemori and

Uses of Persimmon

Fresh Fruit with Slices

Juice (Left) and Dried Fruits (Right) of Persimmon

Face Mask (Left), Soap (Middle) and Shampoo (Right) from Persimmon

Wine (Left), Oil and Cream (Right) from Persimmon

Matsushima, 1985). During fruit maturation astringency levels fall in astringent cultivars and normally disappear completely from non-astringent cultivars provided autumn temperatures are high enough (Kajiura, 1980; Ito, 1980). The soluble tannin concentration of non-astringent cultivars was much lower than that of astringent cultivars (Fei *et al.*, 1994). Analyses of the vacuolar sap revealed that the tannin cells of astringent cultivars contained 10 to 12 per cent (M/V) of tannin as (+)-catechin equivalents whereas the parenchyma cells contained trace amounts of tannins (Yonemori *et al.*, 1997).

Changes in phenolic acid contents during persimmon fruit development were studied in fruits harvested from June to November, 1995 (Ayaz *et al.*, 1997). Salicylic, 4 hydroxybenzoic, vanillic, 3,4-dihydroxybenzoic, syringic, *p*-coumaric, and gallic acids were identified and quantitfied by gas chromatography. The phenolic acid contents, except gallic acid, increased in September, peaked at the end of September, and then decreased rapidly.

2.2.0 Uses

Dried kaki is a traditional feature of the New Year festival season, and a good portion of the Japanese production of kaki fruit is dried in the autumn. Mature but firm fruits are harvested, peeled and sun-dried after threading onto strings. Small fruits may take 18-25 days to dry but large fruits may require 35-40 days and since the late autumn for drying is often short, this may not always be successful. There is considerable risk of contamination of impurities, moulds and discoloration (Kim *et al.*, 1993). In larger commercial operations, fruits are first fumigated with sulphur dioxide and then artificially dried at 32-33°C. Fruits dried by combinations of sun and hot air dried faster than either of the methods alone. However, artificial ventilation resulted in the best quality fruits, followed by sun dried fruits (Lee *et al.*, 1994). The far-infrared method could shorten drying time by approximately 25 days compared with sun drying and the dried fruits were better in quality with respect to colour and appearance than sun drying method (Kim *et al.*, 1993).

The best dried kaki fruits are judged by high sugar content and a thick coating of sugar crystals as an indication of good quality. The texture should not be too mealy and the flesh should be free from stringy fibre. Seedless cultivars are preferred for drying and there should be no remnants of astringency after drying. Hiratanenashi is the best kaki cultivar for drying, but Yotsumizo, Koshu-Hyakume and Yamato-Hyakume are very popular. Ichidagaki and Dojo-Hachiya are also used as dried kaki.

Fresh kaki fruits are often included in particular dishes, especially in various jellies and jams, and the peeled skins are added to pickles to provide a bright orange colour. In Japan, a small amount of kaki is canned in sugar syrup for out-of-season consumption. Green fruits, especially from strongly astringent cultivars, such as Tsurunoko and Tsunagaki, are used in production of tannin of particularly high quality. Several hundred recipes for using persimmon fruit were presented by Griffith *et al.* (1982).

Piretti (1990) reviewed the uses of kaki-tannin as mordant for protecting wood; as treatment of hangover; as antidote against some cobra venoms and bacterial toxins; in pertussis vaccine; as control agent for viral infections in plants, notably against tobacco mosaic virus on tomato. In Japan, astringent persimmons have become an important source of condensed tannins for use as a deproteinizing agent and coating material for wood, paper and fibres (Matsuo *et al.*, 1995).

The various uses of *Diospyros* fruits in Thailand are limited and given in Table 2.

Table 2: Uses of Fruits of *Diospyros* Species in Thailand*

Uses	*Species*
Dyeing	*D. malabarica* var. *siamensis, D. cauliflora, D. rhodocalyx, D. areolata, D. gracilis, D. castanea*
Fish poisoning	*D. wallichii, D. ehretioides, D. toposia, D. montana*
Medicine	*D. mollis, D. montana*
Edible	*D. glandulosa, D. malabarica* var. *siamensis, D. decandra, D. rhodocalyx*

* Utsunomiya *et al.* (1998).

In a brief review, Tchouga *et al.* (2020) indicated the uses of *D. crassiflora* Hiern. According to them, the hard, dense black heartwood of the tree is exploited in Cameroon and other countries for the construction of musical instruments such as the violin, guitar, cello, contrabass, and piano keys (Lemmens *et al.*, 2012; Taylor, 2012). The species is also used locally for medicinal purposes and pharmacological research, and has proven antifungal and antibacterial properties in stem bark extracts (Dzoyem *et al.*, 2007; Tangmouo *et al.*, 2006).

3.0 Origin and Distribution

The persimmon is native to Oriental Asia, most probably China, where it has been cultivated for centuries (Kajiura, 1979). However, most selection and development of the fruit have occurred in Japan, where it is called kaki. Before 1900, kaki was the most important fruit tree in Japan, however, today it is only the fifth most important tree fruit in that country. At the end of the nineteenth century, the persimmon was introduced in many countries in the temperate zone. Commercial plantings exist in the USA, Brazil, Italy, Israel, New Zealand and Australia. Rapid expansion in persimmon plantings occurring in these countries is mainly based on non-astringent cultivars. Because of the low chilling requirement of the non-astringent cultivars, and because higher quality fruit can be produced in warm subtropical regions, further expansion is likely to occur in these regions (George and Nissen, 1985).

Since the emergence of the Australian non-astringent persimmon industry in 1980, it has increased to 1,70,000 trees. Production extends from the north of Queensland down the eastern and southern seaboards to Western Australia (Collins *et al.*, 1995). The first specimens of persimmon cv. Costata were introduced in the southern Ukraine at the end of the 19th century (Kazas and Lobov, 1996). Persimmon was first planted commercially in the state of Santa Catarina, Brazil in 1985 and presently it is cultivated in 6 regions, covering an area of 56.2 ha (Matos, 1993). *Diospyros* species are distributed widely in Thailand (Utsunomiya *et al.*, 1998).

Persimmon (*Diospyros kaki* Thunb.) can produce unisexual (male and female) and bisexual (hermaphrodite) flowers, showing relatively complex sexuality at the species level. According to Guan *et al.* (2020) that in China, there are more than 1000 persimmon germplasms, but most are gynoecious cultivars that bear female flowers only. To date, few germplasms have been reported to generate male flowers, until seven androecious persimmons were recently found in the Dabieshan Mountains at the juncture of Hubei, Anhui and Henan Provinces in central China. They reported a new androecious germplasm located in Jiangxi Province that was cultivated by the first generation of the Luo family in the Song Dynasty nearly for a thousand years ago. Up to now, the millennial androecious germplasm has continuously borne only male flowers and the seeding plant still has strong growth vigor. Investigation of the leaf and pollen morphology showed that this androecious germplasm is similar to 'Male 2' (androecious, discovered in Dabieshan Mountains) but different from two Japanese cultivars ('Taishuu' and 'Zenjimaru'). Polymerase chain reaction (PCR) detection of the sex-linked marker DlSx-AF4S verified that the millennial androecious germplasm carries this sex-linked marker. Moreover, the RO_2/marker, linked Chinese pollination constant non-astringent (C-PCNA) locus, was also present in the millennial androecious germplasm. Additionally, the genetic relationships between SSR markers showed that the millennial androecious germplasm is related to Dabieshan's androecious genotypes and some persimmon cultivars of Chinese origin. According to them their results revealed that the ancient androecious persimmon found in Jiangxi Province is distinguished from previously reported androecious germplasms and has the potential to serve as a breeding resource for selecting sex-linked PCNA persimmons in the future (Guan *et al.*, 2020).

Ercisli and Akbulut (2009), quoting Onur and Onur (1985), opined that Turkey is one of the origin centers of *Diospyros lotus*. The other species (*Diospyros kaki* and *Diospyros oleifera*) have been introduced from other countries at least 200 years ago. *Diospyros oleifera* is more common in the subtropical region (Mediterranean), *Diospyros kaki* in the subtropical (Mediterranean) and warmer temperate region (Black Sea region), however *Diospyros lotus* is more common in the warm temperate region (Northeastern Anatolia).

4.0 Species and Cultivars

4.1.0 Species

The persimmon belongs to the genus *Diospyros* (family Ebenanceae) which includes some 400 species, most of which are tropical or subtropical in requirements (Utsunomiya *et al.*, 1998). The four most important fruit bearing species of the genus are given in Table 3. The American persimmon and the date plum are not grown commercially to the same extent as the Oriental or Japanese persimmon and are occasionally used as rootstock for the Oriental persimmon.

Ng (1978) inferred that *D. kaki* arose directly from *D. roxburghii* (=*D. glandulosa*). Some tropical *Diospyros* species may be closely related to the temperate species (Utsunomiya *et al.*, 1998). In Thailand, 60 *Diospyros* species have been described by Phengklai (1978). Utsunomiya *et al.* (1998) made an extensive survey from 1992 to

1994 on the distribution of *Diospyros* species in Thailand and classified the species according to their habitat (Table 4).

Table 3: Species of *Diospyros* Used Commercially to Produce Fruit*

Species	*Common Names*	*Main Region of Cultivation*	*Uses*
D. kaki	Kaki, Japanese or Oriental persimmon, Yamagaki	Japan, China, Korea	Fresh and processed
D. lotus	Date plum, Ghae tsao, Mamegaki	Asia	Tannin source, rootstock
D. virginiana	American persimmon	North America	Fresh and processed, rootstock
D. oleifera	–	China	Tannin source, rootstock

* Ito (1980).

Table 4: Classification of *Diospyros* Species in Thailand According to their Habitats*

Area	*Species*
Wet	*D. confertiflora, D. toposia, D. sumatrana, D. cauliflora, D. wallichii, D. transitoria, D. areolata*
Wet and dry	*D. castanea, D. montana, D. mollis, D. rubura, D. ehretioides, D. rhodocalyx, D. gracilis, D. variegata*
Cool and wet	*D. glandulosa*
Whole country except for cool area	*D. decandra, D. malabarica* var. *siamenthis*

* Utsunomiya *et al.* (1998)

The characteristic feature of temperate *Diospyros* species is the existence of tannins in the fruit. The tannin is usually contained in edioblasts called tannin cells in *D. kaki* and *D. lotus*. The structure and distribution of tannin cell differ with the cultivar in *D. kaki* (Yonemori and Matsushima, 1987).

Table 5: Classification of *Diospyros* Species According to the Per cent Area Occupied by the Tannin Cells in the Mesocarp*

Per cent Area Occupied by Tannin Cells in Mesocarp	*Species*
Group I (No tannin cells to sparse distribution of tannin cells)	*D. rhombifolia, D. mollis, D. wallichii, D. montana* and *D. ehretioides*
Group II (10 to 30 per cent area occupied by tannin cells)	*D. gracilis, D. oleifera, D. areolata, D. kaki*
Group III (tannin cells in 30 to 50 per cent of the area)	*D. glandulosa, D. cauliflora and D. virginiana*
Group IV (> 50 per cent area occupied by tannin cells)	*D. decandra, D. lotus, D. rhodocalyx* and *D. malabarica* var. *siamensis*

* Utsunomiya *et al.* (1998).

Utsunomiya *et al.* (1998) classified the tropical species available in Thailand into four groups according to the per cent area occupied by tannin cells in the section (Table 5).

The ability of seed formation differs with the species. Many seeds were formed in the fruit of *D. rhodocalyx*, *D. glandulosa* and *D. mollis*, while seedless fruit was often found in *D. decandra*, *D. malabarica* var. *siamensis* and *D. areolata* (Utsunomiya *et al.*, 1998).

Within *Diospyros*, there are three main species cultivated for fruit production: *D. lotus, D. oleifera,* and *D. kaki,* and all three of which produce astringent fruit. According to Guo and Luo (2011), *D. kaki* is hexaploid (2n = 6x = 90), while *D. lotus* and *D. oleifera* are diploid (2n = 2x = 30). *D. kaki* accounts for the greatest number of commercial fruit tree cultivars, and *D. lotus* is widely used as a rootstock. Compared to *D. kaki* and *D. lotus, D. oleifera* exhibits two additional characteristics: the generation of trichomes and the excretion of oils on the fruit surface. Here, the material 'Youshi' (the traditional common name in China for *D. oleifera* Cheng) was used; Youshi means 'oil persimmon' in Chinese (Zhu *et al.*, 2019).

The American persimmon (*D. virginiana*) is a slow-growing, moderately sized tree fruit native to the forests of Kentucky (Pomper *et al.*, 2020). This tree fruit is in the early stages of commercial production with many cultivars selected from the wild. Small orchards of commercially available cultivars are planted in Kentucky. Persimmons are normally dioecious, and female trees require cross-pollination to produce fruit. There are two races of persimmon: the tetraploid (60-chromosome) race is centered in the southern Appalachian region, while the hexaploid (90-chromosome) race generally occupies a range north and west of the tetraploid range. These ranges overlap in Kentucky. Because the ranges overlap, cross-pollination may cause sexual incompatibility, resulting in pollination without fertilization, and therefore seedless fruits of poor quality (Pomper *et al.*, 2020).

Pomper *et al.* (2020) assessed the ploidy level of commercially available American persimmon cultivars and native Kentucky persimmon populations. Leaf samples were collected from 45 cultivars and advanced selections, as well as 45 trees from native populations in Bullitt, Barren, and Franklin Counties. Flow cytometer analysis showed that only four of the selected cultivars were from the tetraploid race: Ennis Seedless, Weeping, Sugar Bear, and SFES; the remaining cultivars were from the hexaploid race. Both hexaploid and tetraploid American persimmon trees were identified in the populations sampled in the Bullitt County locations, but only tetraploid races were found in Franklin and Barren Counties. Because pollen from native trees could result in seedless fruit formation of poor quality when native seedlings are used as pollinizers in commercial production of American persimmon, ploidy level of seedlings needs to be considered (Pomper *et al.*, 2020).

Diospyros crassiflora H. (Ebenaceae), commercially known as the African ebony, is a small forest tree found at elevations less than 1000 m in South Nigeria, Cameroon, Central African Republic, Equatorial Guinea, Gabon, Republic of the Congo, and Democratic Republic of the Congo (Tchouga *et al.*, 2020; Lemmens *et al.*, 2012). It is a shade-tolerant species characterized by a relatively low growth rate

(Bongjoh and Nsangou, 2001). Diospyros crassiflora was listed as vulnerable on the International Union for Conservation of Nature Red List because it is threatened by the destruction of forests for agriculture and grazing, and the commercial logging of other species (Schatz *et al.*, 2019). For now, the species has not been included in extended sylvicultural or agroforestry systems (Tchouga *et al.*, 2020).

4.2.0 Cultivars

The fruits of the Japanese or Oriental persimmon (*Diospyros kaki*), or simply kaki, show wide variation in size, shape and colour, and are broadly classified for horticultural purpose into two major groups: non-astringent and astringent. Each group can further be subdivided, based on their response to pollination (Hume, 1914): pollination constant, no change in flesh colour occurs after pollination; pollination variant, the flesh is light coloured when seedless and dark reddish brown when seeded. When pollination is poor and only one seed is formed, a dark area develops around the seed but the remaining flesh is light coloured. In astringent types (both pollination variant and pollination constant), water soluble tannins, which cause astringency in the flesh, decrease as the fruit softens and becomes edible. Astringency can also be removed by various chemical treatments but these cause variable fruit ripening, some methods producing softening and colour changes more readily than others.

In pollination variant, non-astringent types, the soluble tannins disappear after pollination if enough seeds (usually four or five) form. However, if only one or two seeds are formed, some parts of the fruits will remain astringent. In pollination constant, non-astringent types, the fruit is edible when the flesh is firm but mature, regardless of whether or not pollination has occurred.

4.2.1 Non-astringent Cultivars

More than 1,000 named cultivars of persimmon are known, most of these originating in Japan. Breeding and selection from within seedling populations have also occurred in Australia, New Zealand and Brazil. Over 40 non-astringent cultivars or selections are now being evaluated in various countries. Characteristics of the more important non-astringent cultivars are presented in Table 6. Only one non-astringent cultivar, Fuyu exists, which exhibits differences in fruit shape and size and tree growth charac-teristics. Of the more recently bred cultivars Izu, Suruga and Fuyu Hana (Pomelo 15-7) are now gaining increasing popularity.

In Japan, the most popular non-astringent cultivars grown are Fuyu and Jiro. Both these cultivars occupy about 50 per cent of the plantings. Kajiura (1979) described the more important cultivars grown in Japan.

'Migamjosaeng' (*Diospyros kaki* Thunb.)

It is a new pollination-variant non-astringent type (PVNA) persimmon cultivar developed by Sweet Persimmon Research Institute of Gyeongnam Agricultural Research and Extension Services. This cultivar was selected among the seedlings obtained from the cross between 'Mazumotowase-Fuyu' and 'Nishimurawase' in 1995. The seedlings were cultivated from 1996 to 2001 and 'Migamjosaeng' was selected in 2002. The tree growth habit and fruit characteristics of 'Migamjosaeng'

Table 6: Characteristics of Non-Astringent Cultivars with Potential for Subtropical Regions of the World

Cultivar	Maturity Period	Fruit Characteristics	Tree Characteristics	Comments
Izu	Very early	Orange-red fruit of medium size (200 g). Flat-shaped fruit similar to Fuyu, attractive appearance.	Medium vigour with upright growth habit. Light cropper when young. Some male flowers.	Best of the early non-astringent cultivars. Prone to severe fruit fly attack.
Maekawa Jiro	Mid	Dark orange fruit with few seeds. Excellent quality. Larger than Jiro (250 g). Can set good crops without cross-pollination.	Compact growth habit.	Biennial bearing may be a problem. Flower thinning may be useful to prevent this.
Ichikikei Jiro	Mid-late	Similar to Maekawa Jiro (250 g). Orange-red skin. Flat, square-lobed fruit	Similar habit to Maekawa Jiro, moderate vigour	Good quality fruit but needs adequate pollination for uniform fruit shape. Lobed appearance may be a disadvantage
Jiro	Mid	Large (250 g), orange-red fruit. Round, flattened shape with eight distinct channels. Large calyx. Excellent quality	Vigorous, upright tree. High parthenocarpic ability	Losing favour in Japan due to lobed appearance and stylar-end cracking. Not recommended
Matsumoto Wase Fuyu	Mid	Small- to medium-sized fruit similar to Fuyu. Susceptible to dehiscence	Low vigour with spreading growth habit	Fruit similar to Fuyu but matures about 2 weeks earlier and has a shorter shelf-life. This cultivar needs to be pruned and thinned harder than Fuyu to keep fruit size reasonable
Hana Fuyu	Late	Large-sized (250 g to 300 g) fruit. Moderate quality. Slightly earlier than Fuyu	Medium vigour with upright growth habit	Fruit may be too large for commercial production
Fuyu	Very late	Medium-sized fruit (220 g). A round, flattened shape with four shallow channels from apex. Orange to deep red skin	Medium vigour with spreading growth habit	Good fruit quality and good storage properties. Recommended
Suruga	Late	Small- to medium- sized orange-crimson fruit (200 g), round, dome shaped with definite wrinkles at caylx end. High incidence of dehiscence and misshapen fruit	Semi-dwarf open tree	Matures just before Fuyu but some fruit may hang later. Excellent quality if not picked too early. Appears very susce- ptible to leaf spot disease causing defoliation

Cultivar	Maturity Period	Fruit Characteristics	Tree Characteristics	Comments
Fuyu Hana (Pomelo 152-7)	Mid	Medium-sized fruit (220 g) of excellent appearance, slight dome-shaped.	Weak, open growth	Brazilian cultivar of good fruit quality
20th Century	Mid	Small- to medium- sized fruit, similar to Fuyu	Open growth habit. Low vigour	Fruit and tree appear identical to Matsumoto Wase Fuyu. Recommended for mid-season production. Cultivar may need to be pruned and thinned to achieve reasonable fruit size

were investigated from 2004 to 2007. The tree growth habit was moderate vigor and showed spreading shape. Flowers bloomed only in female flowers. The flowering time was similar to that of 'Nishimurawase' (control) in late May, but the harvest time was about 3 days earlier than 'Nishimurawase' in late September. 'Migamjosaeng' is characterized by its beautiful orange skin color with round flat shape of fruits. Average fruit size was about 160 g. Soluble solids were 16.6°Brix and higher than that of control (14.5°Brix), which contributed to its excellent tastiness. It was noteworthy that occurrence of calyx-end cracked fruits and striped fruit was very rare compared with the control. Therefore, 'Migamjosaeng' trees produced marketable fruits much higher than the control tree. Flesh firmness was slightly high and similar to 'Nishimurawase'(Cheo *et al.*, 2017).

In Italy, the most promising non-astringent cultivars grown are Fuyu, Suruga, O Gosho, Hana Fuyu, Jiro and Kawabata (Bellini, 1979; Bellini and Ragazzini, 1984). In Florida, the most promising non-astringent cutlivars assessed are Ichikikei Jiro, Jiro and Fuyu (Miller, 1984). The most promising non-astringent cultivars tested in California are Fuyu, Hana Fuyu and Twentieth Century (Smith, 1970). Eleven non-astringent persimmon cultivars were evaluated by Miller (1989) for their performances in central and north Florida. The cultivars, listed in ripening order, Izu, Matsumoto Wase, Fuyu, Ichikikei Jiro and Suruga were considered to have the best commercial potential. Cultivar Gailey was the recommended pollinizer.

The cv. Nishimura Wase was introduced to China from Japan. It performed well in the Zhenjiang (Jiangsu) area. The fruits mature in mid- to late September, they are oblate in shape with a glossy orange-yellow/orange-red skin and weigh 138 g on average. The flesh has some brown spots; it is crisp and sweet, with a soluble solids content of 14 per cent, no astringency, and is of good eating quality. Fruits may be kept for 10 days without softening. The trees are productive (Zhou *et al.*, 1997). Ninety six superior cultivars of persimmon have been introduced into Zhejiang Province, China, since 1990. Trials indicated that cultivars such as Japanese Pinghewu, Shandong Red and Boai August Yellow showed better adaptability and economic efficiency than other cultivars, and are suitable for popularization in Lishui (He *et al.*, 2000).

The cultivars Guiboshi, Gosho and Gibrid 27235 (a seedling of the variety Khachia) were selected as the most promising for commercial production in the Shirvan low land area of Azerbaijan. The varieties Gibrid 27235, Geili 18 and Khiakume were recommended as the best for jam production (Kasumov and Semochkina, 1994). The percentage of early ripening cultivars which originated in the northern Japan was higher compared to that in the southern part (Yamada *et al.*, 1994). George *et al.* (1994a) evaluated the growth, yield and fruit quality of two non-astringent cultivars Izu and Fuyu under the subtropical conditions of coastal Australia. Although maximum fruit yield for Izu and Fuyu were high (24 and 37 t/ha respectively) the percen-tages of fruits suitable for export were low and variable ranging from 8.5 to 47 per cent for Izu and from 28 to 44 per cent for Fuyu. Main fruit quality defects were wind rub, petal adherence marks, fruitfly stings, calyx cavity, sunburn and bird damage. Without fruit thinning, both cultivars were susceptible to overcropping in some seasons, which resulted in a high percentage (>40 per cent) of small fruits (<150 g). Most fruits (>80 per cent for Izu and 90 per cent for Fuyu) had Brix levels higher than the recommended export standard of fruits.

4.2.2 Astringent Cultivars

In Israel, the most important astringent cultivar grown is Triumph. Fruits of this cultivar are treated with carbon dioxide at maturity to remove astringency. In California, the most popular astringent cultivar grown is Hachiya (La Rue *et al.*, 1982). In Australia, the most popular astringent cultivars grown are Nightingale and Flat Seedless (George and Nissen, 1982). In more recent testing (George and Nissen, 1988, unpublished data), the cultivar Shakoku produced excellent quality fruit. The cultivars considered most promising in Florida are Giombo, Tanenashi, Fureka and Sheng (Miller, 1984). In Italy, the most promising astringent cultivars are Hiratanenashi, Aizumishirazu, Amankaki, Kaki tipo (Bellini and Ragazzini, 1984). Hundreds of different astringent persimmon cultivars are grown in Japan but only a few produce fruits which are commercially useful. The most important astringent cultivar grown in Japan is Hiratanenashi (Glucina, 1982). If suitable methods of removing astringency from the best quality astringent cultivars could be developed, commercial production of these cutlivars could be encouraged. The quality of treated astringent fruit could be higher than that of non-astringent cultivars.

Xiaofangshi, is a promising *D. kaki* cultivar, which is grown in Nantong district, Jiangsu province, China. It has a compact growth habit (3 m tall at maturity), bears very early (usually in the second year), gives high annual yield with medium-sized fruit and their astringency can be overcome by treatment (Hu *et al.*, 1993). Jiang (1998) reported that Xiandao Seedless is a popular persimmon cultivar grown in Chunan county. Fruits mature in late September weigh 150 g on average and can be stored for up to 2 months at room temperature after harvest. The fruit skin is orange yellow and the flesh orange-red. Trees are self-fertile, very productive, resistant to drought and diseases, and adaptable to infertile soils.

Some 110 Korean native and 31 Japanese cultivars of persimmon were evaluated for 78 characters by Kim and Ko (1997). Refractometric measurement showed that fruit soluble solids content is much higher in pollination constant astringent

Cultivars of Persimmon

Fuyu

Hachiya

Flower and Fruit of Ichikikei Jiro

Flower and Fruit of Izu

Flower and Fruit of Jiro

Cultivars of Persimmon

Flower and Fruit of Maekawa Jiro

Flower and Fruit of Sugura

Flower and Fruit of Yotsumizo

Vanilla Kaki

Fruit Shapes of different Cultivars

(PCA) and pollination variant astringent (PVA) types than in pollination constant non-astringent (PCNA) ones. However, soluble solids content after precipitating soluble tannin with 5 per cent polyethylene glycol did not show any clear difference between cultivar types. Hapcheon-Bansi, Jinyang-Mootbansi and Hiratanenashi were identified as seedless.

5.0 Breeding and Improvement

Historically, *D. kaki* has been confused with *D. roxburghii*, because of their morphological similarities. Latter on Ng (1978) confirmed that there is a critical difference between *D. kaki* and *D. roxburghii* in morphological characteristics of fruit. The chromosome number for *D. kaki* is 2n = 90 (Zhuang *et al.*, 1990) and for *D. roxburghii* is 2n = 30 (Somego, 1978), suggesting that genome duplication occurred during the development of *D. kaki* (Yonemori *et al.*, 1998).

Currently, persimmon is produced worldwide, including in Brazil, Israel, Italy, Spain, Azerbaijan, Uzbekistan, New Zealand, and Australia. In spite of the large worldwide production of this fruit, new persimmon cultivars developed by cross-breeding have been released only in Japan and Korea. New cultivars have been released in Japan since 1959 and in Korea since 2012. Persimmon has been a major fruit crop in Japan for many years, where it ranks fourth in fruit production, after citrus, apple, and Japanese pear (Sato and Yamada, 2016). Oriental persimmon (*Diospyros kaki*) originated in Eastern Asia, and many indigenous cultivars have been developed in China, Japan, and Korea. These cultivars are classified into four groups based on their natural astringency loss on the tree and seed formation: pollination-constant non-astringent (PCNA), pollination variant non-astringent (PVNA), pollination-constant astringent (PCA), and pollination-variant astringent (PVA). PCNA is the most desirable type because the fruit can be eaten without any post-harvest treatment; therefore, one of the goals of our persimmon breeding programs is to release superior PCNA cultivars. The PCNA genotype is recessive to the other three non-PCNA genotypes, and PCNA-type F_1 offspring are obtained exclusively from crosses among PCNA genotypes. Moreover, the number of superior PCNA cross parents have been limited. In the late 1980s, inbreeding depression became obvious, especially in terms of fruit size, tree vigor, and productivity. To mitigate the inbreeding, a backcross program using PCNA [(non-PCNA x PCNA) x PCNA] was started in 1990. This process, however, was inefficient because only 15 per cent of the offspring were PCNA, and all offspring had to be grown to the fruiting stage. Therefore, molecular markers linked to the PCNA locus was developed for discriminating PCNA offspring. A molecular marker linked to Chinese PCNA has also been developed (Sato and Yamada, 2016).

Small breeding and selection programmes are being undertaken in Japan, Brazil, Italy, USA and Australia. A high level of bud mutation has led to several strains of many cultivars. For example, in Australia and New Zealand, there are eight strains of Fuyu currently being grown. These strains have different fruit quality characteristics and different maturity periods. Two major defects associated with many cultivars of persimmons are poor flavour and high susceptibility to blemishing. Both these characteristics can be improved through breeding and selection. Emphasis in breeding programmes should be placed on developing new non-astringent cultivars. The inheritance of non-astringency is complex and under polygene control. However, the recent availability of non-astringent male pollinizer cultivars will significantly reduce the time needed to produce progeny with desirable characteristics.

Pomper *et al.* (2020) presented a brief but an interesting history of American persimmon (*Diospyros virginiana*) improvement ub USA. According to them, the improvement began in the late 19th century with the work of Dr. James Troop at Purdue University (Plumb, 1896; Troop, 1895; Troop and Hadley, 1895). The first named American persimmon cultivar was selected out of the wild in Illinois in 1880 (Miller, 1894). This original cultivar, Early Golden, has served as the female parent of many of the cultivars developed throughout the 20th century. Professor J.C. McDaniel (1973) from the University of Illinois had a strong interest in persimmon and was responsible for selecting the cultivars John Rick and Florence.

The species *D. virginiana* can withstand a temperature of –25°C or below and it has been used in breeding *D. kaki* for frost resistance, *e.g.*, at the Nikita Botanical Garden in the Ukraine, where the variety Rossiyanka 18 and some useful hybrids were bred by interspecific hybridization (Omarov, 1994). In Taiwan, four pollination-constant non-astringent (PCNA) and three pollination-constant astringent (PCA) cultivars were selected for a breeding programme for advanced testing (Wen, 1995). Yamada *et al.* (1994) evaluated 188 cultivars of oriental persimmon native to Japan for fruit ripening time (FRT), fruit weight (FW) and soluble solids content (SSC). High heritability for each fruit was noted. PCA and PVNA (pollination variant-nonastringent) subgroups showed wide variation for each trait. Variations in FRT and SSC of PCNA were characteristically small, which suggest that these cultivars evolved later than the others. There was a week, negative correlation between FW and SSC and a weak, positive one between FRT and FW.

6.0 Propagation

Propagation methods for persimmon have been previously reviewed by Bose *et al.* (1986).

6.1.0 Seed

Seed can either be removed from firm, mature fruit or soft, ripe fruit. Removal from firm, mature fruit is less messy and gives better seed viability. Fruit can be cracked with a rubber hammer to remove seed. With soft fruit, fruit should be pulped and fermented over 2 to 3 days before removing seed. After all the flesh is fermented from the seed, the seeds are washed thoroughly and those floats are discarded.

Gasanov (1968) reported improved germination of *D. lotus* and *D. virginiana* seed after stratification for 60-90 days. In Australia, George and Nissen (1985) found that seed germination of *D. kaki* was high provided the seed is extracted from soft non-decomposed fruit, and planted immediately. No stratification was necessary. Seed was best germinated under warm (28°C) conditions. At this temperature, seeds should take 2-3 weeks to germinate. For long term storage Kotobuki (1978) suggested that seeds should be dried to 45 per cent moisture and stored at 0°C to retain the greatest viability.

Wind rub caused the greatest reduction of marketable yield and this was 5 times higher in Australia despite the windrun in New Zealand being 2.0-2.5 times higher. This was accounted for in the Y-trellising training system adopted by New Zealand compared to free standing trees in Australia (George *et al.*, 1997).

Young seedlings usually take 12 months to be of a suitable size for grafting. Container-grown plants are probably easier to manage and handle than those grown in the nursery beds. Persimmon seedlings are highly susceptible to transplant shock. A suggested method of propagation which avoids this problem is to pre-germinate seed and then raise the seedlings in small containers before transplanting to larger polybags.

6.2.0 Vegetative Propagation

6.2.1 Cutting

In New Zealand, good success has been obtained in rooting leafy summer (December) cuttings taken from forced *D. kaki* seedlings (Glucina, personal communication, 1982). Bases of cutting treated with 1000 ppm IBA (in 50 per cent alcohol) before placement in mistbeds with bottom heat (temperature 23°C) showed high percentage of rooting. Propagation using hardwood cuttings has not been successful. Persimmons could also be propagated by root cuttings but the technique for propagation by this method was not very efficient (Yamada *et al.*, 1976).

Root cuttings from rootstocks of mature trees produced shoots before roots and grew fastest when taken in spring and summer. Roots from young trees regenerated more easily and rapidly than roots from old trees and large root pieces grew better than small pieces. Misting prevented sprouted cuttings from wilting. There were clonal differences in regenerative ability (Yamada *et al.*, 1988). The possibility of vegetative propagation of *D. lotus* using leafy 2-year-old stem cuttings was investigated for difficult-to-root rootstocks in Japan by Sato and Sano (1999). The highest rooting rate (62.5 per cent) and the largest number of roots per cutting (7.8) were obtained with 8000 ppm IBA. In the test of the number of leaf buds per cutting, rooting rates were 75 per cent in the 1-bud, 50 per cent in the 2-bud, 35 per cent in the 3-bud and 15 per cent in the disbudded treatments. Cuttings taken from the bud of a shoot had a higher rooting rate (95 per cent) than those from the middle (60 per cent) and top (10 per cent) portions of a shoot. Defoliation of the leafy cutting greatly affected the rooting rate.

Softwood cuttings of FDR-1, a dwarfing rootstock for persimmon (*Diospyros kaki* Thunb.), show very low rooting percentages when planted under mist irrigation (Tetsumura *et al.*, 2017). Henc the possible use of a fog system in commercial softwood cutting propagation of FDR-1 was tested. Leaf-bud cuttings collected from FDR-1 hedges rooted better under the fog system, which maintained a relative humidity of approximately 100 per cent compared to the mist system, in which humidity often dropped to less than 50 per cent at midday. The amount of water used was lower in the fog system than in the mist system (0.18 vs. 0.89 L/h m^2), thereby affecting leaf SPAD readings. Cuttings planted in early June rooted better than those planted in early July and August, and quick-dip treatment with 6000 mg L^{-1} IBA improved rooting compared with 3000 mg L^{-1} IBA. Cuttings collected from FDR-1 root-suckers rooted better than those from hedges, and almost all rooted cuttings, planted in June, sprouted in April of the following year. Under the fog system a practical rooting percentage of 80 per cent was obtained when cuttings were collected from root-suckers, planted in early June, and quick-dipped in 6000 mg L^{-1} IBA, making it a practicable irrigation system for propagation of FDR-1 leaf-bud cuttings (Tetsumura *et al.*, 2017).

Tetsumura *et al.* (2000) reported that single node stem cuttings of a dwarfing rootstock of the Japanese persimmon (*Diospyros kaki* L.) produced high percent rooting and survival either in late June and July, while significantly low rooting and survival was occurring from the cuttings planted in August.

A practical dwarfing rootstock for the persimmon (*Diospyros kaki* Thunb.), 'MKR1', is normally propagated by cuttings collected from root suckers. However, optimal conditions for propagating leaf-bud cuttings from this variety have not been researched. Thus, several methods influencing survival, rooting and root system structure were investigated (Hejazi *et al.*, 2018). Cuttings planted in late June survived and rooted better than those planted in late July and August, and the rooted cuttings planted in late June also survived winter better.The two different irrigation methods, either a mist system, or a tray with a polyethylene tent (traypolyethylene-tent, or TPT) did not significantly affect survival and rooting. However, the primary root length was longer in the TPT and the root dry weight was heavier under the mist system. Although the rooting medium did not significantly affect the survival of cuttings, the rooting percentage of cuttings planted in perlite was the lowest. The cuttings planted in peat in late June and placed under the mist system had adventitious root initials 17 days after planting and adventitious roots were observed on the cuttings by 22 days after planting. A quadratic regression curve predicted that the highest rooting percentage would result from treatment of 'MKR1' cuttings with approximately 2000 mg·L^{-1} indole-3-butyric acid (IBA), while a different curve predicted that the lowest root number would result from treatment with approximately 1000 mg·L^{-1} IBA. Interestingly, more than 50 per cent of the cuttings treated with a quick dip in 0 mg·L^{-1} IBA successfully rooted. Although the treatment with a higher IBA concentration resulted in greater root system development on the rooted cuttings, treatment with very high IBA concentrations such as 4000 or 5000 mg·L^{-1} caused fading of leaves and dieback of cuttings (Hejazi *et al.*, 2018).

6.2.2 Grafting and Budding

Scion cultivars can either be grafted or budded with success rates in excess of 80 per cent (Mishra, 1982; George and Nissen, 1985). Grafting is generally more successful than budding and should be carried out in September with the start of sap movement. Whip or cleft grafting is suitable. Green grafting in summer has also been successful in New Zealand (Kitagawa and Glucina, 1984). Non-astringent scion cultivars such as Fuyu are propagated by budding (preferably chip) on seedling rootstocks in late February to early March or by whip or cleft grafting in

Propagation of Persimmon

Seed

Cuttings

In-vitro Propagation of Persimmon

Explant initiation (A), shoot elongation (B), in vitro rooted persimmon cultured on 5 mM indole-3-butyric acid (IBA) (C), and acclimatized plant in the greenhouse (D).

***Source*: Palla, K. J., Beasley, R. R. and Pijut, P. M. (2013) *HortScience,* 48(6):747-749.**

late August to early September (Porter, 1987). Budding may be carried out in spring or early autumn. Timing appears to be critical and varies with location. Chip or T budding is suitable provided buds are taken from mature wood. Autumn budding may produce a tree ready for transplanting 6-9 months earlier than one which is grafted (Cesar, 1942). Budwood, taken during deep dormancy, could be stored by wrapping in polyethylene at 2-4°C until required and the highest success rate (97-98 per cent) was obtained by budding in February-March. Summer budding gave only 26-36 per cent success (Omarov and Erokhina, 1991).

Suriyapananont *et al.* (1997a) studied the graft union behaviour of scion Fuyu grafted on *Diospyros lotus* callus originated from cells of cortex, phloem and xylem tissue of both scion and stock and complete tissue connection occurred within 15 days of grafting. On Khiai Rusi stock, Suriyapananont and Suriyapananont (1997b) using Fuyu scion observed that cambium bridge formation started in the stock after 10 days but later in the scion. The formation of vascular connection was complete after 14 days. They (1997b) also reported that callus formation occurred at the interface of the graft between cultivar Xichu and *D. lotus* rootstock, 5-8 days after grafting and vascular connection after 9-10 days. In another experiment using cv. Fuyu on Xichu interstock and *D. lotus* rootstock, Suriyapananont *et al.* (1997b) observed proliferation of cam-bium tissue, callus bridge and new vascular bundle in 4-6 days after grafting. The vascular connection between scion and interstock was complete 6-10 days after grafting. Five grafting methods were compared by Chauhan and Gautam (1986) using cv. Hachiya in 2 experiments and 5 cultivars in a third experiment. Veneer grafting during the second half of September gave the highest success rate (97.67 per cent). Cleft grafting at the crown without a soil cover also gave a high success rate (75.0 per cent) during mid-February and can be used as an alternate grafting method during the dormant period.

Commercial cultivars of Japanese persimmon (*Diospyros kaki* Thunb.) have been propagated by grafting scions onto seedlings of *D. kaki, D. lotus,* and other species. The generally vigorous trees that are grafted on these seedlings tend to grow larger, and thus growers encounter some difficulties in terms of orchard management. Field evaluations of the leading persimmon cultivars in Japan, 'Fuyu' and 'Hiratanenashi', grafted onto cutting-propagated 'MKR1', previously known as "OD-1" or "Rootstock b", showed that the dwarfed trees efficiently produced flowers and fruit without changes in fruit quality (Tetsumura *et al.*, 2010,2015,2019). The developers of 'MKR1' applied to the Ministry of Agriculture, Forestry and Fisheries of Japan for its registration as a new variety, which was granted in March 2015. Since 2015, four dwarfing rootstocks, including 'MKR1', have been offered for field testing at 17 prefectural research institutions in Japan. Regarding FDR-1, a 'Fuyu' persimmon tree grafted on the original seedling was incidentally found to be dwarfed in the orchard at the Fukuoka Agriculture and Forestry Research Center, Japan (Tetsumura *et al.*, 2017). There had been no previous reports about the growth of trees grafted onto clonally propagated FDR-1, partly because micropropagation had been the only method for its vegetative propagation due to the difficulty of rooting cuttings (Tetsumura *et al.*, 2019).

The field performance of 'Taishuu' Japanese persimmon trees grafted onto clonally propagated rootstocks, 'MKR1' and FDR-1, was investigated over 10 years (Tetsumura *et al.*, 2019). These results were then compared with the performance of trees grafted onto seedling stocks (S) and own-rooted trees derived from micropropagation (O-R). 'Taishuu' scions on 'MKR1' and FDR-1 rootstocks grew well initially, but stopped growing taller at seven years after planting. Total shoot length and trunk cross-sectional area (TCSA) increased annually in all trees, while the differences in these parameters between S or O-R trees and 'MKR1' or FDR-1 trees continued to increase year after year. The graft union of trees grafted onto 'MKR1' swelled, and FDR-1 rootstock overgrew the 'Taishuu' scion. Trees grafted onto both 'MKR1' and FDR-1 bore female and male flowers soon after planting, and the percentage of shoots with female flowers relative to total shoots in 'MKR1'- and FDR-1-grafted trees were higher than that in S and O-R trees each year. The numbers of shoots with male flowers varied from year to year in 'MKR1'- and FDR-1-grafted trees, but did not appear to increase yearly. 'MKR1'- and FDR-1-grafted trees almost completely inhibited secondary shoot occurrence in mid-June, and the percentages of dropped fruitlets between the middle of May and the end of July were almost always lower for 'MKR1'- and FDR-1- grafted trees than for S and O-R trees. Cumulative yield efficiencies according to TCSA, canopy area, and canopy volume showed that 'MKR1'- and FDR-1-grafted trees produced fruit most efficiently, although thetotal yields per tree were not significantly different between rootstocks. There were no significant differences in fruit quality between the rootstocks, except for the number of seeds. Concentric cracking and stylar-end cracking occurred in fruit of trees grafted onto 'MKR1' and FDR-1. However, the differences between the rootstock types were not large. The harvest date of 'MKR1' trees was significantly earlier than that of S and O-R trees. In conclusion, 'MKR1' and FDR-1 are both satisfactory dwarfing rootstocks for 'Taishuu', with high yield efficiency and inhibition of early fruit drop (Tetsumura *et al.*, 2019).

6.2.3 Topworking

Topworking should be successful in spring or autumn when sap flow is not high. Techniques similar to those used for other deciduous trees can be used. He *et al.* (1998) standardized a top split method of grafting in persimmon. Healthy scions were selected in mid-late January and kept in warm sand to promote early shooting. Grafting was done from late February to mid-March. Trees (3- to 8-year-old) were grafted at about 1 m height from the ground with 5-10 scions/tree and wrapped with plastic tape. After the union formation, some of the buds were removed to prevent congestion of branches. Branches from scions were trained in May-June to improve the tree architecture. Branches longer than 50 cm were tipped back in summer.

6.2.4 Rootstock

Three rootstock species have been used for the propagation of Oriental persimmon. These are *D. kaki, D. virginiana* and *D. lotus*. *Diospyros kaki* is the preferred rootstock, due to its compatibility with all cultivars (Porter, 1987). There appears to be some incompatibility problems between *D. lotus* and some pollination constant non-astringent cultivars, such as Fuyu (Hodgson, 1939; Bellini and Giannelli, 1982;

George and Nissen, 1985). *Diospyros lotus* also appears more susceptible to crown gall (Camp and Mowry, 1929). *Diospyros virginiana* is used as rootstock for Oriental persimmon in Israel and the USA. It is particularly well adapted to damp, heavy soils and is cold-hardy (Omarov, 1984). However, *D. virginiana* often suckers badly and trees propagated on this are not always uniform is size and vigour (Kitagawa and Glucina, 1984). *Diospyros virginiana* is also highly susceptible to Cephalosporium wilt (Miller, 1984). For these reasons, *D. kaki* is preferred to *D. lotus* and *D. virginiana* as a rootstock for persimmons. Cohen *et al.* (1991) observed the presence of a transmittable biotic factor, both in *D. virginiana* rootstocks and *D. kaki* scions, causing decline only in *D. kaki* scion tissues. Some strains or cultivars of *D. kaki* produce more vigorous and uniform seedlings for use as rootstock than others. For example, the cultivar Saijo produces very uniform seedlings and is used extensively as a rootstock in Japan. Most of the totally sweet scion cultivars were found to be incompatible with date plum persimmon (*D. lotus*) and several wild persimmons from Zhejiang province, but many of the less sweet cultivars were compatible.

Graft compatibility between rootstocks No. 56 and No. 77 and cultivars Maekawa Jiro and Ichikikei Jiro was good (Wang *et al.*, 1996). The 8-year average yields in cv. Khiakume as reported by Omarov (1990) were 65.1 and 46.5 h kg/ha for trees grafted on Caucasian and Viginskaya persimmons respectively. Tsai (1998) reported graft incompatibility in cv. Fuyu.

The rootstock effect of *D. rhombifolia* on the growth of cv. Fuyu was compared with that of seedling rootstock of *D. kaki* (Yamada *et al.*, 1987). Total shoot length, tree height, fresh weight of shoot and root were significantly lower with *D. rhombifolia* as the rootstock. Annual yields and annual shoot and woody shoot weight were also lower on trees on *D. rhombifolia*.

Omarov and Erokhina (1989) studied the root systems of 10-year-old cv. Khiakume trees on *Diospyros lotus* or *D. virginiana* rootstocks by excavation. Root vigour and distribution depended mainly on the biological characteristics of the rootstock. The trees on *D. lotus* were large and had markedly more vigorous root systems (with greater vertical penetration and horizontal spread) than trees on *D. virginiana*. Yields on *D. lotus* and *D. virginiana* averaged 14.5 and 8.6 kg/tree respectively, and the respective mean fruit weights were 193 and 184 g. Fuyu persimmons grafted on 6 rootstocks, *Diospyros kaki* and *D. kaki* var. *silvertris* No. B rootstocks were compa-tible (Liu *et al.*, 2000). Mineral elements (Mn, Ca, P, Mg, B) accumulated more in the leaves of Fuyu persimmon on *D. lotus* rootstocks. The concentrations of soluble sugar and reducing sugar in Fuyu persimmon on *D. oleifera* and *D. glaucifolia* rootstocks exceeded those of Fuyu persimmons on other rootstocks.

7.0 Area and Production

The persimmon is a major horticultural crop in Japan and China. Smaller but expan-ding persimmon industries are being developed in Brazil, Italy, Israel, Korea, USA, New Zealand and Australia. The production of persimmons in some countries, such as Korea and Brazil has almost been doubled in the past ten years. A summary of production and marketing trends in some countries is presented in Table 7.

Table 7: Global Area, Production and Yield (kg/ha) of Persimmon*

Countries	*Area (Ha)*	*Production (Tonnes)*	*Yield (Kg/ha)*
Australia	92	719	78222
Azerbaijan	10474	160092	152850
Brazil	8133	156935	192961
Chile	111	596	53611
China, mainland	852433	3084458	36184
China, Taiwan Province	5239	84301	160910
Iran (Islamic Republic)	1523	22474	147596
Israel	1049	28000	26691
Italy	2468	47615	192929
Japan	19100	208000	108901
Nepal	374	3013	80561
New Zealand	143	2384	166465
Republic of Korea	27203	346679	127440
Slovenia	115	2604	226435
Spain	18601	492320	264674
Uzbekistan	4262	71214	167098

* *Source*: FAOSTAT (2018).

7.1.0 Japan

Production of persimmons in Japan has remained relatively static in the past five years. Sixty per cent of the production is based on non-astringent cutlivars. Japan exports 1-2 per cent of its crop. It is the fourth ranked fruit crop in Japan and produced an estimated 302,000 t in 1994. The main growing areas are Fukuoka, Nara and Gifu Prefectures and the most important astringent and non-astringent cultivars (Yonemori, 1997).

7.2.0 Israel

The persimmon industry in Israel is based on the non-astringent cultivar Fuyu and the astringent cultivar, Triumph, which is treated with carbon dioxide to remove astringency. Ninety per cent of the production is exported to Europe and traded under the name of Sharon fruit.

7.3.0 USA

A marked increase in persimmon plantings has occurred in California towards the beginning of the last decade, based on non-astringent cultivars. The persimmon is considered as an alternative crop in avocado areas affected by phytophthora root rot. A reasonable domestic market has also been created by the influx of Asians into the country.

7.4.0 New Zealand

New Zealand is currently developing an export market for non-astringent persimmon to Japan. Severe calyx cavity and the nutritional disorder green blotch are major problems in some production areas.

7.5.0 Australia

Marked expansion in persimmon plantings has occurred in Australia in the past few years, based mainly on non-astringent cultivars. Export markets to Europe and South East Asia are being developed.

7.6.0 China

D. kaki is the main *Diospyros* species cultivated for edible fruit production in China. Average annual production has been estimated to be 730,000 t (Wang, 1994).

8.0 Soil and Climate

8.1.0 Soil

Although persimmons can grow on a wide range of soil type, well drained sandy loams and loams are best suited for maximum growth and production. Yield is reduced on heavy alluvial soils due to increased fruit drop. Where soils are shallow and poorly drained, mounding to obtain at least 1 m of usable soil depth is recommended. Based on studies with *D. kaki* seedlings, the soil pH range for optimum growth is between 6.0-6.8 (Kumashiro *et al.*, 1957).

Jung *et al.* (1993) observed that approximately 72 per cent of existing sweet persimmon orchard soils in the Yeongnam area of South Korea was located on slopes of 15-60 per cent, of which 54 per cent were north facing. The predominant soils were well drained clay loams of which 31 per cent were vulnerable to drought and 60 per cent were poorly drained. Both were considered unsuitable for persimmon orchards.

8.2.0 Climate

The persimmon is readily adaptable to a wide range of subtropical and warm tempe-rate climates. Trees are deciduous and enter a rest period.

8.2.1 Temperature

Night temperature maintained at 20 ± 2°C advanced anthesis by 10 days relative to the control and reduced the frequency of embryosac degeneration (Fukui *et al.*, 1990). Fruit set of hand-pollinated flowers of non-astringent Australian cultivars at 17/12° and 27/17°C (day/night) was similar and averaged 53.6 and 54.9 per cent respectively. In contrast, the fruit set of hand-pollinated flowers at 27/22° and 32/27°C were significantly reduced by 58 and 77 per cent, respectively compared with non-pollinated flowers. Fruit set of non-pollinated flowers at 17/12°C was much higher than at higher temperature, which might be ascribed to a combination of increased flower size and reduced vegetative growth under these conditions. Shoot starch concentrations at anthesis increased as temperature decreased (George *et al.*, 1994c). However, most cultivars appear to have little chilling requirement (100-200 h below 7.2°C) to satisfy rest and ensure uniform bud-break. The trees, when dormant, will tolerate fairly low minimum temperatures with some cultivars surviving temperatures as low as –15°C (Kitagawa and Glucina, 1984). The hardiest species during the coldest period of the season was reported to be *D. lotus* followed by *D. kaki*, and the least hardy was *D. taitoensis* (which is found in subtropical Taiwan). The lowest temperatures that the plants could sustain without injury were –10°C for *D. taitoensis*, –15°C for *D. kaki* and –20°C for *D. lotus* (Leng *et al.*, 1993). An apparent difference has been noted in cold acclimatization capacity between the species. *D. kaki* seems to have evolved the ability to acclimatize the low temperature by controlling the enzyme systems concerned with production of phenolic substances (Leng *et al.*, 1995). However, freeze damage may occur in some seasons with alternating warm cold cycles (Sharpe, 1966). Occurrence of frost causes browning of twigs. The *D. kaki* cultivars Denglongshi, Hroguantoushi and Hyakume were rated as most frost hardy (Wang and Yang, 1995). Non-astringent cultivars require warmer conditions for fruit maturation than the astringent types. For the non-astringent cultivars, Kajiura (1980) and Chujo *et al.* (1972, 1973) in Japan indicated that the best fruit quality was produced in regions where average temperature in the autumn months ranged between 16 and 22°C and the amount of sunshine recorded during the growing season exceeded 1,400 h. The cvs. Hiratanenashi and Tone Wase, which maintained higher photosynthetic rates during summer, seemed to be more adaptable to the warm climate than other cultivars (Iwasaki and Yoshida, 1996). Leng *et al.* (1997) analysed the relationship between frost hardiness and geographical distribution of Chinese, Japanese and Korean persimmons. Almost no relationship between frost hardiness and place of origin could be established. Analysis of the levels of 4 polyphenol compounds in the same twigs during a pre-hardening period of 12 days at 4°C and the subsequent cold treatment suggested a relationship between an increase in levels of flavans and hardening of twigs.

Attempt has been made to elucidate the relation between cell division in the root tips of Japanese persimmon cv. Nishimurawase and soil temperature. The cell division-temperature relationship obtained from August to October indicated that at 21°C soil temperature the mitotic index of the roots decreased. The proportion of white root tips increased as the soil temperature exceeded 19°C. The appearance of white root tips was related to the end of shoot growth or anthesis (Fukui *et al.*, 1994). Low soil temperature (7°C) extended the period of dormancy, while the highest percentage bud burst (48.3 per cent) was noted at a root temperature of 13°C (Mowat, 1995). Cooling the soil during summer resulted into early maturation of fruit with better quality (Matsumura *et al.*, 1998). Fumuro and Utsunomiya (1999) observed that soil warming to 20°C by circulating warm water pipes hastened the times of bud burst and anthesis. Warming the soil increased shoot growth and accelerated maturation and TSS accumulation of fruits, but the yield was reduced (12.3 compared with 14.0 kg/tree) because of low fruit set and small fruit size.

If non-astringent cultivars are grown under cooler conditions, the fruits may not completely lose their astringency by the time they reach maturity, fail to mature properly and have low sugar content (Glucina, 1982; Chujo, 1982; Kitagawa and Glucina, 1984). A minimum Brix between 14o and 16o at harvest is required. Mowat *et al.* (1997) studied the chemical composition of fruits from orchards in a warm temperature (North Island, New Zealand) and a warm subtropical climate (Queensland, Australia). Mature fruits grown in the warm subtropical region had significantly higher soluble solids concentration and lower soluble tannins content than fruits grown in the warm temperate region. Internal fruit quality attributes appeared more sensitive to micro-climatic differences than the external attributes.

8.2.2 Light

Overcast sky at flowering (irradiance < 10 MJ m^{-2} day^{-1}) reduces fruit set. At irradiances of 13.8 and 5.9 vs 19.8 MJ m^{-2} day^{-1}, fruit set of open pollinated flowers was reduced by 53 and 100 per cent respectively. In contrast, fruit set of hand pollinated flowers was reduced by only 21 and 40 per cent respectively (George *et al.*, 1993). In a preliminary experiment, the effects of plastic cover on fruit maturation and quality were inves-tigated by Mason *et al.* (1992). Control trees were left uncovered. The growth of terminal shoots began earlier in the covered block and trees reached 50 per cent full bloom 7 days before control. Fruits under cover were heavier, redder and sweeter through- out the season and at harvest as well. In the second season covered fruits were less astringent at harvest than control fruits.

The photosynthetic characteristics of persimmon cv. Damopan were studied by Wang *et al.* (1993). The light saturation point was 65 klx and the light compensation point 1.5 klx. The optimum temperature was 25°C. The changes in net photosynthetic rate (Pn) in a day followed a double-peak curve with a minimum at noon; the higher peak at 10.00 h and the lower peak at 18.00 h. The photosynthetic rate also varied with leaf position on the shoot and season.

Asakuma *et al.* (2018) carried out studies to determine the developmental responses of Japanese persimmon seedlings (*Diospyros kaki* Thunb.) to variations in photoperiod and light conditions, and to develop an effective method for scion propagation using light-emitting diode (LED) irradiation for nursery stock production, shoot length and number of nodes increased after an irradiation treatment of 20/h/day compared with that of 12/h/day. Shoot length under red LED light was significantly greater than under blue LED light in the 20/h/day treatment. The number of nodes also increased under red LED light compared to blue LED light. The number of leaf primordia was significantly larger in grafts in late July compared to late June. The survival rate of scions and the shoot lengths were greater for grafts in late July compared to late June. The number of buds in grafts in late July was significantly higher than in late June. Based on these results, we propose that twice grafts per year, consisting of a bark graft in February and a green graft in July, is a better alternative to the conventional once graft in Spring because of the higher survival rate of scions and their accumulation of larger numbers of buds. This alternative method achieves up to a 6-fold improvement in scion propagation rate compared to the once graft approach (Asakuma *et al.*, 2018).

9.0 Cultivation

9.1.0 Planting and Plant Density

Trees are best planted out in July or early August, when dormant. Planting out after bud-break often results in severe transplant shock. Root systems of container grown plants should be disturbed as little as possible during transplanting. Foliar fertilization may reduce the severity of transplant shock. Size and depth of the planting hole should be sufficient to accommodate the spread of roots and to keep bud union 75 mm above ground level.

Fumura *et al.* (1997) described improved method of growing of cultivars Ronewase, Nishimurawase and Shinsyu by planting one-year-old trees on raised beds at 1 m spacing in a sideless plastic house. The trees were trained to a central leader and 3 years after planting the frames of the beds were removed and soil was mounded around the sides of the beds. Precocious cropping with high yields were obtained 3 years after planting. Covering soil surface with aluminized plastic film enhanced fruit colouration and increased total soluble solids.

Planting densities for persimmon depend on cultivar, rootstock and soil type. The planting densities suggested should be taken as a guide only. Dwarf cultivars, such as Ichikikei Jiro can be closely planted at 5.0 x 2.5 m (800 trees/ha), semi-dwarf cultivars, such as Fuyu can be planted at 5.0 x 3.0 m (660 trees/ha) and vigorous cultivars need to be more widely planted at 6.0 x 4.5 m (370 trees/ha). In Azerbaijan, the yields/ha increased with tree density. The highest yields were obtained with trees at 5 x 5 m, followed by 8 x 5 m (Gasanov, 1993). Trees within rows may need to be thinned 7 to 10 years after planting. Higher planting densities may be feasible provided tree size is controlled by using growth retardant, such as paclobutrazol.

9.2.0 Canopy Management

Cultivars differ markedly in vigour and growth habits. Some cultivars are dwarf and highly precocious, others are vigorous upright and late producers. The dwarf and semi-dwarf cultivars are best trained to a modified central leader system (George and Nissen, 1984). Tiers of branches should be well spaced (0.5 m) to avoid fruit rub. The more vigorous, upright narrow crotch angle types may be better trained to a vase or palmette system. Since the flowers of *D. kaki* are borne on current season's wood, heavy pruning reduces crop setting by forcing excessive vegetative growth. Before the beginning of cropping (usually 3 years) trees can be pruned twice a year, when dormant, that is in mid-August, and in the peak of the growing period, usually late November to early December. Once trees begin cropping, cropping load naturally slows vegetative growth and only light dormant and summer pruning are then nece-ssary. Summer pruning of mature trees may thicken laterals and increase fruit size and colour. With the more vigorous, less precocious cultivars, techniques, such as cincturing and limb spreading may be benefical to achieve higher and earlier fruit production. The application of growth retardants, such as paclobutrazol will help in widening crotch angles and make training easier.

Dormant pruning (heading back) of long unbranched shoots in February forced new shoots which fruited well in the following year, whereas heading back short shoots produced vigorous laterals which set little fruit. On the other hand, laterals arising from summer branches which were not headed back during the previous season set more fruits than laterals arising from short, unpruned spring branches (Hasegawa *et al.*, 1991a). The best results, in terms of shoot formation, fruit drop, fruit set and photosynthetic productivity, in cvs. Khaikume and Tsuri-noki were obtained with reduction of crown diameter to 2.5-5 m, and for cv. Zendzhi-maru with reduction of tree height to 3-3.5 m without reducing crown diameter (Gasanov, 1991). Removing shoots from the shaded parts of trees and encouraging the growth of shoots exposed to sunlight increased the proportion of fruits with good colour (Takano *et al.*, 1991).

9.2.1 Trellising

Because of the high yield capabilities of persimmons and the need to provide structural support to prevent limb breakage, various trellising systems including Tatura (Chalmers and Van den Ende, 1975), Lincoln (Dunn, 1974), Ebro-espalier (Anon, 1981) were evaluated in Australia and New Zealand. In Italy, persimmons have already been successfully trained onto a palmette system (Bellini, 1973). Because of the limited research data currently available no recommendation can be made at this point of time on the most suitable trellising system to use.

Persimmon trees in dense orchards have generally been trained to a modified-leader (ML) system or some variants therefrom. However, the ML takes long to fill the space and to be productive, and frequently deteriorates light penetration into the canopy as tree height increases. Low light level within the canopy inhibits flower bud initiation (Choi *et al.*, 2015) and reduces fruiting and fruit quality (Yakushiji *et al.*, 1997).

Applicability of Y-trellis system to high density persimmon was evaluated for 7 years after planting (Choi *et al.*, 2015). Two year old trees of 'Uenishiwase' (*Diospyros kaki*) were planted at a row spacing of 6 m in spring of 1996. Tree spacing within a row was 3 m for modified-leader (ML) and 1.5 m for Y-trellis. It was apparent from 2000 that the increase in trunk cross-sectional area (TCSA) was less in the Y-trellis than in the ML trees. The TCSA of the Y-trellis trees was 19 per cent smaller in 2002 than that of the ML trees. On the other hand, the trees spreaded more in Y-trellis trees. Due to higher planting density and wider tree-spread of the Y-trellis, the cumulative yield per hectare from 1998 to 2002 was 29.4 tons for the Y-trellis trees, a 1.6-fold increase over the ML trees. Light penetration was 10.2 per cent more in the Y-trellis trees. However, there was no consistent difference in fruit characteristics between the two training systems of young trees in this experiment. Despite a higher occurrence of water sprouts in Y-trellis trees on a scaffold limb or land area basis, the results of this study indicated the potential benefits of Y-trellis system for early production of persimmon (Choi *et al.*, 2015).

9.2.2. Pruning

A study was conducted to evaluate the effect of branch and tree thinning on growth, yield and fruit quality in 8-year-old non-astringent persimmon (*Diospyros kaki* L.f.) trees in a high-density orchard over 6 years in Korea (Kim *et al.*, 2017). Adjacent trees in the rows were removed in 2008 (first-year thinning) or in 2010 (gradual thinning). Trees in the second treatment also had 25 per cent of the main scaffold branches removed in the first two seasons. Control plots were left unthinned. First-year thinning elevated leaf macro-nutrient concentrations, increased the production of short fruiting shoots (less than 20 cm) per tree, decreased water sprout development and weight of pruned biomass, and increased fruit set compared with values recorded on control trees and with gradual thinning. First-year thinning gave higher photosynthetic photon flux in the centre of the canopy and better fruit quality. There were generally only small differences in total yield across the three treatments in individual years, but much higher total marketable yield from 2010 to 2013 in the first-year thinning plots. First-year thinning improved overall tree productivity, and the gross returns were 80 per cent higher than those of control from 2008 to 2013 when the extra cost of the thinning was taken into account (Kim *et al.*, 2017).

Persimmon trees, having no dwarf rootstocks of commercial value, tend to grow high. Lowering tree height by heavy dormant pruning has been a routine practice for efficient management of many persimmon orchards (Choi *et al.*, 2011). In densely planted orchards, heavy pruning is an inevitable practice to restrict tree size. This practice, in turn, causes vigorous shoot growth resulting in an excessive crowding of the canopy. Poor fruit set and excessive supply of nitrogen also stimulate the occurrence of vigorous shoots. Tree crowding not only hinders orchard operations, but deteriorates fruit quality as the tree interior becomes heavily shaded. Shoots of vigorous and succulent growth make the tree susceptible to anthracnose (*Colletorichum gloeosporioides*) when humidity within the canopy is high (Choi *et al.*, 2011). Summer pruning is one of many options to alleviate the problems of crowding, ensuring adequate light penetration into the canopy and controlling excessive shoot growth. Choi *et al.* (2011) reviewed the effect of summer pruning in persimmon (*Diospyros kaki* Thunb.) with regard to its positive and negative aspects on tree growth and fruit production. In order for this practice to be of a significant value, a clear understanding is needed to appreciate the fact that so many variables are involved and much remains to be substantiated. Major reasons for summer pruning of persimmon are to improve fruit quality by enhancing light penetration into the tree canopy and to restrict vegetative shoot growth. Summer pruning generally suppresses tree growth even though it elevates leaf activity. Positive effects of summer pruning on skin color, soluble solids, and appearance of fruits are observed mainly in those orchards where the trees are heavily dormant-pruned to lower tree height and to secure space in high density plantings. Secondary shoot pruning and topping could also improve fruit quality, increasing flower bud formation of remaining shoots. Summer-pruning effects are highly dependent on its severity and timing to affect tree growth, shoot regrowth, reserve accumulation, and fruit quality (Choi *et al.*, 2011).

9.3.0 Irrigation

In most regions, irrigation is considered essential for successful persimmon production. This applies even in coastal regions receiving high yearly rainfall.

Dry periods during fruit growth will reduce the size, quality and number of fruits carried to maturity. The spring-early summer period is the most important period in determining the yield and quality of the crop. High levels of soil moisture are required to stimulate leaf growth and flowering. Water stress during early summer may increase fruit drop (Banks, personal communication, 1988). Fruit set of hand-pollinated fruits was reduced by water stress (at 32 per cent soil moisture content). Stomatal conductance and CO_2 assimilation rate were not adversely affected by water stress until available soil moisture fell below 50 per cent (George *et al.*, 1995b).

Water-use reaches a peak in mid-summer due to high daily evaporation. Much of this water-use may be replenished by seasonal rain but irrigation will be needed to ensure that a uniform soil moisture status is maintained during dry periods. Uniform soil moisture levels which prevent over or under watering are important in preventing 'line cracking' of fruit.

During fruit maturity, the amount of water used by plants starts to decrease although good soil water status is still necessary. Water stress during this period can cause premature leaf drop which may affect the maturing fruit by reducing sugar levels and increasing susceptibility to sunburn. 'Line cracking' can still be a problem if soil water supplies are allowed to fluctuate too widely. Uniform level of moisture should be maintained by applying irrigation on a regular schedule related to plant water use.

Persimmons have a natural leaf drop in dormancy period because of their deciduous nature. Creation of water stress to achieve dormancy is not necessary. However, care should be exercised particularly with late maturing cultivars, *e.g.*, Fuyu, because dormancy may begin while fruits are still maturing on the trees. In these cases, irrigation must be continued to allow normal fruit development. Once harvesting is over, irrigation can be reduced to minimal amount required to keep trees alive during dormancy.

The most suitable irrigation system to use with persimmons is an under tree minisprin-kler system. Rates and timing of irrigation can be scheduled according to Class A pan-evaporation or tensiometer readings.

Persimmon cultivation is gaining importance in many semiarid areas of the world. Intrigliolo *et al.* (2018a) reviewed the water requirements of persimmon and responses to variations in water supply and quality. It is concluded that persimmon water needs can be particularly high during summer, when the crop coefficient can reach values close to or even above 1.0. To cope with water scarcity, regulated deficit irrigation (RDI) has been shown as a useful technique for reducing fruit drop in persimmon trees. Late RDI strategies during fruit ripening can accelerate fruit color development resulting in earlier harvesting but with lower fruit weight. This chapter also provides important information on persimmon tree responses to salinity, along with methods to alleviate the detrimental impacts of low-quality irrigation waters on tree performance (Intrigliolo *et al.*, 2018a).

Continuous plant water status monitoring is crucial in order to improve irrigation management. The noninvasive Yara ZIM-probe was assessed for detecting plant water stress in Persimmon trees (*Diospyros kaki* L.f.). The probe measures the pressure transfer function (*P*p) through a patch of an intact leaf, which is inversely correlated with the turgor pressure (Martinez Gimeno *et al.*, 2017). This technology was evaluated in two parallel experiments involving either distinct watering regimes or rootstocks with different drought tolerance [*Diospyros lotus* (L) and *Diospyros virginiana* (V)]. Concomitant measurements of midday stem water potential (stem) and trunk diameter variations were taken throughout the experiments. *P*p was highly correlated with stem. Persimmon leaves exhibited the inversed *P*p curve phenomena under water stress, which enabled the association of a particular range of stem to each of the three leaf turgor states defined. Persimmon trees with no sign of initial or total inversion ensured Ψstem above –0.8 MPa, values considered of a well-watered Persimmon tree. Yara ZIMprobe readings as well as stem and trunk diameter variation measurements pointed L as a more sensitive rootstock to drought than V. In conclusion, results showed that the Yara ZIM-probe can be used to continuously monitor water status in Persimmon trees although further research would be needed to ensure their feasibility for scheduling irrigation (Martinez Gimeno *et al.*, 2017).

9.4.0 Manuring and Fertilization

In orchards where 'green spot' disorder has been observed, soil pH should be adjusted to the upper end of the range to increase calcium availability and decrease quantities of plant available manganese (Clark, 1985). The average height of non-astringent cv. Fuyu grown in waste mushroom compost was almost similar to that of plants grown in commercial compost but significantly greater than that of plants grown in soil (Nee *et al.*, 1994). Young trees upto 3 years of age require an application of complete fertilizer (11:4:14) before bud-break and 3-4 lighter follow-up applications of urea or nitram at monthly interval in the peak growing period. Once trees begin cropping an application of complete fertilizer is applied 4-6 weeks before harvest. Rates of fertilizer to apply need to be adjusted according to cultivar, crop load and leaf nutrient levels. Application of 150 g urea/tree in spring, followed by repeated application (3-5) during summer of 0.1 per cent urea solution as a foliar feed, if possible, combined with spraying of micronutrients were beneficial (He *et al.*, 1998). A 10 per cent reduction in N fertilizer inhibited N uptake, especially in roots whereas it had little influence on the N levels in the trunk, shoots and leaves (Fukui *et al.*, 1999).

An average yield of 31.2 t/ha was reported by Lin *et al.* (1997) in China for 10 years. Each year after harvest, 125 kg compost, 5 kg phosphate and 0.5 kg potassium sulfate were applied to each tree, and in the vegetative season, supplementary fertilizers were applied 3 times and foliar fertilizers applied at regular intervals. Trunk girdling was performed at the beginning of full bloom. The effects of N and irrigation on fruit quality in container and field-grown persimmon cv. Maekawajirou were investigated by Matsumura (1999). For plants in containers, the highest quality fruits were produced from containers which received 3-5 litres water per day and 10 g N per year. Application of N until August also improved fruit quality. Fruit quality was also influenced by fertilizer type.

Matsubara and Hosokawa (1999) studied the effect of mycorrhizal fungi on the growth of seedlings of Japanese persimmon by innoculating plants with: *Gigaspora margarita* (GM); *Glomus aggregatum* (ga); *Glomus fasciculatum* (gf); *Glomus mosseae* (gm); or *Glomus* sp. R-10 (gr). Fifteen weeks after inoculation, plant height showed significantly higher values in fungus-inoculated plants (except gm) than in non-inoculated ones; the effect was more pronounced in GM-, ga- and gr-inoculated plants than in gf-inoculated one. The number of leaves and leaf area were superior in fungus-inoculated plants compared with non-inoculated ones.

There is little information in the literature about the water and nutrients needs of persimmon trees and the response of this crop to variations in water and/or fertilizer supply. This fact is most likely because persimmon cultivation has been traditionally important in areas of Korea and China where the precipitation is often abundant and irrigation is not crucial (Intrigliolo *et al.*, 2018b). The available information in the literature about the nutrient requirements of persimmon trees it has been generally focused on studies conducted on non-astringent varieties cultivated in Asian countries such as South Korea, China and Japan, as well as Australia and New Zealand. In particular, the nutritional studies conducted in Australia by George *et al.* (2005) on Haya and/or Fuyu (non-astringent) varieties are those exposed with greater detail. In Spain, studies conducted by Pomares (2014a, 2014b, 2015) to know the nutrient crop demand of persimmon trees have been mainly focused on the astringent variety "Rojo Brillante". These studies have made it possible to determine the nutritional needs of persimmon trees depending on the fruit production.

9.4.1 Organic Manure

Nitrogen fertilization is one of the most important elements for growth, reproductive behavior and ensure economic yield with a good quality of Costata persimmon trees (Choi *et al.*, 2013). Meanwhile, mineral nitrogen fertilizers cause the accumulation of harmful residual substances like NO_3 and NO_2 which have deleterious effects on soil, water and atmosphere pollution, and reflected on animal and human health (Ennab *et al.*, 2018).

A study was conducted during three seasons to study the effect of replacing mineral nitrogen fertilization by organic nitrogen and biofertilizers on Costata persimmon trees, under calcareous soil conditions (Ennab *et al.*, 2018). The treatments consist of T1 (100 per cent mineral nitrogen), T2 (75 per cent mineral nitrogen + 25 per cent organic nitrogen), T3 (50 per cent mineral nitrogen + 25 per cent organic nitrogen + biofertilizer), T4 (25 per cent mineral nitrogen + 50 per cent organic nitrogen + biofertilizer) and T5 (100 per cent organic nitrogen). Rice straw compost was used as a source of organic nitrogen and N2-fixing free living bacteria namely Azotobacter chroococcum and Azospirillum lipoferum as a source of biofertilizers. The obtained results can be summarized as follows: Treatment of 50 per cent mineral nitrogen + 25 per cent organic nitrogen with biofertilizers gave the highest values of vegetative growth in terms of shoot number/branch, shoot length, leaf numbers per shoot and leaf area of Costata persimmon trees compared to mineral nitrogen fertilizers only. Moreover, treatments included organic and biofertilizers were raising leaf content values of chlorophyll, N, P, K, Ca, Fe, Mn and Zn especially in the second season. As for yield, treatment of 50 per cent mineral nitrogen + 25 per cent organic nitrogen + biofertilizer and 75 per cent mineral nitrogen + 25 per cent organic nitrogen produced the highest yield in terms of number of fruits/tree, weight (kg/tree) and total yield (ton/feddan). All treatments led to improving properties of fruit quality especially treatment of 50 per cent mineral nitrogen +25 per cent organic nitrogen +biofertilizers. Also, treatments containing organic and biofertilizer led to reducing nitrate and nitrite values in fruit juice comparing with 100 per cent mineral nitrogen fertilizer. Based on the study, Ennab *et al.* (2018) recommended: (50 per cent mineral nitrogen +25 per cent organic nitrogen +biofertilizers). This treatment gave the best growth, yield, fruit quality and nutritional status of Costata persimmon trees.

9.4.2 Tissue Analysis and Nutrient Requirement

Leaf nutrient levels and sampling procedures for persimmons have been established by Sato *et al.* (1954), Nakamura *et al.* (1972), Clark (1985) and Clark and Barrett (1985). In Japan, persimmon leaf samples are collected about 2 months prior to harvest. The youngest fully expanded leaves are selected from non-fruiting laterals. The ideal time for leaf sampling appears to be at anthesis, as leaf nutrient concentrations at this stage were found to be more stable than at later sampling dates.

Changes in the concentration and content of macro- and micronutrients in the youn-gest mature leaves on fruiting and non-fruiting shoots of the cultivar Fuyu were recorded by Clark and Smith (1990a). They found no major differences between concentrations in leaves sampled from fruiting or non-fruiting shoots at individual harvest throughout the season. Concentrations of Cu, N, P and S declined during early growth, but reached values which remained constant by mid-season. Develo-ping fruit had a pronounced effect on all leaves, not just those closest to the fruit. Significant quantities of K, N and P (18-63 per cent of the maximum accumulated) were remobilized from the leaves in the 12 weeks prior to leaf-fall, and were sufficient to meet all of the N and K, and about 70 per cent of the P requirements of the fruits during this period. Clark and Smith (1990b) also reported changes in the concentration and quantities of macro- and micro-nutrients in fruits of the cultivar Fuyu. The concentration of all nutrients in the fruit (including skin, flesh and seeds) declined during growth. Concentrations of Ca, Cu and Fe in the skin were consistently greater, and those of P and S consistently smaller than corresponding concentrations in the flesh throughout the season. Concentrations of N, K, Ca, Mg, Mn and Zn in the flesh only exceeded those in the skin during early fruit development. In mature fruit tissues, the highest concentrations were found in the seeds for all nutrients except Ca, which was more concentrated in the skin.

A major advantage of leaf sampling at anthesis, prior to root flushing, is that any detected nutrient deficiencies can be corrected in the current growing season (George *et al.*, 1995a). Leaf nutrient concentrations for the major non-astringent cultivar Fuyu are presented in Table 8.

Table 8: Optimum Nutrient Concentrations of Fuyu Leaf Tissue*

Nutrient Element	*Concentration*
N	2.51
P	0.14
K per cent	2.42
Ca	1.31
Mg	0.45
Fe	81.8
Mn ppm	118
B	32.4

* Kim *et al.* (1993).

Leaf analysis showed that N, P and K concentrations decreased through the season, while Ca and Mg increased (Liu and Wang, 1989; George *et al.*, 1994b). Kaplankiran *et al.* (1998) determined the levels of leaf N, P, K, Na, Ca, Mg, Zn, Mn and Cu for 4 persimmon cultivars, Persimmon Seedless, Fuyu, Tipo Kaki and Tesil Hurma. The trees were grafted on persimmon seedling rootstocks and had been planted. Tesil Hurma had a significantly higher leaf N content (2.41 per cent) compared with Fuyu (1.73 per cent). Persimmon Seedless had significantly higher Ca and Mg contents (1.42 per cent and 0.58 per cent, respectively) compared with Fuyu (1.03 per cent and 0.44 per cent respectively). Other nutrient levels did not differ significantly between cultivars.

Micropropagated persimmon trees were planted in plastic cloth pots containing an artificial medium (sabatsuchi, bark, rockwool and perlite) (Fukui *et al.*, 1997). The trees were grown for 3 years with 2 fertilizer regimes (1 litre liquid feed every week or 1 litre every other week) after which the tissues were analysed for K, Ca and Mg during June, August, October and December. Leaves had a high K content upto August and this declined subsequently, but fallen leaves in November had high K contents. Roots had a low K content in August after which levels increased. K contents in shoots and trunk were at a constant low level throughout the year. Ca content increased after June in leaves and trunk and after August in the roots. Mg content increased after August in roots. The roots were the major storage organ of K, Ca and Mg.

9.4.3 Nutritional Disorders

Many cultivars are susceptible to fruit drop and calyx cavity if nitrogen levels are too high when fruits are sizing. Nitrogen rates should be reduced if these problems are evident. Clark and Kajiura (1986) and Clark and Smith (1985) described the nutritional disorders of persimmon. The most widely occurring nutritional disorder is green blotch. With this disorder the skin, particularly around the apex, stops developing and remains green as the fruits continue to mature (Clark and Kajiura, 1986). Comparisons between affected and healthy orchards in Japan indicate that the manganese concen-tration in the leaves of trees in bearing affected fruits

(600-2000 ppm Mn) is typically 2-4 times greater than normal (300-500 ppm Mn). Spray application of calcium carbonate to young fruit may reduce the disorder (Iimuro *et al.*, 1980).

9.5.0 Weed Control

Mulching around the plant aids in controlling weed growth. Weeds growing through the mulch and those in the tree line may be controlled with a knockdown herbicide, such as paraquat.

9.6.0 Mulching

Investigations by Lim *et al.* (1997) on different mulching materials revealed that soil temperature was highest under black polyethylene film but lower under rice straw mulch. Weeding efficacies were 91 and 67 per cent respectively, with the 2 materials. The organic matter content of surface soil and root weight in the second year after mulching was increased by mulching with rice straw and rice hull. Fruit quality was not greatly influenced by mulching materials, but yields were increased by mulching with rice straw and rice hull.

10.0 Flowering, Pollination, Fruit Growth and Development

10.1.0 Flowering

Flowers are produced on new growth in the spring period. This new growth emerges from one-year-old wood. Three types of flowers are possible: female (pistillate), male (staminate) and hermaphrodite (perfect) flowers. In addition, persimmon trees may be monoecious, dioecious, or rarely polygamous (Kitagawa and Glucina, 1984). However, trees of the most important cultivars bear only female flowers (the stamens are sterile). These occur singly and are relatively larger. They can easily be distinguished from the male flowers by their large, dark green, four-lobed calyx. The male flowers (the ovary is non-functional) are usually axillary and occur in two- to three-flowered clusters on small, weak shoots, and are much smaller than in the female flowers. Hermaphrodite flowers can occasionally be found among the normal male flowers; usually they occupy the centre position in the cluster of three. These hermaphrodite flowers are intermediate in size and characteristics between the male and female flowers. The female flowers can set fruit parthenocarpically (*i.e.*, without seeds). However, in practice, pollination is desirable to reduce natural fruit drop and improve fruit quality (size, even shape, and uniform skin colouration), especially of non-astringent cultivars.

Patterns of floral differentiation in the monoecious type Japanese persimmon cvs. Hanagosho and Kaki-yama-gaki were studied by Yonemori *et al.* (1993). In both the cultivars, the pistillate and staminate floral primordia started to differentiate in early June, and differentiation progressed until August, when the sepal primordia in pisti-llate flowers and petal primordia of staminate flowers had become evident. The buds then entered a quiscent, overwintering state. Sex differentiation in persimmon was associated with previous history of the current season's shoot. Male flowers were formed predominantly on shoots of a branch that had borne male flowers. Likewise, female flowers formed predominantly on branches of had produced pistillate flowers (Yonemori *et al.*, 1992; Yonemori *et al.*, 1993). A weak correlation could be discer-nible between thicker branches and female flowers, but the previous year's flower type and bud position was more influential than branch size in determining sex expression (Yonemori *et al.*, 1992).

In general, the persimmon cvs. Fuyu and Jiro produce only female flowers. However, shoots of individual trees of both cultivars were found bearing male flowers when they were top-worked onto other stocks. These results indicate that the shoots with male flowers are morphologically and enzymatically identical to Fuyu and Jiro, except that the bud sports are staminate-sporadic (Yakushiji *et al.*, 1995). The leaf carbohydrate level at the beginning of flower-bud initiation and N level in 1-year-old wood at the time of flower bud differentiation are important factors which determine the number and quality of flowers. To obtain the optimum yield of good quality fruit annually in cv. Maekawa Jiro, a leaf: fruit ratio of 40 is required, provided N and carbohydrate levels are also optimal (Ooshiro and Anma, 1998).

10.2.0 Pollination and Fruit Set

Cultivars of persimmon are known to have problems with low fruit set and fruit fall. Planting of adequate numbers of pollinizer trees are essential for adequate fruit set (George *et al.*, 1993; Piazza and Intoppa, 1996). The few cultivars which bear pisti-llate and staminate flowers are used as pollinizers. The most important pollinizers used are Gailey, Akagaki and Omiyawase. In Japan, two or three different pollinizers are included in the orchard to ensure that good overlap of flowering occurs. One pollinizer tree should be interplanted with every 8 to 10 trees of the commercial culti-vars. Cultivars recommended as pollinizers for cv. Fuyu include Johongsi, Nishimurawase and Zenjimar. However, it was concluded that local cultivars of Korea Republic A, B and C were most suitable pollinizers for Fuyu than the above mentioned ones (Kim *et al.*, 1996). Fruit set of cvs. Fuyu and Tanenashi was higher with cv. Gailey and Nishimurawase pollen than with non-pollination (Woodburn and Andersen, 1996). Persimmon cultivars fall into 2 groups regarding fruit development. Fruits of cultivars belonging to group I (including cvs. Tabaute, Pomelo, Rubi, Coracao de Boi, Fuyu, Jiro and Fuyuhana) develop in the same way with or without pollination, whereas fruits of group II (including cvs. Rama Forte, Giombo, Chocolate and Luiz de Queiroz) develop a darker flesh colour and lose some of their tannin content when they develop from pollinated flowers (Matos, 1997). Presence of bees must be ensured in the plantation to obtain a higher rate of pollination (George *et al.*, 1993; Piazza and Intoppa, 1996). Initial fruit set has been reported to be the highest in upper parts of the tree canopy, where exposure to irradiance was highest (George *et al.*, 1996). Yamada *et al.* (1987) noted that yearly fluctuations in fruit set were small in pollinated fruits, but large in unpollinated fruits (for example 9-84 per cent in Tenjingosho); these fluctuations occurred in parallel across all cultivars.

Fruit production of the cultivars Fuyu and Saijo is unstable unless they are pollinated. Yamamura *et al.* (1989) ascertained whether gibberellins could substitute for hand pollination. Fruitlets from flowers which were excluded from pollinators during a flowering period were sprayed with 200 mg/l GA_3 and a mixture of GA_4

Flower Types of Persimmon

Flower Buds

Male and Female Flower

American Persimmon (*D. virginiana*)

Diospyroskaki Flowers

'Fuyu' Persimmon Flower

and GA_7 (GA_{4+7}) about 10 days after full bloom. In both cultivars, GA_3 was more effective than GA_{4+7}, and GA_3 significantly increased seedless fruit set. Treatments with GA_3 at 25, 50 or 100 mg/l significantly reduced fruit drop in Saijo. However, even at 25 mg/l GA_3 tended to reduce flowering potential in the following year. In Fuyu, GA_3 was incapable of inducing sufficient fruit set without reducing flowering. Fruit growth was inhibited in seedless fruits of both cultivars, even if they were treated with GA_3. Kitajima *et al.* (1992) obtained different levels of seeded fruit set by (i) hand pollination, (ii) open-pollination, (iii) non-pollination, (iv) bark-ringing of fruiting shoots, and (v) GA_3 application (0-500 mg/l). There was a highly significant correlation between the number of seeds/fruit and the number of seedless fruits abscised. No seedless fruits abscised on ringed shoots until the girdle had healed. On non-pollinated trees seedless fruit set increased as the GA_3 concentration increased. However, more seedless fruits dropped from pollinated trees than non-pollinated trees even at the same GA_3 concentration, indicating a stronger competitive advantage of seeded fruits over parthenocarpic fruits. The effects of growth regulators (5 or 10 ppm BA or 10-100 ppm GA_3) applied during full bloom were investigated over 3 years (Roan and Ni, 1994). The highest fruit set in both cultivars was generally obtained with 100 ppm GA_3.

Mizobutsi *et al.* (1997) treated fifteen-year-old trees of 7 persimmon cultivars (Jiro, Giombo, Regina, Costata, Rubi, Rama Forte and Cereja) growing at Araponga, Minas Gerais, Brazil with 7.35 ml hydrogen cyanamide/l plus 8.0 ml mineral oil/l. Control trees were not sprayed. Spraying advanced budbreak and flowering and harvesting dates in all cultivars, but particularly in cultivars Jiro (by 10 weeks), Regina (by 10 weeks) and Cereja (by 9 weeks). Percentage fruit set was similar in sprayed and control trees except for cultivars Jiro and Regina where spraying increased fruit set by 33.8 per cent and 23.5 per cent respectively. Experiments were conducted in Korea to use water shoots as fruit-bearing mother branches on persimmon trees (Choi *et al.*, 1997). Fruit number increased significantly with increasing water shoot diameter. The highest percentage of fruit set was produced when water shoots were bent to a 45o angle. A total fruit weight of 1305 g/shoot was produced with water shoots compared with 839 g for trees with no water shoots. Fruit quality was not affected by the treatments.

Effects of gibberellic acid (GA_3) and artificial pollination on fruit set and growth of Fuyu persimmons were investigated by Choi *et al.* (1999). In 1996, fruit set was 20, 70 and 15 per cent for the GA_3-sprayed, pollinated and control trees, respectively. In 1997, fruit set was 67, 65 and 41 per cent for the GA_3-sprayed, pollinated and control trees, respectively. GA_3-treatment increased the proportion of seedless fruits; seed number was significantly decreased by the sprays. Wally *et al.* (1999) found that application of Biozyme (containing GA_3, IAA, zeatin and trace elements) + 25 ppm GA_3 on persimmon cv. Costata resulted in highest fruit set, retention and yield.

10.3.0 Biennial Bearing, Thinning and Cincturing

10.3.1 Biennial Bearing and Thinning

Biennial bearing may be a problem with some cultivars. Biennial bearing is related to crop load (Hodgson, 1939), seed production (Hodgson, 1939; Bargioni *et al.*, 1980), tree vigour or age (Kanaeko *et al.*, 1979), soil moisture (Griffith *et al.*, 1982) and pollination (Hague, 1911; Hume, 1914; Hodgson, 1939). Disbudding and flower thinning are used in the 'on' year to overcome the problem. In Japan, early fruit thinning within 20 days of full bloom has also been successful in reducing the problem and resulted in an increase in fruit size (Matsumoto and Kuroda, 1982). In New Zealand, it is recommended that one fruit be left on shoots less than 200 mm long and that longer shoots be thinned to two fruits, well spaced along the shoot (Jordan, 1985). Chemical thinning agents, such as NAA and ethephon have been tested in Japan, but are not used commercially (Sugimoto and Fukunaga, 1972; Yamamura *et al.*, 1976; Yamamura *et al.*, 1980).

10.3.2 Cincturing

In Japan, cincturing has been shown to increase precocity of bearing of young vigorous cultivars (Aoki *et al.*, 1977; Glucina, 1981). Cincturing is normally carried out by removing a strip of bark from around the trunk of the tree. The technique is performed during or just after the flowering period (Blumenfield, 1986). Cincturing may have the undesirable effect of retarding tree growth, particularly on cultivars like Izu which are not naturally vigorous (Douglas, 1980).

The effects of girdling trunks of 6- to 9-year-old Japanese persimmon (cv. Nishimura-wase) during early shoot functions were determined by Fumuro (1998). Trunks were girdled 10 cm above ground level in May (23 days before full bloom). Shoot growth inhibition continued into the next season but had little or no effect on yield and fruit quality in the two years. Growth inhibition was caused by a reduction of dry matter production due to the smaller number and areas of leaves and by inhibition of root functions, such as nutrient absorption and accumulation.

10.4.0 Fruit Growth and Development

10.4.1 Fruit Growth

Persimmon fruit exhibits a double sigmoidal growth pattern (Glucina and Toye, 1985) and fruit shows final swell during growth stage III (Nakano *et al.*, 1998). The two active fruit growth periods occur in mid-summer and the early autumn. The fruit has a large calyx which appears to be an important gas-exchange organ for the fruit (Nakamura, 1967). On the contrary, George *et al.* (1994b) reported that fruitsexhibited a normal sigmoidal growth pattern. Fruit weight in cv. Fuyu was poorly correlated with leaf area, leaf number per fruit, shoot length and shoot base diameter of fruiting shoots, indicating that these variables were not useful criteria for determining thinning levels or final fruit size. Incidence of the fruit disorder, calyx cavity, was lowest in the middle canopy. In contrast, average fruit weight, Brix, colour, texture and flavour were only slightly affected by canopy position. Export-quality fruits can therefore be harvested from any position within the tree canopy (George *et al.*, 1996).

Fruit growth can be inhibited by removing the calyx lobe at early growth stages (Kitagawa and Glucina, 1984). Sucrose is the major assimilate transported from leaves to developing persimmons (Zimmermann and Ziegler, 1975). The key enzyme for sugar metabolism in persimmon is thought to be acid invertase (Zheng and Sugiura, 1990). The high sucrose to hexoses ratios in fruit induced by calyx lobe removal might be due to a decline of acid invertase activity, since acid invertase activity was low in fruit in which the calyx lobes were removed at stages I and II (Hiruno *et al.*, 1995). Yonemori *et al.* (1995) recorded growth inhibition of persimmon fruit by calyx lobe removal in the middle of growth stage 1 (13 July) or at the beginning of stage II (24 August), but no inhibition was observed in the treatment at growth stage III (2 October). Lobe removal at stage I and II resulted in fruit weight reduction of 44 per cent and 76 per cent of the control respectively. Coincident with growth inhibition of the treated fruits, the ratio of sucrose to hexoses (glucose and fructose) increased significantly in the flesh, indicating an unsuitable gradient for further import of the sucrose into the sink organs. Growth inhibition was not related to changes in the activity of endogenous GA-like substances. However, a significant decrease of ABA content after calyx lobe removal at growth stage I or II may suggest its involvement in growth inhibition in persimmons. Growth and development of calyx-lobes-removed fruitlets of cv. Fuyu were retarded so that final fruit weight was only a third of the control (Yamada *et al.*, 2000). Plant growth regulators including auxin (NAA), cytokinin (BA) and gibberellin (GA_3), when applied singly or in combination 4 times at weekly intervals, could not alleviate the growth inhibitory effect of removal of calyx lobes. Sugar content at harvest was lower in the calyx-lobes-removed fruits.

Fruit respiration in persimmon increased rapidly with the onset of growth stage III and maintained a high level until harvest. It was also revealed that fruit growth was markedly reduced, when fruit respiration was inhibited by removal of calyx lobes and sealing the scars with 'vaseline' (Yonemori *et al.*, 1996). The inhibition of fruit respiration seemed to reduce the sink strength of the fruit, which in turn suppressed fruit growth. It was, therefore, concluded that the high respiration during growth stage III in persimmon fruit is important for maintaining sink strength for the final swell of fruit (Nakano *et al.*, 1998). It has been suggested that ethylene is not involved in the respiration rise and that this respiration rise is different from that in climacteric fruits (Nakano, 1999).

Taira and Itamura (1989) reported that flesh firmness and soluble tannin content peaked in late June and July, respectively and then declined, while the total sugar content increased throughout the development. The time of initiation of peel coloura-tion and the rate of colouration varied considerably with the year, resulting in harves-ting dates varying from mid-October to early November.

The effects of controlled temperature during growth stage III of cv. Hiratanenashi at 14, 22 and 30°C on growth and ripening were reported by Sugiura *et al.* (1991). Fruit growth and ripening were greatly accelarated at 22°C compared with control fruits grown at ambient air temperature (range 9.3-28.5°C). At harvest (30 days after treatment), fruits kept at 22°C were much heavier, more deeply coloured and softer than the controls. In contrast, keeping the fruit at 30°C delayed the onset of rapid fruit expansion and ripening. At harvest, however, the fruits, though still green, were about the same size of the controls. With the exception of rapid chlorophyll degradation, keeping the fruit at 14°C had little effect on fruit growth and ripening.

Little difference in sugar content and composition was found between different temperature treatments and controls.

Fruit growth and development in persimmon can be regulated by the application of bioregulators. The growth regulator N-2-chloro-4-pyridyl-N′-phenylurea (KT-30) at 10 to 20 ppm improved percentage of fruit set over control (40 per cent) with 20 ppm producing the best fruit set (>80 per cent). The effect of KT-30 on fruit quality characters varied depending on the cultivar (Hasegawa *et al.*, 1991b). Foliar application of paclobu-trazol (2 g/l) to the terminal 4 nodes of each shoot shortly before and again after anthesis significantly increased fruit weight but did not increase fruit set (George *et al.*, 1995a). Synthetic cytokinins controlled persimmon fruit shape, size and quality (Itai *et al.*, 1995; Sugiyama and Yamaki, 1995). Pre-flowering application of N-2-chloro-4-pyridyl-N′-phenylurea (100 ppm) on cvs. Tonewase and Saijo produced flatter and heavier fruits by promoting transverse growth than other treatments, which was followed by thidiazuron at 100 ppm (Itai *et al.*, 1995). The fruits also had lower sugar concentrations, especially of reducing sugar, than control fruits, because of retarded ripening process. Spraying of CPPU [forchlorfenuron] at 5-100 mg/l increased fruit set, suppressed the sunken fruit apex and delayed fruit coloration (Sugiyama and Yamaki, 1995).

Trees in an orchard at Tsukuba were sprayed with 50 and 250 mg/l *n*-propyldihydro-jasmonate (PDJ) at 131, 138 and 145 days after full bloom (DAFB). Harvest date was advanced between 5 and 15 days, fruits were soft and peel colouring was enhanced by treatment with PDJ. The best treatments were 250 mg PDJ/l at 131 or 138 DAFB and 50 mg PDJ/l at 131 or 145 DAFB. ABA levels were lower in tree and fruits than in control (Gemma *et al.*, 1997).

The plant growth regulators paclobutrazol (PBZ) and gibberellic acid (GA_3) were applied to orchards of cv. Triumph in Israel in order to extend the marketing season of the crop (Ben-Arie *et al.*, 1997). Trees were either drenched with 2.0 g PBZ/tree or sprayed with 0.5, 1.0 and 2.0 g PBZ/tree and with 50 mg GA_3/l. Paclobutrazol, when applied to the roots in the spring, advanced ripening by 2-3 weeks without any detrimental effect on fruit size but it accelerated post-harvest deterioration of the fruit. GA_3, when sprayed at least 10 days prior to harvest, delayed fruit ripening on the tree and also reduced post-harvest deterioration. The post-harvest deterioration of fruits by PBZ could be inhibited by subsequent treatment with GA_3 thus allowing rapid ripening with extended shelf-life. There was a minimum period of one week after GA_3 application before harvesting PBZ-treated fruits, to maintain fruit firmness and not cause a delay in fruit coloration.

Embryo-sac development and fertilization processes of Japanese persimmon cv. Fuyu were histologically investigated by Kitajima *et al.* (1993) to clarify the cause of variations in seed production. The percentage of abnormal embryo sacs was low in trees which produced many seeds per fruit and, conversely, it was high in trees which produced a few seeds per fruit. It was concluded that the variations in seed formation among the trees probably resulted from differences in the frequency of abnormal embryo sacs.

10.4.2 Fruit Drop

Severe fruit drop may occur with some cultivars (Kaneko *et al.*, 1979; Bargioni *et al.*, 1980). Fruit drop may be related to a number of causes including excessive fruit set, lack of pollination, water stress, excessive nitrogen application and insect damage (Nagasawa *et al.*, 1968; Kanaeko *et al.*, 1979; Bargioni *et al.*, 1980; George and Nissen, 1984). Effect of pollinizer on fruit drop of cv. Fuyu persimmon was evaluated by Lee *et al.* (1998). Fruit drop rates with the pollinizers *D. lotus*, wild persimmon, Nishimurawase, Mikado, Bang Gosho, Zenjimaru and control (no pollinizer) were 77.3, 17.1, 22.9, 32.0, 34.0, 13.4 and 65.9 per cent, respectively. A negative correlation was observed between seed number and fruit drop rate. It was concluded that Fuyu persimmon required a pollinizer in order to reduce fruit drop rate and wild persimmon was considered suitable. In a study of 17 cultivars in which flowers were hand-pollinated or bagged to protect against cross-pollination, fruit drop was confined to the early developmental stages in all except Okitsu 15, which showed a heavy late drop (Yamada *et al.*, 1987).

In Israel, spray application of gibberellic acid at full bloom has been shown to reduce fruitlet abscission (Blumenfield, 1986). Tape containing 0.1 per cent IBA, applied to the apices of Hiratanenashi fruits, 10 days after flowering, reduced abscission during the first 35 days after flowering from 37 per cent to 19 per cent (Maotani *et al.*, 1989). A single spray of GA_3, L-canaline or IAA reduced fruit drop to 6.7, 33.3 and 60 per cent respectively compared with 95.4 per cent for the control, but combination of GA_3 and IAA did not reduce fruit drop (Yakushiji and Hase, 1991).

Suzumi *et al.* (1988) also reported the abscission rate of persimmons increased after desiccation or flooding of the soil and defoliation or shading of the trees. Fruits having 3 or more seeds showed delayed abscission. Abscission was inhibited by applying 500 ppm NAA in lanolin to the apex of the fruits, but NAA applied to the terminal bud promoted abscission. They also observed that pre-bloom spraying of AVG [aminoethoxy vinyl glycine] at 500 ppm prevented fruit drop (Suzulsi *et al.*, 1988a). Fruit drop of sweet persimmon in Luotian county in China, a leading persimmon growing area, could be controlled by spraying 2 per cent borax during flowering period (Xu *et al.*, 2000).

The effect of girdling and different concentrations of growth regulators on fruit drop was investigated in persimmon cv. Costata (El-Shaikh *et al.*, 1999). Gibberellic acid (GA_3) at 20 ppm plus girdling at full flowering decreased fruit drop and consequently increased yield. Girdling treatment alone during full flowering significantly increased final yield, while growth regulator application plus girdling gave the best yield compared with control trees. Girdling significantly improved fruit weight and size, while total soluble solids (TSS) level was only slightly affected.

10.4.3 Parthenocarpy

Parthenocarpic fruit set in cvs. Fuyu and Hiratanenashi depended not only on position but also on the presence of a seeded fruit on the same shoot. The variations in setting patterns of parthenocarpic fruits were similar in both the cultivars (Kitajima *et al.*, 1993a). It was concluded that seeded and parthenocarpic fruits on different shoots compete for assimilates, beginning in the latter half of fruit growth stage I, but that it might be possible to produce good quality, seedless Fuyu fruits under ordinary orchard conditions, provided pollination was prevented for all flowers (Kitajima *et al.*, 1993b). Increases in IAA and GA-like substances in the fruitlet of parthenocarpic persimmons during flowering may result in continued growth, but the interaction between elevated ABA and reduced IAA and GA-like substances in the fruitlet may induce its absci-ssion (Kojima *et al.*, 1999). Applications of GA_3 and IAA induced parthenocarpic fruit set in *D. lotus*. Maturation of seedless fruits was earlier than that of the seeded fruits. GA_3 caused a decrease in the fresh mass and size of fruits, while IAA induced an increase in the fresh mass and size in parthenocarpic fruits (Kadioglu and Atalay, 1999).

11.0 Pests, Diseases and Disorders

11.1.0 Pests

Insect pests of persimmons vary considerably in different regions of the world. The major pests include mealy bugs, fruit flies, thrips, spotting bugs and twig girdlers and borers (George and Nissen, 1984; Kitagawa and Glucina, 1984; Miller, 1984). Of all the fruit trees grown in Korea, persimmon ranks the second in terms of acreage. Diseases and insect pests are important factors affecting persimmon production (Son *et al.*, 2009). Questionnaires to sweet persimmon growers indicate that the most serious insect pests is stink bugs (73 per cent of the answers) followed by scale insects (19 per cent). Most growers spray chemicals six to nine times a year to control those pests. Among scale insects, the grape-myrtle scale *Eriococcus lagerstroemiae*, horned wax scale *Ceroplastes pseudoceriferus*, and elongated cottony scale *Phenacoccus aceris* are of major concern. However, they seldom become serious in commercial orchards because of insecticide sprays. Machine oil spray during winter is necessary against the scales. Three species of stink bugs are important: brown marmorated stink bug *Halyomorpha halys*, brown-winged green bug *Plautia stali*, and bean bug *Riptortus clavatus*. They feed on fruits making them unmarketable. A Japanese gall-forming thrips, *Ponticulothrips diospyrosi*, found for the first time in Korea in 2000, is spreading throughout the southern part of Korea. Acreage of persimmon orchard damaged by the thrips was 446 ha in 2005. It feeds on young fruits and leaves of both sweet and astringent persimmons. There are other minor insect pests such as smaller clearwing moth *Synanthedon tenuis* and two-spotted spider mite *Tetranychus urticae* (Son *et al.*, 2009).

11.1.1 *Eriococcus lagerstroemiae*

The first occurrence of the larval stage is reported to be from late June to early July; the second from mid- to late-August and the third from mid- to late-September. Each adult female lays on an average 229.3 eggs. The larvae begin emerging from late April and there are three generations/year. Overwintering eggs are sheltered in bark with atleast 28.7 per cent alive. Two applications of lime sulphur [calcium polysulfide] and machine oil in late June and late August gave 97.8 per cent and 96.8 per cent control, respectively (Kwon *et al.*, 1995).

11.1.2 Hemipterous Insects

Pentatomid yellow brown stink bug (*Halyomorpha halys*) and coreid bug (*Riptortus clavatus*).

Symptoms of damage to persimmon fruits by Hemiptera include concave surfaces or a dark blue colour like black-eye. Fruit surface on cutting reveals spongy structure. In severe attack the fruit becomes soft and red like ripe fruit. Most serious damage occurs from late August to late October. The species *H. halys* shows peak occurrence in early August, and *R. clavatus* in late August. The *H. halys* can be controlled by spraying ethofenprox or henthoate in mid- or late-August, while deltamethrin or fenthion are effective against *R. clavatus* (Chung *et al.*, 1998).

Orthotydeus californicus and *O. caudatus* (tydeid mite) occasionally contaminate persimmon fruit. Field trials were conducted in New Zealand on persimmon to evaluate avermectin, mancozeb, mineral oil, permethrin plus pirimiphos-methyl, pirimiphos-methyl, propargite or tau-fluvalinate to reduce the pest populations (Tomkins *et al.*, 1997). The most effective treatments, early season applications of mineral oil or avermectin, showed promise for use in a pest management system being developed for export persimmons.

In recent years, an increase in the damage from mealybugs (*Pseudococcus citriculus*) in persimmon fruits was observed (Izhar, 1999). The mealybug settles under the sepal and the connection between fruit and leaves. It sucks the fruits and the honeydew causes black knots. Ants are the main transferring factor of the mealybugs and they protect them against predators and parasites. Dorsan (chlorpyrifos) 5 per cent and diazinon 4 per cent applied around the trunk gave excellent results against ants.

11.2.0 Diseases

11.2.1 Crown Gall (*Agrobacterium radiobacter* var. *tumefaciens*)

Symptoms are large galls or swellings at the crown often with smaller, nut-sized galls on large roots. Control involves not planting affected nursery trees and not using *D. lotus* as a rootstock as this species may be more susceptible to crown gall.

11.2.2 Dematophora Root Rot (*Dematophora necatrix*)

In Israel, Dematophora root rot is a major cause of tree loss. *Diospyros kaki* is less susceptible than *D. virginiana* (Szteinberg and Jabareen, 1986). Vegetative propagation of resistant rootstocks is recommended to maintain resistance.

11.2.3 Cercospora Leaf Spot (*Cercospora kaki*)

Symptoms are small, dark brown, angular spots on leaves. Some cultivars are more resistant than others.

11.2.4 Circular Leaf Spot (*Mycospharella nawae*)

Symptoms are small, round, brown spots on the leaves.

Pests of Persimmon

Stink Bug Infestation

Scale Insects Infestation

Long Tailed Mealybug Infestation

Diseases of Persimmon

Cercospora Leaf Spot Infection

Circular Leaf Spot Infection

Bitter Rot or Anthracnose Infection

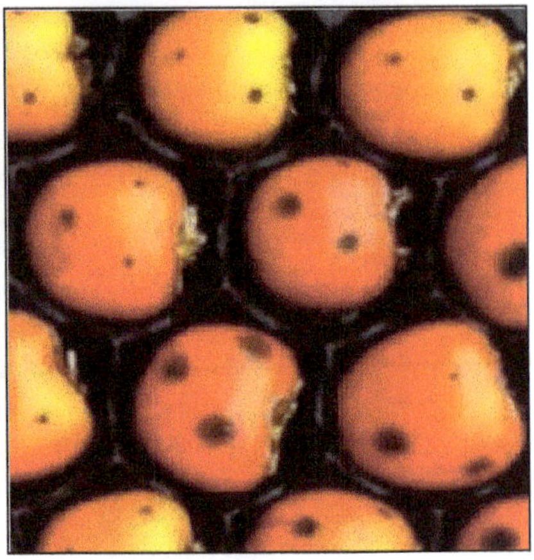

Fruit Rot Infection

Crown Gall Infection

11.2.5 Bitter Rot or Anthracnose (*Gloesporium kaki*)

Anthracnose infects both fruits and shoots, but large differences in resistance occur between cultivars. For example, Fuyu and Izu are very susceptible to the disease. When anthracnose infects the fruit it causes the latter mature rapidly and to drop.

Persimmons are also susceptible to a number of disorders which appears on the fruit surface. Black stains are classified according to the pattern of their visual symptoms. Protuberance and mist stains, referred to as dotted stains, are caused by fungus species: *Heptothyrium pomi* fly speck, *Glomerella* and *Alternaria*. Spots attributed to *Alternaria* sp. are smaller and flatter than those caused by *Glomerella* which form small, scattered protuberances 1-2 mm in diameter. Both types of spotting can be distinguished from the small, black, sunken spots left by copper sprays. Other types of black stains are described as cloud or line disorders. Line stains are thought to be caused by expansion of the fine splits on the fruit skin which arise because of unequal growth of the internal and external tissues during fruit enlargement. Oxidation of polyphenolic compounds in cells in the cracked region or entry of surface organisms in water cause the cracks to blacken. Successful physiological (application of gibberellic acid to delay the onset of ripening) and physical (controlled atmosphere storage) measures to control black spot disease of persimmon caused by *Alternaria alternata* was reported by Ben-Arie *et al.* (1992). *Glomerella cingulata* is prevalent on persi-mmon nursery stock and trees at Quxian county, China. The disease becomes prevalent during late April to early May and lasts until fruit maturity. Heavy rainfall and long rainy days cause severe infection but high air temperature and low humidity do not promote infection. The disease was controlled by spraying with Bordeaux mixture, chlorothalonil or mancozeb (Jia *et al.*, 1997).

Three treatments with GA_3 (20 µg/ml) applied during fruit development at 30, 20 and 10 days before harvest, reduced decay due to black spot disease caused by *A. alternata*. GA_3 apparently affects decay development by enhancing resistance of the fruits (Perez *et al.*, 1995). Modified atmosphere packaging of persimmon fruits resulted in the accumulation of acetaldehyde to a level of 80 µg/ml, ethanol to level of 900 µg/ml and CO_2 up to 30 per cent and storing of fruits in this atmosphere at –1°C for 4 months reduced the incidence of black spot disease caused by *A. alternata*. It is suggested that the increasing concentration of O_2 during storage is the principal factor in the inhibition of black spot disease development (Prusky *et al.*, 1997).

Botrytis cinerea was identified as the cause of leaf blight of persimmon in Taiwan and vinclozolin, eprodione and vinclozolin + metiram effectively controlled leaf blight in field trials (Tung *et al.*, 1994). Among *Diospyros kaki* cultivars grown in the Black Sea coast area of Russia, Seedless, Dzhiro, Khachia and Zendzhi-maru were rela-tively resistant to *Botrytis diospyri* (1-4 per cent infection). Seedless, the most resistant, also gave the highest fruit yields (up to 20.9 t/ha) during the study period (Ostasheva and Omarov, 1988).

11.2.6 Post-harvest Diseases

The most serious post-harvest disease is *Penicillium* (Kitagawa and Glucina, 1984). However, *Penicillium* can not invade the fruit unless an injury occurs to the skin. In general, damage caused by fruit softening during transportation and storage is a more serious problem than damage from decay. Experimental results suggest that hot water treatment (47°C for 90 or 120 min, 50°C for 20 or 30 min and 54°C for 20 min) shows potential for both disinfestation and maintaining quality of fruits during cold storage (Lay-Yee *et al.*, 1997).

Lee *et al.* (1997) studied the effects of preharvest application of MGC-140 (mixture of N, P, K, Mn, B, ethylene, diamine, acetic acid, Fe, $CuSO_4$, $ZnSO_4$ and $(NH_4)2MoO_4$; 800, 1600 or 3200 mg/l) and GA_3 (50 mg/l) on persimmon cv. Fuyu fruit quality during cold storage at 1°C for 114 or 169 days. It was observed that GA_3-treated fruits were the mos firm, regardless of storage period, but the TSS content tended to decline with storage period. The content of total pigments present in stored fruit peels was lowest in GA_3 treated fruits. The content of cryptoxanthin decreased but that of zeaxanthin increased with storage period.

Holding of fruits in 99 per cent CO_2 for 20 h at 20°C followed by holding them in a normal atmosphere at 20°C for 3 days resulted into drastic reduction in tannin concentration in cvs. Thiene and Lycopersicon, but remained the same in cv. Kaki Tipo fruits. In general, CO_2-treated fruits were slightly less firm than control fruits and CO_2 treatment had no effect on pulp colour (Vidrih *et al.*, 1994a). Ishimaru *et al.* (1998) reported that fruits of cv. Tonewase exposed to ambient air or 100 per cent CO_2 for 24 h softened equally fast and ripened within 6 days. Slightly immature fruits exposed to air or 100 per cent CO_2 for 24 h ripened as above, whereas those exposed to 100 per cent CO_2 for 17 h before CO_2 concentration was reduced were still firm after 2 weeks in storage.

11.3.0 Disorders

11.3.1 Calyx Cavity

Calyx cavity (syn. dehiscence) can be a serious problem in persimmons. Symptom is a space or cavity that occurs directly beneath the calyx of the fruit. The cavity becomes a habitat for mealy bugs and fungal growth. Some cultivars are more susceptible than others. The incidence of calyx cavity appears to be less on trees which have heavier crop loads and where fruits have been pollinated (George and Nissen, 1985; Jordan, 1985). Control measures recommended in New Zealand include the avoidance of excessive N and K fertilization, especially in late spring/summer and close to harvest; thinning early in the season to enhance calyx growth; optimising pollination toproduce more than 3 seed/fruit (Glucina, 1987). Sites with deep, fertile and free-draining soils are likely to encourage the disorder, as are areas with high autumn rainfall. Planting of trees on preformed ridges or mounds or into plastic is suggested (Glucina, 1987).

11.3.2 Skin Russeting

Causes are not totally known, but may be due to excessive nitrogen and high relative humidity during fruit maturation (Sugimoto and Yasui, 1982).

Disorder of Persimmon

Calyx Cavity

11.3.3 Calyx-end Cracking

The problem may occur with some cultivars. Causes are not known. However, there was a positive correlation between the severity of cracking and seed number is some genotypes. Generally, cracking was minimal in fruits with no seeds, regardless of whether flowers were bagged or open pollinated. Some genotypes also showed a correlation between fruit weight and cracking (Yamada *et al.*, 1991).

11.3.4 Sujika

Sujika, a disorder producing white lines on the rind of fruit was studied by Kimura *et al.* (1989). Sujika was more frequent on light-fruiting than on heavy-fruiting trees and was not correlated with numbers of seeds/fruit. The rate of sujika was temporarily reduced about 10 days after ringing in mid- or late-June; ringing in mid-June was particularly effective. When fruits still attached to the tree were covered with plastic film the incidence of sujika increased to 70 per cent, compared with 40 per cent for the controls. A night temperature of 18-20°C produced more sujika than one at 15°C. Sujika is concluded to be closely related to fruit enlargement.

12.0 Harvesting and Yield

12.1.0 Harvesting

Persimmons mature between early autumn-winter depending on the cultivar. Harvesting for any one cultivar extends over several weeks with usually 5 to 6 picks necessary to ensure optimum maturity at each harvest time. Fruits are harvested when they are well developed, fully coloured and sugar levels have exceeded 14o Brix (preferably over 16o Brix). Brix levels can be checked using a hand-held refractometer. In Japan, colour charts are used as guides to determine when fruits have reached maturity and can be harvested (Yamazaki *et al.*, 1981; Suzuki *et al.*, 1981). Persi-mmon fruits show 2 types of post-harvest skin discolouration, blackening and browning, of which skin browning (chocolate coloration) is a more severe storage disorder since affected fruits eventually loose their firmness and flavour (Lee *et al.*, 1993; Shin *et al.*, 1994). If fruits are harvested too early they will lack colour, sweetness and flavour. If left on the tree past the optimum picking date, they will have a very short shelf-life. Fruits of astringent cultivars are inedible at harvest and must be treated to remove astringency. Pekmezci *et al.* (1998) concluded that harvesting of cvs. Hachiya and Fuyu, grown in Turkey, was more suitable in mid-October than in September.

Fruit maturing time (FMT), fruit weight (FW) and soluble solids content (SSC) were evaluated by Yamada *et al.* (1995a) in 15 persimmon cultivars at two locations in Japan and China. FMT and FW were significantly affected by location and cultivar, whereas SSC was not affected by location. Coefficients of correlation between the 2 locations were 0.79**, 0.89** and 0.52* for FMT, FW and SSC respectively. The cultivars evaluated at Akitsu, Japan on average, matured 18 days later and fruits weighed 70 g more than did those at Meixian, China. They further reported the differences in the mean and range of FMT and FW were small between the cultivar groups of the two origins. The difference in the mean SSC was also small, although the range of SSC was larger in the Chinese than in the Japanese cultivars (Yamada *et al.*, 1995b).

Fruits are harvested by clipping, leaving the calyx and a short stem attached to the fruit. The fruits should be handled with care so as to avoid bruishing. Fruits are best placed in padded plastic crates or polystyrene containers for transport to the packing shed. Persimmons are very delicate fruits which blemish easily. Dark blemishes are highlighted to well-coloured fruit. Machine polishing is not normally done.

12.2.0 Yield

Dwarf and semi-dwarf cultivars are precocious bearers, and begin cropping 2 to 3 years after planting out. Mature trees of dwarf and semi-dwarf cultivars, such as Fuyu are capable of producing 49 kg of fruit per tree (30 t/ha). Yield appears to be highly correlated with total hours of sunshine (Kanaeko, 1977). In Israel, a combination of both cincturing and spray application of gibberellic acid at full bloom increase yield between 50 and 500 per cent (Blumenfield, 1986).

13.0 Post-harvest Management

Fumigation of fruits with methyl bromide at 30 or 40 g/m3 for 2 hours delayed softe-ning and extended shelf-life for 2 weeks when stored at 20°C (Kawakami *et al.*, 1989). The effects of defoliation and foliar application of gibberellic acid (GA) on the softening of Japanese persimmon cv. Hiratanenashi fruits exposed to ethanol fumes for 3-4 days after harvest were investigated by Itamura *et al.* (1994). Fruits harvested from a tree that had shed leaves early due to declining vigour softened 5-10 days before those from a healthy tree, whereas fruits from a tree growing in a plastic greenhouse, which shed leaves even later in the season, were the slowest to soften. Mature fruit on shoots which were defoliated on 21 July produced ethylene and softened rapidly upon harvest. Spraying bearing shoots on 7 September with 50 ppm GA had no effect on ethylene production, but the treatment did inhibit fruit

Fruit Bearing of Persimmon

Fruit Tree

Persimmon Orchard

Unripe Fruit

Ripe Fruit

Ripe Fruit on Tree

Harvesting and Drying of Persimmon

Hand Picking

Harvesting with Fruit Picker

Harvesting in Orchard

Natural Sun Drying at Home

softening. These results suggested that leaves on bearing shoots were important in the regulation of flesh softening of harvested, mature, ethanol-treated fruits and it was proposed that GA, originating in the leaves and transported to the fruits, inhibited fruit-softening by blocking ethylene action. Changes in 20 physiological and biochemical post-harvest parameters of persimmon cv. Huoshi fruits were analyzed by Tian *et al.* (1994) using biomathematical methods. The results indicated that firmness was positively linearly correlated with TSS, crude cellulose, tannin, protopectin, GA and cytokinin contents; negatively linearly correlated with membrane permeability and ethylene and ABA contents; and not linearly correlated with PG (polygalacturonase) or IAA contents, pH or respiration.

Physiological changes during sun drying of fruit after peeling were recorded by Taira *et al.* (1988). In the astringent cultivars, Hiratanenashi and Yamagatabenigaki, astringency was lost in 10 and 20 days, respectively, the difference probably being due to differences in acetaldehyde accumulation. Soluble tannin content decreased with acetaldehyde accumulation in both cultivars. The non-astringent cv. Jiro, which had no soluble tannins, also accumulated ethanol and acetaldehyde during drying. Total sugar content increased during drying although sucrose decreased and was absent by 40 days.

Experimental evidences suggest that preharvest spray of bioregulators can regulate the quality and shelf-life of fruits. Pre-harvest sprays of gibberellic acid (Kitagawa *et al.*, 1966; Yasunobu, 1971; Testoni, 1981; Ben-Aire *et al.*, 1986) and post-harvest fungicidal dips may be beneficial in extending storage life.

Lee and Toshiaki (1991) treated persimmon cv. Fuyu fruits treated with 100 ppm GA_3, 2000 ppm daminozide, 100 ppm GA_3 + 1000 ppm daminozide, 50 ppm GA_3 + 2000 ppm daminozide, or 2500 ppm CPTA [2-(4-chlorophenylthio)-triethylamine hydrochloride] at 32-39 days before harvest, then harvested and stored at low temperature. Treatment with 100 ppm GA_3 or 2500 ppm CPTA gave the greatest fruit firmness at harvest but firmness decreased more during storage. The highest content of reducing sugars at harvest was with 100 ppm GA_3 +1000 ppm daminozide, while non-reducing sugars were highest with CPTA or 100 ppm GA_3. During storage, reducing sugars increased while non-reducing sugars decreased. Apart from the decrease in firmness during storage, CPTA was the most satisfactory treatment. The maximum storage life of GA_3-treated persimmon fruits (cv. Triumph) in modified atmosphere (MA) packs of 0.08 mm low-density polyethylene was 8 weeks when packed under vacuum, and 20 weeks when packed under 100 per cent N2 (Ben-Arie *et al.*, 1991). The fruits were harvested from trees which had been sprayed with 50 ppm GA_3 10-14 days before harvest. The storage life could be extended to 7 months by storing GA_3-treated fruit in air at –1°C for 3-4 months. The accumulation of ethanol and acetaldehyde in the MA packs after several months' storage caused off-flavours to develop and brown discoloration of the peel and flesh.

13.1.0 Ripening, Storage and Astringency Removal

13.1.1 Ripening

Ripening of persimmon fruits at 18-21°C using ethylene at 10 µl/l for 24 hours in 85 to 90 per cent RH was recommended by Wills *et al.* (1989). Foliar sprays of ethrel

(25 mg/l) at the beginning of stage II hastened fruit maturation, reduced endogenous GA-like activity and increased fruit coloration, softening, sugar accumulation and the contents of reducing sugars and endogenous ABA (Zheng *et al.*, 1991). Ethylene is thought to be an activator of the fruit softening mechanism in persimmon cv. Hiratanenashi fruits. Ethylene evolution either preceded or coincided with a rise in respiration rate, independent of fruit maturity (Itamura *et al.*, 1991). Ethylene evolution and internal ethylene concentration in fruits were highly correlated, regardless of changes in the weather (Suzuki *et al.*, 1987). Treatment of fruits of cv. Fuyugani at a late stage of development (1 month before ripening) with 100 ppm ethephon promoted pigmentation of the fruits and reduced firmness, while GA_3 at 100 ppm delayed the pigmentation process and maintained or increased fruit firmness (Zhang, 1988). The effects on ripening of ethephon at 100 ppm, daminozide at 2000 ppm + GA_3 at 50 ppm and paclobutrazol at 1000 ppm were studied by Lee and Kim (1991). Fruit firmness gradually decreased and sugar concentration increased up to 32 days after treatment (and in the untreated controls), with ethephon producing the softest fruits. The firmest fruits were obtained with daminozide + GA_3. The only treatment to significantly reduce sugar concentration was paclobutrazol. Persimmon fruits treated with 2,4-D at 2000 ppm and benlate at 3000 ppm ripened after 11 and 9 days respectively (Aki *et al.*, 1994). Tian *et al.* (1994) concluded that ripening was depended on the synergistic effects of abscisic acid (ABA) and ethylene, especially ABA. Experimental evidences indicate that polygalacturonase is not responsible for the metabolism of pectins during persimmon fruit ripening (Cutillas-Iturralde *et al.*, 1993).

Changes in cell wall components during ripening and softening of persimmon fruits were thoroughly analyzed by Kang *et al.* (1998a, 1998b and 1998c). During ripening and softening of fruits of cv. Fuyu the content materials soluble in phenol:acetic acid:distilled water increased, and residual cell wall material decreased. In fractions soluble in phenol: acetic acid:distilled water high molecular uronic acid polymer (UAP) of >73 kDa increased during ripening from green stage to maturity. Degradation of high molecular UAP occurred during ripening and softening in the 0.05 M CDTA-soluble fraction. They also reported that cell wall hydrolase activity increased significantly in mature fruits during ethylene treatment. Ethylene biosynthesis was influenced by external spermidine (Spd) (Gao and Zhang, 1999). Ethylene produc-tion decreased in fruits treated with Spd at 10 mM; in such fruits softening was delayed. Ethylene production increased in fruits treated with Spd at 1 and 0.1 mM, and firmness declined. This increase in ethylene production was inhibited by CoCl2 (an inhibitor of ethylene biosynthesis).

ABA. Experimental evidences indicate that polygalacturonase is not responsible for the metabolism of pectins during persimmon fruit ripening (Cutillas-Iturralde *et al.*, 1993).

Changes in cell wall components during ripening and softening of persimmon fruits were thoroughly analyzed by Kang *et al.* (1998a, 1998b and 1998c). During ripening and softening of fruits of cv. Fuyu the content materials soluble in phenol:acetic acid:distilled water increased, and residual cell wall material decreased. In fractions soluble in phenol: acetic acid:distilled water high molecular

Sorting, Grading and Packaging of Persimmon

uronic acid polymer (UAP) of >73 kDa increased during ripening from green stage to maturity. Degradation of high molecular UAP occurred during ripening and softening in the 0.05 M CDTA-soluble fraction. They also reported that cell wall hydrolase activity increased significantly in mature fruits during ethylene treatment. Ethylene biosynthesis was influenced by external spermidine (Spd) (Gao and Zhang, 1999). Ethylene produc-tion decreased in fruits treated with Spd at 10 mM; in such fruits softening was delayed. Ethylene production increased in fruits treated with Spd at 1 and 0.1 mM, and firmness declined. This increase in ethylene production was inhibited by $CoCl_2$ (an inhibitor of ethylene biosynthesis).

13.1.2 Storage

Firmness of persimmon fruits stored at room temperature (10-20°C) decreased linearly during storage and reached a minimum at 4 weeks. Fruit crude fibre, propectin, tannin and protein contents, and pectin-esterase activity decreased with softening, but soluble pectin content and cellulase and polygalacturonase activities increased. Ethylene evolution and respiration rate attained peak after five weeks of storage which was followed by complete softening of fruits (Tian *et al.*, 1991). After removing astringency persimmon fruits cv. Tonewase grown under cover softened more rapidly (at 20°C) than fruits grown in the field. The concentration of total hemicellulose in the cell walls decreased during softening of fruits grown in the open (Itamura *et al.*, 1995).

Bejanidze *et al.* (2019) studied chemical composition, the regularity in the quantity changes of the bioactive substances and technical-economical persimmon parameters of the persimmon fruit during its storage and technological processing; to develop a highly profitable and efficient composite production technology for a new assortment of products and high quality beverages. Organoleptic and chemical tests of 12 persimmon fruit varieties were made. The content and composition of sugars, polyphenols, pectin, and tanning substances, were defined; and the relation of these parameters to the fruit maturity degree, the dynamics of their change in the ripening and storage process were determined. In addition, we determined the content of sugars (inverted, total, saccharose), tanning substances (free, total), polyphenols and nitrogen through a spectrophotometric method, dry solids through a refractometric method, pectin (total, soluble, protopectin) and acidity through a potentiometric method, and cellulose through a chlorite method. The feasibility of the persimmon fruit industrial processing was established. Persimmon fruit of Khachia and Khiakume varieties were proved to be the best raw material for the industrial production of a fruit alcohol and beverages. It should be noted here that persimmon fruit must be harvested in November when they have a dark orange colouring and their dry solids content reaches 18 per cent, inverted sugars make up 16-16.5 per cent, and the acidity is 0.12 per cent. For concentrate production, persimmon must be harvested in October when the fruit are yellow and their tanning and colouring substances reach 1.24 per cent, and pectin substances make up 0.77 per cent. The fruit sugars are glucose and fructose, generally in equal quantities. Tanning substances, leicoantocianes, in particular, are responsible for the tart taste; the lower their content is, the less the tart taste is felt, moreover, ripe fruit are less tart in flavour which is due to the tanning substances getting bound with high molecular substances. All the 12 persimmon varieties cultivated in Georgia can be used as raw material, including substandard and overripe fruit that are not marketable and do not have a long shelf life (Bejanidze *et al.* (2019).

Researches in Japan and Italy have shown that some non-astringent persimmon cultivars, such as Fuyu can be cool-stored at 0°C for up to 2 months (Testoni *et al.*, 1981; Kitagawa and Glucina, 1984). In New Zealand (Anon., 1987), however, it has been shown that the cool storage-life of Fuyu persimmons may be less than 1 month at 0°C because of increased risk of chilling with longer storage period. Temperatures between 4o and 5°C cause more severe chilling damage than that of 9°C. Some cultivars may exhibit internal breakdown and post-harvest diseases (Guelfat-Reich and Ben Aire, 1981). Fruits stored at 18°C were soft and overipe after 2 weeks, while fruits stored at 1°C changed little before 6 weeks (Lyon *et al.*, 1992). On the basis of experimental results, it is recommended that Florida Fuyu persimmons are stored at 68oF if rapid ripening is required, or at 32oF for fruits to be shipped to distant markets (Sargent *et al.*, 1993). Collins and Tisdell (1996) suggested that storage losses might be reduced by not attempting to store certain fruits, based on initial measurements of colour and mass. Cold storage of fruits at 1°C and 90 per cent RH for 80 days resulted into decrease in TSS, total sugar and vitamin C contents, while invert sugar decreased initially and then increased. Carotenoid content of fruits increased during storage. Firmness of all fruits decreased during storage with fruits losing their marketable quality after 40 to 60 days (Turk, 1993). Total carotenoid content initially increased then decreased after storage for 2 months at 4 or 10°C (Yang, 1996a).

Controlled atmosphere storage enables some cultivars, such as Fuyu to be stored for 5-6 months (Kitagawa and Glucina, 1984). It has been suggested that ripening may be delayed by CA storage (Yang, 1996b). Total carotenoids content increased up to 158 per cent of the initial value after 4 months of CA storage (Yang, 1996a). Persimmon fruits cv. Fuyu could be stored for 5 months in 9 per cent CO_2 + 2 per cent O_2 (Chung and Son, 1994).

Collins and Tisdeli (1995) studied quality, storage and shelf life attributes and export potential of fruits of cultivars Fuyu and Suruga before and after exposure to temperatures of 0, 5, 10 and 20°C, followed by 7 days at 20°C to simulate shelf life. Fuyu fruits were increasingly damaged by more than 7 days at 10°C or below, with 5°C being especially injurious. Suruga could withstand these temperatures for at least 42 days at 0 or 10°C, but 5°C did produce some chilling injury (CI). This disorder was associated with a loss of yellow colour (the b scale, blue top yellow) but no other physical or compositional characteristic was consistently associated with the inci- dence of chilling injury. Suruga was less prone to CI than Fuyu, and thus may have greater possibilities for export.

Modified atmosphere storage for up to 5 months was also achieved by sealing fruit in polyethylene bags and holding at 0°C (Tarutani, 1960; Tarutani, 1965; Anon., 1987). In an investigation on storage, Zuthi and Ben-Arie (1989) found that Japanese cultivar Fuyu, which has a limited storage life of up to 6 weeks in regular air storage maintained excellent quality for 18 weeks under modified atmosphere conditions. The fruit was packed after harvest in 0.06 or 0.08 mm polyethylene (PE) bags (6 fruits per 20 x 30 cm bag) and placed in storage at 0°C. Better fruit quality was achieved in the 0.08 than in the 0.06 mm PE bags, but in both instances the fruit was of superior quality compared with non-packed control fruit. Two varieties of astringent persimmon 'Cheongdo-Bansi' and 'Sangju-Dungsi', were stored at 0-3°C in sealed PE film bags (Hong *et al.*, 1993). Changes in soluble solids content, tannin content, and fruit hardness of the persimmons, the CO_2 in the bags, and disorders were investigated during storage. Sealing in 0.8 mm PE film bags reduced softening of stored persimmons but did not change their soluble solids content. After 44 days of storage in 0.08 mm PE film bags, Cheongdo-Bansi was sweet enough to eat. The CO_2 in PE film bags (0.05 mm, 0.08 mm) for both varieties was highest after 7 days of refrigerated storage and then remained constant at a level of 5-6. No storage disorder developed in either variety until after 64 days of storage.

The commercial storage life of persimmon fruits cv. Fuyu, grown in Israel, could be extended from 6 to 18 weeks at 0°C by modified atmosphere packaging (MAP) in a low-density polythene (LDPE) film. MAP retarded fruit softening and inhibited development of peel and flesh disorders (Ben-Arie and Zutkhi, 1992). Use of polyethy-lene bags (0.06 mm thick) improved fruit hardness and rind appearance as well as maintaining eating quality (Nam *et al.*, 1992). Freshly harvested Fuyu fruits held in insecticidal atmospheres (ICA) of 0.25 per cent O_2 showed higher external injury scores than air-stored fruits; however, all injury scores were <1 (1=slight injury). Informal sensory evaluations indicated no loss in quality as a result of the ICA treatments (Mitcham *et al.*, 1997). Fruits of cvs. Hachiya and Fuyu could be stored in bags with ethylene absorbent for 4 months without appreciable loss of quality (Pekmezci *et al.*, 1998). Because of the reduced acid invertase activity, Matsui and Kitagawa (1988) concluded that the ethylene absorbent should be used irrespective of storage tempe-rature, as it conserved the sucrose content of the fruit. They also reported that the ethylene absorbent which delayed fruit softening had only a small effect on polygalacturonase PG activity.

Brackmann *et al.* (1997) in an experiment on storage of fruits of cv. Fuyu and Rama Forte recommended that flesh firmness was generally highest and skin browning was lowest in both cultivars stored in 40 micron MA packs at 0°C. Fuyu fruits could be stored for 3 months in CA storage with 16 per cent O_2 + 15 per cent CO_2 after transfer to 13.5°C and the fruits were still firm and had the lowest percentage of skin browning and no visible storage decay. Non-astringent Fuyu persimmon fruits were stored in a controlled atmosphere (CA, 4 per cent O_2 + 10 per cent CO_2 or 6 per cent O_2 + 10 per cent CO_2), in a modified atmosphere (MA, packaged in 0.06 mm polyethylene film) or air for 28 weeks at 0°C (Park, 1999). Flesh firmness was consistently highest in fruits stored in either of the CA treatments. The occurrence and severity of fruit skin blackening dramatically increased after 12 weeks of storage and was most severe in the MA and air treatments. Polyphenol oxidase (catechol oxidase) activity in the tissue and skin significantly increased after 10 weeks of storage and was highest in the MA and air treatments.

Modified atmosphere packaging (MAP) of persimmon fruit resulted in the accumu-lation of acetaldehyde to a level of 80 μg/ml; ethanol to a level of 900 μg/ml; and CO_2 up to 30 per cent. When fruits were stored at –1°C for 4 months in such atmospheres, the incidence of black spot disease, caused by *A. alternata*, was reduced. The effects of each of these gases were examined by Prusky *et al.* (1997) to determine their individual involvement in the inhibition of *Alternaria* development during storage. Based on analysis of the effect of concentration versus time for each gas accumula-ting in MAP, it was suggested that the increasing concentration of CO_2 during storage was the principal factor in the inhibition of black spot disease development.

Kim *et al.* (1989) investigated the causal factors of skin browning during storage. The occurrence of skin browning increased with cultivar in the order Suruga <Fuyu <Daean Dangam <Jiro, and with longer duration of storage. The skin polyphenol concentration in cv. Fuyu was 4.3 mg/100 g at harvest, 3.9 mg after 40 days of storage and 1.9 mg after 80 days of storage and was correlated with severity of browning. Skin moisture content at harvest was closely correlated with browning. Drying in the shade for 2 days after harvest markedly decreased browning during storage. Browning was less for fruits packed in 0.06-mm-thick bags. Increasing the rate of applied K fertilizer increased fruit K concentration and decreased browning. Spraying with Bordeaux mixture increased skin Cu content which was correlated with increased browning. Lee *et al.* (1999) investigated the effects of modified atmosphere (MA, varying the concentrations of O_2 and CO_2) storage in PE film bags on the development of storage disorders (blackening and flesh browning) in persimmon cv. Fuyu. Blackening occurred gradually in all treatments during 90 days of storage. Flesh browning occurred rapidly between 30 and 60 days of storage in some MA treatments. Blackening was observed mainly at the scar or along the crack of the cuticle layer, although it occurred all over the fruit surface without any regional specificity. Flesh browning always began at the top portion of fruits then expanded towards the equatorial zone. High concentrations of O_2 promoted blackening (incidence of 5 and 16 per cent at O_2 concentrations of 1 and 4 per cent respectively). Flesh browning was only observed in MA treatments with high CO_2 concentrations and low (<1 per cent) O_2 concentrations. Acetaldehyde and ethanol were detected in the atmosphere around fruits with flesh browning. High concentrations of CO_2 promoted flesh browning.

Persimmon (cv. Fuyu) fruits were stored in controlled atmosphere (CA) storage or in modified atmosphere (MA) storage at –0.5°C and 96 per cent RH for 100 days and then held for 4 days under simulated shelf-life conditions (14°C). CA storage in 15 kPa O_2/15 kPa CO_2 maintained the highest flesh firmness after 4 days' shelf-life. Fruits from both CA-storage treatments showed the lowest decay (37-47 per cent of fruits with symptoms of pathogenic decay consisting of lesions >5 mm in diameter) and skin browning incidence after simulated shelf-life. Iprodione treatment reduced decay (from 69 per cent in MA-stored fruits without iprodione treatment to 47 per cent) but caused skin browning. Ethylene absorption reduced skin browning, but did not affect flesh firmness. The effect of fruit size on the incidence of skin blackening and flesh browning during MA storage of Fuyu persimmons was investigated by Lee *et al.* (2000). Harvested fruits were graded to 5 sizes and the average weight of each grade was from 281.3 g to 150.7 g. Five fruits of each grade were packed in 5 different PE film bags and stored at –0.5 or 0°C. Lower O_2 and higher CO_2 concentrations were detected in PE film bags containing larger fruits. In general, skin blackening occurred in a high concentration of O_2 environment, and flesh browning occurred in a low O_2 concentration environment.

Fumigation of fruits with methyl bromide at 30 or 40 g/m3 for 2 hours delayed softening and extended shelf-life for 2 weeks when stored at 20°C (Kawakami *et al.*, 1989).

13.1.3 Astringency Removal

Various methods have been developed on removal of astringency from astringent cultivars, however, most of these result in partial softening of the fruit (Sato *et al.*, 1962; Rouhani *et al.*, 1975). Treatment of the fruit with carbon dioxide has been the most successful of the techniques developed to date. Variations of the techniques have been described by Eaks (1967), Gazit and Adato (1972), Guelfat-Reich *et al.* (1975), Matsuo and Ito (1977) and Glucina (1981). The most important commercial astringent cultivar treated by this method is Triumph which is grown extensively in Israel (Ben-Aire and Guelfat-Reich, 1976; Pesis and Ben-Aire, 1984; Pesis *et al.*, 1986). The effects of interactions between soluble pectin and soluble tannins on the astringency of persimmon fruits were studied by Taira *et al.* (1997). The astringency of both fresh juice and a solution of tannins purified from fruits of the astringent cultivar Hiratanenashi were significantly reduced by the addition of soluble pectin. In intact fruits, a reduction in the total pectin content without a rise in water-soluble solids during flesh softening, suggested the possible formation of pectin tannin complexes, which might reduce astringency when the fruit was eaten, with almost no acetaldehyde accumulation. The results suggested that a complex formation between pectin and soluble tannins might be a factor, other than acetaldehyde, in reducing astringency in persimmons.

Soluble tannin content, the development of tannins and parenchymatous cells were studied in 3 fruit types (2 non-astringent and one astringent) of persimmon (*Diospyros kaki*) in China (Fei *et al.*, 1996). The non-astringent fruit types (pollination constant and pollination variable) exhibited different methods of astringency removal during development. In the first type, astringency was removed during the early stages of fruit development through development of the parenchymatous cells and cessation of enlargement of the tannin cells. In the second type, astringency was removed at the pre-mature stage through coagulation of the tannin cells just before ripeness. Neither of these mechanisms operated in the astringent fruit type. Comparison of acetaldehyde production by flesh discs was made to elucidate cultivar differences between Denkuro and Hiratanenashi persimmons with respect to the ease of removing their astringency (Taira *et al.*, 1994). Soaking of flesh discs of immature astringent fruits in buffer containing ethanol caused more acetaldehyde production in Hiratanenashi than in Denkuro. The results indicated that Hiratanenashi converted ethanol to acetaldehyde more efficiently than did Denkuro, whereas Denkuro metabolized pyruvate more readily to acetaldehyde than did Hiratanenashi.

Ben-Arie and Sonego (1993) reported that astringency was removed from Triumph persimmons by immersion in water at 40°C for 5 h or 60°C for 1 h. Similar treatments at 20°C and 80°C had no effect on astringency reduction. Subsequent application of 80 per cent CO_2 for 48 h was effective for 20°C-treated fruits but not for 80°C-treated fruits, which became non-astringent only after exposure to acetaldehyde. Extending the 60°C-treatment beyond 2 h resulted in recurrence of astringency, which was not reduced by subsequent CO_2 treatment. Astringency also recurred in CO_2-treated non-astringent fruits when exposed to high temperatures. The disappearance and recurrence of astringency correlated with content of methanol-soluble tannins.

Astringency could be removed in 6 and 4 to 6 days by placing the fruits in a 1-litre container with 5 cm^3 of 10 and 30 per cent ethanol respectively (Taira *et al.*, 1991). Combine treatment of ethanol and CO_2 caused removal of astringency from cv. Hiratanenashi fruits more quickly than either of the treatments alone because more ethanol and acetaldehyde accumulated in the fruits. The loss of astringency was almost entirely due to the effects of CO_2 (Taira *et al.*, 1992). Astringency disappeared faster in fruits treated with ethanol than in those treated with acetaldehyde. Ethanol treatments caused a slight loss of astringency during the first two days, followed by a rapid loss over the next 2 days. The results indicated that ethanol removed astringency first by a non-enzymatic method of acetaldehyde breakdown, and second by free water being bound to hydrophilic surfaces of breakdown products and of cell wall polysaccharides, thus increasing the osmotic pressure (Fukushima *et al.*, 1991). Treatment of fruits with acetaldehyde vapour, post-harvest treatment with ethanol vapour, and freezing and thawing, resulted in tannins with low, intermediate and high solubility respectively (Taira *et al.*, 1999). It was suggested that complete removal of astringency of Kaki Tipo fruits require 72 h of CO_2 (100 per cent) treatment instead of normal 20 h (Vidrih *et al.*, 1994).

Effects of CO_2 on ethylene production in persimmon fruit were investigated by Ishimaru *et al.* (1999). Immature and mature Japanese persimmon cv. Tonewase fruits were treated with 100 per cent CO_2 for 24 h (control) or 100 per cent CO_2 for 17 h and gradual reduction to 0 per cent CO_2 for 7 h to remove their astringency; a third group was left untreated. The levels of 1-aminocyclopropane-1-carboxylic acid (ACC) and ACC synthase and ACC oxidase activities in their extracts were analyzed. The results indicated that inhibition of ethylene evolution with 100 per cent CO_2 was reduced to 0 per cent might be attributed to the blockage of S-adenosylmethionine conversion to ACC.

In 2 seasons (1993 and 1995) mature fruits of cv. Monpei were treated with ethanol in the presence of 95 per cent CO_2 for 3-5 h at 40°C followed by sealed incubation at 40°C (post-heating) and open incubation at 20°C (Shimomura,1997). Astringency was removed for 3-7 days, but the fruit remained unripe for a longer period. Application of 5 ppm ethylene for 12 h when fruits had lost most astringency decreased fruit firmness and peel colouring giving an optimum taste for 4-8 days after post-heating. Treatment with ethylene after removal of astringency had little effect and although dipping ethe-phon at 50 or 500 ppm promoted ripening, it also produced red spotting on the skin. Treating astringent persimmons with 3 to 5 nl C_2H_4/ml, following the application of ethanol to reduce tannins, decreased fruit firmness and increased colour, producing high-quality fruits in a short period (Kato, 1990).

Kato (1987) described method of removal of astringency in large scale. Fruits were packed in well-ventilated plastic containers which were stacked in a 142 x 142 x 125 cm de-astringency chamber heated to 30°C as ethanol was added. When 4.0 ppm ethylene was introduced (as well as ethanol), astringency removal and ripening were hastened and flesh texture was improved. If the de-astringency treatment was applied for 3 days (± ethylene), the market value of the treated fruits equalled that of fruits treated for 2 weeks by a conventional method. Toye *et al.* (1987) also reported that exposure of the fruits to 90-100 per cent CO_2 for 24 h at 30°C was effective in removing astringency. Rate of astringency loss was slowed considerably when treated fruits were subsequently stored at 0°C, (3 weeks to lose astringency at 0°C, compared with 4-5 days at 20°C). Fruits treated with CO_2 and stored for 4 weeks at 7°C developed chilling injury. Delaying treatment with CO_2 until after 3 weeks' storage at 0°C impaired complete astringency removal. Fruits treated with CO_2 had a storage life of 4 weeks at 0°C, with 5 days of subsequent shelf-life at 20°C.

Astringency was removed by (1) post-harvest alcohol treatment, (2) post-harvest CO_2 treatment or (3) on-tree alcohol treatment, followed in each case by storage of the fruit at 2 or 20°C (Taira *et al.*, 1987). On the basis of appearance after storage at 20°C, treatment (3) was considered best, followed by (2) and then (1). Characteristics of astringency removal in persimmon cv. Saijo fruits treated with 80 per cent CO_2 and ethanol vapour (EtOH) were investigated in relation to acetaldehyde production (Tamura *et al.*, 1999). The results demonstrated that ethanol vapour treatment caused the removal of astringency in Saijo fruits, but ethanol vapour might act independently of the accumulation of acetaldehyde in the tissue.

The effect of astringency removal treatments (ethanol treatment, CO_2 treatment, and calcium oxide suspension dipping method) on fruit quality and softening was inves-tigated by Shiesh *et al.* (1999) in cv. Hiratanenashi. Fruits treated with CO_2 and ethanol softened quickly, whereas those treated with calcium oxide softened slowly. Astringency decreased faster by CO_2 at higher temperatures. The time required to reach <2 of the astringency index were 2, 3, 5 and >5 days at 35, 30, 25 and 20°C, respectively. Flesh firmness significantly decreased following CO_2 treatments at 30, 25 and 20°C. CO_2 treatment of fruits at 30°C for 48 h followed by storage in air at 15°C for 24 h produced non-astringent fruits of excellent quality without abnormal softening.

14.0 Biotechnology

14.1.0 Tissue Culture

14.1.1 Micropropagation

In vitro propagation of persimmon is still in the early stages of development with limited success being achieved in producing rooted plantlets with a small range of cultivars (Yokoyama and Takeuchi, 1981; Cooper and Cohen, 1985; Sugiura *et al.*, 1986). Shoot tips (1 mm) of cv. Nishimurawase from dormant axillary buds exhibited best shoot growth on MS medium with half-strength N supplemented with 10^{-5} M zeatin. Single-node cuttings of elongated shoots subcultured on the above medium showed a 5-fold proliferation rate over 6 weeks (Fukui *et al.*, 1989). Dormant bud explants of the cvs. Jiro and Kurogaki produced shoots on modified MS medium with half the normal NO_3– concentration and supplemented with 5 mg BA/l. Jiro produced more shoots than Kurogaki. Shoot cuttings rooted at various frequencies when kept in the dark for 9 days on medium containing 250 mg IBA/l (Murayama *et al.*, 1989). Kim *et al.* (1992) obtained success in shoot tip culture of cvs. Fuyu, Ichikikei Jiro and Nishimurawase on half-strength MS medium supplemented with 5 mg BA/l. They also noted that half-strength MS medium supplemented with 0.2 per cent activated charcoal was effective for root induction when supplemented with 2-5 mg IBA/l. High frequency adventitious bud formation from callus cultures initiated from leaf primordia and maintained for > 4 years has been described by Tamura *et al.* (1992). Callus could be induced from leaf explants of astringent persi-mmon on half-strength MS medium supplemented with IAA at 1 mg/l and BA at 0.1 mg/l in the dark (Gondo *et al.*, 1999). It was observed that modified MS medium supplemented with zeatin at 0.5-2.0 mg/l + IAA at 0.05-0.10 mg/l or zeatin at 2 mg/l alone were the most suitable for proliferation and growth of persimmon buds and rooting percentages were markedly higher in the presence of IAA than IBA (Kong *et al.*, 1998). Roots of microshoots of cv. Jiro, when cultured on solidified MS medium, produced shoots spontaneously. The percentage of root segments forming adventitious shoots increased with increasing segment length (Tetsumura and Yukinaga, 1996).

Shoot explants of Nishimurawase survived for 30 weeks at 10°C on medium con-taining 15 g sucrose/l, while those of Fuyu survived for only 12 weeks at the same temperature (Fukui *et al.*, 1995).

Axillary and apical buds from adult persimmon (cv. Giombo) trees were cultured on half strength MS medium supplemented with various concentrations and combinations of zeatin, NAA, IAA and/or kinetin. The best shoot development was recorded from apical buds on media supplemented with 10 µM zeatin or 10 µM zeatin + 0.05, 0.10 or 0.20 µM IAA. Rooting occurred in cultures transferred to MS medium supplemented with 0.20 µM IBA and kept for 16 days in the dark (Salomao *et al.*, 1999). A protocol for *in vitro* establishment of shoot tip culture, included the use of nodal segments of dormant shoot and decontaminated by dipping in benomyl (2 g/l) for 16 h, 70 per cent ethanol for 1 min and 2.5 per cent NaOCl for 30 min. The shoot tips were removed (2-mm length) and cultured on MS/$2NO_3$ solid medium supplemented with 20 µM BA. Over 90 per cent of explants were established without contamination and >50 per cent of the shoot tips developed into shoots (Biasi *et al.*, 1999).

Several factors influence the acclimatization and field performance of Japanese persimmon micropropagules, such as, optimum temperature (28°C), continuous lighting rather than under a 16-h photoperiod, enhanced photosynthetic photon flux, transplanting after root induction, culture on the rooting medium for 20 days (Tetsumura *et al.*, 1993).

Ivanova *et al.* (2019) investigated the *in vitro* morphogenic capacity of vegetative buds and leaf cuttings in persimmon cultivars 'Zolotistaya', 'Nikitskaya Bordovaya' and 'Ukrainka' bred in Nikita Botanical Gardens. Shoot segments with buds were surface-sterilized with 70 per cent ethanol (1 min), 0.3-0.4 per cent Cl_2 (15 min) and 1 per cent Thimerosal (10 min), and contamination frequency decreased to 15-20 per cent. The explants in the culture vessels were kept in a growth room at 25±1°C under 16-h photoperiod supplied by cool-white fluorescent lamps giving 37.5 µmol m^{-2} s^{-1}. After 4-6 weeks of growing on MS medium supplemented with 13.30-22.20 µM BAP and 0.49-1.28 µM IBA, about 1-2 microshoots per explant were obtained for all the studied cultivars. Multiplication rate up to 3-6 adventitious microshoot per explant were increased by subculture. Addition of TDZ to culture medium allowed callus formation and microshoot differentiation in leaf cuttings from in vitro-grown persimmon microshoots. Calli were formed on MS medium subculture with 3, 6, 9 and 12 µM TDZ. Adventitious buds and microshoots were regenerated after 4-6 weeks from small and protuberant structures formed on the green callus surface. Histological analysis demonstrated that 6.0 µM TDZ stimulated the formation of numerous areas of active meristematic cells and apical meristem cells. Leaf primordia in epidermal and subepidermal layers of callus in cultivar 'Ukrainka' were formed. By indirect regeneration from morphogenic callus, about 7-8 microshoots per explant were obtained for the cultivar 'Zolotistaya', 4-6 – for 'Ukrainka' and 2-3 – for 'Nikitskaya Bordovaya'. Subculture on MS medium supplemented with 1.0 µM TDZ was favorable for microshoot elongation (Ivanova *et al.*, 2019).

Tchouga *et al.* (2020) made a first report on micropropagation of *Diospyros crassiflora* micropropagation, which paves the way for rapid ebony multiplication to respond to needed conservation efforts. *D.crassiflora* is known as African ebony. In vitro culture of nodal segments from seedlings was performed in half-strength Murashige and Skoog (MS) medium supplemented with either zeatin (0.0, 2.3, 4.6, 9.1, 13.7, 18.2, 22.8, and 27.4 mM) or 6-benzylaminopurine (BAP) (0.0, 2.2, 4.4, 8.8, 13.3, 17.8, 22.2, and 26.6 mM). After 12 weeks, all media allowed shoot budbreak. Shoots displaying the greatest budbreak were observed with 22.8 mM zeatin and 22.2 mM BAP. The shoots were then transferred to a medium supplemented with indole-3- butyric acid (IBA) and phloroglucinol (PG) at different concentrations for root induction. Root induction was observed on the shoots initially induced in the medium with BAP, but not in those grown in the mediumwith zeatin. In half-strengthMSsupplemented with 396.5 mM PG plus 14.2mM IBA, the formation of a single tap root was observed on 79 per cent of shoots, with an average root length of 2.8 ± 0.18 cm. Rooted plants were successfully acclimatized to the greenhouse, with a 50 per cent survival rate (Tchouga *et al.*, 2020).

14.1.1.1 Performance of Micropropagated Plants

Growth of micropropagated Japanese persimmon trees cv. Nishimurawase during the initial 3 years after field establishment in Japan was compared with that of grafted trees on seedling stocks (Tetsumura *et al.*, 1998). Judging from the mean length of annual shoots/tree and the yearly increases in height, trunk diameter, and top and root dry mass, the grafted trees on seedling stocks grew poorly during the first and second growing seasons. Micropropagated trees raised in pots did well after field establishment. These trees had more fine, thin, middle and large roots.

In contrast, grafted trees on seedling stock had one large taproot, which died back to some extent after field establishment, with few fine roots.

Tetsumura *et al.* (1999) also reported that shoot growth, flowering and fruiting of micropropagated (M) and grafted (G) Japanese persimon trees Nishimurawase were compared over 7 years. Overall, M trees established better without transplanting shock and grew more vigorously, as indicated by changes in tree height and trunk cross-sectional area. There was generally only a small effect on female flowering between the tree types and no significant effect on male flowering. The cumulative yield of M trees was about 20 per cent higher than that of G trees because of higher yields from M trees in the last 2 years, but the difference was not significant.

14.1.1.2 Protoplast Culture

The protoplasts isolated enzymatically from calli derived from leaf-primordia of adult plants of cv. Jiro were cultured on modified KM8P medium with 10 μM NAA, 1 μM zeatin, 150 mg/l ammonium nitrate and 0.5 M glucose (optimum concentrations). The microcalli obtained showed the most vigorous growth on a modified KM8P medium with 150 mg/l ammonium nitrate and supplemented with 1 μM zeatin plus either 1 μM IAA or 1 μM NAA. On this medium, most microcalli grew to more than 5 mm in diameter in 6 weeks. It was demonstrated that plantlets could be obtained from the protoplast-derived calli (Tamura *et al.*, 1993). Tao *et al.* (1991) also reported regeneration of plantlet from protoplast.

14.1.1.3 Somatic Embryogenesis

Primordial leaf explants of cv. Jiro produced yellowish-white callus in the dark on MS medium with half-strength nitrate and supplemented with 10.0 μM zeatin + 1.0 μM IAA. On transfer to MS medium supplemented with 10.0 μM zeatin + 1 μM IAA under a 12-h photoperiod, 24 per cent of the calli produced adventitious buds. Further, shoot growth occurred on the same without IAA (Tao *et al.*, 1988). Calli from leaf-primordia of *D. kaki* showed low organogenetic potential with only 10-20 per cent of calli forming adventitious buds after no or 1 subculture. However, less than 65 per cent of the calli formed adventitious buds after 34 subcultures (Tamura *et al.*, 1992). Benelli *et al.* (1999) reported 100 per cent callus induction of cultured persimmon shoot-tips (cv. Kaki tipo) after 60 days' culture in the dark on MS medium modified with half-strength NH_4NO_3 and KNO_3 and containing 1 μM IAA and 5 or 10 μM zeatin. Callus growth was consi-derably higher in explants cultured on the medium containing the higher zeatin concentration. The highest percentage of adventitious shoot formation (60 per cent) was obtained when calli from the high concentration zeatin medium were transferred into the light on MS medium containing 10 μM zeatin and 0.1 μM IAA.

14.1.1.4 Somatic Hybrids

Tamura *et al.* (1995) also isolated protoplasts from leaf-derived callus of cv. Jiro and Suruga and subjected them to electrofusion followed by their culture on modified KM8P medium. The calli were regenerated and the nuclear DNA contents revealed that 22 out of 127 calli were somatic hybrids. Nine of these 22 calli formed shoot buds. RAPD analysis confirmed that these were somatic hybrids between Jiro and Suruga with 2n =180 which was twice the diploid chromosome number (2n = 90 with x = 15) of the parental plants.

14.2.0 Biochemical and Molecular Markers

Isoenzyme banding patterns of glucose-6-phosphate isomerase (GPI), malate dehydrogenase (MDH) and phosphoglucomutase (PGM), are employed for distinguishing Japanese persimmon cultivars. Tao and Sugiura (1987) observed that with both leaf GPI and MDH, intracultivar polymorphism was absent. Isoenzyme phenotypes were more varied with GPI than with MDH. When both enzyme systems were taken together, 18 cultivars were uniquely discriminated and the rest could be classified into 22 groups of 2 to 18 cultivars each. In order to distinguish cultivars of Japanese persimmon, Sugiura *et al.* (1988) studied pollen isoenzyme systems of 19 cultivars of *D. kaki* and 1 and 2 clones of *D. lotus* and *D. virginiana*, respectively, by starch gel electrophoresis. Eight enzymes gave from 4 to 20 different banding patterns, of which GPI and PGM, both having a variety of well-resolved banding patterns, were thought to be the most useful enzymes for discriminating between the cultivars studied. Thirteen out of 19 cultivars were distinguished by either GPI or PGM, while all cultivars could be discriminated by using both enzymes. By combining the enzyme systems, Tao *et al.* (1989) could be able to identify 121 cultivars and the remaining 42 were divided into 17 sets of 2 to 4 cultivars each. Though GPI and PGM banding patterns helped to identify 31 trees, but the analysis presented some problems due to large number of bands of both isoenzymes and the polyploid nature of the plant (Parfitt *et al.*, 1991).

Analysis of RFLP variation in amplified cpDNA indicated that three temperate zone species *D. kaki*, *D. lotus* and *D. virginiana*, were monophyletic with one subtropical species, *D. ehretioides*, suggesting a close evolutionary relationship among them. An immediate common progenitor for *D. kaki* and *D. virginiana* is suggested from cpDNA homology and the polyploidized karyotypes of both species (Yonemori *et al.*, 1998). The effectiveness of RAPD analysis of cultivar identification of per-simmons was evaluated by using 10 base primers. A protocol of reproducible RAPD analysis for persimmon was established by determining suitable buffer and optimum concentrations of DNA polymerase MgCl2, primer and dNTP. It might be possible to identify persimmon mutants by this method. In addition, polymorphisms among 11 *Diospyros* species were observed by RAPD, using OPA-10 primer. The same method revealed little polymorphism among intraspecific levels of *D. kaki* or *D. lotus* (Luo *et al.*, 1995).

Research on persimmon biotechnology has resulted in development of techniques for micropropagation, somatic embryogenesis and genetic transformation. Molecular characterization of germplasm using RAPD and RFLP has been standardized. With this development it may be possible to introduce genes for controlling fruit flesh browning and also for controlling ripening using polygalacturonase genes and controlling ACC synthase gene. Developing lines tolerant to chilling injury would be another area of interest. Cloning non-astringent gene would aid in developing better varieties (Sanyal, 2001).

The pollination-constant, non-astringent (PCNA) type of persimmon is ideal for production because its fruits lose astringency at harvest regardless of seed formation. The PCNA trait in Japanese persimmons is controlled by a single locus, *AST*, and is recessive to the non-PCNA trait (Kono *et al.*, 2016). Because cultivated persimmon is hexaploid, only the homozygous genotype with six recessive alleles is PCNA. A region tightly linked to *AST* has been used as a DNA marker for breeding. Three non-PCNA (A) alleles have been reported. Kono *et al.* (2016) showed that the region linked to *AST* is highly polymorphic and includes microsatellites. By analyzing the size of PCR-amplified fragments, we distinguished 12 different A alleles from 14 non-PCNA cultivars and a Chinese PCNA 'Luotian-tianshi.' Then, using A fragment size, we assessed A allele inheritance in six non-PCNA x PCNA populations by analyzing segregation of each A allele in a population and segregation of progeny genotypes. By using A allele segregation analysis, we were able to estimate the copy number of each A allele in five non-PCNA parents but not in 'Amahyakume.' By analyzing progeny genotype segregation, we were able to estimate the 'Amahyakume' genotype. Our approach can be used not only for the selection of PCNA individuals in populations, but also for estimation of the copy number of A alleles in a possible non-PCNA parent. This would enable us to select non-PCNA parents with fewer A alleles, which would segregate more PCNA individuals in crosses with PCNA cultivars.

It is crucial to develop a rapid technique for identifying sexuality in the seedling stage of persimmon (*Diospyros kaki* Thunb.), and the elimination of male progeny has been regarded as an important strategy for enhancing breeding efficiency. Hence, Zhang *et al.* (2016) studied phenotype characterization and genotyping of the male linked *OGI* marker using 205 accessions, including persimmon cultivars, F_1 progeny and nine related *Diospyros* species. All persimmon cultivars displayed consistent results regarding *OGI*amplification and sex phenotype. A total of 143 F_1 progeny were derived from 11 crosses, among which 95 individuals had flowered. In the flowering full sib families, the amplification of the *OGI* marker in agreement with the sex phenotype was obtained in 85 plants (89.5 per cent). The segregation of *OGI* in 'Huashi 1' x 'Luotian Tianshi' and 'Huashi 1' x Male 3 F_1 populations fit a 1: 1 ratio. Furthermore, high *OGI* transferability was observed in nine related species. Overall, the results indicated that the *OGI* locus could be used to distinguish male from female persimmon plants at an early stage.

Kanzaki *et al.* (2010) have developed molecular markers linked to the AST locus that controls fruit astringency type in persimmon; however, these markers are not feasible for practical application to breeding programs since they are not fully effective for discriminating the pollination constant and non-astringent (PCNA) genotype from the non-PCNA genotype in a progeny derived from 'Kurokuma'. They also developed new sequence characterized amplified region (SCAR) markers that enable easy and reliable selection of the PCNA genotype from breeding populations. Genomic regions adjacent to probe 5R, which showed polymorphic fragments between PCNA and non-PCNA genotypes in restriction fragment length polymorphism (RFLP) analysis, were isolated from the genomic libraries of 'Nishimura-wase', 'Jiro', and 'Kurokuma'-derived offspring. The isolated genomic regions were characterized and 3 insertion/deletion mutations were observed between ast- and AST-linked regions. Several primers were designed in the flanking region of Indel-3 and, in multiplex PCR, it was shown that using 2 forward primers, AST-F and PCNA-F and a reverse primer, 5R3R, is the most useful and reliable primer set. The AST linked 220-bp fragment proved to be a common marker of 'Kurokuma'-, 'Nisimura-wase'- and 'Aizumishirazu'- derived progenies. This multiplex PCR is considered the most practical tool for marker-assisted selection (MAS) and can enhance and accelerate progress in persimmon breeding.

14.3.0 Genome Sequencing

Within the large order Ericales, a heterogametic male (XY) sex determination system has evolved independently in at least two genera, *Diospyros* and *Actinidia* (Akagi *et al.*, 2014, 2018). *Diospyros* had evolved a Y-encoded pseudogene called *OGI*, which produces small-RNA, which in turn repress the autosomal feminization gene, *MeGI* (Akagi *et al.*, 2014, 2020).Most angiosperms bear hermaphroditic flowers, but a few species have evolved outcrossing strategies, such as dioecy, the presence of separate male and female individuals. The mechanisms underlying dioecy in diploid persimmon (*D. lotus*) was investigated and Akagi *et al.* (2014) found that male flowers are specified by repression of the autosomal gene *MeGI* by its paralog, the Y-encoded pseudo-gene *OGI*. This mechanism is thought to be lineage-specific, but its evolutionary path remains unknown. In a later study, Akagi *et al.* (2020) developed a full draft of the diploid persimmon genome (*D. lotus*), which revealed a lineage-specific whole-genome duplication event and provided information on the architecture of the Y chromosome. We also identified three paralogs, *MeGI*, *OGI* and newly identified *Sister of MeGI* (*SiMeGI*). Evolutionary analysis suggested that *MeGI* underwent adaptive evolution after the whole-genome duplication event. Transformation of tobacco plants with *MeGI* and *SiMeGI* revealed that *MeGI* specifically acquired a new function as a repressor of male organ development, while *SiMeGI* presumably maintained the original function. Later, a segmental duplication event spawned *MeGI*'s regulator *OGI* on the Y-chromosome, completing the path leading to dioecy, and probably initiating the formation of the Y-chromosome. These findings exemplify how duplication events can provide flexible genetic material available to help respond to varying environments and provide interesting parallels for our understanding of the mechanisms underlying the transition into dieocy in plants (Akagi *et al.*, 2020).

Fu *et al.* (2016) sequenced the complete chloroplast genomes from *D. kaki, D. lotus, D. oleifera, D. glaucifolia* and *D. 'Jinzaoshi'* and found that *D. kaki* was more closely related to *D. oleifera*, whereas *D. lotus* exhibited a closer relationship with *D. glaucifolia*. Kanzaki (2016) proposed that the commercialized persimmon (*D. Kaki*) originated from southern China through polyploidization from diploid ancestors. The complete genome sequence for Diospyros would be tremendously useful for clarifying these issues. Zhu *et al.* (2019) obtained chromosome-scale genome assembly for 'Youshi' (*Diospyros oleifera*, 2n= 2x = 30), the diploid species of persimmon, by integrating Illumina sequencing, single-molecule real-time sequencing, and high-throughput chromosome conformation capture techniques. The assembled D. oleifera genome consisted of 849.53 Mb, 94.14 per cent (799.71 Mb) of which was assigned to 15 pseudo chromosomes, and is the first assembled genome for any member of the Ebenaceae. Comparative genomic analysis revealed that the D. oleifera genome underwent an ancient γ whole-genome duplication event. We studied the potential genetic basis for astringency development (proanthocyanidin biosynthesis) and removal (proanthocyanidin insolublization). Proanthocyanidin biosynthesis genes were mainly distributed on chromosome 1, and the clustering of these genes is responsible for the genetic stability of astringency heredity. Genome-based RNA-seq identified deastringency genes, and promoter analysis showed that most of their promoters contained large numbers of low oxygen-responsive motifs, which is consistent with the efficient industrial application of high CO_2 treatment to remove astringency. Using the *D. oleifera* genome as the reference, SLAF-seq indicated that 'Youshi' is one of the ancestors of the cultivated persimmon (2n = 6x = 90). Our study provides significant insights into the genetic basis of persimmon evolution and the development and removal astringency, and it will facilitate the improvement of the breeding of persimmon fruit (Zhu *et al.*, 2019).

14.4.0 Genetic Transformation

Tao *et al.* (1994) used *Agrobacterium rhizogenes* wild-type strain A4 harbouring agropine type Ri-plasmid (P RiA46) to induce crown galls in *D. kaki* cv. Fuyu, Jiro and Nishimurawase. The tumours thus obtained were cultured on modified MS medium. Plants were regenerated on MS medium containing zeatin and IBA. PCR and Southern blot analysis revealed that some of the regenerants from tumour-derived calli were Ti-DNA transformants, but some non-transformants were also regenerated. A few phenotypic alterations, such as dwarfness and decrease in rooting ability were also observed in the transformants.

Gus gene was easily detected on calli from cutivars Kaki tipo and Jiro, following bombardment with the pcGU delta-o plasmid and using the β-glucuronidase histological assay. The highest levels of expression were observed when calli from cv. Jiro were bombarded with gold particles at a gas pressure of 810 kPa (Benelli *et al.*, 1999).

Tao *et al.* (1997) succeeded in developing insect resistant *D. kaki* plants by transformation with *A. tumefaciens*. The leaf-discs from the leaves of shoot cultures were co-cultivated with *A. tumefaciens* carrying the binary plasmid vector pDU92.710. The T-DNA region in the plasmid contained the kanamycin resistance gene (*nptII*), β-glucuronidase gene (*uidA*) and a synthetic reconstruct of *cry IA* (*c*) encoding the insecticidal crystal protein fragment of *Bacillus thuringiensis* (*Bt*) ssp. *Kurstaki HD-73*. Of the 720 infected leaf discs, 17 putative transformed callus lines showing kanamycin resistance were obtained which in 8 weeks produced 15 shoot-buds resulting into 11 shoots. Out of thee shoots 10 showed GUS activity, and 8 of them showed stable integration of T-DNA into *D. kaki* genome as confirmed by PCR and Southern blot analysis. Western analysis confirmed that *cryIA*(*c*) gene was expressing in the plants. Many of these lines showed highly significant mortality of the test insects, *Plodia* and *Monema* species as compared to the control plants.

14.5.0 *In vitro* Conservation

At 2°C shoot-tip explants of cvs. Fuyu and Nishimurawase pre-conditioned on medium containing 60 g/l sucrose showed high levels of viability. However, for storage at 10°C pre-conditioning on medium containing 15 g/l sucrose gave the best results (Fukui *et al.*, 1994).

15.0 References

Akagi T., Henry, I.M., Ohtani, H., Morimoto, T., Beppu, K., Kataoka, I. *et al.* (2018) *Plant Cell., Amer. Soc. of Plant Biologists*, **30:** 780–795.

Akagi T, Henry IM, Tao R, Comai L. (2014) *Science*, **346:** 646–650.

Akagi T, Shirasawa K, Nagasaki H, Hirakawa H, Tao R, Comai L, *et al.* (2020) *PLOS Genetics* **16(2):** e1008566. https: //doi. org/10.1371/journal.pgen.1008566.

Aki, M.A., Eid, M.A. and Ahmed, F.F. (1994) In: *Post-harvest physiol. Pathol. tech. hort. commodities: recent advances* (Eds. A. Ait-Oubahou and M. El-Otmani), Agadir, Morocco, pp. 17-23.

Anonymous (1981) *Orchardist of New Zealand*, **54**: 147-148.

Anonymous (1987) *Growing Today* (*New Zealand*), Feb., 1987, pp. 33-35.

Aoki, M., Tanaka, K. and Okada, N. (1977) *Res. Bull. Gihci-Ken Agric. Res. Centre*, **9:** 119-130.

Asakuma,H., H.Chijiwa and M.Shiraishi. (2018) *Sci. Hort.*, **240**: 232-238.

Ayaz, F.A. and Kadhoglu, A. (1998) *Turkish J. Bot.*, **22**: 69-72.

Ayaz, F.A., Kadhoglu, A. and Reunanen, M. (1997) *J. Agric. Food Chem.*, **45**: 2539-2541.

Bargioni, G., Pisani, P.L., Ramina, A. and Castelli, F. (1980) *Revista della Ortoflorofruticoltura Italiana*, **63**: 81-92.

Bejanidze, I., T. Kharebava, N. Alasania, N. Didmanidze and N. Davitadze. (2019) *International Conference On Innovations In Science And Education*, March 20-22, 2019, Prague, Czech Republic

Bellini, E. (1973) *Informatore Agrario*, **29**: 12286-12303.

Bellini, E. (1979) *Informatore Agrario*, **35**: 6921-6930.

Bellini, E. and Giannelli, G. (1982) *Informatore Agrario*, **38**: 19027-19044.

Bellini, E. and Ragazzini, D. (1984) *Informatore Agrario*, **40**: 57-59.

Ben-Aire, R. and Guelfat-Reich, S. (1976) *J. Amer. Soc. Hort. Sci.*, **101**: 179-181.

Ben-Aire, R., Bazak, H. and Blumenfield, A. (1986) *Acta Horticulturae*, **179**: 807-813.

Ben-Arie, R. and Sonego, L. (1993) *J. Food Sci.*, **58**: 1397-1400.

Ben-Arie, R. and Zutkhi, Y. (1992) *HortScience*, **27**: 811-813.

Ben-Arie, R., Perez, A. and Prusky, D. (1992) *Phytoparasitica*, **20**: 155S-158S.

Ben-Arie, R., Zhou, H.W., Sonego, L. and Zutkhi, Y. (1997) *Acta Horticulturae*, **436**: 243-250.

Ben-Arie, R., Zutkhi, Y., Sonego, L. and Klein, J. (1991) *Post-harvest Biol. Technol.*, **1**: 169-179.

Benelli, C., Bellini, E., Giordani, E. and Lambardi, M. (1999) *Adv. Hort. Sci.*, **13**: 82-86.

Biasi, L.A., Carvalho, D.C.de, Andrade, A. de and Zanette, F. (1999) *Revista Brasileira de Fruticultura*, **21**: 279-283.

Blumenfield, A. (1986) *Alon Hanotea*, **40**: 539-544.

Bongjoh, C.A. and M. Nsangou. (2001). In: *"Sustainable Forest Management of African Rain Forest"*, Part II Symposium. The Tropenbos Foundation, Wageningen, The Netherlands, p. 112–124.

Bose, T.K., Mitra, S.K. and Sadhu, M.K. (1986) In: *Propagation of tropical and subtropical horticultural crops*, Naya Prokash, Calcutta, India, pp. 357-361.

Brackmann, A., Mazaro, S.M. and Saquet, A.A. (1997) *Ciencia Rural*, **27**: 561-564

Camp, A.F. and Mowry, H. (1929) *Fla. Ag. Exp. B.*, **205**: 527-562.

Cesar, H.P. (1942) *Rev. Agric.*, S. Paulo, **17**: 269-275.

Chalmers, D.J. and Van Den Ende, B. (1975) *Victorian Horticultural Digest* (Aust.), **65:** 12-14.

Chauhan J.S. and Gautam, D.R. (1986) *Proc. National Symp. on Temperate Fruits, 15-18 March, 1984*, Himachal Pradesh Agricultural University, Solan, India, Dr. Y.S. Parmer University of Horticulture and Forestry, pp. 115-118.

Choi, S.T., D.S. Park, K.P. Hong and S.M. Kang. (2011) *Adv. Hort. Sci.*, **25**(3): 164-169

Choi, S.; D. Park; G. Ahn; S. Kim, and T. Choi (2013) *Kor. J. Hort. Sci. Technol.*, **31(6):** 706 – 713.

Choi, S.T., Kang, S.M., Ahn G.H., Park, D.S., Shon, G.M. and Rho, C.W. (1999) *J. Korean Soc. Hort. Sci.*, **40**: 581-584.

Choi, S.T., W.D. Song, G.M.Shon, C.W. Rho and S.M. Kang. (2015) *J Kor. Soc. Hort. Sci.*, **46**(4): 255–259

Choi, Y.W., Cho, D., Lee, Y.M., Soh, J.H., Yun, C.H., Choi, S.S. and Shin, J.O. (1997) *Acta Horticulturae*, **436**: 267-274.

Chujo, T. (1982) *Memoirs of the Faculty of Agriculture*, Kagawa University, No. 37, pp. 1-63.

Chujo, T., Hashimoto, T. and Ashizawa, M. (1973) *Kagawa Daigaku Nogakuleu Gakuzyutu Hokoku*, **25**: 25-34.

Chujo, T., Kataoka, M., Yamauchi, S. and Ashizawa, M. (1972) *J. Japanese Soc. Hort. Sci.*, **41:** 339-343.

Chung, B.K., Kang, S.W. and Kwon, J.H. (1995) *RDA J. Agril. Sci., Crop Protection*, **37**: 376-382.

Chung, D.S. and Son, Y.K. (1994) *RDA J. Agril. Sci., Farm Management Agril. Eng. Sericulture & Farm Product Utilization*, **38**: 692-698.

Clark, C.J. (1985) In: *Proc. Ruakura Horticultural Conference*, Ministry of Agriculture and Fisheries, New Zealand, pp. 15-16.

Clark, C.J. and Barrett, L.J. (1985) *Growing Today* (New Zealand), August, pp. 30-31.

Clark, C.J. and Kajiura, I. (1986) *Growing Today* (New Zealand), April, pp. 47-50.

Clark, C.J. and Smith, G.S. (1985) *Southern Horticulture* (New Zealand), **18**: 21-23.

Clark, C.J. and Smith, G.S. (1990a) *Scientia Hort.*, **42**: 85-97.

Clark, C.J. and Smith, G.S. (1990b) *Scientia Hort.*, **42**: 99-111.

Cohen, Y., Gur, A., Barkai, Z., Blumenfeld, A. (1991) *Scientia Hort.*, **48**: 61-70.

Collins, R.J. and Tisdeli, J.S. (1995) *Post-harvest Biol. Technol.*, **6**: 149-157.

Collins, R.J. and Tisdell, J.G. (1996) *Post-harvest Biol. & Tech.*, **7**: 351-357.

Collins, R.J., Mowat, A.D. and George, A.P. (1995) *Acta Horticulturae*, **409**: 67-74.

Cooper, P.A. and Cohen, D. (1985) *Comb. Proc. Intern. Plant Prop. Soc.*, **34**: 118-124.

Cutillas-Iturralde, A., Zarra, I. and Lorences, E.P. (1993) *Physiol. Plant.*, **89**: 369-375.

Daood, H.G., Biacs, P., Czinkotai, B. and Hoschke, A. (1992) *Food Chem.*, **45**: 151-155.

Douglas, J. (1980) *New Zealand J. Agric.*, November, pp. 11-15.

Duangjai, S., B. Wallnöfer, R. Samuel, J. Munzinger and M.W. Chase. (2006) *Am. J. Bot.* **93**, 1808–1827.

Dunn, J. (1974) *Span*, **17**: 112-113.

Eaks, I.L. (1967) *J. Amer. Soc. Hort. Sci.*, **91**: 818-825.

El-Shaikh, A.R.A., Khalil, B.M. and Hamza, A.Y. (1999) *Egyptian J. Agric. Res.*, **77**: 1707-1724.

Ennab,H.A., M.A.M. Soliman and G. B. Mikhael. (2018) *J. Product. & Dev.*, **23(1):** 39-59.

Ercisli, S. and M. Akbulut. (2009) *Acta Hort.*, **833: 35-38.**

Fei, X.Q., Zhou, L.H. and Wang, J.F. (1996) *Forest Res.*, **9**: 27-31.

Fei, Z.Q., Wang, J.F., Zhon, L.H., Gong, B.C. and Wu, K.Y. (1994) *Forest Res.*, **7**: 106-110.

Fraser LG, Tsang GK, Datson PM, De Silva HN, Harvey CF, Gill GP, *et al.* (2009) *BMC Genom.* **10:** 102.

Frodin, D. G. (2004) *Taxon* **53:** 753–776

Fu, J. M. *et al.* (2016) *PLoS ONE* **11**, e0159566

Fukui, H., Ohba, H. and Nakamura, M. (1995) *Intern. Plant Propagators' Society: Combined Proceedings*, **44**: 245-248.

Fukui, H., Ohba, H. and Subhadrabandhu, S. (1997) *First annual meeting, International Plant Propagator's Society: Combined Proc.*, **44**: 245-258.

Fukui, H., Sugiyama, M. and Nakamura, M. (1989) *J. Japanese Soc. Hort. Sci.*, **58**: 43-47.

Fukui, H., Suzuki, Y., Ogawa, E., Hirata, K., Matsubara, Y.I. and Nakamura, M. (1999) *J. Japanese Soc. Hort. Sci.*, **68**: 578-584.

Fukui, H., Umeda, H., Mochizuki, K. and Nakamura, M. (1994) *J. Japanese Soc. Hort. Sci.*, **63:** 291-297.

Fukushima, T., Kitamura, T., Murayama, H. and Yoshida, T. (1991) *J. Japanese Soc. Hort. Sci.*, **60**: 685-694.

Fumaro, M., Ueda, K. and Okisima (1997) *Acta Horticulturae*, **436**: 275-283.

Fumuro, M. (1998) *J. Japanese Soc. Hort. Sci.*, **67**: 219-227.

Fumuro, M. and Utsunomiya, N. (1999) *J. Japanese Soc. Hort. Sci.*, **68**: 1146-1148.

Gao, M and Zhang, J.S. (1999) *Acta Horticulturae Sinica*, **26**: 364-368.

Gasanov, Z.M. (1968) *Subtrop. Kul'tury*, **5**: 104-110.

Gasanov, Z.M. (1991) *Subtrop. Kul'tury*, **3**: 71-76.

Gasanov, Z.M. (1993) *Sadovodstvo i Vinogradarstvo*, **2**: 17-18.

Gazit, S. and Adato, I. (1972) *J. Food Sci.*, **37**: 815-817.

Gemma, H., Ogata, Y. and Iwahori, S. (1997) *Acta Horticulturae*, **436**: 189-194.

George, A.P. and Nissen, R.J. (1982) *J. Agri. Anim. Sci.*, **39**: 149-158.

George, A.P. and Nissen, R.J. (1984) *Australian Horticulture*, January, pp. 28-41.

George, A.P. and Nissen, R.J. (1985) *Queensland Agric. J.*, **111**: 133-140.

George, A.P., Collins, R.J. and Mowat, A.D. (1997) *Acta Horticulturae*, **436**: 171-177.

George, A.P., Collins, R.J. and Nissen, R.J. (1994a) *Australian J. Exp. Agric.*, **34**: 267-275.

George, A.P., Collins, R.J. and Rasmussen, T.S. (1994b) *J. Hort. Sci.*, **69**: 937-946.

George, A.P., Nissen, R.J. and Collins, R.J. (1994c) *J. Hort. Sci.*, **69**: 225-230.

George, A.P., Nissen, R.J. and Kruff, H.C.B.De (1995a) *Acta Horticulturae*, **409**: 117-120.

George, A.P., Nissen, R.J., Collins, R.J. and Haydon, G.F. (1995b) *J. Hort. Sci.*, **70**: 807-816.

George, A.P., Nissen, R.J., Collins, R.J. and Rasmussen, T. (1996) *J. Hort. Sci.*, **71**: 217-226.

George, A.P., Nissen, R.J., Collins, R.J. and Rasmussen, T.S. (1995c) *J. Hort. Sci.*, **70**: 477-484.

George, A.P., Nissen, R.J., Morley-Bunker, M.J. and Collins, R.J. (1993) *J. Hort. Sci.*, **68**: 447-454.

Glucina, P.G. (1981) *Orchardist of New Zealand*, **54**: 381-388.

Glucina, P.G. (1982) *Annual Journal*, Royal New Zealand Institute of Horticulture, **10**: 18-27.

Glucina, P.G. (1987) *Orchardist of New Zealand*, **60**: 161-163.

Glucina, P.G. and Toye, J.D. (1985) In: *Proc. Ruakura Horticultural Conference - Persimmon*, Ministry of Agriculture and Fisheries, New Zealand.

Gondo, M., Tanaka, N., Tanaka, T., Shimomura, K, Nakanishi, F. and Ishimaru, K. (1999) *Phytochem.*, **51**: 879-881.

Griffith, E., Griffith, M.E. and McDaniel, J.C. (1982) *Persimmons for everyone*, North American Fruit Explorers, USA, pp. 1-145.

Guan,C.,P.Zhang, M.Wu, M.Zeng, S.Chacha, X.Ruan, R.Wang and Y.Yang. (2020) *Scientia Horticulturae* **269**, 27 July 2020, 109392 https: //doi.org/10.1016/j. scienta.2020.109392.

Guelfat-Reich, S. and Ben-Aire, R. (1981) *Institute for Technology and Storage of Agricutural Products*, Special publications No.197, Division of Scientific publications, Bet Dagan, Israel, p. 81.

Guelfat-Reich, S., Ben-Aire, R. and Metal, N. (1975) *J. Amer. Soc. Hort. Sci.*, **100**: 95-98.

Guo, D. L. & Luo, Z. R (2011) *Genet. Mol. Res.* **10**, 1060–1068.

Hague, S.M. (1911) *Bot. Gaz.*, **52**: 34-44.

Hasegawa, K., Imafuji, K. and Nakajima, Y. (1991a) *J. Japanese Soc. Hort. Sci.*, **60**: 9-18.

Hasegawa, K., Kuge, N., Minura, T. and Nakajima, Y. (1991b) *J. Japanese Soc. Hort. Sci.*, **60**: 19-29.

He, X.Y., Chen, L.M., Yan, B.Q. and Xie, J.Q. (2000) *J. Zhejiang For. Sci. and Technol.*, **20**: 26-29.

He, X.Y., Xie, J.Q., Ma, J.Q., Chen, L.M. and Yan, B.Q. (1998) *J. Zhejiang Forestry Sci. Tech.*, **18**: 52-54.

Hejazi, Z., S. Ishimura, C. Honsho and T. Tetsumura. (2018) *The Horticulture Journal* **87**(2): 184–192.

Hirano, K., Yonemori, K. and Sugiura, A. (1995) *J. Amer. Soc. Hort. Sci.*, **120**: 75-77.

Hodgson, R.W. (1939) *Proc. Amer. Soc. Hort. Sci.*, **37**: 338-339.

Homnava, A., Payne, J., Koehler, P. and Eitenmiller, R. (1990) *J. Food Quality*, **13**: 85-95.

Hong, Y.P., Lee, J.S. and Kim, Y.B. (1993) *J. Agric. Sci. Farm Management, Agric. Engineering, Sericulture, & Farm Products Utilization*, **35**: 755-760.

Hu, G.Q., Zheng, P., Huang, Y.X., Gu, J. and Cai, H.D. (1993) *J. Fruit Sci.*, **10**: 186-187.

Hume, H. (1914) *J. Heredity*, **5**: 131-138.

Iimuro, S., Ono, Y., Sugimoto, Y., Fukunaga, S. and Kuroda, K. (1980) *Bull. Nara Agric. Exp. Stn.*, No. 11, pp. 131-138.

Intrigliolo, D., F.Visconti and L.Bonet. *et al.* (2018a). In: Tejero, I.F.G. and V.H.D. Zuazo (Eds.). *Water Scarcity and Sustainable Agriculture in Semiarid Environment Tools, Strategies, and Challenges for Woody Crops*, 2018,Academic Press. Pages 149-177.

Ishimaru, M., Yamamoto, T. and Chachin, K. (1998) *J. Japanese Soc. Hort. Sci.*, **67**: 812-814.

Ishimaru, M., Yamamoto, T., Morioka, A., Ueda, Y. and Chachin, K. (1999) *J. Japanese Soc. Hort. Sci.*, **68**: 890-896.

Itai, A., Tanabe, K., Tamura, F., Susaki, S., Yonemori, K. and Sugiura, A. (1995) *J. Hort. Sci.*, **70**: 867-873.

Itamura, H. Fukushima, T., Kitamura, T., Harada, H., Taira, S. and Takahashi, Y. (1994) *J. Japanese Soc. Hort. Sci.*, **62**: 967-875.

Itamura, H., Kitamura, T., Taira, S., Harada, H., Ito, N., Takahashi, Y. and Fukushima, T. (1991) *J. Japanese Soc. Hort. Sci.*, **60**: 695-701.

Itamura, H., Tanigawa, T. and Yamanura, H. (1995) *Acta Horticulturae*, **398**: 131-138.

Ito, S. (1962) *Bull. Hortic. Res. Stn. Japan, Ser. B(1)*, pp. 1-16.

Ito, S. (1980) In: *Tropical and Subtropical Fruits, Composition, Properties and Uses* (Eds. S. Nagy and P.E. Shaw), AVI Publishing, Inc., Connecticuit, USA, pp. 442-468.

Ivanova, N.N., Kuzmina, Ò.N. and Mitrofanova, I.V. (2019) *Acta Hortic.* 1254: 13-18

Iwasaki, N. and Yoshida, A. (1996) *J. Japanese Soc. Hort. Sci.*, **64**: 757-762.

Izhar, J. (1999) *Alon Hanotea*, **53**: 490-492.

Jia, K.F., Cheng, Y.W. and Wang, L.Z. (1997) *J. Zhejiang Forestry College*, **14**: 45-49.

Jiang, L.H. (1998) *China Fruits*, **4**: 55.

Jordan, D. (1985) *Growing Today*, October, pp. 12-17.

Jung, Y.T., Yun, E.S., Kim, J.K., Son, I.S., So, J.D. and Jo, Y.K. (1993) *RDA J. Agric. Sci., Soil & Fertilizer*, **35**: 245-251.

K. S.Cheo, K.S., J.Y.Son, Y.O. Park, S.T. Choi, G.H.Ahn and *e.g.* Kim (2017) Korean J. Breeding Science.**49**: .410-413

Kadioglu, A. and Atalay, F. (1999) *Bio. Plantarum*, **42**: 155-157.

Kajiura, I. (1979) *J. New Zealand Tree Crops Assoc.*, **4:** 55-65.

Kajiura, I. (1980) *Orchardist of New Zealand*, April, pp. 98-101.

Kanaeko, M. (1977) *Research Bull. Aichi-Ken Agric. Res. Centre*, Japan, **13**: 131-136.

Kanaeko, M.Y., Yamamoto, G. and Imagawa, H. (1979) *Higashi-Mikawa District Hort. Abstr.*, Japan, **51**: 3091.

Kang, I.K., Chang, K.H. and Byun, J.K. (1998a) *J. Korean Soc. Hort Sci.*, **39**: 46-50.

Kang, I.K., Chang, K.H. and Byun, J.K. (1998b) *J. Korean Soc. Hort Sci.*, **39**: 51-54.

Kang, I.K., Chang, K.H. and Byun, J.K. (1998c) *J. Korean Soc. Hort Sci.*, **39**: 55-59.

Kanzaki, S. (2016) *J. Jpn. Soc. Food Sci. Technol.* **63:** 328–330.

Kanzaki,S., T. Akagi, T. Masuko, M. Kimura, M. Yamada, A. Sato, N. Mitani, N. Ustunomiya and K. Yonemori (2010) *J. Japan. Soc. Hort. Sci.* **79**(2): 150–155.

Kaplankiran, M., Demirkeser, T.H. and Toplu, C. (1998) *Acta Horticulturae*, **441**: 295-298.

Kasumov, G.K. and Semochkina, L.G. (1994) *Sadovodstvo i Vinogradarstva*, **3:** 18.

Kato, K. (1987) *J. Japanese Soc. Hort. Sci.*, **56**: 92-100.

Kato, K. (1990) *HortScience*, **25**: 205-207.

Kawakami, F., Nishikawa, S. and Moku, M. (1989) *Res. Bull. Plant Prot. service*, No. 25, pp.

Kazas, A.N. and Lobov, E.M. (1996) *Sadovodstvo i Vinogradarstvo*, **2**: 18-19.

Kim, B.S.,M. Gu, K.C. Cho, S.K. Jung and H.S. Choi (2017) *The Journal of Horticultural Science and Biotechnology*, **92**(4): 432-438

Kim, J.C., Chae, Y.S. and Kang, S.M. (1996) *J. Korean Soc. Hort. Sci.*, **37**: 680-685.

Kim, J.G., Chang, H.S., Kim, Y.B. and Kim, J.H. (1993) *RDA J. Agric. Sci., Farm Management, Agril. Engineering, Sericulture & Farm Products Utilization*, **35**: 766-770.

Kim, M.S., Lee, H.C., Kim, S.B., Kim, J.H., Song, N.H., Kim, M.J. and Shin, K.C. (1993) *RDA J. Agric. Sci.*, Soil & Fertilizer, **35**: 311-320.

Kim, M.S., Oh, S.D. and Son, D.S. (1999) *J. Korean Soc. Hort. Sci.*, **40**: 79-82.

Kim, T.C. and Ko, K.C. (1995) *J. Korean Soc. Hort. Sci.*, **36**: 669-677.

Kim, T.C. and Ko, K.C. (1997) *Acta Horticulturae*, **436**: 77-83.

Kim, T.C., Koh, G.C., Shin, Y.U., Son, D.S. and Kim, Y.S. (1992) *Res. Rep. Rural Dev. Admn., Biotech.*, **34**: 30-36.

Kim, Y.S., Jeong, S.B., Son, D.S., Lee, K.K. and Lee, U.J. (1989) *Res. Rep. Rural Develop. Admin. Hort.*, **31**: 62-72.

Kitagawa, H. and Glucina, P.G. (1984) *Persimmon Culture in New Zealand*, Science Information Publishing Centre, Wellington, New Zealand, pp. 1-74.

Kitagawa, H., Sigiura, A. and Sugiyama, M. (1966) *HortScience*, **1**: 56-60.

Kitajima, A., Akua, H., Yoshioka, T., Entani, T., Nakano, M. and Ishida, M. (1992) *J. Japanese Soc. Hort. Sci.*, **61**: 499-506.

Kitajima, A., Kuramoto, Y., Ohoka, K., Nakano, M. and Ishida, M. (1993a) *J. Japanese Soc. Hort. Sci.*, **62**: 317-325.

Kitajima, A., Ohshita, Y., Nakano, M. and Ishida, M. (1993b) *J. Japanese Soc. Hort. Sci.*, **62**: 505-512.

Kitajima, A., Uchida, T. and Ishida, M. (1993) *J. Japanese Soc. Hort. Sci.*, **62**: 327-335.

Kojima, K., Shiozaki, K., Koshita, Y. and Ishida, M. (1999) *J. Japanese Soc. Hort. Sci.*, **68**: 242-247.

Kong, X.S., Zhang, M.X., Zhang, Y.M. and Wang, Z.Y. (1998) *J. Fruit Sci.*, **15**: 232-238.

Kono, A., Kobayashi, S., Onoue, N. and A.Sato (2016) *Mol Breeding* **36,** 56, https: / / doi.org/10.1007/s11032-016-0480-0

Kotobuki, K. (1978) *Ibaraki-Ken*, pp. 36-42.

Kumashiro, K., Fukuda, S. and Harada, Y. (1957) *Studies Inst. Hort.* (Kyoto, Japan), **8**: 47-49.

Kwon, T.Y., Park, S.D., Choi, B.S. and Kwon, Y.J. (1995) *Korean J. Applied Entomology*, **34**: 295-299.

La Rue, J., Opitz, K.W. and Bentel, J.A. (1982) *Growing Persimmons*, Division of Agricultural Science, University of California, Leaflet 21277.

Lay-Yee, M., Ball, S., Forbes, S.K. and Woolf, A.B. (1997) *Post-harvest Biol. & Tech.*, **10**: 81-87.

Lee, S.D., Lee, M.H., Lee, H.U., Cho, J.K., Lee, Y.S. and Shim, K.H. (1994) *RDA J. Agril. Sci., Farm Management Agril. Eng. Sericulture & Farm Product Utilization*, **36**: 699-704.

Lee, S.K., Shin, I.S. and Park, Y.M. (1993) *Acta Horticulturae*, **343**: 300-303.

Lee, Y.J., Lee, Y.M., Kwon, O.C., Jeong, S.J., Lee, Y.B., Cho, Y.S., Park, Y.M. and Kim, T.C. (2000) *J. Korean Soc. Hort. Sci.*, **41**: 71-74.

Lee, Y.M. and Kim, C.C. (1991) *J. Korean Soc. Hort. Sci.*, **32**: 173-177.

Lee, Y.M. and Kim, C.C. (1994) *J. Korean Soc. Hort. Sci.*, **35**: 233-240.

Lee, Y.M. and Toshiaki, C. (1991) *J. Korean Soc. Hort. Sci.*, **32**: 66-72.

Lee, Y.M., Jang, S.J. and Lee, Y.J. (1997) *J. Korean Soc. Hort. Sci.*, **38**: 157-161.

Lee, Y.M., Kwon, O.C., Cho, Y.S., Park, Y.M. and Lee, Y.J. (1999) *J. Korean Soc. Hort. Sci.*, **40:** 585-590.

Lee, Y.M., Park, H.T. and Lee, Y.J. (1998) *J. Korean Soc. Hort. Sci.*, **39**: 533-536.

Lemmens, L., D. Louppe, and A.A. Oteng-Amoako. (2012) *Ressources vegetales de l'Afrique tropicale.*Fondation PROTA,Wageningen, TheNetherlands.

Leng, P., Itamura, H. and Yamamura, H. (1993) *J. Japanese Soc. Hort. Sci.*, **61**: 795-804.

Leng, P., Itamura, H. and Yamamura, H. (1995) *Environment Control in Biology*, **33**: 43-48.

Leng, P., Okamori, T., Itamura, H. and Yamamura, H. (1997) *Acta Horticulturae*, **436**: 109-118.

Lim, H.K., Lim, K.H., Kwong, O.D., Kim, B.S., Kim, W.S. and Kim, H.W. (1997) *J. Agro-Environment Sci.*, **39**: 27-31.

Liu, Y., Huo, G.H., Liu, S.J., Xiao, D.X. and Luo, L.S. (2000) *Acta Agriculturae Universitatis Jiangxiensis*, **22:** 54-57.

Liu, Y.J. and Wang, W.J. (1989) *Acta Horticulturae, Sinica*, **16**: 109-113.

Luo, Z.R., Yonemori, K. and Sugiura, A. (1995) *J. Japanese Soc. Hort. Sci.*, **64**: 535-541.

Lyon, B.G., Senter, S.D. and Payne, J.A. (1992) *J. Food Sci.*, **57**: 693-695.

Maotani, T., Suzuki, A., Nishimura, T., Kumamoto, O., Oshima, K. and Yamanaka, Y. (1989) *J. Japanese Soc. Hort. Sci.*, **58**: 557-562.

Martinez Gimeno, M.A., M. Castiella, S. Rüger, D. S. Intrigliolo and C. Ballester. (2017)*Irrig Sci* **35**: 159–167.

Mason, K.A., Glucina, P.G. and Mason, P.J. (1992) *Orchadist of New Zealand*, **65**: 40-42.

Matos, C.S. (1993) *Agropecuaria Catarinense*, **6**: 38-41.

Matos, C.S. (1997) *Agropecuaria Catarinense*, **10**: 5-7.

Matsubara, Y.I. and Hosokawa, A. (1999) *J. Soc. High Tech. Agric.*, **11**: 281-287.

Matsui, T. and Kitagawa, H. (1988) *J. Japanese Soc. Hort. Sci.*, **57**: 697-701.

Matsui, T. and Kitagawa, H. (1991) *Tech. Bull. Fac. Agric., Kagawa Univ.*, **43**: 45-50.

Matsumoto, Y. and Kuroda, K. (1982) *Bull. Nara Agric. Exp. St.*, Japan, **13**: 9-20.

Matsumura, H. (1999) *J. Soc. High Tech. Agric.*, **11**: 259-266.

Matsumura, H., Niikawa, T. and Noguchi, Y. (1998) *J. Soc. High Tech. Agriculture*, **10**: 185-193.

Matsuo, T. and Ito, S. (1977) *Plant Physiol.*, **18**: 17-25.

Matsuo, T., Kamigama, N. and Saitoh, M. (1995) *Acta Horticulturae*, **398**: 285-296.

McDaniel, J. (1973) *Fruit Var. J.* **27**: 94–96.

Miller, E.P. (1984) *Proc. Fla. Sta. Hort. Soc.*, **97**: 340-344.

Miller, E.P. (1989) *Proc. Fla. Sta. Hort. Soc.*, **102**: 199-202.

Miller, S. 1894. *Garden Forest.* **7:** 278–279.

Mishra, K.A. (1982) *Scientia Hort.*, **17**: 125-127.

Mitcham, E.J., Attia, M.M. and Biasi, W. (1997) *Post-harvest Biol. & Tech.*, **10**: 155-160.

Miyabayashi, T. (1941) *J. Japanese Soc. Hort. Sci.*, **12**: 143-154.

Mizobutsi, G.P., Bruckner, C.H., Salomao, L.C.C. and Neves, J.C.L. (1997) *Revista Ceres*, **44**: 547-556.

Mowat, A.D. (1995) *Acta Horticulturae*, **409**: 137-140.

Mowat, A.D., George, A.P. and Collins, R.J. (1997) *Acta Horticulturae*, **436**: 195-202.

Murayama, H., Tao, R., Tanaka, T. and Sugiura, A. (1989) *J. Japanese Soc. Hort. Sci.*, **58**: 55-61.

Nakamura, M. (1967) *Res. Bull., Fac. Agric., Gifu University, Japan*, No.2.

Nakamura, M., Kawase, I., Matswra, K., Kani, N. and Okayaki, A. (1972) *Res. Bull., Fac. Agr. Gifu Univ.*, Japan, **33**: 37-49.

Nakano, R. (1999) *Scientific Reports of the Faculty of Agriculture*, Okayama Univ., **88**: 131-139.

Nakano, R., Yonemori, K. and Sugiura, A. (1998) *J. Hort. Sci. Biotech.*, **73**: 341-346.

Nam, M. H., Jung, Y.T. and Yang, E.S. (1992) *Res. Rep. Rural Dev. Admn. Farm Management Agril. Eng. Sericulture and Farm Product Utilization*, **34**: 70-76.

Nee, C.C., Chou, W.Y. and Chen, S.L. (1994) *J. Agric. & For.*, **43**: 39-44.

Ng, F.S.P. (1978) *Malaysian Forester*, **41**: 436-438.

Omarov, M.D. (1984) *Subtrop. Kultury*, No. 4, pp. 94-97.

Omarov, M.D. (1990) *Subtrop. Kul'tury*, No.1, pp. 105-108.

Omarov, M.D. (1994) *Sadovodstvo i Vinogradarstvo*, **3**: 17-18.

Omarov, M.D. and Erokhina, A.I. (1989) *Subtrop. Kul'tury*, No.3, pp. 78-82.

Omarov, M.D. and Erokhina, A.I. (1991) *Sadovodstvo i Vinogradarstvo*, **2**: 37-38.

Onur, C. and S. Onur. (1995) *Proc. II National Horticulture Congress*, 3-6 October, Adana, Turkey, Cukurova University Publishing, pp. 587-590.

Ooshiro, A. and Anma, S. (1998) *J. Japanese Soc. Hort. Sci.*, **67**: 890-896.

Ostasheva, N.A. and Omarov, M.D. (1988) *Subtrop. Kul'tury*, No.3, pp. 143-145.

Parfitt, D.E., Yonemori, K., Ryugo, K. and Sugiura, A. (1991) *Fruit Var. J.*, **45**: 107-113.

Park, Y.S. (1999) *J. Korean Soc. Hort. Sci.*, **40**: 83-87.

Pekmezci, M., Erkan, M. and Gubbuk, H. (1998) *Acta Horticulturae*, **441**: 279-286.

Perez, A., Ben-Arie, R., Dinoor, A., Genizi, A., Prusky, D. (1995) *Phytopath.*, **85**: 221-225.

Pesis, E. and Ben-Arie, R. (1984) *J. Food Sci.*, **49**: 869-899.

Pesis, E., Levi, A., Sonego, L. and Ben-Arie, R. (1986) *Alon Hanotea*, **40**: 1149-1156.

Phengklai, C. (1978) *Thai Forest Bull.*, **11**: 1-103.

Piazza, M.G. and Intoppa, F. (1996) *Informatore Agrario*, **52**: 33-35.

Piretti, M.V. (1990) *Fitoterapia*, **62**: 3-13.

Plumb, C. 1896. *Garden Forest*. **9:** 442–443.

Pomper,K.W., J. D. Lowe, S. B. Crabtree, J. Vincent, *et al.* (2020) *HortScience*, **55 (1)**: 4–7.

Porter, G. (1987) *Comb. Proc. Intern. Plant Prop. Soc.*, **37**: 148-150.

Prusky, D., Perez, A., Zutkhi, Y. and Ben-Arief, R. (1977) *Phytopathology*, **87**: 203-208.

Roan, S.F. and Ni. W.T. (1994) *Special Publication - Taichung District Agricultural Improvement Station*, No. 33, pp. 241-252.

Rouhani, I., Bassiri, A. and Shaybany, B. (1975) *J. Hort. Sci.*, **50**: 73-99.

Salomao, L.C.C., Vieccelli, J.C., Siqueira, D.L.de, Otoni, W.C. and Siste, C.E. (1999) *Revista Ceres*, **46**: 267-277.

Sanyal, D. (2001) In: *Biotechnology of Horticultural Crops*, Vol. I. (Eds. V.A. Parthasarathy, T.K. Bose, P. Das, S.K. Mitra and S. Mohandas), Naya Prokash, Culcutta, pp. 331-340.

Sargent, S.A., Crocker, T.E. and Zoellner, J.J. (1993) *Proc. Fla. Sta. Hort. Soc.*, **106**: 131-134.

Sato, A. and M. Yamada (2016)*Breeding Science* **66:** 60–68.

Sato, K., Ishikara, M. and Harada, R. (1954) *Bull. Nat. Inst. Agric. Sci. Hiratsuka Ser. E.*, Japan, **3**: 169-186.

Sato, T., Ito, S. and Shimura, I. (1962) *Bull. Hort. Res. Stn.*, Japan, Ser. B, No. **1**: 48-56.

Sato, Y. and Sano, K. (1999) *J. Fac. Agric.*, **35**: 105-110.

Schatz, G.E., I. Lowry, J.M. Onana, T. St_evart, and V. Deblauwe. (2019) *Diospyros crassiflora. The IUCN Red List of threatened species 2019.*E.T33048A2831968. 14 Sept. 2019. <http: //dx.doi.org/10.2305/IUCN.UK.2019-1.RLTS. T33048A2831968. en>.

Senter, S.D., Chapman, G.W., Forbus, W.R.Jr. and Payne, J.A. (1991) *J. Foof Sci.*, **56**: 989-991.

Sharpe, R.H. (1966) *Proc. Fla. St. Hort. Soc.*, **79**: 374-379.

Shiesh C.C., Lin, H.L., Lin, R.Q., Chen, S.J. and Feng, S.P. (1999) *J. Chinese Soc. Hort. Sci.*, **45**: 273-280.

Shimomura, M. (1997) *Acta Horticulturae*, **436**: 215-233.

Shin, I.S., Lee, S.K. and Park, Y.M. (1994) *J. Korean Soc. Hort. Sci.*, **35**: 155-164.

Smith, R.L. (1970) *Fruit Var. J.*, **24**: 5-8.

Somego, M. (1978) *Malaysian For.*, **41**: 51-52.

Son, J.K., J. E.Yun and C.G. Park. (2009) *Acta Hortic*. 833: 325-330.

Sugiomoto, Y. and Fukunaga, S. (1972) *Bull. Nara Agric. Exp. St.*, Japan, **4**: 1-7.

Sugiomoto, Y. and Yasui, A. (1982) *Bull. Nara Prefec. Agric. Exp. Sta. Kashihara, Nara*, Japan, **13**: 1-8.

Sugiura, A., Tao, R. and Tomana, T. (1988) *Scientia Hort.*, **36**: 67-77.

Sugiura, A., Tao, R., Murayama, H. and Tomana, T. (1986) *HortScience*, **21**: 1205-1207.

Sugiura, A., Zheng, G.H. and Yonemori, K. (1991) *HortScience*, **26**: 574-576.

Sugiyama, N. and Yamaki, Y.T. (1995) *Scientia Hort.*, **60**: 337-343.

Suriyapananont, S., Suryapananont, V. and Tuntawiroon, O. (1997) *Acta Horticulturae*, **436**: 305-313.

Suriyapananont, V. and Suryapananont, S. (1997a) *Acta Horticulturae*, **436**: 321-329.

Suriyapananont, V. and Suryapananont, S. (1997b) *Acta Horticulturae*, **436**: 331-338.

Suriyapananont, V. and Suryapananont, S. (1997c) *Acta Horticulturae*, **436**: 315-320.

Suzuki, A., Iwanaga, H., Murakami, Y. and Maotani, T. (1988) *J. Japanese Soc. Hort. Sci.*, **57**: 167-172.

Suzuki, A., Murakami, Y. and Maotani, T. (1988a) *Bull. Fruit Tree Research Station*, A (Yatabe), Japan, **15**: 31-39.

Suzuki, A., Murakami, Y. and Maotani, T. (1988b) *Bull. Fruit Tree Research Station*, A (Yatabe), Japan, **15**: 41-49.

Suzuki, A., Murakami, Y., Maotani, T., Iwanaga, H. and Imaoka, A. (1987) *J. Japanese Soc. Hort. Sci.*, **56**: 236-241.

Suzuki, K., Yamayaki, T., Murase, S., Miyagawa, H., Nogata, T., Mitobe, M. and Morita, A. (1981) *Bull. Fruit Tree Research Station*, Yatabe, Ibaraki, Japan, **8**: 85-100.

Szteinberg, A. and Jabareen, H. (1986) *Phytoparasitica*, **14**: 240.

Taira, S. and Itamura, H. (1989) *Bull. Yamagata Univ. Agril. Sci.*, **10**: 903-910.

Taira, S., Funakoshi, N. and Watanbe, S. (1988) *Yamagata Agric. Forest. Soc.*, **45:** 1-4.

Taira, S., Kubo, Y., Sugiura, A. and Tomana, T. (1987) *Engei-Gakkai Zasshi J. Japanese Soc. Hort. Sci.*, **56**: 215-221.

Taira, S., Matsumoti, N. and Ono, M. (1999) *J. Japanese Soc. Hort. Sci.*, **68**: 83-88.

Taira, S., Oba, S. and Watanabe, S. (1992) *J. Japanese Soc. Hort. Sci.*, **61**: 437-443.

Taira, S., Ono, M. and Matsumoto, N. (1997) *Post-harvest Biol. Technol.*, **12**: 265-.

Taira, S., Onodera, K., Itamura, H. and Watanabe, S. (1994) *J. Japanese Soc. Hort. Sci.*, **62**: 897-902.

Taira, S., Satoh, I. and Watanabe, S. (1991) *Agric. Hort.*, **66**: 1401-1404.

Takano, S., Nishino, S. and Kuroda, K. (1991) *Bull. Nara Agric. Experiment Stn.*, **22**: 29-33.

Tamura, F., Tanabe, K., Itai, A. and Hasegawa, M. (1999) *J. Japanese Soc. Hort. Sci.*, **68**: 1178-1183.

Tamura, M., Tao, R. and Sugiura, A. (1992) *HortScience*, **27**: 1048.

Tamura, M., Tao, R. and Sugiura, A. (1993) *Japanese J. Breeding*, **43**: 239-245.

Tamura, M., Tao, R. and Sugiura, A. (1995) *Plant Sci. Lim.*, **108**: 101-102.

Tao, R,M., Dabdejarm A.M., Uratso, S.L., Vail, P.V. and Tebbets, J.S. (1997) *Acta Horticulturae*, **311**: 153-159.

Tao, R. and Sugiura, A. (1987) *HortScience*, **22**: 932-935.

Tao, R., Ito, J. and Sugiura, A. (1994) *J. Japanese Soc. Hort. Sci.*, **63**: 537-541.

Tao, R., Murayama, H., Moriguchi, K. and Sugiura, A. (1988) *HortScience*, **23**: 1055-1056.

Tao, R., Tamura, M., Yonemori, K. and Sugiura, A. (1991) *Plant Sci. Lim.*, **79**: 119-125.

Tao, R., Tetsumura, T. and Sugiura, A. (1989) *Memoirs College of Aric., Kyoto Univ.*, **135**: 31-42.

Tarutani, T. (1960) *J. Hort. Assoc.*, Japan, **29**: 212-218.

Tarutani, T. (1965) *Memoirs of the Faculty of Agriculture, Kagawa University*, Japan, **19:** 1-54.

Tchouga, A.O., V. Deblauwe, S.A. M. Djabou, *et al.* (2020) *HortScience*, **55(4):** 424–428.

Testoni, A. (1981) *Sperimeqtale per al Valorizy-azione Tecnologia dei Prodotti Agricolo*, **10**: 63-70.

Testoni, A., Bellini, E. and Gorini, G. (1981) *Informatore Agrario*, **37**: 16087-16098.

Tetsumura, T. (2000) *Acta Hortic.* **517:** 167–174

Tetsumura, T. and Yukinaga, H. (1996) *HortScience*, **31**: 463-464.

Tetsumura, T., S.Ishimura, C.Honsho and H.Chijiwa. ((2017) *Sci. Hort.*, 224: 150-155.

Tetsumura, T., S. Haranoushiro, T. Marume, C. Torigoe, T.Omori, Y. Kurogi, Y. Uchida and C. Honsho. (2010) *J. Japan. Soc. Hort. Sci.*, **79:** 327–334.

Tetsumura, T., S. Ishimura, T. Hidaka, E. Hirano, H. Uchida, Y.Kai, S. Kuroki, Y. Uchida and C. Honsho. (2015) *Sci. Hortic.* **187:** 87–92.

Tetsumura, T., Tao, R. and Yukinaga, H. (1993) *J. Japanese Soc. Hort. Sci.*, **62**: 533-538.

Tetsumura, T., Tao, R. and Yukinaga, H. (1999) *J. Hort. Sci. Biotech.*, **74**: 251-253.

Tetsumura, T., Yukinaga, H. and Tao, R. (1998) *HortScience*, **33**: 751-753.

Tetsumura,T., S.Ishimura1, T. Takita *et al.* (2019) *The Horticulture Journal* **88 (1):** 57–66.

Tian, J.W, He, P.C. and Xu, M.X. (1994) *Acta Horticulturae, Sinica*, **21**: 217-221.

Tian, J.W., Xu, M.X. and He, P.C. (1991) *Plant Physiol. Commun.*, **2**: 109-111.

Tomkins, A.R., Lut, T., Brown, N., Wilson, D.J. and Thomson, C. (1997) In: *Proc. of the Fiftieth New Zealand Plant Protection Conference*, Lincoln Univ., Canterbury, pp. 414-419.

Toye, J.D., Glucina, P.G. and Minamide, T. (1987) *New Zealand J. Exp. Agric.*, **15**: 351-355.

Troop, J. 1895. *Garden Forest.* **8**: 75–76.

Troop, J. and O. Hadley. 1895. *Trans. Indiana Hort. Soc.*, **35**: 116–123.

Tsai, C.T.J. (1998) *Chinese Soc. Hort. Sci.*, **44**: 11-28.

Tung, B.K., Huang, C.J., Tzen, S.L. and Tsay, J.G. (1994) *Plant Protection Bull.* (Taipei), **36**: 53-63.

Turk, R. (1993) *Acta Horticulturae*, **343**: 190-194.

Utsunomiya, N., Subhadrabandhu, S., Yonemori, K., Oshida, M., Kanzaki, S., Nakatsubo, F. and Sugiura, A. (1998) *Economic Bot.*, **52**: 343-351.

Vidrih, R., Hribar, J., Plestenjak, A. and Simcic, M. (1994a) In: COST94. *The post-harvest treatment of fruit and vegetables: quality criteria* (Eds. J. Hribar, D.S. Johnson, H. Bohling, J. Plaza, De La, K.E. Haffner, E. Hohn and Z. Somogyi), Brussels, Belgium, pp. 151-156.

Vidrih, R., Simcic, M., Hribar, J. and Plestenjak, A. (1994) *Acta Horticulturae*, **368**: 652-656.

Wally, A.S.M., Mokhtar, H. and El-Fakharanay, E.M. (1999) *Egyptian J. Agric. Res.*, **77**: 805-815.

Wang, R. Z.,Y. Yang and G.C. Li. (1997) *Acta Hortic.*, **436:** 43–50.

Wang, R.Z. and Yang, Y. (1995) *Acta Horticulturae*, **403**: 249-255.

Wang, W.J., Liu, Y.J. and Wang, Y.H. (1993) *Acta Horticulturae*, **20**: 105-110.

Wang, Y.Y., Li, B.J., Zhou, B.Y., Liu, G.H. and Zheng, X.M. (1996) *Acta Horticulturae Sinica*, **23**: 110-114.

Wang, Z.L. (1994) *Annual Rep. – Northern Nut Growers' Assoc.*, **85**: 154-155.

Wen, I.C. (1995) *J. Agric. Res., China*, **44**: 49-58.

Wills, R.B.H., McGlasson, W.B., Graham, D., Lee, T.H. and Hall, *e.g.* (1989) In: *Post-harvest BSP Professional Books*, Oxford, p. 174.

Woodburn, K.R. and Andersen, P.C. (1996) *HortScience*, **31**: 218-221.

Wu, Z. X. (2005) *Openings* **7**: 74–79

Xu, Y.H., Zhang, L.T. and Zhou, J.S. (2000) *South China Fruits*, **29**: 37.

Yakushiji, H. and Hase, Y. (1991) *Bull. Fruit Tree Res. Stn.*, **19**: 49-59.

Yakushiji, H., K. Morinaga, and S. Ono. (1997)*Acta Hort.* **436**: 165–168.

Yakushiji, H., Yamada, M., Yonemori, K., Sato, A. and Kimura, N. (1995) *J. Japanese Soc. Hort. Sci.*, **64**: 41-46.

Yamada, M., Imada, Y., Utsunomiya, D., Ohsaki, H., Takahashi, C., Tanaka, C. and Mizutani, F. (2000) *Bull. Experimental Farm College of Agriculture, Ehime Univ.*, **21**: 11-14.

Yamada, M., Kakutani, M., Yamane, H. and Yoshinaga, K. (1991) *Bull. Fruit Tree Res. Stn.*, **20**: 1-11.

Yamada, M., Kurihara, A. and Sumi, T. (1987) *J. Japanese Soc. Hort. Sci.*, **56**: 293-299.

Yamada, M., Sumi, T., Kurihara, A. and Yamane, H. (1988) *Bull. Fruit Tree Research Station, E.* (*Akitsu*), Japan, No. 7, pp. 9-30.

Yamada, M., Wang, R., Yamane, H. and Sato, A. (1995a) *J. Japanese Soc. Hort. Sci.*, **64**: 221-226.

Yamada, M., Wang, R.Z., Yamane, H., Sato, A. and Hirakawa, N. (1995b) *J. Japanese Soc. Hort. Sci.*, **64**: 227-233.

Yamada, M., Yamane, H., Sato, A., Hirakawa, N. and Wang, R. (1994) *J. Japanese Soc. Hort. Sci.*, **63**: 485-491.

Yamada, M.T., Sumi, T. and Kurihara, A. (1976) *Abstr. Japan Soc. Hort. Sci.*, Autumn meeting, pp. 150-153.

Yamamura, H., Matsui, K. and Matsumoto, T. (1989) *Scientia Hortic.*, **38**: 77-86.

Yamamura, H., Naito, R. and Mocheda, K. (1976) *J. Japanese Soc. Hort. Sci.*, **45**: 1-6.

Yamamura, J., Naito, R. and Shibata, Y. (1980) *Bull. Fac. Agr.*, Shimane Univ., Japan, **14**: 4-8.

Yamazaki, T., Suzuki, K., Murase, S. and Otake, S. (1981) *Bull. Fruit Tree Research Station*, Yatab, Ibaraki, Japan, **8:** 79-84.

Yang, Y.J. (1996a) *J. Korean Soc. Hort. Sci.*, **37**: 544-547.

Yang, Y.J. (1996b) *J. Korean Soc. Hort. Sci.*, **37**: 787-790.

Yasunobu, Y. (1971) *Bull. Kanagawa Hort. Exp. St.*, Japan, **19**: 45-50.

Yokoyama, T. and Takeuchi, M. (1981) *J. Japanese Soc. Hort. Sci.*, **49**: 557-562.

Yonemori, K. (1997) *Acta Horticulturae*, **436**: 21-32.

Yonemori, K. and Matsushima, J. (1985) *J. Japanese Soc. Hort. Sci.*, **54**: 201-208.

Yonemori, K. and Matsushima, J. (1987) *J. Amer. Soc. Hort. Sci.*, **112**: 812-817.

Yonemori, K., Hirano, K. and Sugiura, A. (1995) *Scientia Hortic.*, **61**: 37-45.

Yonemori, K., Itai, A., Nakano, R. and Sugiura, A. (1996) *J. Amer. Soc. Hort. Sci.*, **121**: 676-679.

Yonemori, K., Kameda, K. and Sugiura, A. (1992) *J. Japanese Soc. Hort. Sci.*, **61**: 303-310.

Yonemori, K., Kanzaki, S., Parfitt, D.E., Utsunomiya, N., Subhadrabandhu, S. and Sugiura, A. (1998) *Genome*, **41**: 173-182.

Yonemori, K., Oshida, M., Fukuda, F. and Sugiura, A. (1997) *HortScience*, **32**: 694-697.

Yonemori, K., Sugiura, A., Tanaka, K. and Kameda, K. (1993) *J. Amer. Soc. Hort. Sci.*, **118**: 293-297.

Zhang, J.S. (1988) *Shanxi Fruit Trees*, **3**: 31-32.

Zhang, P.X.,S.C. Yang, Y.F. Liu, Q.L. Zhang, L.Q. Xu and Z.R. Luo (2016) *Plant breeding* **135:** 721-727.

Zheng, G.H. and Sugiura, A. (1990) *J. Japanese Soc. Hort. Sci.*, **59**: 281-287.

Zheng, G.H., Yonemori, K, Hirano, K. and Sugiura, A. (1991) *Acta Horticulturae Sinica*, **18**: 193-197.

Zhou, Y.H., Mi, L, Shu, Z.L. and Yang, J.S. (1997) *J. Fruit Science*, **14**: 183-184.

Zhu, Q.G., Y. Xu, Y. Yang, C.F. Guan, Q.Y. Zhang *et al.* (2019) *Horticulture Research* **6**: 138.

Zhuang, D., Kitajima, A., Ishida, M. and Sobajima, Y. (1990) *J. Japanese Soc. Hort. Sci.*, **59**: 289-297.

Zimmermann, M.H. and Ziegler, H. (1975) In: *Transports in Plant. I. Phloem transport* (Eds. M.H. Zimmermann and J.A. Milburn), Springer-Verlag, Berlin, pp. 480-503.

Zuthi, J. and Ben-Arie, R. (1989) *Hassadeh*, **70**: 410-414.

⑤

AONLA

V.A. Parthasarathy, P.N. Bajpai and H.S. Shukla

1.0 Introduction

The aonla (*Emblica officinalis* syn. *Phyllanthus emblica*), an important minor fruit and a crop of commercial significance, is quite hardy, prolific bearer and highly remunerative even without much care. It belongs to the family Euphorbiaceae and is known as *amla, amlaki, amali, ambala, amalakamu* and *nelli* in different parts of India. It has been grown and known in India for last more than 3500 years. In fact, it finds a special mention in ancient Indian text 'Ayurveda' by Sushruta (1500 BC-1300 BC). the father of ancient medicine (Wali *et al.*, 2015).

2.0 Composition and Uses

2.1.0 Composition

The fruit is highly nutritive (Table 1) and it is the richest source of vitamin C among fruits except Barbados cherry (Asenjo, 1953). The ascorbic acid and other constituents are well retained in dried aonla fruits (Srivastava and Srivastava, 1964). An easy way to prepare aonla *murabba* with high vitamin C content has been reported by Gupta and Bopaiah (1986). The variations in ascorbic acid content have been recorded by various workers. The aonla powder is superior to synthetic vitamin C in treating deficiencies. The stability of ascorbic acid and presence of astringency in aonla fruit may be assigned to the presence of polyphenols or leucoanthocyanins (Sastry *et al.*, 1956). Hanif *et al.* (1966) noted marked antioxidant effect of gallic acid present in aonla fruit. A comparative evaluation of different products revealed that dried aonla (flakes) had maximum nutritive value followed by *chayvanprash*, aonla preserve (*murabba*), pickle and brined aonla fruits (Naik and Chundawat, 1993). It also contains calcium, phosphorus and iron. It is a fair source of vitamin B (30 mg/100 g) and nicotinic acid (0.2 mg/100 g). The aonla fruit has calorific value of 59 per 100 g of fruit and due to its antiscorbutic, diuretic, laxative, alternative and antibiotic properties; it is regarded as high value having diverse medicinal, industrial and other applications (Wali *et al.*, 2015).

Table 1: Composition of Aonla Fruits

Moisture	81.2 per cent	Phosphorus	0.02 per cent
Protein	0.5 per cent	Iron	1.2 per cent
Fat (ether extract)	0.1 per cent	Calorific value	59/100 g
Mineral matter	0.7 per cent	Vitamin B	30 mg/100 g
Fibre	3.4 per cent	Nicotinic acid	0.2 mg/100 g
Carbohydrates	14.0 per cent	Vitamin C	600 mg/100 g
Calcium	0.05 per cent		

2.2.0 Uses

The fruits are made into preserves (*murabba*), sauce, candy, dried chips, tablets, jellies, pickles, toffees, powder, *etc.* It is valued as an antiscorbutic, diuretic, laxative, alterative (Nadkarni, 1927) and antibiotic (Ray and Majumder, 1976). One or other parts of the plant could be used in treating chronic dysentery (Chopra *et al.*, 1958), bronchitis, diabetes, fever (Drury, 1873), diarrhoea (Dalzell and Gibson, 1861), jaundice, dyspepsia, cough (Burkill, 1935) and in tanning and dyeing industries. The literature regarding its diverse medicinal, industrial and other applications has been reviewed (Anon., 1952; Morton, 1960). The properties and nature of inhibitors of potato virus in the plant extract (Verma *et al.*, 1969), pharmacological activities of phyllemblin isolated from fruit pulp (Rao and Siddiqui, 1964), protective effects of fruit extract in myocardial necrosis (Tariq *et al.*, 1977) and antiviral activity (Singh *et al.*, 1983) have been reported.

In a brief review, Singh *et al.* (2015) describes that the Indian gooseberry under *Pyllanthus emblica* (Aonla or amla), syn: *Emblica officinalis* is in cultivation in India due to its therapeutic, medicinal and neutraceutical properties since time immemorial. It is an important minor fruit crop having commercial significance. The crop is quite hardy, prolific bearer and highly remunerative. It is said to be indigenous to tropical south-east Asia. It is also reported to be the native of India, Ceylon, Malaysia and China. This is a fruit of subtropical region but its cultivation in tropical climate is quite successful. Soon after the fruit set in spring, the fruits remain dormant through summer without any growth. This makes it highly suitable fruit crop for arid region. Aonla is not very exacting in its soil requirement and grows well in sandy loam to clay soils in India. Due to great tolerance to salinity and sodicity it is well adapted to arid conditions of Rajasthan. To safeguard the existing diversity of P. emblica and to achieve sustainable development based on the use of available genetic wealth, promotion and conservation of this species is of immense importance (Singh *et al.*, 2015).

According to Singh *et al.* (2015), the fruit is used for diverse applications in healthcare, food and cosmetic industry. It has been well studied for immunomodulatory, anticancer, antioxidant and antiulcer activities (Dnyaneshwar *et al.*, 2006). In India, it is predominantly used as a cure for liver disorders (Kirtikar and Basu, 2003). Besides this, the fruit extract has shown significant anticancer activity (Mahata *et al.*, 2013) antioxidant supplement to promote endothelial function and to restore wound healing competency (Chularojmontri *et al.*, 2013) and management of type 2 diabetes mellitus (Usharani *et al.*, 2013).

3.0 Origin and Distribution

Aonla is said to be indigenous to tropical south-eastern Asia, particularly in central and southern India (Firminger, 1947). It is also reported to be the native of India, Sri Lanka, Malaysia and China. It thrives well throughout tropical India and is met with wild or cultivated in the region extending from the base of the Himalaya to Sri Lanka and from Malacca to South China. The tree is also common in the mixed deciduous forests of India ascending to 1,350 metres on the hills (Anon., 1952). It is more popular in India and is commercially cultivated in Uttar Pradesh (Bajpai, 1963; Ram, 1983).

According to Wali *et al.* (2015), naturally growing trees of aonla have been reported from India, Sri Lanka, Cuba, Puerto Rico, Hawaii, Florida, Iran, Iraq,

Uses of Aonla

Aonla Fruit with Slices

Murabba

Jam

Candy

Jelly

Uses of Aonla

Aonla Powder

Aonla Chutney

Aonla Pickle

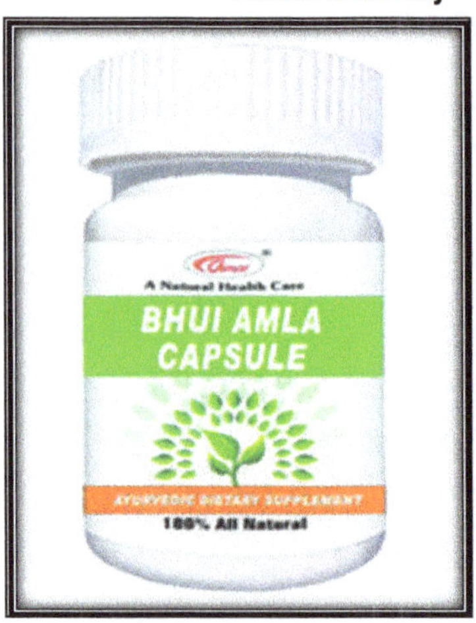

Aonla Capsule

Aonla Chips

Java, West Indies, Trinidad, Singapore, southern Thailand, Pakistan, Malaysia and China and Panama Canal regions. However, its cultivation is more common in India, particularly in the state of Uttar Pradesh where it is cultivated in the districts of Pratapgarh, Raibareli, Jaunpur, Sultanpur, Banda, Kanpur, Agra and Mathura districts. The major concentration of aonla cultivation is in Pratapgarh district. The natural distribution of wild aonla is found on the Himalayas, Chota Nagpur, Bihar, Orissa, West Bengal, North Circars, Deccan, Karnataka and in Western Ghats.

The History of domestication has been interestingly narrated by Wali *et al.* (2015) as follows. "In India, the homeland of aonla, domestication was first started in Varanasi (earlier known as Benaras) district of Uttar Pradesh with the initiative of Maharaja of Kashi. Banarasi, a superior genotype was selected from the wild aonla trees available in large number in the nearby Vindhyan hills. Authentic information regarding its cultivation dates back to 1881-82 in the Partapgarh district of Uttar Pradesh. The ailing state owner of the district (King) was advised for regular consumption of aonla fruit in one way or other. As per information available, few aonla trees were introduced from Varanasi and few from Gujarat. Those brought from Varanasi were named as 'Banarasi' and those brought from Gujarat were known as Francis and later on as *Hathijhool* (because of its drooping branches). A seedling of Banarasi, with prolific bearing and flat fruits was named as Chakla and now it is known as Chakaiya".

4.0 Species and Cultivars

4.1.0 Species

The genus *Phyllanthus* (Greek for leaf flower) comprises about 350 (Hooker, 1973) or even 500 species (Bailey, 1917), mostly shrubs, some herbs or trees. With the revision of the genus, *Phyllanthus emblica* has been placed under *Emblica* as *Emblica officinalis* (Anon., 1969). There are other species which are also used for pickling *etc.*, as *Phyllanthus acidus*, popularly known as Otaheite gooseberry, star gooseberry, or country gooseberry; *Emblica fischeri* (syn. *Phyllanthus fischeri*) and *Phyllanthus longiflorus.*

Pathak (2003) listed 28 species of *Phyllanthus* which are distributed around the world as given below.

Phyllanthus emblica	*P. acutissimus*
P. albidiscus	*P.amarus*
P. angkorensis	*P. chamaepeuce*
P. clarkei	*P. collinsae*
P. columnaris	*P. elegans*
P. emblica	*P. geoffrayi*
P. gracilipes	*P. kerrii*
P. lingulatus	*P. mirabilis*

P. oxyphyllus	*P. pachyphyllus*
P. pulcher	*P. reticulates*
P. roreus	*P. sikkinensis*
P. virgatus	*P. acidus*
P. emblica	*The Phillipines*
P. acidus	*P. javanicus (Ceremai)*

4.2.0 Cultivars

There has been no standardization of cultivars of aonla and they are mostly known on the basis of size (Janakiamal and Raghavan, 1957), colour (Singh *et al.*, 1967) or after the names of places. The cultivars mainly classified according to their colour are green-tinged, red-tinged, pink-tinged, white-streaked, and Bansi Red. Singh (1974) reported that the important cultivars grown in the district of Pratapgarh (Uttar Pradesh) are Banarasi, Chakaiya and Pink-tinged, which can be grown successfully under North Indian conditions.

Banarasi

Tree has upright growing habit with three branchlet per node. Fruits large in size, flattened oblong; skin smooth, yellowish, segments raised in three parts; flesh moderately fibrous, soft, semi-transparent. It has less number of female flowers and has self-incompatibility, hence it is a shy bearing. The fruits contain high amount of ascorbic acid.

Chakaiya

The tree has spreading habit. It bears profusely and has 4.03 female flowers per branchlet. Fruits small to medium, flattened; skin smooth, rosette, greenish in colour; flesh fibrous and hard.

Francis

The branches have dropping habit. Moderate in bearing. Fruits large, flattened oval; skin smooth, greenish yellow; flesh soft, nearly fibreless and has moderate keeping quality and is highly susceptible to fruit necrosis.

Laxmi-52

This is being cultivated in the private orchard as superior chance seedling of Francis *i.e.* Laxmi-52. Its tree has semi-erect growth and branches do not droop like its parent Francis. Fruit size large, 40-60 g, diameter 4.0-4.5 cm with 6 ridges. In early part of fruit growth, fruit colour is light pink which disappears on full development. Leaves are longer, broader, dark green than other aonla cultivars. It is mid season maturing (mid November to December) and is free from necrosis with yielding potential of 2-2.5 q/tree (after 10 years). Owing to larger fruit size and attractive colour, it fetches higher price in the market. The fruit has been found suitable for preparation of segments in syrup, candy and preserve.

Species of Aonla

Emblica officinalis

Phyllanthus acidus

Cultivars of Aonla

Kanchan

Krishna

Chakaiya

Cultivars of Aonla

Neelum-NA-7

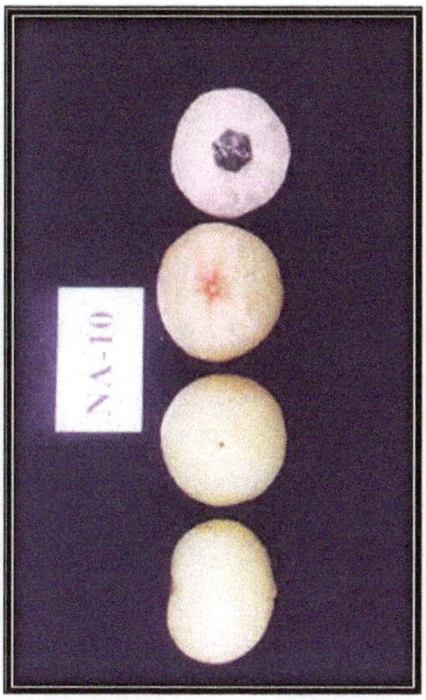

Balwant-NA-10

Francis

Kanchan

Krishna

Chakaiya

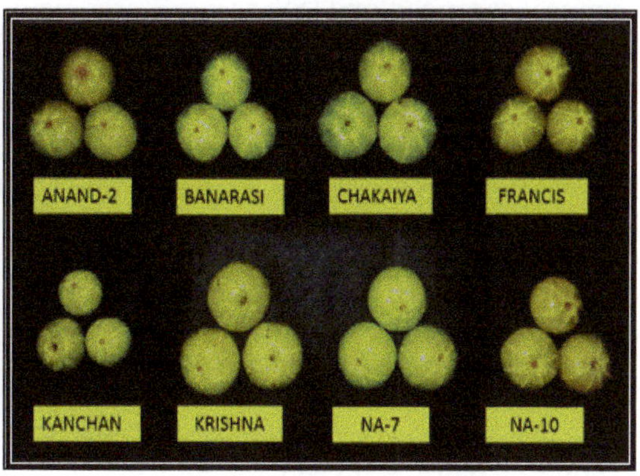

Different Varieties

NA 4 (Krishna)

This is probably a chance seedling of Banarasi cultivar. The cultivar bears moderately. Fruit medium to large in size, flattened, conical, angular, basin papillate in shape; skin smooth, yellowish in colour with red blush on the exposed surface; flesh fibreless, hard and semi-transparent.

NA 5 (Kanchan)

It is believed to be a chance seedling of Chakaiya. The tree has spreading habit. It bears profusely. Fruits small to medium, flattened oblong; skin smooth, yellowish in colour, ideally suitable for preparation of pickles.

NA 6

A seedling selection from Francis, precocious, prolific and regular bearer, no incidence of necrosis, ideal for preparing processed products.

NA 9

Early maturing chance seedling from Banarasi; fruits large, suitable for preservation.

NA-10

Chance seedling developed from cultivar Banarasi, locally known as 'Agra Bold'. Semi-tall and semi-spreading growth habit with deep pink and long inflorescence. Fruits are round and brown in colour, growth initiates early in season, at pea stage appears pinkish yellow, which disappeared gradually later on. Fruit is flattened round with moderate size. Fruit skin rough, yellowish green with pink tinge. Flesh is slightly fibrous, whitish green, soft, juicy and highly astringent. It is the earliest maturing variety with good keeping quality. Mildly susceptible to fruit necrosis

Naik and Chundawat (1993) evaluated some newly evolved aonla cultivars for their physico-chemical attributes. It was observed that the cultivars Anand-2 followed by Anand-1 excelled over Desi in respect of fruit weight, stone weight, pulp/stone ratio, pulp thickness, fruit size and number of fruits per kg weight. However, the cv. Desi was found superior to newly evolved cultivars in respect of chemical composition (TSS, ascorbic acid, acidity, tannin and moisture).

The fruit weight and chemical composition of some aonla cultivars are presented in Table 2.

Other varieties recommended for cultivation is BSR-1 (Bhavanisagar), Gujarat Aonl 1 and Goma Aishwarya (released by CIAH, Bikaner).

5.0 Breeding and Improvement

Aonla improvement has received a lot of attention in recent years. Emphasis has been to identify/develop selections with high percentage of female flowers, good bearing, good skin colour and suitability for processing (Chadha, 1998).

Genetic diversity in aonla was explored by Rai *et al.* (1993). Thirty three genotypes were analysed for physico-chemical traits of fruits. Wide range of

Table 2: Fruit Weight and Chemical Composition of some Aonla Cultivars

Cultivars	*Fruit Weight (g)*	*Total Soluble Solid (°Brix)*	*Total Sugar (Per cent)*	*Acidity (Per cent)*	*Vitamin C (mg/100 g)*	*References*
Banarasi	38.25	12.9	8.56	2.4	636.8	
Chakaiya	26.60	10.7	9.78	1.94	474.5	Singh *et al.*
Desi	12.90	14.8	6.96	2.35	458.9	(1987)
Harpharori (stargoose berry)	12.10	13.5	7.52	2.49	598.0	
Francis	41.5	12.0	4.50	1.70	385.0	
NA 4 (Krishna)	40.0	11.5	4.90	1.40	475.0	Pathak *et al.*
NA5 (Kanchan)	32.0	10.0	4.0	1.45	500.0	(1985)

variation in fruit length, diameter, average weight, number of segments per fruit, TSS, acidity and vitamin C was recorded. Variability was also noted in fruit colour, shape, apex, base, cavity, ridges on fruit surface, tannin and fibre content in pulp. In the character association analysis, fruit length and diameter were positively correlated with average weight. However, acidity was negatively correlated with fruit length, diameter and average fruit weight. A seedling tree (MR/RD-89/9) was found to be promising due to its bigger sized fruits of attractive golden yellow colour having field tolerance to necrosis. A few superior clones (MR/RD-89/17, -30 and -32) from vegetatively propagated elite plants of cv. Banarasi were identified for direct adoption on the basis of profuse bearing, better fruit quality and attractive colour.

Screening of 10 accessions under field condition for resistance to rust revealed that the entry E0.5 was resistant, while E0.2 was highly susceptible. Rest of the accessions were either moderately susceptible or susceptible to the disease (Suriachandra Selvan *et al.*, 1993). Occurrence of xenia effect for fruit size and weight were reported from the crosses Chakaiya x Krishna, Francis x NA7, Banarasi x NA9, Kanchan x NA6 and NA6 x NA9 (Srivastava and Pathak, 1993).

5.1.0 Cytogenetics

Systematic cytogenetic efforts have not been carried out in aonla and crop improvement has been confined to merely selection of desirable traits from the wild as well as seedling populations of the cultivated genotypes. Basic chromosome number of *E. officinalis* is 2n = 26, whereas Perry (1943) reported somatic chromosome number in aonla to be 2n = 28, while variants possessing chromosome numbers from 2n = 23 to 104 have been observed (Amal and Raghvan, 1957).

Phyllanthus emblica L. has 2n=104. It is varied from the previous recorded (2n = 28, 98 and 104). B chromosomes were found in *P. pulcher* Wall. ex Muell. Arg. (Soontornchainaksaeng and Chaiyasut, 1999).

5.2.0 Constraints in Breeding

Pathak (2003) reviewed the status of Aonla and reported that varietal improvement in aonla had been confined mostly to selection of promising genotypes from the existing seedling tubes (Pareek, 2001; Rai *et al.*, 1993; Dhandar and Shukla, 2003). The major constraints in improvement of aonla are (Pareek, 2001; Pathak *et al.*, 1993):

- Highly heterozygous nature
- Perennial growth
- Long juvenile phase
- Complex taxonomy
- Cross and self-incompatibility
- Chromosomal aberrations
- Higher number of staminate flowers in the determinate shoots
- Tiny flower size

5.3.0 Breeding Objectives

Some trees bear small bitter drupes while others produce large and less bitter drupes. The fresh fruit market prefers the larger and less bitter drupes. Therefore, there is a grower driven selection towards the large and less bitter drupe bearing trees. However, small and bitter drupe types are immensely important to maintain the *P. emblica* diversity and mostly to preserve its medicinal value. The grower driven selection leads to uprooting of the small drupe bearing trees causing genetic erosion, threatening the use of it in indigenous medicine (Mawalagedera *et al.*, 2014 a). Pathak (2003) suggested the following areas to be considered as breeding objectives:

- To develop dwarf tree stature for medium density orchards
- To develop precocious and prolific bearing cultivars
- To develop excellent fruit quality for processing (for value added products)
- To develop/varieties with resistance to biotic stress, *viz.*, pests and diseases
- To develop cultivars with resistance to abiotic stresses like frost, heat, drought and humid soils

6.0 Propagation

6.1.0 Seed

Aonla has long been raised from seeds but the plants are not true-to-type and produce small sized fruits, inferior in quality. Seedlings are raised from seeds and used as rootstock. Bajpai (1969b) reported that the seeds attain full maturity by February when they should be extracted and sown as early as possible for getting high percentage of germination. Bhujbal (1975) obtained the highest percentage of germination (92.5) and seedlings with best root system from seeds soaked in 500 ppm GA_3 for 24 hours. Shreekumar and Aravindakshan (1984) obtained 68 to 85 per cent germination by drying the fully ripe fruits in the sun for 2 to 3 days, when they split open within 48 to 60 h and release the seeds, the seeds were dried in the sun and sown immediately. Mechanical isolation of the seeds by breaking open the stones gave only to 38 per cent germination.

The seeds of *P. emblica* exhibit a long dormancy period. The observations indicated that the natural sexual propagation potential of *P. emblica* is very poor and the seed behaviors as semi orthodox (Pushpakumara *et al.*, 2007). Under natural conditions, *P. emblica* seeds are dispersed either through ruminants (Prasad *et al.*, 2006) or by passive means. But there is a high tendency towards passive means of dispersal. Therefore, under natural conditions, this semi orthodox behavior and long-term dormancy aid the seeds to remain viable until the seeds come into contact with the proper substratum for germination. But in terms of agronomy, this is a major drawback in implementing large scale *P. emblica* plantations (Pushpakumara *et al.*, 2007; Mawalagedera *et al.*, 2014b). Also in setting up breeding programs, long-term dormancy causes a big challenge in germinating seeds to create populations. Thus, there is a need to develop a method to break up the seed dormancy artificially. In order to do so, the methods available to break the dormancy of seeds have to be studied extensively for P. emblica (Mawalagedera *et al.*, 2014b). The seeds of *P. emblica* are semi orthodox and exhibit a long dormancy period hindering the natural sexual propagation. Therefore, it still remains as an underutilized fruit crop in Sri Lanka due to its lack of quality planting material and poor propagation techniques. Long term dormancy also causes a big challenge in germinating seeds to create populations in breeding programs. In order to promote *P. emblica* from its underutilized status, what seems most feasible is to develop a method to break up the seed dormancy artificially. In order to do so, the methods of breaking the dormancy of *P. emblica* seeds have to be studied. Therefore, the aim of this study was to identify a method to break up the seed dormancy of *P. emblica*. The seeds were extracted from 21 trees belonging to three different districts in Sri Lanka. The selected viable seeds were subjected to four different pre treatments: none treated seeds (*i.e.* control), seeds scarified, seeds scarified and treated with 1 per cent gibberellin and seed coat removed and followed by seeds treated with 1 per cent gibberellin. From the four treatments, seed dormancy was overturned with a germination percentage of 43 per cent by the seed pre treatment where the seeds were scarified and treated with 1 per cent gibberellin and no other pre treatment methods were successful in breaking the dormancy. This suggests that the natural germination potential of *P. emblica* seeds is very low and it can be overridden by seed scarification and gibberellin pre treatment (Mawalagedera *et al.*, 2014b).

Higher levels of sodicity reduced the percentage of germination, days taken for germination and survival of seedling (Table 3), which might be due to reduced availability of water resulting from high osmotic concentration in soil (Tewari *et al.*, 1999). In seed germination studies, the seeds scarified with concentrated sulphuric acid for 30 seconds were found to be the best one with very high per cent germination (80.39), taking least number of days for 50 per cent germination (12.27) and highest vigour index (1223) (Nayaka, 2006).

Table 3: Effect of Sodicity on Seed Germination, Seedling Growth and Survival of Aonla*

ESP Levels	*pH*	*Seed Germination (Per cent) (Germinators)*	*Seed Germination (Per cent) (Earthen pans)*	*Days taken for 20 per cent Seed Germination in Earthen Pans*	*Seedling Length (cm) 110 Days After Sowing in Earthen Pans*	*Per cent Survival of Seedling at 150 Days After Sowing*
10.00	7.1	35.75	38.75	9.50	10.97	34.25
14.26	8.6	28.0	25.00	11.00	9.35	19.25
32.72	8.9	10.00	19.75	14.44	7.43	8.75
47.35	9.9	0.00	6.75	0.00	0.00	0.00
64.44	10.2	0.00	4.25	0.00	0.00	0.00
77.74	11.6	0.00	0.00	0.00	0.00	0.00
CD (0.05)	2.48	1.78	0.88	0.61	3.01	

Angular transformed values are used for statistical analysis.

*Tewari *et al.* (1999)

6.2.0 Vegetative Propagation

In order to overcome the disadvantages of seed propagation, multiplication of superior types of aonla has been suggested by adopting vegetative methods (Hayes, 1957; Singh *et al.*, 1967; Srivastava, 1964, 1965; Gangwar*et al.*, 1975; Amin, 1978).

6.2.1 Cutting

Rajan and Ram (1982) recorded 87.3 per cent rooting in cuttings using IBA at 15,000 ppm with bottom heat (33°C).

6.2.2 Budding

Of the various methods of vegetative propagation, budding has been found to be most practical and shield budding is the commercial method of aonla propagation (Singh, 1974). Singh (1952) reported that one-year old seedling with a girth of about 1 cm should be shield budded in early June with healthy and plump buds from new growth. Shield budding gave a success of 70-81.1 per cent (Singh *et al.*, 1967; Gangwar *et al.*, 1975). However, even higher percentage of bud-take was recorded with forkert and patch methods (Srivastava, 1964, 1965; Moti *et al.*, 1976).

Pathak *et al.* (1991) conducted a trial to find out the best method and optimum duration of budding for large scale multiplication of aonla plants. Patch budding showed better response followed by modified ring budding (Table 4). It was concluded that budding should be performed from mid-May to mid-August for better success which may be continued up to September with appreciable success. Saroj *et al.* (2000) reported that seedlings for rootstock in perforated polybags were ready for budding in 6 months and patch budding resulted in 90 per cent success.

Table 4: Per cent Bud Sprouting and Survival with different Methods of Budding*

Methods	Per cent Bud Sprouted	Per cent Bud Survival
Modified ring	79.70	67.40
	(63.22)	(55.18)
Patch	83.30	73.02
	(65.92)	(58.95)
Shield	21.26	15.30
	(27.44)	(23.04)
C.D. at 5 per cent	1.40	1.80

Angular transformed values are given in parenthesis.

*Pathak *et al.* (1991)

6.2.3 Grafting

Amin (1978) recorded 70 per cent success following softwood grafting on the terminal shoot of the rootstock raised *in situ* for a year or more. It would be advisable to raise seedling rootstocks *in situ* and bud them with superior types, particularly in dry areas where mortality of budded plants is usually high. Propagation studies revealed that, season of grafting had profound influence on graft success. Maximum success per cent of 87.50 was recorded in November grafting followed by 66.69 per cent in case of December. Least success was recorded during June grafting (16.52 per cent). Influences of rootstock age on success of grafts were significant. The higher per cent successful grafts (62.88) were observed in eight months old rootstocks followed by four months (56.25 per cent) old rootstocks (Nayaka, 2006).

The most important factor is the proper selection of mother plant which is highly fruitful and the buds should be taken from a branch which has good number of female flowers, otherwise the plants will be unfruitful owing to the appearance of a large number of male flowers.

6.2.4 Top Working

Old trees or inferior type can be rejuvenated and easily changed into superior type by resorting to working. Singh (1953) reported that inferior trees can be headed back to a height of 1.2 metres from the ground during March and the stumps produced can be shield budded in early June with scions of improved cultivars. Frost damaged aonla plants were also rejuvenated by T budding (Gangwar*et al.*, 1975).

Top working can be easily adopted in rejuvenated trees to upgrade the old and senile plantations of seedling progeny with superior commercial cultivars (Singh and Mishra, 2007). Top working involves two steps (i) beheading of the tree to be top worked and (ii) budding with a elite material on the new flushes emerging from beheaded tree. The plants are headed back during December – January to an extent of 2.5 to 3.0 m above the ground level. Four to six shoots from the outer directions on main limbs should be allowed to develop. During June- July, in the monsoon period, grafting with the desired scion variety is one. Since aonla is self incompatible, *i.e.* the pollens of same tree/variety cannot fertilize its own ovary, the production from mono-culture orchards without appropriate polliniser varieties, suffers adversely as a result of problem of fruit set. Hence, polliniser variety is budded on developing shoots of pruned trees to strengthen pollination process and enhancement of fruit set and productivity. Budding with mixed varieties results in better yield. The best combination is NA-6 with NA-7; NA-7 with NA-10 and Kanchan with Krishna (Singh and Mishra, 2007).

7.0 Area and Production

The data on the area under aonla cultivation are not available. Although, it is found growing in different states throughout tropical India even up to elevations of 1,500 metres in South India (Singh, 1979), it is more popular in Uttar Pradesh, where it is largely cultivated in commercial orchard in Azamgarh, Pratapgarh, Varanasi and Bareilly districts. In view of its diverse uses, its cultivation is increasing fast and the tree is becoming popular with the orchardists. The production details of major aonla growing states are given in Table 5.

Table 5: Aonla Production in Top 10 States in India*

State	*Production ('000 tonnes)*	*Share (Per cent)*
Uttar Pradesh	384.32	35.79
Madhya Pradesh	302.18	28.14
Tamil Nadu	152.87	14.24
Gujarat	81.90	7.63
Chattisgarh	43.29	4.03
Assam	17.76	1.65
Bihar	14.92	1.39
Maharashtra	12.25	1.14
Jammu & Kashmir	12.10	1.13
Rajasthan	11.19	1.04

* NHB (2017-18).

8.0 Soil and Climate

8.1.0 Soil

Aonla can be grown in light as well as heavy soils except very sandy one. However, well-drained fertile loamy soil is the best. The trees are also adapted to dry conditions and can grow in moderately alkaline soils. Cultivar variation in salt tolerance has been reported. Aonla seedling and Chakaiya exhibited salt tolerance up to ESP 43.5 and 35.0 and ECe 9.7 and 9.0 mmhos/cm respectively (Dikshit, 1987), while Kanchan and Francis could tolerate up to ESP 35.0 and ECe 10.0 mmhos/cm (Shukla, 1990). Salinity caused excessive accumulation of Cl in leaves, while sodicity resulted in accumulation of Na in toxic amounts in all the cultivars (Shukla *et al.*, 1993).

Propagation of Aonla

Patch Budding

T-Budding

T-Budding

Cleft Grafting

Training and Pruning of Aonla

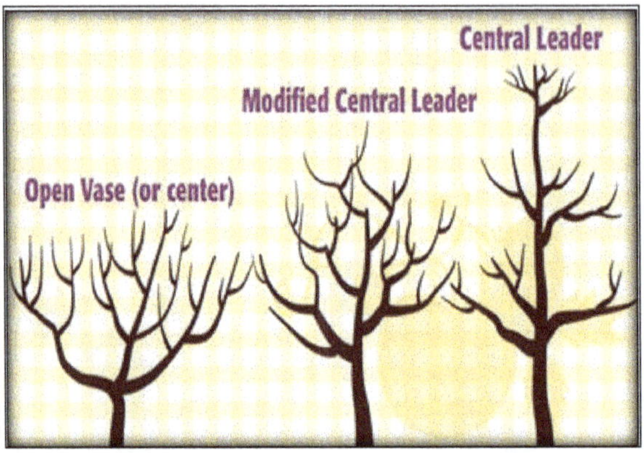

Aonla Training System Diagram

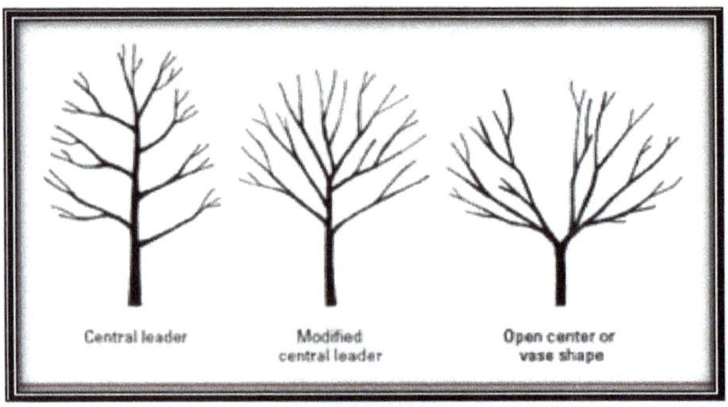

Modified Central Leader System in Aonla

Aonla Shoot Pruning

8.2.0 Climate

Aonla is a subtropical fruit but its cultivation in tropical climate is quite successful. The tree is not much affected either by hot wind or frost, however, older trees are damaged by frost (Gangwar*et al.*, 1975). The young plants should be protected from hot wind during May-June and from frost during winter, at least up to the age of 3/4 years under North Indian conditions. The mature trees can tolerate freezing temperature as well as a temperature as high as 46°C (Shanker, 1969). The warm season appears conducive for the initiation of floral buds (Bajpai, 1963). It can be grown in semi-arid, subtropical region (Pathak, 1991) with an annual rainfall of 350-500 mm (Vashishtha, 1991).

9.0 Cultivation

9.1.0 Planting

Prior to planting, the field should be deeply ploughed, harrowed and levelled. The pits, 1 m3 should be dug during May-June at appropriate distance and filled with surface soil mixed with 10-15 kg of decomposed farmyard manure. Singh (1991) reported that filling of pit with normal soil + 25 kg FYM + insecticide (chlorpyriphos) for preventing termite attack resulted into maximum percentage of plant survival, average height and trunk girth. Healthy grafts or budded plants are planted during rainy season, preferably in early monsoon in July at a distance of 9 m (Singh *et al.*, 1967) or 11 m (Singh, 1974) each way in square system. In dry areas where mortality after transplanting is usually high, the seedling rootstocks can be raised *in situ* at appropriate distance for budding with superior clone. Use of gypsum (75 per cent) in alkaline soil (pH 9.1 and ESP 30.6) was reported to be beneficial for obtaining maximum growth of budded plants (Singh and Gaur, 2000).

By adopting double hedgerow system of planting in *aonla* the productivity per unit area can be enhanced to nearly double over conventional square system of planting under the rainfed condition in semi-arid ecosystem (Singh *et al.*, 2011).

9.2.0 Training and Pruning

As the branches of aonla trees often break off carrying heavy crop load due to brittle nature of wood, the plant should be trained to develop a low headed one and the main branches constituting the foundation framework should be made to arise on the trunk within 0.75 m from the ground. The plant should be trained to modified leader system. The framework should be developed by encouraging the growth of four to six well-spaced branches with fairly wide angle.

The pruning of the bearing plants can be done after the termination of the crop each year. While pruning, dead, diseased, broken, weak, crossing branches and suckers appearing from rootstock should be removed.

9.3.0 Irrigation

Established aonla orchards generally do not require irrigation in normal soils. No irrigation is required during rainy and winter season. Irrigation at 15-20 days interval is desirable in dry summer particularly during early years of orchard establishment under wasteland conditions (Wali *et al.*, 2015).

The young plants require watering during summer months at fortnightly interval, particularly till they have fully established. Watering of mature, bearing plants is also necessary from April to June at bi-weekly interval to obtain higher fruit set and reduce fruit drop. Irrigation during October to December at 20 days' interval helps in better development of fruits.

The effect of irrigation and mulching on growth of aonla (*Emblica officinalis* [*Phyllanthus emblica*]) cv. NA 7 was investigated (Shukla *et al.*, 2000). The treatment comprised 3 irrigations (daily drip irrigation, alternate day drip irrigation and conventional surface irrigation by basin method) and different types of mulching materials, *i.e.* 200 gauge black polyethylene, 8-cm thick farmyard manure (FYM), paddy straw, grass, and un mulched control. Plant height, canopy spread and stock girth were significantly better under alternate day drip irrigation over conventional method. Among mulching treatments, black polyethylene was the most effective mulch material however, among organic mulches paddy straw was the best for the same parameters. Leaf nutrient content (N, P, K, Ca and Mg) was maximum in alternate day drip irrigation and minimum under conventional method. Among mulching treatments, the maximum P, K, Ca and Mg values were found in FYM whereas, the level of nitrogen was more in paddy straw (Shukla *et al.*, 2000).

Shukla and Suhail (2009) studied the effect of drip irrigation and mulches on physical, chemical and biological properties of soil, water use efficiency and productivity of aonla cv. Narendra Aonla-10 (NA-10). There were four irrigation regimes in terms of IW/CPE ratios of 1.0, 0.8, 0.6 and 0.4 and three types of mulches, namely black polythene, paddy straw and control (no mulch). The results indicated that the bulk density of soil was significantly low, while hydraulic conductivity was the highest with the use of paddy straw mulch. Drip irrigation at 1.0 IW/CPE significantly reduced the soil pH, exchangeable soil percentage (ESP) and electrical conductivity (ECe). Mulching with paddy straw enhanced soil organic content, while it reduced soil pH, ESP and ECe. Maximum fungal population in soil was recorded at 0.60 IW/CPE, while earth worm and bacterial populations were higher at 1.0 IW/CPE. Among different mulches, paddy straw mulching resulted in the highest fungal, bacterial and earthworm populations. Maximum water use efficiency and aonla fruit yield were recorded with 0.6 IW/CPE. Among mulches, black polythene was found to be the best for higher water use efficiency and fruit yield. Thus, it helps in cutting down the water requirement of aonla to 0.6 IW/CPE (60 per cent evaporation replenishment) with drip irrigation and black polythene mulching for optimum fruit yield. However, paddy straw mulch was better for maintaining optimum soil health (Shukla and Suhail, 2009).

The Aonla (*Emblica officinalis* Gaertn.) is being cultivated as a rain-fed tree crop in the Shivaliks of Punjab. Soil moisture stress is one of the major constraints for early orchard establishment in the region. Vashisht *et al.* (2018) studied the optimum level of irrigation with and without polythene mulching for the establishment of aonla in the water scarcity situations. During a typical drying period, a total number of 13, 9 and 5 irrigations based on IW/Pan-E ratio of 0.7 (I0.7), 0.5 (I0.5) and 0.3

(I0.3) were applied to aonla plants. The respective improvement in stock girth and plant height was 67.5 and 82.4 per cent higher by irrigating the plants more frequently at I0.7 than at least frequent regime of I0.3. Respective increase of 43.5 and 58.1 per cent in stock girth and plant height was observed with polythene mulch over no mulch after two years of plantation. The stock girth and plant height was significantly higher by 32.0 and 48.27 per cent by irrigating the aonla plants at I0.5 with mulching than irrigating at I0.7 without mulching. This resulted in saving of 20 cm of irrigation water (four irrigations) on net area basis during a typical drying period. With this saving of water, additional area of the orchard may be established (Vashisht *et al.*, 2018).

9.4.0 Manuring and Fertilization

Very little is known about the manuring and fertilization of aonla as no schedule based on experiments has so far appeared. In a sand culture study, Pathak *et al.* (1993) diagnosed the nutrient deficiency symptoms in aonla (Table 6). Concentrations of 0.85, 0.065, 0.208, 0.435, 0.205 per cent or lower of N, P, K, Ca and Mg respectively, in leaves, corresponded to their deficiency symptoms.

Table 6: Nutrient Deficiency Symptoms in Aonla*

Nutrient	*Type of Leaf*	*Symptoms*
N	Older leaves	Yellowing of leaves
P	Older leaves	Tip and marginal yellowing
K	Older leaves	Yellowing and scorching at tip and slightly on the margin of leaves
Ca	Younger leaves	Yellowing, burning, scorching and curling of tip
Mg	Younger leaves	General yellowing in half portion of leaf

* Pathak *et al.* (1993).

Singh *et al.* (1967) observed that each tree should be given 3-4 kg superphosphate. Shanker (1969) suggested a schedule of fertilization, the young plant should be given 15-20 kg of well-rotten farmyard manure and the mature tree 30-40 kg per year during September-October. In addition, application of 30 g N for each year of age in plant up to 10 years and afterwards 680-900 g of N/plant/year should be provided. Every mature tree should also be fertilized with 1 kg superphosphate and 1-1.5 kg muriate of potash. The above fertilizers should be given in two equal split doses to mature bearing tree, once during September-October and again during April-May after setting of fruit. The plants need to be irrigated after fertilizer application. The dose of manure and fertilizer, recommended by the Narendra Dev University of Agriculture & Technology, Faizabad, UP is application of 10 kg FYM, 125 g N, 50 g P_2O_5 and 50 g K_2O per one-year-old plant. The dose should be increased every year in the same proportion upto the age of 8 years (1000 g N, 400 g P_2O_5 and 400 g K_2O/tree) after which the fixed dose should be applied each year, in two splits during September-October and April-May (Kaulgud, 1996).

As regards nutritional status, Garg and Khanduja (1976) observed from leaf analysis for N, P, K, Na, Ca and Mg of plants, raised in normal and alkaline soils, that

N and P were most affected by soil pH. Three months old shoots should be sampled from the middle portion of the indeterminate shoots during July (Pareek, 1997).

The maximum TSS (12.35 per cent), reducing (3.05 per cent), non-reducing (2.58 per cent) and total sugars (5.63 per cent), Vitamin C (560.5 mg/100g of fruit pulp) and fruit yield (99.04) kg/tree) recorded with treatment of $CuSO_4$ (0.4 per cent) + $ZnSO_4$ (0.25 per cent) + Borax (0.25 per cent) whereas minimum acidity (1.37 per cent) and stone weight(1.86g) were also recorded with foliar feeding of $CuSO_4$ (0.4 per cent) + $ZnSO_4$ (0.25 per cent) + Borax (0.25 per cent). On the basis of present finding, it can be concluded that foliar feeding of $CuSO_4$ (0.4 per cent) + $ZnSO_4$ (0.25 per cent) + Borax (0.25 per cent) gave best result for the production of maximum fruit yield and better quality of aonla fruits (Mishra *et al.*, 2017).

9.5.0 Mulching

Mulching helps in conserving soil moisture, checks the weed growth, improves the soil physical and chemical properties and adds organic matter (in case of organic mulches). Rao and Pathak (1996) compared the effects of different types of mulches on soil properties in aonla orchard. It was found that use of locally available organic mulches (paddy straw, sugarcane trash and grasses) was quite effective in improving the soil physical properties *i.e.*, porosity, hydraulic conductivity and field capacity. Organic mulches were also effective in reducing the ECe, ESP and cation contents in tree basin and in improving the organic carbon content, available nitrogen and potassium contents. It was concluded that use of paddy straw over a longer period as mulch needs to be popularized in aonla orchards particularly on salt affected soils.

The effect of different organic mulches such as paddy straw, maize straw, rice husk, grasses, *subabul* loppings on soil properties, earthworm population, growth, yield and fruit quality attributes was studied. Various organic mulches significantly increased the soil moisture status at various soil depths. Paddy straw mulch recorded the maximum soil moisture content followed by maize straw and grasses, among the different evaluated organic mulches. The findings of the study indicated that paddy straw mulch followed by maize straw and grasses had given favourable results with regards to soil moisture and physico-chemical attributes of soil and earthworm population in basin soil. Paddy straw mulches were found to be more effective in producing maximum growth than rest of the mulching treatments. Plants treated with paddy straw mulch recorded highest yield 09.00 kg/plant and 42.15 kg/plant during 2005 and 2006. TSS, total sugar, total phenols and vitamin C were observed maximum with paddy straw followed by maize straw mulch. Among the evaluated organic mulches, straw had showed better response followed by maize straw. In view of the superiority over other organic mulches, use of paddy straw as mulch material over long period needs to be popularized for increasing the productivity of aonla orchards in rainfed condition under semi-arid ecosystem (Singh *et al.*, 2010).

9.6.0 Intercropping

It is better to follow clean culture in aonla plantation except during the rains when leguminous crops, *viz.*, pea, moong, cowpea, *etc.*, can be grown up to 8 years when the plants usually remain in non-bearing stage and their root system and top growth are not much developed.

Traditionally, efforts have been made by the native people, obviously to achieve sustainable production ensuring ecological stability to grow several species together, e g pearl millet, [*Pennisetum glaucum* (L.) R. Br. emend. Stuntz], mothbean [*Vigna acontifolia* Jacq. Marechal], cluster bean (*Trigonella foenum graecum* L.) chickpea (*Cicer arietinum* L.) between the natural stands of khejri [*Prosopis cineraria* (L.) Druce] and boradi (*Ziziphus rotundifolia* Lam.) (Awasthi and Pareek, 2008). However, no scientific data are available. Hence, Awasthi *et al.* (2009) conducted a study on intercropping under arid conditions of Bikaner in newly established 'NA7' aonla (*Emblica officinals* Gaertn). Mothbean (*Vigna acontifolia* (Jacq.) Marechal) grown during rainy (*kharif*) season was a common crop in rotation with winter (*rabi*) crops, i e fenugreek (*Trigonella foenum-graceum* Linn.), chickpea (*Cicer arietinum* L.), mustard [*Brassica juncea* (L) Czernj. & Cosson] and cumin (*Cuminum cyminum* L). Growth parameters in terms of plant height, stem girth, canopy spread and canopy volume of aonla was recorded to be significantly more with intercrops compared with its sole plantation. Higher grain and straw yield were recorded in mothbean–chickpea (497, 1 250 kg/ ha) and mothbean-fenugreek (465, 1 161 kg/ha) crop sequence. Amongst the winter (*rabi*) crops, grain yield of fenugreek, chickpea, mustard and cumin were higher by 28.05, 38.11, 19.96 and 36.50 per cent, respectively, when grown in association with aonla compared to its sole crops. The highest net profit (Rs 28 260/ha) was obtained from mothbean–cumin cropping system, followed by mothbean–chickpea (Rs 25 024/ha) cropping system. Mothbean–chickpea intercropping with aonla supplemented 22.01, 5.00 and 27.90 kg/ha nitrogen, phosphorus and potassium through crop residues, followed by mothbean– fenugreek crop sequence.

An intercropping study was carried out with different leguminous based crop sequences as intercrop in six year-old aonla orchard planted at 6 m x 6 m spacing under rain fed farming situation to identify suitable intercrops for higher economic returns (Negi *et al.*, 2013). The intercropping sequences included in the study were: aonla+green gram (var Samrat)-mustard (var Pusa Tarak); aonla+cowpea (var Kashi Kanchan)-mustard (var Pusa Tarak); aonla+okra (var Kashi Pragati)-gram (var JG-16); aonla+cowpea (var Kashi Kanchan)-lentil (DPL-62) and sole aonla. The results of the two years of investigation revealed that leguminous crop based intercropping sequences in young aonla orchard not only improved the plant growth and fruit yield of aonla but also increased the net returns from the land. Among the four intercrop sequences, cowpea-mustard and okra-gram that gave maximum additional income of Rs 32,990.76 and Rs 32,909.71/ha and benefit: cost ratio (2.16 and 2.11) were found most profitable intercropping sequences in young aonla under rain fed fruit based production system. The sole crop of aonla recorded a net profit of only Rs 22,488.50/-under similar situation (Negi *et al.*, 2013).

From the experimental findings obtained in an investigation (Singh and Singh, 2015) by growing Elephant foot yam (*Amorphophallus*) cultivar Suran as an intercrop in aonla plantation with various sources of nitrogen, it was concluded that the growing suran as intercrop was found most to be suitable under aonla plantation when it was provided with vermicompost and mustard cake as organic sources of nutrients along with urea (Singh and Singh, 2015). Among the different cropping systems, the most economically viable and feasible cropping system was found to be (i) T8 (Aonla + Suran + 25 per cent N_2 from Mustard cake + 25 per cent N_2 from Vermicompost + 50 per cent N_2 from Urea) (ii) T2 (Aonla + Suran + 25 per cent N2 from Vermicompost + 75 per cent N_2 from Urea), (iii) T3 (Aonla + Suran + 50 per cent N_2 from Vermicompost + 50 per cent N_2 from Urea) and (iv) T5 (Aonla + Suran + 25 per cent N_2 from Mustard cake + 75 per cent N_2 from Urea), which were estimated on the basis of growth, yield and quality parameters of intercrop as well as aonla, and soil fertility status, orchard fertility, gross income, net income and cost benefit ratio (Singh and Singh, 2015).

10.0 Flowering, Pollination, Fruit Growth and Development

10.1.0 Flowering

Bajpai (1965) reported flower bud differentiation in Banarasi aonla in the first week of March. The development of male and female gametophyte was normal. The flowers commence opening from the last week of March and the blooming period lasts for 3 weeks. Male flowers appear in clusters in the axil of leaf all over the branchlet while female flowers on the upper end of a few branchlets only (Bajpai, 1957). Bajpai (1968) reported sex ratio (male to female) to be 307.9:1 and 197:1 in two successive years, indicating marked variation in the expression of sex. Aonla genotypes NA6 followed by NA7 had low sex ratio, while Banarasi and NA9 showed high sex ratio. Shoots projected at south had higher number of female and male flowers per branchlet and the number was lowest in west direction. The number of female flowers per branchlet was higher at basal portion of the indeterminate shoot (Srivastava and Pathak, 1993).

The maximum number of male flowers opens between 6 and 7 pm and dehiscence of anthers occurs soon or about 10-15 minutes after anthesis. The female flowers open in stages and it takes 72 hours to open completely and the stigma becomes receptive on the third day of anthesis.

Aonla has unisexual flowers with male-biased sex ratio, peak flower anthesis and anther dehiscence took place in the evening in March-April. The longevity of a male flower was about 60 h and of a female flower, it was 120 h. Pollination by the biotic pollen vectors seemed to be an essential requirement for the successful fruit set in the flowers of this plant (Saini and Sihag, 2020).

10.2.0 Pollination and Fruit Set

Aonla is pollinated by wind as well as by bees the pollen grains are light and are produced in abundance (Wali *et al.*, 2015). Although actual fertilization could not be traced, zygote was observed within 24 to 36 hours of pollination (Bajpai, 1968).

It was believed that there was no self-incompatibility in aonla and the cause of poor fruit set (12.18-18.33 per cent) may be due to a high percentage of staminate flowers (Bajpai, 1968). However, later studies have indicated that many varieties are self incompatible (Singh and Mishra, 2007). An increase in fruit set with hand pollination (18.00-27.74 per cent) indicates the need of pollinating agents for better

Intercropping, Fruit Growth and Development in Aonla

Aonla Fruit Growth

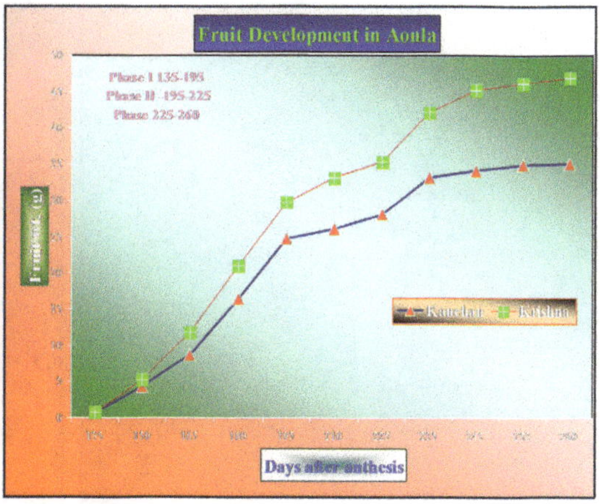

Graphical Representation of Fruit Development

Flowering and Pollination in Aonla

Anthesis of Aonla Flowers

Dehiscence Time of Aonla Flowers

Receptive Stage of Ovary of Aonla Flowers

Fruit Set Stage

setting. Srivastava and Pathak (1993) achieved maximum fruit set under open pollination followed by hand pollination and bagging. Combinations that were found to have better degree of compatibility determined by fruit set were Chakaiya x Francis, Francis x Krishna, NA7 x Krishna, Banarasi x Kanchan, Kanchan x NA6, NA6 x NA8 and NA9 x NA8. Shete and Karale (2000) studied fruiting of 10 aonla cultivars. Fruit set ranged from 43.58 per cent in Modibagh Selection to 81.90 per cent in Francis. Fruit drop continued in July, August and September and final fruiting percentage ranged from 19.23 per cent in Modibagh Selection to 49.52 per cent in Francis.

10.3.0 Flower and Fruit Drop

Bajpai (1968) recorded flower and fruit drop at three stages. The first drop is the highest as 70 per cent of the flowers drop off within three weeks of flowering due to degeneration of the egg apparatus and lack of pollination. The second drop occurs from June to September due to lack of pollination and fertilization. The third drop con-sists of fruits of various stages beginning from third week of August until October probably due to embryological and physiological factors.

The bearing aonla trees often suffer from heavy fruit drop which significantly lowers down the yield. Fruit drop often causes poor fruit set in majority of the cultivars and may result in substantial crop loss in susceptible ones. Among the major causes accounting for fruit drop are self-incompatibility, inadequate pollination, nutritional deficiency, water stress, insect-pest and disease infestations and hormonal imbalances (Allemullah and Ram, 1990 and Singh *et al.*, 2008). Based on the premise that suboptimal biosynthesis and poor translocation of phytohormones may be a major cause of fruit drop, a number of studies have been carried out to assess the potential of plant growth regulators (PGRs) in overcoming this problem in different fruit crops (Singh and Singh, 2015).

Application of plant growth substances on bearing trees of aonla cultivar Narendra Aonla-6 significantly increased fruit retention (Singh and Singh, 2015). The minimum fruit drop was recorded with the use of NAA (15 ppm) + Thiourea (0.1 per cent) followed by 2,4-D (10 ppm) + Thiourea (0.1 per cent) and GA_3 (50 ppm) + Thiourea (0.1 per cent). The maximum increase in size, weight and volume of fruits and the maximum fruit yield were noted on trees sprayed with NAA (15 ppm) + Thiourea (0.1 per cent). The improvement in fruit quality parameters (TSS, ascorbic acid, sugars and acidity) was highest with the application of GA_3 (50 ppm) + Thiourea (0.1 per cent) which showed nonsignificant difference with NAA (15 ppm) + Thiourea (0.1 per cent). Results revealed that foliar application of NAA (15 ppm) + Thiourea (0.1 per cent) or GA_3 (50 ppm) + Thiourea (0.1 per cent), twice during mid-May and mid-July, may effectively overcome the problem of fruit drop leading to higher yield of quality fruits (Singh and Singh, 2015).

An investigation was carried out during 2006 and 2007 to study the effect of boron (0.1, 0.2 and 0.3 per cent), zinc (0.2, 0.4 and 0.6 per cent) and copper (0.1, 0.2 and 0.3 per cent) along with a control on fruit drop, physical parameters and yield of aonla fruits cv. Banarasi (Singh *et al.*, 2012). All the characters studied were significantly improved by application of different micro-elements and their levels showing varying degree of their efficacy. The minimum fruit drop (56.84 and 50.22 per cent), maximum length of fruit (4.01 and 4.10 cm), breadth (4.31 and 4.35 cm), weight (46.85 and 47.34 g) and pulp content (44.66 and 45.16 g) were obtained under the foliar spray of zinc. Among the three concentrations, the higher level proved most effective in respect of all the characters. Zinc at its higher concentration proved the best treatment in improving the yield of aonla (Singh *et al.*, 2012).

10.4.0 Fruit Growth and Development

Kishore (2017) defined codes and phenological stages of Indian gooseberry (aonla) according to the extended BBCH (Biologische Bundesantalt, Bundessortenamt und Chemische Industrie) scale using three-digit numerical system. A total of 8 principal growth stages, *viz.*, bud development, leaf development, shoot development, reproductive development, flowering, fruit development, fruit maturation and senescence and 43 secondary growth stages have been described. The phenology of aonla becomes complex due to the exhibition of phyllanthoid branching and heterogamy. In order to simplify the phenological description, stages of bud development for determinate shoot (branchlet) and indeterminate shoot, and developmental phases of male and female flowers have been defined and described separately as per the extended BBCH scale. The sequential progression of principal growth stages of aonla indicated temporal variation in growth pattern as well as overlapping of some of the phenophases. The extended BBCH scale for aonla has significant importance in crop improvement, germplasm characterization and crop management strategies like management of fruit drop, nutrient and pests, and harvesting of fruits at optimal maturity. The study will not only helpful in assessing the impact of climate on phenology especially on yield attributing parameters but also in evaluating the adaptation potential of genotypes under different agro-climatic conditions.

After the fruits have set, the embryo lies in dormant condition and ovary does not exhibit any symptom of external growth until middle of August. The diameter and volume of the fruit increase rapidly thereafter, and the maximum growth is achieved by November after which there is not much increase in size. The growth of the fruit is due to the enlargement of the cells of the mesocarp while endocarp cells form the hard stone cells (Bajpai, 1968). Anatomical changes in developing aonla fruits have also been reported by Bajpai (1969). Balasubramanyan and Bangarusamy (1998) determined the maturity standard of anola fruits and reported that fruit length and diameter increased rapidly up to 75 days from fruit set and then slowed up to 90 days when there was no further increase. Fruit volume increased up to 75 days after fruit set. Fruit colour changed from green to yellow at 120 days after fruit set, TSS increased steadily up to 120 days after fruit set at which time there was an increase in reducing sugar content. Ascorbic acid increased steadily up to maturity.

10.4.1 Growth Substances in Aonla Fruit

Ram and Rao (1976) reported 6 cytokinin like substances in aonla fruits showing growth after dormancy. They, however, did not find any relationship between cytokinin content and dormancy of the fruits. Further studies on naturally occurring gibberellins in aonla fruits indicated negligible gibberellin activity during

Fruit Bearing in Aonla

dormant stage while higher activity was linked with higher rate of fruit growth and in the rapidly growing fruits 6 gibberellin-like factors were identified, five of which behaved like gibberellin A1, A3, A4, A7 and A9. The young developing seeds appeared to be the major centres of synthesis of auxins, gibberellins and cytokinins in the fruit after its dormancy break (Ram and Rao, 1978, 1981).

11.0 Pests, Diseases and Physiological Disorder

11.1.0 Pests

Recently the insect pests of aonla have been reviewed by Lakra (1999).

11.1.1 Bark Eating Caterpillar (*Inderbelatetraonis; I. quadrinotata*)

It makes tunnel in the main trunk or branches. Trees of all ages are attacked by this pest, but the incidence is more severe in neglected orchards. The larva constructs a typical loose irregular webbing of silken threads, entangling its excretory pellets, pieces of bark and frass of varying colour (Lakra, 1999). Feeding of larva results into deterioration of the vitality of tree and reduction of yield. The pest can be controlled by spraying 0.03 per cent endrin or injecting kerosene oil or petrol in the holes and plugging them with cotton or wet soil during September-October and February-March.

Lakra (1999) suggested a comprehensive control measure against the pest such as: i) keeping the orchard clean, ii) killing the larva by inserting a thin iron spike into the hole, iii) spot application (in September-October) of 10 ml monocrotophos 36 WSC or 10 ml methyl parathion 50 EC or fenitrothion 50 EC diluted in 10 litres of water, iv) injecting into each hole 5 ml emulsion of 2 ml dichlorvos 76 EC or 5 ml methyl parathion 50 EC or 30 ml endosulfan 35 EC or 40 g carbaryl 50 WP in 10 litres water or 10 per cent kerosene-soap emulsion (1 litre kerosene + 100 g soap + 9 litre water).

The entomogenous fungi, *Aspergillus candidus* and *Gibulella* sp. cause considerable larval mortality at some places during rainy season (Lakra, 1999).

11.1.2 Shoot Gall Maker (*Hypolamprus stylophora; Betousa stylophora*)

The young caterpillars bore into shoots and reach the pith in August-September. The damaged tissue develops into a gall like structure. Heavy galling on young plants may prove fatal (Lakra, 1999). Pruning of the affected parts and spraying of 2 per cent parathion have been suggested to kill the larvae.

11.1.3 Pomegranate Butterfly [*Deudorix isocrates (=Virachola isocrates)*]

This is a major pest of pomegranate but occasionally attacks aonla. The caterpillars bore into the developing fruits and feed on hard yet green seeds without eating at all flesh/pulp portion but otherwise destroy the fruits completely (Lakra, 1999). The damaged fruits should be collected and destroyed. Spraying of 0.5 ml phosphamidon 85 WSC, or 1 ml monocrotophos 36 WSC, or 1.3 ml dimethoate 30 EC per litre keeps the pest under control.

Pests of Aonla

Bark Eating Caterpillar

Fruit Sucking Moth

Shoot Gall Maker

Hairy Caterpillar

Aonla Web Insect

Web Insect

Pomegranate Butterfly

Diseases and Disorder of Aonla

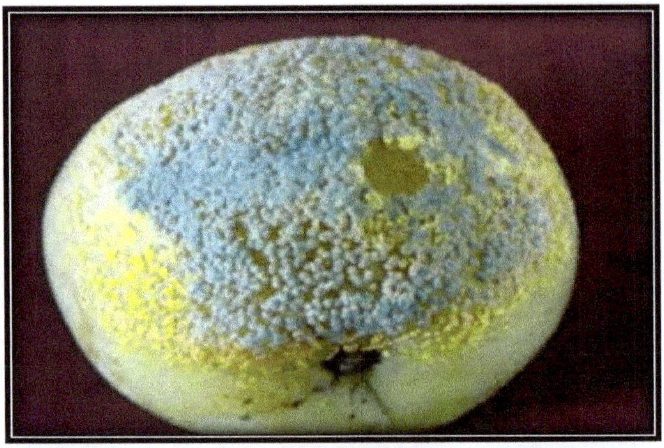

Blue Mould Infection

Rust Infection

Fruit Necrosis

11.1.4 Fruit Sucking Moth (*Achaea janata*)

Fruit sucking moth has been reported to attack aonla (Lakra, 1999). The caterpillar feeds on leaves. The adult moth sucks sap from fruit by making a puncture which is utilized by other moths and gets enlarged serving as an entrance for other pathogens. All the alternate hosts and damaged fruits should be destroyed. Generation of smoke in orchard in the evening keeps the moth away particularly during ripening period. Use of poison baits or spraying of carbaryl (2 g/l) or dichlorvos (0.8 ml/l) protects the fruits from the attack of this pest.

11.2.0 Diseases

11.2.1 Rust

It is caused by *Ravenelliaemblicae*. At Saharanpur, India, Desi cultivar was found severely affected while Banarasi and Chakaiya were relatively free from it (Nirvan *et al.*, 1969). Brown pustules are formed both on leaves and fruits and finally these pustules become dark brown. Spraying with dithane Z-78 at 0.2 per cent proved effective. Aonla rust was effectively controlled by elosal (sulphur), sulphur dust and ultra sulphur (Tyagi and Pathak, 1987).

11.2.2 Blue Mould

It is caused by *Penicillium islandicum* (Setty, 1959). Brown patches with water-soaked areas are formed and the fruit is ultimately covered with bluish green pustules. Control by storage hygiene and treatment of fruits with weak borax or sodium chloride solution have been suggested.

Chahal and Singh (1969) also isolated *Penicillium oxalicum* and *Aspergillus niger* from rotting aonla fruits.

11.2.3 Physiological Disorder

Ram *et al.* (1976) reported internal fruit necrosis, a disorder associated with boron deficiency. It could be controlled with three sprays of borax at 0.6 per cent at fortnightly interval, commencing from early September.

12.0 Harvesting and Yield

12.1.0 Harvesting

Aonla plants start bearing quite late, usually after 8 to 10 years when raised from seeds. The budded plants start fruiting early *i.e.*, after 4 or 5 years. The fruits are light green at first, when they mature, the colour becomes dull, greenish-yellow or rarely brick-red (Naik, 1949). Bajpai (1969) observed that the best time of harvesting aonla fruits is February when the fruits have maximum vitamin C content. If the fruits are allowed to remain on the tree till the next flowering takes place, it gives a false impression that these fruits are from the current season flowering. In South India, fruits are found throughout the year at some places.

Aonla fruits are ready to harvest during November-December. Maximum ascorbic acid content is observed in mature fruits, while immature fruits are acrid and low in ascorbic acid content and minerals (Wali *et al.*, 2015). Fully developed fruits are harvested. Delayed harvesting results in heavy dropping of fruits particularly in Francis and Banarasi cultivars. It also adversely affects the following years bearing. Harvesting should be done in early or in late hours of the day. The method of harvesting is hand picking of individual fruits. Individual fruits are plucked by climbing on the tree with the help of pegged bamboo or ladder. Harvesting should be done in early or in the late hours of the day. A budded/grafted aonla tree starts bearing third year onwards after planting, whereas a seedlings tree may take 6-8 years. Vegetatively propagated plants attain full bearing within 10-12 years and may continue to bear for 60-75 years of age under well managed conditions. An aonla tree may bear 1 to 3 quintalfruits per tree, giving 15-20 tons/ ha (Wali *et al.*, 2015).

12.2.0 Yield

A full-grown grafted aonla tree with good bearing habit yields from 187 to 299 kg fruits per year (Singh *et al.*, 1967). At Allahabad aonla showed a yield of 1.8 and 1.6 quintals from a 10 year-old tree of Chakaiya and Banarasi cultivars, respectively. Singh (1974) reported an average fruit yield of 200 kg per grafted tree.

13.0 Post-harvest Management

Sanjeev Kumar and Nath (1993) reported that physiological weight loss, decay and decrease in vitamin C content were reduced in *E. officinalis* cv. Chakaiya fruits stored in a zero-energy cool chamber (designed to provide lower temperatures and higher humidity than ambient conditions, but without using any power) compared with fruits stored in ventilated polyethylene bags at room temperature. In the zero-energy cool chamber, fruits could be stored for up to 12 days with acceptable minimum decay and quality loss, compared with 4 days at room temperature. In a trial on the effect of post-harvest application of chemicals on shelf-life of aonla cv. Chakaiya, Singh and Kumar (1997 b) observed that post-harvest dipping of fruits for 10 minutes in kinetin (150 ppm) was most effective in reducing the physiological loss of weight in storage (at 18.5°C ± 5.8°C and 67 per cent RH) followed by gibberellic acid (25 ppm). Total soluble solids and acidity increased continuously and ascorbic acid content decreased irrespective of treatments during storage (Tables 7 and 8). Yadav and Singh (1999) reported that 2 pre-harvest sprays of 1 per cent calcium nitrate + 0.1 per cent topsin-M, prolonged shelf-life up to 20 days compared with 10 days in control at ambient temperature. Calcium nitrate-treated fruits had higher concentration of Ca than control fruits. Treatment with topsin-M and bayleton controlled *Penicillium* spp. for 10 days and *Aspergillus niger* for 20 days and extended the shelf-life.

Thokchom and Mandal (2018) made a brief review on the post-harvest research on Aonla. According to them that It is much esteemed for making pickles, preserves, jellies, supari, candy, churan and blended beverages of aonla juice with lime and ginger (Tandon *et al.*, 2003). Chakaiya is a seedling selection having tall upright growth habit. It bears profusely and is regular bearer. Fruit is small to medium, flattened, smooth skin, greenish in colour. Flesh is fibrous and hard. It is a late maturing variety suitable for pickles and other products. The fruit is highly perishable in nature and it is available for a short period. Its storage life

Harvesting, Marketing and Processing

Table 7: Effect of different Storage Conditions on Acidity (per cent) in Aonla cv. Chakaiya*

Storage Condition	*Days of Storage*									*Mean*
	0	*3*	*6*	*9*	*12*	*15*	*18*	*21*	*24*	
Room temperature	2.16	2.28	2.32	2.35	2.63	2.50	2.52	2.55	2.56	2.40
Zero-energy	2.16	2.20	2.26	2.31	2.36	2.40	2.46	2.46	2.48	2.34
Modified storage	2.16	2.19	2.25	2.30	2.34	2.39	2.44	2.47	2.48	2.43
Zero-energy + Modified storage	2.16	2.21	2.26	2.30	2.35	2.41	2.47	2.51	2.53	2.35
Mean	2.16	2.22	2.27	2.31	2.37	2.42	2.47	2.49	2.51	

C.D. at 5 per cent level, storage condition =N.S., period = 0.087, storage condition x period = N.S.

* Singh and Kumar (1997).

Table 8: Effect of different Storage Conditions on Ascorbic Acid (mg/100 g) in Aonla cv. Chakaiya*

Storage Conditions	*Days of Storage*									
	0	*3*	*6*	*9*	*12*	*15*	*18*	*21*	*24*	*Mean*
Room temperature	553.5	542.7	530.6	515.8	492.7	459.7	432.2	389.9	357.2	474.9
Zero-energy	553.5	549.4	539.0	530.7	501.4	460.4	420.7	375.0	324.6	472.7
Modified storage	553.5	551.0	546.5	537.1	512.7	485.7	455.8	422.5	396.1	495.6
Zero-energy + Modified storage	553.5	549.6	546.5	524.4	498.0	475.6	451.2	417.0	381.7	488.6
Mean	553.5	548.2	540.6	527.0	501.2	470.3	440.0	401.1	364.9	

C.D. at 5 per cent level, storage conditions = 6.47, period =9.71, storage condition x period = 19.43.

*Singh and Kumar (1997).

in atmospheric conditions after harvesting is only 5–6 days (Pathak *et al.*, 2009). Therefore, it needs immediate marketing and utilization. To have good return to avoid market glut it becomes essential to store the fruits for a considerable period. Appropriate post-harvest treatment, storage and processing methods can curtail the post-harvest losses to 30 per cent (Goyal *et al.*, 2008) and make the fruit available for longer (Singh *et al.*, 2009). Dip treatment in plant growth regulators like GA_3, NAA and calcium compound chemicals ($CaCl_2$) play a great part for short period storage of various fruits (Gangwar *et al.*, 2012; Thokchom and Mandal, 2018).

Thokchom and Mandal (2018) assessed the influence of various post-harvest treatments on storage behaviour of aonla fruits. Freshly harvested physiological mature fruits of aonla cv. Chakaiya of uniform size, shape, colour, free from disease and bruises were given various post-harvest treatment of gibberellic acid (GA_3: 50 ppm, 100 ppm), naphthalene acetic acid (NAA: 20 ppm, 30 ppm) and calcium chloride ($CaCl_2$: 1, 1.5 per cent) and were air dried and kept in 5 ply corrugated boxes (5 per cent ventilations) with newspaper lining and stored in room temperature. The fruits were analysed for various quality attributes at different storage intervals up to 12 days. The results revealed that physiological loss in weight (4.94, 7.06 and 9.04 per cent) and spoilage percent (0, 0 and 15 per cent) of the fruits remain minimum after 4, 8 and 12 days respectively in $CaCl_2$ (1 per cent) treated fruits during storage. Whereas, TSS (9.28, 10.82 and 1 2.46 oB), acidity (2.31, 2.08 and 1.89 per cent), ascorbic acid (475.06, 456.70 and 413.13 mg 100 g^{-1}) and total sugar (7.86, 8.20 and 8.62 per cent) of the fruits were highest after 4, 8 and 12 days respectively in GA_3 (100 ppm) treated fruits during storage. This study shows that the storability and quality of aonla fruits can be favourably influenced during storage when treated with calcium chloride and GA_3 (Thokchom and Mandal, 2018).

Freshly harvested mature fruits of aonla cultivars, namely, NA-7, NA-10, Chakaiya and a wild aonla were dipped in GA_3 100 mg L^{-1}, 200 mg L^{-1}, 300 mg L^{-1}, MH 100 mg L^{-1}, 200 mg L^{-1}, 300 mg L^{-1} and $CaCl_2$ 10 g L^{-1} 20 g L^{-1}, 30 g L^{-1}solutions (Tarula *et al.*, 2015). Fruits were surface dried, packed in nylon net bags and kept in room temperature (15.5±4°C) and RH 62.7 per cent. Among the different post-harvest treatments, it was found that GA_3 200 mg L^{-1} was best in reducing the physiological loss. No pathological loss was observed in all the treatments up to 14 days of storage. The maximum TSS content was recorded with GA_3 treatment. GA_3 300 mg L^{-1} treatment gave better retention of ascorbic acid and malic acid during the storage of aonla (Tarula *et al.*, 2015).

Growth regulators and fumigants were applied as post-harvest dip to aonla cv. Chakaiya to study their effect on spoilage and biochemical changes (Singh and Kumar, 2000). Fully mature fruits were stored at room temperature (18.5°C and relative humidity 67 per cent). Diphenyl checked decay loss up to 18 days of storage. The maximum decay loss was seen in fruits treated with GA_3 25 ppm. Chlorophyll and carotenoids contents were not affected by any of the treatments. The fungi responsible for decay were mainly Aspergillus spp., Penicillin, spp. and *Collelotrichum* spp. (Singh and Kumar, 2009).

Physically injured and fresh aonla fruits of cultivar 'Desi' were treated separately with $CaCl_2$, $Ca(NO_3)_2$, wax and tap water (control) and stored at ambient conditions (13.3°C, 65.6 per cent RH) for 3 weeks to study changes in their post-harvest physicochemical characteristics (Pathak *et al.*, 2009). The length to width ratio of fruits varied from 0.90 to 0.93 and is not affected by storage period, fruit condition and post-harvest treatment. Physiological loss in weight of injured aonla fruit (extent of injury 1.1 to 20.0 per cent), total soluble solids (12.1 to 17.5°Brix) and the ascorbic acid (932.1 to 751.1 mg/100 g) were affected by post-harvest treatments and storage period. However, the fruits coated with wax did not have adverse effect on their physicochemical characteristics up to 3 weeks. Visually they were not spoiled till 3 weeks. It is therefore, recommended that physically injured fruit could be stored with wax coating for 3 weeks (Pathak *et al.*, 2009).

13.1.0 Storage

The fruit availability period of aonla is very short; hardly 2 to 3 months. Therefore, storage of fruits at an appropriate temperature is essential to extend the availability period and to stabilize the price in the market. Being an underutilized fruit crop, very few studies have been done on low temperature, modified atmosphere and controlled atmosphere storage of this valuable fruit. Refrigerated storage helps in enhancing the storage life of fruit. The five cultivars, *i.e.*,'Banarasi', 'Krishna', 'Francis', 'Chakaiya' and 'Kanchan', were evaluated for their shelf life in the zero energy cool chamber and the results indicated that the cultivars 'Francis', 'Chakaiya' and 'Kanchan' could be stored for up to 18 days, and 'Banarasi' and 'Krishna' for up to 12 days in the zero energy cool chamber (Nath *et al.*, 1992).

Aonla fruits can be stored at 0 to 1.7°C and 85 to 90 per cent RH for 8 weeks (Pantastico, 1975). Singh and Kumar (1997a) observed that the decay loss of aonla cv. Chakaiya was minimum (26.56 per cent) in modified storage condition (boxes containing fruits were packed in polyethylene bags) on 24th day of storage whereas it was maximum (48.7 per cent) in zero energy chamber. The pathogens identified for causing decay of aonla fruits were *Aspergillus* spp., *Colletotrichum* spp. and *Penicillium* spp.

13.2.0 Packaging

Different packaging materials such as nylon net, perforated polyethylene bags, ventilated corrugated fibre boxes, gunny bags, wooden crates, *etc.*, can be used for prolonging the storage life of Fruit (Kore *et al.*, 2013). The comparative performance of different packaging materials was studied in the cultivars 'Chakiya' and 'Francis' packed in gunny bags, fibre carton, pigeon pea branch baskets and wooden boxes (40-kg capacity), all lined with newspaper or polythene. Studies showed that wooden boxes with polyethylene lining resulted in lower weight loss and bruising than other packaging materials (Kore *et al.*, 2013). Packaging of fruits in perforated polyethylene bags was effective in reducing physiological loss in weight, retention in moisture and acceptable physico-chemical quality. The shelf life of aonla fruits was extended up to 15 days at room temperature when fruits were treated with 6 per cent Waxol (an edible wax coating) + [4×10^5 µL Cycocel Chlormequat Chloride (CCC)$\cdot L^{-1}$] and packed in perforated polyethylene bags. This treatment combination recorded the maximum score for marketability among the tested treatments (Dhumal *et al.*, 2008).

13.3.0 Value Addition

The post-harvest losses in aonla vary from 30 per cent to 40 per cent due to its perishable nature, which reduces the market value. Value addition through processing would be the only effective tool for economic utilization of increased production of aonla in the future. Processing not only reduces the post-harvest losses but also provides higher returns to the growers. A number of products such as murabba, pickle, candy, juice, squash, jam, jelly, powder, *etc.*, are prepared from aonla fruits (Kore *et al.*, 2013).

A study was carried out to detect the changes in colour and quality attributes of aonla juice during storage after pasteurization at different temperatures (Bhattacharjee *et al.*, 2011). After extracting juice from aonla cv. Chakaiya, it was pasteurized at five different temperatures and preserved with 500 ppm SO_2 in PET bottles under ambient conditions. Juice was periodically analysed for colour and chemical characters up to 9 months of storage. Though the contents of ascorbic acid and polyphenols in juice decreased with increase in storage period, the effect of pasteurization temperature was not significant. High degree of browning was observed in juice heated at higher temperatures (90 and 95 °C) as compared to lower temperatures (75 and 80 °C) throughout the storage period as indicated by increase in NEB values. The degree of browning was further confirmed by higher negative numerical values of whiteness index in Hunter's scale for intensity of colour. HPLC data indicated that content of gallic acid in juice decreased initially but increased sharply as the storage period prolonged. Higher amount of gallic acid was detected after 9 months of storage in juice pasteurized at 95 °C than in juice heated at 75 °C. The contents of kaempferol and caffeic acid decreased throughout the storage period irrespective of pasteurization temperature. Though least browning was observed in juice pasteurized at 75 °C, but microbial growth was observed after 9 months of storage. Hence, pasteurization temperature of 80 °C was found optimum for preservation of aonla juice under ambient conditions (Bhattacharjee *et al.*, 2011).

Value added dried Indian gooseberry (aonla) shreds were prepared using aonla fruits of cv.'NA-7'. Two blanching methods (hot water and potassium metabisulphite (KMS) at 0.1 per cent) and two drying methods (solar and hot air oven drying) were tried for the production of aonla shreds (Prajapati *et al.*, 2011). Common salt, black salt and ginger juice were mixed for enhancing sensory quality of the product. The best product was obtained with KMS blanching and drying in solar dryer with added common salt at 3 per cent. The most acceptable product had ascorbic acid content 298.3 mg/100 g, tannin 2.4 per cent, acidity 2.6 per cent, reducing sugar 3.0 per cent, non-reducing sugar 21.0 per cent and total sugar 24.0 per cent. The recovery was 8.0–8.5 per cent (Prajapati *et al.*, 2011).

Though, preserve (murabba) is most popular, delicious and high valued product from aonla, its preparation is a cumbersome process involving exhaustive pricking of fruits resulting into loss of most of the nutrients (Jain *et al.*, 1983). Often, fruits break due to over pricking. There are reports that 25-93 per cent losses occur in vitamin C and other nutrients during processing of aonla into preserve (Sethi and Anand, 1983; Kalra *et al.*, 1988; Tandon *et al.*, 2003). Hence, it has been felt to have a product where pricking of fruit may not be required and most of the nutrients are retained.

Aonla segments-in-syrup, an alternative to preserve, was prepared from fresh as well as stored fruits and quality was evaluated (Nayak *et al.*, 2011). Fruit of aonla cv. Chakaiya were packed in CFB boxes using newspaper as lining material and stored under ambient conditions (16-25°C and 60-70 per cent RH). The fruits were withdrawn at 0, 3, 6 and 9 days of storage for quality analysis and the product was prepared after each withdrawal by blanching the fruits in 2 per cent alum solution for 10 minutes, separating the segments and dipping them overnight in successively increasing concentration of sugar syrup (55-750B) and packed in 720B sugar syrup.

The physiological loss in weight, total soluble solids, titratable acidity and tannins increased while the content of ascorbic acid decreased in stored fruits. These losses were more pronounced in the fruits stored for 9 days. The product prepared from fresh and 3 days stored fruits have very little differences in nutritional attributes. The maximum ascorbic acid content (113mg/100g) and organoleptic quality (8.0) was observed in the product prepared from fresh fruits followed by the product prepared from 3 days stored fruits (103mg/100g and 7.6). It was concluded from the study that good quality product could only be prepared from fresh and 3 days stored fruits (Nayak *et al.*, 2011).

13.3.1 Dehydration

The aonla fruits (whole fruit, pricking, splits, segments) were subjected to pretreatments like blanching, osmotic dehydration with salt (2 per cent) and sugar (40 per cent) in different experiments before drying to obtain a product with better keeping quality (Gudapaty *et al.*, 2010). An LPG based drier (CRIDA drier) with capacity to dry 50 kg of fresh Indian gooseberry (aonla) was used. Nutritional quality and rehydration characteristics of CRIDA drier dried products were higher and free from contamination. Drying time was shortest for blanched and osmotically dehydrated segments dried in CRIDA drier and the product had better vitamin C retention, rehydration characteristics and sensory acceptability compared to sun or cabinet drier dried product. The additional expenditure spent on gas in CRIDA drier is compensated by reduced labour cost and higher price for the better quality product. Alternate energy sources like biogas and biomass can be used as fuel in the CRIDA drier (Gudapaty *et al.*, 2010).

14.0 Biotechnology

14.1.0 Tissue Culture

14.1.1 Micropropagation

Very few reports are available on micropropagation of aonla (*E. officinalis*). Verma and Kant (1996) induced shoot proliferation from nodal stem explants after 4 weeks of culture on modified MS medium supplemented with BA (3.0-5.0 mg/l) and NAA (0.5 mg/l). However, bud break was observed in only 8-10 per cent of explants, and a maximum 3-4 shoots per explant were produced. Leaching of phenolics resulted in the loss of almost 90 per cent of the cultures. Shoot proliferation was only observed on explants collected and inoculated in February-April or August-October. Regenerated shoots were elongated on hormone-free MS medium and subsequently rooted on half-strength MS medium supplemented with IBA (3.0 mg/l) and sucrose (1.5 per cent). Kant *et al.* (1999) obtained high frequency regeneration of *E. officinalis* from juvenile and mature explants. Callus formation was achieved from juvenile explants (cotyledonary nodes, shoot apices, cotyledons, hypocotyl and roots) on MS medium containing 2,4-D (2.0 mg/l) and ascorbic acid (25 mg/l). Caulogenesis was observed when this callus was subcultured on MS medium supplemented with BA and/or kinetin (2.0-4.0 mg/l). Direct regeneration from mature nodal explants at the rate of three to four shoots per explant occurred on MS medium supplemented with BA (5.0 mg/l) and IAA (0.5 mg/l) along with antioxidants. Rooting of the shoots was achieved on half MS medium fortified with IBA (2.0 mg/l) and sucrose (1.5 per cent).

In vitro bud induction from the shoot of aonla was dependent on season of explanting and position of node in the mother tree. The shoots collected during August to November showed better establishment in cultures, rapid bud induction and growth of microshoots followed by shoots collected during April toJuly. Slightly green and moderately hard (10-15nodes) shoots gave highest in vitro bud induction and proliferated profusely than the soft (1-10 nodes) or very hard, brown (20-30) nodal segments (Mishra and Pathak, 2001). Nodal segments from mature tree as explants were used with use of growth regulators (Kinetin and GA_3) at various concentrations for shoot proliferation. MS medium supplement with Kinetin at 2.0 mg/l with GA_3 0.5 mg/l resulted in more number of shoots (3.28) and less number of days taken for bud initiation (7.07). All the concentrations of IBA and NAA used individually for in vitro rooting of micro shoots failed to induce rooting. However combination of IBA (2.0 mg/l) and NAA (2.0 mg/l) has resulted in callus induction at cut end of in vitro shoots (Nayaka, 2006).

Nodal segments of *Emblica officinalis* were cultured on twenty different fortifications of MS medium. The basal MS medium fortified with 4.0 mg.l^{-1} BAP and 0.5 mg.l^{-1} NAA was found to be more responsive for shoot proliferation efficiency (47.65 per cent), number of shoot(s) per explant (3.20) and average shoot length (1.43). Rooting was high on MS basal medium supplemented with 2.0 mg.l^{-1} IBA and 0.5 mg.l^{-1} BA (17.20 per cent) in vitro (Patidar *et al.*, 2010).

In vitro shoot proliferation from nodal explants has been observed by using MS, B… and WPM media. Amongst them, MS medium has been observed the best for shoot proliferation (63.3 per cent), while minimum phenol leaching (63.3 per cent) was observed on WPM medium. To enhance the shoot regenerative potential, explants were cultured on MS medium supplemented with various cytokinins, *viz.*, BAP, Kn and TDZ. MS medium supplemented with BAP (4.44 µM /L) was observed the best among all cytokinins. As shoot proliferation depended upon the balance of cytokinins and auxins, and BAP was found ideal for shoot proliferation, additive effect of auxins – IAA, IBA and NAA was assessed at 4.44 µM /L conc. of BAP. Explants when cultured on MS + BAP (4.44 µM/L) with IBA (2.46 µM /L)) showed maximum percentage of shoots (85 per cent) with minimum callus formation (15 per cent). Higher concentration of auxins suppressed the caulogenesis and induced callus formation (Goyal and Bhadauria, 2008).

Regeneration of aonla *in vitro* through shoot bud explant excised from mature plant was tried. Around three cm long shoot apices excised during August-November from mature plants of aonla gave best *in vitro* bud induction on Murashige and Skoog (1962) medium supplemented with kinetin(3.0 mg/litre), GA_3 (1.5 mg/litre) and glutamine (50.0 mg/litre). *In vitro* rhizogenesis was induced on 1/2 MS medium supplemented with NAA 2.0 mg+IBA 10.0 mg/litre. Micropropagated plants were acclimatized on Soil+Sand+FYM (1:1:1) fortified with half strength MS nutrient salt and 1.0 mg/litre paclobutrazol (Chandra *et al.*, 2006).

14.1.2 Somatic Embryogenesis

Sehgal *et al.* (1994) cultured endosperm of *E. officinalis* on MS supplemented with 2,4-D (1 mg/l) + kinetin (1 mg/l) which produced continuously growing compact callus. Subculture of callus on MS supplemented with BA (0.2 mg/l) + IAA (0.1 mg/l) resulted in the production of shoot buds, shoots and embryoids. Roots were formed when embryoids were transferred to filter paper bridges immersed in liquid MS containing NAA (0.02 mg/l).

Verma and Kant (1999) achieved a high frequency plantlet regeneration from callus cultures of *E. officinalis*. The effect of initial explant used (cotyledonary nodes, cotyledons, shoot apices, hypocotyl or roots) on shoot morphogenesis in callus cultures was studied. Regenerative callus was raised from different explants on MS medium supplemented with 2,4-D (2.0 mg/l) and ascorbic acid (25 mg/l). Shoot bud differentiation was achieved in callus cultures on MS-medium supplemented with BA (2.0 mg/l), kinetin (2.0 mg/l), citric acid (25.0 mg/l), ascorbic acid (25.0 mg/l) and PVP (25.0 mg/l). The regenerated shoots elongated on hormone-free medium and were subsequently rooted on half-strength MS-medium fortified with IBA (2.0 mg/l) and sucrose (1.5 per cent).

Effect of NAA, BAP used singly and in combinations was studied in seedling derived root explant of *Emblica officinalis*. It was observed that BAP support shoot proliferation and multiplication. BAP used alone was found to be better than NAA alone or combination with BAP. Effect of subculturing was also observed in root-shoot junction explant of this important medicinal plant. It was noticed that with each progressive subculturing, percentage shoot proliferation increases. Within one week all explants produced callus. Differentiated micro shoots produced roots when inoculated on half strength MS media supplemented with 2.0 mg/l IBA (Gour and Kant, 2009).

Shoots regenerated adventitiously on epicotyl segments from *in vitro* seedlings of *Emblica officinalis* var. 'Kanchan'. Epicotyls derived from 2-week-old aseptic seedlings were most responsive and produced a maximum number of 303 shoots per explant in Murashige and Skoog (1962) medium (MS) augmented with 8.8 μM N^6-benzyladenine (BA) + 1.425 μM indole-3-acetic acid (IAA). Shoots readily elongated in MS lacking growth regulators and rooted in half-salt-strength MS (1/2 MS) supplemented with indole-3-butyric acid (IBA) or α-naphthalene acetic acid (NAA). The highest rooting response was recorded in 1/2 MS containing 14.7 μM IBA. Plantlets were acclimatized inside the green house and 80 per cent of the plantlets survived on transfer to garden soil (Nayak *et al.*, 2010).

A simple and one step reproducible protocol was developed for induction of high frequency somatic embryogenesis from *in vitro* derived juvenile leaf tissues of *Emblica officinalis*. Murashige and Skoog (MS) medium supplemented with twelve combinations of auxins and cytokinins involving 2, 4 – dichlorophenoxy acetic acid (2, 4 - D), indole 3- acetic acid (IAA), 6 – benzylaminopurine (BAP) and kinetin (KIN) were tested at various concentrations (Thilaga *et al.*, 2013). Highest percentage of callus (67.5 per cent) was obtained on media containing 0.45 μM 2, 4- D in combination with 22 μM BAP. Three distinct callus types were induced in various

Micropropagation in Aonla

In vitro Shoot Development from Nodal Segment

Somatic Embryogenesis from Leaf Explants

combinations of media within 60 days of culture. Maintenance of proliferating callus on their respective media led to high frequency somatic embryogenesis. Of the three distinct types of callus, type III and type II callus recorded 33.8 per cent and 8.3 per cent of somatic embryogenesis respectively. Somatic embryos were matured within two weeks when treated with 3.78 µM ABA. Matured embryos were transferred to MS medium fortified with 0.46 µM KIN for germinating into healthy plantlets. About 90 per cent plantlets transferred to *ex vitro* condition were acclimatized for field establishment (Thilaga *et al.*, 2013).

14.2.0 Markers

Reports on DNA markers are very few in Aonla as compared to other major fruit crops.

14.2.1 DNA Isolation

Isolation of high quality DNA from leaf tissues of *P. emblica*, primarily due to high acidic nature as it contains highest amount of vitamin C as compared to any other plant species (Nagarajan *et al.*, 2011). DNA isolation from leaf tissues of *P. emblica* have been described and reported that reduction of pH to extreme acidic level during DNA isolation process was one of the major factors responsible for degradation of DNA. However, the problem in obtaining high quality DNA remains unsolved. Nagarajan *et al.* (2011) reported a reliable and effective protocol for isolation of high quality genomic DNA from leaf tissues of *P.emblica* without degradation of DNA by stabilizing the pH of buffer with NaOH at different stages in the process of DNA isolation. Modified Cetyltrimethylammonium bromide (CTAB) protocol for DNA isolation was developed from leaf tissues of *Phyllanthus emblica* for obtaining high quality genomic DNA. Fresh leaves of three different maturity were analyzed for yield and quality of DNA. Acidity was determined in three different maturity of leaves *viz.* tender, intermediate and mature and their influence on DNA quality was determined. Drastic reduction of pH was the primary cause for poor quality of DNA. However, high quality DNA isolation was achieved by stabilizing the pH by addition of NaOH during different stages of DNA isolation process. The present protocol yielding high quality intact DNA for genetic fingerprinting as well as for amplification of chloroplast genes for molecular analysis (Nagarajan *et al.*, 2011).

14.2.2 Genetic Diversity

In recent years RAPD technique has been used to study the genetic diversity in different sets of P. emblica (Selvi *et al.*, 2008; Chaurasia *et al.*, 2009; Rane *et al.*, 2012; Singh *et al.*, 2014). Besides this some molecular techniques such as ISSR Li and Zhao, 2007) comparative studies of RAPD and ISSR markers (Raut *et al.*, 2010), development of Sequence Characterized Amplified Region (SCAR) marker from the RAPD amplicon (Dnyaneshwar *et al.*, 2006), SCAR and AFLP markers for distinguishing species of the genus Phyllanthus (Bandopadhyay and Raychaudhuri, 2013) and chloroplast *trnL* (UAA) intron identified as candidate gene (Sangeetha *et al.*, 2010).

Investigations on the use of RAPD markers enable to estimate genetic variability among commercially cultivated varieties. This study also enabled us to distinguish these varieties using a set of four decamer primers, which was otherwise difficult by using morphological markers. Cluster analysis revealed three different groups of varieties directly associated to their place of origin. RAPD markers were also able to differentiate varieties of same origin or even selection from same parents (Chaurasia *et al.*, 2009).

The genetic diversity in four populations of *P. emblica* sampled from dry-hot valleys of Yunnan was studied using ISSR inter-simple sequence repeats markers (Li and Zhao, 2010). Based on 12 primers, 135 clear and reproducible DNA fragments were generated, of which 115 were polymorphic, accounting for 85.19 per cent. The coefficient of genetic differentiation GST equalled 0.1222. Such a high level of genetic diversity and low level of population genetic differentiation might result from the breeding system of this species. There was a lack of significant association between genetic and geographical distances r=0.19798, P=0.6513 0.05 among *P. emblica* populations. The results suggest some important recommendations about sampling methods to be used in ecological reforestation of dry-hot river valleys (Li and Zhao, 2010).

14.2.3 Mutant Identification

The effect of different doses of gamma rays in amla variety BSR-1 at molecular level was evaluated. The gamma ray exposure on amla var. BSR-1, RAPD method was applied. The result of this study indicated that gamma ray radiation altered the genetic makeup of amla var. BSR-1 (Selvi *et al.*, 2008).

14.2.4 Species Identification

Random amplified polymorphic deoxyribonucleic acid (RAPD), amplified fragment length polymorphism (AFLP) and sequence characterized amplified region (SCAR) markers were employed for proper identification of five medicinally important species from the genus *Phyllanthus*, thus allowing these species to be used in herbal drug industry straight after bringing them from the field. RAPD data segregated the herb *P. urinaria* from the remaining species, whereas AFLP data successfully distinguished all the herbs, *i.e.*, *P. amarus Schum.*, *P. fraternus Webster* and *P. urinaria L.* A SCAR marker for one of the herbaceous species, *P. urinaria*, was also developed. In this study, AFLP came out as a better polymorphic marker than RAPD (Bandopadhyay and Raychaudhuri, 2013).

The morpho-physiological characterization proved insufficient to distinguish studied varieties of *P. emblica*. Ten RAPD primers exhibited an average of 56.18 per cent polymorphism in banding pattern and 83.1 per cent PIC. They can be used as powerful markers to reveal genetic diversity in *P. emblica*. Present study validates the utility of ITS rDNA region being third generation molecular marker as a reliable indicator of phylogenetic interrelationships, especially for ITS regions as DNA barcode at higher levels can serve as an additional approach for identification and genetic cataloguing of P. emblica germplasm for crop improvement (Singh *et al.*, 2015).

To understand relationship among different popular aonla cultivars fingerprinting using RAPD markers was undertaken (Singh *et al.*, 2014). DNA was isolated from young leaves of ten aonla cultivars. Six RAPD markers (OPA-7, OPA-8, OPZ -4, OPZ-14, OPY-1and OPY-2) were used for the analysis for genetic variation. Out of six primers, OPZ-14 gave most distinct banding pattern for cv. Banarasi. Dendrogram represented four major groups, which was further divided into sub-groups IA and IB. Sub-group IA had NA-04 and NA-10, while IB had cv. Francis. Cluster IIA represented NA-5, NA-6 and NA-9, while IIB represents NA-7 and Chakaiya. Banarasi and Anand represented by group III and IV respectively (Singh *et al.*, 2014).

14.2.5 QTL Markers and Markers for Traits

The genomic diversity of the *P. emblica* germplasm in Sri Lanka has not been studied in detail (Mawalagedera *et al.*, 2014a). Therefore, the present study was conducted to molecularly characterize the *P. emblica* germplasm in Sri Lanka. Young tender leaves from 66 trees of *P. emblica* were collected from various parts of the country for DNA assessment. Genomic DNA was extracted using DNeasy Plant mini kit. Six microsatellite (SSR) markers and one SCAR marker developed for *P. emblica* were used in the PCR amplification. The PCR products were size separated by using 6 per cent denaturing polyacrylamide gel electrophoresis. A binary scoring matrix was constructed on presence and absence of alleles in each single locus and a dendrogram based on genetic similarities was constructed. For each allele, allele frequency, heterozygosity and polymorphic information content were calculated. Cluster analysis was conducted based on marker data and association analysis was carried out to detect putative QTL associated with drupe traits. The dendrogram based on seven DNA marker alleles, showed two major clusters at 14.67 per cent molecular similarity coefficient. A total of 51 alleles were detected in the sampled *P. emblica* germplasm with the six SSR markers and the SCAR marker used. Out of the six SSR markers, four were polymorphic, one was monomorphic and the other one did not produce clear bands. Heterozygosity of the tested markers ranged from 0. 44. to 0.89. Allele at 176 bp of the marker Phyll112 showed a significant association (Mawalagedera *et al.*, 2014a).

P. emblica Linn. is severely infested by *Betousa stylophora* Swinhoe, known as shoot gall maker (SGM). This pest tunnels the shoots of seedlings and actively growing branches of trees and develops gall, leading to stunted growth, unusual branching and death of actively growing shoots. Studies have revealed that trees possessing smooth bark were free from the attack of this pest than those with rough bark surface (Thilaga *et al.*, 2017). Unfortunately, this character is not detectable either at seedling stage or during early growth of trees in the orchard. RAPD genetic fingerprinting of trees possessing smooth and rough bark revealed distinguishable and highly reproducible DNA banding pattern between the two genotypes. Of the 20 RAPD primers tested, five of them produced distinguishable RAPD bands between rough and smooth barked genotypes of P. emblica. Trees with smooth bark produced five unique RAPD bands with molecular weight ranging from 350 bp to 1500 bp and those with rough bark produced six RAPD bands (350 bp–650 bp) to utilize these DNA bands as potential DNA marker for screening tolerant genotypes of this crop against SGM (Thilaga *et al.*, 2017).

14.2.5.1 Chalcone Synthase

The two decisive enzymes in flavonoid biosynthetic pathway are chalcone synthase (CHS) and chalcone isomerase (CHI), wherein the former carries the first committed step of the pathway and later is involved in isomerization of chalcone (Singh *et al.*, 2019). Full-length cDNA sequence of both the genes (*PeCHS* and *PeCHI*) from *P. emblica* (L.) were cloned and sequenced. The 390 and 209 amino acid long polypeptides of *PeCHS* and *PeCHI* are coded by 1514 bp nucleotide (nt) (ORF 1173 bp) and 843 bp nt (ORF 630 bp) sequences respectively. Computational analysis revealed the deduced protein of *PeCHS* contained four CHS protein family specific conserved motifs including residues of active sites and other signature sequences. Four conserved amino acids Thr-48, Tyr-106, Asn-113, and Thr-190 at active sites were identified in *PeCHI*. Both *PeCHI* and *PeCHS* were expressed in E. coli BL21 using pQE-30 UA vector. Expression analysis was carried out in different developmental stages *i.e.* leaf, flower and fruits. Expression of *PeCHS* was maximum in mature fruit while PeCHI expression was highest in young leaves. With respect to fruit, *PeCHI* expression lower down first and then increases with the maturation of fruit whereas *PeCHS* expression increases gradually with the development of fruit (Singh *et al.*, 2019).

14.2.5.2 Type III Polyketide Synthase Mutants

Chalcone synthase (CHS) is a plant-specific type III polyketide synthase that catalyzes the synthesis of flavonoids. Native CHS enzyme does not possess any functional activity on N-methylanthraniloyl-CoA, which is the substrate for acridione/quinolone alkaloid biosynthesis. Here, we report the functional transformation of chalcone synthase protein from *Emblica officinalis* (E°CHS) to quinolone and acridone synthase (ACS) with single amino acid substitutions. A cDNA of 1173 bp encoding chalcone synthase was isolated from *E. officinalis* and mutants (F215S and F265V) were generated by site-directed mutagenesis. Molecular modeling studies of E°CHS did not show any active binding with N-methylanthraniloyl-CoA, but the mutants of E°CHS showed strong affinity to the same. As revealed by the modeling studies, functional analysis of CHS mutants showed that they could utilize p-coumaroyl-CoA as well as N-methylanthraniloyl-CoA as substrates and yield active products such as naringenin, 4-hydroxy 1-methyl 2(H) quinolone and 1,3-dihydroxy-n-methyl acridone. Exchange of a single amino acid in E°CHS (F215S and F265V) resulted in functionally active mutants that preferred N-methylanthraniloyl-CoA over p-coumaroyl-CoA. This can be attributed to the increase in the relative volume of active sites in mutants by mutation. Moreover, metabolomic and MS analyses of tobacco leaves transiently expressing mutant genes showed high levels of naringenin, acridones and quinolone derivatives compared to wild-type CHS. This is the first report demonstrating the functional activity of E°CHS mutants with N-methylanthraniloyl-CoA and these results indicate the role of phenylalanine in altering the substrate specificity and in the evolution of type III PKSs (Aiswarya *et al.*, 2016).

14.2.6 Expressed Sequence Tag (EST) Markers

To date, very few genomic simple sequence repeat (SSR) markers have been reported for *P. emblica* (Pandey and Changtragoon, 2012), and only one (Phyll_68) of the six genomic SSRs was found to be polymorphic in the three studied populations of *P. emblica* (percentage of polymorphic loci = 16.67 per cent;), which are insufficient for further genetic studies. With advances in high-throughput sequencing technologies, especially de novo transcriptome sequencing, expressed sequence tag–simple sequence repeat (EST-SSR) markers can be rapidly mined at a lower cost (Liu *et al.*, 2018). They are also increasingly being used for the evaluation of genetic relationships, because they are codominant highly polymorphic, and are well distributed throughout the genome (Bouck and Vision, 2007 cited by Liu *et al.*, 2018). Furthermore, compared to genomic SSRs, EST-SSRs are more transferable across taxonomic boundaries (Ellis and Burke, 2007; Liu *et al.*, 2018). Therefore, developed 20 EST-SSR markers for *P. emblica.* These new markers have a higher level of genetic diversity than the loci previously reported in Sri Lankan populations (Mawalagedera *et al.*, 2014 cited by Liu *et al.*, 2018). We then evaluated their transferability to two sympatric species in the Phyllanthaceae: *P. reticulates* Poir. and *Leptopus chinensis* (Bunge) Pojark. Based on the transcriptome data of *P. emblica,* 83 EST-SSR primer pairs were designed; 52 primer pairs were successfully amplified, with 20 showing polymorphisms in 90 individuals from three populations of *P. emblica.* The number of alleles per locus varied from 11 to 44. The observed and expected levels of heterozygosity for the 20 loci ranged from 0.240 to 0.868 and 0.754 to 0.933, respectively. Cross-species amplification was successful for all 20 loci in each of the two related species, *P. reticulatus* and *Leptopus chinensis*. These markers will be valuable for studying the population genetics and for mining genes of *P. emblica*, and may be useful for studies of related species (Liu *et al.*, 2018).

14.2.7 Transcriptomics

Despite the high medicinal value, the genomic information of *P.emblica* is still very and till 2016 only71ESTs were available in the National center for Biotechnology Information(NCBI) database (Kumar *et al.*, 2016). The inadequate genomic/transcriptomic data was a major bottleneck in understanding various molecular mechanisms and biosynthetic pathways including flavonoids and vitamin C biosynthesis in *P.emblica*. Earlier, Kumar and Singh (2012) developed a method for RNA isolation and cloned *flavonoids 3 hydroxylase* (*F3H*) gene of flavonoids biosynthetic path way using primers based PCR approach (Kumar *et al.*, 2015). However, complete information about biosynthetic pathways and other molecular details using that approach could not be obtained. Hence, to gain further insight in to metabolic and molecular networks of Aonla,the *de novo* transcriptome study was initiated with foremost emphasis to investigate the candidate genes involved in flavonoids and vitamin C biosynthesis (Kumar *et al.*, 2016).

Kumar *et al.* (2016) developed in-depth transcriptome information of *P.emblica* using Illumina Hiseq 2000 platform and characterized. A total of 31,285,965 high quality reads were assembled into 91,288 contigs with the N50 value 358. Out of them, 47,267 contigs were functionally annotated using BLASTX search against

NCBI-non-redundant (NR) protein database. Further, 31,366 contigs showed similarity with various gene ontology (GO) terms, and 1299 were related to different enzymes and biosynthetic pathways. The transcripts related to each gene involved in flavonoids and vitamin C biosynthesis was identified. Several *cytochrome P450s* (*CYPs*) and *glucosyl transferases* (*GTs*) genes involved in flavonoids biosynthesis and various other metabolic pathways were also documented. Further, 6510 transcription factors and 4420 EST derived simple sequence repeat(SSR) markers were also predicted. This study enlightened various characteristic features of *P.emblica* genome, and provided an important resource for future molecular and functional genomics studies (Kumar *et al.*, 2016).

15.0 References

Aiswarya, G., Mallika, V., Mur, L.A.J. *et al.* (2016) *Plant Cell Rep.*, **35:** 2077–2090.

Allemullah, M. and Ram, S. (1990) *Indian J. Hort.*, **47**: 270-77

Amal, E.K.J. and Raghwan, R.S. (1957) *Proc. Indian Acad Sci.*, **47:** 312-14.

Amin, R.S. (1978) *Curr. Sci.*, **47**: 468-469.

Anonymous (1952) *Wealth of India-Raw materials*, Vol. **3**, CSIR, New Delhi.

Anonymous (1969) *Wealth of India-Raw materials*, Vol. **8**, CSIR, Nes Delhi.

Asenjo, C.F. (1953) *Bull. De colegio de Quimicos de, Puerto Rico*, **10**: 8-17.

Awasthi, O. P. and Pareek, O.P. (2008) *Range Management and Agroforestry*, **29** (2): 67–74.

Awasthi, O.P., Singh, I.S. and More, T.A. (2009) *Indian J. Agric. Sci.*,**79** (8): 587–91.

Bailey, L.H. (1917) *The Standard Cyclopedia of Horticulture*, The MacMillan Co., London.

Bajpai, P.N. (1957) *Indian J. Hort.*, **14**: 99-102.

Bajpai, P.N. (1963) *Ph.D. thesis* submitted to Agra University, Agra.

Bajpai, P.N. (1965) *Agra Univ. J. Res.*, **14**: 167-186.

Bajpai, P.N. (1968) *Hort. Adv.*, **7**: 38-67.

Bajpai, P.N. (1969a) *Hort. Adv.*, **8**: 15-16.

Bajpai, P.N. (1969b) *Hort. Adv.*, **8**: 17-21.

Balasubramanyan, S. and Bangarusamy, U. (1998) *South Indian Hort.*, **46**: 347-348.

Bandopadhyay, S. and Raychaudhuri, S.S. (2013) *Plant Biosyst.*, **147**: 12–20.

Bhattacherjee, A.K., D.K. Tandon, A. Dikshit, A. *et al.* (2011) *J. Food Sci. Technol.*, **48:** 269–273.

Bhujbal, B.G. (1975) *Res. J. Mahatma Phule Agric. Univ.*, **6**: 73-75.

Bouck, A., and Vision, T. (2007) *Mol. Ecol.*, **16:** 907–92

Burkill, I.H. (1935) *A Dictoinary of the Economic Products of the Malay Peninsula*, Vol. **1.**, Crown Agents for the Colonies, London.

Chadha, K.L. (1998) *Indian J. Hort.*, **55**: 265-296.

Chahal, D.S. and Singh, N. (1969) *FAO Plant Prot. Bull.*, No. 17, p. 67.

Chandra, R., M. Mishra and R. Pati. (2006) *Proc.National symp.production, utilization and export of underutilized fruits with commercial potentialities*, Kalyani, Nadia, West Bengal, India, 22-24 November, 2006. pp.111-116

Chaurasia, A.K., Subramaniam, V.R., Bal Krishna and Sane, P.V. (2009) *Physiol. Mol. Biol. Plants*, **15(2):** 169-173.

Chaurasia, A.K., V.R. Subramaniam, B. Krishna and P.V. Sane. (2009) *Physiol Mole Biol Plants* **15:** 169–173

Chopra, R.N., Chopra, I.C., Handa, K.L. and Kapur, L.D. (1958) *Chopra's Indigenous Drugs of India* (2nd ed.), U.N. Dhur and Sons Pvt. Ltd., Calcutta.

Chularojmontri,L., Suwatronnakorn, M., Wattanapitayakul S.K. (2013) *Altern Med* http: / / dx.doi.org / 10.1155 / 2013 / 720728. Accessed 5 Dec 2013

Dalzell, N.A. and Gibson (1861) *The Bombay Flora*, Education Society Press, Byculla.

Dhandhar, D.G. and Shukla, A.K. (2003) Paper presented in Nat. Seminar on production and utilization of aonla, organized by All India Aonla Growers Association, Salem, Tamil Nadu, India.

Dhumal S.S., Karale, A.R., Garande, V.K., Patil, B.T., Masalkar, S.D. and Kshirsagar, D.B. (2008) *Indian J. Agric. Res.* **42** (3): 189–194.

Dikshit, S.N. (1987) *Ph. D. thesis* submitted to Narendra Dev Univ. of Agric. & Tech., Faizabad, India.

Dnyaneshwar, W., Preeti, C., Kalpana, J. and Bhushan, P. (2006) *Biol Pharm Bull* **29:** 2313–2316.

Drury, H. (1873) *The Useful Plants of India* (2nd ed.), Wm. H. Allen & Co., London.

Ellis, J. R., and Burke, J. M. (2007) *Heredity*, **99:** 125–132.

Firminger, T.A. (1947) *Firminger's Manual of Gardening for India*, Thacker Spink Co. Ltd., Calcutta.

Gangwar, R.P., Singh, D. and Chundawat, B.S. (1975) *Haryana J. Hort. Sci.*, **4**: 150-151.

Gangwar, S., Shukla, H.S., Katiyar, D. and Pandey, V. (2012) *HortFlora Research Spectrum* 1(3), 253–258.

Garg, V.K. and Khanduja, S.D. (1976) *Sci. Hort.*, **5**: 243-247.

Gour, V.S. and Kant, T. (2009) *J. Indian bot. Soc.*, **89**: 34-36

Goyal, D. and Bhadauria, S. (2008) *Indian J. Biotechnol.*, **7:** 394-397.

Gudapaty, P.,S. Indavarapu, G.R. Korwar, A.K. Shankar, R.K. Adake, V. Bandi and Gupta, G.K. and Bopaiah (1986) *Indian Hort.*, **31**: 15.

Hanif, M. (1966) *Pakistan J. Sci. Res.*, **18**: 61-63.

Hayes, W.B. (1957) *Fruit Growing in India*, Kitabistan, Allababad.

Hooker, J.D. (1973) *The Flora of British India*, Vol. V, Periodical Experts, Delhi.

Jain S. P, V.K. Tripathi, H.B.Ram and Singh, S. (1983) *Indian Food Pack.*, **37(6)**: 85-90.

Janki Amal, E.K. and Raghavan, R.S. (1957) *Proc. Indian Acad. Sci. Sec. B*, **47**: 313-314.

Kalra, C.L. (1988) *Indian Food Packer*, **38:** 67

Kanchu, S.R. (2010) *J. food sci. technol.*, **47:** 541–548.

Kant, U., Verma, B. and Kishor, P.B.K. (1999) *Proc. Symp. Plant Tissue Culture and Biotechnology: Emerging Trends*, Hyderabad, India, 29-31 January, 1997, pp. 106-111.

Kaulgud, S.N. (1996) *Proc. Production Technology of Arid and Semiarid Fruits*, MPKV, Rahuri, Maharashtra, pp. 109-115.

Kirtikar K.R and Basu B.D. (2003) *Indian medicinal plants,* 2nd edn. Oriental Enterprises, Dehradun, pp 1606–1608

Kishore, K. (2017) *Sci. Hort.*, **225:** 607-614.

Kore, V.T., H. L. Devi And Kabir1, J. (2013) *Fruits,* **68:** 255–266.

Kumar A, Kumar S, Bains S, Vaidya V, Singh B, Kaur R, Kaur J and Singh, K. (2016) *Front. Plant Sci.*, **7:** 1610-1622.

Kumar, A. and Singh, K. (2012) *Mol. Biotechnol.*, 52: 269–275.

Kumar, A., Singh, B., Kaur, J., and Singh, K. (2015) *J. Plant Biochem. Biotech.*, 24: 453–460. doi: 10.1007/s13562-014-0296-0

Lakra, R.K. (1999) *Haryana J. Hort. Sci.*, **28**: 1-11.

Li, Q and J. Zhao. (2007) *Biodivers Sci.*, **15(1):** 84–91

Li, Q.M. and J.Zhao (2010) CBA: 633247.

Liu, X., H. Ma, T. Li, Z. Li, Y. Wan, X. Liu, and Fu, L. (2018) *Applications in Plant Sciences* **6**: e1169.

Mahata S, A.Pandey, S.Shukla,A. Tyagi, S.A. Husain, B.C.Das and Bharti, A.C. (2013) *Nutr Cancer*, **65:** 88–97.

Mawalagedera S.M.U.P, G. A. D. Perera and, Sooriyapathirana, S. D. S. S. (2014b) *Open J. Forestry*, **4**: 38-41.

Mawalagedera,S.M.U.P., P. Janaththani, S.W.M.B. Dunuwille, G.A.D. Perera, C.K. Weebadde, D.S.A. Wijesundara and S.D.S.S. Sooriyapathirana. (2014a). *Ceylon Journal of Science* (*Bio. Sci.*) **43:** 31-46,

Mishra, M. and Pathak, R.K. (2001) *J. Appl. Hort.*, **3:** 103-104.

Mishra, S.M., D. Ram, A. Pandey and Meena, A. K. (2017) *Int. J. Curr. Microbiol. App. Sci* (2017) **6**: Doi: http: //dx.doi.org/10.20546/ijcmas.

Morton Julia, F. (1960) *Econ. Bot.*, **14**: 119-128.

Moti, L. Dhar and Chaturvedi, O.P. (1976) *Punjab Hort. J.*, **16**: 33-38.

Nadkarni, A.K. (1927) *Indian Materia Medica*, S.K. Bijur, Bombay.

Nagarajan, S., M.Steephen, K. Murugan1, R.R. Nair, S.Thilaga P. Alagar and Ganesh, D. (2011) *Iranian J. Biotechnol.*, **9:** 307-313.

Naik, A.G. and Chundawat, B.S. (1993) In: *Golden Jubilee Symp. Horticultural Research – Changing Scenario, Bangalore, Hort. Soc. India*, May 24-28, 1993, Abst. No. 18.6, p. 337.

Naik, K.C. (1949) *South Indian Fruits and Their Culture*, P. Varadachary& Co., Madras.

Nath V., I.S.Singh and, Kumar S. (1992), *Narendra Deva J. Agric. Res.*, **7**(1): 117–120.

Nayak, P. D. K. Bhatt, D. K. Shukla and Tandon, D. K. (2011) *Res. J. Agril Sci.*, **43** (2): 252 – 257.

Nayak, P.P.R.Behera, M.Thirunavoukkarasu and Chand P.K. (2010) *Sci. Hort.*, **123:** 473-478.

Nayaka, D.R.G. (2006) *Ph.D. Thesis* submitted to University of Agriultural Sciences. http: / /krishikosh.egranth.ac.in/handle/1/81625

Negi, R. S., R. Singh, K.S. Baghel and Bose U.S. (2013) *Environ. Ecol.*, **31:** 1288-1290.

Nirvan, R.S., Upadhaya, J. and Prasad, B. (1969) *Hort. Adv.*, **8**: 1-3.

Pandey, M., and Changtragoon S. (2012) *Amer. J. Bot.*, **99:** e468–e469.

Pantastico, Er. B. (1975) *Post-harvest Physiology, Handling and Utilization of Tropical and SubtropicalFruits and Vegetables*, AVI Publishing Co., Westpoint, Conn.

Pareek, O.P. (1997) *Indian Hort.*, **42**: 42-46.

Pathak R.K.,H.Om and Dwivedi, R. (1989) *Indian J. Plant Genet. Resour.*, **2:** 84–86.

Pathak, P. K., Preeti Dwivedi and Sunil Kumar (2009) *J. Food Sci. and Technol. (Mysore)* **46**: 283-285.

Pathak, R.A., Dixit, S.N. and Pathak, R.K. (1993) In: *Golden Jubilee Symp. Horticultural Research – Changing Scenario, Bangalore*, Hort. Soc. India, May 24-28, 1993, Abst. No. 6.19, p. 148.

Pathak, R.K. (1991) *Lecture delivered in the Summer Institute on Fruit Production and Utilization in Wasteland*, Narendra Dev Univ. of Agric. & Tech., Faizabad, June 25-July, 12, 1991.

Pathak, R.K. (2003) *Status Report on Genetic Resources of Indian Gooseberry-aonla (Emblica officinalis Gaertn.) in South and Southeast Asia.* IPGRI Office for South Asia, Pusa Campus, New Delhi 110 012, India, p. 99.

Pathak, R.K., Ojha, C.M., Dwivedi, R. and Hori Om (1991) *Indian J. Hort.*, **48**: 207-212.

Patidar, D.K., M.K. Tripathi, R. Tiwari, B.S. Baghel and Tiwari, S. (2010) *J Agril. Technol.*, **6**: 245-256.

Perry, B.A. (1943) *Amer. J. Bot.*, **30**: 527-543.

Prajapati, V.K.,P.K. Nema and Rathore, S.S. (2011) *J Food Sci Technol.*, **48:** 45–52

Pushpakumara, D. K. N. G., and Heenkenda, H. M. S. (2007) Chapter 6. Nelli (Amla) Phyllanthus emblica L. In: D. K. N. G. Pushpakumara, H. P. M. Gunasena,

& V. P. Singh (Eds.), *Underutilized fruit trees in Sri Lanka*. Volume 1. World Agroforestry Centre (pp. 180-221). New Delhi: South Asia Office.

Rai, M., Gupta, P.N., Rana, R.S. and Dwivedi, R. (1993) In: *Golden Jubilee Symp. Horticultural Research – Changing Scenario, Bangalore, Hort. Soc. India*, May 24-28, 1993, Abst. No. 1.15, p. 8.

Rajan, S. and Ram, S. (1982) *Progressive Hort.*, **14**: 190-191.

Ram, S. (1983) *Aonla (Emblica officinalis* Gaetrn) *uses, botany and culture*, Directorate of Experiment Station, G.B. Pant Univ. of Agri. & Tech. Pantnagar.

Ram, S. and Rao, T.R. (1976) *New Phytol.*, **79**: 441-448.

Ram, S. and Rao, T.R. (1978) *New Phytol.*, **81**: 513-517.

Ram, S. and Rao, T.R. (1981) *New Phytol.*, **81**: 499-513.

Ram, S., Dwivedi, T.S. and Bist, L.D. (1976) *Progressive Hort.*, **8**: 5-12.

Rane, A.A., B. Kag, V.Hegde and Sathyanarayana, B.N. (2012) *Int J Curr Res.*, **4:** 32–36

Rao, M.R.R. and Siddiqui, H.H. (1964) *Indian J. Expt. Biol.*, **2**: 29-31.

Rao, V.K. and Pathak, R.K. (1996) *Indian J. Hort.*, **53**: 251-254.

Raut GR, Senapati SK, Aprajita S (2010) *Czech J Genet Plant Breed* **46:** 135–141.

Ray, P.K. and Majumdar, S.K. (1976) *Econ. Bot.*, **30**: 317-320.

Saini, R. and Sihag, R.C. (2020) *Int. J. Agrcl Res.*, **15:** 1-8.

Sangeetha N,S. MercyS,M. Kavitha, D. Selvaraj, R. Sathishkumar and D. Ganesh. (2010) *Plant Genetic Res.*, **8:** 191–197

Sanjeev Kumar and Nath, V. (1993) *J. Food Sci. Technol.*, **30**: 202-203.

Saroj, P.L., Vishal Nath and Vashishtha, B.B. (2000) *Indian J. Hort.*, **57**: 300-304.

Sastry, L.W., Satyanarayana, M.N., Srinivasan, M., Subramanian, N. and Subrahmanyan, V. (1956) *J. Sci. Indus. Res.*, **15**: 70-80.

Sehgal, C.B., Khurana, S. and Abbas, N.S. (1994) *Adv. Agric. Res. India*, **2**: 1-19.

Selvi, B.S., V. Ponnuswami and P.S. Kavitha (2008) *Advances in Natural and Applied Sciences*, **2 (3):** . 129+. *Gale Academic OneFile*, Accessed 12 Aug. 2020.

Sethi, V., and Anand, J.C. (1983) *Indian Food Pack.*, 37**(6):** 64-67

Setty, K.G.H. (1959) *Curr. Sci.*, **28**: 208.

Shanker, G. (1969) *Indian Hort.*, **13**: 9-11.

Shete, M.B. and Karale, A.R. (2000) *J. Maharashtra Agric. Univ.*, **24**: 322-323.

Shukla, A. K., R.K. Pathak, R.P. Tiwari and Vishal Nath (2000) *J. Appl. Hort.*, **2:** 37-38.

Shukla, S. K. and Suhail M. (2009) *J. Trop Forestry* **25 (3/4):** 18-26.

Shukla, S.K. (1990) *M. Sc. (Ag.) thesis* submitted to Narendra Dev Univ. of Agric. & Tech., *Faizabad*, India.

Shukla, S.K., Pathak, R.K. and Singh, G.N. (1993) In: *Golden Jubilee Symp. Horticultural Research – Changing Scenario, Bangalore,* Hort. Soc. India, May 24-28, 1993, Abst. No. 6.18, p. 147.

Singh,D., Hemlata Bharti and Singh, K.M. (2014) *Indian J. Hort.,* **71**(1): 117-119.

Singh A.K., S. Singh, V.V. Rao B.G. Bagle and More, T.A. (2010) *Indian J. Hort.,* **67:** 124-128.

Singh A.K., S. Singh, V.V. Rao B.G. Bagle and More, T.A. (2011) *Indian J Hort.,* **68:** 461- 465.

Singh, A. and Singh H.K. (2015) *J. AgriSearch* **2:** 20-23

Singh, A.K. and Gaur, G.S. (2000) *Orissa J. Hort.,* **28**: 41-44.

Singh, Gorakh. and Mishra, R. (2007) In: *Aonla Rejuvenation.* PFDC, ICAR-CISH, Lucknow, India, p. 16.

Singh, I.S. (1991) *Lecture delivered in Summer Institute on Fruit Production and Utilization in Wasteland,* Narendra Dev Univ. of Agric. &Tech., Faizabad, June 24-July 12, 1991.

Singh, J.K., J.Prasad, J., Singh, H.K., Singh, A. (2008) *Plant Archives* **8:** 911–913.

Singh, K., A.Kumar, M. Kajal, M. *et al.* (2019) *J. Plant Biochem. Biotechnol.,* **28,** 105–113

Singh, L.B. (1952) *Sci. and Cult.,* **15**: 345-346.

Singh, L.B. (1953) *Indian J. Hort.,* **10**: 59-61.

Singh, P. C., R.S. Gangwar, and Singh, V.K (2012) *HortFlora Res. Spectrum,* **1(1):** 73-76

Singh, R. and Kumar, S. (1997a) *Haryana J. Hort. Sci.,* **26**: 9-12.

Singh, R. and Kumar, S. (1997b) *Haryana J. Hort. Sci.,* **26**: 16-19.

Singh, R. and Kumar, S. (2000) *Haryana J. Hort. Sci.,* **29:** 178-179.

Singh, R.N. (1974) *Indian Hort.,* **19**: 17-18.

Singh, Ranjit (1979) *Fruits,* National Book Trust, New Delhi.

Singh, S., Krishnamurthy, S. and Katyal, S.L. (1967) *Fruit Culture in India,* ICAR, NewDelhi.

Singh, S.K. and Singh, P.K. (2015) *New Agriculturist,* **26**(2): 357–363.

Singh, S.K., P. R. Meghwal and Rakesh Pathak (2015) *Proc. Natl. Acad. Sci., India, Sect. B Biol. Sci.,* **85(4):** 971–978.

Singh, V.K., George, C.K., Gupta, K.P. and Gupta, B.N. (1983) *Sci. and Cult.,* **49**: 354-356.

Soontornchainaksaeng, P. and Chaiyasut, K. (1999) *Cytologia* **64:** 229-234.

Sreekumar, K. and Aravindakshan, M. (1984) *Evergreen,* **13**: 19.

Srivastava, A.K. and Pathak, R.K. (1993) In: *Golden Jubilee Symp. Horticultural Research – Changing Scenario, Bangalore,* Hort. Soc. India, May 24-28, 1993, Abst. No. 1.55, p. 24.

Srivastava, R.P. (1964) *Indian Hort.*, **8**: 15-16.

Srivastava, R.P. (1965) *Sci. and Cult.*, **31**: 149-150.

Srivastava, R.P. and Srivastava, R.K. (1964) *Sci. and Cult.*, **30**: 446-447.

Suriachandra Selvan, M., Anbu, A. and Radhakrishnan, J. (1993) In: *Golden Jubilee Symp. Horticultural Research – Changing Scenario, Bangalore,* Hort. Soc. India, May 24-28, 1993, Abst. No. 2.26, p. 49.

Tandon D.K., R.C. Yadav, S. Sood, S.Kumar and Dikshit A. (2003) *Indian Food Packer* **57:** 147-152

Tandon, D.K., Kumar, S., Dixit, A., Yadav, R.C., Sood, S. (2003). Paper presented at the National Seminar on Problems, Prospects and Utilization of aonla, Salem, Tamil Nadu, 8-10th July, 2003

Tariq, M., Hussain, S.J., Asif, M. and Mahan, M. (1977) *Indian J. Expt. Biol.*, **15**: 485-486.

Tarula, H., J.M.S. Rawat, K.K. Singh, V. Rawat and Singh, B. (2015) *HortFlora Research Spectrum*, **4(1):** 17-21.

Tewari, R.K., Pandey, D. and Pathak, R.K. (1999) *The Hort. J.*, **12**: 77-80.

Thilaga, S., Muthaiah, J.V.L., Alagarsamy, P., Nair, R.R. and Ganesh, D. (2013) *Australian J. Crop Sci.*, **7:** 1480-1486.

Thilaga, S., Nair, R.R., Kannan, M.R. and Ganesh, D. (2017) *J Gen. Engg Biotechnol.*,**15:** 323-330.

Thokchom, R. and Mandal, G. (2018) *Int. J. Bio-res. Stress Management* 9**(2):** 269-273

Tyagi, R.M.S. and Pathak, R.K. (1987) *Indian J. Mycol. Plant Path.*, **17**: 218- 219.

Usharani, P.,N. Fatima and Muralidhar, N. (2013) *Diabetes Metab Syndr Obes.*, **6:** 275-284.

Vashisht, B.B., C.B. Singh and N. Biwalkar. (2018) *J. Soil Water Conserv.*, **17(1):** 98-101.

Vashishtha, B.B. (1991) *Lecture delivered in Summer Institute on Fruit Production and Utilization in Wasteland,* Narendra Dev Univ. of Agric. & Tech., Faizabad, June 24-July, 12, 1991.

Verma, B. and Kant, U. (1996) *J. Phytol. Res.*, **9**: 107-109.

Verma, B. and Kant, U. (1999) *Adv. Plant Sci.*, **12**: 21-25.

Verma, V.S., Raychaudhury, S.P. and Khan, A.M. (1969) *Biol. Plant.*, **11**: 384-387.

Wali, V.K., P. Bakshi, A.Jasrotia, B. Bhushan and Bakshi, M. (2015) In: *Aonla,* Directorate of Extension, SKUAST-Jammu Main Campus, Chatha, Jammu-180 009 (J&K).

Yadav, V.K. and Singh, H.K. (1999) *J. Appl. Hort.*, **1**: 118-121.

6

TAMARIND

S. I. Hanamashetti and V.A.Parthasarathy

1.0 Introduction

Tamarind (*Tamarindus indica* L.) belongs to the family Leguminosae, sub-family Caesalpinae. It is a native of tropical Africa. It is distributed throughout the tropical countries of the world with the largest concentration in India. Tamarind is a hardy tree which grows well under warm climatic conditions of tropics and subtropics. It is drought tolerant and found growing in sandy soil. Further, it also performs well in deep soils and tolerates poor or rocky terrain. In view of its ability to withstand heavy winds, it is preferred as a wind break

Tamarind trees are generally raised on roadsides, in backyards, on the bunds of field and in wastelands, which are either auctioned or utilised by the owners at the time of harvest. In India, tamarind has been in commercial demand. Presently, only seed propagated plantations are largely available in the country. However, in recent times, under social forestry schemes, there has been an effort to raise tamarind in both private and government lands. All the parts of tamarind have some use. In India, the fruit is used mainly for culinary purposes, while in other countries; the fruit is processed into nectar, fruit punch, juice, crystallised fruit and concentrates. The pulp is used to season many foods, *viz.*, chutneys, curries, preserves, confectioneries, ice creams and syrups. The pulp possesses some medicinal value. Hence, it is used in paper sizing, colour printing in textiles, tanning of leather, dressing of homemade baskets and as glue for wood. The seeds also yield amber coloured oil suitable for industrial use (NAS, 1979). The tamarind seed powder commercially known as tamarind kernel powder (TKP), is used as a creaming agent in manufacturing of rubber and latex and as a substitute for cereal starch. Wood is used for agricultural implements, tool handles, wheels, mallets, *etc.* It is suitable for printing blocks, tent pegs and furniture (Gupta, 1993). It has the tremendous potential in agro-forestry programmes (Dagar *et al.*, 1995).

2.0 Composition and Uses

2.1.0 Pulp

Tamarind is a nutritious fruit with a variety of uses. The pulp constitutes 30-50 per cent of ripe fruit (Purseglove, 1987, Shankaracharya, 1998). The shell and fibre account for 11-30 per cent and the seed about 25-40 per cent (Chepman, 1984; Shankaracharya, 1998). Tamarind pulp on an average (the pod is composed of shell) contains15 to 25 per cent, pulp (flesh) 45 to 65 per cent, seeds 25 to 35 per cent and fibre 10 to 15 per cent. The seeds contain testa (seed coat), 25 to 30 per cent and kernel (endosperm) 70 to 75 per cent. The ripe fruit contains 63 to 69 per cent pulp due to the presence of high moisture in it. The pulp content comes down to 45 to 55 per cent after removal of the shell and drying.

The fruit pulp contains mainly tartaric acid, reducing sugars, pectin, tannin, fibre and cellulosic material. The edible portion of the ripe pod, before harvest contains moisture 63.3 to 68.6 per cent, tartaric acid 8.4 to 12.4 per cent, sugars 23 to 30 per cent and shows a pH around 3.15 (Tables 1 and 2) (Duke, 1981; Lewis and

Table 1: Composition of Edible Portion of Ripe Fruit/Pod of Tamarind

Constituent	*Percentage*
Moisture	62.50 – 69.20
Proteins	1.40 – 3.30
Fat /oils	0.27 – 0.81
Sugars, total	21.40 – 30.85
Sucrose	0.10 – 0.80
Cellulose	1.80 – 3.20
Ash	1.20 – 1.72
Tartaric acid, total	8.40 – 12.40
Total acidity as tartaric acid	17.10 – 18.40
pH	3.15
Pentoses	4.20 – 4.80

* Duke (1981).

Table 2: Proximate Composition of Dried Tamarind Fruit*

Constituent	*Percentage*
Moisture	15.00 – 30.00
Proteins	2.00 – 8.79
Fat /oil /lipid crude	0.50 – 2.53
Carbohydrates, total	56.70 – 70.70
Fibre, crude	2.20 – 18.30
Tartaric acid, total	8.00 – 18.00
Reducing sugars	25.00 – 45.00
Total ash	2.10 – 2.90
Pectin	2.00 – 4.00
Cellulosic residue	19.40
Albuminoids	3.00 – 4.00
Total available carbohydrates	41.77
Alcohol insoluble solids	22.70
Water insoluble solids	22.70
Non-reducing sugars	16.52
Total sugars	41.20
Starch	5.70
Tannin (mg)	600.00
Ascorbic acid a (mg)	3.00 – 9.00
β-carotene equivalent (μg)	10.00 – 60.00
Thiamine (mg)	0.18 – 0.22
Riboflavin (mg)	0.07 – 0.09
Niacin (mg)	0.60

* Wealth of India (1966), Duke, (1981), Macmillon (1974), Ishola *et al.* (1990).

Neelakantan, 1964). The dried pulp of commerce contains moisture 15 to 30 per cent, tartaric acid 8 to 18 per cent and reducing (invert) sugars 25 to 45 per cent of the reducing sugars. Present, about 70 per cent is glucose and 30 per cent is fructose.

2.1.1 Colour of Tamarind Pulp

During storage, the brownish red coloured pulp becomes darker and after about a year, is almost black. This is perhaps due to the onset of Maillard reaction, since free amino acids and reducing sugars are present in the pulp. The pulp also becomes soft and sticky as pectolytic degradation takes place and moisture is absorbed, especially in humid weather (Lewis and Neelakantan, 1964). The anthocyanin pigment, chrysanthemin is responsible for the colour of pulp in the red variety of tamarind and the common variety contains leuco cyanidon. The anthoxanthin pigments lutein and apigenin are present to the extent of about 2 per cent in the tamarind leaves. The fruits have low anthoxanthin content, while the flowers contain only xanthophylls. The seed testa contains leucoanthocyanidin (Lewis *et al.*, 1957; Lewis and Johar, 1956; Lewis and Neelakantan, 1964). The mineral contents of the pulp, seed, kernel and testa are given in Table 3. The pulp is rich in calcium and potassium, while the kernel is rich in magnesium, potassium, calcium and zinc and poor in vitamin A and C (Leung and Flores, 1961).

Table 3: Mineral Contents of Tamarind Pulp, Seed, Kernel and Testa

Mineral (mg/100 g)	*Pulp*	*Seed*	*Kernel*	*Testa*
Calcium	81-466	9.3-786.0	120	100
Phosphorus	86-190	68.4-165.0	—	—
Magnesium	72.03	17.5-118.3	180	120
Potassium	62-570	272.8-610.0	1020	240
Sodium	3.0-76.7	19.2-28.8	210	240
Copper	21.83	1.6-19.0	—	—
Iron	1.3-10.9	6.5	80	80
Zinc	1.06	2.8	100	120
Nickel	0.52	—	—	—
Manganese	—	0.9	—	—

Bhattacharya *et al.* (1993), Ishola *et al.* (1990), Marangani *et al.* (1988).

The most outstanding characteristic of tamarind is that its acidity which is mostly due to tartaric acid (2, 3-dihydroxy butanedioic acid, $C_4H_6O_6$, a dihydroxy dicarboxylic acid), ranging from 12.2 to 23.8 per cent, which is uncommon in other plant tissues (Ulrich, 1970). It is an unusual plant acid, which is formed from the primary carbohydrate products of photosynthesis and once formed, cannot be further used in the plant due to the absence of the necessary enzymes. Tamarind contains more of tartaric acid than other sour fruits such as grapes, grape fruit, and raspberries. The tartaric acid is synthesised in tamarind leaves (Kewis *et al.*, 1961; Patnaik, 1974). It is high in young leaves and decreases with age and has been reported to show seasonal variations (Bueso, 1980). Tartaric acid is translocated from leaves to the fruit during ripening (Lewis & Neelakantan, 1964). The content of tartaric acid, however, does not decrease during fruit ripening, indicating that it is not utilised in fruit development, but during this time, reducing sugars increase to 30-40 per cent giving the sour fruit sweeter taste. As the acidity does not disappear with ripening, but is more or less matched with increasing sugar levels, tamarind is known to be simultaneously the most acidic and sweetest fruit (Lewis and Neelakantan, 1964a; Coronel, 1991).

Different authors have reported variations in the tartaric acid and sugar contents of tamarind pulp. In Thailand, the tartaric acid content varied from 2.5 to 11.3 per cent and the sugar from 5.0 to 40.0 per cent. According to Feungchan *et al.* (1996), the tartaric acid content of sweet tamarind in Thailand was as low as 2.0 to 3.2 per cent and the sugar was as high as 39.1 to 47.7 per cent. The sweetness of the pulp is determined by the sugar content and the sugar content of the sweet tamarind types in Thailand (Table 4) varies between 39.00 to 47.71 per cent, whereas the sugar content of sour tamarind is only 4.80 to 38.94 per cent. The sweet and sour types have been reported to contain 6.6-9.8 per cent tartaric acid (tree), 6.7-1.4 per cent tartaric acid (combined), 36.4-38.2 per cent invert sugars and 2.4-4.4 per cent pectin (Anon., 1976).

Table 4: Chemical Composition of Sweet Tamarind Pulp from Thailand

Cultivar	*Tartaric Acid (Per cent)*	*Sugar (Per cent)*	*Acid/Sugar Ratio*	*Fibre (Per cent)*
Sithong	3.18	41.07	1: 13	1.44
Piyai	2.01	47.19	1: 23	1.97
Praroj	2.70	43.09	1: 16	0.88
Srichompoo	2.39	42.52	1: 18	1.96
Kru-in	2.70	39.06	1: 14	1.02
Jaehom	2.98	47.71	1: 16	1.33
Pannanikom	2.34	47.71	1: 20	1.10

* Feungchan *et al.* (1996).

In tamarind pulp, the free amino acids identified are proline, serine, beta-alanine, phenylalanine and leucine. These amino acids are present in higher quantities in the ripe fruits than in immature fruits, indicating the accumulation of free amino acids during the maturation and ripening of tamarind. Tamarind is valued mostly for its fruit and pulp, which is used for a wide variet of domestic and industrial purpose (Kulkarni *et al.*, 1993). Tamarind is not a desert fruit, although the sweet tamarind is often eaten fresh directly from the pod. The pulp is usually removed from the pod and used to prepare juice, jam, syrup and candy. In Asia, however, the immature green pods are often eaten by children and adults and dipped in salt as a snack. Fully grown but still unripe fruits are also eaten in the Bahamas, known as 'Swells'. They are roasted in coal, the skin is then peeled back and the sizzling pulp is dipped in wood ash and eaten (Morton, 1987).

Uses of Tamarind

Nectar (Left) and Fruit Punch (Right) from Tamarind

Juice (Left) and Sauce (Right) from Tamarind

Concentrate (Left) and Crystallized Fruit (Right)

Tamarind Jelly

Tamarind Paste

Tamarid Jam

Tamarind Icecream

Tamarind Sherbet

Tamarind Kernel Powder

More commonly, the acidic pulp is used as a favourite ingredient in culinary preparations, such as curries, chutneys, sauces, ice cream and sherbet in countries where the tree grows naturally (Dalziel, 1937; Eggeling and Dale, 1951; Little and Wadsworth, 1964). In Sri Lanka, tamarind is widely used in cuisine as an alternative to lime and also in pickles and chutneys (Jayaweera, 1981). It is also considered a great delicacy. Immature tender pods are used as seasoning for cooked rice, meat and fresh and delicious sauces for duck, water fowl and geese. In East African countries, the pulp is cooked and made into a porridge called 'uguli' with sorghum or maize flour or dissolved to make a sweet drink.

Tamarind pulp is often made into a juice infusion or brine. In Ghana, a bitter infusion of the pods is used for cooking cereals and is often added to the water in which yams are soaked to detoxify them. In India, the juice is used to preserve fish, along with vinegar. Tamarind is used similarly in Sri Lanka and many other Asia countries (MacMillan, 1943). The juice is also an ingredient of Worcestershire and other barbecue sauces (NAS, 1979). Tamarind drink is popular in many countries. There are many different recipes in some African countries and the pulp juice is mixed with wood ash to neutralise the sour taste of the tartaric acid but the common method is to add sugar to make a pleasantly acid drink. In Ghana, the pulp is mixed with sugar and honey to make a sweet drink 'Jugo' and 'fresco de tamarindo' are favourite tamarind drinks in South America (FAO, 1988). Most of the producing countries manufacture drinks commercially.

In Philippines, Sri Lanka and Thailand, the fruit pulp after removing the fibres, is mixed with sugar, wrapped in paper and sold as toffees. The pulp is also used to make sweet meats with sugar called 'tamarind balls' (Purseglove, 1987) or 'bengal'. Similarly in India, the pulp is eaten raw and sweetened with sugar (Lotschert and Beese, 1994). The pulp is said to improve appetite, it is available commercially in tablet form in Thailand for the reduction of excess weight (Bhumibhamon, 1999). Tamarind pulp alone or in combination with lime juice, honey, milk, dates, spices or camphor is used as a digestive, even for elephants and as a remedy for biliousness and bile disorders.

Tamarind has been used in the treatment of a number of ailments, including the alleviation of sunstroke, datura poisoning and the intoxicating effects of alcohol and ganja (*Cannabis sativa* L.). It is used as a gargle for sore throats, dressing of wounds (Benthal, 1933; Dalziel, 1937; Eggeling and Dale, 1951; Chaturvedi, 1985) and is said to aid the restoration of sensation in cases of paralysis. In Brazil, both the fruit pulp and the leaf are regarded as purgative, diaphoretic and emollient. It is also said to aid in the care of malarial fever (Timyan and Bwa 1996). In south-east Asia, the pulp is prescribed to counteract the ill effects of over doses of chaulmoogra (*Hydnocarpus anthelmintila* Pierre), which is given to treat leprosy and in Mauritius, the pulp is used as a treatment of rheumatism.

2.2.0 Seed

The seed comprises the seed coat or testa (20-30 per cent) and the kernel or endosperm (70-75 per cent) (Coronel, 1991; Shankaracharya, 1998). The seed is a good source of protein and oil. Whole tamarind seed and kernels are rich in protein and the seed coat is rich in fibre and tannins. According to Ishola *et al.* (1990), the seed also contains 47 mg/100 g of phytic acid, which has minimal effect on its nutritive value. It also contains 14 to 18 per cent albuminoid taninins located in the testa. According to Purseglove (1987), the seeds contain 63 per cent starch and 4.5 to 6.5 per cent of semi-drying oil. Both pulp and the seeds are good sources of protein, oil and calcium (Ishola *et al.*, 1990).

The seeds are an important source of proteins and valuable amino acids (Shankaracharya, 1998). The seed is rich in cystine and methionine but threonine and tryptophon are limiting. Tamarind has a very good balance of essential amino acids except for the limiting amino acids, threonine and tryptophan the value of other amino acids is higher than the FAO reference protein (FAO, 1970). The content of sulphur amino acids, 3.5 per cent is equally divided between methionine and cystine and is usually high for legumes, while its high lysine content is similar to that of many food legumes, such as soybean, chickpea, groundnut and cowpea (FAO, 1970; de Lumen *et al.*, 1986, 1990).

The production is high in tamarind and as the seed constitutes over 40 per cent of the pod, a high protein yield can be harvested from the seeds. In India, seed kernels were used during the period of food scarcity in the status of Tamil Nadu, Andhra Pradesh and Madhya Pradesh and elsewhere, either alone or mixed with cereal flours (Shankaracharya, 1998). The seeds are also eaten after removal of the testa after roasting or boiling. The roasted seeds are claimed to be superior to peanuts in flavour (Prati and Delrosario, 1913) and are also used as substitute or adulterant to coffee. Sometimes they are made into flour for bread, Indian chapattis and cake making (Purseglove, 1987). The presence of tannins and other colouring matter in the testa make the whole seed unsuitable for human consumption. Therefore, the testa has to be separated from the kernels by boiling or roasting. Otherwise, such side effects as depression, constipation and gastro-intestinal disorders may result (Anon., 1976).

Seeds give amber coloured oil, free of smell and sweet to taste, which resembles linseed oil. It could be used for making varnishes, paints and burning oil lamps. The oil is said to be palatable and of culinary quality (Morton, 1987). Tamarind seed has shown a relatively high level of *in vitro* protein digestibility. The proteins have a favourable amino acid composition and could supplement cereals and legumes poor in methionine and cystine. They also contain small amounts of anti-nutritional factors such as tannins, phytic acid, hydrogen cyanide and trypsin inhibitor activity. Seeds could be used as a cheaper source of protein to alleviate protein malnutrition, which is widespread in many developing countries (Sidduraju *et al.*, 1995). The seeds are difficult to grind due to the hard testa, which can slightly eased by boiling, followed by decortication. This may account for its low utilisation in livestock feed. In feeding trials, the performance of chickens, however, has been unsatisfactory. The chicks fed on tamarind diets had higher water intakes than the controls and the pancreas weight and intestinal and coecal lengths were also increased (de Lumen *et al.*, 1986). Panigrahi *et al.* (1989) suggested that the poor utilisation of tamarind seed meal by chickens may be due to the indigestible nature of the polysaccharide, rather than its tannin content, or possibly the presence of other toxins. However, tamarind seed meal could be used in the diet of growing pigs, replacing upto 75 per cent of maize without affecting nutrient utilisation (Rao *et al.*, 1989). Other studies show that tamarind seed meal and 10 per cent molasses could completely replace maize in the diet of growing pigs (Reddy *et al.*, 1986).

The seed is usually powdered and often made into paste for the treatment of most external ailments. Powdered seeds have been used to treat boils and dysentery (Jayaweera, 1981; Rama Rao, 1975). Seed powder has been externally applied on eye diseases and ulcers. Boiled, pounded seeds are reported to treat ulcers and bladder stones and powdered seed husks are used to treat diabetes (Rama Rao, 1975). The seed can also be used orally, with or without cumin seed and palm sugar, for treatment of chronic diarrhoea and jaundice. For external appearance, the testa of the seed macerated with vinegar or lime juice is used on the face to prevent formation of pimples and in Indonesia, oil extracted from the seed is used as a hair dressing.

Tamarind seed has many uses as a feed for cattle and pigs, as a valuable remedy in diarrhoea and dysentery (Rao and Srivastava, 1974), as a base in cosmetics in the pharmaceutical industry, as a curative against rheumatism (Anon. 1955) and as a soil stabiliser. The major industrial use, however, is in the form of tamarind kernel powder (TKP) as a sizing material in the textile and jute industries (Lewis and Neelakantan, 1964) and also as an additive in the paper industry as a replacement for starches and galactomannans (Glicksman, 1986). It is frequently used, more effectively in combination with other gums such as guar and alginate.

2.2.1 Seed Kernel Oil

Andriamantena *et al.* (1983) extracted the oil with hexane and a mixture of chloroform and methanol, the yield was 6.0-6.4 per cent and 7.4-9.0 per cent, respectively. The major fatty acids were palmitic, oleic, linoleic and eicosanoic. The lipids contained a relatively large proportion of unsaturated fatty acids with linoleic acid (36-49 per cent) in the highest concentration. Other major fatty acids are oleic acid (15-27 per cent) and palmitic acid (14-20 per cent) (Singh, 1973).The seed kernel is rich in phosphorus (68.4-165 mg/100 g), potassium (273-610 mg/100 g) and magnesium (17-118 mg/100 g). The macronutrients, calcium, magnesium, potassium and phosphorus, however, were low in comparison with other cultivated legumes (Bhattacharya *et al.*, 1994).

2.2.2 Seed Testa

In the production of Tamarind Kernel Powder (TKP) or the jellose, large quantities of testa (seed coat) are left as a residual by-product. The hull (testa) and kernel weigh 28.6 per cent and 71.4 per cent of the whole seed weight, respectively. The testa contains 21.6 per cent crude fibre and 7.4 per cent ash. The presence of tannin (20-24 per cent) and other colouring matter in the testa makes the whole seed unsuitable for human consumption. The seed testa contains leucoanthocyanidin pigment. It contains 38 to 40 per cent of water solubles of which 80 per cent is mixture of tannins and colouring matter. Though the seed testa is very rich in tannins (20-24 per cent), the tannins obtained will be black in colour and highly polymeric due to the roasting of seeds for the removal of testa (Sundara Rao *et al.*, 1997). Further, the tannins impart black colouring to the leather.

2.3.0 Leaves and Flowers

The leaves are used as a vegetable. They contain tartaric acid and maleic acid, the latter is found in excess and increases with the age of the leaves. The leaves are good source of calcium, however, the presence of oxalic acid may affect the nutritive value (Anon., 1976).

The leaves are also used as fodder for domestic and wild animals, including elephants. According to Kaitho *et al.* (1988), crude protein content of the fodder tends to vary with the locality and season. The composition of tamarind leaves as a browse species collected in Ethiopia has been reported by Kaitho *et al.* (1988).

In India, leaves are made into a dish called 'Chindar'. The seedlings are also eaten as a vegetable. Young leaves of tamarind are used as a seasoning vegetable in some Thai food recipes because of their sourness and specific aroma (Coronel, 1991). In Zimbabwe, the leaves are added to soup and the flowers used in salads. The children in Gambia mix the acid leaves with gum from fig trees to make a 'chewing gum' (Sozolnoki, 1985). Flowers are an important nectar resource for honey bees in south India. The honey is of golden yellow colour and slightly acidic in flavour (NAS, 1979; Sozolnoki, 1985).

The leaves are used to rear silk worms, *Anephe* sp. in India, *Hypsoides vuilletti* in West Africa. The silk produced is very fine and considered superior for embroidery. The leaves, flowers and the fruit contain acid and are employed as auxiliaries in dye, especially along with safflower. The leaves and flowers are also useful as mordents in dyeing. A yellow dye derived from the leaves colours wool red and turns indigo dyed silk to green (Anon., 1976; Salim *et al.*, 1998). Mature leaves are used as a bleaching agent in the preparation of young leaves of 'buri (*Corypha alata*) for hat making in the Philippines (Brown, 1954). A sweetened decoction is used in the Philippines as a cough remedy and filtered juice is used for conjunctivitis. Young leaves are reported to cure other eye infections, sprains and wounds.The tender leaves and flowers of tamarind are eaten as a vegetable. The leaves contain oxalic acid (196 mg/100 g) and tender leaves show a calcium /oxalate ratio of 1: 1 at pH 4.5 indicating that the leaves are good sources of calcium, but the presence of oxalic acid may affect their nutritive value. Total tartaric acid in leaves disappear after three months of leaf formation. Malic acid in the young leaves is negligible and the quantity will be 1.5 per cent in the leaves during winter. The leaves and flowers contain high amounts of moisture (70-80 per cent) and rich in phosphorus and calcium (Anon. 1976; Lewis and Neelakantan, 1964).

2.4.0 Twigs and Bark

Natural vegetable dyes have gained considerable importance all over the world due to their eco-friendly nature. The tamarind bark and leaves contain tannin, with the bark reported to contain up to 70 per cent which has been used in the tanning industry. It is used for tanning hides and dyeing. Galls on the young branches are also used for tanning (Morton, 1987). In Zambia, bark tannins are used in the preparation of ink and for fixing dyes (Storrs, 1995). The bark is also burnt to make ink in many other African countries. Tamarind twigs are sometimes used as

'Chewsticks' and the bark is used as a masticatory, alone or in place of lime with betelnut.

2.5.0 Host of Lac Insect

The tamarind tree is a host for the lac insect, *Kerria lacca* that deposits a resin on the twigs. This produce could be harvested and sold as stick-lac for the production of varnishes and lacquers. This is not considered as an important product and tamarind growers often cut the resinous twigs and discard them.

2.6.0 Wood

Tamarind wood is used for making furniture, wheels, mallets, rice pounders, mortars, pestles, ploughs, well construction, tent pegs, canves, side planks for boats, cart shafts and axles and naves of wheels, toys, hubs, oil presses, sugar presses, tools and tool handles, turnery, *etc.* (Coates-Palgrave, 1988; Troup, 1909). Heartwood is considered as a very durable timber and used in furniture making as it takes on a good polish (Jayaweera, 1981). However, due to its hardness, it is difficult to work and liable to crack in seasoning (Dalziel, 1937; Eggeling and Dale, 1951, Bolza and Keating, 1972; Allen and Allen, 1981; Chaturvedi, 1985). The tree tends to become hollow in the centre as it gets old and is therefore difficult to get a tamarind plank of any width. For this reason, its value in house construction is limited. In North America, tamarind wood has been traded under the name of '*Madeira Mahogany*' (NAS, 1979). The wood makes excellent charcoal which is valued for making gun powder (NAS, 1979; Chaturvedi, 1985; Purseglove, 1987) and was used as major fuel for producer gas (gasogen) units that powered Indian cars and trucks during World War II. In Malaysia, although the trees are seldom felled, they are frequently lopped to obtain firewood (Morton, 1987). The ash is used to remove hair from animal hides (Dalziel, 1937; Irvine 1961, Salim *et al.*, 1998) and can be mixed with fruit pulp for cleansing and brightening brass and copper vessels.

3.0 History, Origin and Distribution

3.1.0 History

Cultivation of tamarind is reported in Egypt as early as 400 BC. The tree is also mentioned in the Indian Brahmasamhita scriptures between 1200 and 200 BC and in Buddhist sources from around the year 650 AD (Salim *et al.*, 1998). Tamarind has therefore been known by the people of Asia for a very long time. Arab and Persian traders are thought to have encountered the tree while trading in India, from where they carried the seeds to south-east Asia (Coates-Palgrave, 1988). According to NAS (1979), Marco Polo recorded in 1298, that Arab traders also made tamarind an important commercial item in Medieval Europe. The name tamarind is derived from the Arabic *Tamar-u'l-Hind* because the dark brown pulp of the fruit was thought to resemble dried dates. It was therefore called the *Tamar–i-hind* or the 'date of India'. According to Coates-Palgrave (1988), this inspired Linnaeus in the 18th century to name the tree as *Tamarindus indica*. In the 18th and 19th centuries, New England sea captains also commonly carried preserved tamarind from the Caribbean to Boston (NAS, 1979).

Tamarind is considered to have originated in Madagascar (Von Maydell, 1986; Hocking, 1993). It is now cultivated throughout semi-arid Africa and South Asia, where it has naturalised in several regions. It has been extensively planted in Bangladesh, India, Myanmar, Malaysia, Sri Lanka, Thailand and several African, Australian, Central American Europe during the middle ages. Marco Polo mentioned it in 1298, but it was not until 1563 that its true source was known. The new England sea captains, who traded with the West Indies in the 18th and 19th centuries, frequently brought preserved tamarind fruits to Boston from Jamaica and other islands. Tamarind fruit was at first thought to be produced by an Indian palm. Its name *'amlika'* in Sanskrit indicates its ancient presence in the country (Mishra, 1997). In Burma, it is reported as the commonest village tree in the dry zone (Troup, 1921). Commercial plantations are reported in Central American countries and in north Brazil (Sharma and Bhardwaj, 1997).

3.2.0 Origin and Diversity

Most evidence indicates that Central Africa is the origin of tamarind. The species name *'indica'* gives the illusion of Indian origin (Morton, 1987). Troup (1921) has placed it in Abyssinia. It is indigenous to the drier savannahs of tropical Africa from Sudan, Ethiopia, Kenya and Tanzania, Westward through sub-sahelian Africa to Senegal (Brandis, 1921; Ridley, 1922; NAS, 1979). The high genetic diversity is an indication that evolution has occurred over a long time. A high level of genetic diversity is also represented in the south and south-east Asian regions where tamarind is thought to have been introduced (Brenan, 1967; NAS, 1979) and where it was certainly naturalised a long time ago (Simmonds, 1984; Purseglove, 1987; Coronel, 1991). Tamarind is not presently grown commercially in Africa, but steps have been taken in Asia to evaluate the genetic diversity in the region and develop new varieties.

3.3.0 Distribution

Tamarind occurs widely throughout tropical Africa, where it is frequently planted as a shade tree (Storrs, 1995). Tamarind is now widely spread throughout semi-arid south and south-east Asia (Gamble, 1922; Chaturvedi, 1985). It is presently cultivated in home gardens, farm lands, on roadsides, on common lands and on a limited plantation scale in India and Thailand where the species is more economically important.

Tamarind is now produced commercially in Mexico and is widespread in the States of Chiapas, Calima, Guerro, Jalisco, Oaxaca and Verecruz, covering an area of 4,440 ha (10,000 acres). From Mexico, tamarind was probably introduced throughout tropical America and the Caribbean islands. The dates of introduction are unknown, however, it is thought to have occurred in Hawaii in 1797 where it persists today in small numbers across the islands (Morton, 1987). Tamarind is also found growing throughout Caribbean islands including Jamaica, Cuba, the greater and lesser Antilles and the Dominican Republic. It is common along roadsides, around houses, on hill sides and in dry coastal regions (Little and Wardsworth, 1964). In Puerto Rico, tamarind is produced on a commercial scale. There are also commercial plantations in Brazil and other Latin American countries, but a few trees are found through a natural regeneration. In Guatemala, however, the Montague is said to have so many large tamarind trees growing in one area. The mild, unusual tropical climate of south Florida is also suitable for tamarind and it is grown widely as a shade and fruit tree along roadsides, in dooryards and in parks (Morton, 1987).

4.0 Species and Cultivars

4.1.0 Classification and Cultivars

Tamarind exhibit huge variation for economically important traits ((Benero *et al.,* 1974 and Gunasena and Hughes, 2000) Variation for fruit characteristics, such as length of pod, pod weight, seed number, pod colour and sweetness of pulp is observed. The variation on sweetness and colour of the pulp in tamarind germplasm was recognised as early as 1881 by Von Mueller. The composition of pulp also varies widely and the pulp is usually better developed in Indian than that in African tamarinds (Purseglove, 1987). Considerable variation in fruit yield (from 150-200 kg/tree/year or 800 kg/tree/year) of tamarind have been reported in different countries showing the potential for improvement of fruit production (Jambulingam and Fernandes,1986, NAS, 1979 and Pathak *et al.,* 1992).

The origin of sweet tamarind has been attributed to a point mutation. It is also assumed that the sweetness in tamarind is a rare trait, which may be governed by recessive genes (Feungchan *et al.,* 1996a). Occasionally, isolated branches on a tree may bear sweet fruits, while the other branches have the normal sour type. Grafted plants from these trees are maintained to serve as sources of bud sticks for large scale propagation and distribution. Bud spots are also commonly found in the Philippines. These are isolated branches bearing sweet fruits from normally sour tamarind trees. These bud spots will provide ample opportunities for genetic improvement of tamarind.

4.1.1 Cultivars

Two major types of tamarind are recognised based on the sweetness of the fruit pulp. These are 'sweet fruit' and 'sour flavoured' types. Sometimes the cultivars are differentiated by pulp colour, which may be red or brown. The red coloured pulp is considered to be of superior quality due to its sweet taste. Mostly, unselected tamarind cultivars are grown and these vary in fruit size and degree of tartness. Some of the major tamarind growing countries have selected improved cultivars from among the natural populations as shown in Table 5.

Azhakiamanavalan and Vadivel (1997) reported the performance of PKM-1 tamarind released during 1992. PKM-1 performed better with superior yield pod characters compared to local and registered a mean yield (4 years from 1988-1991) of 263.6 kg per tree as against 165.0 kg per tree in local cultivar. The yield is 26.3 tonnes per hectare as against 16.5 tonnes in the local at the spacing of 10 x 10 m. PKM-1 tamarind exhibits precocity in bearing. Flowering in September-October, matures in February-March and harvest can be done during March-April under Periyakulam conditions.

Types and Cultivars of Tamrind

Manila Sweet

Brown Tamarind

Red Tamarind

Table 5: Selected Tamarind Cultivars Grown in some Countries

Country	*Cultivars*
Thailand	Muen Chong, Sri Tong, Nam Pleung, Jae Hom, Kun Sun, Kru Sen, Nazi Zad, Sri Chompoo
The Philippines	Cavite, Batangas, Bulacan and Laguna
India	
Maharashtra	Prathisthan, Yogeshwari (Red flesh)
Tamil Nadu	Periyakulam (PKM-1), Urigam
Karnataka	Dharwad Tamarind Selection-1 (DTS-1)
	Dharwad Tamarind Selection-2 (DTS-2)

At College of Agriculture, Pune, studies were initiated in 1974-75 on fruiting behaviour and yield potential of 120 tamarind trees planted at college campus in 1907. The tree number 38 showing regular bearing habit coupled with high yielding potential. At Fruit Research Station, Himayatbagh, Aurangabad, 'Pratisthan' and Number 263 stand were released for commercial exploitation. At Ambajogai, a unique red fleshed tamarind tree was located which is released as Yogeshwari (Karal *et al.*, 1997).

Keskar *et al.* (1989) reported an improvement of tamarind by selection. A survey to locate superior strains of tamarind was taken in two districts of western Maharashtra during 1988. Various qualitative and quantitative parameters were studied in 14 seedling types. The study revealed wide variation for TSS, acidity, weight of fruit, pod size, shape and colour of endocarp membrane. The overall study indicated that types like T-9, T-10, T-11, T-12 and T-13 being superior in fruit quality are selected for propagation and for further investigation.

Karale *et al.* (1999) studied the association of fruit characters in tamarind. The results indicated that pod shape could not be given a consideration for selecting a desirable genotype in tamarind since fruit length, breadth and weight are measures of fruit size; the larger the fruit, the more is the pulp weight. To have higher pulp weight, the genotype with higher fruit weight and less seed weight and shell weight should be the basis for the preliminary selection for large number of tamarind trees. These desirable characters were observed when survey was conducted in Maharashtra State in Satara, Solapur and Ahmednagar areas.

University of Agricultural Sciences, Dharwad released two varieties during 1996 based on the clonal progeny performance, *viz.*, Dharwad Tamarind Selection-1 (DTS-1) and Dharwad Tamarind Selection-2 (DTS-2) (Hanamashetti, 1996) (Tables 6,7a,7b).

Hanamashetti (2006) compared clonal and seedling progenies for four years. With respect to seedling progenies, higher mean pod yield was recorded in NTI-19 (DTS-1) (5.79 kg/tree) followed by NTI-14 (DTS-2) (3.29 kg/tree), NTI-62 (2.57 kg/tree) and NTI-5 (2.17 kg/tree). This indicates that seedling progenies can yield higher compared to the use of bulk seeds of unknown origin in case of tamarind (Hanamashetti, 2006).

Table 6: Performance of Dharwad Tamarind Selection-1 (DTS-1) and Dharwad Tamarind Selection-2 (DTS-2)*

Variety	*Pod Yield per Plant (kg)*		*Number of Pods per kg*
	4th Year	*10th Year*	
DTS-1	2.50-3.00	56.00	70.00
DTS-2	2.00-2.50	44.20	75.00

* Hanamashetti (1996)

Table 7a: Pod Characteristics of DTS-1 and DTS-2*

Character	*DTS-1*	*DTS-2*
Mean pod length (cm)	12.16	11.69
Mean pod width (cm)	2.93	2.04
Mean pod thickness (cm)	1.60	1.49
Mean pod weight per pod (g)	16.02	14.15
Number of seeds/pod	4.50	6.20
Seed weight/pod (g)	3.93	4.75
Pulp weight/pod (g)	5.81	4.87
Shell weight/pod (g)	5.40	3.71
Vein weight/pod (g)	0.88	0.83

* Hanamashetti (1996).

Table 7b: Fruit Characters of DTS-1 and DTS-2*

Character	*DTS-1*	*DTS-2*
Per cent of pulp	36.27	34.41
Per cent of seed	24.53	34.56
Per cent of shell	33.70	26.24
Per cent of vein	5.50	4.79
Hundred seed weight (g)	83.20	77.82
Per cent of tartaric acid	17.03	14.73

* Hanamashetti (1996).

Hanamashetti and Sulikeri (1997) found two types of plant growth habit in tamarind genotypes. Hanamashetti (1996) studied the genetic diversity in tamarind through multivariate analysis. The first canonical vector, pulp weight accounted 99.23 per cent towards divergence. Forty genotypes of tamarind were grouped into 10 clusters. Cluster VII, VIII and IX were the largest with 5 genotypes each, while cluster V was the smallest with two genotypes. Cluster mean for 8 quantitative fruits of forty genotypes were calculated. The present study indicates that pulp weight was the most important factor which was responsible for discrimination in the genotypes studies.

Patil (2004) surveyed the Dharwad and Gadag districts of Karnataka. The pod yield recorded from different plus trees ranged between 250 kg per tree in tree number 108 (Balchosur, BLH-3) to 800 kg per tree in tree number 92 (Mulgund, MLG-1).

Biradar (2001) evaluated 21 tamarind genotypes and found pod yield per tree was high in ATI-7 (330.00 kg) followed by ATI-1 (320.50 kg) and ATI-6 (257.50 kg). Two types of pod shape were were observed, *viz.*, straight and semi-curved. Plant height (21.15 m), crown size (20.25 m) and diameter (0.95 m) were maximum in ATI-1 and ATI-7, respectively. Pod weight varied from 9.84 g to 16.44 g, pulp weight from 3.67 g to 6.27 g, tartaric acid from 13.25 to 16.40 per cent.

Biradar and Hanamashetti (2001) found correlation between plant height and plant spread in east to west (0.651), north to south (0.626) and crown size (0.663). Number of pods per 900 cm^2 area was correlated with pod weight (0.402). Pod weight correlated with pulp weight (0.760), shell weight (0.748), vein weight (0.722) and seed weight (0.401). Similar attempts have been made to assess the correlation among the pod characters by Shivanandam and Thimmaraju (1988), Samiullah (1993) and Challapilli *et al.* (1995).

Biradar *et al.* (2002) studied 30 tamarind genotypes. There was wide variation in growth parameters, yield and quality. Among the genotypes studied, HTI-4 and HTI-3 found to be more promising with respect to higher yield and quality, respectively. Higher yield was obtained in HTI-4 (324 kg/tree), pod length and width was higher in HTI-3 (13.75 cm and 2.75 cm, respectively).

Wide genetic variability is observed among the traits studied. Higher values of genotypic and phenotypic coefficient of variation was observed for pod length, mear fruit weight, seeds per pod, pod girth, flesh content and ascorbic acid content. Heritability was higher for mean fruit weight, flesh content, solubility in water, pod girth and seeds per pod, while genetic advance was higher for mean fruit weight, flesh content and pod girth. These estimates indicated the scope for tamarind improvement by resorting to selection (Richard *et al.*, 1998). Ganachary (2005) made the survey and evaluation of different tamarind genotypes at two different locations in Karnataka. Mastan *et al.* (1997) studied the variability in fruit characteristics in tamarind existing in the districts of Chittoor and Anantapur in Andhra Pradesh. Fifty-two selected trees were evaluated. Among them, the largest fruit is found in Anantapur district in Akkampalli (32 cm). The heaviest weighing fruit (63.56 g) was with colour of pulp almost all light brown, colour of endocarp almost all white, yieldwise Begarlapalli (1500 kg) is the highest. In Chittoor district, the largest fruits are in Kurapalle-1, Kurapalle-2, Kurapalle-3 and in weight also, they are superior. Yield wise Chengalapuram-1 (1500 kg), Chengalpuram-2 (1200 kg) are found to be superior.

4.2.0 Genetic Resources

A large variability in germplasm exists in tamarind in terms of acidity of the pulp, pod bearing ability and pod size in countries of Africa, South and south-east

Asia. Therefore, in order to incorporate such traits, germplasm collections have to be made for each region, extending from the African continent to the Indian sub-continent, the Philippines and Indonesia in the south-east Asian region. In recent years, less acidic, sweet tamarind types have been selected in the Philippines and Thailand and are becoming popular. General characters useful for differentiating varieties of tamarind have been documented (Table 8)

Table 8: General Characters Useful to Distinguish Tamarind Cultivars

Plant Part	*Character*
Flower	Colour of petals (dark pink to whitish cream), flowering pattern (early-late)
Pod	Pod form, length, breadth, curvature, shape, pod size, pod weight, colour of shell
Pulp	Colour (red/whitish), pulp: shell ratio, real pulp value, pulp yield /pulp recovery per cent, fibre content, ease of pulp extraction
Seed	Colour, size, weight, number of seeds per pod, seed: pulp ratio
Yield	Total yield of pods, alternate /regular bearing habit
Biochemical	Sweetness of pulp (ratio of tartaric acid: sugar: protein and mineral constituents, amino acid composition
Others	Resistance to salinity, drought degraded soils, water logging, high pH, low pH, grazing disease and pests tree form

4.2.1 Genetic Resources in different Countries

4.2.1.1 India

Selection of tamarind proverances, has, however, been reported in India. Selections such as Prathisthan from Maharashtra and Periyakulam (PKM-1) and Urigam from Tamil Nadu are well established and preferred by farmers, but its performance in Karnataka (Bellary district) is very poor (Hanamashetti, 2006). There is prefuse flowering in case of PKM-1, but no fruit set. Periyakulam (PKM-1) is a sour type, a clonal selection from a local variety of a village, namely Endapalli near Periakulam. It is an early bearer and the grafts come into bearing from the third year against the fifth year in the local cultivars. It has a pulp recovery percentage of 39 per cent compared to the local cultivars which is 28 per cent. The yield is also higher in the Periyakulam cultivar with 263 kg per tree against 167 kg per tree in the local cultivar. The Horticultural Research Station of the Tamil Nadu Agricultural University, Periyakulam, released there varieties.

Studies were also undertaken to select high yielding varieties based on their flowering pattern. In mid and late flowering trees, cross-pollination was enhanced while the early flowering trees were mostly self-pollinated under natural conditions. The study also indicated that mid and late flowering trees should be selected for planting to obtain high yields and was suitable for selection for breeding and improvement (Usha and Singh, 1994).

Recent development in biochemical and molecular markers could be effectively used to identify the genetic variations within and between tamarind population. These modern techniques although very useful have to be used in combination with traditional methods as the farmer will not detect adaptive genetic variation. Variation in tamarind is poorly documented. The majority of accessions are not duplicated elsewhere and are prone to destruction by fire and hurricanes while natural populations are also being depleted due to various reasons. Thus, it is required to develop a national and regional collaborative approach for germplasm conservation.

4.2.1.2 Thailand

Fruits with red and brown coloured pulp have been found in the accessions from Thailand. In Thailand, there are more than 50 sweet tamarind cultivars under cultivation. They include Muen Chong, Sri Tong, Nam Pleung, Jae Hom, Kun Sun, Kru Sen, Nazi Zad and Sri Chompoo. A cultivar commonly grown by the Thai farmers is known as Makham Waen (Morton, 1987). Pod specimens were analysed in Thailand to formulate and criterion for primary screening. The formula derived is expressed as Real value of pulp (RV) and has been used as the criterion for selection.

It was also found that excellent accessions had RV values of over 21 and the number of pods was about 13 to 15 per kilogram. Most of the best accessions were found in the provenances along the Mae Kong River, but the role of the river on the quality parameter has not been fully understood. Sweet tamarind accessions were also collected from the same region. It was found that based on the above RV value, 51 outstanding accessions have been selected for further evaluation (Feungchan *et al.*, 1996a).

The percentage of tartaric acid and sugar which are the main ingredients for taste and flavour also varies with accessions (Feungchan *et al.*, 1996e). Several accessions from Thailand have been selected for improvement for industrial purposes. One accession (No. 86-2-23-017) has given the highest content of tartaric acid upto 11.2 per cent and has potential for commercialization. Accession (No. 86-13-008) had a sugar content of 39.94 per cent and could be valuble for fructose production (Feungchan *et al.*, 1996e). Another accession (No. 87-2-01-035) gave the highest concentrated tamarind flesh which is suitable for making sauces, drinks and confectionary.

4.2.1.3 Philippines

Variation has been observed in fruit characters in the Philippines. Some are oblong and short, while others are long and curved. Sweet and sour types are found in different parts of the country. Several trees bearing sweet fruits have been found in Cavite and Laguna. The University of the Philippines at Los Banos has selected and identified sweet varieties, namely Cavite, Batangas, Bulacan and Laguna (Webster, 1921). These varieties have long plump fruits with their sweet pulp. Some of the sweet types were reported to have pods 7 to 10 cm long and 2 to 3 cm wide, weighing 20 to 30 g. These fruits contain thick sweet pulp with 6 to 7 seeds (Villanueva, 1920). Forty-six accessions of tamarind germplasm are presently available in the Plant Breeding Institute, Los Banos, Philippines.

The varietal characteristics of tamarind lines introduced from the Philippines have been evaluated at the US Department of Agriculture, Subtropical Horticulture

Research Station at Miami, Florida (Knight, 1980) and the Station has developed a variety called 'Manilla Sweet'. It is likely that there will be more cultivars selected based on commercial demand in the future.

5.0 Breeding and Improvement

5.1.0 Cytology

Karyological study of 10 varieties of sweet tamarind (*Tamarindus indica*) in Petchaboon province based on chromosome studies of root tips showed that all varieties possessed the same diploid chromosome number of 24 (x = 12) (Anjirawaroj *et al.*, 2008). The average length of metaphase chromosomes ranged from 1.955 to 2.970 µm. The karyotypes of all 10 varieties were divided into 3 groups as follows. Group 1, including Srichomphoo, Prakaithong, Srithongbao and Intapalum comprised 5m + 6sm + 1st pairs with the arm number (NF) of 46 and a pair of satellites on subtelocentric chromosomes. Group 2, including Khunti, Fagdab, Praroj, Muenjong and Srithong possessed 5m + 5sm + 2st pairs, NF = 44. For each variety of Khunti, Fagdab, Praroj and Srithong, there was a pair of satellites on the first subtelocentric chromosomes. Muenjong, on the other hand, consisted of a pair of satellites on the second subtelocentric chromosomes. Group 3 contained only one distinct variety, namely Sangartit whose karyotype comprised 11m + 1sm pairs, NF = 48 with a pair of satellites on the eighth metacentric chromosomes. In this study, satellited chromosomes were found in all 10 sweet tamarind varieties and were regarded as their chromosome markers. A phylogenetic tree based on parsimony analysis of the ITS region revealed very close genetic relationships among all 10 varieties of the sweet tamarind. Nonetheless, the results of the molecular study confirmed the distinct position of Sangartit which was well corresponded to both the morphology and karyotype (Anjirawaroj *et al.*, 2008).

Tamarind is diploid with 2n=26 in somatic cells. Thirteen bivalents were recorded.

Considerable variation that exists in *Tamarindus indica* species offers immense scope for genetic improvement. Fruit yield and foliage production are the two important characteristics that deserve close attention in this species. Trees with increased fruit pulp yield or foliage production would be high economic value. Through hybridisation, such characteristics of selected individuals can be combined. For set of efficient controlled crossing techniques for improving tamarind, some aspects of its reproductive biology is a prerequisite for hybridisation programmes.

5.2.0 Phenology

Leaf flushing starts by late April and it takes 2 to 3 weeks for developing into fully grown leaves. Formation of reproductive buds is either simultaneous with leaf flushing or sometimes a little later. Tamarind has a mean flowering period of 60-80 days. Heavy flowering and fruiting occurs only in alternate years. In a given inflorescence, only one flower opens every alternate day. Flowers start opening by 20.00 hours and are completely open by 01.00 hour.

5.2.1 Vegetative Phenology

Terminal vegetative shoots are produced annually which bear flowers only in the next immediate flowering season. Two types of terminal shoots production could be observed, *viz.*, shorter terminal shoots and long shoots.

5.2.2 Reproductive Phenology

In an inflorescence, flowers bloomed only every alternative days. Production of flowers significantly varied between clones. Clones with longer vegetative terminal shoots were found to produce more flowers. Style and ovary length also varied between the clones.

5.3.0 Floral Biology

Chavan *et al.* (1999) reported studies on flowering phenology, floral biology and phenotypic variability for floral traits in tamarind. Studies on flowering, phenology, time of anther dehiscence and phenotypic variability for floral traits, *viz.*, inflorescence length, floral buds/inflorescence and blooming pattern was made in tamarind. Flowering was seen between last week of March to fortnight of June. Anther dehiscence was between 9.00 am and 11.00 am. Inflorescence length, floral buds/inflorescence and floral dry weight had high cv. majority of trees (64 per cent) were found to bloom one floral bud, at random, on every alternate day, while 33 per cent trees has a regular blooming of at least one floral bud per day and only two trees (3.18 per cent) were found to bloom more than one floral bud per day. It requires 25 days for completion of blooming of buds in any inflorescence.

Nagarajan *et al.* (1999) studied the phenology and controlled pollination in tamarind considerable vegetative and reproductive phonological variations were observed. Open-pollination fruit setting was between 1 per cent and 2 per cent. Controlled pollinations indicate that tamarind is a predominantly outcrossing species with extremely low level of selfing. Apomixis was absent. Variation in pollen size was observed and pollen sterility was very low. Fruits showed colour dimorphism.

Flowers are cream yellow in colour, bisexual, herkogamous, five sepals, five petals with odd petal being showy, three stamens fused at the base with filaments incurred towards the ovary base style simple, stigma is unbranched and pappillate, ovary is superior with 12 to 14 ovules. Ovary base has numerous uniseriate hairs with copious nectar. Pollination is mostly by honey bees, which is most common among leguminosae members (Arroyo, 1978). Insect visitation peaks between 0800 hours and 1100 hours. Anthesis starts at 2000 hours and flowers are completely open by 0200 hours. Anther dehiscence is only by 0830 hours in the morning. Stigma is receptive nearly for 48 hours with peak respectively on the day of anthesis.

5.3.1 Pollen Biology

Pollen sterility was found to be very low in all clones. Under ambient conditions (37-40°C), pollen viability was 88 per cent until 3 days, while pollen stored in 4°C remained 97 per cent viable upto 100 days. Pollens were produced in two distinct sizes (40-42 μm and 22-25 μm), of which the count of smaller pollens increased late flowering seasons, possibly due to deficient nutrition (Muller, 1979).

5.4.0 Breeding System

Fruit setting in open-pollination ranged from 1 per cent to 2 per cent among the clones. In controlled cross-pollination, fruit set was more than 75 per cent in all clones, in contrast only negligible fruit set could be observed in self-pollination, controlled pollinations clearly indicated that tamarind is a self-incompatible species. No fruit set could be observed in apomixis treatments. Understanding the biology of reproduction in trees is important before attempting their genetic improvement. Such basic studies are crucial especially in lesser understood tropical species (Zobel and Talbert, 1984). Reproductive biology studies help in estimating the genetic variations (Costich, 1995) and also they reveal about the quality and quantity of seeds produced by a species (Nagarajan *et al.*, 1996).

Short and rapid flowering periods, delicate floral organs, difficulty in harvesting pollen, poor conrolled crossing techniques (Bawa and Webb, 1984) and genetic relatedness (Haber and Frankie, 1982) make hybridisation a other related studies difficult among tropical trees. Though generally plants show increased fruit set in crosses (Johnston, 1991; Young and Young, 1992), it has not been so in most tropical trees (Egenji, 1976; Bawa *et al.*, 1985), this can be overcome if the phenology is properly understood. A simple manipulation in number of flowers crossed per inflorescence and their proximity to the maternal axis seems to determine successful fruit setting. It has also been observed that early initiated fruits abort in very low levels (Aker and Udovic, 1981; Udovic and Aker, 1981) and develop better due to position effect (Bawa and Webb, 1984; McNeilage, 1991).

Very low fruit setting noticed in tamarind under open pollination is not a rarity, such a condition is quite common in many tropical trees (Bawa, 1974). This is severe especially when mono culture is practiced, it can be overcome by introducing captive bee hives (Moncur *et al.*, 1995), while specialised anther arrangement makes pollinator interaction effective (Harder and Barrett, 1993; Nilsson, 1988) herkogamy promotes and strongly out crossing breeding system on tamarind under natural conditions. High fruit setting in controlled crosses and negligible fruit set in selfing are ample evidences of self-incompatibility (Mukherjee *et al.*, 1968), however, observed a slightly increased selfing in one genotype. Colouration in reproductive organs, a qualitative trait has been earlier recorded in temperate trees (Copes, 1972; Steinhoff, 1974; Sturgeon and Mitton, 1980; Parris and Mitton, 1985) in comparison no such inheritance has so far been reported in tropical trees. Fruit colour dimorphism in tamarind could be of immense breeding value and can be effectively used as a morphological marker in progeny testing programmes. Tamarind is an out crossing species, low fruit setting observed in natural conditions is mainly due to pollinator limitation. Unique floral adaptations in this species promotes out crossing under natural conditions.

The differences in seed reserve are common within species which may affect seedling establishment and seedling evaluation based on trait variation (Ghaffaripour *et al.*, 2017). Phenotypic trait variation may occur as a result of plant adaptation or the fitness consequences of seed reserve. Ghaffaripour *et al.* (2017) investigated seed reserve variation of tamarind to enable generalization about seedlings' growth interaction. Three provenances from Cuba (Granma), Mali (Faragouaran) and Thailand were selected for this experiment. Seeds were sampled from three populations that differ in seed mass and three seed mass groups within populations which were arranged in a complete factorial experiment as nested design. The seedlings of different populations were grown in a greenhouse under recommended cultural practices in order to determine the above-mentioned relationships. Seedling traits differed significantly among populations, seedlings from Thailand had the greatest total seedling biomass, leaf mass and stem mass, but seedlings from Mali showed a faster ontogeny. Significant variations among seed mass groups within populations could also be detected. We therefore suggest that seed reserve may contribute to construction of trait variation but other mechanisms such as trade–off were observed between seedling traits to enhance their survival and growth based their seed mass. It indicated a weak relationship between seed mass and biomass and there is inherent diversity among populations that corresponded to the alternative life-history adaptations which gives opportunities for selecting high quality planting material (Ghaffaripour *et al.,* 2017).

5.4.1 Breeding System and Pollinators

Tamarindus indica is a cross-pollinated, entomophilous and protogynous species. Bees are the most common pollinators. Insect visitation is found throughout the day and is at maximum during 6.30 to 9.30 am. Under optimal natural open-pollination conditions, fruit set varies between 1 and 6 per cent. Generally, controlled self-pollination results in extremely low levels of fruit set, while crossing yields 70 to 90 per cent success. This indicates that tamarind is self-incompatible (Nagarajan *et al.,* 1999).

Tamarind exhibits herkogamy (anthers and stigma in two different planes), a floral adaptation to avoid self-pollination and promote outcrossing insects belonging to six orders visit the flowers. Among these, the order Hymenoptera (bees and wasps) constitute 60 per cent of the species, Lepidoptera (moths and butterflies) 26.6 per cent, Hemiptera (bugs) 5 per cent, Diptera (flies) 5 per cent, Coleoptera (beetles) and Thysanoptera (thrips) 1.7 per cent each. The most potential pollinators among them are bees like *Allodape* sp., *Apis cerana, Nomia westwoodi* and *Parevaspis* sp.

5.4.2 Pollen Viability

Under natural conditions, pollen fertility is upto 90 per cent and remains viable for 48 hours under room temperature (27-30°C). If stored at 4-5°C, it lasts for more than three months. Use of freshly collected pollen in controlled crossing always results in good fruit set. Pollen viability can be estimate by germinating pollen in modified Brewbaker and Kwack medium supplemented with 10 to 20 per cent sucrose. Pollen tubes growing twice the diameter of the pollen can be scored as germinated. Yet another simple way of estimating viability is with Alexander's stain in which viable pollens stain pink and aborted ones stain green or sometimes may remain colourless. Enzymatic testing can also be done by culturing pollen in sucrose gradients of 1 per cent fluorescein-di-acetate. Under UV light, viable pollen fluoresce brightly when treated with this compound.

5.4.3 Stigma Receptivity

Tamarind being a protogynous species, the stigma is receptive one day prior to anthesis and remains so, till a day after anthesis. The ideal time for crossing is from 10.00 to 2.00 hours on the day of anthesis.

5.4.4 Pollination and Fruit Set

Nicodemus *et al.* (1997) studied the reproductive biology of tamarind and its implications on yield improvement. Tamarind is cross-pollinated mostly by honey bees and red ants. Though flowering is profuse in tamarind, the fruit set under natural conditions is low ranging from 1 to 2 per cent. However, when flowers were artificially cross-pollinated, the fruit set went upto 88 per cent under experimental conditions. Limited fruit set was observed under controlled selfing experiments, indicating that tamarind is highly self-incompatible. It also showed preferential outcrossing when different clones were cross-pollinated under controlled conditions. The fruit set in tamarind was found to be increasing when the number of pollinators visiting the trees was high. It was also found that sufficient number of pollen grains was absent on the stigma of 52 per cent of dropped flowers. These factors suggest that pollen limitation is one of the reasons for low fruit set in tamarind under open pollinated conditions. Planting a large area with ramets of single or two clones is not advisable since all the ramets of a clone are genotypically same. There will be a severe shortage for pollen from unrelated genotypes leading to increased selfing. The ramets of different clones may be thoroughly mixed and planted to ensure maximum outcrossing.

The flowering behaviour of clones was not found to be different when planted outside the region of their occurrence. The flowering period of the clones is constant though the flower opening time varied between locations. When clones are assembled from different parts of the country in commercial plantations or seed orchards; there will be synchronisation of flowering for optimum out crossing.

Very little is known about pollination and fruitset in tamarind. Thimmaraju *et al.* (1977) reported that tamarind is protogynous, entomophilous and highly cross-pollinated. The nectiferous flowers attract honey bees to collect nectar and pollen so they presumably assist in cross-pollination. Self-pollination also results in seed set (Coronel, 1991). Studies on the floral biology of tamarind are few. Thimmaraju *et al.* (1977) reported that flower bud development takes about 20 days. Peak anthesis takes place at 6.00 am and peak anther dehiscence at 10.30 a.m. Maximum stigma receptivity is on the day of flower opening.

Lokesh *et al.* (1997) studied the evolutionary significance of style orientation in tamarind and possible evolutionary constituents for low reproductive success. It is investigated that in tamarind, the style orientation determines the reproductive success. The florets in tamarind possess three petals of which two are arranged side ways as wing petals and the middle one carry nectar (nectar guide). The three anthers are placed just above the nectar guide. It is observed that a bee landing on to a flower is also at random and lands either on left or right petal more often than the nectar guide. Hence, these two levels of randomisation will act as two independent factors and put the reproductive success theoretically at 50 per cent only. It is documented by earlier workers that in spite of profuse flowering in tamarind, the fruit set was very poor, resulting in large scale abscission of flowers as well as fruits during various stages of development. The main causes of poor fruit set might be the failure of fertilisation, sterility or higher percentage of defective pollen grains or slow growth of pollen tube or its early degeneration (Prasad, 1963; Usha, 1986; Laxmi, 1991; Thimmaraju *et al.*, 1977, 1989). In tamarind, the style is oriented towards one of the side petals left or right, in respect of nectar guide. To understand the floral complexities coupled with bee behaviour in shaping the reproductive success, it is necessary as commercial plantations are being taken up by more farmers and organisations.

The inflorescence arises in small terminal drooping racemes, 5 to 28 cm long on the season's growth. Flowers are 2 to 2.5 cm in diameter, zygomorphic, irregular and hypogynous with four sepals and three petals. Androecium with three fertile stamens, filaments united for about half their length. Curling upwards under petals, anthers are transverse, reddish brown and dehise longitudinally hynoecium consists of superior ovary with 10 to 13 ovules. Pistil is oblique, curving upwards, green, sparsely hairy and longer than stamens. Stigma is sticky, small and clavate (Thimmaraju *et al.*, 1977; Usha, 1986). The chance of reproductive success is even lower than 10 per cent pollen reaching stigma has clearly indicated that nearly 54 per cent stigma do not receive at least one pollen. This may be because of pollinator limitation or inability of pollinators to touch the stigma and bring about effective pollination, while 15 per cent stigma received only one pollen. Thus, an evolutionary constraint is imposed on tamarind flowers both at the level of style orientation and at pollinator visitation.

5.5.0 Other Possible Evolutionary Restraints

i) *Self-incompatibility*: Self-pollen will not be effective and outbreeding is obvious. This could be a major restraint because under the conditions of low pollinator activity, the male success (pollen reaching stigma) may be hampered. This situation may be aggravated with protogynous nature (Bawa, 1980).

ii) *Style orientation*: the style in tamarind is either oriented towards right or left side away from the anthers and it is indeed at random. It is further investigated that the bee landing on to a flower is also at random and lands either on left or right petal. Hence these two levels of randomisation will act as two independent factors and put the reproductive success theoretically at 50 per cent only. This necessarily underlines that 50 per cent of flowers could never get the pollen.

iii) *Floral complexities*: A wide variability exists for floral components *viz.*, inflorescence length, number of buds per inflorescence, phenology, nectar content, *etc.* Hence individual trees differ in floral display for pollinators.

iv) *Pollinator movement*: Pollinators visit flowers of the same tree quite often promoting geitanogamy. This behaviour of bees although may be helpful in increasing the male success. The female reproductive success (flower development into fruits) could be low (Leppik, 1988).

v) Lack of potential pollinator: The gap between stigma and the wing petals where a pollinator is forced to land is large. Hence most of the bees (*Apis mellifera, A. cruna, A. florea, etc.*) fail to make essential contact with stigma except *A. dorsata* and xylocopa bees. Eventhough tamarind plants are visited by a large number of pollinators, many are inefficient in bringing about successful pollination.

vi) *Floral predators*: Several floral predators inhibit flowers of tamarind. They damage flowers extensively especially the nectar guide. Hence these flowers may not be visited by the bees thus reducing the reproductive success.

vii) *Phenology*: Flowering phenology of individual species is linked to the abundance of pollinators and competition for those pollinators with other plant species owing to a wide variability for floral traits even inter-tree competition cannot be precluded. Both will limit the seed production.

Large scale plantations, if not supplemented by naturalised bees, may face acute shortage of pollinators resulting in low fruit set thus ultimately incur heavy loss by the farmers or private organisations. The thumb rule could be isolated and self compatible tree.

6.0 Propagation

Tamarind is generally propagated through seeds. It does not breed true to type due to heterozygosity since the flowers are cross-pollinated. Prolonged juvenile phase is one of the problems associated with tamarind propagation (Karle *et al.*, 1997). Therefore, vegetative propagation of superior genotypes is necessary for shortening juvenile phase, production of uniformly growing trees and quality yielders. Non-conventional methods of propagation, *viz.*, tissue culture techniques are gaining momentum recently.

6.1.1 Seed Propagation and Seedling Growth

Seeds should be collected from high yielding trees in March-April by shaking the branches or by hand picking. The pedicels that hold the fruit to the tree are very strong and the pods should, therefore, be removed by clipping in order to avoid damage to the fruit (NAS, 1979). The pods should be dried in the sun and the seeds removed from the pulp manually. Seeds are washed and dried in the shade store in well-ventilated gunny bags or paper bags in a cool place.

The seeds are exalbuminous in the month of April and consist of an outer hard brown testa. The seeds are normally sown directly in sand and germination commences within 5 to 10 days. The young plant grows best in porous soil in shade and reported to be very sensitive to frost. Generally, no pre-treatment is found necessary for germinating the seeds (*Oza, 1978*). The acid (H_2 So_4) scarified seeds.

Some, *et al.* (1990) scarified seeds using 7 per cent H_2SO_4, washed and dried and then stored in sealed containers for 52 weeks at 4°C. The germination percentage after 20, 28 and 52 weeks was satisfactory, but showed little improvement over untreated seed (Some *et al.*, 1990). Seeds soaked in cow urine and cow dung solution increased germination (Sankaranarayanan *et al.*, 1994). At times the ends of the seeds are sliced off to enhance the germination. Germination was more in heavy seeds compared to that of light seeds due to rich nutrient contents in the former. Seedlings attain plantable size of 30 cm and above within 3 to 4 months. Seedlings are raised in the nursery and are transferred to deep bamboo baskets or other deep containers. Seedlings are planted in pits of size 30 cm^3 at 15 m x 15 m. spacing. In pure plantations, spacing is 5 m x 5 m and avenues linear spacing of 10 to 15 m. is adopted. Two years old transplants are better than one year old.

6.1.2. Seed Germination

The behaviour of tamarind seed is orthodox (*Riley, 1981*, Hong *et al.*, 1996). Some studies (FAO, 1988) indicate that germination can be improved by storing the seed for 6 months before planting. Fresh seeds retain viability for at least six months when kept at ambient temperature in dry conditions. Under field conditions, viability is more than a year when the seeds are well dried, mixed with sand and kept in air-tight containers. Seeds could be stored for several years in air tight packs at 10°C and 7 to 15 per cent moisture content (Albrecht, 1993; Hong *et al.*, 1996).

The germination of fresh or well preserved seed is vary from 65 to 75 per cent. Coronel (1991) stated that germination might vary from 30 to 70 per cent epigeal. The radicle emerges from one end of the seed and descends rapidly. The hypocotyl elongates arches slightly and raises the cotyledons enclosed in the testa, above the ground (Troup, 1921). Soaking in cold water for 24 to 48 hours has been reported to hasten germination (Troup, 1921; Von Cerlowitz, 1986). In Malawi, seeds thoroughly cleaned and soaked in water overnight resulted in more than 80 per cent germination (Prins and Magehembe, 1994). Masano (1994) reported that soaking in hot water, reduced germination to 22 per cent. Magehembe (1995) reported that seed germination could be enhanced by sowing immediately after extraction. Seed dressing with carbendazim, Mancozeb, ethyride, coper oxychloride and an organo-mercury fungicide 3 of seed have been recommended for control of seed borne fungal pathogenes (Purohit *et al.*, 1996).

The best medium for seed germination is sand or soil mixed with cow dung. Chattopadhyay and Mohanta (1988) reported that seed germination could be encouraged by using cow dung or cow dung and sand in the propagation medium. However, a nursery potting mixture containing three parts of top soil, one part of sand and one part of compost can be successfully used for germinating tamarind seed. Seeds may be sown in deep polyethylene nursery bags in order to accommodate the tap root without causing and distortion and abnormality.

Seedlings grow rapidly in the early stages and produce a long tap root which may attain 30 cm or more within two months of germination (Troup, 1921). Seedlings are 80 cm before being transplanted to the field.

Propagation of Tamarind

Seed Propagation

Seed Germination in Tamarind

Propagation by Cutting

Propagation by Air Layering

Propagation by Tissue Culture

6.2.0 Vegetative Propagation

Vegetative propagation is preferable to seed propagation as seed propagation does not produce true to type progenies. Mascarenhus *et al.* (1987) reported that vegetative propagation through rooting of cuttings was unsuccessful for *Tamarindus indica*. Swaminath *et al.* (1990) reported an effective technique for propagation through mature stem cutting. Srivasuki *et al.* (1990) reported the superiority of tamarind cuttings over middle or basal cuttings for rooting of tamarind. IBA at 1000 ppm was found to enhance rooting in 10-15 days. Navaneetha *et al.* (1990) reported that semi-hardwood cuttings produced 19.6 per cent rooting followed by softwood cutting with 11.25 per cent in 1000 ppm IBA and more than 48 per cent of them survived. Tamarind can be propagated vegetatively by many methods. Proven methods include stem cuttings, air layering, patch budding or grafting on to seedling rootstocks.

6.2.1 Cutting

The easiest and the cheapest method of propagating tamarind is by stem cuttings. Although vegetative propagation through rooting of stem cuttings was reported to be unsuccessful by Mascarenhas *et al.* (1987), a number of other reports have claimed an efficient response. A technique using softwood terminal cuttings has been developed Srivasuki *et al.* (1990). Shoots bearing new flushes of fully turgid leaves are collected and immediately dipped for 10 seconds in 1000 ppm of indole butyric acid (IBA) and in 50 per cent isopropyl alcohol. They were planted in vermiculite /perlite (1:1) and placed in a mist propagator with 70 to 80 per cent humidity. Use of IBA was found to increase the rooting besides reducing the time taken for root initiation.

Soft or semi-soft stem cuttings (15 to 20 cm) excised from 1 to 2 years old branches is rooted (Swaminath *et al.*, 1990). The cuttings are wrapped in moist cloth after removal from the tree and dipped in IBA (1000 ppm) and Planted in a sand bed in a mist chamber. Bud and root initiated after 20 days leaves are formed after about 45 days. Softwood cuttings are better than semi-hardwood and hard wood cuttings.

6.2.2 Air Layering

Air layering or 'gootee' has been attempted in tamarind. Navaneetha *et al.* (1990, 1991) (Nachegowda, 1997) have reported positive influence of hormone (IBA, NAA, IAA) treatment on rooting. Pradhan (1997) standardised air layering technique using two to three years old branches of pencil thickness who also reported that the success of air layer depended an etiolation medium, age of branch, and season. Navaneetha *et al.* (1991) further confirmed these observations effect of host genotype on success of air layering is reported. Wet saw dust/coir fibre was used to cover the portion and then a polythene paper was wrapped around the etiolation medium and tied on both ends by a string. As regards etiolation medium, saw dust proved superior to coir fibre. Root length and number was more (14 and 10.3 cm, respectively) in the beginning of the rainy season (mid-June to mid-July), whereas it was lowest (5 and 5.8 cm) end of the rainy season (mid-September to mid-October).

Girdling and etiolation increases the percentage of total sugars, reducing and non-reducing sugars, lowers the starch content and raises the percentages of soluble and total nitrogen. This results in a lower C: N ratio, higher level of the rooting cofactor and lower levels of indigenous inhibitors compared with non-girdled etiolated shoots. Hanamashetti and Sulikeri (1997) studied the genotypic response to air layering in tamarind. There was significant difference among the genotypes in respect of root parameters. The survival per cent of layers varied from 11.11 per cent to 100 per cent. Significantly high survival of layers was observed in NTI-60, 61, 31 and 15.

6.2.3 Grafting

Patil (2004) studied the response of genotypes to air layering. In air layering, the range of rooting varied from 60.00 in NTI-82 to cent per cent in NTI-77, SMG-13 and PKM-1 and the survival per cent ranged from 16.66 per cent in NTI-1 to 64.28 per cent in NTI-62. Ramirez *et al.* (1986) has complied various grafting methods employed for tamarind. Purushotham and Rao (1990) and Amin (1978) studied the prospects of veneer and softwood grafting for the propagation of tamarind, which yielded 49 per cent and 68 per cent graft success, respectively. Approach grafting was found to be superior than ordinary grafting (Swaminath and Ravindran, 1989). Kulwal *et al.* (1997) reported softwood in tamarind. Grafting performed in March-April on to 8 months old seedlings of 22 to 32 cm height and 0.3 to 0.4 cm diameter was highly successful. Similar results were observed by Sathisha *et al.* (1997). In another study in green wood cleft grafting, maximum graft success was recorded during May (80 per cent) followed by April (76.7 per cent) and June (70 per cent) and least success was observed in February month of grafting (26.7 per cent) (Nachegouda, 1997).

6.2.3.1 Approach Grafting

Approach grafting is a reliable method and upto 95 per cent success can be obtained (Swaminath and Ravindran, 1989). Root stock should be two year old healthy, vigorous Scion should be selected from genetically superior mother trees for grafting purpose. Graft union could take 90 to 120 days for Stability and establishment. Nadagoudar and Basavanneppa (1997) found approach grafting as on ideal propagation method due to various advantages Such as ease, cost effectiveness, and higher establishment rate. Seasonal influence on success of approach grafting has been reported Biradar (2001) Higher per cent of graft success was recorded during second fortnight of June (88.88 per cent) followed by second fortnight of May (77.77 per cent) and first fortnight of November. Patil (2004) further confirmed these observations.

6.2.3.2 Veneer Grafting

For veneer grafting, 6 months old rootstocks of uniform size is grafted defoliated Scions. Veneer grafts are made 8 cm high on the rootstock and immediately after the stock are cut above the graft union. This method is reported to give about 50 per cent success (Amin, 1978; Purushotham and Narasimharao, 1990).

6.2.3.3 Softwood Grafting

For softwood grafting, the rootstock seedlings are defoliated and their tops cut off at 15 cm high immediately before grafting. A vertical downward cut is made in the centre of the cut stem to about 4 cm depth and the scion sticks are cut into wedge shapes, inserted into the stock and wrapped using 2 cm wide 200 gauge polyethylene ribbon. Softwood grafting was shown to be the best grafting method in terms of successful unions and survival rates (Navaneetha *et al.*, 1990).

The correct spacing of mother plants is important to produce shoots for commercial production of grafted plants. Feungchan *et al.* (1996c) reported that a spacing of 2 m x 2 m is suitable for multiplication purposes. The age of the rootstocks is important for success. Scions pre-conditioned for 30 days prior to grafting on to 6 and 9 months old rootstocks resulted in 69 to 72 per cent success at 60 days. Grafting success has been attributed to the fact that rootstock of this age contain a higher proportion reducing sugars to total sugars than at other ages (Kulwal *et al.*, 1997, Rao *et al.*, 1991, Satisha *et al.*, 1997). In Thailand, a dwarf rootstock has been identified based on morphological characteristics, such as internode length and leaf area. These characters are correlated with the number of stomata on the leaves. Attempts to reduce canopy size by intergeneric grafting an other leguminous species have not been successful. The application of growth inhibiting chemicals has also had no dwarfing effect (Feungchan *et al.*, 1996b).

Age and maturity of rootstock on success of softwood grafting has been reported Biradar (2001). Higher per cent of sprouting (93.33 per cent) was recorded when grafting was done on six months old rootstock followed by 5 months old rootstock (43.33). Patil (2004) studied the age of rootstock for softwood grafting. Higher per cent of success and survival was noticed in softwood grafting done at the age of 7, 8 and 9 months old rootstocks. Genotypic response to softwood grafting was done. The graft-take range was 13.33 per cent to 90.00 per cent in NTI-5 to NTI-19. Nine genotypes possessed significantly higher per cent of graft-take. Kulwal *et al.* (1997) reported on softwood grafting for the propagation of tamarind. Sowing of seed during July to October resulted in seedling of the desired height amenable for grafting as root stock whereas early sowing leads to seedlings with long tap root are not suitable for grafting.

6.2.3.3.1 Preparation of Scion Shoots

Defoliated scion shoots are normally preferred for grafting. The defoliation can be either natural or man made prior to grafting. Seasonal influence on the success of graft has been reported. There was absolutely no success when grafting was performed between July to December. Best period for grafting of tamarind by softwood technique is between leaf fall and new growth.

Multiplication of tamarind by softwood grafting technique – a scion detached method is quite simple. Grafting on 8 months old stock seedling with a height in range of 22 to 32 cm and diameter between 0.3 and 0.4 cm performed in March-April was highly successful.

6.2.3.4 Cleft Grafting

In a green wood cleft grafting studies, grafting during May month recorded maximum graft success (80 per cent) followed by April (76.7 per cent) and June (70 per cent) months of grafting and least success was in February month of grafting (26.7 per cent). This technique can be very well adapted for the large scale multiplication of tamarind clones. Shinde *et al.* (1997) claimed to have produced thousands of grafts using wedge graftings which were being made available to the farmers.

6.2.4 Patch Budding

Budding is one of the clonal methods of propagation, wherein only a single bud with a piece of bark is used as the scion material. The plant that grows after successful union of the stock and bud is known as budded plant. Pathak *et al.* (1991) reported success in patch budding (96 per cent) and modified ring budding (94 per cent). Furthermore, they also opined that patch budding could well be the fittest means of large scale multiplication for orcharding in tamarind.

7.0 Area and Production

Being a non-forest crop, no official estimates are available on area and production. It is also difficult to make a realistic estimate of the area under this crop due to scattered nature of cultivation. However, rough estimates are available on production of tamarind. As per one estimate, production was about one lakh tonnes in 2017-18 (Table). Tamarind cultivation is popular in the Sates of Tamil Nadu, Andhra Pradesh, Karnataka, Orissa and Kerala (Jambulingam and Fernandez, 1986; Anon., 1997; George and Rao, 1997). At present, tamarind is cultivated in 54 countries around the world. The major areas of production are in the Asian and American countries.

The major commercial product is the fruit. Product diversification is taking place slowly in a few countries, such as Thailand, Indonesia, Philippines and some African countries. Although the potential is high, this under-utilised species has not reached the level of commercial exploitation. India is the world's largest producer of tamarind products. It is particularly abundant in the Indian States of Madhya Pradesh, Bihar, Andhra Pradesh, Karnataka, Tamil Nadu, West Bengal and Maharashtra (Table 9). Tamarind is not grown on a plantation scale but trees in patches are common in the villages in many States. In some parts of India, it is naturally regenerated on wastelands and forest lands. India has been exporting processed tamarind pulp to western countries. India has been exporting processed tamarind pulp to western countries, mainly the European and Arab countries and more recently to United States of America. The Annual export to the US exceeds 10,000 tonnes earning about 100 million Indian rupees (US $ 2,165,000).Tamarind products are exported to around 60 countries

Fresh tamarind is exported to 33 countries of which Pakistan, UAE, UK, Bangladesh, Japan and France are importa. Tamarind dried is exported to a 56 countries, during the last five yeas. The important importers are UAE, Saudi Arabia, Syria, Egypt, UK and YAR. Exports to Saudi Arabia and UAE were notably higher than other countries. Recently, Thailand has become a major producer of tamarind with sweet and sour varieties in production. Thailand has an expanding horticultural industry and tamarind is considered one of the major crops, similar in importance to guava and citrus. Thailand is particularly prominent due to the availability of the sweet tamarind types. In the south and north-east of the country, small sweet tamarind orchards have been established by small holders to produce fresh fruits. Research programmes to improve the varieties, cultural practices, processing and storage have commenced, as the potential for both local consumption and export has been recognised. Product development has also be taken place and toffees and many other sweet preparations are sold in the market. Mexico and Costa Rica also produces tamarind. Interestingly the original native of tamarind, Africa is not a major player in the production and marketing of tamarind.

Table 9: Production of Tamarind in India (2017-18)

Sl.No.	State	Production (000 Tonnes)	Share (Per cent)
1	Tamil Nadu	44.55	45.39
2	Kerala	36.47	37.15
3	Maharashtra	8.77	8.93
4	Madhya Pradesh	7.85	8.00
5	Others	0.52	0.53
	Total	98.16	

National Horticulture Board (NHB).

8.0 Soil and Climate

Through tamarind can grow on a variety of soils, the growth is seen best in deep loamy alluvial soils. It is also found in red loam, red sandy, clay loam, clayey and sandy clay soils. It can tolerate slightly alkaline, saline soils poor, degraded eroded, gravelley, saline and alkaline soils.

8.1.0 Climate

The tree is found scattered in dry deciduous forest, mostly in flat or undulating ground or on gentle slopes from sea level to about 1000 m. Through it prefers a mean annual rainfall of 500 to 1500 mm, grows even with 350 mm annual rainfall. (FAO, 1988; Jama *et al.*, 1989; Hocking, 1993). It is adapted to 0-47°C. It has been reported to grow from altitude of 0 to 2000 m (Roti-Michelozzi, 1957; Dale and Greenway, 1961; Brenan, 1967, FAO, 1988, Jama *et al.*, 1989).

8.1.1 Rainfall

In areas where rainfall is very low, the trees are usually located near the water table or along water courses. Tamarind can be grown in very dry areas with supplementary irrigation and can withstand up to 6 months without rainfall. The maximum annual rainfall which tamarind can tolerate is upto 4000 mm with adequate drainage (Duke and Terrell, 1974). Tamarind grows under wet tropics does not flower (Allen and Allen, 1981). Dry weather is important for flower initiation and if heavy rains occur during flowering, tamarind does not bear fruit. As experienced in the wet tropics of north-west India and south Malaysia, despite good growth the trees do not produce any fruit (Morton, 1987). Dry weather is important for fruit development and regardless of total annual rainfall. It produces fruit when subjected to long annual dry period (Allen and Allen, 1981; Von Maydell, 1986). This is Adaptation of turmeric in dray place is due to its deep and extensive root system (Coronel, 1991). Tamarind is usually evergreen but may shed its leaves in very dry conditions during the hot/dry season (Morton, 1987).

8.1.2 Temperature

Tamarind thrives best at annual temperature ranging from 33 to 37°C and a minimum of 9.5 to 20°C. Older trees are more resistant to extremes of temperature than young trees and can withstand temperatures ranging from -3°C-47°C. Tamarind is more cold tolerant than mango, avocado and litchi (Verheji and Coronel, 1991).

8.1.3 Wind

Tamarind tree is resistant to strong winds, and can tolerate violent typhoons and cyclones due to its deep root system (Von Maydell, 1986; Von Carlowitz, 1986), often known as the hurricane-resistant tree (NAS, 1979). It has strong and pliant branches and a deep extensive root system, which anchors it to the ground (Coronel, 1986). Another explanation for the isolated growing habit of the tree is its tolerance to harsh environments.

8.2.0 Soil

The tamarind tree can be grown in a wide range of soils (Chaturvedi, 1985; Sozolnoki, 1985). Sozolnohi (1985) and Galang (1955) have reported that without any cultivation it can flourish in poor soils and on rocky terrain. In India, it tolerates sodic and saline soils where it grows in ravines and on degraded land (Anon., 1991). Older plants are more resistant to salinity than seedlings and have been found growing close to the sea (Pongskul *et al.*, 1988). In north-east Thailand, tamarind has been reported to establish naturally in areas with recently salinised soils (Nemoto *et al.*, 1987). Dwivedi *et al.* (1996) concluded during pot experiment that tamarind could grow in soil containing upto 45 per cent exchangeable sodium. The tree does not tolerate water logging (Relwani, 1993; Vogt, 1995). The optimum pH for tamarind is 5.5 to 6.8, which is slightly acidic (FAO, 1988) though it grows well in alkaline soils (Singh *et al.*, 1997, Rao *et al.*, 1999). It has been suggested (Jansen, 1981) that its association with ant-hills and termite moulds may also be due to a preference for slightly concentrated lime content in the soil. In Africa, the tree is reported to grow near anti- hills or termite mounds, which indicates its preference for well-aerated soils (Dolziel, 1937; Eggeling and Dale, 1951; Irvine, 1961, Allen and Allen, 1981).

9.0 Cultivation

9.1.0 Planting

Tamarind can be grown as pure orchard crop, as one of the one of the mixed crops in agro-forestry species or as a hedge row tree. There are several methods available for establishment of the trees. Transplanting nursery grown plants in the field can have advantages over planting seed at stake. This, however, incur much higher transport cost, is more costly and the newly planted trees need more care and attention than do field grown seedlings. A technique for large scale multiplication for use in roadside plantings has been described by Swaminath (1988). In this method, nursery seedlings are pricked out and transplanted into 12 m x 0.6 m x 0.6 m trenches for 7months before planting in the main field. The trenches are filled with tank silt to 10 to 15 m from top. The seedlings are planted 60 cm apart and watered daily. After 3 months, the watering is done every three days and for the following 3 months, only once a week. The trenches are weeded and off shoots are removed to induce apical dominance. After about 7 months, when the height is about 3 meters and girth 6 to 8 cm, watering is stopped and the seedlings are dug up with a ball of intact earth. They are then tied with paddy straw and kept in a nursery and watered until they have recovered from the uprooting shock. When a new flush of leaves appears, they are then ready for transplanting on roadsides.

Nursery produced plants can be transplanted to the field at about 12 to 14 months by which time they are about 80 cm tall. Under optimum growing conditions, they may achieve this height in 4 to 6 months. In India, seedlings ranging from 40 cm to 2 m in height are commonly planted (Jambulingam and Fernandes, 1986). However, it is difficult to handle these older seedlings as their taproots may have grown deep into the soil. When seedlings or grafted plants are transplanted in the field, their size may range from 40 cm to 2 m tall.

9.1.1 Spacing

The spacing between plants varies between 5 m and 13 m. Spacing may be even wider on fertile soils where the trees grow larger. In a pure plantation, the final spacing should be at least 13 x 13 m square or in the triangle. There can be an advantage at higher density (about 5 x 5 m) followed by thinning two or three times to the appropriate final spacing. This allows further selection in the field and reduces the weeding costs in the early years. The final spacing should be closer when vegetatively propagated plants are used, as they do not attain the height or spread of seed propagated trees. In Central, Eastern and North-eastern Thailand, the recommended tree to tree spacing in sweet tamarind orchards is 10 to 12 m (Yaacob and Subhadhrabhandu, 1995). In the Philippines, the spacing recommended for seedlings is 16 to 18 m and for asexually propagated plants, 8 to 10 m. In Sri Lanka, unproductive scrab forest lands have been successfully planted with tamarind by the Janata Estate Development Board at a spacing of 3 x 3 m in lines (Rodrigo, 1992).

In parts of India, tamarind is established at 8 m x 8 m, 8 m x 12 m or 12 m x 12 m (Jambulingam and Fernandes, 1986). The seedlings are planted at a spacing of 5 x 5 m, which becomes 10 to 15 m and 10 to 15 m after thinning twice of thrice. Tamarind is frequently planted along land borders, canal banks, on field boundaries, in village groves or in home gardens as individual trees.

9.1.2 Time of Planting

The ideal time for field planting is beginning of the rainy season, particularly in secondary dry regions. This will reduce the need for frequent watering until the plants are finally established in the soil. If irrigation is available, field establishment may be undertaken at any time of the year, even in the dry season. It is advisable to provide partial shade to the newly established plants if planting is carried out during the dry periods. In Philippines, the planting time is from May to June and in Sri Lanka, from April to June, to coincide with the onset of the rainy season (Gunasena, 1997). Transplanted trees should be protected from cattle and other animals. Regular weeding around the plant is also essential for early establishment and good growth.

9.2.0 Management

Growth of tamarind is normally fast in the early years. The juvenile phase generally lasts 4 to 5 years. Growth slows down in later years. Attempts have been made to increase the growth rate by vesicular arbuscular mycorrhizal (VAM) fungi (Reena and Bagyaraj, 1990). Tamarind seedlings inoculated with VAM fungi had increased nutrient uptake and growth rates.

9.2.1 Pruning and Training

Tamarind is a compact tree and produces symmetrical branches. Young trees should be pruned to allow 3 to 5 well spaced branches to develop into the main scaffold structure of the tree. In the Philippines, young trees are pruned in the early stages of growth to train 2 to 3 lead branches and to remove the very low branches, thus developing a desirable frame. Bearing trees require very little pruning other than maintenance pruning to remove dead, weak and diseased branches and water sprouts (Salim *et al.*, 1998). In a closely planted orchard regular pruning is needed to rejuvenate fruiting wood as well as to regulate the size of trees.

9.2.2 Intercropping

Tamarind is one of the dry zone agro-forestry tree species. The basic resources for crop production are soil, water and solar energy. Radiation and precipitation are the two important atmospheric inputs. One of the most important factors that decide the feasibility of multiple cropping is light requirement of component crops and the extent of interception by their canopy. Thus, the competition for light often determines the success of a component crop in the tree based multiple cropping systems.

Intercropping can be practised in order to obtain additional income during the early stages of tree growth and until the trees reach bearing age. The intercropping period is limited to 3 or 4 years for vegetatively propagated trees, which commence bearing during this age. It is good practice to plant short season crops which grow in the rainy season. In Thailand's Central delta, intensive cropping is practiced in the plantations of the sweet cultivars, which are always vegetatively propagated and transplanted. Intercropping can be extended upto 5 or 6 years for the slower maturing seedling propagated plantations. In Central America, Mexico and Brazil where there are more than 4000 hectares of well maintained orchards, intercropping is practiced in the early stages. Cover crops are useful to control competition from weeds as well as to conserve moisture. There is a wide choice of intercrops which can be grown depending on the soil, climate and the local market demand.

The long pre-bearing period, shrinking and sufficient inter-space cultivable land has offered scope for growing of variety of annual crops in the inter-space of tamarind plantations. The success of tamarind based intercropping system depends very much on the selection of compatible crop combination for a given agro-climatic condition. Ginger, turmeric and coleus are grown as intercrop under partial shaded conditions.

Intercropping of ginger, safed musli and coleus in tamarind plantation of eight years old recorded significantly higher yield compared to sole cropping in open area (Arunkumar, 2005). Coleus recorded significantly more tuber yield in intercropping (240.33 g/plant) compared to sole cropping (92.14 g/plant).

Higher fresh ginger rhizome yield was recorded under intercropping than sole cropping. Similarly, intercropping of safed musli under tamarind recorded higher fresh weight of tuber (36.73 g/plant), number of tubers (13.13/plant), width of tuber (0.79 cm), dry weight of tuber (8.81 g/plant), dry tuber yield (9.76 q/ha) compared to sole cropping (27.53, 11.6, 6.73 g/plant and 7.04 q/ha, respectively).

Jambulingam *et al.* (1997) studied the tamarind in farm lands of Tamil Nadu. Occurrence in farm lands broadly falls into three categories: (i) Occasional tree in the farm, raised largely for shade for farm animals. (ii) Random planting in rainfed areas of Tamil Nadu. The economic return from this might go as high as Rs. 1000/- per tree, which is equivalent to the profit from one hectare of dryland crop. (iii) Systematic planting with suitable spacing along with annual crops. Such as cereals, pulses (cowpea, black gram) and groundnut as intercrop. The size of such a plantation is ranging from 0.4- 10.0 ha.

The factors indicated for such loss are shade and the tartaric acid released by the decaying leaf litter. Through remuneration in the period up to 3 years, such an intercropping option has been found to be counter productive beyond fourth year as yield loss up to 60 per cent in reported. Economic benefits of intercrops in tamarind can be realized up to fifth year only.

However, the benefit from the tree crop could be derived from the seventh year when it started bearing the fruits and thereafter there was a progressive increase in the pod yield, culminating in a B:C ratio of 1.41 during the tenth year. It is generally opined that the tree start stable production between 15 and 20 years.

An irrigated tamarind plantation adopting an 8 x 8 m espacement was studied in Tamil Nadu (Coimbatore). The amount of inputs directed to this crop is manifold and the farmers donot opt for any intercropping.The benefit cost realized from irrigated tamarind is twice that of rainfed crop. High yield obtained from the irrigated crop is attributed to high input technologies. Regular watering, plant protection measures and manuring have all contributed to early yield of the crop.

9.3.0 Fertilizers and Manures

Trees are known to grow well even without fertilizer application, due to their deep and extensive root system. In India, 25 g urea and 25 g single super phosphate per is recommended, although response of the tree is reported to be poor (Gupta and Mohan, 1990). In Application of 5 kg of farmyard manure at time of planting followed by 5 kg of farmyard manure and 5 kg of neem cake per tree in months of March and April, is a common practice urea at 100 to 200 g per tree. In the Philippines, 100 to 200 g per tree of ammonium sulphate about a month after planting and an equal amount at the end of the rainy season. The amount of fertiliser is gradually increased as the trees grow. When the trees begins to bear fruits, about 500 g of a complete fertiliser containing high amount of nitrogen and potassium is applied per tree twice a year. A full bearing tree should receive 2 to 3 kg of NPK complete fertiliser each year. In Thailand 100-200g urea per tree is recommended particularly for sweet tamarind.

Ammonium sulphate or urea: 2 split does when there is sufficient moisture.

i) Mid-August and

ii) Mid-September

Micronutrients: Iron for trees in alkaline soils.

Fertiliser dose for full grown trees

FYM – 40-50 kg; N – 1000 g; P_2O_5 – 500 g; K_2O – 500 g.

9.3.1 Nitrogen Fixation

Very little information is available on association of nitrogen fixing bacteria in tamarind which was earlier thought to be a non-nodulating species (Allen and Allen, 1936, 1981). However, some evidence suggests that it does form a symbiotic association with rhizobium (Postgate, 1979, Athar and Mahmood, 1982, Quiniones, 1983) enabling it to fix atmospheric nitrogen (Quiniones, 1983, Ding *et al.*, 1986). Rhizobium from the Philippines have been described as gram negative with short and long rods (Quiniones, 1983). Rhizobia inoculation did not yeild nodule formation in Thailand Toroprayoon *et al.* (1996). Chiemsombot *et al.* (1996) further consequently reported that the roots of tamarind did not develop nodule-like structures no rhizobium association. The acetylene reduction assay for did not demonstrate any symbiotic nitrogen fixation in tamarind roots and nitrogen fixing bacteria were absent in the roots. The most outstanding feature noted in the roots were finger-like projections called 'microvilli', whose function is unknown, but it has been suggested that they assist in water absorption.

Yoneyama *et al.* (1993) studied the natural abundance of 15 N (delta 15 N) in terms harvested from tropical legumes in Brazil and Thailand. The 15 N values were the ability of tamarind to fix atmospheric N needs further investigation.

9.3.2 Mycorrhizal Association

Rajasekhar *et al.* (1997) reported association of vesicular arbuscular mycorrhizal fungi *Glomus* and *Acaulospora* in tamarind in semi-arid tropics at Andhra Pradesh. The rhizospheric soils harboured 256 to 518 spores per 100 g dry soil, which was identified as *Glomus fasciculatum, G. geosporum, G. mosseae* and *Acaulospora laevis*. Vijaya *et al.* (1997) studied the effect of *Glomus fasciculatum* on growth and nutrition of tamarind in field condition. Pre and post planting mycorrlizal application was found to increase shoot height, dry matter and uptake of phosphorus, potassium, calcium magnesium zinc, manganese, copper and iron. Pre planting was found more effective than post planting.

Mohan and Mishra (1997) reported endomycorrhizal association in tamarind from arid ecosystem. The results revealed the prevalence of VAM fungal colonisation in root segments and propagules in rhizosphere soils of various samples screened. Different species of the VAM fungal genera of *Glomus* and *Gigaspora* were isolated. Among them, the genus *Glomus* was the most predominant one. The results revealed that all the inoculated seedlings had better plant height and biomass than uninoculated control seedlings.

Tamarind seedlings inoculated with 13 vesicular arbuscular mycorrhiza (VAM) fungi showed increased leaf number, plant height, stem girth, biomass, phosphate and zinc content. The number of VAM spores in the soil, percentage root colonisation and external hyphae measured by soil aggregation were also higher. Tamarind responded positive to inoculation with *Gigaspora margarita* (ICRISAT) and *Glomus fasciculatum* (Reena and Bagyaraj, 1990). Maksound *et al.* (1994) also observed significant increase in growth parameters and leaf P and K when seedlings inoculated with VAM. Pre planting application of inoculum to the pits is effective method of introduction of the VAM fungus.Tamarind responded poorly to inoculation with three arbuscular mycorrhizal fungi (AMF) - *Glomus fasciculatum, G. mosseae* and *Gigaspora margarita* (Selvaraj *et al.*, 1996).

9.4.0 Irrigation

Through not practiced Irrigation is good for establishment in the early stages of growth, especially during the dry seasons (Yaacob and Subhadrabhandu, 1995). In dry areas, techniques rain water harvesting can be adopted as it encourages growth and fruiting (Chundawat, 1990). Irrigation is reported to benefit flowering and fruiting in tamarind.

10.0 Flowering, Fruit Growth and Development

10.1.0 Vegetative Structures

Tamarind is a long lived, magnificent, large, evergreen or semi-evergreen tree, upto 30 m tall (Allen and Allen, 1981; Chaturvedi, 1985; Jambulingam and Fernandes, 1986). Maximum tree diameter at breast height (dbh) of 4.1 m was recorded in Sri Lanka (Gamble, 1922).The trunk can be upto 8 m in circumference with a spreading rounded crown, upto 12 mm diameter. Tamarind is a long lived tree and individuals have been reported with ages of 150 years in Hawaii and 200 years in Sri Lanka (Allen and Allen, 1981; Anon., 1991). The tree has poor form, the bole forking at about 1 m from the ground and often multi stemmed, though it has fairly good coppicing ability. The trunk is light grey to brown, very rough and scaly, the inner bark is 1 to 2 cm thick and green white in colour. The branches are often crooked, thick, wide spreading and drooping at the ends to form a dense crown (Storrs, 1995). A blood-red coloured gum exudes from the bole and branches when they are damaged. The wood is differentiated into bright yellow sap-wood and brown-violet heartwood, it is dense, very hard and close grained and takes a fine polish. It is resistant and durable but susceptible to termite attack.

Flowering, Pollination and Fruit Bearing in Tamarind

Flowering (Left) and Pollination (Right) in Tamarind

10.2.0 Root

In most locations, tamarind produces, a deep tap root, except in poorly drained or compacted soils, and has an extensive lateral root system (Chaturvedi, 1985; Chaturvedi *et al.,* 1986). The main root is wiry and flexuose, while the lateral roots are numerous, moderately long and distributed down the main root. In the seedling, the hypocotyls are distinct and thicker than the root, 3.8 to 8.9 cm long, slightly compressed and finely to tomentose. The cotyledons are sessile 1.3 to 1.8 cm x 0.95 to 1.3 cm in size, plano-convex, thick, fleshy and unequally ovate, orbicular or obovate with the apex rounded and the base projecting about 0.3 cm behind the point of insertion (Troup, 1921).

10.3.0 Leaves

These are alternate, even and paripinnately compound, 7 to 15 cm long with a pulvillus at the base and 6 to 20 pairs of opposite leaflets. The leaflets are 12 to 32 x 3 to 11 mm in size, narrowly oblong, unequal and entire, rounded or slightly emarginated at the apex and rounded at the base, they are glabrous, glaucous beneath and darker green above with reticulate venation and are sessile to sub-sessile. The petiole is upto 1.5 cm long with both the petiole rachis being glubrous or puberious, reticulate venation and the midrib are conspicuous above and below. The apex is rounded to almost square, slightly notched and asymmetric with a tuff of yellow hairs. Stipules are present, 0.5 to 1.0 cm long, falcate, acuminate and pubescent, falling very early (Salim *et al.,* 1998). A prominent scar is seen after leaf fall. Tropism is observed in tamarind the leaflets which fold after dark. This is attributed to lupeol, which is synthesised in the presence of light and degraded in its absence during the night (Ali *et al.,* 1998).

10.4.0 Inflorescence and Flowers

The inflorescence racemes are small, 5-10, 13 cm long, terminal and lateral drooping and often panicled. The flowers are bisexual and 2 to 2.5 cm in diameter, pedicels are about 5 mm long, nodose and jointed at the apex. The bracts are fragrant, zygomorphic, nearly as long as the flower bud and are deciduous. There are two boat shaped bractioles, 8 mm long and reddish in colour. Four sepals, up to 1.5 cm long, unequal, ovate and pink, cream or pale yellow. Five petals in total, the posterior and lateral ones are large and showy, slightly exceeding the calyx, they are obovate to oblong in shape, white, cream, pale yellow or pinkish in colour, streaked with red, the anterior ones are much reduced. There are 3 fertile stamens, the filaments are connate and alternate with bristle-like staminodes to form a sheath, open above and inserted perigynously on the anterior part of the mouth of the calyx tube. Anthers are transverse, reddish brown and dehisce longitudinally. The ovary is superior and unicarpellary on a sheath coming from the posterior part of the calyx tube, it has up to 18 ovules and is oblique, curving upwards and is green with a long, hooked style (Purseglove, 1987).

10.5.0 Fruit

The fruit is pendulous, pods are 5 to 10 (14) cm x 2 cm, oblong or sausage-shaped, curved or straight with rounded ends. The shell is light greyish or scurty brown and minutely scaly, indehiscent, often irregularly constructed between seeds, brittle and easily broken if pressed. It is filled with firm soft pulp surrounding the seed cavities. The outer surface of the pulp has 3 tough, woody, branching fibrerous cords from the base to the apex and the pulp is thick and blackish brown in colour (Coronel, 1991; Purseglove, 1987). The indehiscent pods, ripen about 10 months from flowering and may remain on the tree until the next flowering period (Benthall, 1933; Chaturvedi, 1985; Rama Rao, 1975).

10.6.0 Seed

Pods contain 1 to 10 seeds, each 3 to 10 x 1.3 cm, irregularly shaped, flattened, rhomboid, with the centre of each flat side of the seed marked with a large central depression. The seeds are very hard, shiny, reddish or purplish brown, non-arillate exalbuminous. They are embedded in the pulp, lined with a tough parchment like membrane and joined to each other with tough fibres (Purseglove, 1987).

There are about 720 seeds per kilogram of fruit (Hong *et al.*, 1996). However, this number may vary from 700 to 1000 seeds per kilogram depending on the conditions under which the trees are grown (Little and Wadsworth, 1964; Chaturvedi, 1985, Von Carlowitz, 1986).

11.0 Pest and Diseases

11.1.0 Insect-Pests

Tamarind trees are affected by more than 60 insect pests in India (Table 10). The estimated loss due to pests in tamarind is more than one crore rupees in Kerala (Joseph and Ommen, 1960). Among the various insect pests recorded from india, *Achaea janata* Linn, *Caryedon serratus* Oliv, and *Aonidiella orientalis* (New Stead) are hihgly destructive (Ranjeet Singh, 1997).

Table 10: List of Insects Association with *Tamarindus indica* Linn.

Sl.No.	Family	Species Name	Nature of Attack	Status
		Order: Coleoptera		
1.	Anobiidae	*Lasioderma serricorne* (Fab.) (Cigarette bettle)	Beetle and larva bore in dry seed	Minor
2.	Anobiidae	*Lasioderma* sp.	Beetle and larva bore in dry seed	Minor
3.	Anthribidae	*Araecerus suturalis* Boheman	Grub and beetle bore into fruit and feed within	Minor
4.	Anthribidae	*Autotropis modesta* Conspersa		Minor
5.	Bostrychidea	*Rhizopertha dominica* Fabricius	Grub bore in seed and feed	Minor
6.	Bruchidae	*Caryedon serratus* Oliv. (Tamarind bettle)	Beetle and larva bore in seed	Major
7.	Curculionidae	*Sitophilus linearis* (Herbest) (Tamarind seed borer)	Beetle and larva bore in seed	Minor
8.	Curculionidae	*Sitophilus oryzae* Linn.	Beetle and larva bore in seed	Minor

Sl.No.	Family	Species Name	Nature of Attack	Status
9.	Curculionidae	*Myllocerus blandus*	Adult feed on foliage	Minor
10.	Curculionidae	*Myllocerus discolour*	Adult feed on foliage	Minor
11.	Curculionidae	*Myllocerus viridinus*	Adult feed on foliage	Minor
12.	Curculionidae	*Amblyrhinus poricolles*	Adult feed on foliage	Minor
13.	Melolonthidae	*Holotrichia insularis* Brenske	Grub feed on root (white grub)	Minor
14.	Melolonthidae	*Schizonycha ruficollis* Fabricius	Grub feed on root (white grub)	Minor
15.	Tenebrionidae	*Tribolium castaneum* Herbst.	Larva bore in seed	Minor
16.	Tenebrionidae	*Alphitobius laevigatus* (Fab)	Grub feed on fruit	Minor
17.	Tenebrionidae	*Echocerus maxillosus* Fab.	Grub feed on fruit	Minor
18.	Tenebrionidae	*luoma* sp.	Grub feed on fruit	Minor
Order: Lepidoptera				
19.	Gelechiidae	*Holcocera pulverae* Meyrick	Damage the fruit	Minor
20.	Gracillariidae	*Oecadarchis* sp.	Caterpillar bore in ripened fruit	Minor
21.	Psychidae	*Chaliodes vitea* Hampson	Larva feed on leaves	Minor
22.	Psychidae	*Pteroma plagiophleps* Hampson	Larva feed on leaves	Minor
23.	Cochlidiidae	*Thosea aperiens* Walker (slug caterpillar)	Larva feed on leaves	Minor
24.	Eucosmidae	*Cryptophlebia illepidua* (Butler)	Larva attacks the fruit and bores in seed	Minor
25.	Eucosmidae	*Laspeyresia palamedes* Meyrick	Larva bore in tender buds	Moderate
26.	Galleriidae	*Corcyra cephalonica* Stainton. (rice moth)	Caterpillar bore in seed	Minor
27.	Phycitidae	*Phycita orthoclina* Meyrick	Larva attack the fruits and destroy the seed	Minor
28.	Phycitidae	*Ephesia cautella* (Walker) (Fig moth)	Larva attack the fruits and destroy the seed	Minor
29.	Phycitidae	*Aphomia gularis* Zeller	Caterpillar attack ripened pods	Minor
30.	Phycitidae	*Assaria albicostalis* alker	Larva bore inside the fruit, feed on pulp	Minor
31.	Pyraustidae	*Dichocrocis punctiferalis* (Guenee) (Castor capsule borer)	Larva damage the fruits	Minor
32.	Lycaenidae	*Virachola isocrates* Fabricius (Anar butterfly)	Larva attack fruit and bore in seed	Minor
33.	Geometridae	*Thalassodes quadraria* Guenee	Larva feed on leaves	Minor
34.	Noctuidae	*Eublemma angulifera* Moore	Larva damage the flowers	Major
35.	Noctuidae	*Achaea janata* Linn.	Feed on flowers and foliage	Major
36.	Notodoutidae	*Stauropus alternus* Walker (crab caterpillar)	Infest the foliage	Minor

Sl.No.	Family	Species Name	Nature of Attack	Status
Order: Hemiptera				
37.	Membracidae	*Leptocentrus obliquis* Walker (cow bug)	Nymphs and adults such the sap from tender shoots	Minor
38.	Membracidae	*Otinotus oneratus* Walker (cow bug)	Nymphs and adults suck the sap from tender shoots	Minor
39.	Membracidae	*Oxyrhachis tarandus* Fabricius (cow bug)	Nymphs and adults suck the sap from tender shoots	Minor
40.	Aleyrodidae	*Acaudaleyrodes rachispora* (Singh) (White fly)	Suck sap from tender shoots	Minor
41.	Aphididae	*Toxoptera aurantii* (Fonscolombe) (black citrus aphid)	Nymph and adults suck the sap from tender shoots and leaflets	Minor
42.	Margarodidae	*Drosicha stebbingi* (green)	Nymph feed on tender foliage	Minor
43.	Margarodidae	*Perissopneumon tamarindai* (green)	Nymph feed on tender foliage	Minor
44.	Pseudococcidae	*Nipaecoccus viridis* (Newstead) (mealy bug)	Nymph and adult suck the cell sap from leaflets	Minor
45.	Pseudococcidae	*Planococus lilacinus* (Cockerell) (mealy bug)	Nymph and adult suck the cell sap from leaflets	Minor
46.	Diaspidae	*Aonidiella orientalis* (Newstead) (Oriental yellow scale)	Nymph and adult suck the cell sap	Major
47.	Diaspidae	*Aspidiotus destructor* (Signoret) (coconut scale)	–do–	Minor
48.	Diaspidae	*Aspidiotus tamarindi* Green	–do–	Minor
49.	Diaspidae	*Cordiococcus castilleae* (Green)	–do–	Minor
50.	Diaspidae	*Pinnaspis temporaria* Ferris	–do–	Minor
51.	Diaspidae	*Pinnaspis strachani* (Cooley)	Nymph and adult suck the cell sap	Minor
52.	Diaspidae	*Hemiberlesia lataniae* (Signoret)	–do–	Minor
53.	Diaspidae	*Howardia biclavis* (Comstock)	–do–	Minor
54.	Diaspidae	*Unaspis articolor* (Green)	–do–	Minor
55.	Coccidae	*Saissetia olaea* (Bernard)	Feed on sap of foliage	Minor
56.	Coccidae	*Chionaspis acuminate*	Feed on sap of foliage	Minor
57.	Kerriidae	*Kerria lacca* (Kerr.) (Lac insect)	Nymph suck the sap from tender shoot	Minor
58.	Lacciferidae	*Laccifera lacca* Kerr. (Lac insect)	Feed on sap of twigs, branches and stems	Minor
Order: Thysanoptera				
59.	Thripidae	*Ramaswamiachiella subnudula* Karny	Feed on flowers	Minor
60.	Thripidae	*Scirtothrips dorsalis* Hood (chilli thrip)	Feed on flowers)	Minor
61.	Phlacothripidae	*Halpothrips ceylonicus* Schmutz.	Feed on flowers	Minor

11.1.1 *Achaea janata* Zinn. (Lepidoptera; Noctuidae)

The pest is distributed in all the castor growing areas, namely, Andhra Pradesh, Orissa, Tamil Nadu, Bihar, Madhya Pradesh and Maharashtra. The caterpillar is 55 to 60 mm long and 5 to 7 mm broad, which moults 4 to 5 times before pupation. The full grown larva is dark and is marked with prominent blue-black yellow and reddish stripes and has a pair of reddish process and also a dorsal hump near the head end of the body. The pupa measures about 2.00 cm in length and 0.7 to 0.8 cm in breadth. The adult is pale reddish brown moth with a wing expanse of 4 to 5 cm. The wings are decorated with broad zig-zag markings, a large pale area and dark brown patches. The moth is active at dusk and lays eggs scattered on the tender leaves. A female can lay up to 450 eggs during its life span. The caterpillars feed voraciously on leaves and cause damage to flowers and flower buds of tamarind, causing deflowering which results in heavy losses in fruit setting.

Management

Light trap can be used for suppressing the adult population. Chari and Muralidharan (1985) recommended neem leaf extract @ 10.0 per cent or neem seed kernel suspension @ 2.0 per cent egg and larval parasites. *Trichogramma* sp, and *Microplitis inaculipennis* Schiz, *Rhogas percurrens* Lyle, *Tetrastichus ophisae* Crawp and *Euplectrus leucostomus* Rohwer have been doccumented. *A janata* are susceptible to nuclear polyhederosis virus (NPV) (Ahmed *et al.*, 1988).

11.1.2 *Caeyedon serratus* (Olivier) (Coleoptera; Bruchidue)

It is a large bruchid, 5 to 6 mm long, and probably the most injurious. The bruchids lay eggs on fully matured ripened fruits and on hatching, the grub bore into the fruit by microscopic hole on the pericarp of the fruits. Grub enters into the seeds and feed on the endosperm, without damaging the pulp There are four instars of the grub. The fully matured grubs pupate in a papery cocoon within the seed or protrude out of the seed or fruit. Adult emerge out through the circular opening made by the grub. Fecundity ranges from 11 to 45 eggs with an average of 36 eggs per female (Shivalingaswamy *et al.*, 1991). Total development period varies from 46 to 64 days on fruits while 43 to 80 days in seed. Shivalingaswamy *et al.* (1991) reported infestation ranging from 15.0 to 57.5 per cent.

Moisture content of 15 to 20 per cent and the temperature at 28 to 32°C are very favourable for the development of the bruchid. The moisture content of the seed should be below 8 per cent to prevent infestation. Seeds stored in small containers can be protected by mixing with wood ash, calcium carbonate, precipitated chalk, or staked lime. The proportion is about one per cent for chalk to 15 per cent for slaked lime. This treatment prevents oviposition, but does not stop the development of the bruchid inside the seed already infested. Bags containing tamarind should be dusted with Malathion 50 EC @ 5 per cent and the receptacles or godown should be disinfested with Malathion 50 EC 5 per cent. The stored tamarind may be fumigated with methyle bromide (10 g/m for 4 days) or aluminium phosphide @ 5 tablets per tonne for 8 days.

11.1.3 *Aonidiella orientalis* (New stead) (Hemiptera; Diaspidae)

The oriental scale is one of the sucking pests of tamarind. The scale is flat, circular to oblong and varies in colour from yellow to pale brown to dark reddish brown. The crawlers upon hatching start feeding on soft tissues such as young shoot, green bark leaf and fruits and forms large colonies. Breeding is continued round the year with several overlapping generations. In severe infestation, young 1 to 2 year old saplings are killed, and infestation may lead to die back in shoots. Spray with 0.04 per cent diazinon or monocrotophos is recommended (Butani, 1978). The affected plants are removed and destroyed along with the insects in the initial stage of the attack. Theantly Cardet *et al.* (1998) recommended the use of neem oil and ground nut oil.

11.2.0 Diseases

Several fungal and bacterial diseases have been reported in tamarind, though none of them are known to cause severe damage. The major diseases reported are leaf spot (*Bartolinia robillaridides*, Tassi, *Exosporium tamarindi* Syd., *Hendersonia tamarindi* Syd., *Pestalotia poonensis* V Rao, *Phyllosticta tamarindicola* V Rao, *P. tamarindine* Chandra and Tandon, *Prathiguda tamarindi* Mutappa, *Xanthomonas tamarindi* Cook, *Sphaceloma* spp. and *Stigmina tamarindi* D.C, and *Oidium* spp.), a stooy mould (*Meliola tamarindi* Syd.), stem diseses (*Fracchirea indica* Talde), root and wood rot (*Ganoderma lucidum* (Leyss), Korst and Lenzites Palisoti Fr.) stem rot (*Pholiota gollani* P. Henn.), trunk and root rot (*Stereum nitidulum* Berk), collar rot (*Phytophthora nicotiannae* var. *nicotianne*, Siddaramiah *et al.*,1980), stem canker (*Hypoxlon nectrioides* Speg.) and a bark parasite (*Myriangium tamarindi* Tendulkar) (Mukherji and Bhasin, 1986; Siddaramaiah and Kulkarni, 1982, Morton, 1987; Parrotta, 1990). In Karnataka State, India, stony fruit disease caused by the fungal pathogen *Pestalotia macrotricha* Syd. (Lokesh and Shetty, 1991) makes the fruits hard and stony with fibrous structures.

11.2.1 Fungal Diseases

Pestalotia sps. was reported to cause stony fruit in tamarind. Seeds, fleshy mesocarp, fruit wall and stalks were found affected. Root rot caused by *Ganoderma lucidum*, is reported to affect crown region of trees (ICFRE, 1993). Minor fungal diseases caused *Sclerotium rolfsii* Sacc. (pre- and post-emergence rot in seedlings, *Cercospora tamarindii*, (seedling defoliation). *Oidium sp.* (Powdery mildew) were reported in nurseries. Chemicals like sulphur dust, Karathane (0.1 per cent) and calcium (0.1 per cent) were found to be effective in controlling the disease (ICFRE, 1993)

11.2.2 Bacterial Diseases

Bacterial leaf spot of tamarind caused by *Xanthomonas tamarindi* has been reported from India and Sudan. The bacteria has been reported as an obligate parasite, causing small and rounded spots, at first pale brown with yellow margin later turning jet black, on the leaves of tamarind (ICFRE, 1993). In addition to diseases caused by fungi and bacteria, trees of tamarind have also been found to be parasitised by angiosperm parasites probably belonging to loranthaceae (ICFRE, 1993).

12.0 Harvesting and Yield

Tamarinds may be left on the tree for as long as 6 months after maturity so that the moisture content will be reduced to 20 per cent or lower. Fruits for immediate processing are often harvested by pulling the pod away from the stalk which is left with the long, longitudinal fibers attached. In India, harvesters may merely shake the branches to cause mature fruits to fall and they leave the remainder to fall naturally when ripe. Pickers are not allowed to knock the fruits off with poles as this would damage developing leaves and flowers. To keep the fruit intact for marketing fresh, the stalks must be clipped from the branches so as not to damage the shell. A mature tree may annually produce 33150-225 kg of fruits, of which the pulp may constitute 30 to 55 per cent, the shells and fiber, 11 to 30 per cent, and the seeds, 33 to 40 per cent (Morton, 1987).

One of the most important operations in a packing line is sorting (including sizing). Such an operation requires several parameters to be quickly identified and managed at the same time. These include variety, maturity, color, shape, size, and defects. In order to develop an efficient machine vision experimental sorting system for sweet tamarind pods in Thailand, based on image processing techniques, Jarimopas and Jaisin (2008) used relevant sorting parameters,namely, shape (straight, slightly curved, and curved), size (small, medium, and large), and defects. The variables defining the shape and size of the sweet tamarind pods were shape index and pod length. A pod was said to have defects if it contained cracks. The experiment involved the use of pods from two sweet tamarind cultivars: "Sitong" and "Srichompoo". The sorting system involved the use of a CCD camera which was adapted to work with a TV card, microcontrollers, sensors, and a microcomputer. Analysis was performed with image processing software. Analysis of variance was computed with regard to the variables of shape, size, and defects, and took into account variations in the control factors of belt speed, pod orientation, and spacing. The results showed that the three control factors did not significantly affect shape, size, and defects at a significance level of 5 per cent. The averaged shape indexes of the straight, slightly curved, and curved pods were 51.1 per cent, 61.6 per cent, and 75.8 per cent, respectively. Pod length was found to be influenced by size and cultivar, with Sitong and Srichompoo pods ranging from 10.0 to 14.0 cm and 8.5 to 12.4 cm, respectively. The vision sorting system could separate Sitong tamarind pods at an average sorting efficiency (Ew) of 89.8 per cent, with a mean contamination ratio (_CR) of 10.2 per cent at a capacity of 1517 pod/h. Respective figures for Srichompoo pods were Ew; 94.3 per cent; _CR, 5.7 per cent; and capacity, 1491 pod/h. The contamination ratios met the export standards mandated by the Thai agricultural commodities and food codification.

13.0 Post-harvest Management

13.1.0 Hydration Drying

The seeds were hydrated at room temperature for 24 hours followed by drying at 40 to 100°C for 5 to 9 hours. White TKP can be obtained by drying at 80°C for 6 hours and the hardness of the kernel is also low.

Harvesting, Packaging and Storage of Tamarind

13.1.1 Hydration–Shade drying–Roasting

The seeds were first hydrated at room temperature for 24 hour, followed by drying in the shade for 24 hours and roasting for 125 to 175°C for 3 to 8 minutes.

13.2.0 Storage of Pulp

The brownish-red color of the tamarind pulp turns darker and within one year, it becomes almost black. (Shankaracharya, 1998).Reaction of sugars with amino acids through Maillard reaction is responsible for the non-enzymatic browning in food and agricultural products. Lysine is the major amino acid that contributes to this type of reaction as it has a free ε-amino group that can readily react with reducing sugars (Ajandouz and Puigserver, 1999). The formation of dark colored products in glucose-lysine solutions at pH values ranging between 3 and 9 have been reported to be due to the alkaline degradation of sugar mainly and to a lesser extent to the interaction between lysine and glucose. Obulesu and Bhattacharya (2011) studied the cause of color changes during developmental stage, as well as during storage of tamarind pulp. Tamarind pulp in unripe condition (green pod) shows polyphenol oxidase activity up to 105 days of maturity of the fruit or until it ripens. Thereafter due to ripening there is a marked increase of reducing sugars and available lysine leading to Maillard reaction. This inhibits enzymatic browning in the ripe pulp during subsequent storage.

Qualitative and quantitative changes in the tamarind pulp upon storage in reported. Upon storage the moisture content decreased from 30 per cent to around 20 per cent (Babu *et al.*, 1999). The original brown colour was retained upto 4 months storage in metal and black polythene, as compared to pulp even white polythene wrap and plastic containers. The phoenix mat produced black and sticky first month storage, besides mat remained moist due to its pores nature. Ironically, the most prevailing practice of tamarind pulp transport and storage is by the mat structure. Tamarind can be stored upto two to three years in cold storages without loss of quality and quantity (colour, weight and sourness) (Tables 11 and 12)

Table 11: Per cent Storage Loss in Weight of Tamarind Pulp

Structure	*Months of Storage*			
	1	*2*	*3*	*4*
Metal	0.0	0.0	0.57	1.53
Poly-sheet white	0.0	0.1	0.07	0.38
Poly-sheet black	0.0	0.0	0.06	0.18
Phoenix mat	0.0	0.0	2.22	1.35
Plastic container	0.0	0.0	0.05	0.17

Among the storage pests *Sitophilus oryzae* and *Ephistia cautella*, the incidence of *Ephistia* was high. Among the five different structures tried, plastic containers and black polythene were found to be best for pest control (Table 13).

Table 12: Physical Changes during Storage of Tamarind in different Containers

Structure	*Months of Storage*			
	1	*2*	*3*	*4*
Metal	Brown	Brown	Brown	Brown
Poly-sheet white	Brown	Brown	Brown	Slightly black, sticky
Poly-sheet black	Brown	Brown	Brown	Slight brown, sticky
Phoenix mat	Slight black, sticky	Slight black, sticky	Slight black, sticky	Black, sticky
Plastic container	Brown	Slight black	Slight black, sticky	Slight black, sticky

Table 13: Presence of Insect Pests during Storage of Tamarind

Structure	*Duration of Storage (Months)*			
	1	*2*	*3*	*4*
Metal	0.0	0.0	4.00 (E.A.)	0.33 (E.A.)
Poly-sheet white	0.0	0.66 (E.A.)	1.33 (E.A.)	0.00
	0.0	0.6 (E.P.)	0.33 (E.L.)	
	0.0	0.33 (E.P.)	0.66 (E.P.)	
Poly-sheet black	0.0	0.0	0.33 (E.A.)	0.0
Phoenix mat	0.0	4.0 (E.L.)	4.00 (E.A.)	0.0
Plastic container	1.00 (S.O.)	0.0	0.0	0.0

From the above observations, it is indicative that black polythene wrap for tamarind storage is better than other structures used.

13.3.0 Dehulling

The traditional dehulling of ripe fruit is unhygienic and time consuming. The Dehuller has been developed at the PHT centre, UAS, Bangalore to improve efficiency and eliminate drudgery. The unit is more efficient when fed with large sized curved fruits (hulling per cent of 79.48), while it is low in small size fruits (58.47 per cent hulling). The hulling capacity is 500 kg per hour.

Preparation of tamarind produce for marketing requires additional labour required to produce tamarind pulp (Nageswara Rao, 1997). Ramakumar *et al.* (1997) conducted research on development of a tamarind dehuller and short-term storage of pulp. The study clearly showed that the tamarind dehuller developed is energy and time efficent compared to traditional method.

13.4.0 Tamarind Seed Expulsion

Mechanical seed expelleris nearly 11 times faster than the traditional seed expulsion. The trials showed that during seed expulsion, the fibre could not be removed from the pulp and hence it had to be manually removed. As a result, an additional time of 113 minutes was required to remove the fibre from the pulp. This particular operation could not be mechanised and hence manually removed.A comparison of mechanical and manual seed removal of tamarind fruit showed that the newly developed tamarind seed expeller is 11 times faster than the manual separation and has a capacity of 35 kg per hour at 22 per cent pulp moisture content. During the peak season, the processors/traders often complain about the acute shortage and many a times, non-availability of skilled labourers.Hence a systematic process operations in a sequence is essential to speed up the acvities to achieve maximum operational efficiency. The bulk trials conducted with the seed expeller reveal that the cost of operation worked out to Rs. 0.65 per kg. The cost benefit of squeezing of dehulled fruits is 12.36 times beneficial when compared to manual method of opening up of the fruit.

13.5.0 Value Addition

Tamarind fruit has been used as a raw material for manufacture of several products, *viz.*, tamarind juice concentrate (TJC), tamarind pulp powder (TPP), tamarind kernel powder (TKP), tartaric acid, pectin, tartrates and alcohol.It is exported as fresh, dry and paste. Export of tamarind seed is also taking place both in unground and ground forms.

Among the products, tamarind paste is expensive it is the concentrated form of tamarind pulp. Small quantities of tamarind oleoresin was also exported. Quantity and value of export tamarind oleoresin during these years are presented in Table 19. Oleoresin has the highest unit rate as it is the purest form retaining all the original characteristics of the spices. Tamarind products are exported to around 60 countries. Such as Egypt, France, Germany, Japan, The Netherlands, Pakistan, Switzerland, Syria, Sri Lanka, Saudi Arabia, USA, UK, UAE and Yaman and Arab Republic.Tamarind fresh is exported to 33 countries, of which Pakistan, UAE, UK, Bangladesh, Japan and France are important.

13.5.1 Various Products from Tamarind

Tamarind fruit has been used as a raw material for the manufacture of several products like tamarind juice concentrate (TJC), tamarind pulp powder (TKP), tamarind kernel powder (TKP), tartaric acid, pectin, tartarate and alcohol.

13.5.1.1 Processing and Product Development

The tamarind pulp or tamarind juice concentrate, powder and paste is used for souring curries, sauces, chutneys and other beverages. The pulp is also used as in dyeing and tanning as well as for polishing and cleaning metalware (Lewis and Neelakantan, 1964; Anon., 1982). The CFTRI, Mysore has developed processes for the manufacture of juice concentrate and pulp powder. The concentrate contains about 13 per cent total tartaric acid, 50 per cent invert sugars and 30 per cent moisture and the yield of the concentrate is nearly 75 per cent of the pulp used. Formulae for preparing spice sauces and beverages from the pulp and the extract have been reported (Nagaraja *et al.*, 1975; Lewis *et al.*, 1973). The fruit pulp is converted into a free flowering powder by drying at low temperature adding certain permitted dilutants and anti caking agents. The product is granular, free-flowing but hydroscopic and hence requires air tight container.

An integrated process for production of pectin (yield 2.31-3.70 per cent), tartrate (11-12 per cent) and alcohol (11-13 per cent) and another process for the extraction of tartrates with acidified ethanol and subsequent extraction of pectin have been developed. But factors like low price of imported tartrates and limited demand for pectin have to considered before a commercial venture (Lewis *et al.*, 1957, 1964).

Mixing the fruit with a minimum quantity of water and passing through a pulper removes seeds, fibre and cellulosic material. Drum-drying of this soft homogeneous pulp and upon compression blocks of pulp is made. The Indian standards specifications are available for the tamarind juice concentrate (IS-5955-19700 and the pulp (IS-6364-1979) (Lewis and Neelakantan, 1964; Anon., 1982).

13.5.1.2 Pectin, Tartrates, Tartaric Acid and Ethanol

Tartaric acid and products like ethanol, lactic acid and citric acid have been prepared from tamarind (Lewis *et al.*, 1954). The pulp is repeatedly extracted with boiling water and the filtered extract is cooled to separate potassium bitartrate. The supernatant is concentrated under vacuum and the pectin is separated by the addition of alcohol. The filtrate, after recovering alcohol, is treated with lime to precipitate calcium tartrate. The remaining sugars are fermented with yeast and alcohol is recovered. The recovery of about 2.5 per cent pectin in addition to 12 per cent tartaric acid and 12 per cent alcohol. For the isolation of tartaric acid, the use of unripe green pods has been suggested, as they contain most of the acid in the free form.

13.5.2 Tamarind Pulp

Tamarind pulp is can important ingredient for souring food products like sauces, chutneys, sambar, rasam and some beverages. The fruit pulp is the important raw material for the manufacture of tamarind pulp concentrate, pulp powder, paste, tartaric acid, tartrates, soft drinks, *etc.* The pulp is made into preserve and syrup. 'Jugo' or 'Fresco' de Tamarinda is a favourite beverage in many Latin American countries. The pulp is also used as an auxillary in drying a tanning and for polishing a cleaning metal where the pulp of the fruit is used extensively in the local confectionery industries in several developing countries. The pulp has been used in the Indian system of medicine as a refrigerant, carminative and laxative. The pulp is used as an antiscorbutic (The Wealth of India, 1976; Lewis and Neelakantan, 1964; Duke, 1981).

13.5.3 Tamarind Seed

Tamarind seeds have many uses. Such as feeds for cattle and pigs, as a cosmetics in the pharmaceutical industry, as a curative against rheumatism and as a soil stabiliser (Anon., 1955). The major industrial use, however, is in the form of tamarind kernel powder (TKP), as a sizing material in the textile and jute industries. By using 10 per cent TKP as a binder, saw dust fuel briquetles can also be made. The polysaccharide (Jellose) is a good substitute for fruit pectins in the manufacture of jams, jellies and marmalades. High grade adhesive can be prepared from tamarind seed kernel (The Wealth of India, 1976). Bhattacharya *et al.* (1983) have reported the use of TKP (particularly its carbohydrate function) in food products such as jelly and marmalades. Rao and Subramanian (1984) and Marangoni *et al.* (1988) have attempted production of protein concentrates or meals and studied the functional properties of the kernel proteins.

The presence of tannins and other colouring matter in the testa makes the whole seed unsuitable for human consumption. The kernels are separated from the testa either by roasting or by soaking in water. The testa has to be completely removed before using the kernels for food purposes.

13.5.4 Tamarind Kernel Powder (TKP)

It is prepared by decorticating the seed and pulverising the creamy white kernels. The decorticated seed is ground and stored. The powder tends to deteriorate during storage under humid conditions. Hence storage a dry moisture proof container is important. Mixing with 0.5 per cent of sodium bisulphite before packing will prevent enzymatic deterioration. The TKP will turn rancid and brown on storage. Defatting can be done to increase the storage ability and colour (Sivarama Reddy *et al.*, 1979). The general characteristics of a good tamarind seed its characteristic aroma that should be free burnt or other undesirable odour. It should have good keeping quality and be free from any insect pests, fungal growth or extraneous materials.

In India, TKP is used as a source of carbohydrate for the adhesive or binding agent in paper, textile, weaving and jute industry (Anon., 1976; Shankaracharya, 1998). The sizing property of TKP is due to the presence of upto 60 per cent of the polysaccharide. A high grade adhesive from tamarind seed kernels could be prepared by roasting the seeds at 110°C for 15 minutes, testa removed powdered and passed through a 180 mesh sieve before making porridge by boiling in 200 per cent hot water. Then it is mixed with 5 per cent glucose and 12 per cent sodium bicarbonate and is ready for use (Devadasa and Gothandapani, 1993). Rao and Subramanian (1984), Marangeri *et al.* (1988) and Bhattacharya *et al.* (1994b) have studied the functional properties of the kernel protein for its use, as food.

13.5.4.1 Food Products made from Kernel

Three products (fortified bread, biscuit and jelly) were made using TKP and these products were characterised for their quality by subjective and objective methods. Breads can be made with incorporation of TKP (0-20 per cent) following standard bread making procedure. The ingredients for biscuit making included wheat flour, salt, ammonium bicarbonate, sodium carbonate, milk powder, oil, sugar, shortening and TKP. The dough was allowed to ferment for 2 hours followed by baking at 215 to 220°C for 10 minutes. Results indicated that TKP fortification upto 15 per cent is possible in biscuit without sacrificing the quality of the product.

13.5.5 Pectin

Polysaccharides obtained from tamarind seed kernels from mucilaginous dispersions with water and possess the characteristic property of forming gels with sugar concentrates like fruit pectins. However, unlike fruit pectin, tamarind polysaccharide can form gels over a wide pH range, including neutral and basic conditions. Fruit pectins undergo degradation on boiling and fall to one-third of the original value after one hour of boiling. Therefore tamarind polysaccharide can be useful as a gel formation agent and may be substituted for fruit pectins. Tamarind polysaccharide does not contain galacturonic acid and methyluronate and is therefore not regarded as a true pectin, it is termed 'jellose' (Rao, 1948).

Jellose is prepared on a large scale by adding TKP to 30 to 40 times its weight of boiling water containing citric acid and tartaric acid at a concentration of 0.2 per cent. It is then stirred vigorously and boiled for a further 30 to 40 minutes. The resultant solution is kept overnight for setting and the supernatant liquid is siphoned off and concentrated under vacuum, passed through a filter press and then dried in a drum drier. The resultant product is pulverised in a ball mill. A good sample of jellose should have a relatively viscosity of 5.0 per cent at 35°C in 0.5 per cent solution Jellose is much cheaper than corn starch and is required in smaller quantities.

A device for the extraction of polysaccharides from seed kernels has been developed and patented by Gordon (1968). As the pulp also contains pectin, an integrated process has been developed for the production of pectin, tartarates and ethanol. In this process, the pulp is repeatedly extracted with boiling water and the filtered extract is cooled to separate potassium bitartarate. The supplement is concentrated under vacuum and the pectin is separated by the addition of alcohol. The filtrate after recovering the alcohol, is treated with lime to precipitate calcium tartarate. The remaining sugars are fermented with yeast and alcohol is recovered. The recovery is about 25 per cent pectin in addition to 12 per cent tartaric acid and 12 per cent alcohol, which makes the process economic and attractive. Krishna (1955) reported a process for the extraction of tartarates with acidified ethanol and subsequent extraction of pectin. After removing the alcohol, the residual syrup could be used for edible purposes. Tamarind pectin resembles apple pectin and has the following composition.

13.5.6 Jelly

Jelly can be made using TKP as a substitute for pectin. Jelly was made with TKP, sugar, citric acid, permitted preservative, flavour and colour. The use of TKP and pectin together is another option. Hence, TKP can replace pectin in making jelly. The added advantage of TKP is that it is nutritious and is low in cost, and hence the import of pectin can be reduced. Raghuveer (1997) made the market survey on tamarind products with special emphasis on tamarind paste in Andhra Pradesh. The intrinsic value of raw tamarind can be further enhanced through value addition activities and there is a good market for these processed products both in domestic as well as international circuits. Deseeding of tamarind has traditionally been the easiest and simplest way of enhancing its value.

13.5.7 Tamarind Paste

The preparation of tamarind into paste involves cleaning with water to remove impurities and then passing it through pulverizes. It is estimated that 20 per cent loss occurs during the conversion of ripe tamarind into paste. The demand for ready made tamarind paste is increasing in the recent times. Tamarind powder on the other hand, has not received the same extent of acceptability. The paste after opening the packet has to be necessarily transferred to a dry bottle and stored in the refrigerator to ensure its quality.

13.5.8 Tamarind Juice Concentrate

Tamarind juice concentrate (TJC) is a convenient product as it is easy to dissolve and reconstitute in hot water (Anon., 1982a). It can also be stored for long periods. TJC is prepared by extracting cleaned pulp with boiling water using the counter current principle, where dilute extracts are used for extracting fresh batches of the pulp. Using this process, an extract is obtained containing 20 per cent soluble solids. The extract is separated from the pulp by sieving and is concentrated under vacuum in a forced circulation evaporator. When the soluble solids reach 68 per cent, the pulp is placed in cans or bottles. It sets like jam on cooling. The yield of the concentrate is about 75 per cent of the pulp used. The quality of the product is assured by using recently harvested fruits free from insect and rodent damage (Lewis *et al.*, 1964a).

The tamarind concentrate or tamarind juice concentrate (TJC) is a convenience product and it is easy to disperse and reconstitute well in hot water. The concentrate is hygienic and can be stored well for longer periods.

The process for the manufacture of tamarind concentrate has been developed by the Central Food Technological Research Institute (CFTRI), Mysore and several firms are producing the concentrate on commercial scale based on this process (Anon., 1982). Jaleel *et al.* (1980) used fungal pectic enzyme for the production of tamarind concentrate. The pulp was extracted with hot water followed by squeezing through cloth. Pectin enzyme concentrate (PEC) was used for depectinising the pulp and after 15 hours the juice was extracted. Treatment with PEC gave a clear juice, which on concentrate gave acceptable non-viscous and free flowing products. From 5 kg pulp, 3.46 kg of free flowing concentrate was obtained with the following characteristics. Brix 75°, acidity 18.1 per cent and pH 2.0. The tamarind concentrate or tamarind juice concentrate is a convenience product, disperse and reconstitute well in hot water. The process for the manufacture of tamarind concentrates has been developed by CFTRI, Mysore.

13.6.0 Changes in Chemical Composition and Overall Acceptability of Concentrate during Storage

Ramteke *et al.* (1990) reported that the tamarind concentrate contains the mellow flavoured potassium acid tartarate, besides free tartaric acid, the main acid in the tamarind fruit. The concentrate is free of fibre, seed, *etc.*, and dispress readily in hot water. It retains the natural tang and purity of tamarind and eliminates the cumbersome process of soaking, squeezing and filtering. Besides this, the concentrate is a valuable semi–finished product for the production of respective fruit juices, beverages, jams, jellies and fruit juice powder.

13.7.0 Beverage

Formulae for preparing spiced sauces and beverges from the pulp have been reported by Girdharilal *et al.* (1958). Tamarind extract as a replacement for phosphoric acid, citric acid and other acids in soft drinks is described by Zablocki and Pecore (1995). Beverges containing the extract with a lower pH have improved shelf life. Tamarind fruit can be used as a raw material for the preparation of wine like beverages (Benk, 1987; Latino and Vega, 1986; Sanchez, 1985). The processing techniques for the preparation of canned tamarind pulp and tamarind (pulp) soft drinks have been reported by Bueso (1980). Process for making drinks, syrup, juice, liquor and solid extracts based on tamarind were developed by Mellon (1974). Wine and vinegar production from tamarind fruit is reported by Maldonado *et al.* (1975). Studies were conducted on the preparation and shelf life of tamarind and sour sop – tamarind soft drinks by Benero *et al.* (1974). Tamarind drinks with 9 to 12 per cent pulp concentration and 21.5°Brix and sour sop tamarind blended drinks at 10 to 14 per cent and 15°B and 17°B were proposed. Canned and stored at 85°F canned tamarind drinks can be preserved for one year and blended drinks for about 10 months, blended drinks of 17°B being preferred in terms of acceptability.

Kotecha *et al.* (1995) reported that the production cost of ready to serve (RTS) beverage was Rs. 2.28 per 250 ml. The production cost of RTS beverage, syrup and concentrate were worked out on the basis of cost of raw materials, chemicals added and yield of juice or pulp and products obtained. The data revealed that the cost of production of RTS beverage, syrup and concentrate were Rs 18.76, Rs. 36.48 and Rs. 70.95 per litre. These costs did not include rent and transport charges, sale commission, local taxes, interst on capital, depriciation on equipments and buildings, *etc.* Bawne (2000) reported that the cost of production of RTS beverage, syrup and concentrate of tamarind were Rs. 3.3, Rs. 5.1 and Rs. 19.2 per 250 ml.

13.8.0 Tamarind Pulp Powder

Tamarind pulp powder (TPP) is prepared by concentrating, drying and milling the pulp into a powder form. It is one of the convenience food products produced commercially by several manufacturers in India. Depending on the manufactory process, wide variations in the physico-chemical characteristics are reported. Manjunath *et al.* (1991) reported that an average total solid content of tamarind pulp powder varied from 18.6 to 25 per cent acidity from 8.7 to 11.1 per cent with an average value of 9.9 per cent (as tartaric acid). The moisture content ranges from 3.5 to 8.8 per cent. Among the minerals, the calcium and potassium contents were high and ranged from 74 to 143 mg and 23.8 to 27.7 mg. Starch is the major ingredient in tamarind pulp powder (20-41 per cent). One small portion of the fruit pulp is made into a refreshing drink after dissolving in water and squeezing by hand. After removing the extraneous matter, water is added to dilute the drink as required. Tamarind pulp is enjoyed as a refreshing drink and beverage in most of the Southern and Central American and Asian countries. Tamarind drinks in polypacks are commercially available in Thailand, Indonesia and in many countries of Africa. In Mexico and Guatemala, the drink is carbonated. It has a high content of vitamin B (Thiamine and Niacin) as well as small amounts of vitamin C (ascorbic acid) and carotene (Sozolnoki, 1985). The ascorbic acid in tamarind fruit juice contains about 5.1 to 6.1 mg/100 ml.

The pulping operation consists of breaking the shells by hand and agitation in water to separate the pulp and the seeds. The industrial method used in India for the manufacture of tamarind juice concentrate is by extraction of all the soluble from the fruit pulp in boiling water (Nagaraja *et al.*, 1975). The recommended concentration of tamarind pulp in syrup is between 20 and 24 per cent, so as to produce a beverage with distinctive flavour and acidity (Bueso, 1980). Such syrup contains 56.7 per cent total solids, 43.8 per cent reducing sugar and total acidity of 1.11 per cent as tartaric acid. A clarified tamarind juice is one in which the structural and colloidal phases are broken to prevent sedimentation which can be done by adding gelatine (Baragano de Mosqueda, 1980). In this process, 0.12 to 0.15 per cent gelatine is agitated with the pulp and left to stand for 10 to 15 days at 10°C. The colloidal particles will precipitate during this period and are removed by decantation. The juice maintains good colour and flavour and is transparent. Bueso (1980) also prepared a blended tamarind and a sour sap (*Annona muricata*) drink, which had commercial acceptance. The tamarind drink contained 9 to 12 per cent pulp and was adjusted to 21.5°Brix. The blended drink had 10 to 14 per cent pulp and 15° and 17°B. The blended drink had 10 to 14 per cent pulp and 15 and 17°B. The drink after pasteurisation at 185 to 190°F was canned.

Zablocki (1995) described a process for using tamarind extract as a replacement for phosphoric acid, citric acid and other acids added to soft drinks. Latino and Vega (1986) and Sanchez (1985) have used tamarind pulp for making wine like beverages. Central Food Technological Research Institute has developed a processing the preparation of tamarind powder which is easy flowing in nature and retains its original colour and flavour characteristics for periods upto 6 months. The product is much superior to the available market samples. A method for preparation of tamarind powder has been patented. The general characteristics of good quality tamarind powder are that (i) it will have flavour characteristics of good tamarind pulp which dissolved in water and will be free from burnt and any other undesirable flavour (ii) it will have a good keeping quality and (iii) it will be free from fungal growth, live or dead insects and insect fragments (Manjunath *et al.*, 1991).

13.9.0 Adhesive

TKP, when dissolved in water containing boric acid and phenol as preservative, gives a very good paper adhesive (The Wealth of India, 1976). Xyloglucan of tamarind seed (tamarind gum) is available as a food additive for improving the viscosity and texture of processed foods.

13.10.0 Polysaccharide (Jellose)

The polysaccharide in TKP do not contain galacturonic acid or methyl uronate groups, Jellose forms the best gel when its concentration is 0.7 to 0.9 per cent.

13.11.0 Tamarind Pickles

Pickles are hot, spicy and salty-sour taste, and can be preserved for several months. Preservation is due to the presence of salt, increased acidity and spices.

When making pickles, fresh mature tamarind fruits are selected and soaked in clean water for 12 hours and the pulp is separated. For each kilogram of tamarind, one kilogram of sugar is added and boiled. Stirring the mixture continuously, then mixtures of spices are added.

These are ground into a paste and added to the boiling tamarind and mixed thoroughly. The product is then packed into presterilised glass jars or plastic bags and allowed to cool at room temperature. The pickle can be stored in small polythene bags or in clean jars and sealed. It will keep well if stored in a cool place. Care should be taken to prevent air bubbles becoming trapped in the pickle and to ensure that a layer of oil covers the product.

13.12.0 Other Products

Tender green fruits of tamarind are used to prepare tamarind pickles as they are very acidic. Tender fruits can be a raw material for the manufacture of natural tartaric acid. The ripened fruit pulp is used in the manufacture of mixed fruit jams, sauces and tamarind paste (Goce *et al.*, 1993). The tamarind pulp is also canned for long use. The tamarind juice is concentrated after mixing with spices, sugar and salt and converted into a preserve. Tamarind juice extract is also used in certain confectionary products like tamarind candy (Sadasivam *et al.*, 1979; Girdharilal *et al.*, 1958; Gowramma *et al.*, 1968).

Kotecha and Kadam (2003) studied the effect of storage tempearature and period in the chemical composition of RTS beverages, syrups and concentrates prepared from tamarind during storage. Tamarind fruits (pods) of local cultivars, at full maturity stage, were obtained from local market. Various steps in the extraction of pulp and juice from tamarind fruits are outlined. The juice was utilised in the preparation of RTS beverages, syrups and concentrates.

13.12.1 Tamarind Sauce

Process is developed by Kulkarni (2004). Tamarind sauce was prepared using different ingredients like tamarind pulp, sugar, salt, garlic, cloves, cardamom, black pepper, cinnamon, chilli, powder, preservative (sodium benzoate) and the receipe was standardised with different levels of sugar, salt.

13.12.2 Jam

For making tamarind jam, ripe fruits are shelled and the pulp is boiled for 10 minutes. The pulp is then drained and separated from the seeds. For every cup of pulp, two cups of sugar are added. The mixture is then cooked and constantly stirred while boiling until it becomes thick in consistency. The resulting jam is cooled, packed in dry, sterilised jars and sealed.

13.12.3 Syrup

Tamarind syrup is made by boiling immature fruit pulp until it is soft and then strained through cheese cloth. To every cup of juice, a half teaspoon of baking soda is added. The mixture is boiled down to one-half the original quantity, removing the rising serum in the process. The juice is again strained and for every cup obtained, a quarter cup of sugar is added. The mixture is boiled again for 20 minutes. The cooled syrup is poured into sterilised bottles and sealed.

13.12.4 Candy

Sweetened tamarind fruit is made by peeling whole ripe fruits and pouring boiling (62°C) sugar syrup over the fruits and placing in a deep enamel basin. Boiling syrup is prepared by mixing three parts sugar with one part water. After soaking for three days, the fruits are drained of the old syrup and again covered with freshly prepared syrup. The process is repeated until the fruits are sweet enough. The sweetened fruits are arranged on bamboo racks, dried in sun and covered with a wire screen to keep off flies. Final drying is done in an oven at a low temperature. After cooling, the dry fruits are wrapped in cellophane.

13.12.5 Champoy

Tamarind fruits may also be processed into balls or 'champoy', which is a common tamarind product in the Philippines. To one cup of pulp with seeds are added two cups of boiled and mashed sweet potato (*Ipomea batatas*), two cups of sugar, one eighth cup of salt and one cup of water. The mixture is cooked in a moderate fire with constant stirring until it becomes thick and may be shaped into balls. The balls are rolled in sugar and wrapped in cellophane.

13.12.6 Ade

This is a refreshing tamarind drink prepared in the Philippines and some tropical American countries by mixing ripe pulp with sugar and water until the desired taste is attained. The sample home method of making ade is to shell the fruits, place in 2 to 3 litres of water and allow to stand for a short time, then add a tablespoonful of sugar and shake vigorously. For a richer beverage, the shelled tamarind should be covered with hot sugar syrup and allowed to stand for several days and finally diluted to the required strength with iced water and strained. Sometimes spicy seasonings, such as cloves, cinnamon, ginger, pepper or lime slices are added to improve the taste (Morton, 1987). In Brazil, Ade is prepared by covering the shelled fruits in cold water for 10 to 12 days, then the seeds are removed and strained and a cup of sugar is added to every two cups of pulp. The mixture is boiled for 15 to 20 minutes and placed into glass jars topped with paraffin and diluted as desired. It is a popular drink in the tropics of America and it is bottled in carbonated form in Guatemala, Mexico and Puerto Rico.

13.13.0 Quality

Quality parameters are important for commercialization of tamarind based products. It is due to high moisture level, seeds, fibre and rind contents. It is not rare that tamarind is adulterated with foreign matter which are of organic and inorganic nature. Poor Quality are due to poor post-harvest management practices including processing.

For dry tamarind, there are three grades, such as special, Grade A and Grade B under Agmark rules based on the percentage of rind, fibre, moisture and insect damage, compared with other agricultural comodities. A higher level of moisture content varying from 15 to 20 per cent is allowed in various grades from special to Grade B. Agmark has also prescribed grade specifications for seed extracted tamarind. There are five grades such as special, Grade A, Grade B, Grade C and Non-specified for this product. Being seed extraction, a difficult process, allowance of 5 to 20 per cent seed is allowed to grades special to Grade C. There are also quality specifications for undecorticated and decorticated tamarind seeds and powder.

Tamarind pulp is prepared removing the seeds from the matured fruits and pressed to form a cake. It is generally free from rind, seed and fibre. Tamarind seed content should not exceed 3 per cent for special, 5 per cent for Good, 7 per cent for Fair and 10 per cent for Average quality tamarind pulp. A higher moisture content of 20 per cent is allowable for pulp as per Agmark grades. Tamarind concentrate is a jam like product resulting from the concentration under vacuum of hot water extract of soluble solids of tamarind pulp. Also, use of pectin enzyme concentrate (PEC) in the process aids better extraction by depectinising the pulp (Jaleel *et al.*, 1980). The concentrate is clean and easily dispersible in hot water. Because of the superior quality, the concentrate is found wide acceptance in Indian homes and also in the preparation of various food products commercially. The minimum total soluble solids in tamarind concentrate should be 65 per cent.

13.14.0 Tamarind Related Patents

Several tamarind related patents have been obtained since 1974. Eight were issued in India and these related to the preparation of TKP as a sizing agent, tanning material, preparation of gum, preparation of juice and health care applications. The US Patent And Trademark Office (USPTO) granted 30 patents between 1978 and 1997. Eight of those 30 patents are owned by Merck and Co. and relate mainly to the utilisation of TKP as a gum and binding agent (Table 14).

Table 14: Patents on Processes, Products and Uses of Tamarind

Patent Granted/Application Field	*Assignee/Applicant*	*Year*
Patents Granted (India)		
Method of preparation of tamarind powder	Chankrapani, S. and Chankrapani J.	1974
A new process for the preparation of TKP	CSIR	1974
A simple process for obtaining a good tanning material from tamarind seed testa	CSIR	1978
Purification of tamarind gum by air classification	General Mills Chemicals Inc.	1979
A enzymatic process for the preparation of tamarind juice concentrate	CSIR	1985
Patents Application (India)		
Process for making tamarind pickles	Dilip Shanthnaran Dahanukar	1995
A process for preparing tamarind extract in the form of paste/ jam	Shoki Kobayashi	1996
Patent granted/application field	Assignee/Applicant	Year

Patent Granted/Application Field	*Assignee/Applicant*	*Year*
Patents granted/application field	Assignee/Applicant	Year
A manufacturing process for tamarind paste and concentrate	Yelantinaga	1996
A new process for recovery of tartaric acid and other products from tamarind pulp	CSIR	1996
Patents Granted (USPTO)		
Purification of tamarind gum	General Mills Chemicals Inc., USA	1978
Use of TKP as an anti-migrant	Merck and Co. Inc., USA	1982
Clarified tamarind kernel powder	Merck and Co. Inc., USA	1984
Process for separating polysaccharides from tamarind seeds	Shikibo Ltd., Japan	1990
Methods for preparing tamarind oligosaccharides	Lafayette Applied Chemistry Inc., USA	1995
Beverages using tamarind extract and methods of making such beverages	Nutra Sweet Co., USA	1995

Source: TIFAC bulletin – intellectual property rights (IPR), Vol. 3, No. 9, 1997.

Dutt (2001) surveyed the world R&D efforts on tamarind-based innovations as reflected in the United States Patent and Trademark Office (USPTO) and European Patent Office (EPO) free patent search databases. It enlists 62 patents found at both the databases. A majority of the patents are held by corporations. Since 1992 there is a slowing down and inconsistency in patenting activity the world over. Innovation trends during the last ten years, from 1992 to 2001, suggest that food and beverages, health care, and industrial applications have been the focus of the patenting activity. It must be pertinent to mention that the traditional Indian medicine has mentioned a lot of uses for Tamarind in the texts of Charak Samhita. Efforts must be made to validate and patent those uses.

13.15.0 Marketing and Economics

Nageswara Rao (1997) made the economic analysis of income and employment of tamarind trees in rural areas in Andhra Pradesh. Total cultivation cost of tamarind per hectare was Rs. 14,247 for grafts and Rs. 12,579 for seedlings. Tamarind trees come up well in different agro-soil-climatic conditions with minimum investigation and highly suited to marginal and less fertile soils under dry land conditions. They are considered boon for dry land farmer as they provide stable income even when all other crops fail for one reason or the other. Tamarind yields can be obtained even in the third or fourth year only by using grafts or improved seedlings and adopting good management practices. Tamarind is highly suited to development of wastelands. It is a multipurpose plant and every part of it has economic value. The yields and prices are highly stable compared to all other field crops and ultimately result in stable returns. Jeeva *et al.* (1997) opined that tamarind is a promising tree for wasteland development. The production of tamarind in India is mainly concentrated in the States of Tamil Nadu, Andhra Pradesh, Karnataka, Orissa and to a certain extent in Kerala. Tamarind always prefers a warm climate and well adapted to rainfed conditions but can not tolerate continued damp weather or frost condition. Rains during fruit ripening adversely affect the yield and quality.

While producton is larger in south India, its demand is spread all over the country. Hence, tamarind is moved from the villages to assembling markets through brokers/commission agents or directly by the wholesale merchants. Distribution of tamarind within the country is done after the consignment is arrived in the terminal markets. The cost at various stages of marketing including the profit margin of the village merchants, fee or commission of the commission agents/brokers and profit margin of the wholesale merchants both in the assembly markets and terminal markets. Stockists and retailers are added to the consumer's price. Among various markets, Mumbai has significant place for tamarind and its products. In many of the consuming markets, the retail price is over Rs. 30/- per kg for fresh tamarind. Recently the sweet tamarind pods are sold in the market at Rs. 200 per kg. An intersting feature of the consumer price is that much of the fluctuations in the wholesale prices are not reflected and more or less a steady upward trend is maintained. The producer always gets only very low price. Average price realisation is about Rs. 10 per kg at the farm gate. In other words, in the consumers price the share of the producer is hardly 33 per cent. In a non-perishable commodity like tamarind fresh, the share of the producer in the consumer's price is very low. It is mainly because of the unscientific marketing practices and the urge of retailers for higher profits.

There are various levels of merchant associations, such as regional associations in Andhra Pradesh based in Guntur, Visakhapatnam, Vizianagaram and Hyderabad and State level association is located in Hyderabad and all India level association of tamarind merchants is based in New Delhi.

14.0 Biotechnology

14.1.0 Tissue Culture

Micropropogation, Can complement conventional propagation methods, as it can potentially augment multiplication of selected trees on large scale. Micropropagation from hypocotyl and shoots of seedlings is reported (Mascarenhas *et al.*,1987). Shoot and root was induced in Ms and whites medium respectively. (White, 1963) cytokinins and auxins. Koppa and Nataraga (1990) *In vitro* shoot proliferation from seedling explant is reported. Splitt Stoesser and Mohammed-Yasseen (1991) observed multiple shoot regeneration from cotyledon nodes when thidiazuron was used instead of BAP. Jaiswal and Gulati (1992) reported high frequency regeneration from excised cotyledon on MS medium containing BAP where as Rooting was achieved with IAA Who further reported multiple shoot regeneration from shoot tips and nodal explants with BAP. Similarly multiple shoot generation and rooting of micro-shoots was observed in tamarind seedling explants by Rao *et al.* (1997). MS containing 0.5 mg per IBA, 0.5 mg per 1 KN and 0.1 mg per 1 IAA was sufficient for promoting 2-3 multiple shoots as documented in their reports. Rhizogenesis was facilitated on MS containing 0.2 mg of IAA, IBA and NAA. Ganga and Balakrishnamoorthy (1997b) optimized high frequency regenerations from seedling explants of genotype 'Urigum'. Shoot bud differentiation, the earliest response and the maximum number of multiple shoots were recorded from These explants an MS supplemented with 2.5 mg of GA. Induction of micro shoots from auxiliary buds (Balakrishnamoorthy and Ganga, 1997a, 1997b) and *in vitro* rhizogenesis of these micro-shoots of genotype Urigam is also reported (Ganga and Balakrishnamoorthy, 1997a). MS at half strength plus 0.5 mg of 1AA and 1BA recorded maximum rhizogenic response. Tissue culture plants showed good growth in height, branching habit, spread of branches, early flowering and fruiting cycle (ICFRE, 1993; Mishra, 1997). Tissue culture plants started flowering earlier than seed plants (3.7 m as against 8 m) have recommended planting. Spacing recruitment for TC plants is 9x9 m instead of 12.5 x 12.5 m.

14.2.0 Micropropagation

Shoot tips, cotyledons and its nodes have been successfully used as explants for tissue culture (Splittstoesser and Mohamed, 1991). Jaiwal and Gulati (1991) reported 100 per cent shoot bud differentiation in 4 months when the adaxil surface of the entire cotyledon (excised from 12 days old seedlings) was in contact with MS medium containing BAP (5 x 10^{-6} M) Regenerated plants could be established in the soil with 70 per cent success. The explants were cultured in a MS (Murashige and Skoog, 1962) medium supplemented with thidiazuron (TDZ)/LBA (indole butyric acid) or BA (butyric acid)/NAA and incubated at 28°C for 5 weeks under an 18 hours photoperiod. The shoot tips of tamarind regenerated only one shoot, while the cotyledonary nodes produced multiple shoots. The shoot tips of tamarind regenerated only one shoot, while the cotyledonary shoot nodes produced multiple shoots. All shoots were rooted in MS with IAA (indole acetic acid) (5 x 10^{-6} m) and the micropropagated plants were successfully transferred to the soil (Jaswal and Gulati, 1992). Shoot tips from *in vitro* grown seedlings have been regenerated in to plantlets in MS medium containing IBA or IAA. Higher concentration of growth hormones, (2 x 5 mg/l) induced better rooting, but a media supplemented only with coconut water or BAP supported growth of shoots, but not that of roots (Koppa and Nataraja, 1990).

Response to axillary bud proliferation was on MS medium plus BA and GA, while the greatest rhizogenic response was on MS medium at half strength plus 0.5 mg per litre each of IAA and IBA. Tissue culture plants have been reported to show better growth in height, branching habit, spread of branches, early flowering and fruiting cycle with the start of flowering at an average height of 3.7 m. (Mishra, 1997 cited in Rao *et al.*, 1999). Although micropropagation have shown promise, none of them have attained the stage of commercialisation. Balakrishnamurthy and Ganga (1997) reported that the axillary buds collected and cultured during the month of May recorded the best results with respect to the per unit bud break (82.50), number of days taken for bud break (9.33), per unit response to multiple shoot induction (22.50), number of shoots produced per culture (6.40) and length of the micro-shoot (1.27 m).

The axillary buds collected and cultured during the months of December, January, February, March, September and October did not record any response. Similar result is also reported by Prabhakaran *et al.* (2003) who found the explants collected during the month of June (T_6) showed the least contamination (7.32 per cent) and the highest bud break percentage (84.14). The contamination was the highest (77.00) and the bud break was the lowest (45.39). Among the five different genotypes of tamarind tested for their suitability for multiple shoot induction, the genotype Urigam exhibited the highest percentage of bud break (73.63), with more number of shoots per bud (1.40) and the shoot length was also the highest (3.34 cm) followed by PKM-1, Pollachi,-2, Asanoor-H-1 and Salem-144.

A protocol for in vitro regeneration of plants via adventitious bud formation from mature embryo axis of tamarind was standardized (Mehta *et al.*, 2000). Explants consisting of longitudinal section of embryo axis with attached cotyledon were cultured in the Murashige and Skoog (MS) medium with various combinations and concentrations of a-naphthaleneacetic acid (NAA), 6-benzylaminopurine (BAP) and sucrose. Induction of adventitious shoot buds was achieved in the cut surface of the axis when cultured in a medium containing 2.69 mM NAA, 44.39 mM BAP and 4 per cent sucrose. A medium consisting of 0.91 mM zeatin, 2.22 mM BAP, 0.41 mM calcium pantothenate and 0.40 mM biotin supported the differentiation of the buds to form elongated shoots. The shoots developed roots in a half strength MS medium with 2 per cent sucrose following a 72 h treatment with auxin mixture in the dark. On transfer to soil 24 plants survived (Mehta *et al.*, 2000).

Germination of tamarind seeds in medium containing thidiazuron (TDZ) resulted in induction of nodular protrusions in and around the cotyledonary node meristem(Mehta *et al.*,2005). The structures developed radially in well-defined circles and subsequently spread towards the cotyledonary bridge and also in the proximal part of the hypocotyl. The structures developed into shoots on transfer to medium devoid of growth regulators. Histological studies revealed that the protrusions initiated from the nodal meristem and extended to the non-meristematic region between the two meristems and also in the proximal part of the hypocotyl in seedlings germinated in 9.08 mM TDZ. Newly formed cell layers and less-differentiated meristematic protrusions were also seen. With the increase in the distance from the meristem, the buds were less differentiated; in the proximal part of the hypocotyl only the multiple layers of meristematic cells were noted. With extension of the period of incubation, the TDZ-induced meristematic activity extended laterally in circles towards the neighboring region. The radial spread of the meristematic activity from the center of the nodal meristem was also evident at 18.16 mM TDZ. From the pattern of the morphogenic development and the histological studies it may be hypothesized that in tamarind, TDZ influences the existing meristems specifically. Subsequently de novo organogenesis is triggered in the neighboring cells (Mehta *et al.*,2005).

14.3.0 Somatic Embryogenesis

Reports on *in vitro* morphogenesis in tamrind are limited, due to its recalcitrant and callogenic nature (Mehta *et al.*, 2004). To overcome these limitations, an attempt was made to induce meristematic activity in seedling explants. Seedlings were germinated in medium with or without thidiazuron (4.54, 9.08, 13.12, 18.16 µ*M*). This growth regulator restricted the differentiation of the apical meristem to form shoots.

It triggered proliferation of the meristematic tissue at the cotyledonary node and a large number of meristematic buds appeared in a ridial pattern around the node. The meristematic activity extended to the junction of the epicotyl and hypocotyl, giving rise to buds in the form of protuberances in all sides of the junction. These buds differentiated to form shoot primordia and subsequently to shoots in medium devoid of growth regulators. Plants developed by micrografting of these shoots on seedling-derived rootstocks survived in soil.

The effect of varying concentrations of 6, furfurylamino purine (Kinetin), 6, Benzylamino purine (BAP) (1 Naphthalene acetic acid), was evaluated for primary callus induction (Mowobi *et al.*, 2016). Their roles on shoot induction were also investigated. Different concentrations of the three samples were investigated to study their effects on callus induction and shoot induction. The seeds of Tamarind were cultured in Murashige and Skoog (MS) media containing different concentrations of the three hormones. The results revealed that 0.2 mg/l of Kin, 0.5 mg/l of BAP and 0.1 mg/l of NAA were suitable for callus induction and 0.05 mg/l of Kin and 0.1 mg/l of BAP were also suitable for shooting using the leaves while cotyledon was more responsive to callus induction as compared to leaves and roots (Mowobi *et al.*, 2016).

14.4.0 Markers

14.4.1 Genetic Diversity and Phylogenetic Analysis

The genetic diversity and genetic relatedness of 36 tamarind genotypes were studied using amplified fragment length polymorphism (AFLP) markers (Ahmed *et al.*, 2011). Twelve primer pairs were used for the screening of genotypes based on the number and quality of polymorphic fragments produced. These primer combinations produced 702 bands, 627 (89.3 per cent) of which were polymorphic. The mean Nei's gene diversity index was 0.4868, and the Shannon's information index was 0.6802, indicating the abundant genetic diversity within the tamarind population. Among the tamarind genotypes, the dissimilarity coefficient ranged from 4.4 to 14. Cluster analysis grouped all tamarind genotypes into two major clusters (A and B) at 37 linkage distance and two minor clusters each. Cluster A included the genotypes predominantly with brown to dark brown fruit pulp and dark green leaves. However, the straight to semi-curved fruits were sub grouped in sub-cluster 'A1' and semi-curved to curved fruits in sub-cluster 'A2'. Cluster B contained a collection of genotypes predominantly characterized with trees of orthotropic growth and semi-curved fruit shape. A good correlation of these amplified fragment length polymorphism-based groupings with their morphological traits was observed. We found moderate genetic diversity in these tamarind genotypes. The use of AFLP markers and the level of genetic variability detected within Southern Indian tamarind germplasm suggested that this is a reliable, efficient, and effective marker system that can be used for diversity analysis and subsequently in tree breeding programs. Nyadoi *et al.* (2010) characterized variations in tamarind diameter at breast height (DBH), crown section area (CSA) and height (morphological characters) in different environment (temperature, rainfall, soils, habitats of East Afric altitudes and latitudes) and mapped the variations in relation to established mitochondria haplotypes. Results show that tamarind mean height, DBH and CSA increased from Island to mainland Phytocoria respectively < 5 m to 14 m, 61 cm to 189 cm and 28 m³ to 229 m³ and were influenced (P<0.05) by altitudes, phytocoria and habitats. Soil exchangeable potassium also influenced (P = 0.03) tamarind DBH while CSA further varied (P<0.05) between the Somalia Maasai and other mainland phytocoria. Correlation (R) between tamarind morphological variations and environment was 59.5 per cent and interestingly, the variations coincided with haplotypes diversity in populations, suggesting in homogeneous environment, morphological variants could be useful as markers for germplasm and conservation unit selection. Further correlation analysis is however needed to ascertain level of precision with which morphological (DBH, height and CSA) variants will be useful as selection markers in tamarind conservation programmes where these traits are of interest.

Sarmiento *et al.* (2017) investigated the genetic diversity of 32 tamarind plus trees using inter-simple sequence repeat (ISSR) markers. Eighty four loci were examined using 12 markers, with a mean number of 4.42 loci per primer; 8 loci (9.52 per cent) were monomorphic and 76 (90.48 per cent) polymorphic, revealing genetic variability among the individuals. Polymorphic information content (PIC) values varied from 0.29 (ISSR_808) to 0.93 (ISSR_HB12), whereas the marker index ranged from to 26.4 (ISSR_814) to 62.5 (ISSR_17899A). Primers ISSR_HB11, ISSR_836, ISSR_842, ISSR_848, ISSR_860, ISSR_17899A and ISSR_17899B were useful to discriminate the grouping of the accessions according to their PIC values. Ward cluster analysis grouped accessions into two major groups with five subgroups with 46 per cent similarity according to Jaccard distance. The genotypes from Loja, Manabí and Guayas provinces were grouped in the first cluster; while only individuals from Manabí located in the other group, indicating major diversity in the latter province. Genotypes T1-ECUM-001 and T1-ECUM-002 presented 76 per cent similarity, while T1-ECUM-008, T1-ECUM-010, T1-ECUM-012, T1-ECUM-017 and T1-ECUM-018 shared 60 per cent. All materials from Loja grouped with 65 per cent similarity. Other genotypes clustered with similarity of 54 per cent. The cophenetic correlation coefficient (0.634) showed a good fit between the data matrix and the dendrogram results. A reasonable degree of diversity was found among tamarind genotypes.

Genetic relatedness and genetic diversity among 13 Indian collections of Tamarind were evaluated by Gangaprasad *et al.* (2013). Twenty eight selected arbitrary primers were used for characterization using RAPD that generated 131 fragments, of which 116 (88.54 per cent) were polymorphic. Two genotypes, NTI62 and NTI84 were distinguished by unique band specific to them. The genetic similarity coefficient values suggested a wide genetic base for genotypes considered for the study. Cluster analysis based on unweighted pair group method using arithmetic average (UPGMA) clearly indicated that genotypes did not cluster according to their site of collection. This could be attributed to highly cross pollinating nature, small distribution area and that most tamarind genotypes grown in India are from seed source (Gangaprasad *et al.*, 2013).

The genetic diversity of 3 tamarind samples, random exploitation of enhanced polymorphic DNA (RAPD) markers was reported (Kamalambigeswari *et al.*, 2020). Tamarind plants were sampled in 3 environmental zones covering domestically known fruit morphotypes. On 3000 fruits and flowers, 12 morphological descriptors were evaluated. Univariate and canonical analyses carried out on morphological descriptors revealed essential similarities and verified the difference between morphotypes as viewed by indigenous people. However, the assessment of variance components showed significant differences within morphotypes, indicating a great heterogeneity within seeds historically categorized as belonging to an equal morphotype. Quantitative descriptors should, therefore, be coupled with domestically perceived qualitative characteristics (pulp style and color) in order to cause a bunch of strong morphological discrimination. The differences discovered were significantly linked to ecological factors. The size and mass of fruits and seeds are cared for with increased wetness and decreased with aridity. Results also stated that fruit mass could be an intelligent pulp output expression, although its predictive strength varied among morphotypes. Outputs from the variance component assessment recommended that if any genetic tests were not finished, germplasm assortment should be performed by testing a mild variety of forests per morphotype to ensure that a big variety of genetic diversity is captured. 10 Random modified polymorphic DNA (RAPD) primers were used to evaluate tamarind trees ' genetic variety. At the molecular stage, genotypes that were strongly linked morphologically were discovered to be unrelated. A significant amount of intra-population variation reported in the donation research that could be used in interbreeding programs to introgress the intriguing characteristic of concern with effectiveness (Kamalambigeswari *et al.*, 2020).

14.4.2 DNA Bar Codes for Thai Tamarind

Srisuvoramas *et al.* (2015) investigated *Tamarindus indica*.L. chloroplast rbcL gene for ribulose-1,5-bisphosphate carboxylase/oxygenase, large subunit as the molecular marker to identified the cultivar level of *T. indica*.L. in Thailand. The partial rbcL gene of 19cultivars were sequences approximately 611-712 bp. The sequence alignment and Phylogenetic Tree analysis were generated by Clustal W and NJ methods. The result of nucleotide sequence alignment of partial rbcL gene for all cultivars were divided into 19 groups. The different of these groups similar to the different of local name of Tamarind cultivars in Phetchabun Province, Thailand. Base on these results indicated that 19 cultivar DNA Barcodes can supported the local opinions act as the scientific data to identification of tamarind cultivars in Thailand (Srisuvoramas *et al.*, 2015).

15.0 References

Ahmed, A.Q., Yahya, A., Papanna N. and Simon, L. (2011) *Int. J. Fruit Sc.*, **11**(1): 1-66.

Ahmed, S.I., Pankajam, S. and Vijaychandarn, S.N. (1988) *Proceedings of 75th Indian Congress*, Pune.

Ajandouz, E.H. and Puigserver, A. (1999) *J. Agril. Food Chem.*, **47:** 1786–1793.

Ali, M.S., Ahmad, V.U., Azhari and Usmanghani, K. (1998) *Fitoterapica*, **69**(1): 43-46.

Allen, O.N. and Allen, E.K. (1936) *Soil Science*, **42**(2): 87-91.

Allen, O.N. and Allen, E.K. (1981) Madison, W.S., Univ of Wisconsin Press, p. 812.

Allen, O.N. and Allen, E.K. (1981) W.S., University of Wisconsin Press, p. 852.

Amin, R.S. (1978) *Curr. Sc.*, **47**: 468-469.

Amin, R.S. (1978a) *Indian J. Hort.*, **35**: 105-108.

Andriamantena, R.W., Artuad, J., Gaydofn, E.M., Iatrides, M.C. and Chavalier, J.L. (1983) *J. American Oil Chem. Soc.*, **60**(7): 1318-1321.

Anjirawaroj, S., Rangsiruji, A. and Donsakul, T. (2008) *SWU Sci. J.* **24:** 184.

Anonymous (1955) *Indian Farming*, **5**: 21-24.

Anonymous (1955) *Indian Farming*, **8**: 21-24.

Anonymous (1976)In: *The Wealth of India* (raw materials series), Council of Scientific and Industrial Research, New Delhi, **10**: 114-122.

Anonymous (1982a) *Some Recent Developments*. Central Food Technological Research Institute, Mysore, India.

Anonymous (1982b) *Indian Food Industry*, **1**: 43-44.

Anonymous (1984) *Profodil Bull.*, **19**: 24-26.

Anonymous (1987) The wealth of India.

Anonymous (1989) Karnataka Forest Department, Silviculturist, Northern Zone, Dharwad, Karnataka, pp. 41-48.

Anonymous (1991) Daseanayake, M.D. and Fosberg, F.R., Oxford and IBH Publishing Co. Pvt. Ltd., New Delhi, **7**: 96-97.

Anonymous (1997) *Area and Production of Spices in India and the World*. Spices Board, Cochin, Kerala.

Arunkumar, B. (2005) *M.Sc. Thesis*, University of Agricultural Sciences, Dharwad.

Athar, M. and Mahmood, A. (1982) *Pakistan J. Bot.*, **14**: 36.

Azhakiamanavalan, R.S. and Vadivel, E. (1997) *Proceedings of National Symposium on Tamarindus indica* L., Tirupati, pp. 7-9.

Babu, C.K., Ramakumar, M.V. and Ranganna, B. (1999) *Technical Bulletin No. UAS/PHT/99/33*, GKVK Campus, Bangalore.

Balakrishnamoorthy, G. and Ganga, M. (1997) *Proceedings of National Symposium on Tamarindus indica* L., Tirupati, Andhra Pradesh, Forest Department of Andhra Pradesh, pp. 98-100.

Balakrishnamoorthy, G. and Ganga, M. (1997a) *Proceedings of National Symposium on Tamarindus indica* L., Tirupati, Andhra Pradesh, Forest Department of Andhra Pradesh, pp. 110-112.

Baragano-De-Mosqueda, M. (1980) In: *Tropical and Subtropical Fruits*. (Eds. Nagy, S. and Shaw, P.E.). AVI Publishing, Westport, Conn, pp. 375.

Bawa, K.S. (1980) *Annu. Rev. Ecol. S.*, **11**: 15-40.

Bawa, K.S. and Webb, C.J. (1984) *American J. Bot.*, **71**: 736-751.

Bawne, P.R. (2000) *M.Sc. (Agri.) Thesis*, Mahatma Phule Krishi Vidyapeeth, Rahuri.

Beeson, C.F.C. (1941) Manager of Publication, Government of India, New Delhi, pp. 767.

Benero, J.R., Collazo-De-Rivera, A. and Georage, L.M.I. (1974) *J. Agril Univ.*, Puerto Rico, pp. 58-99.

Benk, E. (1987) *Alcoholic Industries*, **100**: 87-129.

Bennet, S.S.R., Durai, A., Nicodemus, A., Gireesan, K., Nagarajan, B. and Varghese, M. (1997) *Proceedings of National Symposium on Tamarindus indica* L., Tirupati, pp. 47-50.

Benthall, A.P. (1933) Thacker Spink and Company, p. 513.

Benthall, A.P. (1933) Thacker Sprink and Co., Calcutta, pp. 110-112.

Bettencourt, E., Mazekamp, T. and Perry, M.C. (1999) International Plant Genetic Resources Institute, Rome.

Bhat, S.G. (1966) *Indian Oil Soap J.*, **32**: 53-57.

Bhattacharya, S. (1997) *Proceedings of National Symposium on Tamarindus indica* L., Tirupathi, Andhra Pradesh, pp. 162-168.

Bhattacharya, S., Bal, S. and Mukherjee, R.K. (1994) *Food Chem.*, **49**: 1-9.

Bhattacharya, S., Bal, S., Mukherjee, R.K., and Bhattacharya, S. (1993) *J. Food Engg.*, **18**: 77-79.

Bhattacharyya, P.K., Bal, S. and Mukherji, R.K. (1994) *J. Food Sc. Technol.*, **31**(3): 372-376.

Biradar, S. (2001) *M.Sc. (Hort.) Thesis*, University of Agricultural Sciences, Dharwad.

Biradar, S.R. and Hanamashetti, S.I. (2001) *J. Plantation Crops*, **29**(3): 64-65.

Biradar, S.R., Hanamashetti, S.I. and Patil, P.B. (2002) *My Forest*, **38**(1): 17-20.

Bolza, E. and Keating, W.G. (1972) Melbourne, Australia, Commonwealth Scientific and Industrial Research Organisation, Division of Building Research, p. 700.

Brandis, D. (1921) Constable and Co. Ltd., London, **13**: 52-253.

Brenan, J.P.M. (1967) In: *Flora of Tropical East Africa*. (Eds. Milne-Redhead, E. and Polhill, R.M.). London, Agents for Overseas Governments and Administrations, pp. 151-153.

Brown, W.H. (1954) *Technical Bulletin*, Bur Printer, Manila.

Bueso, C.E. (1978) *Abstracts of Papers ACS 176*, AGFD 45.

Bueso, C.E. (1980) In: *Tropical and Subtropical Fruits*. (Eds. Nagy, S. and Shaw, P.E.). AVI Publishing Westport, Conn., pp. 375.

Butani, D.K. (1978) *Pesticides*, **12**(11): 34-41.

Challapalli, A.P., Chimmad, V.P. and Hulamani, N.C. (1995) *Karnataka J. Agril. Sci.*, **8**(1): 114-115.

Chapman, K.R. (1984) Complied by Page, P.E. Queensland Department of Primary Industries, Information Series Q 183018, Brisbane, pp. 83-86.

Chari, M.S. and Muralidharan, C.M. (1985) *J. Entomol. Res.*, **9**(2): 243-245.

Chattopadhyay, P.K. and Mohanta, S.K. (1988) *South Indian Hort.*, **36**(6): 324.

Chaturvedi, A.N. (1985) *Uttar Pradesh Forest Bulletin*, Lucknow, India, Government of India Press, **15**(1): 286.

Chaturvedi, A.N., Bhat, D.N., Misra, C.M. and Singh, S.L. (1986) *Forest Bulletin*, No. 53.

Chavan, R., Lokesha, R. and Nayak, S.N. (1999) *Mysore J. Agril Sci.*, **12**(1-4): 55-59.

Chiemsombat, P., Navitoom, K., Jiwa Jinda, S., Klinkong, Krai-Rod, A. and Feungchan, S. (1996) *Thai J. Agril. Sci.*, Special issue, **1**: 68-92.

Chundawat, D.J. (1990) Arid Fruit Culture. Oxford IBH Publishers, New Delhi, pp. 158-163.

Coates-Palgrave, K. (1988) C.S. Striuk Publishers, Cape Town, pp. 278-279.

Corner, E.J.H. (1945) Tamarindus. Singapore Government Printing Office, **1**: 404.

Coronel, R.E. (1986) *Processing Fruits of the Philippines*. 2nd Edition, University of Philippines, Los Banos.

Coronel, R.E. (1991) In: *Plant Resources of South-East Asia*, Wageningen; Pudoc. No. 2, Edible fruits and nuts. (Eds. Mkerheij, E.W. and Coronel, R.E.). PROSEA Foundation, Bogor, Indonesia, pp. 298-301.

CSIR (1976) Raw materials, Vol. 10, Council of Scientific and Industrial Research, New Delhi, India, pp. 118-122.

Dagar, J.C., Singh, N.J. and Singh, G. (1995) *J. Trop. For. Sci.*, **7**(4): 623-624.

Dale, I.R. and Greenway, P.J. (1961) Buchanan's Estate Limited, p. 654.

Dalziel, J.M. (1937) Crown Agents for Overseas Governments and Administrations, p. 612.

De-Lumen, B.O., Becker, R. and Reyes, P.S. (1986) *Anim. Feed Sci. Tech.*, **22**(4): 285-293.

De-Lumen, B.O., Becker, R. and Reyes, P.S. (1990) *J. Agril. Food Chem.*, **34**: 361-364.

De-Pauda, L.S., Lugod, G.C. and Pancho, J.V. (1978) *Los Banos Technical Bulletin*, College Laguna, **11**: 3.

Devadas, C.T. and Gothandapani, L. (1993) *The Hindu*, pp. 3.

Ding, M.M., Yi, W.M. and Liao, L.Y. (1986) *Nitrogen Fixing Trees Research Reports*, **4**: 9-10.

Duke, J.A. (1981) *Handbook of legumes of World Economic Importance*. Plenum Press, New York, pp. 228-230.

Duke, J.A. and Terrell, E.E. (1974) *Introduction Taxon*, **23**(5-6): 759-799.

Dutt, B. (2001) *J. Intellectual Property Rights*, **6: 215-226.**

Dwivedi, K., Ali-Wahid and Pathak, R.K. (1996) *Annals Agril. Res.*, **17**(4): 447-449.

Eggeling, W.J. and Dale, I.R. (1951) *The Indigenous Tree of the Uganda Protectorate Entabbe*, Uganda. The Government Printer, p. 491.

FAO (1970) *National Studies FAO/UN*, Rome, **24**: 50-52.

FAO (1988) *Technical Notes*. FAO-SIDA Forestry Paper, **34**: 165-167.

FAO (1998) *A Resource Book for Promoting the Exploitation and Consumption of Food Plants in Arid, Semi-arid and Sub-humid Lands of Eastern Africa*. Paper, **42**: 478-483.

FAO (1999) *Eleventh Session*, 29 September-October FORGEN/99/5.

Feungchan, S., Yimsawat, T., Chindaprasert, S. and Kippowsong, P. (1996c) *Thai J. Agril. Sci.*, Special issue, **1**: 42-44.

Feungchan, S., Yimsawat, T., Chindaprasert, S. and Kitpowsone, P. (1996b) *Thai J. Agril. Sci.*, Special issue, **1**: 34-41.

Feungchan, S., Yimsawat, T., Chindaprasert, S. and Kitpowsong, P. (1996e) *Thai J. Agril. Sci.*, Special Issue No. **1**: 28-33.

Feungchan, S., Yimsawat, T., Chindaprasert, S. and Kitpowsong, P. (1996a) *Thai J. Agril. Sci.*, Special issue, **1**: 1-11.

Frankel, O.H. and Hawkes, J.G. (1975) *Crop Genetic Resources: Today and Tomorrow*. Cambridge University Press, Cambridge.

Galang, F.G. (1955) *Fruits and Nuts Grown in the Philippines*. Malabon, Rizal AIA Printing Press.

Gamble, J.S. (1922) *A Manual of Indian Timbers*. London Sampson Low, Marston and Co., p. 866.

Ganachary, V. (2005) *M.Sc. Thesis*, University of Agricultural Sciences, Dharwad.

Ganga, M. and Balakrishnamoorthy, G. (1997a) *Proceedings of National Symposium on Tamarindus indica* L., Tirupati, Andhra Pradesh, Forest Department of Andhra Pradesh, pp. 105-106.

Ganga, M. and Balakrishnamoorthy, G. (1997b) *Proceedings of National Symposium on Tamarindus indica* L., Tirupati, Andhra Pradesh, Government of Andhra Pradesh, pp. 107-109.

Gangaprasad, S.,R Rajkumar, R L Ravikumar M H Savita, K Krishnamurthy, S Hittalama. (2013) *J. Spices and Aromatic Crops*, **22:** 55-61.

George, C.K. and Rao, Y.S. (1997) *Proceedings of National Symposium on Tamarindus indica* L., Tirupati, Andhra Pradesh, Government of Andhra Pradesh, pp. 156-161.

Ghaffaripour, S., Van den, N. and Samson, B. R. (2017) *Scientia Horticulturae* **222**: 145-152.

Girdharilal, L., Das, D.P. and Jain, N.L. (1958) *Indian Food Packer*, **12**: 13-16.

Glicksman, M. (1986) In: *Food Hydrocolloids*. (Ed. Glicksman, M.), Vol. III, CRC Press, Boca Raton.

Goce, S.P., Ombico, M.F.T. and Raymundo, L.C. (1993) *Agriculturist*, **76**: 103-112.

Gordon, A.L. (1968) *Tamarind Extract*. United Stats Patent Office Patented on August 29, 1968.

Gowda, N. (1983) *Thesis Abstract*, Haryana Agricultural University, **9**(3): 276-277.

Gowramma, R.V., Radhakrishnaiah Setty, G. and Sastry, M.V. (1968) *Indian Food Packer*, **22**: 12-15.

Grieve, M. (1995) Tamarind: Modern herbal home page (http: //www.botanical.com/botanical/mgmh.html). Electric Newt, pp. 1-2.

Gunasena, H.P.M. (1997) *Proceedings of the Regional Workshop on domestication of agroforestry trees in southeast Asia. Yogyakarta*, Winrock International and ICRAF, Indonesia, pp. 49-53.

Gunasena, H.P.M. (1999) *National Workshop on Domestication of Agro-forestry Trees in Southeast Asia*. A Publication of Winrock International in Collaboration with ICRAF, Yogyakarta, Indonesia, pp. 49-53.

Gunasena, H.P.M. (1999) Paper presented at the *National Workshop on Fruits for the Future*. 4th November, Gannoruwa, Peradeniya, Sri Lanka.

Gunasena, H.P.M. and Hughes, A. (2000) *Tamarindus indica* L. International Centre for Underutilised Crops, Southampton, United Kingdom.

Gupta, G.N. and Mohan, S. (1990) *Indian Forester*, **116**(8): 622-630.

Gupta, R.K. (1993) *Multipurpose Tree for Agro-Forestry and Wasteland Utilization*. Oxford and IBH Publication Co., pp. 482-484.

Hanamashetti, S.I. (1996) *Ph.D. Thesis*, University of Agricultural Sciences, Dharwad.

Hanamashetti, S.I. (2006) *Report on ICAR Ad-hoc Project*, New Delhi, p. 37.

Hanamashetti, S.I. and Sulikeri, G.S. (1997) *Proceedings of National Symposium on Tamarindus indica* L., Tirupati, pp. 59-68.

Hanamashetti, S.I. and Sulikeri, G.S. (1997) *Proceedings of National Symposium on Tamarindus indica*, Forest Department of Andhra Pradesh, Tirupati, pp. 101-104.

Hasan, S.K. and Ijaz, S. (1972) *Science Industry* (Karachi), **9**(3-4): 131-137.

Hernandez-Unzon, N.Y. and Lakshminarayana, S.L. (1982) *Hort. Sci.*, **6**: 938-940.

Hockin, G.D. (1993) *Family Leguminosae: Caesalpinoideae*. Oxford and IBH Publishing Co., New Delhi, pp. 305-307.

Holden, J.H.W. and Williams, J.T. (1984) *Conservation and Evaluation*. George Allen and Unwin, London.

Hong, T.D., Linnington, S. and Ellis, R.H. (1996) *Handbook for Genebanks No. 4*, International Plant Genetic Resources Institute, Rome.

ICFRE (1993) *Technical Bulletin*, Forest Research Institute, Dehradun, India, pp. 16.

Irvin, F.R. (1961) *Woody Plants of Ghana*. Oxford University Press, London, p. 868.

Ishola, M.M., Agbaji, E.B. and Agbaji, A.S. (1990) *J. Sci. Food Agri.*, **51**: 141-143.

Jaiwal, P.K. and Gulak, A. (1991) *Plant Cell Reports*, **10**(11): 569-573.

Jaiwal, P.K. and Gulati, A. (1992) *National Academy Science Letters*, India, **15**(3): 63-67.

Jaleel, S.A., Narayana, K., Sreekanthaiah, K.R. and Ramanathan, F.K. (1980) *Indian Food Packer*, **34**: 12-13.

Jama, B., Nair, P.K.R. and Kurira, P.W. (1989) *Agro-forestry Systems*, **9**: 17-27.

Jambulingam, R. and Fernandes, E.C.M. (1986) *Agroforestry Systems*, **4**(1): 17-32.

Jambulingam, R., Chellapilli, K.L. and Murugesh, M. (1997) *Proceedings of National Symposium on Tamarindus indica* L., Tirupati, Andhra Pradesh, Government of Andhra Pradesh, pp. 35-39.

Jansen, P.C.M. (1981) Centre for Agricultural Publishing and Documentation (PUDOC), Wageningen Agricultural Research Reports, No. 906.

Jarimopas, B. and Jaisin, N. (2008) *J. Food Engg.*, **89:** 291–297.

Jayachandran, B.K., Bai, M.M., Salam, M.A. and Mammen, M.K. (1992) *Indian Cocoa, Arecanut and Spices Journal*, **15**: 40-41.

Jayaweera, D.M.A. (1981) *Part II. Flacourtiaceae-Lytharaceae*. National Science Council of Sri Lanka, pp. 244-246.

Jordan, C.F. and Gajaseni, J. (1990) *Proceedings of International Conference,* Marburg. Eds. Warner, D. and Muller, P., 8-12 October, pp. 76-81.

Joseph, K.V. and Oommen, P. (1960) *Indian J. Entomol.*, **22**(3): 172-180.

Joshi, P. (1985) *Botanical Bulletin of India*, **27**(1-4): 100-104.

Kaitho, R.J., Nsahlai, I.V., Williams, B.A., Umunna, N.N., Tamminga, S. and Van Bruchem, J. (1988) *Agroforestry Systems*, **39**: 129-144.

Kamalambigeswari, R., Sangeetha, K. and Kanimozhi, S. (2020) *Plant Cell Biotechnol. Molecular Biol.*, **20:** 1255-1258.

Karale, A.R., Kaulgud, S.N. and More, T.A. (1997) *Proceedings of National Symposium on Tamarindus indica* L., Tirupati, pp. 10-12.

Karale, A.R., Wagh, A.P., Pawar, B.G. and More, T.A. (1999) *Maharashtra Agricultural University*, **24**(3): 319-320.

Karikalan, T.V., Divya, M.P. and Gopi, D. (2002) *Indian J. Agron.*, **4**(2): 88-93.

Katho, R.J., Nsahlai, I.V., Williams, B.A., Umunna, N.N., Thamminga, S. and Vanbrochem, J. (1988) *Agroforestry Systems*, **39**: 129-144.

Keskar, B.G., Karale, A.R., Dhawale, B.C. and Choudhar, K.G. (1989) *Maharashtra Journal of Horticulture*, **4**(2): 121-124.

Khan, M.G. (1946) *Indian J. Entomol.*, **8**: 111-115.

Knight, T. Jr. (1980) In: *Tropical and Sub-tropical Fruits*. (Eds. Nagy, S. and Shaw, P.E.). AVI Publishing Westport, Conservation., p. 1.

Koppa, M.S. and Nataraj, K. (1990) *Indian J. Forestry*, **13**(1): 30-33.

Kotecha, P.M. and Kadam, S.S. (2003) *J. Food Sci. Technol.*, **40**(1): 76-79.

Kotecha, P.M., Adsule, R.N. and Kadam, S.S. (1995) *Indian Food Packer*, **49**(4): 5-11.

Krishna, M. (1955) Indian Patent No. 52, Bureau of Indian Standards, New Delhi.

Kulkarni, R.S., Gangaprasad, S. and Swamy, G.S.K. (1993) *Minor Forest Products News*, **3**(3): 6.

Kulwal, L.V. and Tayde, G.S. (1985) Abstracts of paper presented at *Second International Symposium on Mango*, May 85, Bangalore, p. 24.

Kulwal, L.V., Tayde, G.S. and Deshmukh, P.P. (1985) *PKV Res, J.*, pp. 33-36.

Kulwal, L.V., Tayde, G.S., Chaudhari, M.H. and Verma, R.M. (1997) *Proceedings of National Symposium on Tamarindus indica* L., Tirupati, Andhra Pradesh, Government of Andhra Pradesh.

Kumar, V. (1995) *Nursery and Plantation Practices in Forestry*. Scientific Publishers, Jodhpur, pp. 187-215.

Lakshmanan, K.K. and Narayanan, A.S.S. (1994) *J. Econ. Taxon. Bot.*, **14**(1): 171-173.

Latino, S. and Vega, D. (1986) *Boletin Tecnico-Labal*, **7**: 13-17.

Laxmi, D.V. (1991) *M.Sc. Thesis*, University of Agricultural Sciences, Bangalore.

Lefevre, J.C. (1971) *Review of the Literature on the Tamarind Fruits* (English summary), **26**: 687.

Leppik, E.E. (1988) *Floral Evolution in Relation to Pollination Ecology*. International Book and Periodicals Supply Service, New Delhi.

Leung, W.T. and Flores, M. (1961) *Tables for use in Latin America*. National Institute of Health, Bethesda, Md.

Lewis, Y.S. and Johar, D.S. (1956) *Curr. Sci.*, **25**: 325-326.

Lewis, Y.S. and Neelakantan, S. (1964a) *J. Sci. Ind. Res.*, **23**: 204-206.

Lewis, Y.S. and Neelakantan, S. (1964b) *Curr. Sci.*, **15**: 460.

Lewis, Y.S., Dwarakanath, C.T. and Johar, D.S. (1954) *J. Sci. Ind. Res.*, **13**A: 284-286.

Lewis, Y.S., Dwarakanath, C.T. and Johar, D.S. (1957) *Curr. Sci.*, **26**: 394-395.

Lewis, Y.S., Menon, P.G.K., Natarajan, C.P. and Amla, B.L. (1970) *Indian Food Packer*, **24**: 18-20.

Lewis, Y.S., Neelakantan, S. and Bhatia, D.S. (1961) *Curr. Sci.*, **30**: 381-382.

Little, E.L. and Wardsworth, F.W. (1964) Common trees of Puerto Rico and the Virgin Islands. *Agriculture Handbook 249. U.S. Department of Agriculture, Washington, DC*, p. 548.

Lokesh, A.R., Sarala, B.S. and Chavan, R. (1997) *Proceedings of National Symposium on Tamarindus indica* L., Tirupati, Andhra Pradesh, Government of Andhra Pradesh, pp. 212-217.

Lokesha, S. and Shetty, H.S. (1991) *Int. J. Trop Plant Dis.*, **9**(2): 179-181.

Lotschert, W. and Beese, G. (1994) *Tropical Plants*. Collins Photo Guide, Harper Collins Publishers, p. 223.

Macmillan, H.F. (1943) *Handbook of Tropical Plants*. Medicinal Plants. Anmol Publication, New Delhi, p. 362.

Magehembe, J.A. (1995) SADC-ICRAF Agro-forestry Project, Zomba, Malawi. (Eds. Magehembe, J.A., Ntupanyama, Y. and Chirwa, P.W.) pp. 39-49.

Maille, P. (1991) *TRI News*, **10**(2): 13-20.

Maksoud, M.A., Haggag, L.F., Azzay, M.A. and Saad, R.N. (1994) *Ann. Agric. Sci.*, **39**(1): 355-363.

Maldonado, O., Rolz, C., Schneider and De Cabreva, S. (1975) *J. Food Sci.*, **40**: 262-265.

Manjunath, M.N., Sattigeri, V.D., Rama Rao, S.N., Udahrani, M. and Nagaraja, K.V. (1991) *Indian Food Packer*, **45**: 39-42.

Marangoni, A., Alli, I. and Kermasha, S. (1988) *J. Food Sci.*, **53**: 1452-1455.

Marothia, D.K. and Gauraha, A.K. (1992) *Indian J. Mark.*, **2**: 85-91.

Masano, M. (1994) *Bulletin of Penelitian-Hittan*, **563**: 33-42.

Mascarenhas, A., Nair, S., Kulkarni, V.M., Agarwal, O.C., Khushpee, S.S. and Mehta, V.J. (1987) *Cell and Tissue Culture in Forestry*. (Eds. Bonga, J.M. and Durzan, D.J., Martinus Nijhopp, Dordrecht). **3**: 316-325.

Mastan, M., Sivaram Prasad, N.V., Chalama Reddy, K. and Prasada Reddy, B.V. (1997) *Proceedings of National Symposium on Tamarindus indica* L., Tirupati, pp. 26-34.

Meghwal, P.R. (1997) *Indian Hort*, October-December: 26-27.

Mehta, U.J., Sahasrabudhe and Hazra, S. (2005) *In Vitro Cell. Dev. Biol. Plant*, **41:** 240–243.

Mehta, U.J., Barreto, S.M. and Hazra, S. (2004) *In Vitro Cell. Dev. Biol. Plant*, **40:** 279–283. https: //doi.org/10.1079/IVP2003507.

Mehta, U.J., Krishnamurthy, K. V. and Hazra, S. (2000) *Curr. Sci.*, **78:** 231-1234.

Meillon, S. (1974) French Patent No. 2231322 (English Summary).

Mishra, R.N. (1997) *Proceedings of National Symposium on Tamarindus indica* L., Tirupati, Andhra Pradesh, Government of Andhra Pradesh, pp. 51-58.

Mohan, V. and Mishra, S.N. (1997) *Proceedings of National Symposium on Tamarindus indica* L., Tirupati, Andhra Pradesh, Government of Andhra Pradesh, pp. 248-254.

Morad, M.M., El Magoli, S.D. and Sedky, K.A. (1978) *Fette Seifen Austrichmittel*, **80**: 357-359.

Morton, J. (1987) Fruits of warm climates. (Ed. *Miami*, F.L.). pp. 115-121.

Morton, J. (1987) In: *Fruits of warm climates*. (Eds. Morton, Miami, F.L., Julia, F.,) p.115–121.

Mowobi, G.G., Osuji, C., Salisu, A. and Yahaya, F.M. (2016) *J. Environ. Life Sci.*, **1:** : 26-31.

Mubabamba, L. and Sitaubi, L.A. (1995) *Seed Pre-Treatment, Growth and Phenology of Some of the Indigenous Fruit Trees of the Niomboecozone at Nadungu Forestry Reserve, Malawi*. (Eds. Mageheme, J.A., Ntupanyama, Y. and Chirwa, P.W.). pp. 66-72.

Mukherji, K.G. and Bhasin, J. (1986) *A Source Book*, New Delhi, Tata McGraw Hill, p. 486.

Murashige, T. and Skoog, F. (1962) *Physiol. Plant*, **15**: 473-497.

Murthy, K.N. (1997) *Proceedings of National Symposium on Tamarindus indica* L., Tirupati, pp. 20-25.

Nachegowda, V. (1997) *Proceedings of National Symposium on Tamarindus indica* L., Tirupati, Andhra Pradesh, Government of Andhra Pradesh.

Nadagoudar, B.S. and Basavanneppa, M.A. (1997) *Proceedings of National Symposium on Tamarindus indica* L., Tirupati, Andhra Pradesh, Government of Andhra Pradesh.

Nagaraja, K.V., Manjunath, M.N. and Nalini, M.L. (1975) *Indian Food Packer*, **29**: 17.

Nagarajan B., Nicodemus, A., Varghese, M., Sasidharan, K.R., Mahadevan, N.P. and Bennet, S.S.R. (1999) *ICFRE Technical Bulletin*, IFGTB, Coimbatore, India, pp. 1-8.

Nageswara Rao, A. (1997) *Proceedings of National Symposium on Tamarindus indica* L., Tirupati, Andhra Pradesh, Government of Andhra Pradesh, pp. 127-135.

Naidu, C.V., Sivaram Prasad, N.V. and Ramana Reddy, T.V. (1997) *Proceedings of National Symposium on Tamarindus indica* L., Tirupati, pp. 198-211.

NATIONAL ACADEMY OF SCIENCES (1979) Resource for the future. Washington DC, pp. 117-121.

Navaneetha, N., Palniswamy, K.P., Abdul-Khader, Md. and Kumar, N. (1991) *South Indian Hort.*, **39**: 102-105.

Navaneetha, N., Palniswamy, K.P., Abdul-Khader, Md. and Kumar, N. (1990) *South Indian Hort.*, **38**(4): 220-224.

Nemoto, M., Panchaban, S., Vichaidis, P. and Takai, Y. (1987) *Thailand J. Agric. Sci.* (Tokyo, Nogyo Daigaku), **32**(1): 1-9.

Ngognje, J. (1984) *Feeding Habit of Giraffe* (*Giraffa cameloparadalis* Peralta) *in Wata National Park*, Cameroon, Mammalia, **48**(2): 173-184.

Nicodemus, A., Nagarajan, B., Durai, A., Gireesan, K., Sasidharan, K.R., Mahadevan, N.D. and Bennet, S.S.R. (1997) *Proceedings of National Symposium on Tamarindus indica* L., Tirupati, Andhra Pradesh, Government of Andhra Pradesh, pp. 218.225.

Nyadoi, P., Okori, P., Okullo, J. B., Obua, J., Burg, K., Fluch, S., Nasoro, Magogo, Saleh, Haji, Temu, A. B., Jamnadass, R. (2010) *Gene Conserve*, **9**(36): 1-28.

Obulesu, M. and Bhattacharya, S. (2011) *Int. J. Food Prop.*, **14(3):** 538-549.

Osawa, T., Tsuba, T., Watanabe, M., Ohshima, K. and Yamamato, A. (1994) *J. Agric. Food Chem.*, **42**(1): 2671-2674.

Padolina, F. (1931) *Philippines J. Agric.*, **2**(4): 347-355.

Panigrahi, S., Bland, B. and Carlaw, P.M. (1989) *J. Anim. Feed Sci. Tech.*, **22**(4): 285-293.

Parham, J.W. (1972) *Plants of the Fiji Islands*. Revised Edition, Authority of the Government Printer, Sura, p. 102.

Parrotta, J.A. (1990) SO-ITF-SM-30 June, USDA Forestry Service, Rio-de-Janeiro Piedras, Peutro Rico, pp. 1-5.

Pathak, R.K., Ojha, C.M. and Dwivedi, R. (1992) *Indian Hort.*, **36**(3): 17.

Patil, S.S. (2004) *Ph.D. Thesis*, University of Agricultural Sciences, Dharwad.

Patnaik, K.K. (1974) *Biol. Plant*, **16**: 1.

Perham, J.W. (1972) Revised Edition. Authority of the Government Printer, Suva, p. 102.

Pitka, P.M., Singh, P.P. and Srivastava, H.C. (1977) *J. American Oil Chem. Soc.*, **54**: 592.

Pongskul, W., Wanapat, S., Mungporm, P. and Sirimukdakul, S. (1988) *Khon Kaen Agr. J.*, **18**(2): 72-79.

Postgate, J. (1979) *Nitrogen Fixation*. Southampton, The Camelot Press Limited.

Prabakaran, G., Chezhiyan, N. and Jansi Rani. (2003) *South Indian Hort.*, **51**(1-6): 76-82.

Pradhan, K. (1997) *Proceedings of National Symposium on Tamarindus indica* L.,Tirupati, Andhra Pradesh, Government of Andhra Pradesh.

Prasad, A. (1963) *Madras Agric. J.*, **50**: 202-203.

Prati, D.C. and Delrosario, J.I. (1913) *Philipp J. Sci.*, **48**: 59-80.

Prins, H. and Magehembe, S.A. (1994) *Forest Ecol. Manag.*, **64**(2-3): 111-125.

Purohit, M., Jamaluddin, M. and Mishra, G.P. (1996) *Indian Forester*, **112**: 492-495.

Purseglove, J.W. (1987) *Tropical Crops*. Dicotyledons, Longman Science and Technology, pp. 204-206.

Purushotham, K. and Narasimharao, S.B.S. (1990) *South Indian Hort.*, **38**(4): 225.

Quiniones, S.S. (1983) *Nitrogen Fixing Tree Research Reports*, **1**: 12.

Raghuveer, P. (1997) *Proceedings of National Symposium on Tamarindus indica* L., Tirupathi, Andhra Pradesh, pp. 184-187.

Rajasekhar, A., Kumari, P.C., Prasada Reddy, B.V. and Rao, P.S. (1997) *Proceedings of National Symposium on Tamarindus indica* L., Forest Department of Andhra Pradesh, pp. 238-240.

Ramakumar, M.V., Babu, C.K., Subarmanya, S., Ranganna, B. and Krishnamurthy, K.L. (1997) *Proceedings of National Symposium on Tamarindus indica* L., Tirupathi, Andhra Pradesh, pp. 145-150.

Ramarao, M. (1975) *Flowering Plants of Travancore*. Dehra Dun India. Bishen Singh Mahendra Pal Singh, p. 484.

Ramirez, L., Montesinos, A. and Guzman, L. (1986) *Peru Turrialba*, **36**(1): 99-104.

Ramteke, R.S., Eipeson, W.E. and Patwardhan, M.V. (1990) *J. Food Sci. Technol.*, **27**(5): 277-279.

Rao, K.H. and Subramanian, N. (1984) *Proceedings of National Symposium on Protein Foods and Feeds*, Madras, India.

Rao, P.S. (1948) *J. Sci. Ind. Res.*, **68**: 89-90.

Rao, P.S. and Srivastava, H.C. (1974) Tamarind. In: *Industrial Gums*. Academic Press, New York, pp. 370-411.

Rao, V.S., Mary Mathew, K., Pradeep Kumar, K., Salykutty, Joseph, Laxmanan, R., Madhusoodanan, K.J. and Potty, S.N. (1997) *Proceedings of National Symposium on Biotechnology of Spices, Medicinal and Aromatic Crops* (BIOSAP). 24-5 April, Calicut, pp. 39-44.

Rao, Y.S., Mary Mathew, K. and Porry, S.N. (2000) *Indian J. Arecanut, Spices and Medicinal Plants*, **1**(4): 127-145.

Rao, Y.S., Mathew, K.M. And Potty, S.N. (1999) *Indian J. Arecanut, Spices and Medicinal Plants*, **1**(4): 127-145.

Rao, Z.P., Rao, S.S., Reddy, P.M. and Reddy, K.K. (1989) *Indian J. Animal Nutrition*, **6**(1): 81-84.

Reddy, D.V., Prasad, D.A., Reddy, B.S. and Charyulu, S. (1986) *J. Animal Sci.*, **56**(10): 1094-1098.

Reena, J. and Bagyaraj, D.J. (1990) *World J. Microb. Biot.*, **6**(1): 59-63.

Relwani, L.L. (1993) *The Bharathia Agricultural and Industrial Foundation Journal*, **13**(2): 10.

Retrieved from https: //ikprress.org/index.php/PCBMB/article/view/4868.

Richard, R., Thangaraj and Balakrishnamurthy, G. (1998) *Spice India*, pp. 14.

Ridley, H.N. (1922) *The Flora of the Malay Peninsule*. 47. *Tamarindus* Linn., Vol. 1, Polypatalae L., Reeve and Co. Ltd., London, p. 636.

Rodrigo, T.M.A.S.K. (1992) *Sri Lanka Forester*, **19**: 43-45.

Rodriguez, R., Del Valle, N., Arango, W., Torres, R. and Fernandez, M. (1985) *Century Agriculture*, **12**(1): 71-80.

Roti-Michelozzi. (1957) *Tamarindus adumbr*. fl. Aethiop. 6 Webbia, **13**(1): 134-141.

Roy, R.N., Maiti, S.S. and Mondal, C.R. (1987) *Environ. Ecol.*, **5**(3): 467-471.

Sadasivam, R., Thirumaran, A.S., Manimegalai, G. and Neelakantan, S. (1979) *Proceedings of First Convention of Food Scientists and Technologist*. (FSTA 1979, 11, 12 A87I).

Saideswara Rao, Y. (1997) *Proceedings of National Symposium on Tamarindus indica* L., Tirupati, pp. 124-126.

Salim, A., Simons, A., Waruhin, A. and Orwa, C. (1998) *Agro-Forestry Tree Database: A Tree Species Reference and Selection Guide and Tree Seed Suppliers Directory*. International Council for Research in Agro-forestry, P.O. Box 30677, Nairobi, Kenya.

Samiullah, R. (1993) Genetic diversity in tamarind. Abstracts of *Golden Jubilee Symposium on Horticulture Research: Changing Scenario*. Horticulture Society of India, New Delhi, p. 473.

Sanchez, P.C. (1985) *A Lucrative Business*. Research at Los Banos, The University of the Philippines, **3**: 10-13 (FSTS: 86-05H0063).

Sankaranarayan, R., Vijayakumar, M. and Rangasamy, P. (1994) *Indian Hort.*, **38**(4): 15.

Sano, M., Miyata, E., Tamano, S., Hagiwara, A., No, N. and Shivai, T. (1996) *Food Chem. Toxicol.*, **34**: 463-467.

Sarmiento, L., Iris, P. A., Díaz, B., Hugo, Á. and William, V. (2017) *Bioagro*, **29**(3): 153-162.

Satisha, J., Melanta, K.R. and Venkatesha, J. (1997) *Curr. Res.*, University of Agricultural Sciences, Bangalore, **26**: 6-7.

Savur, G.R. (1948) *J. Biochem.*, **199**: 501-509.

Selvaraj, M., Syamala, D., Arumugam, D. and Rao, M.V. (1996) *Indian Forester*, **112**: 1161-1167.

Shankaracharya, N.B. (1997) *Proceedings of National Symposium on Tamarindus indica* L., Tirupati, Andhra Pradesh, Government of Andhra Pradesh, pp. 226-230.

Shankaracharya, N.B. (1998) *J. Food Sci. Technol.*, **35**(3): 193-208.

Sharma, S. and Bharadwaj, R. (1997) *Proceedings of National Symposium on Tamarindus indica* L., Tirupati, Andhra Pradesh, Government of Andhra Pradesh, pp. 4-6.

Shinde, B.N., Shinde, N.N., Sirdeshpande, V.R. and Naik, D.M. (1997) *Proceedings of National Symposium on Tamarindus indica* L., Tirupati, Andhra Pradesh, Government of Andhra Pradesh, pp. 113-114.

Shinde, N.N., Ingle, G.N., Shinde, B.N. and Chavan, S.D. (1997) *Proceedings of National Symposium on Tamarindus indica* L., Tirupati, Andhra Pradesh, Government of Andhra Pradesh.

Shivalingaswamy, T.M., Balasubramanian, R. and Sivaramakrishnan, V.R. (1991) *My Forest*, **27**(2): 112-116.

Shivanandam, V.N. and Thimmaraju, K.R. (1988) *Mysore J. Agric. Sci.*, **22**: 229-231.

Siddaraju, P., Vinayakumari, K. and Janardhanan, K. (1995) *J. Food Compos. Anal.*, **8**: 351-362.

Siddaramaiah, A.L. and Kulkarni, S. (1982) *Indian Forester*, **108**(5): 361-364.

Siddaramaiah, A.L., Desai, S.A. and Bhat, R.P. (1980) *Sci. Cult.*, **46**(10): 358-359.

Sidduhuraju, P., Vijayakumari, K. and Janardhanan, K. (1995) *J. Food Compos. Anal.*, **8**: 351-362.

Simmonds, N.W. (1984) *Evaluation of Crop Plants*. Longmans, London.

Singh, B.G., Vijayachandran, S.N., Mahadevan, N.P. and Santhi, K. (1997) *Proceedings of National Symposium on Tamarindus indica* L., Tirupati, Andhra Pradesh, Government of Andhra Pradesh.

Singh, D. (1963) *Punjab J. Hort.*, **3**: 194-198.

Singh, L.B. and Singh, R.N. (1958) *Indian J. Hort.*, **16**: 168-172.

Singh, M., Singh, A.K. and Singh, A.K. (1989) *Geobios* (India), **16**(2-3): 74-77.

Singh, P.P. (1973) Oxalic acid content of Indian foods. *Chemical Abstracts*, **79**: 135506W.

Singh, R. (1997) *Proceedings of National Symposium on Tamarindus indica* L., Tirupati, Andhra Pradesh, Government Of Andhra Pradesh, pp. 255-261.

Singha, R.K. (1995) *Int. J. Sust. Dev. World*, **2**(4): 278-284.

Sivaramareddy, G., Jaganmohan Rao, S., Achyutaramayya, D., Azeemoddin, G. and Tirumala Rao, S.D. (1979) *J. Oil Tech. Assoc.*, **11**: 91-93.

Some, L.M., Sary, H. and Bellefontaine, K. (1990) *Bois et Forets des Tropique*, **225**: 42-46.

Sozolkoki, T.W. (1985) Hamburg, Federal Republic of Germany. Tiftung Walderhatung in Africa, p. 132.

Splittstoesser, W.E. and Mohamed-Yasseen, V. (1991) *Proceedings of the Inter-American Society for Tropical Horticulture*, 37th Annual Meeting, Chile, 7-12 October, **35**: 6-8.

Srisuvoramas, B., Aiumsamang, S., Piamalung, T., Sansenya, S. and Sripanom, L. (2015) DNA barcodes for cultivar identification of Thailand *Tamarindus indica* L. *International Conference on Science and Technology (TICST)*, Pathum Thani, pp. 148-152. doi: 10.1109/TICST.2015.7369352.

Srivastava, R.P. (1985) *Second International Symposium on Mango*, May 1985, Bangalore, pp. 20-24.

Srivasuki, K.P., Reddy, R.D. and Reddy, K.K. (1990) *Indian Forester*, **116**: 984-985.

Storrs, A.E.G. (1995) *Know Your Trees – Some Common Trees Found in Zambia*. Regional Soil Conservation Unit, Zambia.

Sundararao, V.S., Venkataboopathy, K. and Madhulatha, W. (1997) *Proceedings of National Symposium on Tamarindus indica*, Tirupathi, p. 21.

Sutter, *e.g.* and Barker, P.B. (1985) *Plant Cell Tiss. Org. Cult.*, **5**: 13-21.

Swaminath, M.H. (1988) *My Forest*, **24**(1): 1-5.

Swaminath, M.H. and Ravindran, D.S. (1989) *My Forest*, **25**(4): 57-60.

Swaminath, M.H., Ravindra, D.S. and Mumtaz, J. (1990) *My Forest*, **26**(2): 207-208.

Thimmaraju, K.R., Narayana Reddy, M.A., Swamy, N. and Sulladmath, U.V. (1977) *Mysore J. Agril. Sci.*, **11**: 293-298.

Timyan, J. and Bwa, Y.O. (1996) *Southeast Consortium for International Development*, 1634, 1 Street N.W. Suite 702, Washington DC, 20006.

Toroprayoon, S., Rodtong, S., Feungchan, S., Chindaprasert, S., Yimsawat, T. and Kitpowsong, P. (1996) *Thai J. Agril. Sci.*, Special issue, **1**: 57-67.

Trimen, H. (1894) *A Handbook to the Flora of Ceylon*. Part 11. Connaraceae-Rubiaceae *Tamarindus indica* L. Dalau and Co., 377, Soho Square W. London, p. 114.

Troup, R.S. (1909) Economic Product Series, Superintendent of Government Printing, Calcutta, **1**(1): 253.

Troup, R.S. (1921) *Tamarindus indica* L. Oxford, Clarendon Press, **11**: 263-363.

Troup, R.S. (1921) *The silviculture of Indian trees*. Vol. IV. Leguminosae, pp. 231-225.

Ulrich, R. (1970) In: *The Biochemistry of Fruits and their Products*. Vol. I, (Ed. Hulme, A.C.). Academic Press, New York.

Usha, K. (1986) *M.Sc. Thesis*, University of Agricultural Sciences, Bangalore.

Usha, K. A. and Singh, B. (1994) *J. Crop Improv.*, **23**(2): 199-202.

Verheij, E.M.W. and Coronel, R.E. (1991) Handbook 2, Wageningen, The Netherlands.

VIjaya, T., Prasada Reddy, B.V., Rao, P.S. and Srivasuki, K.P. (1997) *Proceedings of National Symposium on Tamarindus indica* L., Forest Department of Andhra Pradesh, pp. 241-247.

Vijayakumar, M., Shivashankar, G. and Nachegowda, V. (1991) *Crop Res.*, **4**(1): 169-170.

Villanueva, L.B. (1920) *Philippine Agri.*, **9**: 97-110.

Vogt, K. (1995) SOS Sahel International, UK 1, Tolpuddle Street, London, NI OXT, p. 52.

Von Carlowitz, P.G. (1986) *Multipurpose Trees and Shrubs Seed Directory*. International Council for Research Agro-forestry, Nairobi, p. 265.

Von Maydell, H.J. (1986) *Rossdorf: Tz-Verlagsgesellschaft*, pp. 382-387.

Von Maydell, M.J. (1986) Deutsche. Geseuschft Fuer Techhische Zusammenarboit, Eschborn, Germany, pp. 111.

Watt, G. (1893) *Dictionary of Economic Products of India* (silk to tea), **6**(3): 404-409.

Webster, P.J. (1921) *Agri. Rev.*, **14**(3): 211-384.

Webster, P.J. (1924) *Philippine Agri. Rev.*, **17**(1): 46-51.

Wenkam, N.S. and Millar, C.D. (1965) Hawaii Agriculture Experimental Station, University of Hawaii, *Bulletin*, **135**: 1965.

Yaacob, O. and Subhadrabhandu, S. (1995) *The Production and Economic Fruits in South-East Asia*. Oxford University Press, pp. 133-142.

Yadav, S.K. (1986) *J. Trop. Forestry*, **2**(1): 53-58.

Yimsawat, T., Feungchan, S., Chindaprasert, S. and Kitpowsong. (1996) *Thai J. Agril. Sci.*, **1**: 45-47.

Yoneyama, T., Muraoka, T., Murakami, T. and Boonkerd, N. (1993) *Plant Soil*, **153**(2): 295-304.

Zablocki, L.P. (1995) US Patent No. 547791, Nutraswert Co., Deerfield IL, USA.

7

PASSIONFRUIT

C.M. Menzel, C.W. Winks, D.R. Simpson,
B. Ghosh and Ivi Chakraborty

1.0 Introduction

The passionfruit plant (*Passiflora edulis* Sims), of the estimated 500 species of *Passiflora*, belongs to the family Passifloraceae. Within this species, there are mainly three types of passionfruit under cultivation. They are the yellow passionfruit (*Passiflora edulis forma flavicarpa*) which is suited to tropical conditions or the plains, the purple passionfruit (*Passiflora edulis forma edulis*) which grows best under sub-tropical conditions or high altitudes, and the giant granadilla (*Passiflora quadrangularis*). Although several species of *Passiflora* are native to the tropics and subtropics, commercial development of this crop did not occur until some 70 to 80 years after *P. edulis* and its closely related forms were introduced into Hawaii, South Africa and Australia in the late 1800s. Since its initial development, the passionfruit industry in many countries has been noted for its instability and unpredictability. Although high yields are technically feasible in many parts of the tropical and subtropical world, there are marked annual and seasonal fluctuations in production.

The yellow passionfruit is tolerant to many of the soil borne pests and diseases that affect the purple type, and is more prolific, bearing larger, heavier fruit with more juice, which has a higher acid content than the purple type (Joy, 2010). The purple passion fruit is native of tropical America while yellow passion fruit is considered as a mutation of purple variety, or as a natural hybrid between purple and another related species of passion fruit (Strohalm *et al.*, 2007; Watson *et al.*, 2008). Purple variety is mainly grown for fresh juice and its flavour is a vital attribute which make the juice a desirable ingredient for many formulated beverages (Farid *et al.*, 2010). Purple passion fruit is available in the Nilgiris in south India and in various parts of northern and north-east India. In many areas, the vine has run wild. The yellow form was unknown in India until just a few decades ago when it is introduced from Ceylon and proved well adapted to low elevations around Chennai and Kerala. In India, the vines bear throughout the year but two peak periods are, first, August to December and second, March to May. At the latter time, the fruits are somewhat smaller, with less juice. For juice processing, the fruit is allowed to attain a deep-purple colour or pale yellow colour with glossy skin. In India and Israel the fruits are always picked from the vine rather than being allowed to fall.

The most serious limitation of existing commercial types is related to their disease susceptibility. In some countries, there are also problems with low pulp recovery and low juice yield, small fruit size and loss of good skin colour for fresh market during overcast weather. Many cultivars also displayed limited adaptation to climatic and soil extremes. Heavy yield of passionfruit can only be achieved by close attention to vine management. High establishment, growing and harvesting and marketing costs often limit grower returns in developed countries with high labour costs. Good opportunities exist for increasing productivity by breeding and selection of passionfruit cultivars better adapted to diseases and environmental stress.

A general account of passionfruit is provided by Abeysinghe (1973), Whittaker (1972), Fisch (1975), Purseglove (1974) and Knight (1980). Several authors have presented information on passionfruit growing in various countries, *e.g.*, South Africa (Kuhne, 1975), Sri Lanka (Bertin, 1976), Kenya (Lippmann, 1978), Cameroon (Haury, 1979), Hawaii (Pope, 1935; Anonymous, 1979), Zimbabwe (Gilmore, 1983), Australia (Beal and Farlow, 1984) and India (Joy, 2010; Tripathi *et al.*, 2013; Thokchom and Mandal, 2017)

2.0 Composition and Uses

2.1.0 Composition

The nutrient composition of three major commercial species of passion fruit is presented in Table 1. Passion fruits are fair to good sources of provitamin A, ascorbic acid, riboflavin and niacin and have a high mineral content. The chemistry of passion fruit juice has been described in detail by Pruthi (1963) and Chan (1980). The fruits of passion fruit contain low protein (4.6 per cent) and fat (0.6 per cent), but it is rich in fibers (35–90 per cent of dry matter) and compounds with antioxidant activity (Sousa *et al.*, 2011; Hernandez-Santos *et al.*, 2015). More than 110 kinds of chemical constituents have been isolated and identified from the *P. edulis.* Among them, flavonoids, triterpenoids, and carotenoids are the primary type (He *et al.*, 2020). Several bioactive compounds, such as quercetin, luteolin, and cyanidin 3-O-glucoside have been identified (Farid *et al.*, 2010; Ayala-Zavala *et al.*, 2011). These compounds possess high antioxidant activity, and could be used as a source of food ingredients for the development of functional foods (Contreras-Calderon *et al.*, 2011; Martinez *et al.*, 2012; Lopez-Vargas *et al.*, 2013; Correa *et al.*, 2015, 2016). The unique flavour of passion fruit is attributed to several volatile compounds which get deteriorated with the increase in temperature (Strohalm *et al.*, 2007). A cyanogenic glycoside is found in the pulp of passion fruits at all stages of development, but its content is highest in very young unripe fruits and lowest in fallen, wrinkled fruits, the level in the latter being so low that it is of no toxicological significance (Morton,1987).

The composition of passion fruits cultivated in Japan was determined by Kikutani *et al.* (1988). Three types, purple (*Passiflora edulis*), yellow (*P. edulis* f. *flavicarpa*) and the hybrid Tai Non No.1, were analysed. The purple type showed higher levels of protein, fat and ascorbic acid than the others. The yellow type was relatively rich in phosphorus, iron, sodium and potassium. The hybrid, which was best suited to commercial production showed a markedly higher content of vitamin A (370-460 IU per 100 g of edible portion) than the other types.

The chemical composition of the rind (on the dry basis) of the yellow one has moisture (16.85 per cent), crude fibre (25.7 per cent), protein (4.6 per cent) and total ash (6.8 per cent) (Pruthi, 1963). Otagaki and Matsumoto (1958) reported that crude protein content of yellow passion fruit rind varies from 4.6 to 9.2 per cent. On the other hand, Pruthi (1960) and Susheela *et al.* (1961) reported that crude protein content in purple rind vary from 12 to 15 per cent. Morton (1987) has mentioned the use of the rind in cattle and pig feed.

Table 1: Nutrient and Approximate Composition of Passion Fruit per 100 g of Edible Portions*

Composition	*Species*		
	P. edulis	*P. edulis f. flavicarpa*	*P. quadrangularis*
Moisture (per cent)	85.6	84.9	88.0
Protein (g)	9.4	0.7	0.9
Fat (g)	0.1	0.2	0.2
Carbohydrate (g)	13.6	13.1	10.1
Ash (g)	0.3	0.5	0.9
Ca (mg)	3.6	3.8	10.0
P ((mg)	12.5	24.6	22.0
Fe (mg)	0.2	0.4	0.6
Vitamin A (IU)	717	2410	70
Thiamine (mg)	trace	trace	–
Riboflavin (mg)	0.1	0.1	–
Niacin (mg)	1.5	2.2	2.7
Ascorbic acid (mg)	30	20	20

* Chan (1980).

Aside from its unique flavour, the comparatively high acid content of passionfruit juice (2.4-5.0 per cent, depending on the species) is it's most distinctive characteristic and is important in processing and formulation of products containing the fruit. The floral mixed-fruit character of the juice has been attributed to the presence of deep yellow/orange volatile compound, in particular ethyl butyrate, ethyl hexanoate, hexyl butyrate and hexyl hexanoate. These four esters make up to 95 per cent of the volatile oil with hexyl hexanoate as the principal constituent. Differences in the proportion of these substances have been given as an explanation for the difference in favour between the golden and purple passion fruit. Fully mature fruit have the highest proportion of juice, sugar, ascorbic acid and flavour compared with immature fruit in the yellow-green stage and partially mature fruit.

Yield and properties of juice from fruits of purple (*P. edulis* f. *flavicarpa* x *P. edulis*) and yellow (*P. edulis* f. *flavicarpa*) passion fruit, maypops (*P. incarnata*) and tetra-ploid passion fruit hybrids (*P. edulis* x *P. incarnata*) were compared by Senter *et al.* (1993). Weight, proportion of fruit filled, number of seeds and juice recovery were lower in tetraploid hybrids than in *P. edulis*. Acidity and soluble solids were highest in tetra-ploid hybrids and lowest in maypops. Fructose and glucose were the major sugars in purple and yellow passionfruit and sucrose was highest in tetraploid hybrids and maypops. Citric acid was the predominant acid in all samples and was highest in the yellow passionfruit (6.6 mg/g) followed by tetraploid hybrids and purple passionfruit.

2.2.0 Uses

Passionfruit is proved to have analgesic, anti-anxiety, anti-inflammatory, antispasmodic, cough suppressant, aphrodisiac, cough suppressant, central nervous system depressant, diuretic, hypotensive and sedative activities (Joy, 2010; He *et al.*, 2020). Besides, it is traditionally reported to possess anticonvulsant, antidepressant, astringent, cardiotonic (tones, balances, strengthens the heart), disinfectant, nervine (balances/calms nerves), neurasthenic (reduces nerve pain), tranquilizer and vermifuge (expels worms) activities. It may have promising and powerful effects on neurological disorders and chronic diseases such as heart disease and cancer. The Native American Indians, Aztecs and Mayas used *Passiflora* as a remedy for pains and ailments, a tradition which is still continued today. Local markets offer dried passion flowers which are used to brew a pain-killing tea. Dr. Alberto Seabra, a well-known Brazilian homeopath, is quoted as saying "The passion fruit is currently known throughout the world and is used on its own or with other products in different formulas by various laboratories and under many names. It must be recommended for all types of nervous restlessness, hysteria, neurasthenia, and in some cases, obsessive depression" (Anon., 1998)

The crop is grown for extraction of fruit juice, which is often added to other fruit juices to enhance aroma. Because of its strong acidic, aromatic and flavour characteristics, passion fruit juice is generally diluted (1:10) when consumed as nectar by itself or used as a minor constituent when combined with other fruit juice. Juice can also be preserved with sulphur dioxide in cordials with sugar and water. Rind can preserve as powder and pickle (Chakraborty *et al.*, 2011). The fruit is eaten alone or in fruit salads, sherbets, ice cream, jams, cool drinks, liquors and as concentrates (Howell, 1976; Meletti, 2011; Joy and Divya, 2016). As an edible fruit, it contains several components such as acids and sugars, nutrients, and non-nutritive phytochemicals that make passion fruit a tasteful and healthy addition to the diet. It is used for mood disorders (depression, anxiety, stress); insomnia and sleep disorders; headaches, migraines and general pain; stomach problems (colic, nervous stomach, indigestion, *etc.*) and to relieve menstrual cramps and premenstrual syndrome (PMS) and also as a sedative.

The sweet passionfruit (*Passiflora alata* Curtis.) has restricted yield and commercialization; most people do not know it and has only been consumed fresh as a table fruit because of its low acidity (Meletti, 2011). Passion rind is a source of good quality pectin that can compensate with the high cost apple pectins. Isolation and characterization of pectin with good yield, quality characteristics and gelling capacity from the peel of the passion fruit, a locally abundant and unexploited pectin source (Joy and Divya, 2016)

The leaves and juice contain passiflorin, a natural sedative, and tea prepared with leaves has a diuretic effect. It also shows purgative, sedative and anti-inflammatory properties. The seeds act as anthelminthic. It is popularly believed that the tea leaves, besides acting as a sedative, are also an effective antipyretic and help to treat skin inflammation, but these two uses are based on popular beliefs (Joy and Divya, 2016). The passion fruit seeds represent about 6 per cent to 12 per cent of the total fruit weight is considered as industrial waste. Oil can be extracted from the seeds for industrial use such as soaps and cosmetics. Passion fruit oil is yellow, has a pleasant taste and flavour, and has a high content of unsaturated fatty acids. It has a relaxing aromatic substance (passiflorin) that reduces anxiety, improves sleep, and reduces stress and fatigue. Passion fruit oil is rich in linoleic acid, which acts to aid in the restoration of the lipid layer of the skin and provides moisturizing and emollient properties that increase the texture of the skin (Joy and Divya, 2016).

3.0 Origin and Distribution

The edible commercial species of passionfruit originated on the edges of South American rainforests in the Amazon region of Brazil and possibly in Paraguay and northern Argentina. The purple passionfruit (*P. edulis*) is adapted to the cooler subtropics or at high altitudes in the tropics, while the golden passionfruit (*P. edulis* f. *flavicarpa*) is more suited to tropical lowland conditions. The two forms of passionfruit hybridize readily and produce fertile seedlings intermediate in appearance between the parents.

Passion fruit became naturalized in most of the tropical and subtropical world, reaching South Africa, Hawaii, California and Florida by the end of the 19th century and Kenya, Sri Lanka and Fiji by the middle of the 20th Century. It was also taken to Europe by the 17th century where it was grown under glass for its attractive flowers.

4.0 Botany, Species and Cultivars

4.1.0 Botany

The family Passifloraceae includes 550 species in 12 genera and is represented by more fruiting species than probably any other plant family. The most important genus *Passiflora* has about 400 species which are mostly native to tropical America, but include about 40 in Asia, Australia and the South Pacific and one in Madagascar.

Commercial *Passiflora* species are vigorous woody perennial vines with medium to large leaves (8-30 cm long) with toothed margins and usually bearing glands on the petioles (Bailey, 1949; Purseglove, 1974). The bisexual flower is borne solitary at leaf nodes of new growth and is about 4-5 cm in diameter and very conspicuous, colourful and fragrant. It has five whitish sepals and petals and two purplish rows of thread-like rays (corona). The pistil consists of an ovary and tripartite style, each branch terminating in a sticky stigma. Passionfruit species have a protandrous habit with pollen from the anthers ripe before the stigmas are receptive. The parts of the flower supposedly represent the implements of the crucifixion. The fruit is round or oval, about 4-6 cm long and greenish yellow or purple when ripe and smooth. It consists of a thin, shell-like rind within which are numerous small black or brown seeds, each enclosed in a yellow or orange aromatic juicy pulp which has an acidic but pleasant flavour.

4.2.0 Species and Cultivars

The classification of *Passiflora* has been studied in detail by Killip (1938) and their commercial significance reviewed by Knight (1969), Morton (1967), Martin and

Nakasone (1970), Howell (1976) and Fouque (1982). New Species are occasionally added to the genus *Passiflora* (Escobar, 1986).

There are about 600 known species of *Passiflora* now found worldwide (Joy, 2010). Out of which 60 species of *Passiflora* which bear edible fruit, only about ten are well-known outside limited areas where they grow wild or are cultivated commercially. In most countries, passion fruit production is based on cultivars of the yellow or golden passion fruit (*P. edulis* f. *flavicarpa*). The major exceptions are South Africa, Kenya and New Zealand where production is based on lines of the purple passion fruit (*P. edulis*) and in Australia where hybrids between the two forms are exploited. Another type, the giant granadilla (*Passiflora quadrangularis*) is also cultivated rarely. The characteristics of the three types under cultivation are given in Table 2.

Table 2: Characteristics of Three *Passiflora* types under Cultivation

Species	*Common Name*	*Characteristics*
Passiflora edulis Sims f. *edulis* Degener	Purple Passion Fruit/Mountain Sweet Cup	Round to ovoid fruit (5-8 cm long and 4-8 cm diameter), purple in color with tough, waxy smooth rind and orange yellow colored pulpy juice with pleasant and sub-acidic flavored arils.
Passiflora edulis Sims f. *flavicarpa* Degener	Yellow Passion Fruit/Golden passion fruit	Round to ovoid fruit, larger in size (8-10 cm long and 4-10cm diameter) with smooth, glossy, light and airy thick (3-4mm) rind of yellow to light orange pulp having highly aromatic and acidic juice.
Passiflora quadrangularis Medic	Giant Granadilla	Oblong-ovoid fruit of very large size (20-30 cm long and 12-15 cm diameter) thick skin, greenish-white to pale yellow color, whitish to yellowish sweet acid arils having mild flavor.

The origin, world distribution and importance, botany, fruit composition, climatic and soil requirements, cultivars, and breeding and selection of *Passiflora* spp. were reported by Winks *et al.* (1988). The characteristics of tropical American species used as fruiting plants [*P. quadrangularis*, *P. ligularis*, *P. laurifolia*, *P. maliformis*, *P. mollissima* and *P. antiquiensis* (*P. antioquensis*)] or as sources of cold/disease tolerance in breeding programmes (*P. incarnata*, *P. caerulea*), minor tropical species cultivated for their edible fruits (*P. alata*, *P. coccinea*, *P. mixta*, *P. popenovii*, *P. semanii*, *P. serratodigitata*), the fruit composition of *P. edulis* and *P. quadrangularis*. Mac Dougal (1989) described two new species *Passiflora nubicola* and *P. gilbertiana*. *P. nubicola* is closely related to *P. biflora* and shows close morphological similarities. Medagoda and Balasubramaniam (1988) collected forty samples of the genus *Passiflora* from several localities in Sri Lanka and keys were constructed for the identification of species and cultivars based on morphological characteristics. The distribution of cyanogenic glycosides in the leaves, stems, tendrils, floral parts, roots and fruits of the widely cultivated *P. edulis* f. *flavicarpa* was studied. The lowest levels were found in *P. edulis* f. *edulis* (8 ppm HCN) and the highest levels in *P. qua-drangularis* (138 ppm HCN).

Several other tropical American *Passiflora* species have been naturalized in many countries and are grown commercially for fruit production or used as rootstocks

Uses of Passionfruit

Salad (Left), Icecream (Middle) and Concentrate (Right) from Passionfruit

Juice (Left, Middle) and Squash (Right) from Passionfruit

Jam (Left), Nectar (Middle), and Syrup (Right) from Passionfruit

Species of Passionfruit

Passiflora quadrangularis

Passiflora ligularis

Passiflora lauriflora

Passiflora maliformis

Species of Passionfruit

Passiflora mollissima

Passiflora antiquensis

Passiflora incarnata

Passiflora caerulea

Species of Passionfruit

Passiflora alata

Passiflora coccinea

Passiflora mixta

Passiflora popenovii

Species of Passionfruit

Passiflora semanii

Passiflora serratodigitata

Passiflora nubicola

Passiflora gilbertiana

Species of Passionfruit

Passiflora ambigua

Passiflora cincinnata

Passiflora mucronata

Passiflora nigradeniya

Species of Passionfruit

Passiflora riparia

Passiflora serratifolia

Passiflora flavicarpa

Cultivars of Passionfruit

Yellow and Purple Passionfruit

Giant Passionfruit

Kaveri Hybrid

Giant Passionfruit

Banana Passionfruit

or as source of resistance to unfavourable environmental conditions (Tables 2 and 3). Other species which deserve evaluation include: *P. ambigua, P. cincinnata, P. quazumaefolia, P. mucronata, P. nigradenia, P. nitida, P. riparia and P. serratifolia.*

The major commercial cultivars in Australia are Purple Gold, Lacey, E-23 and Black Beauty.

In India no well-established cultivars are regularly available, so plants are customarily grown from seed or cuttings of vines selected for desirable characteristics (Joy, 2010). A form of yellow passion fruit that sets fruit abundantly from self-pollination is often grown from cuttings or seeds. Its fruit is usually smaller than that of other cross-pollinated yellow passion fruits. Seeds of large fruited selections of yellow passion fruit were brought from Hawaii and plants from this source are grown. Locally selected purple passion fruit is vegetatively propagated for commercial production. A hybrid 'Kaveri' has been developed from Central Horticultural Experiment Station, Chettalli, Karnataka, India and it is popular in south India (Kishore *et al.*, 2006). This was a cross between purple and yellow varieties. Each plant bears 40-60 fruits per annum. Fruits ovoid to round and purple dotted. Fruits contain 25-30 per cent juice, 11.5-12.0 percent sugars and 3.0-3.5mg citric acid/100ml. Juice. Plants are tolerant to brown leaf spot, wilt, and collar rot, root-knot nematode and thrips. On an average 25 tonnes of fruits/ha can be harvested.

5.0 Breeding and Improvement

As a group, the cultivated species of *Passiflora*, grows in climates ranging from tropical lowland to the elevated subtropics. Many species, including the cultivated ones, naturalize easily outside their area of origin.

The basic chromosome number is normally 3, 6, 9 or 12, mostly 8 (Beal and Farlow, 1984). The inheritance of new characteristics has been studied in the golden and purple passion fruit (Nakasone *et al.*, 1967). The purple form is susceptible to *Fusarium* wilt, the yellow resistant. A single gene appears to be involved. Fruit skin colour appears to be controlled by a single pair of genes with no dominance as three colour types were found in the F_2 progeny.

Most of the industries around the world are dependent on seedling lines of the yellow or purple passion fruit. A large-fruited yellow passion fruit was selected in Hawaii for its unusually bright orange coloured juice and tolerance to *Alternaria passiflorae* spot and was released as Noel's Special (Ito, 1978). Like most other yellow forms, it requires interplanting with other cultivars for adequate pollination.

In South Africa, the blue passion fruit (*P. caerulea*) is used as a rootstock for the purple passion fruit (*P. edulis*). It is resistant against *Phytophthora* stem rot and *Fusarium* wilt and displays high tolerance to nematodes, water-logging and chilling injury. Purple scions grafted onto *P. caerulea* out yielded those grafted onto the golden passion fruit and purple seedling by 41-74 per cent (Anon., 1984, 1986a,b).

Many of the species in the genus *Passiflora* can be crossed. At least 13 species, including nine with edible fruits have been used in interspecific hybridization. Most breeding has been with yellow and purple forms of *P. edulis* and rootstocks, such as

P. caerulea and *P. incarnata*. A major aim has been to overcome the susceptibility to numerous diseases. Another has been to improve cropping regularity and quality attributes.

An outline of cultivar improvement in Australia since the 1950s is given below as an example of such breeding efforts.

The Australian passion fruit industry is totally dependent on hybrids derived from a breeding programme conducted at Redlands Horticultural Research Station (RHRS) in the late 1950s. Various genotypes from that programme have achieved prominance in the industry including selection 3-1 (Redlands Triangular), 3-26 (Redlands Pink), E-23, Purple Gold (Days Special), Lacey and Black Beauty.

Because of susceptibility to passion fruit woodiness virus (PWV) and especially the severe or necrotic strain (PWV-N), cultivar 3-1 and 3-26 are no longer used. The main cultivars currently used are E-23, Purple Gold and Lacey. Of these, E-23 is the predominant cultivar with Purple Gold, Lacey and Black Beauty used to a lesser degree.

The significant feature of all selections from the RHRS breeding project was their considerable advantage in yield over the purple passion fruit (*P. edulis*) on which the industry was then based. Yield of hybrids was up to five times higher than the purple passion fruit due to their inherent cropping characteristics.

The most serious limitations of existing commercial hybrid cultivars are related to their disease susceptibility. Passion fruit woodiness virus (PWV) has for many years been the most serious disease problem and with the development of the necrotic strain (PWV-N), severe debilitation of vines is common after slow growth periods in the winter months or where cultural management is inadequate.

Because hybrids are vegetatively propagated, there is no opportunity to break the virus cycle. Methods to produce virus-free material have not been found. As the purple passion fruit was true-to-type from seed, the practice of grafting seedling of *P. edulis* onto *P. edulis* f. *flavicarpa* seedlings for commercial production provided virus-free plants. Although *P. edulis* was highly susceptible, commercial production was possible through this break in the virus cycle and only gradual infection in plantations. Chemical control of virus vectors was used but at best only slowed the rate of infection.

Virus-free plants appear a desirable objective in improvement work, either by elimi- nation of virus from existing cultivars if methods can be developed or through new breeding work. However, maintaining virus-free plants during progeny evaluation is not a practical reality.

In recent year, *Alternata* spot (*Alternaria alternata*) has caused serious losses and downgrading of fruit. This disease, either through increased virulence or through conditions which favour its development has posed a more serious problem than woodiness virus. Whereas other fungal diseases have been amenable to control with fungicides, alternata spot has proved a most difficult problem even with more intensive fungicide strategies.

In recent year, processors have experienced problems with low pulp content and low juice yield. This problem is probably related to decline in vine health through increased disease pressure rather than any change in the genotype. Inadequate pollination is another possible factor. Another problem in the cultivar Purple Gold is loss of fruit size of mature vines. Once again this is probably related to pollination or possibly vine vigour associated with nitrogen and potassium nutrition and disease incidence.

A second species, the sweet passion fruit (*P. alata*), is native to the Brazilian plateau and the eastern Amazon region, but is cultivated as a low-intensity crop only in the South and Southeast of Brazil. It is appreciated for its typical aroma and flavor characteristics and can therefore command up to triple the price of the yellow passion fruit at local markets. *P. alata* is a semi-woody perennial climbing vine that produces large attractive, hermaphrodite flowers. In common with *P. edulis*, it is a diploid (2n = 18), out-crossing, self-incompatible species (Bruckner *et al.*, 1993; Suassuna *et al.*, 2003; Ferreira *et al.*, 2010). A recent Brazilian survey on agricultural production showed low productivity of the crop (IBGE, 2017). This is because fruit production is not stable and there is lack of improved varieties that meet the needs of both producers and consumers in terms of quality and yield. In 2017, the Brazilian Agricultural Research Corporation released the first sweet passion fruit cultivar for cropping mostly in the central region of the country for which it was developed.

An evaluation of genotypes belonging to two species of passionfruit was carried out in Brazil to identify high yielding and better fruit quality type (Weber *et al.*, 2017). The genotypes BRS Sol Cerrado, BRS Rubi do Cerrado and Catarina belonging to *Passiflora edulis* have the higher productivity, fruit mass, fruit length, fruit diameter and pulp yield in relation the genotype Urussanga (*Passiflora alata*). The genotype Urussanga has the highest pericarp thickness and soluble solid content, besides the lowest acidity.

Castillo *et al.* (2020) evaluated the agronomic and morphological characteristics of 50 purple passionfruit genotypes including commercial cultivars (8 genotypes), genebank accessions (8), and landraces (34) across two different elevations and agro-ecological sites as a base for germplasm enhancement in the highlands of Colombia. Results indicated that the mid-elevation site produced higher yields (fruit/plant) than the high elevation site, although some landraces were highly productive there. Commercial cultivar and genebank accessions clustered together in a principal component analysis (PCA); while landraces showed high levels of variation in the trait descriptors with five different clusters. Therefore, landraces of purple passion fruit contained greater genetic diversity than commercial cultivars or the genebank, and breeding programs for the crop should use landraces to increase diversity of varieties available to producers and to further expand the crop to new regions, at higher elevations, or with different agro-ecologies.

The research work of development of new passion fruit varieties is in progress at CHES Chettalli (IIHR), IIHR, Bangalore; Banana Research Station, Kerala, India (Tripathi *et al.*, 2013). The performance of some of the elite lines at CHES Chettalli revealed that the fruit yield per vine was highest in PF-7 (10.94 kg) followed by Kaveri (10.24 kg), PF-3(9.07kg) and PF-2 (8.14 kg).The fruit weight was highest in

Kaveri (101.4 g) followed by PF-2(99.5 g) and PF-7 (99.1 g). Fruit girth was highest in PF-7 (7.08 cm) followed by Kaveri (6.35 cm) and PF-3(6.15cm). Fruit length was highest in PF-7 (6.73 cm) followed by Kaveri (6.59 cm), Few accessions were found promising. PF-3 is a seedling selection of Coorg yellow. The fruits are Yellow colour fruit. Average fruit weight is 95 g with 17.5 °Brix TSS and higher yield (9.07 kg per vine). PF-7 is a seedling selection of Thrissur yellow. The fruits are Yellow colour fruit. Average fruit weight is 99 g with 17.2 °Brix TSS and higher yield (10.9 kg per vine).

In passion fruit, like other fruit crops, the main purpose of breeding is to meet the demand for quality (da S Pereira *et al.*, 2017; Cavalcante *et al.*, 2017; Viana *et al.*, 2017). According to Jung *et al.* (2007), sweet passion fruits should weigh 200 to 300 g, be oval in shape, free from apical softening, with a firm skin, rich pulp yield (over 30 per cent), high sugar content and significant yield. However, all these goals are not easily achieved in a few selection cycles, requiring a medium-term program that takes into account correlations between these traits. Breeding for yield and fruit quality traits in passion fruits is complex due to the polygenic nature of these traits and the existence of genetic correlations among them (Chavarrýa-Perez *et al.*, 2020). Therefore, studies focused on crop management practices and breeding using modern quantitative genetic approaches are still needed, especially for *Passiflora alata*.

Although some work has been done on genetic and phenotypic analysis (da S Pereira *et al.*, 2017; de Jesus *et al.*, 2018), studies focused on crop management practices and breeding using modern quantitative genetic approaches are still needed. In this context, understanding the genotype by environment interaction (GEI) has been one of the most important challenges faced by plant breeders, affecting every aspect of the decision-making process in breeding programs. In this context, Chavarria-Perez *et al.* (2020) were able to demonstrate that the use of selection indices (weight of skin, thickness of skin and weight of pulp) could help reduce the magnitudes of correlations between desirable and undesirable traits. These indices allowed the selection of six promising genotypes that were also mostly cross compatible and therefore could be used commercially or for continuing the breeding program. Their results provide a comprehensible view of the genotype by environment interaction and helped to interpret how the sweet passion fruit genotype performs across environments. This research group also conducted breeding programs using modern quantitative genetics to generate information and select genotypes that can lead to the development of varieties with improved fruit quality and yield. To do this, a segregating population (F_1) was developed by crossing two outbred, divergent accessions of *P. alata*. In parallel, this full-sib family (n = 180) was genotyped using molecular markers (Penha *et al.*, 2013) and used to construct a unique linkage map for the species (Pereira *et al.*, 2013). One hundred individuals were then sampled and field-evaluated, and QTL mapping analysis performed to identify loci associated with fruit quality traits. The MSI was also used to select the 30 most promising genotypes (da S Pereira *et al.*, 2017).

Breeding work is also ongoing in Brazil, mostly for yellow passion fruit (Fagne *et al.*, 2014; Junqueira *et al.*, 2006, 2013; Lenza *et al.*, 2009; Moreira, 2009; Roncatto *et al.*, 2014) and China and Kenya for purple passion fruit (Liu *et al.*, 2017; Matheri *et al.*, 2016) with an emphasis on creating hybrids, improving fruit characteristics or developing varieties that resist a number of diseases and pests.

5.1.0 Scion Breeding

The primary aim of the current plant improvement programme in passion fruit is to overcome the passion fruit woodiness virus and *Alternaria alternata* problem in current cultivars. In the breeding strategy there is also the opportunity to select for numerous other characteristics desirable to the industry including cold tolerance and resistance to septoria spot (*Septoria passiflorae*) and brown spot (*Alternaria passiflorae*).

All the commercial hybrid cultivars were derived from crosses between the purple passionfruit and the GP50 strain of the golden passionfruit which has since been lost. Attempts to repeat this success with other strains of golden passionfruit have not been successful.

Self pollinating E-23 was thought to be a useful approach but did not produce useful types. Back-crossing to *P. edulis* was also tried, with some success, but even the best selections did not achieve commercial status. In light of this, the intercrossing of commercial hybrids to recombine the genetic variability that exists in these genotypes appears the most likely source of superior genotypes which combine all the desirable characteristics required in a commercial cultivar. At a later time interspecific hybri-dization was attempted to incorporate greater variability into commercial passionfruit. Greater cold tolerance and resistance to passionfruit woodiness virus were twoobvious opportunities, as well as disease resistance.

By the mid 1970s, passion fruit woodiness virus had become severe in commercial plantings of cultivar 3-1 which ultimately lead to the demise of this cultivar. Subsequent work by plant pathologists identified the emergence of a more severe strain (PWV-N).

Interspecific hybrids between a *P. edulis* hybrid and *P. incarnata* showed vigorous growth and healthy foliage at the end of winter when all other vines were affected by cold and PWV-N. Tests of presence of virus showed that PWV was present indicating a high degree of tolerance. Breeding work with this material has been continued and improvement towards commercial fruit types has been achieved in plants retaining cold and virus tolerance.

Breeding work by the New South Wales Department of Agriculture, using the disease resistant cultivar Tom's Special has shown a high proportion of useful types in the F_1 progeny when crossed with E-23, 3-1 and Lacey. It is hoped that one of these elite progenies will combine resistance to alternata spot with good agronomic performance.

Current scion breeding work at Maroochy Horticultural Research Station (MHRS) includes hybridization between current commercial cultivars, further breeding with *P. edulis* x *P. incarnata* selections and identification of sources of resistance to *Alternaria alternata* for use in new breeding work.

5.2.0 Rootstock Breeding

The industry currently depends on seedlings of *P. edulis* f. *flavicarpa* as rootstocks in commercial production. This rootstock was selected by Grozman and Purss (1958) for its resistance to passion fruit wilt caused by *Fusarium oxysporum*. It has the added advantage of resistance to phytophthora blight and nematodes but is the least cold tolerant of *P. edulis* forms and is very susceptible to PWV-N.

Sources of cold and virus tolerance were identified in the interspecific hybrid between *P. edulis* and *P. incarnata* and incorporation of these features in a new roostock was seen as a useful objective in rootstock breeding. Several selections from the F_1, F_2 and F_3 or progeny from breeding work with this material were tested as seedling rootstocks. The best selection (3-79/F_3) showed vigorous seedling growth, reached graftable size earlier than *P. edulis* f. *flavicarpa* and showed vigorous early growth in the field. However, in trial it performed only slightly better than *flavicarpa* with high variability due to its early generation hybrid status.

Because fusarium wilt resistance is essential in passionfruit rootstocks, the level of resistance in a number of selections was tested against *P. edulis* f. *flavicarpa* and highly susceptible *P. edulis*. The level of resistance in selections varied from resistant to very susceptible with selection 3-19 showing susceptibility midway between the two. For this and other reasons more breeding work is indicated before a commercial rootstock can be released to industry.

Other sources of cold tolerance have been identified in *P. caerulea* which is currently used in South Africa because of poor performance of the golden passionfruit in this industry. Testing of seedlings lines of *P. caerulea* introduced from South Africa has given poor results in Queensland and other lines of *P. caerulea* have been obtained. This rootstock has been tested previously with purple passionfruit but was discarded because of suckering.

New accessions *P. incarnata* have been established for further use in breeding work. Current rootstock breeding work at MHRS includes hybridization of *P. incarnata* with *P. edulis* f. *flavicarpa* and F_3 selection 3-19, and attempts to hybridize *P. edulis* f. *flavicarpa* and *P. caerulea*.

6.0 Propagation

Passionfruit is propagated by seeds, cuttings or grafting. In Hawaii, Brazil, Sri Lanka, Kenya and New Zealand, commercial production is based on seedlings of the yellow or the purple passionfruit. In contrast, in Australia and South Africa where the industry exploits hybrids between the two types of passionfruit or where particular rootstock characteristics are favoured, the main form of propagation is grafting. Protocol for micropropagation of hybrids and species of passionfruit *in vitro* has been standardized.

6.1.0 Seed Propagation

Seed should preferably be extracted from freshly harvested fruit. The seeds are removed from the pulp and juice after fermentation for about 3 days, thoroughly washed, dried and kept in a sealed container in a refrigerator. Such seeds produce germination better than 85 per cent after a period of 3 months.

Nakagawa *et al.* (1991) stored passionfruit seeds for 57 months in (i) paper bags in ambient condition in a laboratory, (ii) paper bags held in a cool chamber or (iii) polyethylene bags also held in a cool chamber. Germination was 60 and 50 per cent for seeds in treatments (iii) and (ii), respectively, at the end of storage. In treatment (i), the percentage germination was zero after 32 months.

Seeds may be sown in flats or pots of sterile soil and kept in a moist place shaded from direct sunlight. Seeds ordinarily germinate in 10-20 days and young plants grow rapidly. Seedlings should be potted individually in small containers as soon as practical after germination. They can be transferred to a permanent location when they are at the 2- to 3-leaf stage. Dehydrated seeds of many *Passiflora* species may require from many months up to two years to germinate. *Passiflora* seed coats are very tough, in nature they are softened by the stomach acids of birds and other animals. This can be mimicked by placing them for 24 hours in milk or citrus juice. Rinse the seeds and plant them immediately in an airy general-purpose seedling soil. Cover them with 5 mm of the same medium and gently press it down a bit, put them in a bright spot at 20°C. Use a hand sprayer to keep the substrate moist constantly. Transplant them to individual pots after they formed a few leaves. Plantlets are grown under shaded conditions and hardened before being transplanted into the field. It takes 6 to 8 weeks from the time of sowing to transplanting in the field.

6.2.0 Vegetative Propagation

6.2.1 Cutting

Passionfruits are also propagated by cuttings. Cuttings from young, newly mature wood with 2-3 internodes may be rooted in about one month and ready for setting out in 90 days. The most desirable cutting material is that portion of the stem from the first fully expanded mature leaf, back to the area of the fully extended branch. The best period to obtain cutting material is when the vines are actively growing, after the summer and winter crops. Cuttings are taken from vigorous, actively growing vines. Start cuttings from woody sections of the plant. A 2-foot (60cm) branch yields 4 cuttings which have 2 growing points each. Cut the branch diagonally 2 cm above the upper and 3 cm below the bottom growth point. Remove the tendril and leaves from the bottom shoot by swiftly rubbing your fingers down the branch. The upper shoot remains intact but cut away half of the leaf to avoid dehydration. Put the bottom of the cutting in cold water and let them drip off in the fridge, then dab them in rooting hormone powder (eg. IBA). Tap of the excess powder and stick them 2/3 in a sandy airy mix. Place them in a bright spot in a humid environment (under clear plastic or in a mini-conservatory), not in direct sunlight. Roots will take about 4 weeks to appear, don't interfere with them. Water the pot from a tray - watering from the top will collapse the airy soil. Then it should be treated similar to seedlings. Plants from rooted cuttings are less vigorous than seedlings and could be planted at a closer spacing than seedlings.

Gonzalez *et al.* (1989) reported rooting up to 99.89 per cent in 20-30 cm long cuttings with 2-3 nodes/cutting, placed in medium in black polyethylene bags, after 7 days from planting.

6.2.2 Grafting

Purple passion fruit is sometimes grafted onto a yellow passion fruit rootstock to alleviate nematode and disease problems affecting the root system of purple passion fruit. Seedlings of both stock and scion should be about 30-40 cm tall and have a stem diameter about that of a pencil thickness when grafted. Scions should be about 8-10 cm long and contain at least 2 nodes. The stock should be cut off 25-30 cm above the soil line. For grafting, a long, slanting cut is made from one side to the other through the base of the scion for about half its length, and a similar cut is made through the stem of the stock. The two cut surfaces are then placed together with cambia aligned and the graft is tied firmly with budding tape. The graft is enclosed in a small plastic bag tied shut below the graft, and placed in a warm, shady location for 10-14 days or until the union takes. Then the bag is loosened to admit air and is removed when scion buds begin to grow. The budding tape is removed before it thickens the growing stem. Grafted plants are ready for transplanting into the field after the graft has callused which normally takes from 4-5 weeks.

In a greenhouse experiment in Bahia, Brazil, *Passiflora edulis* f. *flavicarpa* was grafted (using side wedge grafting or cleft grafting) on *P. alata, P. cincinnata, P. foetida, P. caerulea, P. gibertior P. edulis* f. *flavicarpa* (Lima *et al.*, 1999). Grafting success was highest on *P. cincinnata* and *P. caerulea* (52.6 and 49.4 per cent, respectively after 120 days); no differences between grafting methods were observed.

Passiflora alata, P. giberti, P. edulis f. *flavicarpa, P. caerulea, P. nitida* and *Passiflora* sp. (snake passionfruit) were evaluated as rootstocks for *P. edulis* f. *Flavicarpa* using cleft grafts enclosed in plastic bags (Menezes *et al.*, 1994). With the exception of *P. nitida*, all species, and especially *P. caerulea* which was most compatible, were promising as rootstocks.

7.0 Area and Production

Passionfruit is grown in most tropical and subtropical parts of the world, and it is particularly important commercially in Australia, Hawaii, South Africa and Brazil. Brazil is the world's foremost producer of passion-fruit, with about 90 per cent of the production, followed by Peru, Venezuela, South Africa, Sri Lanka and Australia. Brazilian production is around 478,000 t with a yield of about 13.8 t/ha. The northern and north-eastern regions of the country are responsible for more than 80 per cent of the national production. Brazil, Venezuela, South Africa, Sri Lanka, Australia, Papua New Guinea, Fiji, Hawaii, Taiwan and Kenya account for 80-90 per cent of the world's passionfruit production (Table 3). Production is expanding in Brazil, Fiji, South Africa, Kenya and Sri Lanka, but declining in Papua New Guinea and Hawaii. Other countries where passionfruits are grown commercially are New Zealand, West Samoa, Cook Islands, Malaysia, Israel, Congo, Angola, Ivory Coast, Cameroon, Peru, Colombia and West Indies. Commercial culture in these countries was initially developed to meet domestic requirement, but it is

Propagation of Passionfruit

Plants Obtained from Passionfruit Seeds

Cutting

Cleft Grafting

Tissue Culture from Mature Endosperm (Left) and Leaf Disc (Right)

now estimated that more than half of the world production of passionfruit juice is exported. The principal market is Europe, which imports more than 90 per cent of the juice. However, there are very good prospects in the American, Canadian and Japanese markets. India, for many years, has enjoyed a moderate harvest of purple passion fruit in the Nilgiris in the South and in various parts of northern India. In many areas, the vine has run wild. The yellow form was unknown in India until just a few decades ago when it was introduced from Sri Lanka and proved well adapted to low elevations around Chennai and Kerala. It was quickly approved as having a more pronounced flavor than the purple and producing within a year of planting heavier and more regular crops. The passionfruit vine is cultivated predominantly in small orchards, on average 1.0 to 4.0 ha, and is an important source of income for small to medium producers.

Table 3: Estimated Area under Passionfruit (Hectares) for the Major World Producers*

Australia	320
Papua New Guinea	28
Fiji	709
New Zealand	80
Sri Lanka	800
South Africa	150
Kenya	166
Hawaii	100
Brazil	20,000
Venezuela	100
Peru	1,000
Ecuador	600

* Morton (1967); Abeysinghe (1973); Fisch (1975); Crestani *et al.* (1986); Fajac (1986, 1987).

In 1999, Maldonado reported that Brazil is the main world producers of passionfruits, with over 30000 ha planted, but average yields are low (12000 kg/ha).

8.0 Soil and Climate

8.1.0 Soil

Passionfruit grows successfully in a specific range of soil, but light to heavy sandy loams are the most suitable. Passion fruit has a superficial root system (60 per cent of the roots located within 30 cm of the surface). So it is important that there is no impedance to root growth in the top 60 cm. In general, it is recommended that the soil should be deep, relatively fertile and well drained. Poorly permeable soils with high clay content, subject to flooding, are not recommended. The best soils are sandy clay. For good growth, it is recommended that the soils have neither impermeable, rocky or hardened layers, nor a water table at less than 2 m to avoid the appearance of dry rot. They survive temporary water-logging but die after week of immersion. They have poor tolerance to extremes of pH (optimum of 5.5 to 7.5) and to salinity. In an experiment in pots containing naturally sodic (ESP of 48.0) and artificially sodic (ESP of 12-60) soils for 12 months, Valia *et al.* (1994) observed that there was a 50 per cent reduction in whole plant FW at an ESP levels corresponding to 20, 50 and 65 per cent growth depression were 7.8, 52.0 and 55.8, respectively. The Ca:Na and (Ca + Mg + K): Na ratios showed a marked decrease as ESP increased from 3.82 (in the control) to 12.

They grow on infertile soils, although the yield is reduced without fertilizers. They do best in deep (>60 cm) fertile well-drained loamy soils. On shallow or heavy textured soils they are best planted on ridges.

8.2.0 Climate

Passionfruits are adapted to tropical and subtropical areas with high rainfall. Temperature, relative humidity, light intensity and precipitation have an important influence on the longevity and the yield of the plants, but also favour the incidence of pests and diseases. The purple passion fruit crops do best at higher altitudes above 2000 m in the tropics, while the golden passion fruit and hybrids between the two forms are superior in the lowlands.

Passionfruits are sensitive to frost and are killed or severely injured by prolonged temperatures below freezing, but can withstand light frost. They may survive short periods below -2°C. Young vines and actively growing shoots are more susceptible than mature hardened plants. Species in increasing chilling tolerance are: *P. edulis* f. *flavicarpa*, *P. incarnata*, *P. edulis* and *P. caerulea* (Patterson *et al.*, 1976, 1978). In cooler sites, it is desirable to select the north and east facing slopes (southern hemisphere) to increase temperatures in the orchard. At the present state of knowledge, it is suggested that commercial passionfruit sites should be free of frost.

The temperature range between 21 and 25°C is considered as the most favourable for the growth of the plant, being best between 23 and 25°C, but passionfruit is being successfully cultivated in temperatures between 18°C and 35°C. Lower temperatures slow the growth of the plant and reduce the uptake of nutrients and fruit production, while very high or very low temperatures affect fruit bearing. At intermediate temperatures of 23°C to 28°C, the fruit growth period is 60 days, when the temperatures were lower (23°C) and higher (33°C) the period was 75 days. The germination period of the seeds is shorter in summer time than in the coldest months, when the period is longer.

Passionfruits are limited in their adaptation to extremes of temperatures. Temperatures below 15-18°C restrict vegetative growth and flowering, while temperatures above 30-32°C promote growth at the expense of flowers and fruits. Low temperatures can also reduce pollination, there being no pollen germination below 20°C. Heaviest yield is obtained between 20 and 30°C, a compromise between flowering and excessive vegetative growth.

In a temperature experiment, Menzel *et al.* (1994) determined the 2 critical root zone temperatures and reported 90 per cent of maximum growth at about 20 and 35°C for vine extension, leaf area, node and leaf production, and 20 and 30°C for

Table 4: Tropical American *Passiflora* Species which are Utilised as Fruiting Plants or as Sources of Cold/Disease Tolerance in Breeding Programmes*

Species	*Origin*	*Description of Fruit*	*Environmental Adaptation*	*Cropping Ability*	*Other Uses*	*Areas of Cultivation or Importance*
P. quadrangularis (giant granadilla)	Peru/Brazil	Very large (20-23 cm long), green-yellow skin, oblong shape, white pink flesh with central cavity filled with purple/pink delicately flavoured pulp	Warm wet tropical lowland (>13°C)	Heavy cropping	-	Cultivated throughout tropical world, increased northern Queensland
P. lingularis (sweet granadilla)	Venezuela/ Brazil	Large (7-8 cm diam.), orange/brown to yellow/ tan when ripe, hard shell, ovoid shape, white edible aromatic pulp	Tropical highland (1000-3000 m)	Very vigorous, fair cropping	Ornamental (large attractive flowers)	Cultivated in Mexico, California and central tropical America
P. laurifolia (yellow granadilla or bell-apple)	West Indies/ north-eastern South America	Large (7-8 cm long), ellipsoidal shape, yellow, white spotted, skin, edible white pulp with flavour or pear	Qarm wet tropical lowland	Very vigorous poor cropping (shy flowering self-incompa- tible)	—	Grown in West Indies, coastal Venezuela, Peruvian Amazon, Brazil, Asia and Ceylon
P. maliformis (sweet calabash)	Tropical South America	Medium (5 cm diam.), oval shape, hard green-orange shell, white sweet pulp with flavour of grapes	Warm wet tropical lowland	Fair cropping (flowers self- incompatible), fast-growing, susceptible to	—	Cultivated in many tropical South American countries
P. mollissima (banana passion- fruit)	Andes of tropical South America	Large (7 cm long), ellipsoidal, yellow skin, orange/salmon edible pulp aromatic and sub-acid to sour	Suited to 2000-3000 m in tropics	Fair cropping	Attractive flower	Cultivated in east Columbia, S.E. Peru, W. Bolivia and W.
P. antiquiensis	Columbia	Large (7 cm long) ellipsoidal, yellow skin (pubescent) similar in taste to *P. mollissima*	Fruits at high (>2000 m) in tropics	Fair cropping	Attractive flower	Grown at high altitudes tropical South America, edible fruit
P. incarnata (vine apricot or wild passionfruit)	Subtropical and tropical N. America/ Bermuda	Medium (5 cm diam.) ovoid shape, yellow skin, pulp similar to taste to tart apricot	Suited to warm subtropics, strong perennial vine roots	—	Useful as a source of cold tolerance in breeding programme	Cultivated as a garden specimen in southern USA
P. caerulea (blue passionfruit)	Brazil	Medium (5 cm long), ovate to subglobose shape, yellow skin	Suited to cooler tropics and subtropics	—	Very cold hardy, useful as rootstock, (hybri- dizes readily)	Used as a rootstock in South Africa

Table 5: Some Minor Tropical *Passiflora* Species Cultivated for their Edible Fruit*

Species	*Origin*	*Description of Fruit*	*Environmental Adaptation*	*Other luses*	*Commercial Importance*
P. alata	Peru/Brazil	Large (8-10 cm long), obovoid shape, thin yellow skin, edible pulp similar to *P. quadrangularis*	Tropical lowland	Good source of germplasm for *P. quadrangularis*	Cultivated widely
P. coccinea	N.E. South America	Medium (5 cm long), ovoid shape, stripped and mottled yellow skin, brittle	—	—	Occasionally cultivated
P. mixta	Tropical America	Medium (4-6 cm long), ovoid shape, green to green-yellow skin, pulp similar to *P.* mollissima	Tropical highland	Hybridizes readily	Often cultivated
P. popenovii	Peru/Brazil	Medium (7 cm long), ovoid shape, thin yellow skin, delicately flavoured pulp (sometimes candied)	—	—	Cultivated commercially in Ecuador
P. semanni	Panama to Colombia	Small (4-5 cm long), ovoid shape, edible pulp	Tropical highland	—	Occasionally cultivated in the Andes
P. serratodigitata	West Indies to Brazil	Small (4-5 cm long), ovoid shape, thin brittle yellow skin, white pulp with flavour of guava	Tropical highland	—	Seldom cultivated

* Knigh (1969); Martin and Nakasone (1970).

flower production. Leaf and stem DW were optimum between 18 and 34°C while maximum root-growth occurred at 38°C.

It can grow at altitudes between 100 and 1,000 m. Plantations at lower altitudes last for a shorter period of time than those with higher altitudes. In South Africa, at altitudes between 1,200 and 1,400 m, plantations may be productive for eight years, owing to the longer cycles, implicating a greater longevity.

Relative humidity has a great influence on vegetative development and the phytosanitary state of the passionfruit. Air relative humidity of around 60 per cent is the most favourable for the crop. Elevated temperature, associated with constant wind and low relative humidity, causes a drying out of the tissues by excessive transpiration and impedes the development of the passionfruit. Relative humidity of greater than 60 per cent, when associated with rains, favours the incidence of disease, like citrus scab and anthracnose (black spot) in the aboveground parts of the vine.

Passionfruit may survive drought conditions, but will not grow and crop. Consequently, unless irrigation is available, high well-distributed rainfall in excess of 1200 mm per annum is usually considered essential for commercial passionfruit growing.

Because passionfruit evolved on the margins of tropical rainforests, they are very sensitive to wind damage. Cold southerly and south-easterly winds in Australia restrict vegetative growth and increase the severity of PWV. Strong cyclonic winds cause branch breaking, flower drop and fruit rub and sometimes vine death. Hot-dry northerly and north-westerly winds at flowering dehydrate flowers, while fruits do not set on vines subject to continued strong winds. It is vital for passionfruit to be planted in a protected site with permanent windbreaks, like bamboo, grevillea, pine, hibiscus, eucalyptus and grass species around the block if winds are strong.

It has been suggested that passionfruit is a long-day plant, requiring day lengths in excess of 10.5 h to flower and fruit (Watson and Bowers, 1965; Vallani *et al.*, 1976). In the semi-arid regions of Brazil, with more than 11 hours of day light associated with high temperatures throughout the year, the passionflowers continuously producing fruits throughout the year, as long as there is an adequate supply of water. Light intensity greatly influences the phenological phenomenon of flower opening in the yellow passionfruit. The flowers normally open at 12.00 hrs, immediately following the maximum incidence of photosynthetically active radiation (PAR), and close at 15.00 hours; however when light intensity is lower, they close at 14.30 hrs.

Passionfruit needs a constant supply of water. The demand for water varies from 800 to 1750 mm and must be well distributed throughout the year, preferably with 60 to 120 mm of water each month, by rain complemented when necessary with irrigation. Generally, annual rainfall should be at least 900 mm. It is reported that annual rainfall in passionfruit growing areas of India ranges between 1000-2500 mm. Although the plant withstands droughts relatively well, prolonged drought damages its vegetative development, causing, in severe cases, leaf fall and the formation of smaller and lighter fruits. On the other hand, intense rains in the flowering period also damage production, because they inhibit pollination by diminishing the activity of pollinating insects and causing pollen grains to burst. In regions where the rains occur in specific periods, resulting in shortage for a few months, irrigation is indispensable to guarantee good production and fruit quality.

9.0 Cultivation

9.1.0 Land Preparation

Land preparation aims to improve the soil conditions for root development, by way of increasing aeration and water infiltration and reducing soil resistance to root growth. Land should be prepared thoroughly 2 to 3 months before planting the vines, as the commercial life of vines may be from 3-6 years. A soil analysis should be made at this stage and necessary corrections made before planting. The pH should be in the range of 5.5 to 6.5. Use dolomite if magnesium levels are low. Manual soil preparation starts with clearing existing vegetation and using it as mulch or burning it. Soil preparation is limited to the manual opening of the pits for planting the vines. Mechanical preparation is done by machine, taking care not to remove the superficial layer of soil, which is rich in organic material. This is followed by ploughing and then pit digging or creating of furrows for planting. Surface scarifying may substitute for ploughing for minimum soil preparation. Passionfruits respond to the application of bulky organic manures before planting, especially on sandy soils. An alternative method is to grow and work in a bulky green crop before planting.

9.2.0 Plant Density

Recommended spacing for this crop is 2-4 m between rows and 2.5-5.0 m between plants in the rows (500-1300 vines/ha). Closer spacing provides increased early yield and greater yields later in the life of a block, when the bearing canopy tends to become sparse in places (Araujo Filho *et al.*, 1981; Gurnah and Gachanja, 1984; Manica *et al.*, 1985; Anon., 1986b). Double rows increased yield of the golden passionfruit in Cameroon (Haury, 1979). Research in the purple passionfruit in South Africa did not give similar results (Bester *et al.*, 1978).

Within-row spacings of 2.00, 2.75, 3.50, 4.25, 5.00 and 5.75 m were tested by Manica *et al.* (1987) with a between-row spacing of 2.50 m, giving a plant density range of 695 to 2000 plants/ha. Fruit yield and fruit number decreased linearly with increasing distance between the plants, from 19 to 9 t/ha, and from 240000 to 110000 fruits/ha, at the 2.00 m and 5.75 m spacing, respectively. Average fruit weight (56 g/fruit) was unaffected by spacing. In another experiment on plant density, Manica *et al.* (1994) found that fruit weight/plant and the number of fruits/plant in the first and second years were highest with spacing at 2.00 x 3.25 m (cumulative fruit weight and fruit number for the 2 years were 17.86 kg and 233.25 fruits, respectively). Cumulative yields/ha for the 2 years were highest at the 2.00 x 1.25 m spacing (33.22 t/ha), followed by spacing at 2.00 x 2.25 m, 2.00 x 1.75 m and 2.00 x 2.75 m (33.06, 31.72 and 30.59 t/ha, respectively). Mean fruit weight decreased in the second harvest. At spacing between 1.25 m and 5.00 m (2.5 cm between rows), Kist *et al.* (1995), on the contrary, recorded the highest yields/ha at spacing of 5.00 m (10.97 t/ha) and 2.75 m (10.31 t/ha) but these were not statistically different to yields at the other spacing (7.34-9.31 t/ha). The total numbers of fruits/ha were also highest at spacing of 5.00 m (126210) and 2.75 m (114 530) but not significantly higher than at the other spacing (90590-99600). Yields/plant increased as spacing increased, from 2.80 kg/plant at a spacing of 1.25 m to 13.72 kg/plant at a spacing of 5.00 m. Mean fruit weight was similar at all spacings.

9.3.0 Construction of Trellises

Passionfruits are normally grown commercially on either vertical or horizontal trellises. Horizontal trellises have cross-pieces at the top of each post with 2-4 wires strung horizontally 60 cm (2 ft.) apart along the top of each cross-piece. Vertical trellises consist of heavy posts without cross-pieces, with 2 or 3 wires strung along the row like barbed wire fencing, attached to the posts from the top down at intervals about 30-40 cm (12-16 in.) apart. Trellis wires should be No. 9 or 10 galvanized steel. The posts need to be stout enough to withstand the weight of the vines throughout a season that normally includes the buffeting of strong winds. Ideally they should be long enough to provide a trellis height of 2.0 m, with 45-75 cm (18-30 in.) in the ground. Trellis rows should be oriented north-south (in the southern hemisphere) and east-west (in the northern hemisphere) for maximum exposure to sunlight, and the vines should be allowed to grow together along the trellises to promote cross-pollination. Experiments with the purple passionfruit in Kenya showed that a horizontal trellis with 3 wires did not outyield a single wire trellis over 3 years (Gachanja and Gurnah, 1980).

Large crops are obtained on trellises 7′- 8′ tall. Such a tall trellis with a wire attached near the top of the posts and two other wires each attached near the ends of cross-bars a foot or more below the top of the posts tends to give the best spread of vines and the greatest convenience of working in the planting but is expensive especially for a short-lived planting.

In rocky or gulch areas it is often possible to use standing trees to guy the trellis wires. In some circumstances heavy cables are stretched between trees to support lighter wires at right angles to the cable. In other locations, rather tall trees have been used and the wires guyed to shorter trees or posts, giving a tent-like appearance. However, a standard trellis has the advantage of giving early returns, greater yield, accessibility for fertilization and harvesting; also more leaf surface can be exposed to the light.

In constructing the standard trellis, it is desirable to use at least eight or nine gauge heavy wire. To withstand the weight of the vine and tension on the wire, the wire should be placed on top of the cross-arm or post and stapled securely or holes drilled through the cross-arm and post and the wire strung through the holes. It is then possible to adjust the tension of the wires which is not possible when wires are stapled.

With the yoke trellis, the cross-member at the end of the trellis should be sufficiently sturdy to support the strain that will be placed on it. A 20′ length of trellis 3′ wide is required to support a minimum weight of 150kg of vine and fruit.

Trellising and Irrigation in Passionfruit

Trellising System in Passionfruit

Drip Irrigation in Passionfruit

A study was conducted to evaluate 6 different types of live and non-live stakes with 3 different types of wires as trellising materials for passionfruit vines grown in the mid-country wet zone of Sri Lanka (Heenkenda and Punchikumarihami, 1995). Live staking with gliricidia (*Gliricidia maculata*), used in combination with plain steel wire, gave the highest fruit yield (10.1 kg/vine) and the highest benefit:cost ratio (1.56). Bamboo (*Bambusa vulgaris*), another live stake, used with barbed wire gave similar yields and the second highest benefit:cost ratio. Since the initial cost was very high, concrete and 'L' iron supports gave very low benefit:cost ratio.

9.4.0 Training and Pruning

Plant passionfruit vines in full sun except in very hot areas where partial shade is preferable. The aim of training is to get the vines onto the wires as quickly as possible (Anon., 1967). One or two vigorous leaders are selected and trained to the top wire. The vine can be rather rampant, so it is important to plant it next to a chain link fence or install a strong trellis before planting. The plants can also be trained into an attractive arbor.

Plants in fertile soil extend their stems about 3 m per year. Each year during the annual dry season, the leaves fall off and the twigs die, leaving the main stem and a few important branches alive to rebuild the crown after the rains begin. Because fruiting takes place on new wood, light pruning does not reduce yield. Plants live from 3 to 8 years and do not re-sprout. Commercial stands are managed in vineyards, somewhat like grapes. They are planted in trellised rows 4.5-6 m apart and spaced 3-4.5 m apart within rows. The orchards are replanted every 4 to 6 years. Alternately, vineyards may be established using small trees or bamboo as standards. Fruiting plants can even be grown in pots under glass. The roots normally form mycorrhizal associations and benefit from inoculation with superior strains of fungi.

Some kind of temporary support at each plant is necessary to train it to get a good hold on the wires of the trellis, after which it takes care of itself and needs no other support. The principal objective is to get the vine to the trellis wires in the simplest, quickest and least expensive manner.

If the young vine is supported in an upright position, with a strong wire or light pole to grasp, it usually will grow quickly to the trellis wire with a minimum of lateral branching. A terminal branched portion of bamboo, inverted and hung over the trellis wire, provides an excellent support for the vine and eliminates the necessity for frequent tying.

Four to six laterals may be trained in both directions onto the overhead wire, and the sooner they come to a horizontal position on the trellis the more quickly they will flower and fruit.

Passion vines in their native state clamber up available trees or rocks and spread out to catch the available sunlight. The yellow passion fruit has naturalized in this manner. In cultivation, vines should be trained to cover the wires of the trellis or fence on which they are grown.

Young vines are trained by aiming a growing up toward the top of the trellis and once there, allowing a shoot to grow along each wire in each direction. A 2-wire trellis provides 4 sprouts growing along the trellis away from the vine's trunk. Once started, the vine should be allowed to grow without pruning throughout the season. With self-incompatible forms of yellow passion fruit it is particularly desirable to allow 2 different, cross-fertile vines to grow through each other and intertwine so as to promote heavy fruit production.

The pruning of passionfruit vines is conducted in order to reduce the level of pests and diseases and to encourage new vine growth. Vines should be pruned in late winter when they are not actively growing. All dead and weak wood should be cut out and the vine pruned back to vigorous, well-budded stems so that it can resume healthy, active growth in early spring. Disinfect pruning shears between each pruning to avoid spreading disease from vine to vine. A good time to make the first fertilizer application is after pruning. Heavy pruning should only be performed once per year, after the July to September crop. Since pruning tools are means by which diseases are spread they should therefore be kept clean.

Generally, pruning is done while the vine is dormant and consists of removing any growth that is weak or trailing on the ground, and shortening strong canes by about one-third. Regular pruning is necessary because the fruit is borne on new shoots arising from old canes. This pruning encourages new growth and removes unproductive wood.

A spur should be left at the base of each cane to replace the old cane after it has borne fruit for a year or two. This pruning is done when the vine begins active spring growth. Vigorous vines branch freely and branches that trail on the ground should be removed. Severed stems and pruned branches should be allowed to become dry and brittle so that they can be disentangled easily from the vine.

In some countries (*e.g.*, Kenya, Autralia, New Zealand, South Africa, Sri Lanka and Hawaii), passionfruit vines are pruned to encourage new growth and reduce disease and pest infestation. However, unpruned vines of both the yellow and pruple passionfruit have been found to consistently out yield those which have been pruned (Gachanja, 1975; Gachanja and Gurnah, 1980; Gurnah and Gachanja, 1984). This is because pruning of old growth removes the current season's crop, while pruning of new growth reduces future cropping. If pruning is to be carried out, it should be restricted to the lower parts of the vine hanging on the ground, during winter when the vines are dormant.

9.5.0 Irrigation

Rainfall in the range of 800 to 1,700 mm, regularly distributed throughout the year, is ideal for passionfruit (Borges and Lima, 2007). In areas where the rains are insufficient (below 1200 mm per year) or poorly distributed, irrigation is essential, not only to increase productivity, but also to improve the quality of the fruit, via continuous and uniform production. Water requirement is high when fruits are approaching maturity. If the soil is dry, fruits may shrivel and fall prematurely.

The soil water level influences passionfruit flowering. A lack of moisture results in loss of leaves and fruits, especially at the start of their development, and this affects the production and quality of the fruits. Flower bud development and fruit set are less sensitive to water stress than leaf initiation. Heavy rain during pollination prevents fertilization (Carr, 2013).

Fruit cracking can occur during fruit development when heavy rain follows a period of dry weather. The fruit skin cracks because of abrupt changes in the pressure (turgor) potential in the skin of the fruit, and the sudden expansion of the fruit that follows. As the fruit grows, the cracks become corky. Irrigation can help to reduce the incidence of fruit cracking.

The method most frequently used is localized irrigation, using drip and micro-sprinkler systems (Carr, 2013). Micro-sprinkler promotes a greater wetted area of soil compared to dripping, thus permitting greater volume of root system expansion. The drip irrigation system has been more widely adopted by farmers because it provides moisture and aeration conditions that favour the development and productivity of the plants. Drip irrigation has the advantage of not contributing to the formation of a humid transitory microclimate within the culture orchard, thus helps to decrease the risk of diseases.

Over a three-year period (2005-2007), Srinivas *et al.* (2010) conducted a drip irrigation experiment in Bangalore, India. The two watering treatments compared were 25 per cent and 50 per cent replacement of evaporation from a USWB Class A pan, applied on a daily basis after allowing for rainfall. Seedlings (cv. Kaveri), 40 days old, were planted in the field at a spacing of 3 x 3 m and subsequently trained on to a bower. The first harvest was made 185 days after planting and harvesting continued for another 175 days, at intervals of eight to 10 days. After the first 20 harvests, a total of 361 fruits plant^{-1} had been harvested from the 50 per cent replacement treatment compared with 305 fruits plant^{-1} from the 25 per cent replacement (a reduction of 15 per cent). The corresponding yields of fresh fruit were 23 and 19 kg plant^{-1} (–17 per cent). In the second group of harvests (15 in total) fruit yields totalled 12 and 10 kg plant^{-1} respectively (–19 per cent). Yield differences were mainly the result of differences in fruit number, not size. The yield figures are averaged over four fertilizer treatments. As there Water requirements of passion fruit 593 was no control – 'dry' – treatment, nor a well-irrigated treatment from which to set yield limits, no broad conclusions can be drawn from this experiment. The incremental water productivity between 50 per cent and 25 per cent replacement treatments (assuming that the term water-use means irrigation water applied), equated to 25 kg (fresh fruit) ha^{-1} mm^{-1} (irrigation).

The most common system of irrigation used in Australia is microjet with either one 360o or two 180o jets at either side of the vine with the lateral suspended 30 cm above the ground. The other popular form of irrigation is trickle with two drippers sited 30 to 100 cm from the stem. Good filtration is essential for trickle irrigation and chlorine injection is usually required. Trickle irrigation is not suitable for sandy soils. These two forms of irrigation require about 5 to 6 megalitres of storage per hectare of orchard. The initial design of the irrigation system should allow for the maximum water requirements of a mature orchard.

The amount of irrigation required depends on rainfall, soil type and water usage of the crop. With microjet systems in Australia, a mature vine will need about 200 litres in winter and 350 litre water in summer. In an experiment in polyethylene bags (50 cm x 36 cm), the supply of 10 litres of water a week (2 applications of 5 litres) maintained the substrate water content at a level adequate for crop development. Soil mulch exercised a positive effect only on the dry matter of stems and roots of plants (Aranjo *et al.*, 2000).

A useful method of scheduling irrigation is to use tensiometers which measure soil moisture. Irrigation should commence when a tensiometer at 30 cm depth reads 30 centibars on a clay loam and 10 centibars on a sandy loam. Irrigate until the tensio-meter at 60 cm (installed next to the shallow tensiometer) reads zero centibars, indi-cating the soil in the feeder root zone has reached field capacity. Both tensiometers should be installed halfway between the trunk and the post. Separate tensiometers are required for each soil type and block. The tensiometers should be read early in the morning as they are less reliable once the soil warms up. In a lysimeter-based experiment with the yellow passion fruit in tropical Brazil, *etc.* rates during apical vegetative growth after transplanting averaged 2.8 mm d^{-1}. These increased to 5.0 mm d^{-1} when the side branches were developing, and then to 5.7 mm d^{-1} when the plants were flowering and Water requirements of passion fruit 591 fruiting. When averaged across a number of ancillary treatments, the corresponding Kc values (based on a Penman–Monteith estimate of ETo) were c. 0.57, 0.90 and 1.10, respectively. Irrigating with saline water (4.5 dS m–1) reduced *etc.* rates by 18 per cent during vegetative growth and by 8 per cent when the plants were in flower and forming fruit. Organic mulch reduced total water use (ETc) by about 5 per cent (Freire *et al.*, 2011). In a similar lysimeter study again in Brazil, the values of Kc varied from 0.42 to 1.12 (Silva and Klar, 2002).

It is essential that vines do not suffer water stress at any stage of development as this markedly affects yield. Water stress reduces vegetative growth and flowering and induces premature drop of young fruit. Once fruits have reached a certain size they are less prone to shedding, but may subsequently become stunted in growth and have lower pulp recovery at harvest. Overwatering is more likely to occur in shallow soils or where heavy clay subsoil exists. Water-logging leads to loss of vine vigour and leaf colour, nutrient imbalances and root rots. Excessive watering will also leach nutrients out of the root zone.

Elias Menesis *et al.* (2000) did not find any effect of irrigation levels on incidence of *Fusarium* but optimum level (field capacity) had a significant effect on growth.

9.6.0 Manuring and Fertilization

Various nutrients have different roles in maintaining healthy and productive vines. Nitrogen is the major nutrient and occupies an important position in a nutrition programme. The other major nutrients are phosphorus, potassium, calcium and magnesium.

When the concentration of an available nutrient reaches abnormally low or high levels in plants, characteristic symptoms appear in the leaves, stems, flowers or fruits. Normally, growth and yield are reduced long before visible symptoms appear. The only way to avoid these yield reductions is to monitor the levels of the nutrients in the plant and soil. The nutrition programme in passionfruit begins with adjustment of soil pH and preplant fertilizing in conjunction with soil analysis of the orchard site. Once the vines have begun to grow after planting out, nitrogen fertilizer is applied monthly and complete fertilizer every second month. The aim of early nutrition management is to obtain the greatest vine size and earliest commercial bearing.

Several solution culture studies have described the effect of reduced nutrient supply (N, P, K, S, Ca and Mg) on growth and fruiting of the golden passionfruit (Avilan, 1974; Blondeau and Bertin, 1978, 1980; Marchal *et al.*, 1978, 1980; Primavesi and Malavolta, 1980b). Growth was primarily affected by deficiencies of N, S and Ca, whereas, P, K and Mg had less effect. The symptoms of deficiency for major and minor nutrients in passionfruit were reported by Morales and Muller (1976) and Aguire *et al.* (1977).

There is some confusion on the appropriate rates and time of application of the major nutrients for passionfruit vines once they begin cropping. Suggested rates for mature vines (density of 850 vines per hectare) are 500 kg N, 100 kg P and 500 kg K per hectare per year. Split applications of phosphorus should be made in spring and late summer and monthly applications of nitrogen and potassium between September and May (southern hemisphere). It is possible that excessively high rates of nitrogen may promote vegetative growth at the expense of flowers and fruits, while inadequate amount of potassium would reduce vegetative growth, fruit set, fruit size and pulp recovery. The greatest demand for potassium occurs in the fruit as it matures.

Some authors indicated that passionfruit showed no response to N, P and K fertilizer while others reported an increase in yield by 200 per cent or more (Partridge, 1972; Baumgartner *et al.*, 1978; Souza *et al.*, 1979; Singh *et al.*, 1980). In an experiment on the effect of N, P and K on productivity of passionfruit plants, trained on support received annually N at 0, 300 or 600 g/plant, P at 0, 150 or 300 g/plant and K at 0, 200 or 400 g/plant, in different combinations, as ammonium sulphate, single superphosphate and potassium chloride, respectively. The fertilizers were split into 3 doses applied in March, September and December. The treatments had no appreciable effect on the average yield which ranged from 13.7 t/ha in the control and in one of the NPK treatments to 16.7 t/ha with N at 600 g + P at 300 g + K at 0 g/plant. Foliar analysis showed that the macronutrient content was adequate for the crop (Faria *et al.*, 1991). Menzel *et al.* (1991) studied the effect of different concentrations of nitrogen in sand culture on plant growth, flowering and dry matter production and distribution.

The importance of the macronutrient to passion fruit production showed that best yield was obtained with fertilization of 250-50-80 kg N-P_2O_5- K_2O/ha (Olermo *et al.*, 2017).

Carvalho *et al.* (2000) studied the effect of 50, 250, 450 or 650 g N/plant/year and irrigation to replace 0, 25, 50, 75, 100 or 125 per cent of reference evapotranspiration (ETo). There was no yield response to N at 0, 25, 100 or 125 per cent ETo. With 50 or 75 per cent ETo, yield increased with increasing N rate up to 250 g, then decreased with 75 per cent ETo giving greater yields than

Harprakasarao *et al.* (1988) reported that increasing levels of N (50, 100, 150 or 200 g/vine), P (0, 50 or 100 g/vine) and K (0, 50 or 100 g/vine) were reflected in higher leaf N, P and K status during the flowering stage. There were significant correlation between petiole N and fruit yield and between leaf K and fruit yield. Increased levels of K_2O application significantly increased fruit yield. The intracorrelations between various leaf nutrients indicated that there were synergistic effects of N on P and Fe, P on Fe and Ca on Fe. Antagonistic effects of K on Mg and Ca on Mn were also observed.

The effect of monoammonium phosphate (MAP) and organic matter (chicken manure; 0, 100, 200, 300 or 400 l/m3 of soil) on passionfruit seedling production was reported by Pejxoio *et al.* (1999). The use of MAP or organic matter produced taller seedlings than untreated controls, with more root development, higher dry weights of aerial parts and roots, and larger stem diameters.

In a similar experiment on K fertilization (76, 307, 531 or 764 g K/plant/year, using KCl as the source of K) grown under 6 irrigation regimes (0, 25, 50, 75, 100 or 125 per cent) of reference crop evapotranspiration (ETo) using drip irrigation. Carvalho *et al.* (1999) reported that K application increased yields, up to an estimated maximum of 34.6 t/ha with 434 g K/plant/year. K also increased juice content up to an estimated maximum of 13.8 t/ha with 562 g K/plant/year. Yields were highest (35.1 t/ha) with irrigation at an estimated 85 per cent Eto and fruit weight was greatest (145 g) with irrigation at an estimated 79.4 per cent ETo.

Rodriguez *et al.* (1995) in an experiment concluded that absence of vesicular-arbuscular mycorrhizas (VAM) in the soil (due to sterilization and non-inoculation) had a negative effect on plant development and P uptake in passionfruit. The best results with regard to the indices studied, were obtained with plants receiving P at 50 kg/ha. Of the 3 VAM species, *Acaulospora longula* was the most effective.

The effect of a soil conditioner, consisting mainly of chitosan oligosaccharides, on the flowering and fruiting of pot-grown passionfruit under conditions of high and low N was investigated by Utsunomiya *et al.* (1998). The number of flowers and harvested fruits, fruit weight and juice production was increased significantly by the annual addition of soil conditioner and high N.

The amount of nutrients absorbed by roots, stem, leaves and fruits of golden passionfruit vines after 370 days' in Brazil were as follows (1500 vines/ha): 205 kg N, 18 kg P, 184 kg K, 16 kg Mg, 25 kg S, 295 g B, 198 g Cu, 779 g Fe, 2180 g Mn and 316 g Zn (Haag *et al.*, 1973). Similar data were provided by Fernandes *et al.* (1977) and Primavesi and Malavolta (1980a).

The schedule of manures and fertilizers for passionfruit grown in two premier belts (Southern and northeast) of India is given by Seema and Maiti (2006) in the following manner (Table 6)

In the absence of soil analytical data apply 50 g of micro-nutrients mixture in the planting hole. Zinc and B are the micro-nutrients taken up in largest amounts by the plant, followed by Mn and Fe (Joy, 2010). With Zn deficiency, apply 20 g of zinc sulphate ($ZnSO_4.H_2O$) per plant, and of B, apply 6.5 g of boric acid (H_3BO_3)

per plant. Boron and zinc recommendations for the yellow passionfruit are given by Borges and Lima (2003) in the following manner (Table 7).

Table 6: Schedule of manures and fertilizers for passionfruit grown in India

Manures and Fertilizers	*South India*			*Northeast India*		
	Planting	*2-4 years*	*>4years*	*Planting*	*2-4 years*	*>4 years*
i. FYM (kg/vine)	5	10	15	2	4	6
ii. Nitrogen (g/vine)	25	80	150	20	60	80
iii. Phosphorus (g/vine)	10	30	50	10	40	40
iv. Potassium (g/vine)	25	60	100	10	50	50

Table 7: Boron and Zinc Recommendation for the Irrigated Yellow Passionfruit

Nutrient	*Soil Composition (mg/dm³)*	*Classes of Fertility*	*Nutrient Dose (kg/ha)*
	Hot water		
B	<0.2	Low	2
	0.21-0.6	Medium	1.0
	>0.6	High	0
	DTPA		
Zn	<0.5	Low	6
	0.6-1.2	Medium	3
	>1.2	High	0

Deciding whether to split fertilizer application depends on the texture and the CEC of the soil, as well as the pattern of rainfall. In sandy soils and those with a low CEC, fertilizers should be applied weekly or biweekly. In more clayey soils, fertilizers can be applied monthly or bimonthly, especially when applied to the soil.

With fertigation nutrients can be applied weekly or biweekly, depending on soil texture. Drip irrigation is the most appropriate for fertigation, because nutrients are applied directly to the zone of greatest concentration of active roots. Nitrogen is the nutrient usually applied via irrigation, because it has high mobility in the soil, especially in the form of nitrate (NO_3^-), but care must be taken not to favour loss by leaching. In fertigation, N is applied accordingly to the demand of the plant to reduce losses, especially in sandy soils. Because it is so mobile in soil, N should be applied frequently at three to seven day intervals, except in sandy soils when the interval should be around three days. The recommended quantity should be distributed throughout the period between the first four months of growths, corresponding to the formation phase of the plant and the beginning of the production phase (first year). Solid N fertilizers are available in four forms: ammoniacal (ammonium sulphate), nitric (calcium nitrate), ammoniacal-nitric (ammonium nitrate) and amide (urea), all being soluble in water and adequate for fertigation, including drip irrigation. In general, these N sources behave similarly and can be applied with other nutrients. Their different effect on the soil pH has to be considered.

In general, P is not often applied with irrigation due to the low solubility of most P fertilizers and their tendency to precipitate, causing blockage of the lines and emitters. Phosphoric acid, apart from a risk of corrosion in metallic lines and connections, does not cause problems of emitter blockage, and is applied via irrigation water to promote cleaning of the lines and emitters in fertigation systems. Apart from this, diammonium phosphate (DAP) and monoammonium phosphate (MAP) can be used in fertigation.

Like N, the application of K via irrigation water is viable, because K fertilizers are soluble. When using split applications, it is important to consider its potential loss by leaching in very sandy soils and its adsorption by clay minerals in heavy soils. Potassium fertilizers normally used in fertigation are potassium chloride, potassium sulphate, potassium nitrate and potassium and magnesium sulphate. The application of K with irrigation may be done every six or seven days and it is recommended to continue the distribution throughout the growth of the plant. Starting from the second year, the recommended quantity of K_2O for the production period may be divided between the 5th and the 12th month after the seedlings have been transplanted.

Usually two emitters per plant are placed at a distance of 60 cm between them and each one placed 30 cm from the stem.

The fertilizer rates suggested above are a guide only and should be supported by leaf and soil analysis, with a record of leaf colour, vine vigour, bloom and yield. Rates of fertilizer should be adjusted according to vine response. The ultimate rate depends on vine size, vigour, crop load, cultivar, spacing, climate and soil type.

9.6.1 Tissue Analysis and Nutrient Requirement

Leaf and soil analyses provide a guide to the balance of elements including trace elements. The timing and selection of sample leaves for passionfruit leaf analysis was investigated by Marchal and Bourdeaut (1972); Kotur *et al.* (1983); Marchal (1987).

The first leaf sample should be accompanied by a soil sample (0-30 cm), therefore, a soil sample once every two to three years is usually sufficient. A leaf and soil sample should represent a planting of no more than three hectares (separate samples are recommended for each soil type, block and cultivar). Approximately 20 uniform and well-spread vines are selected for sampling.

Tentative leaf standards for passionfruit have been developed in Australia (Table 8).

A survey of passionfruit orchards in Brazil showed that leaf Ca, Mg, Cu, Fe and Mn were low in many vines (Ferreira Filho *et al.*, 1984). The most common nutrient deficiencies in passionfruit orchards in Queensland are N, K, Mg, B, Fe and Zn (Menzel *et al.*, 1986b). Several foliar applications of micronutrients remove the visual symptoms of deficiency, but effective tissue levels increase only slightly. Soil applications are required for longer-term response. Soil treatment usually corrects the problem for two to three years. However, correction of Zn and B deficiency is sometimes difficult on red loam soils, possibly because of soil fixation.

Table 8: Leaf Nutrient Standard Ranges for Healthy Passionfruit Vines in Southern Queensland*

Nutrients	*Optimum Range*	*Occurrence of Deficiency in Passionfruits Orchards*
N (per cent)	4.75-5.25	Common
P (per cent)	0.25-0.35	Rare
K (per cent)	2.0-2.5	Common
Ca (per cent)	0.5-1.5	Rare
Mg (per cent)	0.25-0.35	Common
S (per cent)	0.2-0.4	Rare
Na (per cent)	0.1-0.2	—
Cl (per cent)	0.6-1.6	—
Mn (ppm)	50-200	rare
Fe (ppm)	100-200	Common
Zn (ppm)	45-80	Common
Cu (ppm)	5-20	Rare

* Anon. (1983).

Menzel *et al.* (1993) studied the leaf nutrient composition in relation to vegetative growth and fruiting. They reported that the concentration of N, P and K in the most recently matured leaf decreased between October and December. Leaf N and K also declined in April. These declines in nutrient concentrations were associated with periods of heavy vegetative flushing and/or fruit set. Tentative standards for the most recently matured leaves in Australian passionfruit orchards during May to August before the commencement of active vegetative growth are 4.25-5.25 per cent for N, 0.15-0.25 per cent for P, 2.0-3.0 per cent for K, 1.75-2.75 per cent for Ca, 0.30-0.40 per cent for Mg, < 2.0 per cent for Cl, 100-200 μg/g for Fe, 100-500 μg/g for Mn, 50-80 μg/g for Zn, 5-10 μg/g for Cu, 40-60 μg/g for B and <800 μg/g for Na. These nutrient standards would also be a reliable guide for purple or golden passionfruits provided sampling during major periods of leaf growth and fruiting were avoided.

Menzel *et al.* (1991) also reported a relationship between the growth of passionfruit hybrid E-23 and N status in sand culture. High N supply (0.5, 2.5, 5, 10, 25 or 50 mM N as NH_4NO_3) compared with no N, increased the concentration of total N in new growth (1.75-7.03 per cent), index leaves (youngest fully expanded leaf plus next 9 older leaves) (1.56-7.98 per cent N), old leaves (1.37-6.86 per cent N), stems (0.72-4.13 per cent N) and roots (1.37-3.60 per cent N). The index or old leaves were the best indicators of vine N status throughout the range from deficiency to toxicity. Flower number increased as N concentrations in the index leaves rose from 1.5 to 8.0 per cent. The critical ranges of N concentration in index leaves using 95 per cent to maximum growth were: 6.0-8.0 per cent N for vine extension, 6.3-7.8 per cent N for the number of nodes, 4.3-6.6 per cent N for the number of leaves,

5.0-7.2 per cent N for leaf area, 4.4-7.1 per cent N for leaf DW and 5.5-8.6 per cent N for stem DW.

Based on the fact that each nutrient has a specific role in the physiological functions of plants, then excesses or deficiencies, *i.e.* imbalances, often result in characteristic symptoms, which permit the identification of the cause of the disorder. To establish the cause of visual symptoms requires knowledge of the symptom and its cause determined in both controlled experiments, which simulate the nutritional disorders systematically, and from soil and plant analysis. The symptoms of nutritional deficiencies are described by Borges and Lima (2003) in the following Table 9

Table 9: Visual Symptoms of Nutrient Deficiency in Passionfruit Leaves

Nutrient	*Leaf Age*	*Leaf Symptoms and Causes*
N	Oldest	Light green and smaller area. Yellowing and premature falling. *Cause:* low composition of organic matter, acidity (lower mineralization), leaching, prolonged drought.
P	Old	Dark green, later yellowing from the edges to the centre. *Cause:* low composition of P in the soil, low pH (lower availability).
K	Old	Progressive chlorosis from the edges to the centre, necrosis and tissue "burn". *Cause:* low composition of K in the soil, leaching and excessive liming.
Mg	Old	Yellowish spots between the veins, wizened lamina curling down. *Cause:* soils low in Mg, acidity and excessive potassium in fertilization.
Ca	Young	Death of apical sprout, interveinal chlorosis and necrosis. *Cause:* low Ca composition in the soil, excessive potassium in fertilization.
S	Young	Chlorotic, yellowish veins on the bottom side of the leaves. *Cause:* low soil S composition, low organic matter content.
Cu	Old	Large and wide leaves, dark green in color and partially shrivelled, thickening of the veins on the upper side, curved downwards. *Cause:* low soil Cu composition, excessive liming and high levels of organic matter.
Mo	Old	Interveinal chlorosis. *Cause:* acidity, excessive sulphate.
B	Young	Plants atrophied necrosis of the terminal sprout. Smaller and shrivelled leaves with waves along the edges. *Cause:* low soil B composition, low organic matter content, excessive acidity, leaching.
Fe	Young	Interveinal chlorosis. *Cause:* Excessive liming, elevated organic matter content, low soil Fe composition and elevated moisture.
Mn	Young	Chlorotic spots between the veins. *Cause:* excessive liming, elevated organic matter content, low soil Mn composition.
Zn	Young	Small leaves, gaunt and pointed lobes, milky white spots with yellow edges *Cause:* low soil Zn composition, excessive liming and phosphatized fertilization.

9.7.0 Weed Control

Passionfruit plantations should be kept free from weeds throughout the year. Weed control can be done manually in the rows and mechanically between the rows. During harvest weed control needs to be especially well done in the rows parallel to the planting lines because the fruits are usually collected from the ground. Mechanical weeding (close to the plant less than one meter) is not recommended in order to prevent damages to the roots, which are largely concentrated within 15 to 45 cm from the stem. Chemical weeding, using selective herbicides eliminates not only the weeds, but reduces operational costs and simplifies the work.

The best pre-emergence herbicides are "diuron" and "bromacil". Paraquat or glyphosate have been used in some plantations at 1.5 to 2.0 l/ha, and applied between the rows, leaving the dead weeds as a form of mulch covering the soil surface. Almedia Lima *et al.* (1999) tested diuron (1.2, 2.4 and 4.8 kg a.i./ha), oxyfluorfen (0.48, 0.96 and 1.92 kg a.i./ha), alachlor (2.8, 5.6 and 11.2 kg a.i./ha) and atrazine + metachlor (3.0, 6.0 and 12.0 kg a.i./ha) for selectivity to passionfruits under greenhouse conditions. Only atrazine + metachlor at 6.0 and 12.0 kg a.i./ha caused severe injuries to the seedlings. The other herbicides were concluded to be promising for use in passionfruit crops under field conditions.

In an attempt to investigate the phytotoxic effects of herbicides applied in post-transplanting of seedlings of yellow passion fruit, Paiva *et al.* (2015) observed that of the 21 herbicides tested, at least four have caused reductions in height gain. They also observed that only the herbicides oxadiazon, fenoxaprop-ethyl, tembotrione, chlorimuron-ethyl did not harm growth and isoxaflutole not intoxicated the seedlings, being the most promising compounds for use in total area. Even though the yellow passion fruit is membership of sweet passion fruit, these results cannot be extrapolated for dealing with different species.

Amara *et al.* (2018) evaluated the phytotoxicity of seven herbicides with different mechanisms of action in seedlings of sweet passion fruit. Chlorimuron-ethyl (ALS inhibitor), glyphosate (EPSP synthase inhibitor), paraquat (photosystem I inhibitor) and atrazine (photosystem II inhibitor) have the higher effect on the variables [initial plant height (IPH) and final plant height (FPH); initial number of sheets (INS) and end number of sheets (ENS); intoxication of plants (INTO); fresh leaf mass (FLM) and dry leaf mass (DLM); fresh stem mass (FSM); dry steam mass (DSM); fresh root mass (FRM) and dry root mass (DRM) and total dry mass (TDM)] analyzed, and they caused some kind of intoxication in the plants of sweet passion fruit. Conversely, haloxyfop-p-methyl (ACCase inhibitor) and fomesafen (protox inhibitor) showed a lesser effect on the sweet passion fruit plants.

10.0 Flowering, Fruit Growth and Development

Growth and development of passionfruit plantlets were studied by Maciel *et al.* (1997a). During growth and development, the plant went through embryonic, juvenile and transition phases before reaching the adult phase. Germination and emergence occurred 10.5 and 15.5 days after sowing, respectively. Germination was phanero-cotyledonary and epigeal. The juvenile phase lasted 80 x 10 days, when the plant grew its first 2 foliar helices, and the phase ended when plants had 10 leaves. During the transition phase, from the 3rd to the 7th foliar helix, several morphological changes occurred; leaf size and shape, internode elongation, appearance of tendrils and the beginning of branching, the plant also took on a creeping habit. The adult phase began when the main axis started flowering (plant was about 195 days old and 8 foliar helices had formed). Growth stages of passionfruits were further analysed by Maciel *et al.* (1997b). During the juvenile phase, plants grew as an orthotropic axis (monopodium of radial symmetry and continuous growth). During the transition phase, the axis adopted dorsiventral symmetry by internode enlargement and secondary leaf orientation giving rise to the climbing form. The adult vine exhibited monopodial, continuous growth and lateral flowering. Maciel *et al.* (1994) described the growth and development of passionfruit vines and morphological variable. The vine behaves as a perennial in which the juvenile phase corresponds to an erect form and the transition and adult phases are represented by a climbing form. The juvenile phase takes place in 80 x 10 days during the formation of the first 10 leaves, equivalent to 2 phyllotaxic turns. The transition occurs during the formation of the 3rd-7th phyllotaxic turn. In this phase, tendrils, trilobed leaves and lateral branches begin to appear. The adult phase is initiated when the first flowers appear on the main axis, after the 8th phyllotaxic turn. This takes place about 195 days from seedling emergence. The vine's architecture is determined by a monopodial axis of continuous growth, lateral flowering and acropetal branching. The vine reaches 4 orders of branching (including the epicotyl axis).

Growth and flowering of passionfruit were also studied by Bautista and Salas (1995). Axillary branching occurred from the base of the distal part of the main stem during the plants' vegetative growth phase. Flowers were formed on the main stem above the 24th node (counting from the first based tendril) and continued to form on lateral shoots. All shoots were potentially fruit-bearing. Vegetative growth and flowering were continuous and simultaneous.

There is a strong relationship between fruit set and the rate of vegetative growth in passionfruit, since flowers are initiated in the leaf axils of new growth. Loss of yield potential inevitably follows a reduction in vegetative growth. Although flowers are borne on new growth, not all nodes may set and carry fruits to maturity. After a certain number of fruits have set along a branch, there is temporary cessation of new growth, flowering and fruits setting. Flowering and fruit set do not commence until the first fruit has begun to mature. This alternation in vine development is thought to be related to competition for carbohydrates between vegetative tissues, flowers and fruits or possibly the translocation of growth/flower inhibitors from developing fruit.

The seasonal patterns of vegetative and reproductive growth were investigated by Menzel *et al.* (1993) in 4 passionfruit orchards over 2 years in subtropical Queensland. Vegetative growth occurred all the year round with major activity in September and October and least in November. Flowering occurred throughout the year but at very low levels in May-August or May-June and November. Fruit set also occurred throughout the year, but was lowest in January and February and

July-September or in May-September. Yield was highest between November and April and lowest in September

The root distribution of 2-year-old passionfruit vines growing in a sandy loam soil was studied by Arenas Ocampo *et al.* (1994). The greatest root density was found at a depth of 0-20 cm and a distance of 51-100 cm from the base of the vine. These results indicated that passionfruit vines tended to have shallow root systems, suiting production in shallow and/or wet soils. Menzel *et al.* (1994) also studied the relationship between shoot and root growth in *Passiflora edulis* f. *Edulis* x *P. edulis* f. *flavicarpa* hybrid E-23 grafted onto *P. edulis* f. *flavicarpa* seedling. Shoot and root growth increased as root zone volume increased from 0.3 to 1.4, 4, 12 and 24 litres. Shoot weight was correlated with root weight, with the plants allocating a smaller proportion of DM to the roots as root weight increased.

The main environmental factors influencing growth and development are temperature, solar radiation, water and nitrogen nutrition (Table 10).

10.1.0 Temperature

Experiments with the commercial hybrids E-23, Lacey and Purple Gold showed that potential yield was restricted by either low temperatures (<15°C) which reduced vegetative growth or by high temperatures (>30°C) which prevented flower production (Menzel *et al.*, 1987) (Table 10).

Table 10: Effect of Temperature on Floral Bud, Open Flower Number (Means of a 6 vines, in X + 1) Transformation of Passionfruit Cultivars (Mean followed by different letters are significantly different (P<0.05)*

Cultivar	*Days/Night Temperature (°C)*			
	15/10	*20/15*	*25/20*	*30/25*
Floral bud number				
Lacey	0.0a	0.0a	0.0a	0.0a
E-23	14.3c	13.6c	0.0a	0.0a
Purple Gold	17.6d	20.6d	8.7b	0.0a
Open flower number				
Lacey	0.0a	0.0a	0.0a	0.0a
E-23	3.9c	3.8c	0.0a	0.0a
Purple Gold	3.0b	12.0e	4.9d	0.0a

*Menzel *et al.* (1987).

Menzel *et al.* (1986b) showed that spring growth in passionfruit was associated with an increase in maximum air temperature, while minimum air and soil temperatures remained low until summer. In the studies by Simon and Karnatz (1983) and Meinke (1986), warm roots (23-28°C) and cool shoots (18-25°C) promoted fruit production. From the same laboratory (Kantharajah, 1985; Besold and Karnatz, 1986) demonstrated that constant (20, 24, 28 or 30°C) and diurnal fluctuating (22/18, 24/16 or 32/38°C) root-zone temperatures had very little effect on flowering and fruiting of passionfruit hybrids.

Table 11: Effect of Soil Water Potential on Vine Extension, Node Number, Leaf Area, and Flower Bud and Open Flower Number of Cultivars E-23 (Means followed by different letters are significantly different $Log_{10(y+1)}$ transformation, P<0.05; Data are the mean increases of eight vines)*

Soil Water Potential	*Wine Extension (cm)*	*Node Number*	*Leaf Area (cm^2)*	*Flower Bud Number*	*Open Flower Number*
–0.0025	191.5	28.4	2250	10.3a	8.9a
–0.01	85.1	15.5	884	2.1c	2.1b
–0.14	89.9	16.3	780	5.06	1.5bc
–1.5	34.9	8.3	259	0.5d	0.5c
LSD (P<0.05)	30.8	3.7	321	–	–

* Menzel *et al.* (1986a).

In another investigation when passionfruits cv. Purple Gold were grown in a greenhouse under semi-controlled conditions, high air and soil temperatures (25°C) increased leaf area, leaf number and shoot length (Meinke and Karnatz, 1990). Grafting also increased these parameters, particularly under low air and soil temperatures (18°C). An air temperature of 18°C increased flower numbers per vine; air and soil temperatures of 18°C and 25°C, respectively, resulted in the highest fruit set. A soil temperature of 25°C also increased fresh fruit weight and seed number. Grafted plants gave the highest fruit dry matter under air and soil temperatures of 18°C and 25°C, respectively. The length of time between flowering and fruit drop was shortest when both air and soil temperatures were 25°C.

In an experiment on the effect of day/night on growth and flowering, temperature, purple passionfruit plants grafted on *P. caerulea* rootstock were grown in a green- house at day/night temperature of 23/18°C, 28/23°C or 33/28°C (Ustunomiya, 1992). Shoot growth was greatest at 33/18°C, no lateral developed but some new flower buds were initiated during the flowering period. Flower development was faster at the higher temperatures.

10.2.0 Light

Irradiance also has a major influence on passionfruit flowering and fruit set (Menzel and Simpson, 1988). Potential production in the hybrid E-23 was reduced for each reduction in irradiance below full sun (20.9 MJ m^{-2} day^{-1}). Low irradiance typically increased shoot elongation and reduced leaf area, plant dry weight and floral initiation and development. In an experiment on intermittent shading of passionfruit, Menzel and Simpson (1989) grew plants of *Passiflora edulis* f. *edulis* x*P. edulis* f. *flavicarpa* cultivar E-23 under various periods of full sun and shade, but with some of the plants receiving similar amounts of total shortwave radiation. Low irradiance treatments (*i.e.*, 90 per cent shade; 1 week full sun: 3 weeks 90 per cent shade; 50 per cent shade; alternate weeks of full sun and 90 per cent shade; and 3 weeks full sun:1 week 90 per cent shade) generally increased vine extension and reduced plant DW and floral bud and open flower numbers, compared with full sun or alternate weeks of full sun and 50 per cent shade. The effects of shading on growth and flowering were stronger with shorter periods of heavy shade (90 per cent) than with longer periods of intermediate shade (50 per cent), when the average irradiances during the experiment were similar (16.5 and 15.4 MJ $m^{-2}day^{-1}$ and 11.2 and 10.0 MJ $m^{-2}day^{-1}$). It was concluded that intermittent periods of heavy shade had residual effects on growth and flowering during periods of full sun. Short periods of heavy shade for as little as 1 out of 4 weeks could reduce flowering and potential yield significantly. These results suggested that increases in plant density which led to inter-vine shading would decrease flowering and fruit set.

10.3.0 Soil Water

Milne *et al.* (1977) showed that irrigation increased passionfruit yield in South Africa upto 95 per cent, with no significant difference between a 7-, 14- or 28-day cycle. Menzel *et al.* (1986a) demonstrated reduced shoot growth and flowering when the soil water potential fell below –0.01 MPa before rewatering. Reductions in growth with water stress were associated with lower leaf water potentials, but not leaf conductances. Very high leaf conductances were obtained on leaves that were visibly wilted.

The impact of soil water deficits on the expansion, functioning and water status of leaves and the opening of flowers of hybrid passionfruit cv. E-23 grafted on *Passiflora edulis* f. *flavicarpa* seedling rootstock was studied in a series of pot experiment (Turner *et al.*, 1996). The leaves of well-watered plants expanded in a sigmoid pattern to about 65 cm^2 over 15 to 18 days. Drying the whole root system stopped leaf expansion after 6 days, but did not change the pattern of leaf growth. Growth resumed after rewatering but the final leaf size was halved.

Effects of nutrient solution concentration on photosynthesis and growth of passion- fruit were investigated by Letchamo *et al.* (1993). The highest biomass production, chlorophyll concentration, net CO_2 assimilation and leaf area were obtained at an EC of 6.8 mS/cm. At 9.0 mS/cm plant, growth was depressed and photosynthetic rate declined sharply. The highest photosynthetic rate (8.45 µml m^{-2} s^{-1}) was recorded at an EC of 6.8 mS/cm and a photosynthetic photon flux of 1600 µml m^{-2} s^{-1}.

10.4.0 Nitrogen Nutrition

Both vegetative and reproductive development was reduced when passionfruit plants were grown in solution culture with little or no added nitrogen. For examples, when the nitrate-nitrogen concentration was reduced to 25 per cent of the control solution, root, stem, leaf and fruit weight were reduced by 68.5, 50.0, 83.7 and 100 per cent, respectively. The role of nutrients on growth and yield has already been discussed in detail under cultivation.

10.5.0 Floral Biology and Pollination

Depending on the season, flower buds usually take 17 to 46 days to develop, while the time from fruit set to fruit maturity is 60 to 90 days (Haddad and Figueroa, 1972; Gachanja and Gurnah, 1978).

The floral biology of passionfruit was described by Sriram and Raman (1961) and Ishihata (1981). The flowers are protandrous, the anthers dehiscing 3-4 hours prior to maturity of the stigma. A study of the floral biology and pollination of yellow passionfruit, conducted in southern Florida by Hardin (1987) showed that honey bees (*Apis mellifera*) were the sole pollinators, and up to 25 per cent of open flowers produced fruit. Percentage fruit set in the natural environment was unknown. Flowering was erratic and each flower underwent a series of daily stylar movement. Nectar contained 46.38 per cent sugar on average and peak production occurred at 16.30 h. Fruit weight, juice volume and percentage sugar were positively correlated with the number of seeds/fruit.

Most cultivars of the golden passionfruit are self-incompatible and must be cross-pollinated from another seedling line to set fruit successfully. If the yellow passionfruit is to be planted, several strains which are known to be cross compatible to each other should be selected. The practice of planting seedlings normally alleviates the problem of incompatibility in most countries when the golden passionfruit is exploited. Self-incompatibility does not appear to occur with the purple passionfruit or hybrids between the golden and purple forms.

Hammer (1988) recorded that although *P. edulis* f. *flavicarpa* was capable of a very small degree of self-compatibility, cross-pollination resulted in increased fruit set, fruit weight, seed set, juice volume and, to a small extent, juice sugar concentration. Large bees such as *Xylocopa* spp. were expected as pollinators; honey bees (*Apis mellifera*) were the only pollinators observed.

Ishihata (1983) showed that the most favourable temperature for pollen germination was between 25°C and 30°C for the purple passionfruit. There was almost no germination at 15°C or 35°C. Pollination was not effective during rainy weather, since the pollen grains burst in contact with water (Ishihata *et al.*, 1984).

Field observations in Hawaii, West Indies and Brazil (Nishida, 1958, 1963; Akamine and Girolami, 1959; Ruggiero *et al.*, 1976; Corbet and Willmer, 1980) showed that insects were required for pollination of passionfruit. The main visitors to passionfruit flowers were the carpenter bee (*Xyloco pasonorina*), the honey bee (*Apis millifera*) and the hover fly (*Eristo lisarvorum*). The carpenter bee was the most efficient pollinator. Bees are attracted to passionfruit flowers, since they have a good nectar supply and heavy sticky pollen. If pollination is poor, hives can be introduced into the orchard (Sale, 1981). Caicedo *et al.* (1993) monitored visits by *Xylocopa* spp. to yellow passionfruit flowers at 3 sites in Colombia. Each bee spent 2.75 h/day visiting flowers, with an average feeding time of 5.4 second/flower. Fruit set from pollinated flowers averaged 45 per cent.

The giant granadilla also needs pollination to ensure fruit set. It requires mild temperatures for normal fruiting and may bloom but set no fruit (or misshapen fruit) during the hottest part of the summer. Hand pollination may be the easiest way to ensure fruit production on a few passion vines growing in the home garden.

Flowering, Pollination and Fruit Growth in Passionfruit

Apis cerena **and** ***Apis mellifera*** **Pollinating Flower of Yellow Passionfruit**

10.6.0 Fruit Set and Fruit Growth

10.6.1 Fruit Set

Development of fruit is more rapid under warm temperatures compared with cool temperatures. Maximum growth of the fruit (mainly of the skin, rind and seeds) takes place in the first 20 to 40 days after pollination. The development of the juice does not occur until just before harvest (Gilmartin, 1958; Ishihata *et al.*, 1987). Because flowers are borne on new growth, a crop from one flush of vegetative growth and fruit setting is harvested over a period of two to three months. Seasonal pattern of vegetative flushing, fruit set and yield in the passionfruit of eight vines from four orchards was studied by Menzel *et al.* (1993).

Liao (1987) studied the fruit set and fruit growth in passionfruit. Fruit set under natural conditions averaged 23.5 per cent, but artificial pollination at the viable flowering stage increased set to 36.5 per cent. The fruits attained their maximum sizes on average 15 days after anthesis. By 18 days after anthesis the fruits no longer increased in size. Most energy was expended on seed development and the formation of the ovary, the edible portion of the fruit. It took 89-91 days for the formation of the fruit from fertilization to maturation. To promote fruit set, heavy applications of fertilizers before bud burst, proper shoot pruning and pollination by honey bees were effective. Nives *et al.* (1999) reported that the base temperature of yellow passionfruit (*Passiflora edulis* f. *flavicarpa*) was determined to be 10°C. The fruit set rate was 19 per cent and 41 per cent for flowers opened in January and April, respectively. For those periods, the degree-day requirements were 867.6 and 861.7°C days, respectively. Ustinomiya (1992) also reported that fruit set was maximum at the lowest temperature (23°/18°C). Fruit development was fastest and fruit juice content was greatest at 28°/23°C. Fruit weight and juice content were lowest at 33°/28°C. Sucrose accumulation in the fruits was greatest at 23°/18°C but glucose and fructose contents were greatest at the highest temperature.

10.6.2 Fruit Growth

Fruit growth (diameter) of purple passionfruit (*P. edulis*) and maypop (*P. incarnata*) followed a sigmoidal growth curve. Passionfruits were larger than either greenhouse-grown or wild maypop fruits. Yellow passionfruit had the lowest percentage of bulb and highest soluble solids concentration (SSC) and greenhouse-grown maypop had the lowest SSC among the 4 groups tested. Purple and yellow passionfruits had lower juice pH than maypop. Yellow and purple passionfruit juice had higher fructose and glucose contents than did maypop juice (Arjona *et al.*, 1991). Enamorado *et al.* (1995) investigated growth and development of yellow passionfruits (*Passiflora edulis* f. *flavicarpa*), grown in Brazil, from anthesis to 80 days after anthesis (DAA). The fruit showed a single sigmoid growth curve with a peak at 21 DAA. Fruit FW and DW increased until 50 and 70 DAA, respectively. Initial fruit growth (until 21 DAA) was mainly associated with peel growth, while advanced fruit growth was primarily due to growth of the pulp (arils and seeds). Accumulation of soluble solids and acids peaked at 63 DAA, followed by a decline of the former coinciding with the initiation of the climacteric rise of respiration and ethylene production.

In an investigation on fruit growth and development, Lederman and Gazit (1993) found that the fruits had a rapid sigmoid growth pattern reaching their final size about 3 weeks after anthesis. The relative fruitlet growth rate was highest (210 per cent /day) in 4- to 7-day-old fruits and daily increases in fruit weight and volume were highest in 11- to 15-day-old fruits. After the short period of rapid fruit growth, fruit weight and size varied little over the next 2 months until ripening and abscission (approx. 85 days after anthesis).

Guadarrama (1996) studied the development of yellow passionfruit grown near Maracay, Venezuela, harvested at various developmental stages. Fruits showed changes in protein, pectin and soluble solids contents during development. Internal and external diameter was related to mesocarp fresh weight in an exponential way.

In order to harvest mature but not ripe fruits and thus extend storage life, developmental stages of fruits were studied by Gammarra Rojas and Medina (1997). Based on FW, DW and volume data, fruit growth followed a single sigmoid pattern. Fruit deve-lopment was characterized by 3 distinct phases; fast growth up to 25 DAA when fruits reached maximum FW and volume; accumulation of assimilates, between 25 and 50 days from ovule fertilization; and physiological maturity, at about 50 DAA. Fruit age was a good index for the developmental stage for all variables evaluated.

Antognozzi and Proietti (1995) applied 20, 60 or 120 ppm CPPU (forchlorfenuron) 2 weeks after full bloom. CPPU increased fruit drop slightly in all conditions; in non-irrigated conditions the 60- and 120-ppm rates slightly reduced yield/tree. It increased the pulp:pit ratio in all treatments. It was concluded that CPPU was very effective in improving fruit size both with and without irrigation and that the best rate was 20 ppm.

11.0 Pests and Diseases

11.1.0 Pests

Several insect pests attack passionfruit vines causing economic losses because of fruit blemishes and/or loss of plant vigour and productivity (Sale and Alexander, 1986b). The major pests are fruit fly (not in all countries), citrus mealy bug, California red scale and passionvine mite. Pests which are sometimes important are fruit spotting bug, green vegetable bug, soft brown scale, passionfruit hopper, aphids, thrips, rutherglen bug and broad mite. A description of the pest complex affecting passionfruit orchards in subtropical Australia is provided by Murray (1976).

11.1.1 Fruit Fly

The most troublesome pests of passionfruit are Oriental fruit fly (*Dacus dorsalis*), found at high elevations; Melon fly (*Dacus cucurbitae*); Mediterranean fruit fly (*Ceratitis capitata*) seem to prefer lowlands and Queensland fruit fly (*Dacus tryoni*) in Australia.

The pest punctures the immature fruit before the rind hardens. As the fruit enlarges, a corky area develops around the puncture. If the fruit is small and undeveloped the damage will be sufficient to cause it to shrivel and fall. Mature fruit can be stung and blemished by the fly, but larvae usually do not develop. The main economic loss from this pest is the downgrading in quality of mature fruit. Quarantine restrictions due to fruit fly make marketing extremely difficult in southern states of Australia and impossible in countries like Japan and the USA.

The pest is most active during warm wet periods from September to April. Stinging is least important in the main crop in mid-summer and more important in the autumn and winter crops.

The main objective in fruit fly control is to destroy the gravid females which usually breed elsewhere but lay eggs in the orchard. An important step, then, is to eliminate nearby overripe fruits on which the adults feed and breed.

Fruit fly adults may be destroyed with various insecticides. One is Malathion 25 per cent wettable powder, which is sprayed at 1.5 kg/ha in 500 l of water. Absolute caution should be exercised in making applications and every precaution taken to do so as safely as possible. Fruit fly adults may be destroyed also by use of bait sprays made with 1.5 kg of Malathion and 0.5 kg of yeast hydrolysate in 500 l of water per hectare. Frequency of application depends on the population. When adults are numerous, applications twice weekly may be necessary when young fruit is present. Trapping offers a reliable indication of fruit fly activity but provides no control. Thorough coverage, especially of sheltered fruit is essential for good control.

11.1.2 California Red Scale (*Aonidiella aurantii*)

California red scale is common on old vines during dry weather in December and January. Older leaders are first infested and as the population increases the scale spreads to young sections of the vine and eventually to the fruit. Severe infestations may reduce plant vigour and kill the vine.

Parasitic wasps are important in the control of red scale when insecticide usage is restricted providing sufficient numbers of wasps are present at the beginning of the outbreak. When infestation of red scale is serious, two sprays of white oil, one in early December and one 4 to 6 weeks later are usually required. For heavy infestations or during warm weather (>30°C) use methidathion. The regular use of pesticide for control of fruit fly often depletes the population of beneficial wasps which parasitise red scale.

11.1.3 Mealy Bug (*Planococcus pacificus*)

Passionfruit mealy bugs aggregate on leaf nodes and under flowers and fruit bracts and dead leaves. They may also encrust developing fruit. Outbreak of mealy bug is more common in late summer and autumn when the growth cycle from egg to adult is shortest (four weeks). If unchecked, an infestation may cause loss of vigour, leaf drop, fruit malformation and eventually vine death. Honey dew excreted by the mealy bug promotes the growth of sooty mould on the leaves and fruit.

Natural control by Ladybirds, wasp parasites and lacewing larvae is important in maintaining mealy bug populations below damaging levels. Ants which tend the mealy bug interfere with natural control by parasites and predators. Consequently, the first approach to control of the mealy bug should be to remove the ants.

Pests of Passionfruit

Queensland Fruit Fly Infestation

Mealy Bug Infestation

Mite Infestation

Leaf Footed Plant Bug Damage

Occasional outbreaks of this pest are best controlled by two sprays of white oil or methidathion combined with white oil 2 to 4 weeks apart. White oil is preferred to as methidathion is highly toxic to the mealy bug's natural enemies. Chemical control is often difficult because of the secluded habit of the bug and the dense foliage of mature vines. Control by natural enemies is the preferred option.

11.1.4 Spider Mite (*Brevipalpus phoenicis*), *Tetranychus telarius*; Broad Mite (*Hemitarsonemus latus*)

The passion vine mite is a common pest of passionfruit throughout coastal Queensland affecting both young and old vines during summer and autumn. Mites are generally most damaging in areas of low rainfall and during prolonged dry seasons. Severe infestations cause leaf drop, bud failure and stunted growth. Effective control is achieved by sprays of wettable sulphur or propargite, 2 to 3 weeks apart.

Biological control of *Pseudaulaca spispentagona* was reported by Liebregts *et al.* (1989) at two sites in Western Samoa 8 months after the *Encarsia barlese* and *E. Diaspidicola* parasitoids were introduced. A model for interaction of *P. pentagona* and parasitoids predicted higher levels for parasitization than that revealed by microscopic examination.

11.1.5 Aphids

Severe damage by aphids usually results from their attacks on young plants. Two aphids, *Myzus persicae* and *Macrosiphum solanifolii* are efficient vectors of the passionfruit woodiness virus. These aphids are present in Hawaii but the virus is not found.

Woodiness disease has been identified in Australia as cucumber mosaic virus 1. It is thought to be transmitted by aphids and is the most serious threat to the purple passionfruit in Australia and Kenya.

In this disease, the leaves become leathery and malformed; the fruits gradually decrease in size and the rind becomes thick and hard and little pulp is produced. Spraying of imidacloprid @ 0.03 per cent or acetamiprid @ 0.15 per cent may be effective against these pests.

11.1.6 Caterpillars

Caterpillars have been destructive of the passionfruit plants, especially attacking new leaf growth. *Heliconius* caterpillars have an appetite for Passiflora and most cause allergic reactions or skin-irritation when picked off by hand. Avoid using chemical weapons but put on pair of gloves and pick them off - some of these insects are more beautiful than your plants and play an important role in the ecosystem. Hairy caterpillars can cause respiratory problems or irritation to exposed skin - wear protective glasses and mask and keep your body covered when removing them. Tortrix moths have caterpillars which are known as leaf-rollers, they fold the leaf to create a protective cocoon whilst munching away the plant. Chlorpyriphos @ 0.25 per cent can be effective for caterpillar control.

11.2.0 Diseases

Several diseases limit the production of passionfruit (Sale and Alexander, 1986b). Basically the disease complex affecting the crop is the same in most countries with one or two substitutions. The major exceptions are that *Phytophthora* and *Fusarium* root rot do not affect orchards in Australia, South Africa, Hawaii and other countries where the industry is based on golden passionfruit rootstocks. A summary of the major diseases in subtropical Australia is provided by Inch (1978).

The major diseases of passionfruit are septoria spot, brown spot, *Phytophthora* blight, *Alternata* spot, woodiness virus and base rot. The first three named diseases can normally be controlled with available fungicides, the fourth can sometimes prove difficult to control with fungicides, while there is no chemical control available for the last two named diseases. *Phytophthora* rot of roots and *Fusarium* wilt (*Fusarium oxysporum*) are not problems in Queensland because of the use of resistant *P. edulis* f. *flavicarpa* (golden passionfruit) rootstocks.

Septoria spot (*Septoria passiflorae*), brown spot (*Alternaria passiflorae*) and alternata spot (*Alternaria alternata*) infect leaves, stems and fruit of passionfruit in Queensland. Extensive leaf fall follows infection with septoria spot, while serious infections of brown spot cause wilting and death of long sections of vines. Leaf lesions of alternata spot are often inconspicuous and the vine itself may appear healthy. The main economic loss associated with these diseases is the downgrading of fruit. Infected fruits are acceptable only for processing.

Fruit infection with septoria spot, brown spot and alternata spot occurs at any stage of growth and is characterized by light-brown blotches on the skin. In septoria spot, these blotches may coalesce to cover much of the fruit. In brown spot, the blotches become sunken and wrinkled as they enlarge covering up to half the fruit. In contrast, young fruit infected with alternata spot have lesions characterized by a tan or brown centre and greasy water-soaked margins. The spots are superficial and confined to the skin and are rarely more than 7 mm wide. The golden passionfruit and hybrid Purple Gold are less susceptible than Lacey or E-23.

Septoria spot, brown spot, alternata spot and *Phytophthora* blight are more common during warm wet weather from spring to autumn. Copper oxychloride or mancozeb every three to four weeks during May to September and every 10 to 14 days during October to April will normally give good control of these diseases. However, when alternata spot is severe, it is essential that vines be sprayed every ten days with iprodione mixed with copper oxychloride at 50 per cent higher concentration than normal. It is important to obtain full spray coverage of fruit and foliage. The addition of stickers to high volume spray is useful in this regard. Pruning of diseased parts of the vine will also reduce inoculum level and improve spray penetration.

Severe infections of yellow passionfruits caused by *Xanthomonas campestris* pv. *passiflorae* were reported from Brazil by Torres and Ponte (1994). The best control and the highest yield were achieved with copper oxychloride 30 per cent (as Coprantol), copper oxychloride 50 per cent (as cupravitverde) and copper oxychloride + maneb

+ zineb (as peprosan) at 15-day intervals. Pruning of the canopy did not improve disease control.

In Australia, base rot produces a soft decay at the base of the stem and upper parts of the main roots with associated gradual decline. Vines may show symptoms of the rot for sometime if very vigorous in growth before dying. There is some evidence that it might be due to *Phythium* species entering the crown through wounds, especially in poorly-drained soils. Zheng and Huang (1997) reported that in China, base rot disease of passionfruit is caused by *Fusarium oxysporum* f. sp. *passiflorae*. The disease develops between March and December. Thiophanate-methyl (70 WP) was shown to control the disease effectively.

Sawant *et al.* (1995) identified the causal organism of collar rot of passionfruit in India as *Rhizoctonia solani* and described the suitable control methods. Of the two mycoparasitic fungi tested *Trichoderma harzianum* controlled the disease while *Gliocladiumvirens* had no effect and of the 6 fungicides tested tolclofos methyl (as rizolex) and carbendazim (as bavistin) effectively controlled the disease.

In Uganda, collar rot caused by *Fusarium solani* was controlled by copper oxychloride applied as a drench at a rate of 60 g/20 l of water. Other fungicides tested were rovral 70 WP (iprodione) and cercobin 50 L (thiophanate-methyl). To address the environmental issue, the use of resistant species was investigated and *Passiflora edulis* f. *flavicarpa* and *P. maliformis* were found to be partially resistant to collar rot, andcould be recommended for use as rootstocks to susceptible *P. edulis* f. *edulis* and hybrid clones (Ssekyewa *et al.*, 1999).

Phytophthora blight (*Phytophthora nicotianae*) causes blackening and death of new growth, defoliation, wilting and collapse of the vine. Large grey-green water-soaked areas appear on the fruits causing them to drop rapidly. Fruits are unmarketable.

Commercial hybrids in Queensland are susceptible to the necrotic strain of the passionfruit woodiness virus (PWV-N). Affected vines have mottled green-yellow puckered leaves, retarded growth and bear small misshaped fruit with abnormally thick rind and small pulp cavity. Affected fruits are unmarketable. Most vines contact the disease, the virus being transmitted from diseased to healthy vines by migrating aphids and on pruning implements. Vines show the greatest symptoms of PWV-N during cool winter and may outgrow the virus during the warmer months.

Careful selection of grafting tips in winter from vines not showing virus symptoms can often reduce the incidence of PWV-N. Once vines are established in the field, it is essential to maintain plant vigour as regular sprays for control of virus vectors is not normally completely effective. The long-term solution to woodiness is through plant breeding. Lines of *P. edulis* x *P. incarnata* show resistance to PWV-N.

Kitajima *et al.* (1997) reported a severe outbreak of a disease in passionfruit (*Passi- flora edulis*) characterized by necrotic lesions in stems and green spots in mature fruits and senescent leaves, usually resulting in the death of the plants, occurred in the orchards of Sao Paulo, Brazil. This condition was associated with heavy infestation by *Brevipalpusphoenicis* and the death of plants was caused by the girdling of the stem by the confluence of the necrotic lesions. In extreme cases, orchards of a few hectares were completely destroyed. Electron microscopic

Diseases of Passionfruit

Phytophthora **Fruit Rot infection**

Fusarium **Wilt Infection**

Septoria **Spot (Left) and** ***Alternaria*** **(Right) Infection in Fruit and Leaf**

Collar Rot (Left) and ***Rhizoctonia solani*** **Root Rot (Right) Infection**

Diseases of Passionfruit

Brown Spot (Left) and Anthracnose (Right) Infection

Woodiness Virus Infection

examination of tissues from stem lesion and green spots of leaves and fruits revealed the presence of particles with round to bacilliform profile contained in membrane-bounded cavities and dense masses in the cytoplasm, similar to those found in tissue infected by the citrus leprosis rhabdovirus (CitLV).

12.0 Production Season, Harvesting and Yield

12.1.0 Production Season

Cropping season of commercial hybrids in southern Queensland is continuous with a major peak in mid-summer and two smaller peaks in late autumn and late winter (Hargreaves, 1979). In contrast, the golden passionfruit has a single extended cropping season over later summer, autumn and early winter. The purple passionfruit produces a heavy summer crop and virtually no winter crop.

In Hawaii, the golden passionfruit produces fruit over 7-8 months from November to July. In Kenya, Zimbabwe and South Africa where the industry is based on lines of the purple passionfruit, there is a major harvest in November to January and one or two smaller harvests in March to April and June to July. Figures quoted for the purple passionfruit in New Zealand indicate that 86 per cent of the crop is harvested between March and June (Sale and Alexander, 1986a).

Data collected by Fullerton (1986) indicated that crosses between a seedling of E-23 (purple x gold hybrid from Australia) and local golden seedling produced a major peak in April to June and two smaller peaks in December to January and February to March.

Season of harvesting influences the quality of fruits. Yellow passionfruits produced in the south of Minas in Brazil, were harvested during 3 growing seasons: GS_1 (May to July, 1995), low temperature and rainfall and high insolation; GS_2 (October to December, 1995), moderate temperature and rainfall and low insolation, and GS_3 (January to March, 1996), high temperature and rainfall and moderate insolation and their physico-chemical characteristics were evaluated by Nascimento *et al.* (1998). The percentages of total sugars and glucose, vitamin C content and the TSS:TA ratio were highest in GS_2 fruits. Sucrose was highest in GS_3; however, sucrose content was always lower than glucose content. Fruits harvested in GS_2 had the best quality.

There are two major seasons of production, June to August and November to January in India (Joy, 2010). The first season may extend from May to October.

12.2.0 Harvesting

The passion fruit vine, especially the Yellow, is fast-growing and will begin to bear within 1 to 2 years. Ripening occurs 70 to 80 days after pollination. A passionfruit orchard may have an economic life span of 3-5 years. Fruits fall after ripening on the vines and are picked up from the ground 2 to 3 times weekly.

Passion fruit being a climacteric fruit and ripening may also take place off the tree; therefore, fully matured fruits (10-11 weeks old) may be harvested and may be kept for ripening. Immature passionfruit do not ripen off the vine. They may lose some acidity, but do not develop full flavour or colour and have a higher concentration of cyanogenic glycosides (a toxin produced by the vine). The usual method of harvesting in Australia is to pick up dropped fruit in 20 litre containers once or twice peak week in winter and daily in summer or in rainy weather to reduce sunburn and spoilage by fungal rots. Harvesting is the most difficult and expensive operation in passionfruit growing. Mechanical harvesting has been used experimentally in southern Queensland, but has not been adopted by the industry (Menzel *et al.*, 1990). Veras *et al.* (2000) suggested that sweet passionfruit (*P. alata*) could be harvested up to 4 days while acid passionfruit (*P. edulis*) need to be harvested up to 8 days before fruit abscission without affecting quality.

12.3.0 Ripening

Purple passion fruit and giant granadilla flower throughout the year but major bloom occur in March-April and July-August. The yellow passion fruit and Kavery major flowering occurs during May-June and during September-October. Fruits of passion fruits mature between 70-80 days of flowering (Thokchom and Mandal, 2017).

Ripening characteristics of purple fruit was investigated by Shiomi *et al.* (1996a). Analysis of the rates of increases in size and weight showed a single sigmoidal curve. Size reached a maximum 20 days after flowering (DAF), whilst fruit weight continued to increase gradually. The respiration rate of fruits left attached to the vine until just before measurement decreased gradually with time but without showing a climacteric peak, whereas ethylene production increased sharply soon after 70 DAF, thereafter remaining constant. After successive detachment of fruits at 40 or more DAF, a respiratory peak was shown, the onset of which occurred quicker the longer the fruit had remained attached.

As fruits matured, soluble solids content steadily increased and reached maximum levels 70 DAF while titratable acidity increased rapidly up to 60 DAF and then decreased up to 80 DAF. Changes in fruit colour from green to purple were attained between 70 and 80 DAF and this corresponded with the commencement of autocatalytic ethylene production. The results indicate that purple passionfruits appear to attain physiological maturity at about 60 DAF and can be harvested with acceptable quality at about 70 DAF for commercial use.

Shiomi *et al.* (1996b) also studied ethylene biosynthesis and the pattern of changes in some chemical constituents in purple passionfruits harvested at the turning stage of ripeness (when the skin colour began to change from green to purple). During ripening, sucrose content decreased while fructose and glucose contents increased. Citric and malic acid contents increased slightly during the early stage of ripening and decreased thereafter. Ethylene biosynthesis in purple passionfruits is regulated mainly by ACC synthase activity, and the sensitivity to ethylene may change after harvest as the fruits ripen.

Enamorado *et al.* (1995) reported that ethephon, applied at 1000 ppm to fruits harvested at 40, 50, 60 or 70 days after anthesis (DAA), advanced both the climacteric peak of CO_2 evolution and the appearance of a yellow fruit colour. Untreated fruits harvested at 40 DAA, however, did not ripen, suggesting that yellow passionfruits

Harvesting, Sorting, Grading and Packaging of Passionfruit

Passionfruit in Bearing Stage

Harvesting

Mechanical Grading

Packaging in CFB Box

grown under Brazilian conditions require more than 40 DAA to reach physiological maturity.

12.4.0 Yield

The economic yield starts after 1-2 year of planting and a healthy plant produces about 150-180 fruits/year. Purple passion fruit produces more fruits than yellow and giant granadilla due to compatibility of pollens (Thokchom and Mandal, 2017).

Yield depends on several factors including cultivar, seasonal weather and vine management. Commercial hybrids in subtropical Queensland yield up to 20 to 25 tonnes per hectare when passionfruit woodiness virus (PWV) is not severe. In contrast, yield upto 30 to 50 tonnes per hectare has been recorded for the golden passionfruit, *P. mollissima* and *P. quadrangularis* in Hawaii, Colombia and northern Queensland. The purple passionfruit is less productive with maximum yields of about 5 to 10 tonnes per hectare where passionfruit woodiness virus is not severe. The yields of passion-fruit species in different countries are presented in Table 12.

Based on physical and chemical fruit and juice characteristics, harvesting 63 days after anthesis, when 40-50 per cent of the fruit peel is yellow, was recommended by Aular *et al.* (2000).

Table 12: Yield of *P. edulis, P. edulis* f. *flavicarpa* and Hybrids in Various Countries

Country	*Species*	*Density (Vines/ha)*	*Yield (t/ha)*	*References*
Brazil	*P. edulis* f. *flavicarpa*	667-1665	2.4-21.6	Baumgartner *et al.* (1978) Souza *et al.* (1979) Araujo Filho *et al.* (1981) Manica *et al.* (1985) Weber *et al.* (2017)
Hawaii	*P. edulis*	1200	5.0-10.0	Akamine and Girolami (1959)
	P. edulis f. *flavicarpa*	1200	25.0-50.0	Abeysinghe (1973)
Australia	*P. edulis* x *P. edulis* f. *flavicarpa*	850	10.0-25.0	Menzel *et al.* (1988)
New Zealand	*P. edulis*	1320	10.0-12.0	Sale and Alexander (1986a)
Fiji	*P. edulis* f. *flavicarpa*	550	13.6-37.4	Partridge (1972)
India	*P. edulis*	110	4.6-7.2	Singh *et al.* (1980)
			10.0-12.0	Tripathi *et al.* (2013)
Sri Lanka	*P. edulis*	1000	4.0-6.0	Abeysinghe (1973)
	P. edulis f.	*flavicarpa*		Bertin (1976)
South Africa	*P. edulis*	1600	18.9-24.4	Bester *et al.* (1978)
Kenya	*P. edulis*	1200-1500	3.0-66.0	Gachanja (1975) Lippmann (1978)
Cameroon	*P. edulis* f. *flavicarpa*	2500	6.0-21.8	Haury (1979)

Average yield for purple-gold hybrids in Australia is about 10 to 15 tonnes per hectare, with more favourable growing conditions in Queensland compared with northern New South Wales accounting for higher productivity. In Maracay, Bautista and Salas (1995) recorded around 18,000 to 23,900 kg/ha yield. Vines may produce commercial crops for 6 to 8 years. However, the average life of vines is usually no more than 3 to 4 years, especially in warm environments where diseases and pests are difficult to control in dense canopies (Menzel *et al.*, 1990). In India, the vines bear throughout the year but two peak periods are, first, August to December and second, March to May. At the latter time, the fruits are somewhat smaller, with less juice.

13.0 Post-harvest Management

13.1.0 Grading and Packaging

Fresh fruit marketed in Queensland must comply to the following standard: fruits mature, half to full ripe but not overripe, large (>40 mm in diameter) with at least 35 per cent pulp recovery (Menzel *et al.*, 1988). Diseased or badly blemished fruits are unacceptable. Fruits are mechanically graded for size, and for colour by hand, and then packed in 8-10 kg cartons. During most of the year, it is only the smooth dark purple well-graded, blemish-free fruits that attract good prices. The demand is for fruits with a count of 120-140 per carton. The fruit count, grower's name and address and agent are displayed on the carton. Some buyers demand fruit in plastic bags. Tray packs offer good presentation but higher packaging costs make it unviable most of the year. There is limited demand for yellow-skinned fruit, but the market is easily oversupplied.

The main characteristics used in grading passion fruit are size, skin colour and uniformity, shape, firmness, and the amount of surface blemishes. Fruit marketed domestically should be clean, firm, free from visible signs of disease, mature, uniformly coloured, and free from damage which detracts from the appearance or edibility of the fruit (*i.e.* bruises, cuts, shrunk or cracks, insect damage, sunburn, *etc.*). At least 90 per cent of the fruits in any lot should have a glossy appearance and not be soft or shrivelled (Joy and Divya, 2016).

Three different grades (Grade 1, Grade 2, and Grade 3) have been established by the National Bureau of Standards for domestic marketing of passion fruit (Table 13).

Table 13: Grade Standard of Passionfruit

Passionfruit Grades	*Features*
1	Fruit will be firm with shiny appearance; signs of shrivelling will be absent, free from visible evidence of insects, disease, and surface blemishes. These fruit types are highest in quality.
2	These passion fruits must be of good quality, although 10 per cent of the fruit in this grade may be soft or show signs of shrivelling, and will not be having a glossy appearance
3	These passion fruit do not qualify for inclusion in the higher grades, but should be clean, mature, free from pests and diseases, will have similar colour, shape, and size

13.2.0 Storage

Passionfruits begin to lose moisture after a few days at room temperature causing shrivelling of the skin and some loss of flavour. In contrast, fruit will store for up to 4 to 5 weeks with minimum loss if fungal rots are controlled and they are stored at 5°C and 80 to 90 per cent relative humidity. Waxes and fungicidal dips have been used in Australia to extend shelf life of fresh fruit by reducing water loss and rotting, but are not common.

Arjona *et al.* (1992) studied the effect of temperature and storage time of yellow passionfruit stored for 15, 30 or 45 days at 5, 10 or 15°C and changes occurring during storage. Fruit's external appearance deteriorated rapidly at 5 and 15°C. Pulp percentage at 5°C increased linearly with storage duration and did not change at 10°C. Pulp percentage at 15°C changed quadratically with storage time, increasing up to 30 day and then decreasing by 45 day. Soluble solids concentration did not change at 5 or 10°C but decreased linearly at 15°C. Sucrose content decreased quadratically at 5°C and linearly at 15°C, but increased linearly at 10°C. Fructose and glucose content decreased quadratically with storage time at 15°C. Glucose content increased linearly at 5 and 10°C, fructose content did not change at these temperatures.

Sanez *et al.* (1991) reported that cold storage for 21 days significantly reduced weight loss (1.36 per cent) compared with ambient conditions (4.45 per cent). The lowest weight loss over 28 days, *viz.*, 2.21 per cent in cold storage, was obtained with fruits packed in non-perforated bags. No differences in titratable acidity or TSS were found between different packing treatments. Both acidity and TSS were higher in cold storage; their highest values were recorded at 14 and 21 days of storage and thereafter acidity, TSS and fruit firmness decreased. In an investigation on storage, fruits of yellow and purple passionfruit cultivars were packed in sealed or perforated low density polyethylene bags (0.025 mm thick) and stored at 10, 20 or 30°C and 80-90 per cent RH for 20 days (Mohammed, 1993). Fruits were sampled at 4-day intervals and the marketable percentage, FW loss, shrivelling, decay, pH, TSS, TA, TSS:TA ratio and flavour assessed. The best results were obtained with storage at 10°C in perforated bags, which gave 61.2 and 50.1 per cent marketable fruits for the purple and yellow cultivars, respectively, after 16 days.

Post-harvest changes in aroma in vine-ripened yellow passionfruit were studied by Narain and Bora (1992) after storage for up to 15 days at 29 x 2°C. Ethyl butanoate and ethyl hexanoate concentrations decreased rapidly after 3 days of storage. It was concluded that ripe yellow passionfruits retain their fresh, characteristic, ester-type aroma for 3 days only after harvest.

Fruits of *Passiflora edulis* f. *flavicarpa*, 5-7 cm in diameter, were treated with auto-citrol wax, tecto-600 (thiabendazole) or sodium hypochlorite, then packed or not in 60 x 40 cm bags, and stored at 6°C for up to 42 days (Gama *et al.*, 1991). The stored fruits showed a reduction in total sugar content and acidity, and an increase in Brix:acid ratio the longer they were stored. The soluble solids and reducing sugar contents remained unchanged during the storage period. Packing in bags reduced weight loss and the fruits stored well for 42 days. Neither wax, thiabendazole nor sodium hypochlorite application had a beneficial effect on fruit storability.

Arjona *et al.* (1994) studied the quality of vine-ripened yellow passionfruits in Styrofoam trays, wrapped with VF-60 plastic film and stored for 15 or 30 days. Wrapping prevented fruit weight loss while maintaining external appearance. Storage time contributed to quality loss of external appearance.

In sweet passionfruit (*P. alata* cv. Dryander), Silva *et al.* (1999) observed that wax coating with Stay Fresh was the best treatment for the conservation of sweet passion- fruit. Precooling and precooling + wax, were not efficient for the maintenance of post-harvest quality. Gomez (2000) found that use of wax and/or a storage temperature of 12°C and 80 per cent RH, extended the shelf life of fruits, reduced weight losses and maintained an adequate external appearance.

Effects of dipping fruits of several cultivars for 3 min in 2-8 per cent $CaCl_2$ 5 days after harvest and the effects of the treatment on the control of storage rots and fruit respiration and composition were studied. The highest concentration caused fruit scald, and 4 per cent $CaCl_2$ generally gave better results than 2 per cent $CaCl_2$. Storage losses were reduced, in some cases substantially (in cv. Tamopan Bolshot from 30 to 9 per cent), the respiratory response also differed with cultivar. Sugar and vitamin C contents were highest in fruits treated at 4 per cent (Treglazova and Fataliev, 1989).

13.3.0 Processing and Value Addition

Passion fruit all over the world is utilized mostly for juice production. There are different methods of extraction of pulp and juice from purple and yellow passion fruits. Several kinds of mechanical extractors are available *viz.* centrifugal extractor, converging cone extractor commonly used in Australia and pass press extractor in Italy (Casimir, 1974). For small scale production the simplest procedure is to give the fruit a pleasant cut by a cutting knife and scoop out the pulp with stainless steel spoon or pitting or scooping knives. This method is still being used in Africa, India and some other South East Asian countries.

Formulation for blended beverages of passion fruit with guava, pine apple, papaya, mango, jackfruit *etc.* and passion fruit concentrates was mentioned by several workers (Brekke, 1973; Chakraborty *et al.*, 2011; Joy and Divya, 2016). The formulations contained added sugar, chemical preservatives and a suspending agent and were intended to be diluted for consumption at the rate of 1 volume of base to 5 volume of water. Passion fruit juice was one of the ingredients for blended tropical fruit nectars that were examined for acceptability and shelf life

Passion fruit rind containing about 50 per cent in the purple and about 50-60 per cent in the yellow colour fruit is considered to be "factory waste". Its chemical composition and nutritive value ensure the possibility of its economic utilization and minimizing the environmental pollution. Chaurasiya (2010) developed attractive candy, preserve and pickle from yellow passion rind. Chakraborty *et al.* (2011) observed that these products could easily be stored for more than a year at ambient temperature with acceptable and microbiologically safe condition

14.0 Biotechnology

The use of protoplast, cell and tissue culture, organogenesis and genetic transformation of *Passiflora edulis* and other *Passiflora* species were reviewed by Lima *et al.* (2000).

14.1.0 Tissue Culture

14.1.1 Micropropagation

Kawata *et al.* (1995) described a method for long-term micropropagation of passion fruit from subcultured multiple shoot primordia. Excised tissues from leaves cultured on MS medium supplemented with 3 per cent (w/v) sucrose, 1 µM BA and 1 µM IBA formed multiple shoot primordia which were transferred to a medium containing 10 µM BA and subcultured repeatedly at 3-week intervals. The subcultured shoot primordia were produced when transferred to a medium containing lower concentrations (0.1 or 1.0 µM) of BA. Subcultured shoot primordia continued to regenerate plants for more than 3 years. Shoots developed and rooted on MS medium without growth regulators and the rooted plantlets were transferred to pots and grown in the greenhouse. Rejuvenation was obtained by this method of *in vitro* culture. Lin *et al.* (1996) cultured terminal buds, internode segments, and leaf-stalks of *P. edulis* on half-strength MS medium supplemented with 2 per cent sucrose and 1-4 mg IBA/l. The most effective rooting (100 per cent) was achieved with terminal buds in 2 or 4 mg IBA/l; rooting occurred within 10-12 days. Shoot induction rate was 100 per cent for internode segments cultured on half-strength MS medium supplemented with 2-3 mg zeatin/l. Root segments and leaf petioles gave 33 and 37 per cent shoot induction, respectively, with 2.5 mg zeatin/l.

Gloria *et al.* (1999) reported direct shoot regeneration from leaf-derived explants *in vitro* in light using BA and rhizogenesis in presence of NAA.

14.1.1.1 Performance of Tissue Cultured Plants

The growth of tissue-cultured plants of cv. Tainong No.1 *P. edulis*x*P. edulis* f. *flavicarpa*, as own-rooted tissue-cultured plants, tissue-cultured scions grafted on the seedling rootstock of *P. edulis* f. *flavicarpa,* and scions grafted on the same rootstocks was studied at the Viticulture Research Station, Taiwan (Huang *et al.*, 1997). Tissue-cultured, own-rooted and tissue-cultured, grafted plants were longer and larger than the grafted plants. Fruit yield from tissue-cultured, grafted plants was 1.5 times more than from tissue-cultured, own-rooted plants and twice that of the grafted plants. Passionfruit woodiness virus and passionfruit mottle virus infection were lowest and of shortest duration in the tissue-cultured own-rooted plants, while infections of grafted plants were highest and longest.

14.1.2 Micrografting

The dissemination of a disease caused by the *Cowpea aphid-borne mosaic virus* (CABMV) (Nascimento *et al.*, 2006) in Brazil has caused reductions in productivity and fruit quality, diminished plantation longevity, and has essentially eliminated the use of vegetative propagation (Ribeiro *et al.*, 2008). A low-cost technique of micrografting (grafting the shoot apices of selected plants onto the stems of young rootstock plants that had been cultivated *in vitro*) was developed to eliminate CABMV in yellow passionfruit plants (Ribeiro *et al.*, 2008) using soil substrates to grow seedlings for rootstock instead of culture medium, and stem apices (up to 400 lm long with from two to four leaf primordia) for grafting under semi-sterile conditions. This technique resulted in a 53 per cent success rate for scion development, with 93 per cent of the plants being pathogen-free. This technique also proved useful for the elimination of viruses in superior genotypes of *P. edulis* (Ribeiro *et al.*, 2008; Andrade *et al.*, 2010), with the possibility of its utilization in other *Passiflora* species (Pereira *et al.*, 2009).

The cv. E 23, a F_1 hybrid between *P. edulis* and *P. edulis* f. *flavicarpa*, is propagated by shoot-tip grafting. Growth of adult buds on intact stem sections was promoted by 1 week of dark incubation on MS basal medium plus 10 µM 2iP, 200 µM adenine sulphate and 17.1 µM IAA and further developed into shoots on MS medium plus 4.9 µM 2iP and 5.7 µM IAA. On contrast, juvenile shoots of E 23, and other *Passiflora* species *edulis* f. *flavicarpa*, *edulis*, *alata*, *caerulea*, *mollissima*, *coccinea*, *herber-tiana* and *suberosa* grew rapidly on MS medium plus 10 µM kinetin and 5 µM IAA. Rapid multiplication was achieved on MS plus 20 µM BA, 10 µM kinetin and 5 µM IAA, and roots initiated on MS plus 5 µM IAA (Drew, 1991).

Ribeiro *et al.* (2015) grafted shoot tips with 2–3 leaf primordia, and 600–800 µm in length, to the hypocotyl of 21–28 day-old rootstock seedlings, and their development was followed for 30 days using scanning electron and visible light microscopy. The success of micrografting was found to depend on the preservation of the vascular tissue of the rootstock and the placement of the scion adjacent to the rootstock phloem. They observed callus formation, which initiates approximately 3 days after micrografting (DAM) through the proliferation of parenchymatous cells of the rootstock cortex, fills the space left by the incision and guarantees adherence and nutrition for the scion during the initial phases of development. At seven DAM, connective cells that develop at the base of the scion produce a junction with the callus. At 10 DAM, differentiation of procambial strands and parenchymatous cells initiates in the callus. At 15 DAM, parenchymatous cells derived from the callus give rise to procambial strands and initiate the differentiation of tracheal elements and the epidermis in the junction region. Vascular connections are established at 20 DAM, promoting the accelerated development of the scion, which, at 30 DAM, shows shoot development. The developmental phases of micrografting in passionfruit plants therefore include: placement of the scion; callus formation by the rootstock; cellular connections; differentiation of the callus; vascular connection; and shoot development.

Meristem culture is a potential regeneration technique for obtaining virus-indexed stock plants for mass propagation of *P. edulis* f. *edulis*. Meristems excised from a greenhouse-grown plant were micrografted onto *in vitro* germinated seedlings. Fifty per cent of the micrografted meristems developed a small leaf 2 weeks after micrografting. Micrografted plants were successfully transplanted in a greenhouse into a standard soil mix in small pots (Biricolti and Chiari, 1994).

14.2.0 Protoplast Culture

New plantlets of passionfruit were regenerated and genetically improved by somatic cell techniques through protoplast culture (D'UtraVaz *et al.*, 1993; Otoni *et al.*, 1995a,b). Protoplast yield and viability varied according to the enzyme mixture used for isolation (Otoni *et al.*, 1995a). Protoplast isolated from newly expanded leaves by enzymatic digestion exhibited maximum frequency when planted at a density of 1.5×10^5 protoplasts/ml in agarose-solidified droplets of KM8P medium containing 250 mg cefotaxime/ml. Protoplasts-derived calli cultured for 30 days on agar-solidified MS medium with 5.0 mg NAA, 0.25 mg BA/l, and additional vitamins regenerated shoots on transfer to MS medium with 1.0 mg/l BA and were rooted in half-strength MS medium with 3.0 mg IBA and 0.5 mg NAA/l, followed by subcultures to hormone-free MS medium (D'UtraVaz *et al.*, 1993).

Plants were also regenerated from dark-green compact callus derived from leaf mesophyll protoplasts after 3 subcultures at 30-day intervals on MS medium supplemented with either 1.0 mg BA + 0.001 mg NAA or 1.0 mg zeatin/l, in both cases containing 3.0 per cent (w/v) sucrose and solidified with 0.7 per cent (w/v) agar (Otoni *et al.*, 1995a). The inclusion of antibiotic cefotaxime proved essential for inducing and sustaining cell division in isolated protoplasts (Otoni *et al.*, 1995b). Dornelas and Vieira (1993) isolated protoplasts from seedling cotyledons of yellow passionfruit (*P. edulis* var. *flavicarpa*) and two related wild species *P.amethystina* and *P. cincinnata*, which divided in culture and produced calli. Shoot regeneration was obtained in MS medium containing 2.0 mg BA/l. Regenerated plants produced roots in half-strength hormone-free MS medium and was transferred to soil after acclimatization.

14.3.0 Tissue Culture

In vitro regeneration in passion fruit (*Passiflora edulis* Sims. f. *flavicarpa* Deg.) has been obtained from shoot apices or nodal segments cultivation (Monteiro *et al.*, 2000; Reis *et al.*, 2003) or from adventitious buds developed from leaf discs (Appezzato-da-Glória *et al.*, 1999; Trevisan and Mendes, 2005), hypocotyl (Faria and Segura, 1997b), or internodal segments (Biasi *et al.*, 2000). Optimized protocols have been established mainly by using different combinations of plant growth regulators, such as BAP and IBA (Kawata *et al.*, 1995), BAP and NAA (Dornelas and Vieira, 1994), BAP and IAA (Faria and Segura, 1997b), and different salt solutions (Faria and Segura, 1997b; Monteiro *et al.*, 2000).

14.4.0 Somatic Hybrid

Somatic hybrid plants were produced following electrofusion of mesophyll protoplasts of *P. edulis* f. *flavicarpa* and *P. incarnata*. These hybrids exhibited morpholo- gical characteristics intermediate between those of the 2 parental species. Further evidence of hybridity was provided by isozyme analysis, ploidy determination and RAPD analysis.

14.5.0 Biochemical and Molecular Markers

Characterization of genetic variation among passion fruit genotypes is imperative for its effective conservation, management of germplasm and efficient utilization in breeding schemes. Genetic diversity of passion fruit has been studied with a number of molecular marker techniques such as Random Amplified Polymorphism, RAPD (Aukar *et al.*, 2002; Bellon *et al.*, 2009; Cerqueira-Silva *et al.*, 2010), Amplified Fragment Length Polymorphism, AFLP (Segura *et al.*, 2002), Simple Sequence Repeats, SSRs [Cerqueira-Silva *et al.*, 2014; Cerqueira-Silva *et al.*, 2015], and Inter Simple Sequence Repeats, ISSRs (Santos *et al.*, 2011). RAPD technique is simple, convenient, and inexpensive, but low stability and poor reproducibility limits its utilization (Roodt *et al.*, 2002). On the other hand, AFLP technology presents good reproducibility and high polymorphism, but is complex, requires multiple steps and shows pseudo-polymorphism when methylation-sensitive restriction enzymes are used (Budak *et al.*, 2004). Further, SSRs are stable, abundant, co-dominant, highly polymorphic and reproducible, but they require sequencing for primer development for each species, which makes the method time-consuming and expensive (Li *et al.*, 2001). Although ISSRs are widely distributed throughout the genome and are highly polymorphic, they have a disadvantage of being dominant marker and thus unable to differentiate heterozygosity at a given locus (Budak *et al.*, 2004).

Do *et al.* (1992) compared chloroplast DNA (ctDNA) of *P. edulis* (including f. *flavi-carpa*), *P. alata*, *P. coccinea*, *P. suberosa* and intraspecific and interspecific hybrids by RFLP analysis using 6 restriction enzymes. According to the fragment patterns generated by BamHI, purple-type passionfruit was characterized by B8 (4.3 kb), B9 (4 kb) and B13 (2.9 kb) fragments, while the yellow-type (f. *flavicarpa*) was characterized by B3a (7.5 kb), B7c (4.7 kb) and B19b (1.8 kb) fragments. All the evidences indicated that the pollen parent might contribute ctDNA to the fertilized egg, but interaction of organelles between both parents led to exclusion of chloroplast genomes. It appeared that yellow passionfruit might have stronger transmission abillity than its related species.

Figardo *et al.* (1998) carried out genetic analysis based on RAPD on 52 accessions representing 14 species of the genus *Passiflora* using 50 random 10-mer primers. Large intraspecific variation was found *P. ligularis* and *P. adenopoda* while *P. edulis* and *P. maliformis* exhibited a low level of intraspecific variation. The clusters based on RAPD markers correlated fairly well with the present classification scheme based on morphological description with two exceptions: subgenus *Passiflora* was split into *P. edulis* and the rest of the members of subgenus *Passiflora* separated by sub-genus *Tacsonia*; and the two species of the subgenus *Decaloba*, *P. adenopoda*and *P. coriacea*, were not clustered together on the dendrogram of genetic relationship.

Some 125 accessions of *Passiflora* belonging to 26 species were characterized by Sanchez *et al.* (2000) using 50 RAPD. Furthermore, 35 accessions were analysed by RFLP of chloroplast DNA. RAPD revealed a high degree of genetic diversity, RFLP indicated high interspecific polymorphism in *P. ligularis* and *P. maliformis*.

A recent study was undertaken to determine the genetic diversity of passionfruit genotypes from major growing regions in Kenya using sequence-related amplified polymorphism (SRAP) markers (Oluoch *et al.*, 2018). Twenty four SRAP primer combinations were screened using three passion fruit genotypes and only seven that displayed polymorphic and stable amplification profiles were used to analyze 22 genotypes. The seven primer combinations amplified a total of 931 clear bands with an average of 133 bands per primer pair, of which 610 (65.5 per cent) bands were polymorphic. The similarity coefficients among the 22 passion fruit germplasms ranged from 0.51 to 1.0 with an average of 0.755. The 22 passion fruit genotypes were classified into two groups by cluster analysis using unweighted pair-group method with arithmetic mean (UPGMA) with 12 per cent similarity. Shannon's diversity index was 0.0934 and Nei's gene diversity index was 0.1370 in the present study. The study findings demonstrate the existence of a considerable amount of genetic variability among passion fruit genotypes grown in different regions of Kenya. This indicates the potential application of these genotypes in breeding programs by exploiting the use of molecular markers for selection of specific agronomic traits.

14.6.0 Genetic Transformation

The success of genetic engineering in controlling viral diseases in some fruit species and the need for minimizing problems associated with CABMV in passionflower orchards have encouraged the development of a research project for the production of transgenic Passiflora plants in Brazil. Initially, *P. edulis f. flavicarpa* cultivars were transformed using *Agrobacterium tumefaciens* and a gene construct containing the coat protein gene (CP) of CABMV (Trevisan *et al.*, 2006), which resulted in the selection of transgenic plants resistant to this virus. These plants were self-pollinated, and R1, R2, and R3 generations were obtained (Monteiro-Hara *et al.*, 2011) and propagated. The results of the study conducted by Correa *et al.* (2015) on sweet passionfruit indicated that the incorporation of CABMV-gene fragments into the Passiflora genome may influence the resistance of these plants to the pathogen. The availability of plants with some level of impairment of virus infection and/or replication is important and these plants may serve as parents for hybridization in breeding programs. In addition, the use of these plants together with other control measures in integrated disease management may be beneficial for maintaining disease intensity below an economic injury threshold (Nutter, 2007).

Mander *et al.* (1994) cocultivated leaf and stem explants with a disarmed strain of *A. tumefaciens* harbouring the co-integrate vector pMON200. Four plants were regenerated from leaf explants on agar-solidified MS medium supplemented with 4.43 µM BA and 86 µM kanamycin sulphate. The plants were rooted by transfer to MS medium supplemented with 14.7 µM IBA and 2.68 µM NAA for 7 days, followed by hormone-free MS medium. Both rooting media contained 172 µM kanamycin sulphate. Rooted plants were potted and grown to maturity. Three of the plants synthesized nopaline and expressed NPTII activity. DNA dot blot and polymerase chain reaction analyses confirmed the presence of the *nptII* gene. Lin (1998) smeared *Agrobacterium rhizogenes* MAFF03-01724 over a stem segment of *P. edulis* inoculated upside-down on MS solid medium. After 4 weeks, tumours broke out of the injured stem segment. Adventitious roots were derived from the tumours the following week. The hairy roots were transferred onto MS medium supplemented with 1 mg BA and 0.1 mg 2,4-D/l where they developed into plantlets.

15.0 References

Abeysinghe, A. (1973) *J. Nat. Agric. Soc. Ceylon*, **9**: 87-111.

Aguire, A.C.D., Lourenco, R.S. and Malavolta, E. (1977) *Solo (Brazil)*, **69**: 30-31,

Akamine, E.K. and Girolami, G. (1959) *Tech. Bull. Hawaii Agric. Exp. Sta.*, p. 39.

Almeida Lima, A.de, Carvalho, J.E.B.de and Caldas, R.C. (1999) *Rev. Bras. Frutic.*, **21**: 379-381.

Amaral, Uirá do, Ferreira, Luiz Leonardo, de Souza, Kelly Martins, Vieira, Emerson Pereira and Barcelos, Michelle Nunes (2018) *J. Agric. Sci.*, **10** (9): 423-429.

Andrade, S.R.M., Ribeiro, L.M., Vieira, L.M., Pereira, W.V.S., Nery, L.A., Fogaca, C.M., Rosa, S.D., Faleiro, F.G., Silva, M.S. and Junqueira, N.T.V. (2010) *Boletim de Pesquisa e Desenvolvimento (Embrapa Cerrados)*, **279:** 6–29.

Anonymous (1967) *Queensland Agric. J.*, **93**: 130-141.

Anonymous (1979) *Circ. Hawaii Agric. Exp. Sta.*, p. 345.

Anonymous (1983) *Plant Tissue Analysis Interpretation Manual*, Consolidated Fertilizers Limited, Brisbane.

Anonymous (1984) *Inf. Bull. Citrus Subtrop., Fruit Res. Inst.*, No. 141, pp. 104.

Anonymous (1986a) *Inf. Bull. Citrus Subtrop. fruit Res. Inst.*, No. 164, pp. 1-5.

Anonymous (1986b) *Inf. Bull. Citrus Subtrop. fruit Res. Inst.*, No. 168, pp. 10-11.

Anonymous (1998) GRAN ANGULAR, "Discover the passionfruit growing in the Chapare" Magazine on *Development in the Cochabamba Tropics*. Number 33, Jan

Antognozzi, E. and Proietti, P. (1995) *Acta Hort.*, **379**: 159-166.

Appezzato-da-Glória, B., Vieira, M.L.C. and Dornelas, M.C. (1999) *Pesqui. Agropecu. Bras.*, **34:** 2007-2013

Araujo Filho, G.C., Fatima Pinheiro Sa, M. and Vasconcelos Lopes, J.G. (1981) *Commun. Tech. Epace*, **8**: 1-6.

Arenas Ocampo, M.L., Evangelista-Lozano, S. and Valdes Estrada, M.E. (1994) *Proc. Interamerian Soc. Trop. Hort.*, **38**: 139-141.

Arjona, H.E., Matta, F.B. and Garner, J.O.Jr. (1991) *HortScience*, **26**: 921-923.

Arjona, H.E., Matta, F.B. and Garner, J.O. Jr. (1992) *HortScience*, **27**: 809-810.

Arjona, H.E., Matta, F.B. and Garner, J.O.Jr. (1994) *HortScience*, **29**: 295-296.

Aukar, A.P.A., Lemos, *e.g.*M. and Oliveira, J.C.O. (2002) *Rev. Bras. Frutic.*, **24**: 738–740.

Aular, J., Ruggiero, C. and Durigan, J.F. (2000) *Rev. Bras. Frutic.*, **22**: 6-8.

Avilan, A.R. (1974) *Agron. Trop.*, **24**: 134-140.

Ayala-Zavala, J. F., Vega-Vega, V., Rosas-Dominguez, C., Palafox-Carlos, H., Villa-Rodriguez, J. A. and Siddiqui, M.W. (2011) *Food Res. Int.*, **44:** 1866-1874.

Bailey, L.H. (1949) *Manual of Cultivated Plants*, MacMillan Co., New York.

Baumgartner, J.G., Malavolta, E. and lourenco, R.S. (1978) *Cientifica*, **6**: 361-367.

Bautista, D. and Salas, D. (1995) *Agronomia Tropical (Maracay)*, **45:** 331-345.

Beal, P.R. and Farlow, P.J. (1984) In: Tropical *Tree Fruits for Australia* (Ed. P.E. Page), Queensland Govt. Printer, Brisbane.

Bellon, G., Faleiro, F., Peixoto, J., Junqueira, P., Junqueira, N., Fonsceca, K. and Braga M. (2009) *Rev. Bras. Frutic.*, **31**: 197-202.

Bertin, Y. (1976) *Fruits*, **31**: 171-176.

Besold, D. and Karnatz, A. (1986) *HortScience*, **21**: 857.

Bester, C.W.J., Van Zyl, H.J. and Etsebeth, J.W. (1978) *Deciduous Fruit Grower*, **28**: 75-77.

Biasi, L.A., Falco, M.C., Rodriguez, A.P.M. and Mendes, B.M.J. (2000) *Sci. Agric.*, **57:** 661-665.

Biricolti, S. and Chiari, A. (1994) *Adv. Hort. Sci.*, **8**: 171-175.

Blondeau, J.P. and Bertin, Y. (1978) *Fruits*, **33:** 433-443.

Blondeau, J.P. and Bertin, Y. (1980) *Fruits*, **35**: 361-367.

Borges, A. L., and Lima, A. de A. (2003) In: Passionfruit. Embrapa Mandioca e Fruticultura, Caixa Postal 007, CEP 44380-000, Cruz das Almas-BA, Brazil.

Borges, A.L. and Lima, A. de A. (2007) In: *Fertilizing for High Yields and Quality: Tropical Fruits of Brazil*, (Ed. A. E. Johnston) Horgen, Switzerland: International Potash Institute, IPI Bulletin 18, pp. 163–178.

Brekke, J. E. (1973) *Res.Pep. Hawaii. Agric. Exp. Sm.*, 198: 1-8.

Bruckner, C.H., Casali, V.W.D., De Moraes, C.F., Regazzi, A.J., and Da Silva, E.A.M. (1993) *Int. Symp. Trop. Fruits.*, **370**: 45–58.

Budak, H., Shearman, R.C., Parmaksiz, I., Gaussoin, R.E., Riordan, T.P. and Dweikat I. (2004) *Theor. Appl. Genet.*, **108:** 328–334.

Caicedo, R.G., Vargas, G.H. and Gaviria, M.J. (1993) *Rev.Colomb. Entomol.*, **19**: 107-110.

Carr, M.K.V. (2013) *Expl. Agric.*, **49** (4): 585–596.

Carvalho, A.J.C.de, Martins, D.P., Monerat, P.H. and Bernardo, S. (1999) *Rev. Bras. Frutic.*, **21**: 333-337.

Carvalho, A.J.C.de, Marting, D.P., Monneratl, P.H. and Bernardo, S. (2000) *Pesqui. Agropecu. Bras.*, **35**: 1101-1108.

Casimir, D.J. (1974) *Ph. D. thesis* submitted to University of New South Wales, Sydney, NSW, Australia.

Castillo, Nohra Rodríguez, Ambachew, Daniel, Melgarejo, Luz Marina and Blair, Matthew Wohlgemuth (2020) *Hortscience*, **55**(6): 768–777.

Cavalcante, N.R., Krause, W., Viana, A.P., Silva, C.A., Porto, K.K.X., Martinez, R.A.S. (2017) *Acta Sci. Agron.*, **39:** 143. https: //doi.org/10.4025/actasciagron. v39i2.31022.

Cerqueira-Silva, C.B.M., Conceição, L.D.H.C.S., Santos, E.S.L., Cardoso-Silva, C., Pereira, A.S., Oliveira, A.C. and Corrêa R.X. (2010) *Genet. Mol. Res.*, **51:** 2421–2428.

Cerqueira-Silva, C.B.M., Jesus, O.N., Oliveira, E.J., Santos, E.S.L., Souza, A.P. (2015) *Euphytica*, **202** (3): 345–359.

Cerqueira-Silva, C.B.M., Jesus, O.N., Santos, E.S.L. and Corrêa, R.X. (2014) *Int. J. Mol. Sci.*, **15:** 14122–14152.

Chakraborty, I., Chaurasiya, A.K. and Saha, J. (2011) *J. Food Sc.i Technol.*, **48** (6): 750–754

Chan, H. J.Jr. (1980) In: *Tropical and Subtropical Fruits: Composition, Properties and Uses* (Eds. S. Nagy and P.E. Shaw), AVI Publishing, Wesport, USA.

Chaurasiya, A.K. (2010) *Ph.D thesis* submitted to Bidhan Chandra Krishi Viswavidyalaya, West Benagal, India

Chavarria-Perez, Lourdes Maria, Giordani, Willian, Dias, Kaio Olimpio Gracas, Costa, Zirlane Portugal, Ribeiro, Carolina Albuquerque Massena, Benedetti, Roberto, Anderson, Cauz-Santos, Luiz Augusto, Pereira, Guilherme Silva, Rosa, João Ricardo Bachega Feijo´, Garcia, Antonio Augusto Franco, Vieira, Maria Lucia Carneiro (2020) *PLoS ONE*, **15**(5): e0232818. https: //doi.org/10.1371/ journal.pone.0232818.

Contreras-Calderon, J., Calderon-Jaimes, L., Guerra-Hernandez, E. and Garcia-Villanova, B. (2011) *Food Res. Int.*, **44:** 2047-2053.

Corebet, S.A. and Willmer, P.G. (1980) *J. Agric. Sci.*, **95**: 655-666.

Correa, Marcelo Favareto, Pinto, Ana Paula Chiaverini, Rezende, Jorge Alberto Marques, Harakava, Ricardo and Mendes, Beatriz Madalena Januzzi (2015) *Eur. J. Plant Pathol.*, **143:** 813–821.

Correa, R. C. G., Peralta, R. M., Haminiuk, C. W. I., Maciel, G. M., Bracht, A., and Ferreira, I. C. F. R. (2016) *Trends Food Sci. Technol.*, 58: 79–95.

da S Pereira, G.S., Laperuta, L.D.C., Nunes, E.S.E.S., Chavarria, L., Pastina, M.M.M.M. and Gazaffi, R. (2017) *Trop. Plant Biol.*, **10:** 18-29.

de Jesus, C.A.S., de Carvalho, E.V., Girardi, E.A., Rosa, R.C.C., de Jesus, O.N. and de Jesus, C.A.S. (2018) *Rev. Bras. Frutic.*, **40:** https: //doi.org/10.1590/0100-29452018968.

Do, Y.Y., Shii, C.T. and Huang, P.L. (1992) *Spl. Pub. Taichung Dist. Agric. Improv. Sta.*, 29, pp. 303-317.

Dornelas, M.C. and Vieira, M.L.C. (1993) *Plant Cell Rep.*, **13**: 103-106.

Dornelas, M.C. and Vieira, M.L.C. (1994) *Plant Cell, Tissue and Organ Culture*, **36:** 211-217.

Drew, R.A. (1991) *Plant Cell Tissue Organ Cult.*, **26**: 23-27.

D'UtraVaz, F.B., Santos, A.V.P.Dos, Manders, G., Cooking, *e.g.*, Davey, M.R. and Power, J.B. (1993) *Plant Cell Rep.*, **12**: 220-225.

Elias Menesis, S., Panesso, I.F.A., Torres, C.G. and Mejia, de T.S. (2000) *ASCOLFI Informa*, **26**: 3-4.

Escobar, L.K. (1986) *Syst. Bot.*, **11**: 88-97.

Fagne, E., Souza, D., Toledo, M., Galape, F., Borges, L., Barbosa, A. and Nietsche, S. (2014) Avaliacxao do desempenho agronomico de tres genotipos de maracujazeiro-azedo sob cultivo irrigado no norte de Minas Gerais. FEPEG, Forum Ensino, Pesquisa, Extensao, Gestao. Unimontes, Brasil.

Fajac, F. (1986) *Fruits*, **41**: 563-573.

Fajac, F. (1987) *Fruits*, **42**: 59-64.

Faria, J.L.C., Colauto, N.M., Manica, I., Stronski, M.D.S. and Appel, H.B. (1991) *Pesqui. Agropecu. Bras.*, **26**: 311-314.

Faria, J.L.C. and Segura, J. (1997) *HortScience*, **32:** 1276-1277.

Farid, R., Rezaieyazdi, Z., Mirfeizi, Z., Hatef, M. R., Mirheidari, M. and Mansouri, H. (2010) *Nutr. Res.*, **30:** 601-606.

Fernandes, P.D., Oliveira, G.D. de, Ruggiero, C. and Haag, H.P. (1977) *Solo* (*Brazil*), **69**: 16-21.

Ferreira Filho, C. de Q., Silva, J.U.B., Trindade, J. and Silva, L.M.S.Da (1984) *Ann. Brazil Cong. fruit*, **7**: 1000-1008.

Ferreira, T.G.T., Penha, H.A., Zucchi, M.I., Santos, A.A., Hanai, L.R. and Junqueira, N. (2010) *Plant Breed.*, **129:** 727–730.

Figardo, D., Angel, F., Grum, M., Thome, J., Lobo, M., Roca, W.M. and Sanchez, I. (1998) *Euphytica*, **101:** 341-347.

Fisch, M.B. (1975) *Calif. Rare Fruit Growers Yearb.*, **7**: 13-55.

Fouque, A. (1982) *Fruits*, **37**: 599-608.

Freire, J.L.O., Cavalcante, L.F., Rebequi, A.M., Dias, T.J. and Souto, A.G.L. (2011) *Revis. Caating.*, **24**(1): 82–91.

Fullerton, R.A. (1986) *Rep. Res. Advisory Committee Totokoitu Res. Sta.*, Rarotonga.

Gachanja, S.P. (1975) *Acta Hort.*, **49**: 219-222.

Gachanja, S.P. and Gurnah, A.M. (1978) *East Afr. Agric. For J.*, **44**: 47-51.

Gachanja, S.P. and Gurnah, A.M. (1980) *J. Hort. Sci.*, **55**: 345-349.

Gama, F.S.N.da, Manica, I., Kist, H.G.K. and Accorsi, M.R. (1991) *Pesqui. Agropecu. Bras.*, **26**: 305-310.

Gammarra Rojas, G. and Medina, M. (1997) *Fruits* (*Paris*), **52**: 19-26.

Gilmartin, A.J. (1958) *Trop. Agric.*, **35**: 41-58.

Gilmore, J.G. (1983) *Zimbabwe Agric. J.*, **80**: 83-92.

Gloria, B.A.da, Vieira, M.L.C. and Dornelas, M.C. (1999) *Pesqui. Agropecu. Bras.*, **34**: 2007-2013.

Gomez, P.K. (2000) *Revista de la Facultad de Agronomia, Universidad del Zulia*, **17**: 1-9.

Gonzalez, G., Rodriguez, J. and Sourd, D. (1989) *Agrotecnia de Cuba*, **21**: 107-109.

Grozmann, H.M. and Purss, G.S. (1958) *Queensland Agric. J.*, **84**: 214-216.

Guadarrama, A. (1996) *Acta Hort.*, **429**: 539-545.

Gurnah, A.M. and Gachanja, S.P. (1984) *Trop. Agric.*, **61**: 143-147.

Haag, H.P., Oliveira, G.D. de, Borducchi, A.S. and Saurudge, J.R. (1973) *Anais Scola Superior Agric. Luiz de Queiroz*, **30**: 267-279.

Haddad, G.O. and Figueroa, R.M. (1972) *Agron. Trop.*, **22**: 483-496.

Hammer, L.H. (1988) *Proc. Fla. Sta. Hort. Soc.*, **100**: 283-287.

Hardin, L.C. (1987) *Proc. Interamerican Soc. Trop. Hort.*, **30**: 35-44.

Hargreaves, J.R. (1979) *Queensland J. Agric. Anim. Sci.*, **36**: 147-150.

Harprakasarao, M., Subramanian, T.R., Ravishankar, H. and Srinivasan, V.R. (1988) *Indian J. Hort.*, **45**: 263-267.

Haury, A. (1979) *Fruits*, **34**: 289-293.

Heenkenda, H.M.S. and Punchikumarihami, S.M.A. (1995) *Trop. Agric.*, **147**: 11-18.

Hernandez-Santos, B., de Los Angeles Vivar-Vera, M., Rodriguez-Miranda, J., Herman-Lara, E., Gabriel Torruco-Uco, J. and Acevedo-Vendrell, O. (2015) *Int. J. Food Sci. Tech.*, **50:** 268-274.

He, X., Luan, F., Yang, Y., Wang, Z., Zhao, Z., Fang, J., Wang, M., Zuo, M. and Li, Y. (2020) *Front. Pharmacol.*, **11, doi**: 10.3389/fphar.2020.00617.

Howell, C.W. (1976) *Proc. Fla. State Hort. Soc.*, **89**: 236-238.

Huang, B.H., Yang, Y.S., Weng, S.W., Chen, S. and Chang, C. (1997) *Special Publication - Taichung District Agricutural Improvement Station*, No. 38, 301-311.

IBGE (2017) In: Produc¸ão agrý´cola municipal: culturas tempora´ rias e permanentes. Rio de Janeiro.

Inch, A.J. (1978) *Queensland Agric. J.*, **104**: 479-484.

Ishihata, K. (1981) *Bull. Fac. Agric. Univ. Kagoshima*, **31**: 25-31.

Ishihata, K. (1983) *Bull. Fac. Agric. Univ. Kagoshima*, **33**: 7-11.

Ishihata, K., Hayashi, M. and Ikeda, M. (1984) *Bull. Fac. Agric. Univ. Kagoshima*, **34**: 9-16.

Ishihata, K., Masuyama, H. and Iwahori, S. (1987) *Bull. Fac. Agric. Univ. Kagoshima*, **37**: 17-28.

Ito, P.J. (1978) *HortScience*, **13**: 197.

Joy, P.P. (2010) In: *Passionfruit Production Technology*. Pineapple Research Station (Kerala Agricultural University), Kerala, India, pp.1-23.

Joy. P. P. and Divya. B. (2016) In: *Post-harvest handling of passionfruit.* Pineapple Research Station (Kerala Agricultural University), Kerala, India

Jung, M.S., Vieira, E.A., Brancker, A. and Nodari R.O. (2007) *Ciência Rural St Maria.*, **37**: 963–969.

Junqueira, N.T., Icuma, I.M., Veras, M.C., Oliveira, M.A. and Santos, J.R. do (2013) *Sci. Technol.*, **6:** 178–183.

Junqueira, N.T., Lage, D.A., Braga, M.F., Peixoto, J.R., Borges, T.A., and Andrade, S.R. (2006) *Rev. Bras. Frutic.*, **28**: 831– 836.

Kantharajah, A.S. (1985) *Dr. Agric. Sci. Thesis*, Technical Univ., Berlin.

Kawata, K., Ushida, C., Kawai, F., Kanamori, M. and Kuriyama, A. (1995) *J. Plant Physiol.*, **147**: 281-284.

Killip, E.P. (1938) *Field Mus. Nat. Hist. Bot. Ser.*, No. 19, pp. 407-408.

Kikutani, N., Momma, K., Iguchi, M., Tomomatsu, T., Ozawa, K., Wada, M., Aoki, M., Tosaka, M. and Urano, M. (1988) *Ann. Rep. Tokyo Metropolitan Res. Lab. Pub. Health*, No.39, pp. 164-167.

Kishore, K. K., Pathak, A., Yadav, D.S., Bujarbaruah, K.M., Bharali, R. and Shukla, R. (2006) *Tech Bull.*, pp. 2-4.

Kitajima, E,W., Rezende, J.A.M., Rodrigues, J.C.V., Wegato, L.G., Piza Junior, C.T. and Morozini, W. (1997) *Fitopatol. Bras.*, **22**: 555-559.

Kist, H.G.K., Manica, I. and Boaro, J.A. (1995) *PesquisaAgropecuariaGaucha*, **1**: 21-26.

Knight, R.J.Jr. (1969) *Proc. Trop. Reg. Am. Soc. Hort. Sci.*, **13**: 265-274.

Knight, R.J.Jr. (1980) In: *Tropical and Subtropical Fruits: Composition, Properties and Uses* (Eds. S. Nagy and P.E. Shaw), AVI Publishing, Westport, USA.

Kotur, S.C., Singh, H.P. and Yadav, I.S. (1983) *South Indian Hort.*, **31**: 32-33.

Kuhne, F.A. (1975) *Calif. Rare Fruit Growers Yearb.*, **7**: 56-71.

Lederman, *i.e.* and Gazit, S. (1993) *Pesqui. Agropecu. Bras.*, **28**: 1195-1199.

Letchamo, W., Xu, H.L., Desroches, B. and Gosselin, A. (1993) *J. Plant Nutrition*, **16**: 2521-2537.

Lenza, J.B., Valente, J.P., Roncatto, G., and Ching, L.A. (2009) *Rev. Bras. Frutic.*, **31**: 831–836.

Liao, J. (1987) *J. Fujian Agricultural College*, **16**: 306-312.

Liebregts, W.J.M.M., Sands, D.P.A. and Bourne, A.S. (1989) *Bulletin Entom. Res.*, **79**: 163-171.

Li, G.Y., and Quiros, C.F. (2001) *Theor. Appl. Genet.* **103**: 455–461.

Lima, D.M. de, Golombieski, E.R. and Ayub, R.A. (2000) *Ciencia Rural*, **30**: 359-363.

Lima, A.de A., Caldas, R.C., Cunha, M.A.P.da and Santos Filho, H.P. (1999) *Rev. Bras. Frutic.*, **21**: 318-321.

Lin, S.Q. (1998) *J. Fujian Agric. Univ.*, **27**: 225-256.

Lin, S.Q., Chen, F.X., He, B.Z. and Su, M.S. (1996) *J. Fujian Agric. Univ.*, **25**: 302-306.

Lippmann, D. (1978) *Cultivation of Passiflora edulis – General Information on Passionfruit Growing in Kenia (Kenya)*, German Agency for Technical Co-operation, Gar-Hammarskjold.

Liu, S., Li, A.D., Chen, C., Cai, G., Zhang, L., Guo, C. and Xu, M. (2017) *Forests*, **8**: 435.

Lopez-Vargas, J. H., Fernandez-Lopez, J., Perez-Alvarez, J. A. and Viuda-Martos, M. (2013) *Food Res. Int.*, **51**: 756-763.

Mac Dougal, J.M. (1989) *Annals Missouri Bot. Garden*, **76**: 608-614.

Maciel, N., Bautista, D. and Aular, J. (1994) *Proc. Interamer. Soc. Trop. Hort.*, **38**: 133-138.

Maciel, N., Bautista, D. and Aular, J. (1997a) *Fruits (Paris)*, **52**: 11-17.

Maciel, N., Bautista, D. and Auiar, J. (1997b) *Fruits (Paris)*, **52**: 93-97.

Maldonado, J.F.M., Cruze Silva, J.A.da, Fernande, S.G., Carvalho, S.M.P. de, Cosga, R.A.da, Oliveira, L.A.de, Sarmento, W.da R.M. and Cunha, H. (1999) *Documentos-Empresa de PesquisaAgropecuaria do Estado do Rio de Janeiro*, **49**: 34.

Manders, G., Otoni, W.C., D'UtraVaz, F.B., Blackhall, N.W. and Power, J.B. (1994) *Plant Cell Rep.*, **13**: 697-702.

Manica, I., Ritzinger, R., Koller, O.C., Riboldi, J., Ramos, R.M. and Rodrigues, A.E.C. (1985) *Fruits*, **40**: 265-270.

Manica, I., Ritzinger, R., Mundstock, E.C., Koller, O.C., Marodin, G.A.B. and Rodrigues, A.E.C. (1987) *Proc. Interamerican Soc. for Trop. Hort.*, **31**: 1-9.

Manica, I., Fioravanco, J.C., Barradas, C.I.N. and Kist, H. (1994) *Pesqui. Agropecu. Bras.*, **29**: 1083-1090.

Marchal, J. (1987) In: *Plant Analysis as a Guide to the Nutrient Requirements of Temperate and Tropical Crops* (Eds. P. Martin-Prevel, J. Gangnard and P. Gautier), Lavoisier Publ. Inc., New York.

Marchal, J. and Bourdeaut, J. (1972) *Fruits*, **27**: 307-311.

Marchal, J., Blondeau, J.P. and Bertin, Y. (1978) *Fruits*, **33**: 681-691.

Marchal, J., Blongeau, J.P. and Bertin, Y. (1980) *Fruits*, **35**: 529-536.

Martin, F.W. and Nakasone, H.Y. (1970) *Econ. Bot.*, **25**: 333-343.

Martinez, R., Torres, P., Meneses, M. A., Figueroa, J. G., Perez-Alvarez, J. A. and Viuda-Martos, M. (2012) *Food Chem.*, **135**: 1520-1526.

Matheri, F., Mwangi, M., Runo, S., Ngugi, M., Kirubi, D.T., Njoroge, A., Mawia, A.M., Kioko, F.W., Lagat, R. and Daniel, N.A. (2016) *Adv. Crop Sci. Technol.*, **4**: 226.

Medagoda, I. and Balasubramaniam, S. (1988) *Trop. Agriculturist*, **144**: 135-144.

Meinke, H. (1986) *M. Sc. Thesis*, Technical Univ. Berlin.

Meinke, H. and Karnatz, A. (1990) *Scientia Hort.*, **43**: 237-246.

Meletti, L.M.M. (2011) *Rev. Bras. Frutic., Jaboticabal*, E: 83-91.

Menzel, C.M. and Simpson, D.R. (1988) *Scientia Hort.*, **34**: 77-88.

Menzel, C.M. and Simpson, D.R. (1989) *Scientia Hort.*, **41**: 83-96.

Menzel, C.M., Simpson, D.R. and Dowling, A.J. (1986a) *Scientia Hort.*, **29**: 239-249.

Menzel, C.M., Simpson, D.R. and Price, G.H. (1986b) *Scientia Hort.*, **28**: 339-346.

Menzel, C.M., Simpson, D.R. and Winks, C.W. (1987) *Scientia Hort.*, **31**: 259-268.

Menzel, C.M., Winks, C.W. and Simpson, D.R. (1988) *Queensland Agric. J.*, **114**: 13-18.

Menzel, C.M., Winks, C.W. and Simpson, D.R. (1990) In: *Fruits - Tropical and Subtropial* (Eds. T.K. Bose and S.K. Mitra), Naya Prakash, Calcutta, pp. 690-721.

Menzel, C.M., Haydon, G.F. and Doogan, V.J. (1993) *J. Hort. Sci.*, **68**: 215-229.

Menzel, C.M., Turner, D.W., Doogan, V.J. and Simpson, D.R. (1994) *J. Hort. Sci.*, **69**: 553-564.

Menezes, J.M.T., Oliveira, J.C.de, Ruggiero, C. and Banzatto, D.A. (1994) *Cietifica Jaboticabal*, **22**: 95-104.

Milne, D.L., de Villers, E.D., Logie, J.M., Bredell, G.S., Barnard, C.J. and Kuhne, F.A. (1977) *Citrus Subtrop. Fruit J.*, **524**: 16-18.

Mohammed, M. (1993) *Proc. Interamerican Soc. Trop. Hort.*, **37**: 85-88.

Monteiro, A.C.B.A., Higashi, E.N., Gonçalves, A.N. and Rodriguez, A.P.M. (2000) *Plant*, **36:** 527-531.

Monteiro-Hara, A.C.B A., Jadão, A.S., Mendes, B.M.J., Rezende, J.A. M., Trevisan, F., Mello, A.P.O.A., Vieira, M.L.C., Meletti, L.M.M. and Piedade, S.M.S. (2011) *Plant Dis.*, **95**: 1021–1025.

Morales, A.A. and Muller, L.E. (1976) *Turrialba*, **26**: 331-336.

Moreira, P. (2009) In: *Tesis pregrado Ingeniero Agronomo*. Universidad Tecnica de Manabi, Facultad de Ingeniería Agronomica. SantaAna-Manabi, Ecuador.

Morton, J.F. (1967) *Proc. Fla. State Hort. Soc.*, **80**: 320-330.

Morton, J.F. (1987) In: Fruits of warm climates, Miami, Florida, pp. 320–328.

Murray, D.A.H. (1976) *Queensland Agric. J.*, **102**: 145-151.

Nakagawa, J., Cavariani, C. and Amaral, W.A.N.do (1991) *Rev. Bras. Sementes*, **13**: 77-80.

Nakasone, H.Y., Hirano, R. and Ito, P. (1967) *Tech. Prog. rep. Hawaii Agric. Exp. Sta.*, No. 161.

Narain, N. and Bora, P.S. (1992) *J. Food & Agric.*, 529-530.

Nascimento, A.V.S., Santana, E.N., Braz, A.S.K., Alfenas, P.F., Pio-Ribeiro, G., Andrade, G.P., Carvalho, M.G., Zerbini, F.M. (2006) *Arch. Virol.*, **161**: 21–34.

Nascimento, T.B. do, Ramos, J.D. and Menezes, J.B. (1998) *Rev. Bras. Frutic.*,**20**: 33-38.

Nishida, T. (1958) *J. Econ. Entomol.*, **51**: 146-149.

Nishida, T. (1963) *Tech. Bull. Hawaii Agric. Exp. Sta.*, 55.

Nives, C.S.V.J., Carvaiho, S.I.C.de and Neves, P.M.O.J. (1999) *Rev. Bras. Frutic.*, **21**: 128-130.

Nutter, F.W. Jr. (2007) In: *General concepts in integrated pest and disease management* (eds. A. Ciancio and K. G. Mukerji), Dordrecht: Springer, pp. 45–79.

Olermo, N. C., Santos, L. A. and Labindao, J. P. R. (2017) *Asia Pac. J. Multidis. Res.*, **5** (2): 125-130.

Otoni, W.C.,Casali, V.W.D., Cecon, P.R., Davey, M.R. and Power, J.B. (1995a) *Revista Ceres*, **42**: 461-468.'

Otoni, W.C., Casali, V.W.D., Cecon, P.R., Moria, J., Power, J.B. and Davey, M.R. (1995b) *Revista Ceres*, **42**: 507-515.

Paiva, M. C. G., Barcellos Júnior, L. H., Pereira, G. A. M., Gonçalves, V. A., Felipe, R. S., Machado, M. S. and Silva, A. A. (2015) *Revis. Brasil. de Herbicidas*, **14**(4): 280-287.

Partridge, I.J. (1972) *Fiji Agric. J.*, **34**: 97-99.

Patterson, B.D., Kenrick, J.R. and Raison, J.K. (1978) *Phytochem.*, **17**: 1089-1092.

Patterson, B.D., Murata, T. and Graham, D. (1976) *Aust. J. Plant Physiol.*, **3**: 435-442.

Pejxoio, J.R., Paiva Junior, M.C.de, Angelis, B.de and Oliveira, J.A. de (1999) *Rev. Bras. Frutic.*, **21**: 49-51.

Penha, H.A., Pereira, G.S., Zucchi, M., Diniz, A. and Vieira, M.L.C. (2013) *Plant Breed.*, **132**: 731–735.

Pereira, W.V.S., Ribeiro, L.M., Vieira, L.M. and Mercadante-Simoes, M.O. (2009) *Pesqui. Agropecu. Bras.*, **44:** 446–453

Pope, W.T. (1935) *Bull. Hawaii Agric. Exp. Sta.*, 74.

Primavesi, A.C.P.A. and Malavolta, E. (1980a) *Anais Scola Superior Agric. Luiz de Queroz*, **37**: 603-607.

Primavesi, A.C.P.A. and Malavolta, E. (1980b) *Anais Scola Superior Agric. Luiz de Queroz*, **37**: 609-630.

Pruthi, J.S. (1963) In: *Advances in Food Research*, Vol. 12 (Eds. C.O. Chichester, E.M. Mrak and G.F. Steward), Acad. Press, New York.

Pruthi, J. S. (1963) *Adv. Food Res.*,**12:** 203-282.

Purseglove, J.W. (1974) *Tropical Crops – Dicotyledons*, Longman Group Ltd., London.

Reis, L.B., Paiva Neto, V.B., Toledo Picoli, E.A., Costa, M.G.C., Rêgo, M.M., Carvalho, C.R., Finger, F.L. and Otoni, W.C. (2003) *In Vitro Cellular and Developmental Biology- Plant*, **39**: 618-622.

Ribeiro, L.M., Peixoto, J.R., Andrade, S.R.M., Fonseca, R.S., Vieira, L.M. and Pereira, W.V.S. (2008) *Pesqui. Agropecu. Bras.*, **43:** 589–594.

Ribeiro, Leonardo Monteiro, Nery, Lays Araujo, Vieira, Lorena Melo, and Mercadante-Simoes, Maria Olývia (2015) *Plant Cell. Tiss. Organ. Cult.*, **123:** 173–181.

Rodriguez, G.A.M., Hurtado, M. and Sanchez de Prager, M. (1995) *Acta Agronomica, Universidad Nacional de Colombia*, **45**: 89-98.

Roncatto, G., Kamel, T. and Saraiva, L. (2014) Avaliacxao de cultivares de maracuja-amarelo enxertadas em diferentes especies de maracuzeiro no estado do acre. *Resultados de Pesquisas en EMBRAPA*, pp. 5273–5276.

Roodt, R., Spies, J.J. and Burger, T.H. (2002) *Bothalia*, **32**: 117–122.

Ruggiero, C., Lam-Sanchez, A. and Banzatto, D.A. (1976) *Acta Hort.*, **57:** 121-124.

Sale, P.R. (1981) *Hort. Produce Practice* (N.Z.), 97.

Sale, P.R. and Alexander, G. (1986a) *Hort. Produce Practice* (N.Z.), 291.

Sale, P.R. and Alexander, G. (1986b) *Hort. Produce Practice* (N.Z.), 321.

Sanchez, I., Angel, F., Fajardo, D., Castillo, M.F., Lobo, M., Thome, J. and Roca, W. (2000) *Proc. Internamer, Soc. Trop. Hort.*, **42**: 267-271.

Sanez M.V., Castro-Barquero, L. and Gonzalez-Calvo, J. (1991) *Agronomia Costarricense*, **15**: 79-83.

Santos, L.F., Oliveira, E.J., Silva, A.S., Carvalho, S.M., Costa, J.L. and Pádua, J.C. (2011) *Biochem. Genet.*, **49**: 540–554.

Sawant, I.S., Sawani, S.D. and Ganapathy, M.M. (1995) *Indian Phytopath.*, **48**: 202-205.

Segura, S., d'Eeckenbrugge, G.C., Bohorquez, A., Ollitrault, P., and Tohme J. (2002) *Genet. Resour. Crop Evol.*, **49** (2): 111–123.

Sema, A. and Maiti, C. S. (2006) *Status and Prospects of Passion Fruit Industry in Northeast India*. Central Institute of Horticulture, Medziphema-797 106, Nagaland, India.

Senter, S.D., Payne, J.A., Knight, R.J. and Amis, A.A. (1993) *J. Sci. Food and Agri.*, **62**: 67-70.

Shiomi, S., Wamocho, L.S. and Agong, S.G. (1996a) *Post-harvest Biol. Technol.*, **7:** 161-170.

Shiomi, S., Kubo, Y., Wamocho, L.S., Koaze, H., Nakamura, R. and Inaba, A. (1996b) *Post-harvest Biol. Technol.*, **8**: 199-207.

Silva, A. A. G. and Klar, A. E. (2002) *IRRIGA.*, **7**(3): 185–190.

Silva, A.P.da, Viettes, R.L. and Cereda, E. (1999) *Scientia Agricola*, **56**: 797-802.

Simon, P. and Karnatz, A. (1983) *Acta Hort.*, **139**: 83-90.

Singh, H.P., Ganapathy, K.M. and Anjaneyulu, K. (1980) *Indian J. Agric. Sci.*, **50**: 260-263.

Sousa, M.S.B., Vieira, L.M. and Lima, A. D. (2011) *Brazil. J.Food Technol.*, **14:** 202-210.

Souza, E.A., Nociti, P.R.H., Ferreira, M.E., Ruggiero, C. and Grisi, C. (1979) *Cientifica*, **7**: 327-330.

Strohalm, H., Dregus, M., Wahl, A. and Engel, K. H. (2007) *J. Agric. Food Chem.* **55:** 10339-10344.

Srinivas, K., Sulladmath, V. V., Palanappan, R. and Venugopalan, R. (2010) *Indian J. Hort.*, **67**(3): 289–292.

Sriram, T.A. and Raman, K.R. (1961) *South Indian Hort.*, **9**: 3-7.

Ssekyewa, C., Opio, A.F., Swinburne, T.R., Damme, P.L.J.van and Abubakar, Z.M. (1999) *Intern. J. Pest Management*, **45**: 173-177.

Suassuna, T. de M.F., Bruckner, H., de Carvalho, R. and Borem, A. (2003) *Theor. Appl. Genet.*, **106:** 298–302.

Susheela, R., Pruthi, J. S. and Lal, G. (1961) In: Nitrogenous substances in passion fruit waste (rind and seeds) *Proc. Symp. Protein*, held at CFTRI, Mysore, India, Aug. 1960, p. 435.

Thokchom, R. and Mandal, G. (2017) *J. Agric. Engin. Food Technol.*, **4:** 27-30.

Torres Filho, J. and Ponte, J.J.da (1994) *Fitopatologia Brasileira*, **19**: 34-38.

Treglazova and Fataliev (1989) *SubtropicheskieKul'tury*, No. 6, pp. 108-111.

Trevisan Flavio and Mendes Beatriz Madalena Januzzi (2005) *Sci. Agric.* (*Piracicaba, Braz.*), **62:** 346-350.

Trevisan, F., Mendes, B.M.J., Maciel, S.C., Vieira, M.L.C., Meletti, L.M.M. and Rezende, J.A.M. (2006) *Plant Dis.*, **90:** 1026–1030.

Tripathi, P.C., Karunakaran, G., Sakthivel, T., Sankar, V. and Senthilkumar, R. (2013) Passionfruit cultivation in India, Central Horticultural Experiment Station, Indian Institute of Horticultural Research, Karnataka, pp.1-13.

Turner, D.W., Menzel, C.M. and Simpson, D.R. (1996) *Scientia Hort.*, **65**: 25-36.

Ustinomiya, N. (1992) *Scientia Hort.*, **52**: 63-68.

Utsunomiya, N., Kinai, H., Maisui, Y. and Takebayash (1998) *J. Japanese Soc. Hort. Sci.*, **67**: 567-571.

Vallani, P.C., Lam-Sanchez, A., Ruggiero, C. and Ferreira, F.R. (1976) *Acta Hort.*, **57**: 233-236.

Valia, R.Z., Patil, V.K., Patel, Z.N. and Kapadia, P.K. (1994) *J. Maharashtra Agric. Univ.*, **19**: 354-357.

Veras, M.C.M., Pinto, A.C. de Q. and Menesses, J.B.de (2000) *Pesqui. Agropecu. Bras.*, **35**: 959-966.

Viana, A.P, de Lima e Silva, F.H., Gloria, L.S., Ribeiro, R.M., Krause, W. and Boechat, M.S.B. (2017) *Euphytica*, **213**: 228. https: //doi.org/10.1007/s10681-017-2020-3

Watson, D.P. and Bowers, F.A.I. (1965) *Hawaii Farm Sci.*, **14**: 3-5.

Watson, R. R., Zibadi, S., Rafatpanah, H., Jabbari, F., Ghasemi, R. and Ghafari, J. (2008) *Nutr. Res.*, **28:** 166-171.

Weber Diego, Nachtigal, Jair Costa, Barreto, Caroline Farias and Malgarim, Marcelo Barbosa (2017) *J. Exp. Agric. Intl.*, **18**(6): 1-8.

Winks, C.W., Menzel, C.M. and Simpson, D.R. (1988) *Queensland Agric. J.*, **114**: 217-224.

Whittaker, D.E. (1972) *Tropical Sci.*, **14**: 59-71.

Zheng, J.X. and Huang, Y. (1997) *J. Fujian Academy of Agric. Sci.*, **12**: 40-43.

Zhu, Xiang-hao, Duan, Zhen-hua, Yang Yu-xia, Huang Xin-hui,Xu Cheng-ling and Huang, Zhi-zhuo (2017) *IOP Conf. Ser.: Earth Environ. Sci.*, **100**, doi: 10.1088/1755-1315/100/1/012080.

8

JACKFRUIT

V.A. Parthasarathy, C. Aswath,

H.N. Samaddar and G. Karunakaran

1.0 Introduction

The jackfruit was not regarded as a commercial fruit crop earlier although it is widely cultivated in southern Asia, the East Indies and other warm areas of both hemispheres. But in view of its high yields and nutritional quality, it is considered as a super food and is gaining lot of importance. It is also found growing in Myanmar, Malaysia and Brazil and is very popular in eastern and southern parts of India. It is the largest among the edible fruits. Its origin is in India at Western Ghats as well as at Eastern Himalayas or may be it had simultaneous origin.

2.0 Composition and Uses

2.1.0 Composition

Tender jackfruit appears in the market in spring and continues until summer as a popular vegetable. Since common vegetables are scarce and costly at that time of the year, jackfruit enjoys a high demand and premium price. Ripe fruit has high nutritive value. The fruits are normally fibrous and are composed of mono-, di-, and polysaccharides (Berry and Kalra, 1987). Zaghlol *et al.* (1983) noted glucose (10.94 per cent), fructose (1.42 per cent), xylose (0.18 per cent), rhamnose (0.44 per cent), arabinose (1.51 per cent), galactose (2.47 per cent), and three other unknown sugars (13.4 per cent) in ripe jackfruits. Jackfruit is an important source of pectin (Krishnarnurthi and Giri, 1949) and contains about 1.9 per cent protein on fresh weight basis. Hossain *et al.* (1979) stated that the fruit contained 2.64-11.77 mg of ascorbic acid/100 g and 250-1740 mg carotene/100 g. The jackfruit latex is reported to possess bacteriolytic activity. The composition of edible portion is detailed in Table 1.

Table 1: Composition of Jackfruit* (per 100 g of edible portion)

Nutrients	*Tender Fruit*	*Ripe Fruit*	*Seed*
Moisture (per cent)	84	77.2	64.5
Carbohydrate (g)	9.4	18.9	25.8
Protein (g)	2.6	1.9	6.6
Fat (g)	0.3	0.1	0.4
Fibre (g)	–	1.1	–
Total mineral matter (g)	0.9	0.8	1.2
Calcium (mg)	50	20	21
Phosphorus (mg)	97	30	28
Iron (mg)	1.5	500	–
Potassium (mg)	246	–	–
Vitamin A (I.U.)	0	540	17
Thiamine (mg)	0.25	30	–
Riboflavin (mg)	0.11	–	–
Vitamin C (mg)	11	–	–
Calorific value	–	84	–

* Anon (1979).

Jack has unique aroma which is favoured by people. In Brazil, according the consistence and size of the fruits, the species are known as "jacadura" (hard jackfruit) and "jacamole" (soft jackfruit). The pulp of the former is harder in comparison with the latter that often is also called butter-jackfruit. On the other hand, the aroma of both fruits is very distinguishable. The hard jackfruit is bigger than the soft jackfruit while the latter is often more sweet and aromatic. In the Amazon region it is cultivated only on a domestic scale (Maia *et al.*, 2004). The aroma volatiles from two fruit varieties of jackfruit (*Artocarpus heterophyllus*) growing in the Amazon were obtained by simultaneous distillation-extraction and analysed by GC–MS. The major components identified in the aroma concentrate of "hard jackfruit" variety were isopentyl isovalerate (28.4 per cent) and butyl isovalerate (25.6 per cent). The aroma concentrate of "soft jackfruit" was dominated by isopentyl isovalerate (18.3 per cent), butyl acetate (16.5 per cent), ethyl isovalerate (14.4 per cent), butyl isovalerate (12.9 per cent) and 2- methylbutyl acetate (12.0 per cent) (Maia *et al.*, 2004).

2.1.1 Aroma of Jackfruit

An aroma extract dilution analysis applied to the volatiles isolated from jackfruit pulp by solvent extraction and solvent-assisted flavor evaporation resulted in the detection of 48 odorants with flavor dilution (FD) factors between 1 and ≥8192 (Grimm and Steinhaus, 2019). Application of gas chromatography-olfactometry to static headspace samples revealed additional five highly volatile odorants. The structures of 44 out of the 53 detected odorants could be assigned. These 44 compounds were quantitated using stable isotopically substituted odorants as internal standards, and their odor activity values (OAVs) were calculated as the ratio of the natural concentrations in jackfruit pulp and the odor threshold values in water. High OAVs were in particular obtained for ethyl 3-methylbutanoate (74000), ethyl butanoate (1800), 3-methylbutanal (1500), and 2-methylpropanal (1400). An aroma model solution based on the natural concentrations of the 35 compounds for which OAVs > 1 had been calculated fully mimicked the characteristic jackfruit pulp aroma (Grimm and Steinhaus, 2019). The same authors (Grimm and Steinhaus, 2020) in a subsequent study, screening the volatiles of cempedak [*Artocarpus integer* (Thunb.) Merr.] pulp for odoractive compounds by aroma extract dilution analysis and gas chromatography (GC)–olfactometry of static headspace samples revealed a total of 55 odorants, among which 47 were identified. Using stable isotopically substituted odorants as internal standards, these compounds were quantitated by GC–mass spectrometry, and odor activity values (OAVs) were calculated as ratios of the natural concentrations in cempedak pulp to the orthonasal odor detection thresholds. An aroma reconstitution model based on the 41 compounds with OAVs >1 in their natural concentrations successfully mimicked the characteristic aroma of cempedak pulp including the pronounced sulfury, oniony note which is intense in cempedak pulp but absent in jackfruit pulp. Further sensory tests finally showed that 2-(methylsulfanyl) propane, 2-(methylsulfanyl) butane, and 2-(methylsulfanyl) pentane are the key compounds responsible for this unique aroma note in cempedak pulp and vitally contribute to the aroma difference between cempedak pulp and jackfruit pulp.

2.2.0 Uses

2.2.1 Fruits

Jackfruits are eaten unripe at 25-50 per cent full size as a vegetable or ripe as a fruit. When used as vegetable, it is peeled, sliced and boiled and then seasoned or mixed with other food. Jackfruit is an integral part of common Indian diet and is freely available in Indian and adjoining continents, its medicinal properties are also mentioned in Ayurveda. The plant is reported to possess antibacterial, anti-inflammatory, antidiabetic, antioxidant and immunomodulatory properties. Jackfruit is an important source of compounds like morin, dihydromorin, cynomacurin, artocarpin, isoartocarpin, cyloartocarpin, artocarpesin, oxydihydroartocarpesin, artocarpetin, norartocarpetin, cycloartinone, betulinic acid, artocarpanone and heterophylol which are useful in fever, boils, wounds, skin diseases, convulsions, diuretic, constipation, ophthalmic disorders and snake bite *etc.* (Om Prakash *et al.*, 2009).

In eastern India, tender jackfruit appears in the market in spring and continues until summer as a popular vegetable. Since common vegetables are scarce and costly at that time of the year, jackfruit enjoys a high demand and premium price.

Ripe fruit has high nutritive value. In India it is called as the poor man's fruit. The ripe fruit also significantly contributes to the nutrition of low income families as a source of vitamins, minerals and calories. It contains a fairly good amount of β-carotene, a precursor to vitamin A. Jackfruit is an important source of pectin and contains about 1.9-2.2 per cent protein on fresh weight basis. Fructose, glucose and sucrose are the major sugars in all parts of the fruit, except in the outer spiny rind, which is devoid of glucose. Capric, myristic, lauric, palmitic, oleic, stearic, linoleic and arachidic acids are the major fatty acids, with varying proportions in different parts of the fruit (Chowdhury *et al.*, 1997).

The seasonal nature of the fruit with short storage life, even under low temperature conditions, necessitates processing of the ripe fruit. A number of products have been developed from unripe as well as ripe fruits.

2.2.2 Seeds

The seeds are boiled or roasted and used in many culinary preparations. The seeds are very rich in carbohydrate and protein. A systematic study of proximate composition and functional properties of jackfruit seed revealed it to be a good source of protein having high solubility near neutral pH. It has low water and fat absorption capacities. Incorporation of jack seed to deep fat fried products revealed that it brings down fat absorption to a remarkable extent with affecting sensory attributes of savoury product (Rajarajeshwari and Prakash, 1999).

Jacalin, the major protein from the jackfruit seeds, is a tetrameric two-chain lectin (MW 65 kDa) combining a heavy α-chain of 133 amino acid residues with a light β-chain of 20-21 amino acid residues. Even in its sialylated form, it is highly specific for the -o-glycoside of the disaccharide Thomsen-Friedenreich antigen (Galbeta 1-3Ga 1NAc). Because of the properties, jacalin is suitable for studying o-linked glycoproteins, particularly human IgAl. Jacalin is strongly mitogenic for

Uses of Jackfruit

Raw Jackfruit Packaging

Beverages from Jackfruit

Powder (Left) and Treat (Right) from Jackfruit

Different Value Added Products from Jackfruit

human CD4+ T lymphocytes, and it has proved useful for the evaluation of the immune status of patients infected with human immunodeficiency virus-1. The abundance of source material for the production of jacalin, its ease of purification and its yield and stability has made it an attractive cost-effective lectin (Kabir, 1998).

2.2.3 Wood

Jack timber is valuable for making furniture as it is rarely attacked by white ants. Both wood and charcoal are excellent raw materials for fuels (Komarayate and Honda, 1994). Chemical analysis of wood of the jackfruit tree revealed a cellulose content of 56.74 per cent, lignin content 28.76 per cent, pentosan content 18.64 per cent, yield of charcoal 38.74 per cent, charcoal calorific value 7108.37 cal/g. Charcoal briquette properties shows 5.10 per cent moisture, 3.06 per cent ash, 25.5 per cent volatile matter, 71.23 per cent fixed carbon, 0.63 g1cm^3 density, 350 kg/cm^3 compression strength and 6487.28 cal/g calorific value (Kornarayati, 1995).

The leaves and skin of the fruit are excellent cattle feed. The latex from the bark contains resins. Sometimes this is used to plug holes in earthen vats and buckets. The skin and core of the fruits, could be used commercially for pectin extraction.

3.0 Origin and Distribution

3.1.0 History and Origin

Jackfruit (*Artocarpus heterophyllus* Lam.) is native to India. Its likely centre of origin is either Western Ghats or Eastern Himalayas. It is likely they have independent origin. Considering the wide genetic diversity in SE Asian nations including Bangladesh and NE Region, the Primary centre of origin in Eastern Himalayas seems tenable with Western Ghats as Secondary centre. The jackfruit is cultivated throughout the tropical lowlands in South and Southeast Asia, parts of central and eastern Africa and Brazil. Major jackfruit producers are Bangladesh, India, Myanmar, Thailand, Vietnam, China, the Philippines, Indonesia, Malaysia and Sri Lanka.

Has Jackfruit has two simultaneous centres of Origin? Western Ghats and Eastern Himalayas? In Assam, Jackfruit has been mentioned since long. There is a village called *Kathalguri* which translates to " area with jackfruit trees". This village is 22km from Nagoan district HQ. Huan Tsang visited Kamarupa,present day Guwahati, in 643 AD during the rule of Bhaskar Burman, the aristocratic King of Assam. He was surprised to see deft planting of Jackfruit trees and methodical cultivation of Jackfruit.

Ramayana does mention about Jack fruit in both NE Himalayas and Western Ghats. In *Valmiki Ramayana - Aranya Kanda* in Prose *Sarga* 73 "Kabandha extols Pampa Lake and details Rama about the course to be adopted to proceed to Mt. Rishyamuka ...."Oh, Rama, have a recourse to westward, and where the trees of Rose-apple, Priyaala, Jackfruit, Plaksha, Banyan, Tinduka, Pipal, Karnikara, Mango, and others. and even trees like Dhanva, Naaga, Tilaka, Naktamaala, blue Ashoka, Kadamba, Karaviira, Agnimukha, Ashoka, red-sandalwood trees and Neem trees will be heart-pleasingly shining forth in full blossom, that alone is an optimistic course for you. (https://tamilandvedas.com/tag/jack-fruit/).

Antiquity of jackfruit is well understood by Sangam literature of oldest language Tamil in First Sangam period which lasted 4400 years ago. Simlarly another tamil poem in Kurunthogai 18 (Period: 500 to 70 AD) also mentions jackfruit. The ancient Sangam Tamil literature talks about the abundance jack in Chera Kingdom (Present Kerala and Western Ghats).

The jackfruit has played a significant role in Indian agriculture for centuries. Archeological findings in India have revealed that jackfruit was cultivated in India 3000 to 6000 years ago. Findings also indicate that Indian Emperor Ashoka the Great (274–237 BC) encouraged arbori-horticulture of various fruits including jackfruit. Varahamihira, the Indian astronomer, mathematician, and astrologer, wrote a chapter on the treatment of trees in his Brhat Samhita. His treatise includes a specific reference on grafting to be performed on trees such as jackfruit (https://karkanirka.org/2008/09/20/kurunthokai18/).

A well-preserved ninth-century AD shipwreck was excavated in 1998-9 off the Indonesian island of Belitung between Sumatra and Borneo. The principal features of the wreck include planks joined by stitching with wadding inside and outside the hull, a sharp bow with little rake, stitched-in frames, through-beams stitched to the hull, removable ceiling timbers, a keelson and stringers, and a composite iron and wood anchor. Ethnographic and iconographic evidence suggests that these are the features of ancient Indian and Arab vessels, an origin which is strongly supported by the timber species identified in the wreck. The timber mentioned by him includes Jackfruit tree timber (Flecker, 2001).

A comprehensive account of Jackfruit in Bangladesh and the origin of jackfruit have been presented by Khan *et al.* (2010). Interestingly, Jackfruit is the national fruit of Bangladesh. According to the review presented by them, the place of origin and wild ancestor of jackfruit is unknown. Jackfruit has been cultivated for millennia and was referred to as early as 300 B.C.E. by Theophrastus (Hort, 1916). It is now so widely cultivated that the region in which it is indigenous and its wild progenitor is unclear. Today it can be found in cultivation at low elevations from the Indian subcontinent through Bangladesh, Myanmar, into southern China, Laos, Vietnam, Cambodia, Thailand, Malaysia, Indonesia, and Oceania. It is also commonly cultivated in the Philippines, and has been introduced throughout Africa and the Neotropics. It has been under cultivation for so long that Jarrett (1959) speculated it would not be possible to identify the wild progenitor. It is, nevertheless, believed to be native to the Indo-Malaysian region (Beddome, 1873; Brandis, 1906; Gamble, 1902; Kanjilal *et al.*, 1940; Talbot, 1911; Wight, 1843). More specifically, jackfruit is often considered to originate from the Western Ghats of India as Wight (1843) reported finding trees growing in primary forest away from human habitation. Barrau (1976) suggested that it originated in Malaysia due to the great diversity of jackfruit cultivars found there. While jackfruit is important and widely cultivated in Malaysia, it was likely introduced there and never found in the wild (Jarrett 1959). Bashar and Hossain (1993) suggest that wild jackfruit relatives inhabit the Andaman

Islands and report that it is found only under cultivationin Bangladesh. However, it is often found in forests in Bangladesh away from human habitation (cf. Khan *et al.*, 2010). In many districts like Tangail, Mymensingh, Dhaka, and adjoining areas, many jackfruit trees grow in the wild near and in forest edges and fallow lands. Many of these trees grow tall and bear relatively small but numerous fruits that are, in most cases, inedible or of very poor quality. The local people call these "jangli kathal" (wild jackfruit). A wide and undocumented diversity of jackfruit has been observed throughout Bangladesh, but little attention has been paid to the documentation and conservation of its genetic resources (Khan *et al.*, 2010). This documentation is crucial to better understand the history of jackfruit domestication. If Bangladesh is a "secondary center of diversity" for jackfruit, this implies that it was domesticated elsewhere and subsequently introduced to Bangladesh, where the crop was then diversified. If there truly are wild jackfruit trees in Bangladesh, then it is possible that Bangladesh is part of a broader center of origin of the crop (a "noncenter," sensu Harlan 1971). On the other hand, perhaps the trees growing in forests and fallow lands that exhibit "wild" characteristics are not truly ancestrally wild, but rather represent naturalized feral populations of trees that in the absence of human selection evolved back toward wild characteristics after the crop was introduced to the area millennia ago (Khan *et al.*, 2010).

Breadfruit (*Artocarpus altilis*) has long been all important staple crop in the Pacific Islands. Principally important as a carbohydrate food source, every portion of the breadfruit tree has yielded materials useful to islanders. There are two species of breadfruit in the Paciny Islands, *A. altilis* and *A. mariannesis.* A third species *A. camansi*, a spiny, seeded breadfruit similar to *A. altilis* is found in New Guinea, Islands of Southeast Asia and the Philippines. It is rarely seen in the South Pacific with the exception of recently introduced trees, and it is not found anywhere in Micronesia (Ragone, 1991). Whereas *A. mariannensis* and *A. camansi* are always seeded, there are both seeded and seedless forms of *A. altilis.* The seeded form of *A. altilis* appears to be indigenous in New Guinea, where it is a dominant member of alluvial forests in low land areas. Outside of New Guinea, *A. altilis* occurs only in cultivation, although long abandoned plantations are often mistaken for wild trees (Ragone, 1995). The distribution of *A. altilis* beyond New Guinea reflects the movements of humans through the Pacific Islands. Seeded *A. altilis* is not found in Micronesia and it has not been distributed beyond Oceania. Seeded and seedless cultivars follow a west to east gradient in the South Pacific. Seeded forms predominate in the western islands of Melanesia, with few-seeded and seedless forms beginning to occur in eastern Melanesia. Seeded and few seeded types outnumber seedless cultivars in Samoa, but seedless cultivars 'predominate in eastern Polynesia. A few seedless cultivars typical of Polynesia are found throughout the northern Pacific Islands from the Marshall Islands to the east to Guam, Yap and Belau in the west (Ragone, 1995).

Breadfruit is an ancient domesticated cultigen, but the long-held supposition that it was first domesticated in the Indo-Malaysian region (Barrau, 1976; Jarrett, 1959) may be inaccurate (Ragone, 1995). It is likely that breadfruit was first domesticated in New Guinea and associated islands such as the Bismarch Archipelago (Ragone, 1991).

The chempedak (*Artocarpus integrifolia*) (syn. *A. polyphema; A. chempeden*) (Rajendran, 1992) is another important species of genus *Artocarpus.* Chempedak is a native of Malaysia and is distributed in Myanmar, peninsular Thailand and Malaysia, the Indonesian islands and western New Guinea. Like jackfruit, it shows high variability in characters (Nakasone and Paull, 1998). The chempedak can be distinguished from jackfruit by having smaller fruit with thinner rind and more juicy flesh, which is darker yellow when ripe (Jarrett, 1959).

Outside Borneo, tarap (*Artocarpus odoratissimus*) is not well known, though in the Philippines scattered trees of tarap can be found in the southern islands. It appears confined to Sabah, Brunei and Sarwak with its greatest diversity in Brunei (Tinggal, 1992). Male and female flowers are borne on terminal branches. Young fruits when damaged exude thick white latex. Ripe fruits are light brown, weighing 1-3 kg and contain 60-200 seeds. The stiff hairs give way to touch, but are sticky and gritty. The whole fruit is an inflorescence with the seeds attached to the central core. Each seed is covered with sweet, juicy white pulp (Osman *et al.*, 1996).

The other *Artocarpus* species which produces edible fruits and have the potentiality to promote are:

Artocarpus anisophyllus

A native of Malaysia, locally known as bintawa. The almost syncarp shows fairly dense spines and is yellowish brown when ripe. The inedible husk can be easily peeled off to reveal the yellowish orange fleshy perianths which are very sweet. Seeds are edible when roasted.

Artocarpus elasticus (Syn. *A. sericicarpus*)

Indigenous to Malaysia, locally known as 'pedalai' or 'terkalong'. The oblong yellowish brown syncarp, 8.5 cm x 5 cm, has long flexible spines. The inedible husk can be easily peeled. Fleshy perianths whitish in colour, sweet in taste display a strong aroma. Seeds are edible when roasted.

Artocarpus rigidus

Known as pala munsoh or 'tempunik' in Malaysia, probably belongs to subspecies *rigidus* (Wong, 1994). Fruit consists of a globose syncarp, measuring 8 cm in diameter, with densely-set, rigid, long spines. When ripe, fruit turns yellowish brown and husk can be easily peeled.

4.0 Taxonomy

The family Moraceae to which *Artocarpus* belongs includes fig and mulberry. *Artocarpus heterophyllus* Lam., belongs to the family Moraceae. This family encompasses about 1,000 species in 67 genera, mostly tropical shrubs and trees, but also a few vines and herbs (Bailey, 1942; Merill, 1912). The Philippines is the home to about 150 species representing 13 of these genera (Merill, 1912). The genus *Artocarpus* contains about 50 species; most are native to Asia and of which about 15 produce edible starchy fruits. The three most important species are the more tropical breadfruit, *A. altilis* (syn. *A. communis; A. incisus*), the jackfruit *A. hetero phyllus* (syn.

A. integer; A. integrifolius) and its close relative, chempedak, *A. integrifolia*, (syn. *A. polyphema; A. chempeden*). The other *Artocarpus* species which produce edible fruits include *A. anisophyllus, A. elasticus* (syn. *A. sericicarpus*), *A. odoratissimus, A. rigidus, A. anisophyllus, A. julvicortex, A. polyphema, etc.*

There is high variability in breadfruit (2n = 27, 28, 56), jackfruit (2n = 56) and chem pedak (2n = 56) characters.

4.1.0 Tree

Jackfruit is a monoecious, evergreen, latex-producing tree, 20-30 m in height. The breadfruit tree grows up to 30 cm. Both jackfruit and breadfruit have straight trunk with thick branches.

4.2.0 Leaves

Leaves are alternate, coraceous, fleshy, entire, simple, stipules leave a scar on the branch. The chempedak has long, \wiry, brown hairs on the leaves.

4.3.0 Flowers

Jackfruit is monoecious, with male and female inflorescences appearing separately on the same tree (Moncur, 1985). During flowering, specialized laterals develop on the trunk or main branches (cauliflory), which are 46.0 ± 8.0 cm in length, initiating 6.6 ± 0.5 leaves, and 2.2 ± 0.6 inflorescences. Female inflorescences occur only on these laterals whereas the males may occur also in the terminal leaf axils or leader and lateral shoots. There appears to be no regular sequence in the incidence of the male and female inflorescences, although there are considerably more males than females.

Each terminal is enclosed by a pair of thick stipules which protect the bud initially but leave a prominent annular scar on the node. The stipules enclose curled leaf primordia and vegetative or reproductive apices. Whereas male and female inflorescences are similar during early development, the female is later distinguished by a thicker peduncle and a large annular disc at the base of the spike. The female is noticeably shorter and wider than early-emerged males at anthesis, but later-emerged males are smaller. The light green female inflorescence is 5-8 cm by 3-5 cm at anthesis. The stigmatic surface is composed of papillae, which become sticky when receptive 1-2 weeks after exsertion; this condition may last for a further 2 weeks. A few interfloral bracts may be present but are usually shed soon after anthesis. Prior to anthesis, the perianth segments remain free, though tightly packed. Later, they enlarge in the centre and fuse but remain free at the lower and upper regions. The lower region containing the seed is fleshy and edible, while the other two regions constitute the outer protective rind. As it grows and becomes heavier, the fruit becomes pendulous.

The male inflorescence is oblong and light green and grows to a length of 5-10 cm by 2-4 cm. It is densely covered with minute flowers, each with a 2-lobed calyx and a single stamen. Each stamen consists of four anthers which split open at anthesis, exposing the masses of spherical pollen grains. Prior to anthesis the surface appears armoured, with spikes and scales which split as the stamen exserts. The male perianth segments are shorter than the female, and though tightly packed are always free. Anthesis commences 2-3 weeks after emergence from the stipules and lasts about 2 weeks during which inflorescence appears yellow. Pollen production is abundant and ranged from 1,50,00,000 to 1,70,00 and 000 per catkin. The percentage of pollen fertility ranged from 89 to 93 and the pollen diameter ranged from 16 to 22 μ (Joseph and Kumaran, 1994).

4.4.0 Fruits

The fruit, a gigantic syncarp, is the largest of all cultivated fruits (Moncur, 1985). Different shapes (obloid, spheroid, high spheroid, ellipsoid, clavate, oblong, broadly ellipsoid, *etc.*) of fruits are found in jackfruit. The rind colour also have wide variations (green, greenish yellow, yellow, reddish yellow, brown, *etc.*). The fruits can have 10500 flakes and seeds. The seeds are 2-4 cm long by 1-2 cm wide each surrounded by a horny endocarp and sub gelatinous exocarp. The various flake shape (spheroid, cordate, twisted, elongated ovate, rectangular, oblong with curved tip, irregular, *etc.*) and seed shape (spheroid, ellipsoid, elongate, oblong, reniform, irregular, *etc.*) have been observed. On an average in ripe jackfruit, the flakes, seeds and rind form 29, 12 and 59 per cent of the bulk respectively (Bhatia *et al.*, 1955).

5.0 Species and Cultivars

Jackfruit belongs to the genus *Artocarpus*, a genus rich in culturally important species including breadfruit (*A. altilis*), dugdug (*A. mariannensis*), and breadnut (*A. camansi*). These three species represent some of the most important traditional subsistence trees of Pacific islands (Elevitch and Manner, 2006). Elsewhere in Southeast Asia and India, champedak (*A. integer*), lakoocha (*A. lakoocha*), marang (*A. odoratissima*), kwai muk (*A. lingnanensis*), and others are important fruit trees, all with culturally important uses, and many with other valuable products, such as timber. The genus *Artocarpus* comprises about 50 species, 11 of which are known to produce edible fruits. Even at the species level, a high degree of genetic variability exists. This is true in the case of jackfruit (*A. heterophyllus*), breadfruit (*A. altilis*) and marang (*A.odoratissimus*). At least two species, jackfruit and champedak (*A.integer*) hybridize freely in nature. Furthermore, some species are graft-compatible. The seedless breadfruit, for example, is graft compatible not only with the seeded strains but also with the kamansi (*A. camansi*) and pedalai (*A. sericicarpus*). The monkey jackfruit (*A. rigidus*) is another promising species because it is very productive, produces perhaps one of the smallest fruits in the genus, has sub-acid taste and possesses good flavour, better than jackfruit, champedak and marang (APAARI, 2012).

One of its wild relative Cempedak (*Artocarpus integer*), also spelt "chempedak", is a tropical fruit from the Moraceae family. It is found in Myanmar, Thailand, Vietnam, Malaysia and Indonesia. Cempedak is similar to the jackfruit in appearance as well as in the way the fruit is used. The pulpy flesh covering the seeds found within the fruit is sought after for its fragrant taste. However, the hard seeds can also be cooked and eaten (http://eresources.nlb.gov.sg/infopedia/articles/SIP 201-2005-02-01.html). Molecular studies have confirmed that the jackfruit and chempedak (*A. integer* Merr.) were monomorphic and therefore indistinguishable for

Species of Jackfruit

Artocarpus heterophyllus

Artocarpus anisophyllus

Arotocarpus altilis

Artocarpus integrifolius **(Chempedak)**

Species of Jackfruit

Artocarpus integer

Artocarpus rigidus

Artocarpus polyphema

Artocarpus elasticus

Artocarpus odoratissimus

all restriction sites. They have many morphological similarities. Do they (*A.Integer and A. heterophyllus*) warrant separate species status?

Another close relative of jackfruit is *A.lakoocha* which can be considered as Progenitor of Jack. According to geographical distribution, the nearest modern species is *A. lakoocha* which is still growing in West Coast from Konkan southwards through Kanara and Malabar to Tranvancore as well as endemic to NE India. This species is also found in the Andaman Islands; Kumaon Eastwards through Bengal, Assam and Chittagong, in plains and hill forests of Myanmar, Bihar, Orissa and Madhya Pradesh (Pearson and Brown, 1932).

Classification of Cultivars

The cultivars are broadly classified on the basis of a few characteristics of the fruit. The classifications were done on the basis of aroma of pulp, yield of trees, and seasonal availability of fruits (APPARI, 2012). However, the other important characters were overlooked and not used for this classification due to other local names of cultivars of jackfruit (Azad, 2000; Haq, 2006). With such wide diversity, it requires a thorough study to classify Jackfruit. Based on aroma: Gulabi (rose scented), Champa (flavour like that of *Michelia* sp.)

Based on Bearing: Hazari (bearing more number of fruits in a tree).

Based on texture of pulp: Vela (Soft pulp)

Based on size: Barka (Small fruited) and Kapa (Large fruited)

Different types of jackfruit have come into commercial cultivation with different local names. There is very little comprehensive or reliable information on any true variety of jackfruit which has been developed through breeding programme. However, the cultivars Rudrakshki and Singapore Jack (Ceylon Jack.) in India are known to breed more or less true to type from seed (Soepadmo, 1992). These cultivars produce fruits weighing 5-20 kg within 2-5 years of planting the seedlings. In most jackfruit producing countries, there are many local names given to clonal selections which are propagated (*e.g.* Gulabi, Chamooa, Hazari in India; Gala, Khaja, Hazari, *etc.* in Bangladesh), but how these are to be distinguished from each other is not yet established. Selections that moved from one country to another were given separate names which caused confusion. In Sri Lanka, a variety called Vela with soft pulp is the same as described by de Costa as Gerisssa in India. Under these two names, there are many subtypes which differ in taste, colour, fibre content and sweetness and they have their local names. Many of the local names relate to attributes rather than places *e.g.* Gulabi (rose scented); Champa (taste like champedak), Hazari (producing large number of fruits) (Samaddar, 1990). Manjunath (1948) also classified jackfruit into two types on the basis of fruit size, (i) Barka - a smaller fruit type and (ii) Kapa - large fruit type. The Barka is locally called Katahali (in Uttar Pradesh) and yields small fruits (2-8 kg each) which are somewhat sour in taste. Kapa is called Kathal in Uttar Pradesh and neighbouring states, yields large fruits weighing up to 40 kg and at ripening has fleshy sweet bulbs (APPARI, 2012) Sreenivasan (1970) reported a local cultivar, Muttam Varikha,with an average weight of 7 kg in Tamil Nadu (India) produced at ground level and with a large fruit weighing up to 55 kg. Singh

(1986) also reported local selected strains, Hairialyalva, 44 Bhadunha, Zarda and Bhusala from Northern India. Morton (1965) reported selected cultivars of jackfruit which included Safeda, Khaja, Bhusila, Bhadaiyan, Handa, T-Nagar Jack and also Velipala a local selection from the forest having large fruits with superior quality large bulbs.

In India, mainly two types, soft flesh and firm flesh are available. Some types with local names like 'Gulabi' (rose-scented), 'Champa' (flavour like that of *Michelia* sp.); Hazari (bearing large number of fruits), Rudrakshki (roundish fruits) are also available. A genotype (No.7) was selected from Uttar Pradesh, with moderate yields, large fruits (> 15 kg) and bulbs (about 20 g), small cylinders, high pulp/cylinder ratio and average fruit quality (Kumar *et al.*, 1996). From eastern India (mainly West Bengal) about 35 types have been identified as superior clones. Wide variability in yield (141200 fruits/tree), fruit weight (2.0-16.0 kg), fruit shape (oblong, roundish, conical); peel colour, pulp colour, number of flakes (30-380/fruit), number of seeds (30-365/fruit), texture of pulp, pulp weight, total soluble solids (15-30 per cent), total sugar (13.027.0 per cent) and acidity (0.10-0.33 per cent) were observed among the genotypes. The growth, bearing (twice a year in some types) and maturity (June-September) of fruits also show variations among the genotypes (Mitra and Maity, 2000).

The Mekarsari Fruit Garden of Indonesia has a rich collection of different *Artocarpus* sp., collected from the various regions in the Archipelago (Tirtawinata, 2000). Two important collections of jack fruit maintained at the garden are cvs. Kandel and Mini.

The jackfruit has been grown in South Florida for 90 years or more, but has never had commercial importance. In recent years, many new superior cultivars have been imported by Fairchild Tropical Garden which offers commercial possibilities. Of the 22 cultivars presently in collection, Black Gold, Golden Nugget, J-31, Honey Gold and Lemon Gold have been the most precocious. The fruits of these cultivars vary greatly in their size, shape, flesh characteristics and taste. All compared with NS-l, the most popular cultivar in South Florida. Galaxy and Honey Gold have shown a tendency for weak growth, with severe dieback. Delightful, Dang Rashmi, Golden Nugget and Black Gold have been the most vigorous. Fruit ripening has been concentrated in July and August in most cultivars but J-31 has ripened in late-May and early-June and Black Gold has ripened as late as mid-September (Campbell and McNaughton, 1994).

The important varieties grown in different countries are summarised in Table 2.

SIDDU (Karunakaran and Singh, 2019)

The novel jackfruit tree (age 35 years) was identified from Mr S S Paramesha Chelur, Tumkur district, also Karnataka. The tree shape is broadly pyramidal. It bears solitary fruits on primary as well as secondary branches with small sized fruits (2-5 kg); weighing 2.44 kg/fruit; fruiting season is May-July. The fruit shape is irregular with 450 fruits/tree; and highly suitable for homesteads and commercial cultivation. Being a small fruit, it suits to family of four members. Each fruit

Table 2: List of Selected Cultivars in different Countries

Country	Cultivars
Bangladesh	Topa, Hazari, Chala, Goal, Koa, Khaja
India	Khujja or Karcha, Ghila, or Ghula, Hazari, Rudrakshi,Gulabi,Hazar, NJT I, NJT 2, NJT 3 NJT 4,Koozha Navarikka or Pazam Varikka,Safeda, Khaja, Bhusila, Bhadaiyan, Handia,Mammoth, Swarna, Everbearer, Rose-scented, Kooli,Varika, Gerissal, Barica, Ghila, Bhadaiyan,Karcha, Rudrakshi, Champa, Safeda, Khaja, Bhusila,Handia, T-Nagar, Jack, Velipala, Ghulabi, Champa,Singapore or Ceylon Jack
Myanmar	Talaing, Kala
Sri j	Vela, Varaka (Waraka), Peniwaraka, Kuruwaraka,Singapore Jack/Ceylon Jack
Philippines	Torres, Mabini, 'EVIARC Sweet', 'Sinapelo',Burabod', 'Cervantes,Gold', and 'MMSU-SROSweet'.
Hawaii	Black Gold, Dang Raismi, Golden Nugget, HoneyGold, and NS1

Source: Updated by authors after Haq (2006).

contains 30 bulbs. The flakes are attractive, sweet and firm, and coppery red colour significantly proved better as compared to the yellow and white colour flaked one. Estimated yield is 1098 kg/tree. Total soluble solids of these varieties is quite high (31.0 0B) which is an important parameter related to fruit quality. Coppery red jackfruit is rich in phytochemicals such as carotenoids (4.43 mg/100g), lycopene (1.12 mg/100g), flavonoids (3.74 mg catechin equivalents/100g), phenols (31.76 mg gallic acid equivalents/100g), vitamin-C (6.48 mg/100 g) compared to yellow and white coloured flakes. Total antioxidant activity is quite high (11.00 and 14.93 mg AEAC/100g in FRAP and DPPH assay, respectively) which indicated potential of its health promoting and functional food components.

SHANKARA (Karunakaran and Singh, 2019)

It is a selection from a 25 year old tree at farmers' field of Shri Shankariah, Chowdlapura village, in Tumkur District of Karnataka. Growth habit is medium tall tree with semi spreading nature and fairly large dark green leaves. It bears solitary fruits on the primary as well as secondary branches. Female flowers are larger than the male and pedicel is thick one. It bears fruits in bunches on the primary branches and also on secondary branches. Fruit is a multiple type with small size fruits (2-5 kg), fruit's skin colour is dark green when mature and tip of spine becomes brownish to black when ripe. The average fruit weight is 2.73 kg and contains 61.40 number of flakes with average flake weight of 17.95 g; flake shape is irregular and thickness is 5.86 mm. The fruit contains the edible, sweet, aromatic, medium crispy bulbs and coppery red coloured flakes. The crop is generally harvested during April-June. Well-managed tree is estimated to yield around 460.12 kg per fruit tree. Total soluble solid variety is quite high (28°B) which is an important parameter related to fruit quality and taste.The farmer variety has deep coppery red colour with high amounts of carotenoids (5.83 mg/100 g) and lycopene (2.26 mg/100 g) as compared to white coloured flakes. Total phenols (37.99 mg Gallic acid equivalents/100g) also higher in coppery red coloured flakes compared to white and yellow coloured flakes. Total antioxidant activity was 10.91 and 9.42 AEAC/100g in FRAP and DPPH assay, respectively, which indicated potential of its health promoting and functional food components.

Palur 1 Jack or PLR 1

Tamil Nadu Agricultural University (TNAU) in India developed this improved variety in 1992. It is a clonal selection from Panikkankuppam local of Panruti taluk, in South Arcot district of Tamil Nadu. It is high yielding with medium height, less spreading and suitable for high density planting. In addition to regular bearing season (March-June), it produces fruits during off season (October- December) also. The annual yield per tree is about 80 fruits weighing around 900 kg. The average fruit weight is 12 kg containing 115-120 flakes. The off season bearing is a unique trait of this variety and the fruits fetch premium market price. The fully ripe fruits have flat stigmatic surface instead of spiny surface. The fruit quality is good with attractive golden yellow firm flakes. The flakes are very sweet with high consumer appeal and good keeping quality. The seeds are also edible with high palatability. TSS is 19°brix.

PPI 1 Jack

Tamil Nadu Agricultural University (TNAU) has developed this improved variety in 1996. It is a clonal selection from Ulagumoodu Local near Pechiparai. It is a medium tall tree, yields 105 fruits per tree per annum (each weighing 17 kg) which accounts for 40.8 per cent more yield than local. The fruit bearing occurs twice annually (April-June and November - December) and produces high quality crisp carpels with high TSS and ascorbic acid content. The flakes are sweet and tasty with pleasant aroma. The maximum bearing takes place on tree trunks. This variety is suitable for commercial planting and home gardens.

PLR (J) 2

Tamil Nadu Agricultural University (TNAU) developed this improved variety during 2007. It is a clonal selection from Pathirakkotai Local. It possesses good quality and bigger size fruits, and highly palatable and edible flakes and fetches more price due to attractive characters and better keeping quality resulting in increased income to the farmers. This is having less incidences of major insect–pests and diseases. This can be planted during June–December. This is suitable for growing in tropical climates of deep well drained loam soils. This variety yields 95–110 fruits/tree/year weighing an average of 1600–1950 kg. This is suitable for growing in tropical climates and deep well drained loam soils.

Swarna

The University of Agricultural Sciences, Bangalore, Karnataka (India), released a jackfruit selection Swarna in 2010. Swarna is a very good yielder that bears fruits in bunches on the trunk, primary branches and also on the secondary branches. The medium sized fruits weigh 6-8 kg each. The fruit lets are golden brown in colour, juicy with a TSS of 25-26 °Brix, with thick flakes and a very thin rind (<1.0 cm).

Cultivars of Jackfruit

Mini

Golden Nugget

Black Gold

Honey Gold

Dang Rasimi

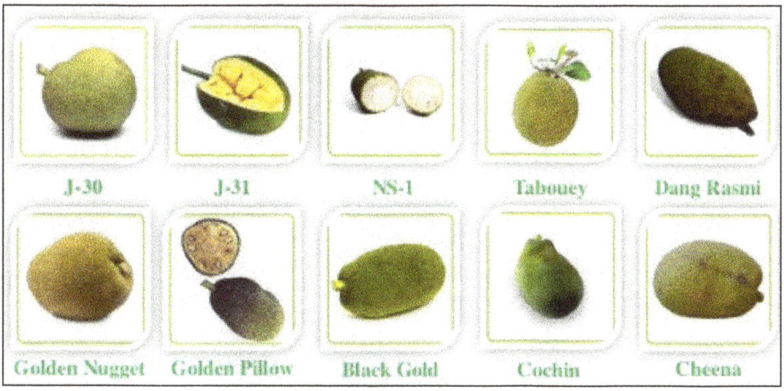

Different Cultivars of Jackfruit

Cultivars of Jackfruit

Gumless Type (Pre-released)

The gumless type developed at the Indian Institute of Horticultural Research (IIHR), Bangalore, India has medium to low latex with medium fruits (6.4-9.0 kg) having oblong fruit shape containing 120-140 flakes/fruit with flakes: fruit ratio of 0.45-0.50. The flakes are light yellow, sweet in taste having TSS of 26-30°Brix has been found promising and is under the process of release.

Konkan Prolific

This variety was released in 2004 by the Regional Fruit Research Station (RFRS), Vengurle of Dr. Balasaheb Sawant Konkan Krishi Vidyapeeth, Dapoli, Maharashtra. It is a seedling selection with a potential yield of 450-550 kg/tree and average yield of 420.56 kg/tree producing 73 fruits/tree/year with individual fruit weight of 5.70 kg, fruit length of 34.4 cm fruit breadth of 15.2 cm, medium sized fruit, oblong shape, firm flesh even in monsoon season, golden yellow colour of flesh, fruit bearing in bunches, spiny surface, flowering in January –February, bearing after 6-7 years of planting. Fruits have TSS of 25 ° Brix and seed weight of 132 g.

Muttom Varikka

A promising jackfruit variety developed by Kerala Agricultural University. This variety of *Artocarpus integrifolia[Artocarpus integer]* bears fruits of good quality, which have crisp, sweet, golden yellow flesh and which keep well TNAU has followed this breeding approach and released some promising varieties like PLR 1, PLR 2 and PPI 1 (Kumar and Kavino, 2011).

PPI-1

A new variety of jackfruit (*Artocarpus heterophyllus*) derived as a clonal selection from the Kazhukupala type was released for general cultivation in Tamil Nadu. The variety is high yielding and comes to first fruiting in 5 years. In trials conducted at Mulagumoodu during 1980-94, PPI1 gave an average yield per tree of 107 fruits weighing 1818 kg, out yielding the local variety PLR1 which gave 76 fruits weighing 865 kg. (Alagiapillai *et al.*, 1996).

Red Flesh Jackfruit

It is a new dry-bulb cultivar selected from natural seedlings of Thailand jackfruit. The variety has the following characteristics: early fruiting and lower fruiting location, flowering year round but mainly from March to October. The fruit development period is 110 to 135 days. Its fruit is oblong, with an average weight of 10.3 kg. The fruits have thick and less latex bulbs. The flake is orange red, crisp and with sweet aroma. The total soluble solids (TSS) content of flake is 18.87 per cent (Wang *et al.*, 2012)

AHPah – 1, 2, and 3

These varieties were developed as selection at Agricultural Research Station, Bangladesh Agricultural Research Institute, Pahartali, and Chittagong during 2013-2014.

AHPah-1 has been found promising for table purpose followed by AHPah-2 and AHPah-3. Minimum days (117) to 1st harvest were observed in AHPah-1. The number of fruits per plant was exceedingly higher (73) in AHPah-1 whereas minimum number (41) was found in AHPah-2. Maximum weight (8.40 kg) per fruit was observed in AHPah-2 and minimum was in AHPah-1(3.40 kg). The highest single fruit length (37.25 cm) was found in AHPah-2 and breadth (27.00 cm) was produced by AHPah-3. Maximum number of bulbs (116) was produced in AHPah-1, whereas minimum (63.00) was in AHPah-3. Maximum weight of bulbs per fruit (4.24 kg) was produced in AHPah-2. Individual bulb weight was higher (54.42 g) in AHPah-2 whereas; the lowest (16.71) was in AHPah-1. Edible portion was higher (69.27 per cent) in AHPah-1 whereas; the lowest (53.43 per cent) was in AHPah-3. The TSS was the highest (21.00 per cent) in AHPah-1. The highest bulb length and breadth was found in AHPah-3. Highest seeds weight (639 g) was produced in AHPah-2. Individual seed weight (8.19 g) was higher in AHPah-2. Therefore, the genotypes can be included in the variety development program after comparing with the already BARI released jackfruit variety.

Salient characters of important jackfruit cultivars growing in different countries are as follows:

5.1.0 Indonesia

Kandel

Fruits are of medium weight (15-25 kg), thick flesh (upto 12 mm), crispy, very sweet and dry flesh. The tree bears fruit all year round, with two big harvests in a year. It was released as a national superior cultivar of Indonesia. Propagation can only be done by approach grafting on rootstocks of the same cultivar. It is highly incompatible with rootstocks from other cultivars. The latex is very sticky, thus inhibits root formation on marcotting, and inhibits cambium contact on budding or wedge grafting. Seeds germinate very easily, even while still intact in the fruit.

Mini

The cv. Mini is a very precocious bearing tree found in the Riau province in Sumatra. It starts to bear fruit just 18 months after sowing the seed. Although the fruit is relatively small (5-12 kg) and the quality varies from one tree to the other, there is potential in using it for breeding programmes. Potted plants bear fruit easily too.

Tabouey

The tree has small, dark green, rounded leaves and forms an open, rounded canopy. The fruits are long, thin and tapered at the stem-end. They are often misshapen and lobed, with a bright yellow colour and irregular, blunt spines. The flesh is firm, pale yellow and has almost no aroma. Its flavour is mild and pleasant.

5.2.0 Malaysia

J-30

The tree is vigorous, forming an open, conical canopy. Fruits are uniform and hang singly on the major limbs. The fruits are dark green and blocky, with blunt spines on the skin, average weight 5-6 kg, with an edible flesh percentage of around 27 per cent. The flesh is deep orange and firm, although the walls of the flesh are thin.

J-31

The tree is moderately vigorous, forming a spreading, open canopy. The fruits are large and irregular in shape with prominent, blunt spines. Average weight of fruit is 12 kg with an edible flesh percentage of 36. The flesh is deep yellow and firm with thick walls. The flavour is sweet and rich with a strong, earthy aroma.

NS 1

It is one of the first superior jackfruit cultivars imported into the Western Hemisphere from Malaysia. The tree is medium vigorous, forming a dense, upright canopy. Production is heavy. The fruits are blocky and dark green colour, with blunt spines on the skin. The fruit weighs 4-5 kg, with a flesh percentage of 34. The blunt spines flatten and open as the fruit ripen. The flesh surrounding each seed is firm, and thick-walled. The colour is a dark orange, with a rich and sweet flavour.

5.3.0 Australia

Golden Nugget

The tree is fast growing, with a distinctive dark green, rounded leaf It forms a dense canopy. The fruits are small (3-3.5 kg), green and rounded, with sharp, fleshy spines on the skin. The spines flatten to a smooth, golden-yellow upon maturity. Excellent edible flesh percentage is 41. The deep orange flesh is soft to medium firm depending on fruit ripeness. The flavour is excellent and the flesh has no fibre.

Black Gold

The tree is vigorous, with a spreading and highly manageable canopy. The fruits are medium-sized, averaging 6.7 kg, with an edible flesh percentage of 35. The exterior of the fruit is dark green and has sharp fleshy spines. The spines do not flatten at maturity, making it difficult to judge the proper harvest time and causing difficulty in handling the fruit after harvest. The deep orange flesh is soft to medium firm, with a strong, sweet flavour and aroma. The flesh is easily removed from the fruit compared with other cultivars.

Honey Gold

The tree has a slow to moderate growth rate, forming a small, open and spreading canopy. The blocky, dark green fruit has small, sharp spines that open and turn a golden yellow upon maturity. The fruit weighs 4.5 kg, with an edible flesh percentage of 36. The flesh is dark yellow to orange and firm, with a sweet, rich flavour. The fruit has a strong, sweet aroma.

Lemon Gold

The tree is moderately vigorous, forming a spreading, open canopy. The fruits are bright green and blocky, with prominent fleshy spines on the skin that flatten at maturity. Fruits average 6 kg and have 37 per cent edible flesh. The flesh is lemon yellow, thick-walled and firm, with a sweet and aromatic flavour.

5.4.0 Thailand

Dang Rasimi

The tree is open, spreading and fast growing. The trees are highly productive. The fruits are bright green to pale yellow and uniform in shape. The skin has sharp spines that do not flatten at maturity. Fruits are medium to large (5-6 kg), with 32 per cent edible flesh. The flesh is deep orange and medium firm to firm depending on fruit ripeness. It has a mild, sweet flavour and a sweet, pleasant aroma.

5.5.0 Bangladesh

Topa

Fruits and seeds are small, flesh soft, low latex, flavour good. Starts bearing in 4-5 years.

Hazari

Produced one thousand fruits, hence the name. Fruits are very small. Flesh soft and sweet.

Chala

Medium-sized fruits. Both flakes and seeds are small. Flesh soft, sweet and fibrous. Bears entirely on the tree trunk.

Goal

Fruits and bulbs both are roundish in shape. Flesh soft, fibrous and very sweet. Precocious bearing.

Koa

Medium-sized fruits. Flesh very soft and sweet. Bulbs are elongated in shape.

Khaja

Medium-sized fruits. Flesh firm and sweet. The fruits are brownish at maturity.

5.6.0 Philippines

Torres

The first jackfruit variety released in the 1980's by IPB (Philippines) using the Farmer Participatory Breeding.

Mabini

This jackfruit released in 200 is a dual-purpose variety intended for both fresh fruit and processing. The selection was released as, a variety called 'Mabini', named after the national hero, Apolinario Mabini, a sublime paralytic who once lived in Talaga, Tanauan City, Batangas, the place where the original jackfruit tree was selected in 2000. This selection was proven outstanding in terms of fruit qualities both as fresh fruit and processing type 'Mabini' is a dual-purpose jackfruit variety with unknown parentage registered with the National Seed Industry Council. The fruit is oblong, weighs 15,433.33 g and measures 451.67 mm long and 290.67 mm wide. The skin and spines are yellow green (RHS 1-B) when ripe, rough textured and thick (11.63 mm). The fruitlet is big (32.70 g) while the flesh is bright yellow orange (RHS 23-A), juicy, smooth, semi-crispy, has strong aroma, moderate latex, no fibers, thick (5.97 mm) and sweet (20.40 Brix). The whole fruit has a medium-sized core (1,100 g) with a diameter of 98.33 mm and relatively large edible portion (50.20 per cent). The seed is oblong with grey-orange (RHS 165-D) testa, weighs 5.27 g, and measures 34.21 mm long and 21.07 mm wide. The tree is 7 m tall, vigorous, evergreen, has semi-spreading growth habit, a regular bearer yielding 50-80 mature and marketable fruits per season, with a peak fruiting season from February to July (Valencia, and Magdalita, 2010).

The other jackfruit varieties that were developed using the Farmer Participatory Breeding approach in the Philippines are: 'EVIARC Sweet', 'Sinapelo', 'Burabod', 'Cervantes Gold', and 'MMSU-SRO Sweet'. The other jackfruit varieties that were developed using the Farmer Participatory Breeding approach in the Philippines are: 'EVIARC Sweet', 'Sinapelo', 'Burabod', 'Cervantes Gold', and 'MMSU-SRO Sweet'.

5.7.0 China

Jackfruit, a world-famous tropical fruit originally planted in ancient India, was introduced to China during Tang Dynasty (618-907). Growing up to 15-20 meters, the trees usually produce fruits ranging from 5 kg to 20 kg (some even exceed 50 kg) on their trunk (https://www.visitourchina.com/blog/detail-174.html)

Jianmila

It is a cultivar belonging to *Artocarpus champeden*, a tropical evergreen fruit crop. It performs well at the area with yearly temperature of 22.4 °C, minimum temperature of 8.0-9.0 °C, yearly rainfall of 2131.4 mm. Fruit mature in September to October, having an elongated round shape with a yellow green skin, weighing 2-4 kg. Flesh is yellow to dark yellow in colour, having a sweet and fragrant flavour, containing a soluble solids content of 25 per cent, starch 1.26 per cent, ascorbic acid of 3.43 g. The fruit skin is easily peeled from the flesh. It could be propagated by seeds or grafting (Wu *et al.*, 2010).

6.0 Breeding and Improvement

The jackfruit is regarded as a minor fruit and it is seldom found in regular plantations. In spite of its high nutritive values, a wide range of preserves and utility of timber and foliage, the tree is a neglected one. The following measures are suggested for improvement of this potential fruit crop.

The most important means is to exploit the wide variation among different types grown. Trees of outstanding merits are to be selected after a thorough survey in the growing region. The selected trees should bear large number of fruits of small to medium size. Big size of fruit should be avoided. Fruits should contain larger quantities of pulp with smaller seeds and thinner skin. Selected clones should be propagated vegetatively and a mother orchard has to be established for preparing planting materials.

6.1.0 Cytogenetics and Breeding

6.1.1 Cytogenetics

Very little information is available on the cytology of *A.heterophyllus* gene pool. Darlington and Wylie (1956) have reported that it is tetraploid with a somatic chromosome number of 56 (2n=4x=56), thus the basic chromosome number (x) is 14. It crosses freely with *A.integer*, *A.lanceaefolius* and *A. rigidus* and Kanzaki *et al.* (1997) believed that these species are close relatives of *A. heterophyllus.*

Character of genome sequence, evolvement biology and the screening of functional genes of jackfruit, determination of the genome size of jackfruits were studied by Yang *et al.* (2015) using Tomato and litchi 'feizixiao' as internal references, and Otto as extract. The fluorescence intensities of the internal reference and jackfruit were measured by flow cytometry (FCM) after the mixed nucleus samples dyed by propidium iodide (PI). Comparing the multiple relationships between the two peak values, the ploidy level and genome size was calculated for different jackfruit lines. The selected jackfruit lines were diploid. The hard jackfruit genome size range from (1 243.75+or-9.21) Mb to (1 540.63+or-3.55) Mb, while the soft jackfruit genome size was about (1 902.87+or-8.08) Mb. They concluded that the genome sizes varied in different jackfruit lines, and the Nuclear DNA contents of soft lines were much greater than that of hard lines. The karyomorphology of 16 species in 13 genera representing Moraceae and Cecropiaceae was investigated in an effort to contribute to a better understanding of chromosome features and evolution in the families (Oginuma and Tobe, 1995). All genera investigated have similar karyomorphology, but differences are found in (1) chromosome features of Interphase nucleus (simple, simple-complex, or complex chromocenter type), (2) basic chromosome number (*x*=13 or 14), (3) size variation (mono-or bimodial), and (4) frequencies of chromosomes with median centromeres (*m*-chromosome) (25–85 per cent) and those with subterminal (or terminal) centromeres (*st*-chromosome) (14–69 per cent). Comparisons with Ulmaceae as an outgroup of the remainder of Urticales suggest that the simple chromocenter type, *x*=14 comprising both *m*- and *st*-chromosomes, and the monomodial karyotype are plesiomorphies in Moraceae and Cecropiaceae. Most of Moraceae and Cecropiaceae retain generalized chromosome features of the order, but have involved a few evolutionary changes in karyomorphology.

6.1.2 Breeding Objectives

In jackfruit, the trees are erratic in bearing and the fruit yield varies every year. Hence, the first objective is to develop a variety that is regular and prolific in bearing high quality fruits (APPARI, 2012). According to Bhatia *et al.* (1955), the edible pulp (bulbs) forms 29 per cent of the ripe fruit, the seeds 12 per cent and the outer rind 59 per cent. The outer rind consists of soft structural matrix (non-edible internal tissues) and fleshy yellow-green spines on the skin. The content of this mass in a fruit needs to be reduced through breeding. Sincere breeding efforts are required to enhance the weight of edible pulp from about 30 per cent to 45 per cent of the fruit. The seeds should constitute about 5 per cent of the fruit and the outer rind along with internal structural matrix should not exceed 50 per cent of the total fruit weight. Usually big size fruits are inconvenient to handle. Thus, to improve yield, number of fruits per tree should be increased instead of getting lower number of bigger sized fruits.

Jackfruit tree has a long juvenile phase (7-8 years). Reducing this period is another important breeding objective. According to Richards (1950), Singapore variety comes into bearing in about 18 months under favourable conditions in Sri Lanka. Nangka Salak is an early bearing variety in Java (Yaacob and Subliadrabandhu, 1995). Such varieties can be used as one of the parents for inducing precocity. Reduction in tree height is desired for high density plantings and also for convenience in cultural operations.

Haq (2006) believed that most important characteristics of a jackfruit ideotype are: (i) easy to manage and requiring less or no pruning for fruit production; in case of timber, long straight trunks for crop diversification and agroforestry systems, (ii) vigorous and prolific plants compatible with one or more rootstocks with early flowering and regular bearing, (iii) good quality fruits with acceptable flesh colour and texture, good flavour and sweetness, (iv) fruits with symmetrical form and acceptable size, (v) resistance to fruit borer, die-back and flood, (vi) wider geographical and environmental adaptation, (vii) wide range of fruit maturity period (viii) off-season fruiting, (ix) long post-harvest life, and (x) high yield.

6.2.0 Genetic Resources Survey, Collection, Conservation and Evaluation

6.2.1 Genetic Diversity

An extensive survey of barhal growing areas was conducted in Kasia block of Kushinagar district in Uttar Pradesh. During the exploration twenty three genotypes having a wide spectrum of variability for various characters were collected from 17 villages. Various tree and fruit parameters were studied with objectives to establish variability in the natural germplasm and to identify elite germplasm among its population. Highest GCV along with high PCV was recorded for tree cross section area (70.51 per cent, 70.57 per cent), pulp weight (47.22 per cent, 47.62 per cent), fruit volume (43.24 per cent, 43.20 per cent) and fruit weight (42.05 per cent, 42.17 per cent). Highest heritability coupled with high genetic advance was noticed for fruit weight (0.994 per cent, 83.45 per cent) followed by fruit volume and pulp weight which ensured that substantial improvement of these traits could be achieved through direct selection whereas selection of specific combinations followed by mating could lead to improvement of traits expressing high heritability but moderate genetic advance (Dwivedi *et al.*, 2010).

Jackfruit is one of the indigenous and underutilized fruit of Rain Forests of India. Naturally Eastern and North Eastern India have the rich genetic base of jackfruit with wide genetic variability of plant growth, fruiting behaviour, fruit physical and quality attributes of fruit. The quality attributes of some jackfruit genotypes of Eastern and North Eastern India was studied by Deb *et al.* (2013). The properly mature fruits of sixteen selected jackfruit genotypes (soft type) of different locations of Eastern and North Eastern states were collected. Quality parameters like total soluble solids (TSS), titratable acidity, total sugar, reducing sugar, ascorbic acid, carotene, dry matter content *etc.* were evaluated. Sensory evaluation was also carried out to judge the acceptance of the selected jackfruit genotypes. TSS of the bulbs of different jackfruit genotypes ranged from 17.35 to 28.12 degrees Brix while titratable acidity from 0.42 to 0.93 per cent. Maximum total sugar and reducing sugar were 17.71 and 8.03 per cent respectively. A range of ascorbic acid content of fruit pulp of different jackfruit genotypes from 5.24 to 10.12 mg/100 g was recorded while it was from 311.2 to 496.7 micro g/100 g in beta -carotene content. The jackfruit selections collected from North Bengal, lower Assam and Tripura were superior with respect to different quality parameters and sensory analysis.

The North Eastern State of Tripura has a great diversity for jack. Physico-chemical characteristics of seventeen types of jackfruit collected from elite clones through an extensive survey has led to the identification of trees for different parameters of qualitative and quantitative characters aspect of fruits (Islam *et al.*, 2014). A wide variation among selected fruits was observed in jackfruit like fruiting season, bearing habit, fruit attractiveness, fruit weight, fruit shape, latex exudation, bulb number, bulb colour, bulb size, pulp juiciness, pulp taste, pulp flavour, flake fibre content, flake thickness, seed weight, total soluble solid (TSS), acidity, sugar and TSS:acid ratio *etc.* The result of the study are to be helpful for understanding the variety and best type are attempting the selection of superior desirable jackfruit genotypes for bring to commercial cultivation.

Bangladesh is often recognized as a secondary center of diversity for jackfruit and is expected to harbor important genetic diversity of this underutilized tree crop (Khan *et al.*, 2010). Unfortunately, genetic erosion is occurring before this rich local resource has been documented and utilized for potential crop improvement. The aim of this study was to carry out a village-based survey and make morphological measurements to document and assess jackfruit diversity across trees associated with a gradient of three locations/habitats (homesteads, public lands, and forest or fallow lands). We also tested the hypotheses that cultivated jackfruit found in homesteads exhibited positive selection pressure for characteristics desirable in the market, and that the tree location reflected its history of origin and human selection. This was accomplished using 28 standardized morphological descriptors and represents the first large-scale assessment (900 trees) of jackfruit diversity in multiple locations (nine villages). Among the descriptors studied, those most closely associated with jackfruit marketability showed a trend for selection among the trees located in homesteads. Both fruits and leaves were larger in plants on homesteads compared to plants in forest/fallow lands, and fruit quality was significantly higher in homestead trees compared to the other two location categories. However, 18.7 per cent and 23.7 per cent of the fruits found on public lands and forest/fallow lands, respectively, were still considered to be of excellent fruit quality and tree vigor was significantly higher in jackfruit in forest/fallow lands compared to the other location categories. The combination of high tree vigor and presence of excellent fruit quality among jackfruit in forest/fallow lands suggests that the "wild" Bangladeshi jackfruit germplasm, which is considered inferior for market and is being negatively impacted, has valuable genetic diversity to contribute to jackfruit cultivation. Furthermore, sampling across a gradient of habitats may enable the detection of possible trends resulting from domestication pressures.

An analysis of genetic divergence for 18 economically important quantitative traits in 44 jackfruit genotypes from different geographical regions of West Bengal, India was carried out using Mahalabonis D^2 statistics (Maiti *et al.*, 2002). The analysis of variance showed significant differences among the 44 genotypes. The genotypes were grouped into 13 clusters. The average intra and inter cluster distance analyses showed that the intra cluster distance was zero for clusters I, IV, V, XI, XII and XIII. The average values of intra cluster distance were highest in cluster X (172.7) and lowest in cluster IX (71.7). The maximum values of inter cluster distance were observed between clusters III and IV (14319.1), followed by clusters II and IV (11744.3), and clusters I and IV (10335.6). The minimum values of inter cluster divergence were recorded between clusters I and II (153.5) followed by clusters XI and X (218.1). The genotypes included in clusters with maximum inter cluster distance were genetically more divergent.

The natural variability existing in physico-chemical qualities of jackfruit bulbs through an extensive survey of coastal zone of Karnataka (situated in the Western Ghats of India), was reported by Jagadeesha *et al.* (2010). A significant variation in physico-chemical characters of jackfruit bulbs was observed among the 30 jackfruit selections surveyed and studied in this zone. A high coefficient of variation was observed for bulb parameters such as flake mass (74.79 per cent), bulb mass (71.17 per cent) and single bulb mass (50.11 per cent). Among the biochemical parameters starch content (67.30 per cent), TSS: Acid ratio (34.69 per cent) and titratable acidity (31.76 per cent) exhibited a wide variation.

Forty four superior jackfruit genotypes were collected from three districts of West Bengal, India based on jackfruit descriptor (2000). These genotypes were analyzed based on 19 important quantitative and 7 qualitative characters (Wangchu *et al.*, 2013). From the analysis of genetic parameters, it could be concluded that characters like weight of fresh flakes without seed, weight of fresh flakes with seed, stalk length, fruit yield per tree, rachis diameter, rachis length, fruit length, shelf life, number of flakes per kg of fruit, flake width, number of seeds per kg of fruit, vitamin C, TSS/acid ratio and sugar/acid ratio could be used as selection criteria for development of effective and productive plant types in jackfruit. Further, these 44 genotypes were grouped into ten clusters for quantitative characters and eight for qualitative characters.

Different cultivars of jackfruit from Rajshahi, Natore, Noagoan and Chapai Nawabgonj areas of Bangladesh were studied (Ibrahim *et al.*, 2013). These cultivars showed variability in morphological fruit characters and nutritional food values. Among them the percentage of pulp (38.60-47.37 per cent), percentage of rind (15.67-20.00 per cent), percentage of skin (14.86-23.68 per cent), percentage of seed (09.46-19.33 per cent), moisture content (63.39-76.62 per cent), TSS (18.80-27.37 per cent), total sugar (11.84-17.01 per cent), vitamin C content (17.82-31.55 mg/100 g) and acidity (0.037-0.075 per cent as citric acid) were estimated. The findings show that, evaluation of these promising cultivars have identified them as superior ones. Lot of Jackfruit trees are grown in Home gardens which serve as a great repository of diversity. The current state of jackfruit diversity within home gardens in seven unions of Chapai Nawabganj District of Bangladesh was investigated in a comparative study on 210 home gardens, interviews of 335 villagers and screening on 1400 jackfruit plants (Sarker *et al.*, 2015). The report highlighted the gross lack of scientific information about home gardens for in situ conservation of jackfruit morphotypes and emphasized the urgent need for a systematic research upon which future strategies for home gardens in Bangladesh can be based. Researchers applied folk classification, ethno botanical and morphological measurement to obtain sustainable design for natural link between conservation and development. 15 landraces and two separate communities were identified by the local farmers. Samples from the selected sites exhibited significant variation for six quantitative characteristics evaluated for phenotypic diversity of jackfruit. On age related measures, only Debinagar union displayed no significant alteration for any of six morphological characters. But all the other unions demonstrated distinct deviations for three (Jhilim and Shahjahanpur), four (Narayanpur and Islampur) and five (Ranihati and Sundarpur) features respectively. Ethnobotanical and morphological assessment clearly distinguished the reduction of jackfruit variation over time among six unions except Debinagar. Present study suggested establishing common names for landraces to protect them from extinction.

Character Association

A study was conducted to assess the correlation among fruit weight and its components and path coefficient analysis to provide adequate genetic background for direct and indirect effect of various characters on fruit weight of jackfruit in West Bengal (Maiti, 2010). The results revealed that genotypic and phenotypic correlation were positively significant between fruit weight of edible part (0.980 and 0.977), fruit and rind weight (0.976 and 0.971), and number of stones and flakes (0.999 and 0.999). The path coefficient analysis indicated that weight of edible part had positive direct effect on fruit weight both at genotypic (0.459) and phenotypic (0.451) levels. Hence, at the time of selection of jackfruit genotypes, one has to put emphasis on the characters like weight of edible part, rind weight, flake and stone numbers.

Trait Specific Genetic Resources

Crack Resistance

Based on overall performance with respect to vegetative growth, yield and quality parameters, jackfruit cracking and non-cracking genotypes studied at Karnataka, indicated that cracking trait jackfruits were more likely to be varietal characters. The investigation also revealed that cracking genotypes contained less pectin when compared with the non-cracking genotypes that may cause cracking due to inelastic nature of the rind (Singh *et al.*, 2011).

Offseason Jack Varieties

Fresh and tender fruits are usually not available after November. Three trees were identified; in the terai region of sub-Himalayan foot hills of Cooch Behar and

Jalpaiguri districts of West Bengal, India, which bear fruits throughout the year (Saha *et al.*, 2011). These trees were studied for three years. On an average, 26 to 31 fruits were harvested per month per tree in the months of April to November. However, it was only 16 to 21 fruits during the months of December to February. Kerala, which lies in the southern most part of Western Ghats, is well known for its diversity in jack fruit. Jack fruit accessions with off season fruiting, beginning from January and extending up to July were identified in the present study (Gomez *et al.*, 2015). Accession AH-6 produced fruits round the year. Wide variation was observed in physico-morphological characteristics like fruit shape, weight, and rind colour and percentage composition of the various constituents. Accession AH-8 recorded the maximum fruit weight (20.13 kg) while percentage of pulp was highest (43.04 per cent) in AH-9. Accession AH-10 had the highest number of bulbs (300). Significant variation in the biochemical constituents was also observed. Total soluble solids, titrable acidity, total sugars, brix: acid ratio exhibited maximum variation among the biochemical attributes. Highest total soluble solids were recorded in AH-5 (30.2 °Brix) whereas total carotenoids were highest in AH-2 which was on par with the check cultivar Muttam Varikka.

Cold Resistance

In China studies were conducted to select the Jackfruit with strong cold-resistant characteristics. Eight field-grown Jackfruit strains (E1 to E8) were studied (Lu *et al.*, 2012).The cold-resistant characteristics among them were contrasted and classified, and some physiological indexes on their cold-resistant characteristics were also measured and correlated with the chilling damage degrees. The results showed that the continuous cold weather could result in chilling damage of Jackfruit strains, and the damage degree depends on their cold-resistant abilities. Of all the eight Jackfruit strains, four of them were first-class chilling damage; three of them were second-class chilling damage; one of them was third-class chilling damage. There were a significantly positive correlation between the value of leaf conductivity and the chilling damage degree, a significant negative correlation between the soluble sugar content and the chilling damage degree, and no significant correlation between the vitamin C content and the chilling damage degree of Jackfruit strains. In conclusion, the Jackfruit strains with strong cold-resistant characteristics were E8, E6, E5 and E3. The physiological indexes of the leaf conductivity and soluble sugar content could be used for evaluating the cold-resistant characteristics of Jackfruit strains.

Insect Resistance

Sixty-five germplasm collections of jackfruit were screened for relative resistance to shoot and fruit borer, *Diaphania caesalis* (Wlk.) under field conditions in Bengaluru (Soumya *et al.*, 2015). A significant variation in infestation of *D. caesalis* was observed among jack germplasm. The result s showed that out of 65 accessions, 4 accessions were found resistant, 26 were least susceptible, 22 were moderatively susceptible and 13 were highly susceptible. The identified resistant accessions form a basis for further studies on morphological and biochemical analysis to understand host plant resistance.

6.3.0 Approaches to Crop Improvement

6.3.1 Introduction

The jackfruit (introduced from tropical Asia) has been grown in South Florida for 90 years or more, but has never had commercial importance. In recent years, many new superior cultivars have been imported by Fairchild Tropical Garden which offers commercial possibilities. Of the 22 cultivars presently in the collection, Black Gold, Golden Nugget, J-31, Honey Gold and Lemon Gold have been the most precocious. These cultivars vary greatly in their size, shape, flesh characteristics and taste. All compare well with NS-1, the most popular cultivar in South Florida. Galaxy and Honey Gold have shown a tendency for weak growth, with severe dieback. Delightful, Dang Rasimi, Golden Nugget and Black Gold have been the most vigorous. Fruit ripening has been concentrated in July and Aug. in most cultivars, but J-31 has ripened in late-May and early-June and Black Gold has ripened as late as mid-Sep. (Campbell and McNaughton, 1994).

6.3.2 Selection

The Farmer Participatory Breeding is a direct approach being used by fruit breeders in the Philippines for developing jackfruit varieties (Magdalita *et al.*, 2011). The breeders exploited the natural variability existing in the jackfruit population consisting of trees grown from seeds. On-site selection and evaluation of desirable jackfruit genotypes was conducted directly in the farmer's and grower's fields in various places in the Philippines. The main objective of the participatory breeding work is to identify and select promising jackfruit trees that could be released as a registered variety. Farmer Participatory Breeding for fruit crops was first recorded at the Institute of Plant Breeding (IPB), College of Agriculture (CA), University of the Philippines Los Baños (UPLB). The first jackfruit variety released in the 1980's by IPB using the Farmer Participatory Breeding is the 'Torres' jackfruit. The 'Mabini' jackfruit is another variety released by IPB and registered to the National Seed Industry Council (NSIC) in 2005. It is a dual-purpose variety intended for both fresh fruit and processing. The other jackfruit varieties that were developed using the Farmer Participatory Breeding approach in the Philippines are: 'EVIARC Sweet', 'Sinapelo', 'Burabod', 'Cervantes Gold', and 'MMSU-SRO Sweet'. The selected jackfruit variety after it has been evaluated by the breeders in collaboration with the farmer-cooperators for three consecutive years is being registered with the National Seed Industry Council (NSIC) of the Department of Agriculture to make the variety valid. At present, the only involvement of the farmer-cooperators in the Farmer Participatory Breeding is in the selection of promising jackfruit trees. In the future, it is expected that a full-blown participation of the farmers in on-farm trials, on-site propagation and nursery establishment and holding of jackfruit contests, shows and fora to address issues on production, pest and disease management, processing and entrepreneurship will be attained to advance the jackfruit industry in the Philippines

6.3.3 Rootstock Selection

Recalcitrant seeds are desiccation and chilling sensitive and are viable for a very short period. Seeds of *Artocarpus heterophyllus* being recalcitrant in nature pose storage problems. Studies were taken up to identify methods to prolong seed viability of the species. Seeds were stored at four different temperatures: Ambient (25+or-2 degrees C), 20 degrees C, 15 degrees C and 0+or-2 degrees C and subjected to germination tests at the end of 1, 2, 3, 4 and 5 weeks of storage. Seeds stored at 20 degrees C retained viability for 5 weeks with 41 per cent germination. Experiments to store seeds with and without bulb revealed that seeds should be extracted from the bulb to ensure effective storage. The viability of seeds can be prolonged to 32 weeks (with 48 per cent germination) when stored at 20 degrees C (Warrier *et al.*, 2009).

Performance of rootstock for grafting success depends on genotype and seed weight. Nivetha and Maheswari (2015) carried out studies to standardize the genotype and seed weight on root stock production. They used five different genotypes of jack collected from Panruti (G1), Virdhachalam (G2), A. Puthur (G3), Kadampuliyur (G4) and Neyveli (G5) with four different seed weight 4 g (W1), 5 g (W2), 6 g (W3) and 7 g (W4). The results of the experiment revealed that among these genotypes, G1 (AH-1) was the promising genotype in achieving maximum value for the characters like germination percentage, shoot length, root length, plant height, vigour index, number of leaves, leaf length, leaf breadth, stem girth, fresh and dry weight of seedling, which was followed by G4 (AH-4). Among the various seed weight tried, the above said characters were favourably enhanced by W4 (7 g) followed by W3 (6 g). Among the various factorial combinations, rootstocks produced from 7 g seed of Panruti collection (G1W4) recorded the maximum values for these growth characters. This was followed by 7 g seed of Kadampuliyur collection achieved (G4W4) the next best value for these characters. Further, after doing soft wood grafting, the characters like graft success and graft survival percentage, days taken for first and last sprouting, number of sprouts, sprouting percentage, number of leaves, leaf area, plant height and stem girth were maximum in the factorial combination of G1W4 and it was followed by G4W4.

6.3.4 Hybridization

The first breeding programme in Jackfruit was initiated in Tamil Nadu at the Fruit Experimental Station at Kallar in 1952 with a view to developing short, compact, multi-branched trees; precocious and productive; bearing large, yellow, high quality fruits, 1/2 in the main season, 1/2 late. The variety Singapore Jack was chosen as the female parent because of its early and late crops and Velipala, a local selection from the forest having large fruits with large carpels of superior quality, and borne regularly in the main summer season was selected as the male parent. After 25 years of testing, one hybrid was rated as outstanding for precocity, fruit size, off-season as well as main season production, and yield excelling its parents (Chellappan and Roche,1982).This was made available as Hybrid Jackfruit. Later organized breeding work was reported from Florida in USA.

6.4.0 Controlled Pollination Technique

The technique followed by El-Sawa (1998) at Florida for her thesis work is given below.

Paper Pollentector (TM) bags (Carpenter Paper Co., Des Moines, Iowa) were placed over the inflorescences, before the inflorescences emerged from the stipules, to ensure no pollen landed on the surface of the flowers prior to treatment. The paper bags were then closed tightly around the peduncle of the inflorescence and tied with a wire label. In 1997, the paper bags were used as described above and additionally secured with parafilm to preclude any space between peduncle and bag that might allow access for small insects or pollen. The parafilm kept the peduncles moist and therefore was changed every other day to prevent rotting of the peduncles. For the cross-pollinated treatments, each bag was removed from the inflorescence for about 1 to 2 minutes for hand pollination. The pollen was obtained from the male parents and rubbed in the palm of a sterilized hand, then gently rubbed against the female inflorescences. For the self pollinated treatments, pollen was obtained from the same trees and applied in the same manner. Hands were washed with 70 per cent ethyl alcohol between pollinations to prevent contamination between the self- and cross-pollinated treatments and between the different cultivars. Hand pollinations were carried out every other day at mid-morning, beginning from the time the stigmas first appeared on the surface of the inflorescences, until all stigmas disappeared. The bags remained on the inflorescences for two weeks after hand pollinations to prevent any contamination. Bags were then removed and fruit left to develop on the trees until maturity (approximately 160 days). The harvested fruit were evaluated for total weight, flesh weight, seed weight, seed number, flesh percentage and edible percentage. Evaluation was carried out by first measuring total weight of the fruit, then opening the fruit and separating the edible flesh from seeds and un-pollinated flowers.

Jackfruit has been under investigation in the Fairchild Tropical Botanic Garden since 1987. The programme has focused on the conservation of genetic resources, and the promotion and development of viable cultivars. An improvement breeding programme was initiated in 1995 in conjunction with studies on the hand pollination of jackfruit. Reciprocal crosses were made among a subset of the core genetic collections. Cultivars were selected for inclusion based on precocity, productivity and fruit quality traits (aroma, edible percentage, flesh firmness, colour and flavour). 'Black Gold', 'Cheena', 'Dang Rasimi', 'Galaxy', 'Golden Nugget', 'Honey Gold', 'Lemon Gold', 'J-30 per cent ', 'J-31', 'NS-1' and 'Tabouey' were used. All cultivars used in the breeding are of Thai or Australian origin. Controlled crosses were made over 2 fruiting seasons, seeds were collected and 15 seedlings of each cross were planted at the USDA-ARS Subtropical Horticultural Research Unit, Miami, Florida, USA. The total seedling population began at 600 trees. Evaluations began in 2003 with the objective of the selection of superior jackfruit cultivars for South Florida estate and patio gardening. Preliminary results demonstrate considerable promise for the selection of improved jackfruit cultivars for tropical America (Ledesma *et al.*, 2005). In an effort to combine desirable characteristics of jackfruit (*Artocarpus heterophyllus*) cultivars, a breeding programme was initiated in 1998 at Fairchild Tropical Garden in Miami, Florida, USA. Using hand pollination of 12 cultivars within the Fairchild Tropical Garden germplasm collection (Black Gold, Cheena (a natural hybrid between jackfruit and A. integer), Cochin, Dang Rasimi, Galaxy, Golden Nugget, Honey Gold, J-30, J-31, Lemon Gold, NS1 and Tabouey), selected crosses of jackfruit cultivars have been made, using each cultivar as both male and female parents, but no self pollination (132 possible crosses). Fifteen seedlings of each cross will be planted in the field and evaluated for fruit size, seed number and weight, flesh to seed ratio, taste, texture and ease of flesh removal. Trees will be evaluated for size and ease of management (Wasielewski and Campbell, 1999).

6.5.0 Future Thrusts

From the foregoing, it is clear that in spite of the importance of this fruit, it is still considered as underutilized or crop of the future. Huge amount of diversity exists and many important aspects of the floral biology and the pollination are poorly studied. Varieties need to be classified and duplicates understood since many of them known by local names. Some of the areas that could be addressed in the form of research are as follows:

a) Origin of Jackfruit has to be scientifically investigated. Even though, it is mentioned in some texts and papers that Jackfruit has originated in Western Ghats, it lacks scientific data since the diversity in Eastern Himalayas and Indo-Malayan region is far greater.

b) Phylogenetic relationships and cross compatibility among the Artocarpus needs to be studied to eliminate the species status of some of the botanical varieties which are crossable with *A.heterophyllus*.

c) Ecosystem studies would be an important area since many of these trees are grown in homesteads.

d) Pollination and floral biology needs to be studied to confirm the presence or absence of incompatibility and mechanism behind the phenomena. The phenomenon of agamospermy needs to be studied.

e) In jackfruit, collecting, characterization, documentation and evaluation from the region of origin and centers of diversity have been sporadic and are far from complete. There is an urgent need to undertake targeted collecting from the Indian sub-continent and other jackfruit growing countries of Southeast Asia to assemble the diverse germplasm and use in the varietal improvement. In terms of genetic resources, collecting of wild species will not be justified unless they are evaluated and used for crop improvement. However, pointers can be derived from data recorded on local knowledge and ethnic uses. Future collecting and genetic diversity studies should focus on specific desirable characters and the germplasm need to be conserved both in *in situ* and *ex situ* conditions. There is a need to develop strategic plans based on the results of research and cooperation between neighbouring countries or regions, for genetic conservation of the jackfruit gene pool, needs to be strengthened. (APPARI, 2012).

f) Farmers participatory Breeding has been proved to be an efficient method for selection and breeding of Jackfruit in Philippines. This method may be further studied by other jackfruit growing countries.

g) In breeding for improvement of qualitative traits, studies on genetic correlations between traits need to be undertaken in order to decide the parents for the hybridization programme. An assessment of the patterns of existing genetic diversity needs to be made. Such a study will help planning for national genetic conservation activities, either in natural ecosystems or on-farm sites, or in well-known, well-described and focused *ex situ* collections.

h) Suitable cultivars/varieties need to be developed to meet the specific needs for fruit, vegetable and timber production as well as for multipurpose use.

i) Information on the mode of inheritance of important characters need to be generated and identification of gene(s) for seedlessness, dwarfness, resistance against fruit fly, fruit-rot and soil salinity need to be initiated on priority.

7.0 Propagation

7.1.0 Seed

The most common method of propagation of jackfruit is by seed. Generally, 4-5 seeds are planted *in situ* so that the taproot can grow undisturbed. However, seedlings can be raised in pots or polybags. After one or two years the seedlings are planted at site.

Germination of seeds deteriorates in storage. Seventy per cent of the seeds germinate within 15 days after extraction. But, after 30 days' storage, the germination declined to 40 per cent (Anon., 1967). Hayes (1953) opined that larger seeds germinate better. Soaking of seeds in water for 24 hours has been found to improve the germination (Singh, 1969). Krishnasamy (1990) suggested that the critical seed moisture content in a cultivar was 39 per cent, at which moisture content approximately 93 per cent of seed germinated. Freshly extracted seed had a moisture content of 62 per cent and 100 per cent germination.

Sinha and Sinha (1968) obtained 76.7 per cent germination by soaking in 25 ppm NAA for 24 hours compared with 60 per cent with 50 ppm NAA and 50 per cent under control. Prakash (1998) recorded maximum germination of seeds from which the outer thin leathery seed coat was removed and soaking in water for 8 hours. Treatment of seeds with biotin (10 ppm) and kinetin (50 ppm) resulted in a significantly higher vigour index. Seedling growth was also better when seeds were soaked in 25 ppm NAA. Sanmugavelu (1971) obtained 100 per cent germination by soaking seeds for 48 hours in GA_3 upto 500 ppm compared with 80 per cent under control. Seedling growth was also better. It was also noted that IAA, IBA, NAA and CPA were less effective.

7.2.0 Vegetative Propagation

A number of vegetative propagation methods have been attempted by different workers. But the plants were not superior to the seedlings although they could reproduce true to the type (Naik, 1952).

7.2.1 Cutting

Mukherjee and Chatterjee (1979) reported 84 per cent success in rooting of cuttings by etiolation and forcing of shoots followed by treatment with 5000 ppm

Propagation of Jackfruit

Seed Propagation

Chip Budding

Cleft Grafting

Veneer Grafting, Whip and Tongue and Cleft Grafting

IBA, and then keeping under a mist. No rooting was obtained without dipping in IBA. Survival of cuttings was 50-70 per cent. Dhua *et al.* (1983) recorded maximum rooting success (90 per cent) from shoots which were etiolated and ringed for 30 days and then treated with IBA at 3000 ppm in combination with ferulic acid at 2000 ppm. Pre-conditioning (especially etiolation) or stock plant treatment with ethephon before taking cutting and treating the cuttings with IBA (5000 ppm) and *p*-hydroxybenzoic acid (200 ppm) at the time of planting was found effective in inducing rooting of semi-hardwood cuttings (Mitra and Bose, 1996).

7.2.2 Layering

7.2.2.1 Air-layering

Sen and Bose (1959) reported 100 per cent success in rooting of air-layers. Treatment with IDA markedly improved the root formation. Pre-treatments of shoots like gird ling, etiolation and treatment with IBA and NAA were found more effective in rooting of layers when applied in combination rather than when applied individually (Lingarajappa, 1982). Nazzem *et al.* (1984) recorded 81.29 per cent rooting in air-layers of one-year old shoots of bearing plants when coconut pith was used as rooting medium. Datta (1988) recorded 100 per cent rooting by pre-treatment of stock plants with ethrel (250 ppm) or cycocel (1000 ppm) and application of IBA at 3000 or 5000 ppm. Alila *et al.* (2000) concluded that the treatment with IBA (5000 ppm) could be used for adventitious root formation in air-layering of jackfruit in combination with etiolation for a period of 30 days.

7.2.2.2 Mound Layering

Alila *et al.* (1997) reported that medium sized stool resulted into more number of bud emergence. IBA at all (500, 1500, 3000 and 4500 ppm) but preferably lower concentration proved better than NAA (250, 500, 750 and 1000 ppm) in rooting of layered shoots. No distinct variations could be established in field survival of rooted shoots as compared to control.

7.2.3 Grafting

7.2.3.1 Inarching

Experiment on inarching conducted at the Fruit Research Station, Tamil Nadu revealed that by using *A. hirsuta* or Rudrakshi as rootstocks, jackfruit could be inarched with 60 to 70 per cent success (Naik, 1952).

Effect of different pre-treatments of scions on success of grafting was investigated by Maiti and Sen (1999). They observed maximum success (60 per cent) by ringing and blanching of scions for 3 days before grafting.

7.2.3.2 Epicotyl Grafting

Epicotyl grafting with mature, plump, terminal scion shoot on germinating jackfruit seedling of about 8 to 10 days by wedge method during April-May gave 50 to 90 per cent success and the successful epicotyl grafts attained saleable size within a year (Gunjate *et al.*, 1982). In the experiment on grafting of jackfruit, Jose and Velsalakumari (1991) reported highest percentage of sprouting (83.30 per cent) and survival (61.67 per cent) for grafting done during June. Epicotyl grafting with 3 to 4-month-old scions and 5-dayold rootstocks resulted in the highest percentage sprouting (29.50 per cent) and the highest percentage survival (21.67 per cent). Moist conditions increased the percentage of success in both kinds of grafts.

7.2.3.3 Softwood Grafting

The efficacy of softwood grafting in promising jack fruit types was evaluated by Madala *et al.* (2018). Out of the five types, Gumless resulted in the early emergence of grafts (28.67 days) along with highest percentage of sprouting (46.90). Maximum survival percentage (75.00 per cent) and more number of leaves (8.33) were observed in Cluster type while Early bearing Varikka resulted in maximum shoot length (8.37 cm).

The response of different jackfruit genotypes to softwood grafting was studied (Sampath *et al.*, 2019).The success varies between genotypes. The difference in the graft success was significant among the different genotypes. The highest graft success was noted in MCR-1 (54.78 per cent) and the lowest was found in MDA-1 (39.24 per cent). The graft survival was highest in MCR-1 (60.42 per cent) and lowest in MNS-1 (35.00 per cent). The number of leaves, number of leaves, graft height and graft girth were significantly different for the different genotypes. The number of sprouts (3.30), number of leaves (8.04), graft height (19.16 cm) and graft girth (8.31 mm) were highest in MCR-1. The number of sprouts (3.30), number of leaves (8.04), graft height (19.16 cm) and graft girth (8.31 mm) were highest in MCR-1. The genotypes MDA-1 had lowest graft success (39.24 per cent), number of sprouts (1.71), number of leaves (5.40), graft height (14.35 per cent) and graft girth (6.40 mm). (Sampath *et al.*, 2019).

7.2.4 Budding

Several methods of budding, tried at different research stations have proved successful.

Forkert budding: Modified forkert budding was found very successful in Java (Naik, 1952).

Chip budding: Success was 41 per cent by chip budding at the Horticultural Research Station, Krishnagar (Samaddar and Yadav, 1970).

Patch budding: Teaotia *et al.* (1963) reported 100 per cent success by patch budding. Patch, chip and T methods of budding were tried during February to September by Singh *et al.* (1982). The maximum success (90 per cent) was obtained following patch method in June and T budding was found least successful (Table 4).

In experiments with patch bud grafts, retaining all or some of the leaves on the rootstock at grafting had no effect on graft take and survival. Tying with single polyethylene strips was recommended as the most economical method (Kelaskar *et al.*, 1991). The grafts became ready for planting one or two years after grafting.

Table 3: Success (Per cent) Obtained by different Methods in Budding*

Method of Budding	Feb.	March	April	May	June	July	Aug.	Sept.	Mean
Patch	50.0	25.00	61.67	66.67	90.00	81.67	63.33	35.0	59.17
Chip	40.00	15.00	45.00	50.00	65.00	60.00	51.67	30.00	44.58
T	35.00	18.33	36.67	48.33	56.67	50.00	40.00	26.67	38.96
Mean	41.67	19.44	47.48	55.00	70.56	63.89	51.67	30.55	47.57

* Singh *et al.* (1982).

8.0 Area and Production

The estimate of actual area under this crop is not available. It is rarely grown in plantations but preferred very much in homesteads and as a shade tree or as a mixed crop. It occupies a considerable area in coffee gardens and in roadside plantations. The largest area under jackfruit in India is in Assam where the area is about 8,000 hectares. In Bihar, the area is approximately 4,000 hectares and in South India the trees occupy an area of about 2,000 hectares. In northern India, jackfruit is grown in foothills of the Himalayas. It grows throughout South India upto an altitude of 2,400 metres (Singh, 1969). Jackfruit also flourishes in Western Ghats.

Table 4: The Jackfruit Production in India*

State	Production('000 tonnes)	Share (per cent)
Tripura	291.59	16.84
Odisha	232.79	13.44
Assam	197.20	11.39
West Bengal	196.76	11.36
Karnataka	193.47	11.17
Kerala	190.14	10.98
Chattisgarh	183.23	10.58
Jharkhand	117.46	6.78
Madhya Pradesh	53.09	3.07
Tamil Nadu	49.73	2.87
Total	**1,705.46**	

* National Horticultural Board (2015-16).

9.0 Soil and Climate

9.1.0 Soil

The jackfruit can grow on a wide variety of soil although it prefers a rich, deep alluvial soil. Soil drainage is of utmost importance. Sub-soil drainage congestion, rise in water table or flood severely damage the trees and may lead to death. Jackfruit trees can be grown on open-textured light soil or lateritic soil provided nutrients are available.

9.2.0 Climate

Warm humid plains are suitable for jackfruit. It flourishes in humid hill slopes also up to an elevation of 1,500 m. Quality of fruits deteriorates in higher altitudes. It also grows well in arid and warmer plains of South India. Cold weather and frost are harmful. But perhaps it is the most cold tolerant among the members of the genus (Ochse *et al.* 1981).

10.0 Cultivation

10.1.0 Planting

Commonly, the square system is followed for planting. Hexagonal system may be followed in less fertile soil. In fertile soil a spacing upto 12 m x 12 m accommodating 70 plants per hectare will be sufficient for this fruit crops. On average soil, trees may be planted at 11 m apart. Higher density of planting can be practised in lighter and poorer soils. Shorter spacing in fertile soil will lead top crowding of trees.

For planting jackfruit, 1 m cube pits are dug at least 10 days before planting. About 30 kg well-rotted farmyard manure and 500 g superphosphate are mixed with the soil of each pit and the pit is refilled. Chloropyriphos may be applied in the pit to avoid insect attack. *In situ* planting of 3-4 seeds per pit leads to stronger plant. But nursing and raising of a large number of plants in this way are difficult. After planting, the soil is pressed firmly to avoid water-logging in pits during rainy season. This is important for the fact that jackfruit cannot withstand water-logging.

The best time for planting grafts or seedlings is June through August. Prolonged dry weather after planting may lead to the death of plants. The taproot should not be disturbed while planting to avoid damage to plants.

10.2.0 Care of Plants

Jackfruit leaves being a favourite feed for goats, the young plants are frequently damaged by +stray goats and cattle unless adequately guarded by providing gabions for about two years. Gabions may have to be replaced after a year. Hand watering of young plants during one or two summers is necessary for assured survival and good growth of plants. In colder regions, protection against frost at least during first few years is safe. Cleaning of basins by spading and ploughing of orchards should be followed as a routine measure. Frequent weeding and mulching are necessary to achieve normal plant growth.

10.3.0 Irrigation

The jackfruit is not normally irrigated. The trees are sensitive to drought. Irrigation during dry periods is considered essential in arid regions for normal growth. In order to economise use of water, ring system may be adopted for irrigation. For young orchards, hand watering is necessary during first 2 to 3 years till the root system has penetrated deep enough. The frequency of irrigation will depend on the soil moisture condition.

The salinity is considered one of the main obstacles to agriculture worldwide, constituting one of the limiting factors to the growth and development of the plants. Mesquita *et al.* (2019) evaluated the effects of salinity of irrigation water in the growth of seedlings two varieties of jackfruits, and with the application of common bovine biofertilizer and chemically enriched bovine biofertilizers in greenhouse conditions, in Brazil. The substrate used was a material collected of the first 10 cm of depth of a Dystrophic Regolitic Entisol. The levels of irrigation water saline of 0.5; 1.0; 2.0; 3.0 and 4.0 dS m^{-1}, in soil without and with common bovine biofertilizer and enriched biofertilizer, were tested with two varieties of jackfruit (jack-soft and jack-hard), and with six replicates. The biofertilizer after dilution in non-saline water (0.5 dS m^{-1}), in proportion of 1:3, where was applied once to 10 per cent of the substrate volume, two days before sowing. The increment of irrigation water salinity inhibited alls variables studied in the jackfruits plants, but with less intensity in the treatments with bovine common biofertilizer (Mesquita *et al.*, 2019).

10.4.0 Manuring and Fertilization

Experimental evidence on nutritional requirement of jackfruit is not available. If growth and production appear to be marginal, manuring will be desirable. Farmyard manure or compost may be applied according to the growth and age of the tree and condition of the soil. The trees, however, need nutrition for regular and good cropping. Application of farmyard manure (37 kg/plant), 750 g ammonium sulphate, 625 g single superphosphate and 300 g muriate of potash per plant per year in the second and third year of planting and in the fourth year an increased dose of 938 g ammonium sulphate, 781 g single superphosphate and 375 g of muriate of potash were found to increase growth, flowering and fruiting (Mukherjee *et al.*, 1983).

For quick growth of trees, manures and fertilizers may be added twice a year, before and after monsoon. It is advisable to apply 80 kg of farmyard manure per tree per year. In addition, according to the nutrient status of the soil and growth of the trees, chemical fertilizers should be applied for desired results.

The fertilizers and manures should be spread in the basin and thoroughly mixed with soil by spading. The fertilizers should be applied in an area upto the leaf-drip around each tree leaving a portion around the trunk.

10.5.0 Intercropping

The jackfruit requires a long time to occupy fully the land provided during planting for future mature trees. It is desirable that the interspace should not be left unutilized. Suitable crops should be grown every year till the trees reach bearing stage. When the soil moisture is not a limiting factor, vegetable crops like okra, brinjal, chilli, tomato and pulses like *kala;*, gram, *etc.*, can be conveniently grown. When the trees come into bearing, pulses like gram and *kala;* can be grown as intercrops. These crops will also improve the nitrogen status in the soil.

11.0 Flowering, Fruit Growth and Development

11.1.0 Flowering

Irrespective of the method of propagation, the tree starts bearing from the seventh or eighth year after planting. The flowers generally start appearing in December and continue up to March and the· fruit ripens in summer. At higher altitude fruit growth may continue up to September. Occasionally, off-season flowering in September October may be noticed.

The tree is monoecious. Large number of flowers is borne on club-shaped rachis. The inflorescence is a spike covered by two spathes. The female spikes are borne on foot-stalks while the male spikes appear both on the footstalks as well as on the terminal branches. Footstalks bearing female spikes are much more vigorous than those carrying only the male or both male and female spikes. Sambamurthi and Ramalingam (1954) noted that footstalks measuring more than 6.5 cm in circumference at the base seemed to produce female spikes only and that the spikes appearing at the beginning and at the end of flowering season were mostly males.

11.2.0 Sex Distribution, Floral Biology and Pollination

11.2.1 Floral Biology

Floral biology of jack was not studied much. A classical piece of research was published by Brantjes (1981) on *A. altilis* (breadfruit) in Brazil. According to him, nectar is a floral attractant, which, in contrast to pollen as attractant, can be produced in both staminate and in pistillate unisexual flowers. In this way nectar enables the evolution of dicliny in entomophilous flowers. In the genus *Artocarpus* the top parts of the perianths of the adjacent pistillate flowers are continuous (or fused), leaving no outside opening in between the flowers (Jarrett, 1977). This will impair access to the nectar if production occurs in a similar way as in the staminate flowers. Therefore, no nectar can be expected in the pistillate inflorescences, and in fact there were never been observed on any of these inflorescences.

What might be its adaptive value? Brantjes (1981) opined the possibility of "insect assisted wind pollination". That means that the bees' activities promote the release of the pollen from the protruding anthers, which pollen grains thereafter are borne by the wind.

Self incompatibility in jackfruit was reported from the controlled pollination studies carried out to study the breeding system of jackfruits in South Florida (El Sawa and Campbell, 1997). The effect of self and cross pollination and elimination of pollination on total weight, flesh weight, seed weight, seed number, flesh percentage and edible percentage of cultivars Dang Rasimi and Cheena were studied. Significant cultivar by treatment interactions were found in seed weight, seed number and edible percentage. In addition to the interaction, the treatments differed significantly in total weight, flesh weight and flesh percentage, while the cultivars differed in total weight and flesh weight. Pollination was necessary for fruit set in jackfruit. Neither cultivar was highly self-compatible, which may be attributed to self-incompatibility or inbreeding depression. However, Dang Rasimi and Cheena produced good quality fruits when crossed with pollen from another cultivar. The control and the crossed treatments yielded similar results for both Dang Rasimi and Cheena in total weight, flesh weight, seed weight, seed number and edible percentage, indicating that jackfruits are out breeding plants.

Occurrence of agamospermy (formation of fertile seeds without fertilization) of *A. heterophyllus* Lam. was reported by Pushpakumara (2006). He carried out three controlled pollination treatments (selling, out-crossing, without pollination) were used. Fruit-set, seed-set, mean seed weight and percentage germination of seeds following the three controlled pollination treatments and open pollination were assessed. He found that *A. heterophyllus* has very high fruit-set and very low seed-set, with a mean of 3 per cent of the flowers in an inflorescence producing seed as a result of open pollination. Pollination did not appear to be a prerequisite for fruit- and seed-set, with 53 per cent of inflorescences isolated with pollen-proof bags producing syncarps. Dissection of these ripened syncarps produced seeds although the number of seeds per syncap was significantly lower than in open pollination. Further, seeds were not found in the inflorescences where the top, bottom or lateral half of stigmas had been shaved, and the rest normally pollinated, in contrast to those inflorescences, which were completely shaved. These results suggest the occurrence of facultative agamospermy in *A. heterophyllus*. The mean seed weight of agamospermically produced seeds was not significantly different from that of open-pollinated seeds. Seed germination started within a week after sowing and was completed in approximately 35 days in seeds collected from all treatments. The overall germination success of agamospermic seeds was high (93 per cent) and not significantly different from other pollination treatments.

The distribution of sex on the jackfruit tree is interesting. On a 16-year-old tree about 2,200 spikes appeared of which 96 percent were males (Samaddar and Yadav, 1982). Sahadevan *et al.* (1950) recorded 3 to 40 per cent female spikes while Srivastava (1961) reported 20 per cent.

The footstalks carrying female spikes appear on the trunk and main branches adventitiously in the central region of the tree. Male spikes appear both in the central and peripheral regions.

Sex of a spike can be easily identified when it is small. The length and diameter of female spike are much more than those of the males. Surface of a young male spike is smooth while in the case of a female flower it is granular.

A single male flower is actually a stamen enclosed in a green leathery tubular perianth. During anthesis, the stamens protrude out of the perianth tube and appear on the surface of the spike. About 4-6 days after opening· from the spathes the first stamen appears. In a couple of days the whole surface of the spike is covered with undehisced yellow anthers or ashy grey dehisced ones held on filaments. After dehiscence, the male spike gradually turns black due to the growth of a mould on it and drops down in another weeks' time. In a study at the Horticultural Research Station, Krishnagar, it was observed that dehiscence mainly took place in the afternoon. The peak period of anthesis was between 2.00 and 3.00 pm and of dehiscence between 3.00 and 4.00 pm (Samaddar and Yadav, 1982). In a female

Flowering and Fruit Set in Jackfruit

Flowering in Jackfruit

Male (Left) and Female (Right) Flower

Pollination

Fruit Set

Flowering to Fruit Set

Jackfruit Orchard

flower the perianth encloses the ovary and style. Creamy white stigmas protrude out of the surface of the female spike, 4-6 days after opening from the spathes. In another 4-6 days the whole surface looks woolly. The receptivity of the stigma continues for about 36 hours (Sambamurthi and Ramalingam, 1954). Pollination and fertilization of a spike are complete within 3-6 days after anthesis. The axis of the inflorescence, the ovaries and perianths all grow simultaneously and develop into a multiple fruit, botanically called a sorosis. The pericarp around each seed and the fleshy perianth are edible.

The jackfruit is anemophilous. Under open pollination the fruit set is about 75 per cent which is improved by hand pollination. Gopinath *et al.* (1983) recorded pollen fertility of94.5 per cent. The best medium for pollen germination and pollen tube growth was 1 per cent agar and 10 per cent sucrose.

The fruits develop during spring and summer. The fruits become ready for harvest in June or early July. If the flowers on all sides are not pollinated the fruit does not develop normally. The size is small and the shape may be irregular in less pollinated fruits. Lack of pollination may lead to the drop of the female spikes.

11.2.2 Pollination

The pollination system of the genus *Artocarpus* still needs study (Sakai *et al.*, 2000). There are only brief reports on drosophilid fly pollination in *A. heterophyllus* (van der Pijl, 1953), moth pollination in *A. odoratissimus* (Momose *et al.*, 1998), and wind pollination of the other species (Corner, 1952; Primack, 1983; Endress, 1994). A previously undescribed pollination system involving a monoecious tree species, *Artocarpus integer* (Moraceae), pollinator gall midges, and fungi is reported from a mixed dipterocarp forest in Sarawak, Borneo. The fungus *Choanephora* sp. (Choanephoraceae, Mucorales, and Zygomycetes) infects male inflorescences of *A. integer*, and gall midges (*Contarinia* spp., Cecidomyiinae, Diptera) feed on the fungal mycelia and oviposit on the inflorescence. Their larvae also feed on the mycelia and pupate in the inflorescence. The gall midges are also attracted by female inflorescences lacking mycelia, probably due to a floral fragrance similar to that of male inflorescences. In flower visitors and pollen load observed on *Contarinia* spp. collected on both male and female inflorescences, *Artocarpus integer* is thought to be pollinated by the gall midges. Although several pathogenic fungi have been reported to have interactions with pollinators, this is the first report on a pollination mutualism in which a fungus plays an indispensable role.

11.2.3 Fruit Set

An experiment on fruit characteristics borne on different bearing position of the tree of six jackfruit germplasm lines was carried out at the Germplasm Centre of Fruit Tree Improvement Program (FTIP), Bangladesh Agricultural University, Mymensingh (Hossain *et al.*, 2012) Fruit bearing habit of jackfruit was cauliflorous *i.e.*, fruits are borne on trunk and branches. They found that fruit growth occurred rapidly up-to 50 to 60 days after fruit set (DAFS) and growth did not differ significantly between the two bearing positions. Both quantitative and qualitative parameters varied with bearing position. Among the physical parameters, length of pedicel (24.83 cm fruit-1), weight of fruit (7.11 kg fruit-1), rachis (0.51 kg fruit-1), flat ribbons (98.31 g fruit-1), rind (3.28 kg fruit-1) and number of flat ribbons (5.61 per fruit) and bulbs (120.44 per fruit) were significantly higher for fruit on the main trunk than primary branches.

11.3.0 Fruit Growth and Development

Female spikes take 90 to 110 days to develop into a mature fruit (Sahadevan *et al.*, 1950). Teaotia and Chauhan (1969) reported that 126 days are needed from spike emergence to the full maturity. Rahman *et al.* (1999) analysed the soft and firm types of jack fruit at 3 stages of maturity (7-8, 10-12 and 14-16 weeks after anthesis). The dry matter content of perianth and seed of the soft and firm varieties increased from 10.0 to 32.0 per cent and 19.0 to 52.0 per cent, respectively, while the ash content decreased from 5.7 to 2.0 per cent and 4.9 to 1.5 per cent, respectively, on a dry matter basis. The free sugars increased with maturity from 1.5 to 10.5 per cent and 1.4 to 5.2 per cent of dry matter for soft and firm varieties respectively. In all cases, varying proportions of glucose, fructose and sucrose were the major sugar constituents. The starch content of the perianth samples increased from 7.8 to 47.0 per cent and from 9.0 to 50.5 per cent, on a dry matter basis for the soft and firm varieties, respectively. However, seed starch content increased up to 65.0 and 59.0 per cent, respectively. Microscopic examination of the samples showed that the perianth contained thin-walled cells packed with starch granules, some organized into distinct clusters. From both the chemical and histological studies, it appeared that the starch content of both perianths of seed of soft and firm varieties gradually increased with increasing maturity. The total dietary fibre varied from 42.0 to 55.0 per cent in the perianth. However, in seeds it changed very little.

12.0 Pests and Diseases

12.1.0 Pests

Although a number of pests are known to attack jackfruit, the most important ones are described below:

12.1.1 Shoot and Trunk Borer

It is often serious in Assam, Uttar Pradesh and Bihar. It bores into the tender shoots and buds. The affected parts should be nipped off and destroyed. Application of carbaryl (Sevin) 50 per cent at the rate of 4 g per litre of water by spray in flowering season is advisable.

The brown weevil (*Ochyromera artocarpi*) is also reported to bore into the tender buds and shoots. They may be controlled by destroying fallen fruits and buds and by collecting grubs and adults.

In addition, mealy bug and jack scale are also found to attack jackfruit which can be controlled by use of a suitable contact insecticide.

12.2.0 Diseases

Fruit rot caused by *Rhizopus artocarpi* is the most common disease of jackfruit in India. The disease causes premature fall of young fruits due to rotting and may result in heavy loss in yield under unusually humid conditions. Spraying of dithane M-45 (0.2 per cent) and bavistin (0.05 per cent) at 15 days interval during fruit growth was found effective in controlling fruit rot (Gupta and Pandey, 1985). *Botrytis cinerea and Rhizopus stolonifer* were also found to cause premature drop of young fruits. This could be controlled by two applications of 0.04 per cent of benomyl at an interval of 15 days (Pandey *et al.*, 1981).

Pink disease, stem rot, male inflorescence rot and soft rot are occasionally noticed in jackfruit. However, these diseases are not serious maladies and can be controlled by pruning the affected parts and using Bordeaux paste at the cut-ends.

13.0 Harvesting and Yield

13.1.0 Harvesting

Tender jackfruits are harvested for use as vegetable during early spring and summer until the seeds harden. The fruit matures towards the end of summer in June. Period of fruit development is February to June. The optimum stage of harvest maturity of jackfruit has been reported to be 90-110 days after the appearance of the spike (Anon., 1975). Harvesting is done by cutting off the footstalks carrying the fruits.

13.2.0 Yield

Bearing starts from the seventh to eighth year when a few fruits may develop. The tree reaches its peak bearing stage within fifteen to sixteen years after planting. Yield is not similar every year. Normally, a tree bears a few to about 250 fruits annually at this stage. The weight of fruits varies widely depending on the type. Individual jackfruit may weigh from about 1 kg to about 20 kg.

14.0 Post-harvest Management

14.1.0 Ripening

In most cases ripening is not a problem. The fruit ripens when the maximum temperature reaches during the end of the summer season. In colder regions, the fruits mature late. Selvaraj and Pal (1989) studied the biochemical changes during ripening of jackfruit. They observed that the fruit firmness decreased considerably as a result of ripening. From harvest maturity to half-ripe stage, little softening was observed. The intensity of fruit softening increased from ½ ripe to ¾ ripe stage and in ripe fruits. A two-fold increase in pulp colour (carotenoid pigments) was observed from harvest maturity to eating ripe stage. Dry matter, alcohol insoluble solids, and starch contents decreased with ripening. Sucrose formed the major sugar followed by fructose and glucose. A three-fold increase in sucrose concentration was observed during ripening. Citric and malic acids were the major non-volatile organic acids identified. The concentration increased at ¾ ripe stage and declined in ripe fruits. The decline was much more for malic acid resulting in increased citric: malic acid ratio in ripe fruits. Asparagine, aspartic acid, glutamine, glutamic acid, glycine, α-aminobutyric acid, phenylalanine and proline formed the major soluble amino acids. Jackfruit also contained very little tanins and their concentration

Pests of Jackfruit

Shoot and Trunk Borer

Fruit Borer

Brown Weevil

Bark Eating Caterpillar

Mealybug Infestation

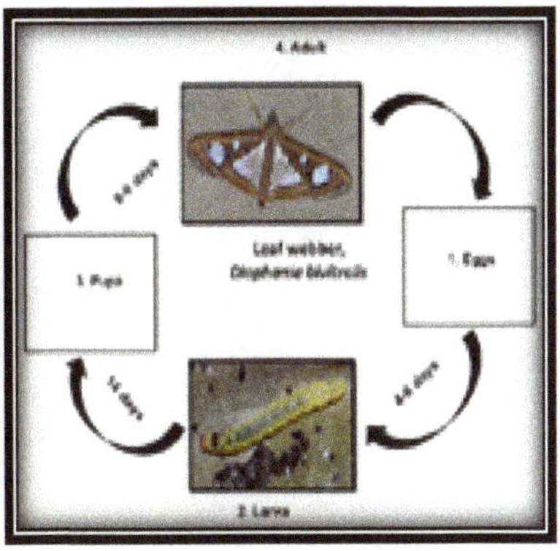

Leaf Webbar Life Cycle

Harvesting, Storage, Transportation and Marketing of Jackfruit

decreased during ripening. Total lipids, total fatty acids, un saponifiable and total sterols increased whereas phospholipids and free acids decreased during ripening. The respiration pattern of bulbs with seeds excised from mature fruit exhibited a typical climacteric rise on ripening. Climacteric peak in respiration was reached three days after harvest, followed by a steady decline until 8th day when fruit reached the eating ripe stage. Biale and Barcus (1970) reported that the fruits respire at the rate of 21-19 ml O_2/kg/hr and 21-32 ml CO_2/kg/hr at 20°C.

14.2.0 Storage

Jackfruit is not normally stored in cold storage. A storage life of about 6 weeks is expected when the temperature is 11.1-12.8°C and humidity between 85 and 90 per cent (Singh, 1972). The initial quality and stage of maturity at harvest are important factors on which the storage life depends. Mathur *et al.* (1952) studied the changes in chemical composition of jackfruit during storage at different temperatures and found that storage at 11.1 to 12.8°C and RH 85-90 per cent for 42 days resulted in decrease in sucrose content from 9.5 to 5.0 per cent while reducing sugar content increased from 2 to 5.98 per cent. Reduction in the level of ascorbic acid content from initial 8.2 per cent to 3.5 per cent was also recorded.

14.3.0 Minimal Processing

Jackfruit is a tropical fruit with high nutritional value and reported health benefits, but its commercialization is hindered by its mass (> 10 kg) and difficulty to peel. Jackfruit bulbs (JFBs) are often sold fresh, washed only with water and packaged in polypropylene containers, which minimizes their shelf life and makes it necessary to consume them shortly after purchase (Anaya-Esparza *et al.*,2018). Minimally processed products (MPPs) are ready-to-eat fresh-cut products, and the technologies used to produce them to allow better utilizing fruits, extending their shelf life and freshness as much as possible. Anaya-Esparza (2018) reviewed the reported advantages and limitations that minimal processing methods exert on JFBs. These methods include low temperatures, anti-browning agents, anti-respiratory substances, texturizing agents, controlled atmospheres, edible coatings, osmotic agents, preservatives, and UV treatments among others. Their effects on physiological and physicochemical parameters, as well as effects on shelf life and sensory attributes are also discussed. Evidence shows that minimal processing techniques are a viable option that can be used on JFBs, which preserve quality and sensory attributes by reducing metabolic reactions like respiration rate (RR) and ethylene production. Shelf life can be extended up to 50 days without changes in sensory attributes, and higher retention of bioactive compounds in comparison with untreated JFBs, but results vary with each specific treatment. Minimal processing technologies are a rapid, efficient, and reliable alternative that retain quality and extend the shelf life of fresh JFBs. With further improvements in preservation and standardization, the commercialization of jackfruit could be brought to bear (Anaya-Esparza *et al.*, 2018).

14.4.0 Starches

Jackfruit is composed of rind, edible yellow flesh, procarp and seeds. Jackfruit seeds which represent 8–15 per cent of the fruit weight (Kittipongpatana *et al.*, 2011 cited by Zhang *et al.*, 2017) have a short shelf life because of their high carbohydrate content (*i.e.*, starch content above 60 per cent, dry basis) and could thus be used as good culture for microorganism (Babital *et al.*, 2007 cited by Zhang *et al.*, 2017). Although they contain a large amount of potentially useful starchy material, the seeds are usually discarded as a waste in jackfruit processing, thus contributing to environmental pollution (Tran *et al.*, 2015). Due to their high starch concentration, the best way to preserve the seeds is to extract the starch, which is used in different industries based on its suitability (Zhang *et al.*, 2017).

The molecular structure of starches isolated from five jackfruits (M2, M3, M4, M8 and X1) and its relationship with physicochemical properties were investigated (Zhang *et al.*, 2017). Although they had uniform amylase (AM) content, the five jackfruit starches displayed different physicochemical properties, including their pasting, thermal, crystal and texture properties. Furthermore, differences in the molecular structure (*i.e.*, average weight-average molar mass (Mw) of amylose and amylopectin (AP) as well as the same AP fine structure) were also found in the five jackfruit starches. The results indicated that jackfruit starch with a larger Mw of amylose and proportions of DP 25–36, DP $\geq$ 37 and chain length had a lower peak viscosity, breakdown, final viscosity, setback and adhesiveness, but a higher pasting and gelatinization temperature, gelatinization temperature range, gelatinization enthalpy and relative crystallinity.*Xiangyinsuo 1 hao* (X1) starch, which originated from Xinglong in Hainan province, China, had special physicochemical properties, which were ascribed to its lower amylopectin Mw, smaller particle size, and perfect amylopectin structure. The results showed that the most important intrinsic factors that could determine the physicochemical properties of starch were its molecular structure, including the Mw of amylose and AP as well as a fine AP structure (Zhang *et al.*, 2017).

14.5.0 Utilisation of Wastes

According to Alves *et al.* (2020), around 2.96 million tons of potential feedstocks are generated annually in the form of jackfruit waste, including peels (JP) and seeds (JS). They studied the suitability of JP and JS as new feedstocks for combustion and their preliminary potential for bioenergy considering their physicochemical properties, bioenergy indicators, combustion behaviors and emission characteristics. The jackfruit wastes presented attractive features for direct application in combustion, such as their composition of 75.1–81.3 wt. per cent of volatile solids, 40.8–41.8 wt. per cent carbon content, a higher heating value of 16.3–17.2 MJ kg^{-1}, and negligible amounts of nitrogen and sulfur. Their ash contents varied from 5.6 wt. per cent to 6.7 wt. per cent, similar to that of other biomass fuels. The bioenergy yield obtained for JP and JS were 2.5 TJ ha^{-1} yr^{-1} and 0.9 TJ ha^{-1} yr^{-1} (dry basis), respectively, which exceed that of well-known bioenergy crops. The use of jackfruit residues as raw material for bioenergy can be highly feasible, given the low CO, CO_2 and SO_2 emissions, and a considerable equivalence with reference fuels. Combustion properties of JS and JP revealed a favorable burning performance. In conclusion, the highlights of this study provide useful information for the valorization and large-scale application of jackfruit wastes and contribute to their establishment as sustainable and renewable sources of bioenergy (Alves *et al.*, 2020).

15.0 Biotechnology

Micropropagation of jackfruit has proved successful. Investigations on other aspects of biotechnology are very scanty.

15.1.0 Micropropagation

In jackfruit, new plantlets were successfully raised from nodal explants and shoot apices cultured on MS medium supplemented with various concentrations of cytokinin for shoot proliferation and sub cultured on similar medium supplemented with different concentrations of auxin for root regeneration (Rajmohan and Mohanakumaran, 1988a,b; Roy *et al.*, 1990). Micropropagation of jackfruit was also successful by using terminal and axillary buds of 40-year-old trees on supplemented MS medium (Roy *et al.*, 1991). Roy *et al.* (1993) propagated successfully the flood tolerant type of jackfruit by using shoot buds as explants. They obtained highest number of shoot buds by using MS medium containing 8.88 μM BA and 2.68 μM NAA. These shoots were found to proliferate through 7 or more subcultures with an average of 10 shoots per transfer. For shoot multiplication, the concentration of BA and NAA were lowered to 4.44 and 0.54 μM respectively and coconut milk (10 per cent) was added. Well developed shoots were then excised and implanted individually on a rooting medium consisting of half-strength MS salts supplemented with 5.37 μM NAA and 4.92 μM IBA. Eighty per cent of the shoots produced roots, and the rooted plantlets were transplanted directly from the culture tube to earthen pots containing sterile sand, soil and humus (1:2:1) and covered with transparent plastic bags. After acclimatization the seedlings were flooded with water for 30 days and 75 per cent of these plants survived. Shoot-tips and nodal segments collected from coppiced shoots on the main trunk of a new jackfruit cultivar bearing fruits twice a year exhibited maximum proliferation on MS medium containing 2.5 mg BA + 0.5 mg NAA/l after 28 days. The shoots were then sub cultured for several times. The number of shoots increased to 40 when the concentration of BA and NAA were lowered to 1.25 and 0.25 mg/l, respectively and 15 per cent (v/v) coconut milk was added to the medium. Rooting of excised shoots was achieved on half-strength MS medium containing NAA and IBA each at 1.0 mg/l. The rooted plantlets were transplanted directly from the tube to pots containing sterile sand, soil and humus (1:2:1) and covered with transparent polythene bags. Seventy five per cent of the seedlings survived after transplanting into the open field (Roy and Roy, 1996). Singh and Tiwari (1996, 1998) used nodal explants of mature trees on modified MS medium supplemented with 18.0 mg BA/l + 0.2 mg IBA/1. The effects of carbon source, type of culture vessel, gelling agents and gibberellic acid on shoot proliferation were studied in jack fruits cv. Singapore (*Adiga et al.*, 1998). Cultures supplemented with sucrose or sugar at 3 or 4 per cent produced more shoots than those supplemented with sucrose or sugar at I per cent. Gibberellic acid (6 mg/l) promoted shoot length.

Amin and Jaiswal (1993) developed a tissue culture technique for rapid vegetative propagation of mature jackfruit trees using apical bud cultures. Shoot-tip cultures were established on MS medium with 5- to 10-mm-long explants dissected from terminal buds of new growth from the trunk. After initial culture of bud explants, 1- to 2-node pieces were taken from the microshoots formed and used to proliferate further axillary shoots from multiplying and maintaining shoot cultures. Benzyladenine and kinetin (4.5-9.0 JIM), either separately or together, supported shoot proliferation.

15.2.0 Embryo Culture

There are some reports on the embryo culture of jackfruit. Chin *et al.* (1988) excised embryos of the recalcitrant *Artocarpus* spp. to determine suitable media and cultural conditions for their development into whole plants. MS medium modified by the addition of 0.17 g NaH_2PO_4/l was found to be a suitable basal medium. The basal medium supplemented with 2 g/l activated charcoal and 1.0 mg/l NAA or IAA in combination with 1 mg/l kinetin, BA or 2iP was found to be suitable where the embryos developed normally to form whole plants.

15.3.0 Markers

15.3.1 Molecular Markers

Using polyacrylamide gel electrophoresis, peroxidase and esterase isozymes Wangchu *et al.* (2011) used leaf tissue samples of 44 selected (superior) jackfruit genotypes,collected after surveying 1500 trees in the districts of Nadia, 24 Parganas (N) and Coochbehar in West Bengal. The selected genotypes were analyzed for isozyme variation for peroxidase and esterase. Polymorphism was observed in both the enzyme systems studied, where several bands were found showing similarities and dissimilarities among the selected types. In the peroxidase assay, 7 loci were identified among the 44 genotypes. The relative mobility (Rm) values of the loci ranged between 0.29 and 0.50. Whereas in the esterase assay, 6 loci were identified and the relative mobility (Rm) values of the loci ranged between 0.22 and 0.65. Moreover, it was observed that genotypes collected from nearby locations did not show similar banding patterns, indicating that forces other than geographical origin, such as exchange of genetic stocks, genetic drift, spontaneous variation, natural and artificial selection, cross pollination and seedling origin of the plant are responsible for the genetic diversity. RAPD markers were used to estimate genetic diversity in 12 high-yielding jackfruit accessions obtained from different locations in southern India (Simon *et al.*, 2007). Marker data were compared with morphological data obtained over three successive seasons. PCR-amplifiable DNA was isolated using the CTAB method and 171 amplified fragments were obtained using 23 random primers. The genetic dissimilarity matrix was calculated based on Squared Euclidian Distances, which revealed a maximum genetic distance of 7.9 per cent between a clone of 'Mottavarica' ('M0'), and 'Chandrahalasu' from distant locations, while the minimum genetic distance (5 per cent) was between the genotypes ('M0') and 'Kerala', indicating their similar geographical origin. Ward's method of cluster analysis grouped all individuals on the dendrogram into two major clusters according to their geographical location. The present study showed low-to-moderate genetic diversity among the 12 jackfruit accessions, which will assist in the identification and management of jackfruit germplasm for breeding purposes. In another similar study, Deivasigamani *et al.* (2009) carried out RAPD analysis on leaf material to assess the authentic identification of three wild varieties (FH 11, FH 1 and NC 3) and two tissue cultured varieties (TCJ 5 and TCJ 2) of Jackfruit using the primer H15 (AATGGCGCAG). A total of 31 bands were obtained by using PCR from these five varieties of jackfruit. So, approximately 6.2 bands were obtained from a single variety. Eighteen bands (58.06 per cent) were monomorphic, 10 bands (32.25 per cent) were monomorphic and unique to all varieties, while the other 3 bands (9.6 per cent) were polymorphic in the five tested varieties.

ISSR (inter-simple sequence repeat) markers were used in China to analyze genetic diversity in 76 accessions of jackfruit (Ye *et al.*, 2009). Among 477 bands produced by 24 ISSR primers, 427 were polymorphic (accounted for 89.52 per cent). The average PIC (polymorphism information content) was 0.23. These 76 jackfruit accessions could be discriminated by the 24 ISSR primers. Genetic distance analysis revealed that, low genetic diversity existed in the jackfruit germplasm studied, and the genetic similarity coefficient ranged from 0.626 to 0.945, with an average of 0.775. Cluster analysis showed that 4 groups could be clustered at a genetic distance coefficient of 0.752. The results indicated that, 2 accessions from Tropical Plant Garden, WJ2 and GSYWJ1, were genetically different from other accessions. Accessions of soft flesh type and firm flesh type could not be discriminated. Accessions collected from different sources could form independent clusters, while accessions collected from different areas of Leizhou peninsula could not form independent clusters. The genetic diversity and genetic relatedness of 50 jackfruit accessions were studied using amplified fragment length polymorphism markers. Of 16 primer pairs evaluated, eight were selected for screening of genotypes based on the number and quality of polymorphic fragments produced. These primer combinations produced 5976 bands, 1267 (22 per cent) of which were polymorphic. Among the jackfruit accessions, the similarity coefficient ranged from 0.137 to 0.978; the accessions also shared a large number of monomorphic fragments (78 per cent). Cluster analysis and principal component analysis grouped all jackfruit genotypes into three major clusters. Cluster I included the genotypes grown in a jackfruit region of Karnataka, called Tamaka, with very dry conditions; cluster II contained the genotypes collected from locations having medium to heavy rainfall in Karnataka; cluster III grouped the genotypes in distant locations with different environmental conditions. Strong coincidence of these amplified fragment length polymorphism-based groupings with geographical localities as well as morphological characters was observed (Shyamalamma *et al.*, 2008). They found moderate genetic diversity in these jackfruit accessions. This information should be useful for tree breeding programs, as part of our effort to popularize jackfruit as a commercial crop.

15.3.1.1 Phylogeny

The phylogenetic relationships among eleven *Artocarpus* species, including two economically important species, the jackfruit (*A. heterophyllus* Lamk) and the breadfruit (*A.altilis* Fosberg), were estimated by comparing 30 restriction-site mutations in an amplified region of chloroplast DNA (cpDNA). Neighbor joining and parsimony analyses were performed on the data to identify the relationships (Kanzaki *et al.*, 1997) The jackfruit and chempedak (*A. integer* Merr.) were monomorphic and therefore indistinguishable for all restriction sites. This confirmed the close relationship between these two species. The breadfruit and *A. elasticus* formed a monophyletic group with strong support from both neighbor-joining and parsimony analyses, indicating the possibility that the breadfruit was derived from *A. elasticus* or its close relatives. The results were in agreement with the conventional classifications in *Artocarpus* species except for the position of *A. chaplasha*.

The genetic relationship of seven *Artocarpus* species was determined through isozyme analysis by Medina *et al.* (2010). Samples were obtained from the field genebank of the National Plant Genetic Resources Laboratory, Institute of Plant Breeding, Crop Science Cluster, College of Agriculture, University of the Philippines Los Banos and at the vicinity of Mt. Makiling. The species include jackfruit (*A. heterophyllus* Lam.), monkey jackfruit (*A. rigidus* Blanco), breadfruit [*A. altilis* (Park.) Forsberg], breadnut (*A. camansi* Blanco), marang (*A. odoratissimus* Blanco), gumihan (*A. sericicarpus* Jarret) and lakoocha or monkey jack (*A. lakoocha* Roxb.). Five presumptive loci were detected for three enzyme systems considered, namely: acid phosphatase (ACPH), alkaline phosphatase (ALP) and esterase (EST). Two loci were observed for ACPH: monomorphic ACPH-1 with Rf value of 0.08-0.13 and a polymorphic ACPH-2. All of the seven species showed SS genotype for ACPH-1. For ACPH-2, both slow (S) and fast (F) moving bands were noted. The S band migrated to Rf value of 0.23-0.30, while the F moving band localized to Rf value of 0.35-0.42. *A. odoratissimus, A. heterophyllus, A. camansi*, and *A. altilis* exhibited SF genotype. Both *A. sericicarpus* and *A. lakoocha* showed SS genotype, while *A. rigidus* showed FF genotype. For ALP, only one locus (ALP-1) with S and F moving bands were noted with Rf value of 0.17-0.22 and 0.27-0.32, respectively. *A. odoratissimus, A. camansi, A. altilis* and *A. lacoocha* exhibited SF genotype. *A. sericicarpus* and *A. rigidus* exhibited SS genotype. Two loci were detected for esterase namely: EST-1, with Rf value of 0.10-0.15 and EST-2 with Rf value of 0.26-0.31. All the species exhibited SS genotype for EST-1, while only *A. camansi, A. altilis, A. sericicarpus* and *A. lakoocha* showed SS genotype for EST-2. From the isozyme data, a dendrogram showing the genetic relationship of the seven species was constructed. At 0.70 similarity index, A. *odoratissimus and A. heterophyllus* were found to be in one cluster. *A. altilis* and *A. camansi* clustered with *A. lakoocha. A. rigidus* did not cluster with any of the other species. Based on leaf and fruit morphology, *A. odoratissimus* and *A. heterophyllus* have elliptic to oval leaf shape, *A. camansi* and *A. altilis* have deeply lobed leaves and round to oval fruit shape while *A. lakoocha, A. sericicarpus* and *A. rigidus* have different leaf and fruit characteristics from the other species.

Fifteen simple sequence repeats (SSRs) were identified in chloroplast sequences from four *Artocarpus* transcriptome assemblies (Gardner *et al.*, 2015). The markers were evaluated using capillary electrophoresis in *A. odoratissimus* (105 accessions) and *A. altilis* (73). They were also evaluated in silico in *A. altilis* (10), *A. camansi* (6), and *A. altilis x A. mariannensis* (7) transcriptomes. All loci were polymorphic in at least one species, with all 15 polymorphic in *A. camansi*. Per species, average alleles per locus ranged between 2.2 and 2.5. Three loci had evidence of fragment-length homoplasy. These markers will complement existing nuclear markers by enabling confident identification of maternal and clone lines, which are often important in vegetatively propagated crops such as breadfruit.

The research on jackfruit trees has been frequently conducted including genetic diversity analysis using RAPD method as reported earlier by many workers. (Chunhai *et al.*, 2005; Simon *et al.*, 2007; Deivasigamani *et al.*, 2009; Anburaj and Sudarmani, 2010; Pusphakumara and Harris, 2007; Palupi *et al.*, 2019).

Palupi *et al.* (2019) assessed variation in jackfruit based on molecular marker RAPD (Random Amplified Polymorphic DNA) of jackfruit in Indonesia. Seven RAPD primers namely A-13, A-27, B-4, H-15, L-1, OPB-1, and OPD-19 generated 70 DNA bands, of which 40 were polymorphic. In this study H-15 primers produced more polymorphic fragments than the other six primers and had the highest percentage of polymorphic fragments, namely 91.7 per cent. Similarity matrix was calculated using coefficient of Jaccard. Unweighted Pair Group Method Using Arithmetic Mean (UPGMA) cluster analysis was performed to develop a dendogram. This data analysis was performed by NTSYS software ver. 2.1. In the present studies, jackfruit tress accessions from the eleven provenances were divided into two clusters with similarity coefficient ranging from 63.33 to 85.71 per cent, provenance of East Java and West Java were the most closely similar (85.71 per cent) based on molecular characters (Palupi *et al.*, 2019).

Use of microsatellites in jackfruit for identification of cultivars, phylogenetic analysis and diversity analysis has been reported.

Use of magnetic beads in SSR analysis of jackfruit genomic DNA of Jackfruit (*Artocarpus heterophyllus* Lam.) was reported by Wang *et al.* (2010). They extracted and digested with restriction enzyme MseI. The fragment of DNA were hybridized with biotins labeled Oligo (GA)10, (AC)10 probes and several other probes, and the hybridized complex was fixed by magnetic beads coated with streptavidin. The selected fragments were amplified using adapter specific primers and cloned into pGEMT-vectors and transformed into E. coli hosts. An SSR enriched genomic DNA library was thus constructed. 83 clones were selected by colony-PCR method and then sequenced. And 19 pairs of SSR primers were found usable (Wang *et al.*, 2010).

Nakintu *et al.* (2020) analyzed the genetic diversity of *A. heterophyllus* trees from 12 districts belonging to three agro-ecological zones and three political regions of Uganda. Ten SSR loci were used to assess the genetic relationship among 200 trees, 197 from Uganda and 3 out-group individuals. All SSR loci were polymorphic with an average of 10.9 alleles per locus. STRUCTURE analysis proposed two genetic clusters: Cluster 1 was composed of samples from Eastern and neighboring Central districts, and Cluster 2 which constituted out-groups and samples from Western and neighboring Central districts. Results of STRUCTURE analysis were confirmed by PCoA. Mbarara District exhibited the highest genetic diversity (He = 0.79, I = 1.71), while Kamuli (He = 0.61, I = 1.08 and Pallisa (He = 0.59, I = 1.12) displayed the lowest genetic diversity despite high abundances of jackfruit trees. Molecular variation was higher within populations than among populations. Moderate and significant genetic differentiation was registered among geographical zones, while varietal differences displayed little insignificant genetic differentiation. Soft and white pulped varieties, considered inferior on the market, harboured private alleles which may be genetically valuable resources (Nakintu *et al.*, 2020).

Witherup *et al.* (2019) tested whether genetic diversity is changing over time and investigated a possible cause—a shift from direct seed planting by tree owners to purchasing seed-propagated saplings from nurseries, a method that has increased in popularity since the 1980s. They measured genetic diversity over time (across both reported tree age and tree size classes) using 13 microsatellite loci for 361 jackfruit individuals collected throughout Bangladesh. They found downward trends in diversity over time (regardless of seedling source), and no change in diversity between owner seed-propagated and nursery seed-propagated sapling trees that were planted since the early 1980s. Jackfruit, long an important crop in South and Southeast Asia, is gaining global popularity. Because it is a long-lived, out-crossing crop, changes in genetic diversity may occur gradually, and locally adapted alleles could be lost in transitioning to commercialized uniform cultivars. It is important to measure and conserve diversity baselines before selection bottlenecks occur in underutilized crops, like jackfruit, on the verge of increased industrialization.

15.3.1.2 Markers Specific to Traits

The edible part of the fruit develops from the perianths, and contains many sugar-derived compounds. However, its sugar metabolism is poorly understood. A fruit perianth transcriptome was sequenced on an Illumina HiSeq 2500 platform, producing 32,459 unigenes with an average length of 1345 nt. Sugar metabolism was characterized by comparing expression patterns of genes related to sugar metabolism and evaluating correlations with enzyme activity and sugar accumulation during fruit perianth development (Hu *et al.*, 2016). During early development, high expression levels of acid invertases and corresponding enzyme activities were responsible for the rapid utilization of imported sucrose for fruit growth. The differential expression of starch metabolism-related genes and corresponding enzyme activities were responsible for starch accumulated before fruit ripening but decreased during ripening. Sucrose accumulated during ripening, when the expression levels of genes for sucrose synthesis were elevated and high enzyme activity was observed. The comprehensive transcriptome analysis presents fundamental information on sugar metabolism and will be a useful reference for further research on fruit perianth development in jackfruit.

Cracking: The RAPD analysis identified OPC-07 as the polymorphic primer showing band size of 810 bp as the polymorphic marker for identification of cracking genotypes of jackfruit. These identified RAPD marker can be used as a tool for identification of cracking free genotypes of jackfruit in the future (Singh *et al.*, 2011a and b).

Kavya *et al.* (2019) studied the genetic diversity present in twenty jackfruit genotypes for different pulp colours. The genotypes were identified for commercial cultivation, based on fruit and flake quality characters *viz.*, fruit weight, fruit rind weight per kg fruit, number of flakes per kg fruit, weight of flakes per kg fruit, individual fresh flake weight and TSS as per the Jackfruit descriptors given by

IPGRI, Rome. The best genotypes identified from the study were Swarna, Lalbagh Madhura, Byrachandra, NSP, Ashoka Yellow and NEL and these can be used for commercial purpose. Molecular diversity analysis was also carried out using 22 SSR primers, out of which six primers (SSR 9, SSR 10, SSR 30, SSR 34, SSR 45 and SSR 48) showed polymorphism among twenty genotypes. The genetic similarity co-efficient ranged from 0 to 0.96, indicating a vast variation in genetic diversity for pulp colour. A dendrogram was constructed by UPGMA analysis revealed three major clusters. The genotypes with cream and yellow colour pulp grouped in cluster I and II, followed by yellow, orange and red in cluster III. These markers effectively segregated the genotypes based on different pulp colours and hence, can be used for both diversity analysis and in breeding applications (Kavya *et al.*, 2019).

15.4.0 DNA Bar Coding

There are two flesh types; soft flesh (SF) (Wela) and fi rm flesh (FF) (Waraka) (*i.e.* highly preferred).There is no strategy available to identify FF plants at nursery or vegetative stages. Hence Karunarathne *et al.* (2018) conducted a study was conducted to assess the applicability of leaf morphological variation and DNA bar-coding to diff erentiate SF and FF bearing jackfruit plants. The syncarp morphology and sensory parameters of SF and FF ripen flesh were also assessed to check the variation with 10 trees from each flesh type. The trees were chosen from SF and FF types and leaves and mature syncarps were collected. The leaf and syncarp morphological variation, pH, soluble solids and ascorbic acid content were measured. The sensory parameters of the flesh types were ranked by employing a taste panel. The DNA bar-coding was carried out for two loci rbcL and ITS to identify the sequence variants. The leaf tip length was longer in SF, ascorbic acid content was higher in FF and all the other parameters were not diff erent ($P < 0.05$). The SF leaves were elliptical whereas FF leaves were obovate. The association analysis revealed that the sweetness and sourness of flesh were inversely associated. The DNA bar-coding for rbcL and ITS showed that rbcL locus was not informative however, ITS locus can be used to characterize the genetic diversity of jackfruit. Three haplotypes were detected within ITS locus where CC haplotype is characteristic to SF and CT or TC is detected in FF enabling marker assisted seedling selection. This is the first reported study attempting to use sequence variants to diff erentiate SF and FF typed trees in jackfruit (Karunarathne *et al.*, 2018).

15.5.0 Genomics

Two of the most economically important plants in the *Artocarpus* genus are jackfruit (*A. heterophyllus* Lam.) and breadfruit (*A. altilis* (Parkinson) Fosberg). Both species are long-lived trees that have been cultivated for thousands of years in their native regions. There are hundreds of breadfruit varieties that are native to Oceania, of which the most commonly distributed types are seedless triploids. Jackfruit is likely native to the Western Ghats of India and produces one of the largest tree-borne fruit structures (reaching up to 45 kg). To-date, there is limited genomic information for these two economically important species (Sahu *et al.*, 2020). They generated 273 Gb and 227 Gb of raw data from jackfruit and breadfruit, respectively. The high-quality reads from jackfruit were assembled into 162,440 sca olds totaling 982 Mb with 35,858 genes. Similarly, the breadfruit reads were assembled into 180, 971 sca olds totaling 833 Mb with 34,010 genes. A total of 2822 and 2034 expanded gene families were found in jackfruit and breadfruit, respectively, enriched in pathways including starch and sucrose metabolism, photosynthesis, and others. The copy number of several starch synthesis-related genes was found to be increased in jackfruit and breadfruit compared to closely-related species, and the tissue-specific expression might imply their sugar-rich and starch-rich characteristics. Overall, the publication of high-quality genomes for jackfruit and breadfruit provides information about their specific composition and the underlying genes involved in sugar and starch metabolism (Sahu *et al.*, 2020).

16.0 References

Alagiapillai, O. A., Premjoshua, J., Kuttalam, S., Subramanian V. and Jayasekhar, M. (1996) *Madras Agric. J.*, **83:** 310-312

Ali, P., Sanyal, D. and Kar, P.L. (1997) *Crop Res.*, **14**: 433-4361.

Alila, P., Chiesotsu, S., Kar, P.L. and Sanyal, D. (2000) *The Hort. J.*, **13**: 9-13.

Alves, J.L.F., Silva, J.C.G., Mumbach G.D. *et al.* (2020) *Renewable energy* **155:** 1328-1338.

Amin, M.N. and Jaiswal, V.S. (1993) *Plant Cell Tissue Organ Cult.*, **33**: 59-65

Anaya-Esparza, L.M., González-Aguilar, G.A., Domínguez-Ávila, J.A. Olmos-Cornejo, J.E., Pérez-Larios, A., and Montalvo- González, E. (2018) *Bioprocess Technol.*, **11**: 1761–1774.

Anburaj, J. and Sudarmani, D.N.P. (2010) *Adv. Biotechnol.*, **10:** 34-36.

Anonymous (1966-67) *Annual Report of Horticultural Research Station*, Krishnagar, Nadia, West Bengal.

Anonymous (1975) *Annual Report*, CFTRI, Mysore, India, pp. 24-25.

Anonymous (1979) Extension Bulletin No. 22, *Indian Institute of Horticultural Research*, Bangalore.

APAARI (2012) Jackfruit Improvement in the Asia-Pacific Region – A Status Report. Asia-Pacific Association of Agricultural Research Institutions, Bangkok, Thailand, p.182

Azad, A.K. (2000) *Ph. D Thesis*, University of Southampton, UK. p.200.

Babitha, S., Soccol, C.R. and Pandey (2007) *Bioresour. Technol.*, **98:** 1554–1560.

Bailey, L.H. (1949) *Manual of Cultivated Plants*. Macmillan Co. New York. 338 p.

Barrau, J. (1976) In: *Evolution of Crop Plants* (Ed. N.W. Simmonds), Longman, London, pp. 201-202.

Bashar, M. A. and Hossain A.K.M.A. (1993) Present Status of Jackfruit in Bangladesh. A Paper Submitted to the International Centre for Underutilized Crops (ICUC), Southampton University, U.K.

Beddome, R. H. (1873) Forester's Manual of Botany for Southern India. Gantz Brothers at the Adelphi Press, Madras.

Bhatia, B.S., Siddappa, G.S. and Lal, G. (1995) *Indian J. Agric. Sci.*, **25**: 303-307.

Biale, J.B. and Barcus, D.E. (1970) *Trop. Sci.*, **12**: 93-104.

Bhatia, S., Siddappa, G.S. and Lal, Giridhari (1956) *Indian J. Agric.*, **25:** 408.

Brandis, D. (1906) Indian Trees. Bishen Singh, Mahendra Pal Singh, Dehra Dun.

Brantjes, N.B.M. (1981) *Acta Bot. Need.* **30 (5/6):** 345-352.

Berry, S.K. and Kalra, C.L. (1987) *Indian Fd. Pack.*, **42**: 62-76.

Campbell, R. J. and McNaughton, R. (1994) *XXXX Annual meeting of the Interamerican Society for Tropical Horticulture*, Campeche, Mexico, 13-18 Nov. 1994. *Proc. Interamer. Soc. Trop. Hort.*, **38**: 201-205.

Campbell, R.J and McNaughton, R. (1994) *Proc. Interamer. Soc. Trop. Hort.*, **38**: 201-205.

Campbell, *R.J.*, El-Sawa, S.F. and Eck, R. (1998) *Fairchild Horticulture Series*, Vol.-2, p. 23.

Chellappan, K. and Roche, J.I. (1982) *South Indian Hort.*, **30**(2): 76-81.

Chowdhury, F.A., Raman, M.A. and Mian, A.J. (1997) *Food Chem.*, **60**: 25-28.

Chunhai, Y., Yingzhi, L. and Feng, F. (2005) *Acta Hort. Sin.*, **32:** 1088-1091.

Corner, E. J. H. (1952) Wayside trees of Malaya II. Government Printing Office, Singapore, Malaysia.

Darlington, A. and Wylie, A.P. (1956) Chromosome Atlas of Flowering Plants. George Allen and Unwin Ltd. London. p.184.

Datta, P. (1988) *M.Sc.* (*Ag.*) *thesis* submitted to Bidhan Chandra Krishi Viswavidyalaya, India. Chin, H.F., Krishnapillay B. and alang, Z.C. (1988) *Pertanika*, **11**: 357-363.

Deb, P., Munsi P.S. and Bhowmick N. (2013) *Proc. II Internatl. Symp. On Minor Fruits and Medicinal Plants for Better Lives 2nd ISMF and MP*, University of Ruhuna, Sri Lanka, 20-December-2013, pp. 68-73.

Deivasigamani, B., Maheswaran, E., Alagappan, K.M. and Kumaran, S. (2009) *Asian J. Exp. Sci.* **23**(3): 605-610.

Deivasigamani, B., Maheswaran, E., Alagappan, E.K.M., and Kumaran, S. (2009) Analysis of genetic diversity on jackfruit using H-15 RAPD marker.*Asian J. Experiment. Sci.*, **23:** 605-610

Dhua, R.S., Sen, S.K. and Bose, T.K. (1983) *Punjab Hort. J.*, **23**: 84-91.

Dwivedi, D.H., Mishra V.and Dwivedi S.K. (2012) *Proc. Internatl. Symp. Minor Fruits and Medicinal Plants for Health and Ecological Security* ISMF and MP, West Bengal,India, 19-22 December, 2012, pp. 117-119.

Elevitch, C.R. and Manner, H.I. (2006) *Artocarpus heterophyllus* (jackfruit) Species Profiles for Pacific Island Agroforestry. www.traditionaltree.org ver.1.1 p.17.

El-Sawa, S. (1998) *FIU Electronic Theses and Dissertations*. 3129. http: // digitalcommons.fiu.edu/etd/3129.

El-Sawa, S. F. and Campbell, R. J. (1997) *Proceedings of the Interamerican Society for Tropical Horticulture*, **41**: 228-235.

Endress, P.K. (1994) Diversity and evolutionary biology of tropical flowers. Cambridge University Press, Cambridge, UK.

Flecker, M. (2001) A ninth-century AD Arab or Indian shipwreck in Indonesia: First evidence for direct trade with China. *World Archaeology* **32 (3):** 335-354.

Gamble, J. S. (1902) A Manual of Indian Timbers. An Account of the Growth, Distribution, and Uses of the Trees and Shrubs of India and Ceylon with Descriptions of Their Wood Structures. Sampson Low, Marston and Company, London

Gardner, E. M., Laricchia, K. M., Murphy, M., Ragone, D., Scheffler, B. E., Simpson, S., Williams, E. W. and Zerega, N. J. C. (2015) *Applications in Plant Sciences* **3:** No.9 pp.15.

Gomez, S. P.M., Joseph, K.B., Sheela, C.V., Sruthi, and Lal L. (2015) *J. Tropical Agriculture* **33(**2 (Part III)**:** 1249-1255

Gopinathan, R., Prasannakumari, S. and Kumaran, K. (1983) *South Indian Hort.,* **31:** 178-180.

Gunjate, R.T., Kokekar, D.T. and Salvi, M.J. (1982) *Proc. Nat. Sem. on Pl. Propagation,* Calcutta, p.87.

Grimm, J.E. and Steinhaus M. (2019) *J. Agric. Food Chem.,* **67:** 258–266.

Grimm, J.E. and Steinhaus M. (2020) *J. Agric. Food Chem.,* **68:** 5838–5846.

Gupta, J.H. and Pandey, I.C. (1985) *Prog. Hort.,* **17**: 361-362.

Haq, N. (2006) Jackfruit, *Artocarpus heterophyllus*, Southampton Center for Underutilised Crops, University of Southampton, Southampton, UK, p.192

Hayes, W.B. (1953) In: *Fruit Growing in India,* Kitabistan, Allahabad, pp. 355-359.

Hossain, M., Haque, A. and Hossain, M. (1979) *Bangladesh J. Agric.,* **4**: 9-12.

Hort, A. F. (1916) Theophrastus Enquiry into Plants. Harvard University Press, Cambridge.

Hossain, A. (1996) *Status report on genetic resources of jackfruit in Bangladesh.* IPGRI Regional Office, Singapore, 30.

Hossain, M.M., Islam M.M. and Rahim, M.A. (2012) *Thai J. Agril. Sci.,* **45:** 203-211.

Hu, L.S., Wu, G., Hao, C.Y., Yu, H. and Tan, L. (2016) *Plant Science.* **248:** 45-56.

Ibrahim, M., Islam, M.S. M.O.H. Helali, A.K.M.S. Alam and Shafique M.Z. (2013) *Bangladesh J. Scientific and Industrial Research,* **48:** 287-292

Islam, S., Hasan, M.A. and Munsi, P.S. (2014) *J. Interacademicia,* **18:** 546-552

Jagadeesha, S.L., Reddy, B.S., Basavaraj, N., Swamy, G.S.K. and Hegde, L.N. (2010) *Karnataka J. Agril. Sci.,* **23:** 293-297

Jarrett, F.M. (1959) *J. Arnold Arboretum Harvard University*, **40**: 329-334.

Jarrett, F. M. (1959) *J Arnold Arboretum* **40: 113–155**

Jarrett, F.M. (1977) *Gard. Bulk, Singapore* **29:** 35-39.

Jose, M. and Velsalakumari, P.K. (1991) *South Indian Hort.*, **39**: 264-267.

Joseph, T. and Kumaran, K. (1994) *Madras Agric. J.*, **81**: 239-241.

Kabir, S. (1998) *J. Immunologica Methods*, **212**: 193-211.

Kanjilal, U. N., Kanjilal, P.C. and Das, A. (1940) Flora of Assam, Published under the Authority of the Government of Assam, India.

Kanzaki, S., Yonemori, K., Sugiura, A. and Subhadrabandhu, S. (1997) *Sci. Hort.*, **70:** 57-66.

Karunakaran, G. and Singh, P. (2019) *Indian Hort.* March-April, 2019 pp.18-19.

Karunarathne, S.I., Ranaweera, L.T., Daundasekera, D.M.K.C. *et al.* (2018) *The J. Agric Sci., Sri Lanka*, **13:** 200-218.

Kavya, K.S., Shyamalamma, and Gayatri, S. (2019) *Indian J. Agric. Res.*, **53**: 8-16.

Kelaskar, A.J., Desai, A.G. and Salvi, M.J. (1991) *Indian J. Plant Physiol.*, **34**: 58-62. Komarayati, S. (1995) *J. Penelltian Hasll Hutan*, **13**: 45-51.

Khan, R., Zerega, N. Hossain, S. and Zuberi, M. I. (2010) *Econ. Botany*, **64:** . 124–136.

Kittipongpatana, O. S. and Kittipongpatana, N. (2011) *LWT - Food Science and Technol.* **44**, 1766–1773, https: //doi.org/10.1016/j.lwt.2011.03.023.

Komarayati, S. and Honda, D. (1994) *J. Penelltian Hasll Hutan*, **12**: 39-41.

Singh, K. (1972) *Farm Information Bull. No.* 71, Directorate of Extn., Ministry of Agriculture, New Delhi.

Krishnamurthi, C.R. and Giri, K.V. (1949) *Proc. Indian Acad. Sci.*, **298**: 155-167. Krishnasamy, V. (1990) *South Indian Hort.*, **38:** 218-219.

Kumar, N. and Kavino, M. (2011) In: *The jackfruit*, pp. 115-127.

Kumar, S., Singh, I.S. and Kumar, S. (1996) *Recent Hort.*, **3**: 55-59.

Ledesma, N., Campbell, R.R.J., El-Saw, S., Wasielewski, J. and Ayala Silva, T. (2005) Selection of jackfruit (*Artocarpus heterophyllus* Lam.) for tropical America.

Lingarajappa, V.G. (1982) *M.Sc. thesis* submitted to USA, Bangalore, India.

Lu, Q.F., Li, Y.Z., Yu, W. and Ye, C.H. (2012) *J. Fruit Sci.*, **29**: 81-85.

Madala A., Rajagopalan, A., Satheeshan, K.N., Meera Manjusha A.V. and Raji Vasudevan, N. (2018) *J. Tropical Agric.*, **56:** 179-182.

Magdalita, P. M., Dayap, F.T. and Valencia, L.D.C. (2011) Farmer participatory breeding and selection. In: The jackfruit 2011 pp.89-113.

Maia,J.G.S., Andrade, E.H.A. and Zoghbi, M.G.B. (2004) *Food Chem.* **85**: 195-197.

Maiti, C. S., Wangchu, L.L. and Mitra, S. K. (2002) *Indian J. Gen. Plant Breed.*, **62:** 369-370.

Maiti, C.S. and Sen, S.K (1999) *The Hort. J.*, **12**: 97-99.

Maiti, C.S. (2010) *Indian J.Hort.*, **67**(2)**:** 169-172

Manjunath, B.L. (1948) Wealth of India. Vol.1 CSIR Publications.New Delhi, 1-240.

Mathur, P.B., Singh, KK. and Kapur, M.S. (1952) *Indian Soc. Refrig. Engineers*, **3**: 40-46.

Medina,.P.A.D.R, Mendioro, M.S. and Cruz. F.S.Jr.dela (2010) *Philippine J. Crop Sci.*, **35:** 16-21

Merrill, E.D. (1912) A Flora of Manila (1976 reprint), Manila: Bu.of Printing, p.369.

Mesquita, F.D.O., Cavalcante L. F. and Jackson de M. Alves. (2019) *J. Agric Sci.*, **11:** 396-407.

Mitra, S.K. and Bose, T.K. (1996) *Scientific Hort.*, **5**: 1-7.

Mitra, S.K. and Maity, C.S. (2000) *Symp. Booklet, Int. Symp. On Tropical and Sub-tropical Fruits,* Cairns, Australia, p. 45.

Momose, K., Hatada, A., Yamaoka, R. and Inoue, T. (1998) *Tropics* **7:** 165-172.

Moncur, M.W. (1985) *Australian J. Bot.*, **33**: 585-593.

Morton, J. (1965) *Proc. Fla. Sta. Hort. Sci.*, **78**: 336-344.

Mukherjee, S.K. and Chatterjee, D.K (1979) *Scientia Hort.*, **10**: 295-300.

Mukherjee, S.K., Rao, D.P., Mukherjee, S., Ray, M. and Mohanty, B.D. (1983) *Indian J. Hort.*, **40**: 150-154.

Naik, K.C. (1952) *ICAR Bull. No. 33*, New Delhi.

Nakasone, H.Y. and Paull, R.E. (1998) *Tropical Fruits,* CABI Publication, U.K., pp. 328-341.

Nakintu, J., Albrecht, C., Müller, C.M. *et al.* (2020) *Genet. Resour. Crop Evol.*, **67:** 605–619.

Narasingha Rao, U. (1952) *ICAR Bull. No. 21*, New Delhi.

Nazeem, PA., Gopikurnar, K. and Kumaran, K. (1984) *Agric. Res. J. Kerala*, **22**: 149-154.

Nivetha, K. and Maheswari, T.U. (2015) *Plant Archives*, **15:** 875-878

Ochse, J.J., Soule, M.J.Jr., Dijkman, M.J. and Welburg, C. (1981) *Tropical and Subtropical Agriculture,* Macmillan Co., pp. 652-655.

Oginuma, K. and Tobe, H. (1995) *J. Plant Res.* **108**: 313. https: //doi.org/10.1007/ BF02344357.

Om Prakash, Kumar, R., Mishra A. and Gupta, R. (2009) *Phcog. Rev.*, **3:** 353-358.

Osman, M., Putch, M. and Mohamad, A. (1996) In: *Underutilized Tropical Plam Genetic Resources: Conservation and Utilization* (Eds. S.S. Quah, R. Kiew, I. Bujang, M. Kusnan, N. Haq and P. de Groot), Penerbit Universiti Pertanian Malaysia, pp. 231-248.

Palupi1, D., Rahayu, S.S.B. and Daryono, B.S. (2019) *SABRAO J. Breed. Genet.*, **51 (1):** 57-67.

Pandey, P.C., Puri, Y.N. and Rehill, P.S. (1981) *Indian J. Forestry*, **4**: 5-7.

Prakash, M. (1998) *Ann. Pl. Physiol.*, **12**: 75-77.

Primack, R. B. (1983) Forester's guide to the Moraceae of Sarawak. Forest Department, Sarawak, Malaysia.

Pushpakumara, D.K.N.G. (2006) *Tropical-Agriculturist* **156:** 87-98.

Pushpakumara, D.K.N.G., Boshier, D.H. and Harris, S.A. (1997) *Trop. Agril Res.* **9:** 11-14

Pushpakumara, D.K.N.G. and Harris, S.A. (2007) *J. Nat. Sci.* **35:** 175-179.

Ragone, D. (1991) *Ph.D. dissertation.* University of Hawaii, Honolulu. Ragone, D. (1995) *Acta Hort.*, **413**: 93-98.

Rahman, MA., Nahar, N., Mian, A.J. and Mosihuzzaman, M. (1999) *Food Chem.*, **65**: 91-97.

Rahman, M.H., Patwary, M.M.A., Barua, H., Nahar, S., and Ahmmed, A.N.F (2016) *The Agriculturists* **14:** 107-111

Rajarajeshwari, H. and Prakash, J. (1999) *The Indian 1. Nutrition and Diet*, **36**: 312-319.

Rajendran, R (1992) In: *Plant Resources of South East Asia.* No.2. *Edible Fruits and Nuts*

Rajmohan, K. and Mohankumaran, N. (1988a) *Agric. Res. J., Kerala*, **26**: 29-38. Rajmohan, K. and Mohankumaran, N. (1988b) *Agric. Res. J., Kerala*, **26**: 169-174.

Roy, S.K and Roy, P.K. (1996) *Acta Hort.*, **429**: 497-502.

Richards, A.V. (1950) *Trop. Agiculturist.*, **105:** 12-13.

Roy, S.K., Hossain, T. and Sahana, B. (1991) *Acta Hort.*, **289**: 145-146.

Roy, S.K., Islam, M.S., Sen, J., Hossain, A.M.E. and Hadiuzzaman, S. (1993) *Acta Hort.*, **336***:* 273-278.

Roy, S.K., Rahman, S.L. and Majumder, R. (1990) *HortScience*, **65**: 335-338.

Saha, B.C., A.D.; Sharad, and R. Barman (2011) *Acta Hort.*, **918:** 785-790.

Sahadevan, P.C., Abdul Samad, A. and Menon, KC.P. (1950) *Madras Agric. J.*, **37**: 497-498.

Sahu, S.K., Liu, M. Yssel, A. *et al.* (2020) *Genes*, **11:** 27; doi: 10.3390/genes11010027

Sakai, S., Kato, M. and Nagamasu, H. (2000) *Amer. J. Bot.* **87(3):** 440-445.

Samaddar, H.N. (1990) Jackfruit. In: Bose TK and Mitra B (eds.) *Fruits of India*, Tropical and Sub-tropical. Naya Prakash, Calcutta, India: 638-649.

Samaddar, H.N. and Yadav, P.S. (1970) *South Indian Hort.*, **18**: 182.

Samaddar, H.N. and Yadav, P.S. (1982) *The Hort. J.*, **1**: 54-60.

Sambamurthi, K and Ramalingam, V. (1954) *Indian J. Hort.*, **11**: 24-29.

Sampath P.M., Swamy, G.S.K., Honnabyraiah, M.K., Shyamalamma, Jayappa, S. J. Suresha G.J. and Tanushree, S. (2019) *Int. J. Chem. Studies* **7(6):** 2680-2682.

Sanmugavelu, KG. (1971) *Madras Agric. J.*, **58**: 97-103.

Sarker, S.R, Roy, P.P. and Zuberi, M.I. (2015) *Gene Conserve.* **14**(57): 1-28

Selvaraj, Y and Pal, D.K (1989) *Fd. Sci. Technol.*,**26**: 304-307.

Sen, P.K. and Bose, T.K. (1959) *Indian Agric.*, **3**: 43-47.

Shyamalamma, S., Chandra, S. B. C., Hegde, M. and Naryanswamy, P. (2008) *Genetics and Molecular Research*, **7**: 645-656.

Simon, L., Shyamalamma, S. and Narayanaswamy, P. (2007) *J. Hort. Sci. Biotechnol.* **82**: 764-768.

Singh, A. (1986) *Fruit physiology and production*. Kalyani publishers,New Delhi, India. 210 p.

Singh, R. (1969) *Fruits*, National Book Trust, New Delhi, p. 115.

Singh, R. and Tiwari, J.P. (1996) *Appl. Hort. Navsari*, **2**: 86-90.

Singh, R. and Tiwari, J.P. (1998) *Indian J. Hort.*, **55**: 213-217.

Singh, S.R., Narayanaswamy, P., Banik, B.C., Shyamalamma, S. and Simon, L. (2011c) *Crop Res. Hisar*, **42**(1/2/3): 157-162.

Singh, U.R., Pandey, T.C. and Upadhyaya, N.P. (1982) *Punjab Hort. J.*, **22**: 103-105.

Singh, S.R, Narayanaswamy, P., Banik, B.C., Shyamalamma, S. and Simon L. (2011b) *Crop Res. Hisar* **42**(1/2/3): 151-156.

Singh, S.R, Narayanaswamy, P., Yumnam, S.S. and Banik, B.C. (2011a) *Asian J. Hort.*, **6**: 199-202.

Sinha, M.M. and Sinha, S.M. (1968) *Sci. and Cult.*, **34**: 372-373.

Soepadmo, E. (1992) In: Plant Resources of Southeast Asia No. 2: Edible Fruits and Nuts. (Eds. E.W.M. Verheij and R.E. Coronel), PROSEA, Wageningen, Netherlands. pp. 86-91.

Soumya, K., Krishnamoorthy, A., Patil, Prakash and Venkatesha, M.G. (2015) *Pest Management in Hort. Ecosystems* **21**: 8-10

Srinivasan, K. (1970) *Agril Res. J. Kerala* **8**: 51-52.

Srivastava, R.P. (1961) *Allahabad Fmr.* **35**: 187-190.

Talbot, W. A. (1911) Forest Flora of Bombay Presidency and Sind, 2. Government of Bombay, Poona.

Teaotia, S.S. and Chauhan, R.R. (1969) *Hort Science*, **1**: 29-37.

Teaotia, S.S., Dayal, K and Asthana, M.P. (1963) *Sci. Cult.*, **29**: 46-47.

Tinggal, S.O.S. (1992) *Acta Hort.*, **1**: 106-111.

Tirtawinata, M.R. (2000) *Symp. Booklet, Int. Symp. Tropical and Sutropical Fruits*, Cairns, Australia, p. 42.

Tran, P.L. *et al.* (2015) *LWT-Food Sci. Technol.*, **62**: 1091–1098.

Valencia, L.D.C. and Magdalita, P.M. (2010) *Philippine J. Crop Sci.*, **35**: 74-79.

Van Der Pijl, L. (1953) *Annals Bogorienses* **1**: 77–99.

Wang, Y.H., Li, Y.Z., Feng, F. and Ye, C.H. (2010) *J. Fruit Sci.*, **27:** 1046-1051

Wang, Z.H., Pan, D.F., Deng, Z.Q., Li, J.G., Ye, Y.X., and Zhu, J.Y. (2012) *J. Fruit Sci.* **29:** 518-519.

Wangchu, L., Singh, D. and Mitra, S. K. (2011) *Acta Hort.*, **890:** 91-101.

Wangchu,.L, Singh, Dheera and Mitra, S.K. (2013) *Genet. Resour. Crop Evol.*, **60**(5): 1749-1762

Warrier, R.R., Singh, B.G., Anandalakshmi, R., Sivakumar, V. Geetha, S., Kumar A.M. and Hegde, M.T. (2009) *J. Agril. Biol.Sci.*, **4**(2): 6-9.

Wasielewski, J. and Campbell, R.J. (1999) *Proc. Fla. State Horticultural Society* 1999

Wight, R. (1843) *Icones Plantarum IndiaeOrientalis*, J. B. Pharoah, Madras.

Witherup, C., Zuberi, M.I., Hossain, S. *et al.* (2019) *Econ Bot.*, **73**: 233–248.

Wong, K.C. (1994) *Japan 1nt. Res. Centre for Agric. Sci. Int. Symp. Series*, No.3, pp. 27-38.

Wong, K.C., Lim, C.L. and Wong, L.L. (1992) *Flavour and Fragarance J.*, **7:** 307-311.

Wu,G., Yang, F.F.,Yan, L. and Tan, L.H. (2010) *South China Fruits*, **5:** 60-61

Yaacob, O. and Subliadrabandhu, S. (1995) The Production of Economic Fruits in South-East Asia, Oxford Univ. Press, Kuala Lumpur, Malaysia. pp. 122-128.

Yang, Z.Y., Wu, C.L., Feng, F., Lu, Q.F.J., Wang, N. and Ye, C.H. (2015) *J. Fruit Sci.*, **32:** 567-571.

Ye, C.H., Wang, Y.H., Li, Y.Z.and Feng, F. (2009) *J. Fruit Sci.*, **26:** 659-665.

Zaghlol, H.A., Elohamy, I.M., Moustafa, S.M.I. and El-Olemy, M.M. (1983) *Sci. Pharma*, **51:** 391-396.

Zhang,Y., Zhang, Y., Xu, F., Wu, G. and Tan, L. (2017) *Sci. Rep.*, **7:** 13423 | doi: 10.1038/s41598-017-13435-8.

9

ACEROLA

V.A. Parthasarathy and S.K. Mitra

1.0 Introduction

Acerola (*Malpighia emarginata* DC), also known as "Barbados Cherry" and "West Indian Cherry", is a shrub found naturally on several Caribbean islands and in Central America and parts of Amazonia (Carrington and King, 2002). Acerola (Malpighia emarginata D.C.) acquired world-wide importance after discovery the high vitamin C level in its fruits (Adeodato *et al.*, 2011). The acerola crop in Brazil began in the mid-twentieth century with the introduction of the first seeds from the Antilles and the United States. In the 1980s, this crop gained great attention due to the high content of ascorbic acid in their fruits, which can be as high as 5g/100 g pulp, corresponding to 80 times the amount found in oranges and lemons (Moraes Filho *et al.*, 2013). Brazil is the world's largest producer of acerola, the Northeast Brazil is responsible for 60 per cent of the national production.

2.0 Composition and Uses

2.1.0 Composition

The fruits contain vitamin C ranging from 900 to over 3000 mg per 100 g when fully ripe. The fruit is also rich in provitamin A (408 IU vitamin A) but is low in B-vitamins (thiamine 0.01 mg, riboflavin 0.04 mg and niacin 0.5 mg). The minerals, calcium (21.4 mg), iron (1.0 mg) and phosphorus (12 mg) are also low. Carotenoid composition has been investigated in acerola fruits (*Malpighia emarginata* DC. syn. *Malpighia glabra* L.) and derived products (Mezadri *et al.*, 2005). In the ripe fruit, four major carotenoids were identified (β-carotene, β-cryptoxanthin, lutein, and violaxanthin) together with other minor carotenoids (neoxanthin, antheraxanthin, neochrome, luteoxanthin, auroxanthin, β-cryptoxanthin-5,6-epoxide, β-cryptoxanthin-5,8-epoxide, *cis*-β-carotene, and *cis*-lutein). An average composition for the ripe fruit has been estimated as follows: β-carotene (536.55 μg/100 g fw), β-cryptoxanthin (417.46 μg/100 g fw), lutein (99.21 μg/100 g fw), violaxanthin (395.33 μg/100 g fw), and total minor carotenoids (197.33 μg/100 g fw). Vitamin A values are similar to those described in tomatoes and some tropical fruits such as guava and papaya. After juice-making, including a pasteurization stage as thermal processing, decreases in carotenoid content were observed as well as progress of *cis*-isomers and structural rearrangement of xanthophylls containing 5,6 epoxide groups. The occurrence of such modifications affected the nutritional value of fruits as well as their antioxidant capability, properties that could be used as a measurement of processing quality. The chemical composition of Acerola fruits are given in Table 1.

2.2.0 Uses

Acerola (*Malpighia emarginata* DC.) is one of the richest natural sources of ascorbic acid and contains a plethora of phytonutrients like carotenoids phenolics, anthocyanins, and flavonoids. There is an upsurge of interest in this fruit among the scientific community and pharmaceutical companies over the last few years. The fruit contains an exorbitant amount of ascorbic acid in the range of 1500–4500 mg/100 g, which is around 50–100 times than that of orange or lemon. Having

Table 1: The Composition of Acerola (100 g fresh weight)*

Water	91.41 g	Zinc	0.10 mg
Energy	32 kcal	Total ascorbic acid	1677.6 mg
Protein	0.40 g	Thiamine	0.020 mg
Total lipid (fat)	0.30 g	Riboflavin	0.060 mg
Carbohydrate, by difference	7.69 g	Niacin	0.400 mg
Fiber, total dietary	1.1 g	Vitamin B-6	0.009 mg
Calcium	12 mg	Folate	14 µg DFE
Iron	0.20 mg	Vitamin A	38 µg RAE
Magnesium	18 mg	Fatty acids, total saturated	0.068 g
Sodium	7 mg	Fatty acids, total polyunsaturated	0.090 g
Sodium	146 mg		
Phosphorus	11 mg		

* Praksah and Baskaran (2018).

a reservoir of phytonutrients, the fruit exhibits high antioxidant capacity and several interesting biofunctional properties like skin whitening effect, anti-aging and multidrug resistant reversal activity. Countries like Brazil, realizing the potential of the fruit have started to exploit it commercially and have established a structured agro-industrial based market. In spite of possessing an enriched nutrient profile with potent "functional food" appeal, acerola is underutilized in large part of the globe and demands greater attention (Prakash and Baskaran, 2018). Besides used as table fruit, acerola fruit is used in preparation of jelly, *sherbet*, juice, salad, *etc.* It was observed that both hot- and cold-pressed frozen juice retained about 85 per cent of the original vitamin C even after 8 months of freezer storage (Miller *et al.*, 1961). Phytochemicals are non-nutrients present in plants, which are known to have diverse biological activities and reduce the risk of many chronic diseases. The major group of phytochemicals includes carotenoids, phenolics, alkaloids, nitrogen-containing compounds, and organosulfur compounds. Acerola is one of the few fruits, which apart from having an exorbitant content of ascorbic acid, also contains a plethora of other phytonutrients like phenolics, flavonoids, anthocyanins and carotenoids in a fair amount. The fruit also contains pro-vitamin A, vitamins B_1 and B_2, niacin, albumin, iron, phosphorus and calcium (Prakash and Baskaran, 2018).

Aparecida De Assis *et al.* (2008) presented a comprehensive review on Acerola detailing importance, culture conditions, production and biochemical aspects. According to them, the demand for acerola has increased significantly in recent years because of the relevance of vitamin C in human health, coupled with the use of ascorbic acid as an antioxidant in food and feed, namely in juices. Acerola fruit has additionally been shown to also contain other significant components, which are likely to lead to a further increase in its production and trade all over the world. The need for careful planning in order to avoid deterioration of biofunctional activity has been highlighted. In particular, it has been shown during the immature green and green stages. Acerola can be a potential useful source for pectin, a macromolecule currently used in the candy industry and in the enrichment of food as dietary fibre.

Uses of Barbados Cherry

Different Cherry Products

Dry Cherry

Cherry Extract

Cherry Powder

Juice with Pulp

Juice

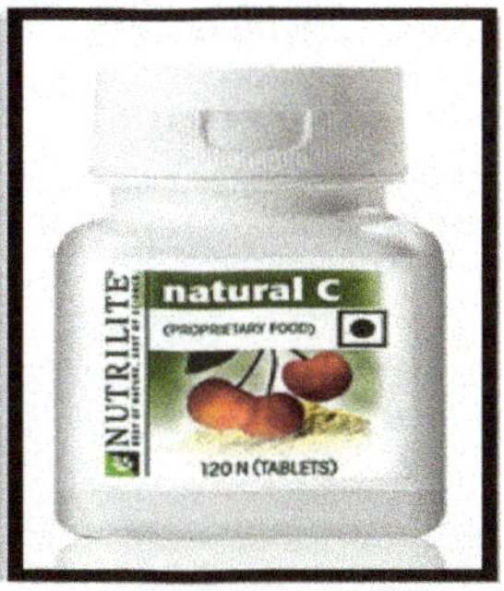

Cosmetics Tablets

Icecream

Fruit Dry Powder

Species and Types of Acerrola

Malpighia punicifolia

***Malpighia glabra* L.**

West Indian Cherry (*Cerise acerola*)

Acerola (Florida Sweet Variety)

Ripe and Unripe Fruits

3.0 Origin and Distribution

It is originally from Yucatan, and can be found in Mexico, Central America, South America and the southern region of Brazil. It is cultivated in the tropics and subtropics throughout the world including the Canary Islands, Ghana, Ethiopia, Madagascar, Zanzibar, Srilanka, Taiwan, India, java, Hawaii and Australia (Kumar *et al.*, 2017). It is believed to be introduced to Florida from Cuba in the 1880s. It is now grown in Caribbean, Latin America, Puerto Rico, Florida and Hawaii. It is also grown in Okinawa (Japan) and Taiwan.

4.0 Species and Cultivars

Acerola belongs to the family Malpighiaceae. The genus *Malpighia* has about 30 species. Acerola (*Malpighia punicifolia*syn, *M. glabra*) is also known as Barbados cherry, West Indian cherry or chercese. Other species producing edible fruits are *M. souzae* and *M. shaferi*.

The acerola is a bushy shrub or small tree attaining a height of 4-5 m with either drooping or upright branching habit. The leaves are small and slightly pubescent. The pink flowers are 1.8-2.0 cm in diameter and are produced in clusters in the axils of the leaves on lateral spurs. There are 10 stamens, 3 styles and a superior ovary (Yamane and Nakasone, 1961).

There are mainly two types of acerola *i.e.*, acid type and sweet type. The acid type acerola fruit (pH 3.1-3.3) contained more vitamin C than the sweet type (pH 3.4-3.6). The acid type is also high yielder. The acid type 'B-17', a selection of Puerto Rico yielded about 8.1-12.1 t/ha with average fruit weight of 9-12 g and 13.3-22.5 g/l juice vitamin C content (Arostegui *et al.*, 1954). In Florida the sweet type selection 'Florida Selection' produces fruits weighing 14 g each, 45.4 kg yield/tree (Ledin, 1958). Puerto Rican 'B-17' is better suited to processing and Florida Sweet is consumed fresh (Gomez *et al.*, 1999).

De Freitas *et al.* (1995) used electrophoretic analysis of the enzyme systems for identifying clones of 14 breeding stocks. According to Cavalcante *et al.* (2007) that, in Brazil, the research has led to the development of varieties, namely, Sertaneja, Olivier, UEL3-Dominga, UEL4-Lígia and UEL5-Natália, UFRPE-7 and UFRPE-8. California Rare Fruit grower (1996) listed the following varieties.

B-17

A common variety. High in vitamin C and subacid to acid.

Dwarf

A low-growing cultivar, to about 2 ft. tall. Grows well in a hanging basket. Can take colder weather than others, to 22° F.

Florida Sweet

A common California variety. Fruit large, 1-1/4 inches in diameter. Skin thick. Flesh very juicy, flavour apple like, semisweet. Vitamin C content about 1,500 to 2,000 mg. per 100 g. Tree erect, with open-type growth and outstanding yields. Originated in Homestead, FL by the Florida Sub-Tropical Engineering Station.

Manoa Sweet

Orange-red fruit of the sweet type. Tree upright, spreading, very productive. Originated in Honolulu, HA by Henry Y. Nakasone, University of Hawaii. Introduced in 1963.

Other named varieties include: Beaumont, Haley, Hawaiian Queen, Maunawili, Red Jumbo, Rehnborg and Tropical Ruby. Varieties other than B-17 and Florida Sweet are not often encountered in Southern California, although several are now available from Pacific Tree Farms, Chula Vista, CA.

In 1956, workers at the University of Florida's Agricultural Research and Education Center in Homestead, after making preliminary evaluations and selections, chose as superior and named the 'Florida Sweet', a clone that was observed to have an upright habit of growth, large fruits, thick skin, apple-like, semi-sweet flavor, and high yield.

The first promising selections in Puerto Rico, on the bases of fruit size, yield and vitamin content, were identified as 'A-l' and 'B-17', but these were later found to be inferior to 'B-15' in ascorbic acid level and productivity. Yields of 10 clones ('A-l', 'A-2', 'A-4', 'A-10', 'A-21', 'B-2', 'B-9', 'B-15', 'B-17', and 'K-7') were compared over a 2-year period (1955-56) in Puerto Rico and 'B-15' far exceeded the others in both years. A horticultural variety in St. Croix, formerly known as *M. thompsonii* Britton & Small, has displayed unusually large leaves and fruits and more abundant flowers than the common strain of Barbados cherry.

Dey *et al.* (2018) described the following varieties.

A. Sweet Varieties

Tropical Ruby

A good yielder variety selected from B-17 seedlings. Trees growth resembles Manoa Sweet with Multiple leader production and requiring certain amount of training to obtain single trunk tree. It grows up to a height of about 15 feet without pruning. Fruit characteristics are similar to those of other sweet varieties type.

Queen

It is a selection from B-17 seedlings. Trees growth habit is upright, spreading, and open. Trees can be trained to a single trunk with minimum afford compared to other sweet varieties. Fruits are resembles to those of other sweet varieties.

B. Sour Varieties

J.H. Beaumont

3B-21 was christened as J.H. Beaumont after the late Dr. John H. Beaumont. It was a clone obtained from seeds. It a compact, low, densely branch growth habit and can easily be trained to a single trunk tree. Both fruit and ascorbic acid production are good. Fruits are large with orange-red colour when fully ripe.

C.F. Rehnborg

This is a high yielding cultivar named after Mr. Carl F. Rehnborg. It was derived from the Miami seeds germinated in Hawaii. Trees have a compact, densely branch growth type and can be easily trained to a single trunk tree. Fruits are large, orange - red colour, turning dark red when fully ripe with low ascorbic acid content.

F. Haley

This cultivar was named after the late Mr. Frederick Haley, the first person to introduce the acerola into Hawaii in large quantities for commercial purposes. It was also derived from Miami seeds. Trees can be easily maintained to a single stem and it has an upright growth habits with branches are mostly of short laterals and are not so spreading. Fruits are medium sized, turning purplish-red when on full maturity. This variety has a wide adaptability especially in the drier areas.

Red Jumbo

It is a selection made from the Miami seedlings. Trees are of single stem with compact growth and well branches, dense foliage and low growing habits. Flowering period extended to January and February and fruit setting percentage and bearing is very high indicating prolific flowering. Fruits are large (9.3 g/fruit), attractive colour turning cherry red to purplish-red upon full maturity.

5.0 Breeding and Improvement

5.1.0 Cytology

Most of the species of the genus *Malpighia* are diploids, *i.e.* 2n=20. The chromosome of Barbados cherry are small about 1.71 to 2.56 μm and metacentric except chromosome 2 which was classified as sub-meta-centric (Mondin *et al.*, 2010). Mondin *et al.* (2010) confirmed the Previous reports indicating that most *Malpighia* species are diploids, including *M. emarginata* with 2n = 20 (Kumar *et al.*, 1981). They described its karyotype was described, providing the identification of the homologues for the ideogram construction. The acerola chromosomes are small (1.71 to 2.56 μm) and metacentric with the exception of chromosome 2 that was classified as submetacentric (Mondin *et al.*, 2010).

5.2.0 Breeding Objectives

Very little work has been done on this crop except in case of Brazil. Breeding studies for acerola (*Malpighia glabra*) improvement aim at obtaining plants that produce fruits with uniform chemical and physical attributes, including high levels of vitamin C, which can provision the market with fresh fruit and frozen pulp. High variability in fruit quality is observed in Brazilian acerola crops, especially those propagated by seeds. In this context, the objective of our research was to evaluate the physical and chemical characteristics of Brazilian acerola genotype fruits (Cavalcante *et al.*, 2007).

In spite of having been introduced into Brazil in the fifties acerola only became a crop of commercial status in the beginning of the 1980s, based on the demand in countries of Europe, Japan and United States, and more recently, of the growing domestic market. The euphoria provoked an unbalanced surge of orchards where the most varied planting techniques possible were used (Paiva *et al.*, 2007). The planting of seedlings obtained through seeds led to a strong segregation in the plant architecture, size, shape and vitamin C content of the fruits and in the yield (Lopes and Paiva, 2002 cited by Paiva *et al.*, 2007). To meet the demands of increased productivity and fruit quality on commercial plantations Embrapa Agroindústria Tropical (Tropical Agroindustry Research Center) developed actions along two lines of research: the population improvement and clonal improvement. In the first phase, plants from seed-based commercial plantations were selected, targeting the establishment of open-pollinated progenies and cloning (Paiva *et al.*, 2007).

Breeding studies for acerola improvement have as an objective to obtain plants that produce fruits with uniform chemical and physical attributes, including high levels of vitamin C, which can provision the market with fresh fruit and frozen pulp (Gomes *et al.*, 2000). The physicochemical parameters are important indicators of maturation and internal and external quality of fruits (Cavalcante *et al.*, 2007), decisive factors for accomplishment of market demands. In this sense, research that studies acerola chemistry and indicates genetic material for fruit production with good external quality and high levels of vitamin C has a high relevance (Cavalcante *et al.*, 2007). Mean yields on such plantations attain between 20 and 50 kg fruits per plant per year (Alves and Menezes,1995).

5.3.0 Genetic Diversity

Carpentieri-Pípolo *et al.* (2000) studied the genetic divergence in acerola and divided divide the 14 genotypes into 3 groups: group I contains 12 genotypes, which corresponds to 86 per cent of the genotypes; groups II and III involve one genotype each, representing 7 per cent of the genotypes. Based on the study of genetic divergence and the most important agronomic character (Vitamin C), the recommended crosses are as follows: genotype AM Mole (group III) with genotypes PR AM, N ° 18, PR 17, PR 16, Eclipse, AM 22 and Dominga (group I), obtaining 7 crossings, totaling 7.7 per cent of the total that could be done; thus, only 7 crossings out of a total of 91 are made, not including the reciprocal, which can be performed in the partial diallel among the 14 genotypes evaluated. The selection of large fruits with small seeds is difficult because of the positive phenotypic and genotypic correlations between the characters fruit weight and seed weight. The selection of large fruits with a larger diameter results in the selection of fruits with a greater amount of pulp.

5.4.0 Selection

5.4.1 Selection Criteria

Cluster analysis and principal components analysis were applied to morphological, yield and fruit quality data from 12 genotypes of *M. emarginata* from Brazil, in order to classify genotypes and identify those with better commercial performance (Gomes *et al.*, 2000). The variables (characteristics) that contributed most to the first main component were average fruit height, average fruit diameter and average weight of pulp in the fruit. The variables contributing most to the second main component were ascorbic acid content per 100 ml juice and canopy diameter. Genotype 5 with 1306.22 mg ascorbic acid/100 ml, 7.27 cm average branch growth, 8.32° Brix and 1.82 x 2.42 cm fruit size was chosen as the most suited for commercial production of fresh fruit for the local market and pulp for the export market (Gomes *et al.*, 2000).

5.4.2 Selection

Paiva *et al.* (2007) gave a graphic history of Acerola breeding in Brazil. According to them an evaluation of the progenies in the first selection cycle consisted of the measurement of the yield and physical-chemical fruit evaluation to determine the vitamin C content, soluble solids content (°Brix) and mean fruit weight (MFW). The proportional gain estimated for the monthly yield, with selection among and within progenies varied from 3.68 to 44.98, respectively, for the months of June and May 1999, while for the other traits the variation in total tritratable acidity and pH was 28.3 to 327.2, respectively. It was expected that the following population would present an over 30 per cent higher vitamin C content than in the plants of the first selection cycle (Paiva *et al.*, 2002).

Further their (Paiva *et al.*, 2007) review indicated that acerola clone selection is the most efficient way of satisfying the immediate demand of varieties, with visible results in the short term (Paiva *et al.*, 2003a). In 2003 Embrapa Agroindústria Tropical released four acerola clones with a view to commercial planting in the region and recommended the joint planting of the clones BRS 235 (Apodi), BRS 236 (Cereja), BRS 237 (Roxinha) and BRS 238 (Frutacor), to avoid genetic uniformity and reduce the vulnerability to pests and diseases in the orchard (Paiva *et al.*, 2003b).

Subsequently, since 1996, plants were selected and genetic parameters estimated in 64 second selection cycle progenies as part of an acerola population improvement program of Embrapa.The variance components, genetic parameters and genetic values were obtained by the individual REML/BLUP procedure. Results indicate that plant selection based on only one fruit harvest is not advisable. The selection encompassed 39 per cent of the progenies (selection among progenies) and 11 per cent of the plants (selection within progenies). This led to the selection of the best 25 progenies for the traits yield, vitamin C, total soluble solids and mean fruit size. In the end, 38 plants were selected and vegetatively propagated to establish a seed orchard to supply commercial seed (Paiva *et al.*, 2007).

6.0 Propagation

Ritzinger *et al.* (2018) have presented a detailed review on the propagation of Acerola. According to them the acerola tree has traditionally been propagated by sexual (seeds) and asexual (grafting and cutting) methods. Chronologically, the cultivation of the acerola tree in Brazil began with the importation of seeds from Puerto Rico, by the Federal Rural University of Pernambuco, in the 1950s, arousing general interest due to the high content of vitamin C in the fruits. However, a significant, though disorganized, growth of commercial plantations, mostly consisting of sexually propagated seedlings, only occurred in the late 1980s and early 1990s. The high genetic variability present in these plantations, with the consequent lack of uniformity and low productivity, motivated the selection of genotypes with good agronomic traits and fruit quality for use as commercial varieties. The first prominent varieties vegetatively propagated for the Brazilian Northeast were 'Flor Branca', 'Okinawa' and 'Sertaneja'.

6.1.0 Seed Propagation

The internal morphology and viability of acerola seeds were studied using tetrazolium (2, 3, 5 triphenyl tetrazolium chloride). Of the tested clones, Florida Sweet had the lowest percentage of seeds with normal embryos (10 per cent) as well as in reaction to seeds without embryos (8 per cent) and the highest percentage of seeds with deformed embryos (81 per cent) (Costa *et al.*, 2003). Clone 07-OS showed a higher percentage of seeds with normal embryos (51 per cent) and a number considered high with no embryos (34 per cent). The other clones showed intermediate values. For all evaluated clones, seeds with normal embryos showed 100 per cent viable embryos. These seeds submitted to the tetrazolium test for a period of 12 hours, presented an intense red color, considered ideal for the positive evaluation of seed viability. These results cannot, however, be taken for prognosis and the calculation of the germination and dormancy rate, just indicating that the seeds are alive (Costal *et al.*, 2003).

The extraction of acerola seeds can be accomplished by the natural fermentation of the fruit pulp (Vichiato and de Vichiato, 2008). The effects of the duration of natural fermentation (0, 48, 96 or 144 h) on the germination of Barbados cherry seeds were studied. Fermentation for 48 h resulted in the highest germination rates. The removal of pulp had greater effects on the enhancement of the germination rate than the duration of fermentation (Vichiato and de Vichiato, 2008).

6.1.1 Seedling Growth and Maintenance

The effects of different substrate compositions and percentage levels of bovine biofertilizer supplied to soil for seedling production of acerola were evaluated by Dantas *et al.* (2014). The levels of bio-fertilizers (0.0, 4.0, 8.0 and 12.0 per cent volume) and two substrates S1= (50 per cent sand, 25 per cent soil, 25 per cent cattle manure); S2 = (50 per cent manure, 25 per cent sand, 25 per cent soil) were evaluated. The application of biofertilizers were applied as basal treatment, 24 hours before sowing and a cover at 32 DAS (days after sowing). The organic manure with biofertilizer, regardless of the substrates, positively influenced the growth and dry matter accumulation of barbados cherry. The addition of bovine biofertilizer to 7.15 per cent by volume, regardless of the substrate used, promoted good quality seedlings (Dantas *et al.*, 2014).

6.2.0 Vegetative Propagation

Acerola is usually propagated by air-layering during spring and summer which rooted in six to eight weeks. Propagation by cutting is also possible. Semi-hardwood leafy cuttings rooted in two months in the mist following IBA treatment at 2500 ppm. Cuttings or air-layered plants are ready for transplanting in six to twelve months. Seeds are used for raising rootstocks for propagation by budding or grafting. Seedlings of *M. suberosa* are considered resistant to nematode and are used as root- stock. Propagation by side-veneer grafting, cleft grafting or shield budding is also in practice.

Propagation of Acerola

Production of Acerola Seedlings.

(A) steps involving the process of fruit pulping, washing, sifting and separation of cores; (B) dry cores containing the seeds and seeds removed from the interior of the cores (C) seedlings emerged in sand bed and (D) plants under development in black polyethylene bags (Ritzinger *et al.*, 2017).

Stages of the process of top wedge grafting in acerola plant ((A) selection, cutting and preparation of the scion and rootstock and wedge-shaped cut at the base of the scion so that it can be fixed in the rootstock; (B) open slit in the rootstock for grafting; (C) the scion with the wedge-shaped cut at the base is introduced into the open slit; (D) wrapping with plastic tape for union between scion and rootstock; (E) the scion is covered with a transparent plastic bag to form a humid chamber and avoid dryness and (F) grafted plant ready for planting (Ritzinger *et al.*, 2017).

Propagation of Acerola

Stages of the Method of Propagation of Acerola Plants by Minicuttings.

(A) new sprouting, ideal for use as minicutting; (B) tray with minicuttings covered with clear plastic; (C) rooted minicuttings; (D) plants developed from minicuttings transplanted to tubes (Ritzinger *et al.*, 2017).

Propagation by Cutting (Left) and Air Layering (Right).

Silva de *et al.* (2010) studied different containers and substrates in the rooting of cuttings of acerola from February to May in Brazil. The cuttings were potted in polyethylene bags with sizes of 8x14 cm (R1), 12x16 cm (R2) and 15 x 21. 5 cm (R3), using six substrates: (S1) 100 per cent of organic compound, (S2) 86 per cent vermiculite and 14 per cent of organic compound, (S3) 86 per cent soil and 14 per cent of organic compound, (S4) 86 per cent large coconut fiber and 14 per cent organic of compound; (S5) 86 per cent fine coconut fiber 14 per cent of organic of compound and (S6) 86 per cent Plantmax® and 14 per cent of organic compound. The rooting of acerola cutting is influenced by substrate. The vermiculite, organic compound and soil proved to be good. The size of container did not influence in rooting (Silva de *et al.*, 2010).

7.0 Area and Production

Acerola is native to the Caribbean islands, Central America and the Amazonian region. More recently, it has been introduced in subtropical areas (Asia, India and South America). Currently, Brazil is the greatest world producer of acerola. Most of the large size plantations are observed, particularly in the Northern region of this country (Aparecida De Assis *et al.*, 2008)

8.0 Soil and Climate

Acerola can be grown on a wide range of soil; however, the soil should be well drained and free of nematodes. A soil pH of 6.0-6.5 is considered most suitable for growing acerola. Landrau and Samuels (1956) observed an increased yield in soil with pH 5.4 with application of lime at 4.5 t/ha.

The plant grows well where the temperature varies between 15-32°C and ceases growth when night temperature falls to 10-15°C. It cannot stand temperature below 5°C. A well distributed rainfall of about 1500-1700 mm per year is satisfactory for growth and fruiting. Although, it grows well at sea level, it is found growing at an altitude of 150 m in Hawaii.

9.0 Cultivation

9.1.0 Planting

Planting is usually done during the rainy season; however, with proper irrigation planting may be done at any time of the year. Before planting, 13 m pits are dug and 15-20 g of complete fertilizers are placed at the bottom of the pits which are refilled with top soil. Trees are planted at 5 m x 4 m spacings. Hedgerow system with 2.44 m within rows and 4.6 m between rows is practiced for clones with compact and upright growth. In some environments, fruit set is a problem which can be partially overcome by planting of several clones.

9.2.0 Training and Pruning

Acerola plants should be trained to a single trunk and 4-6 well distributed scaffold branches should be allowed to grow at 60-90 cm above ground. Normally the plants become bushy making management and harvesting difficult. Pruning should be done every year after harvesting to make the plants more productive.

Pruning encourages axillary branching. Miyashita *et al.* (1964) recorded flower-bud production after 15 to 18 days of pruning.

9.3.0 Irrigation

Irrigation at 10-15 days interval is necessary to maintain production round the year. In tropical areas, flowering and fruiting of acerola may occur up to six times in one year. It has been found to flower or extend new shoot growth following water application (Jackson and Pennock, 1958). However, Michelini and Chinnery (1988) reported that following 15 days drought, irrigation resulted in both vegetative growth and flowering. Growth flushes started approximately 4 days after the first water application, with flower opening beginning approximately 5 days later and anthesis spread over a period of 5-10 days.

Effects of water stress in low season were investigated in acerola trees with five, 10 and 15 days on water stress effect (Adeodato *et al.*, 2011). It was evaluated the buds break dormancy performance, bloom induction in order to increase the yield during the low season that correspond dry season, no use chemicals only water managements. The experiment was conducted at Nutrilite Brazil Farm in Ubajara-CE, with randomized blocks designed with four replications. The treatments were the control treatment and different water stress levels, such as: five days, 10 days and 15 days. Each plot was composed of 12 plants with BV 07 acerola variety, after the each level of stress it has been completed, the irrigations were resumed than maintained next soil field capacity that is 12 per cent. The treatment that had the best yield was 10 days, followed for five days and 15 days, even so the treatment with 15 days of water stress showed results of 34.4 per cent higher when compared to control treatment. Thereby it has been concluded that the water stress management in low season increased the acerola yield (Adeodato *et al.*, 2011).

9.4.0 Manuring and Fertilization

Acerola produces flowers almost round-the-year; hence several applications of fertilizer are required to keep the plants productive. Very little information is available on the nutrient requirements of the tree. Deficiencies of nitrogen (Landrau and Hernandez-Medina, 1959), phosphorus, boron, sulphur and iron were found to reduce growth and yield. Landrau and Hernandez-Medina (1959) reported that maximum fruit yields and ascorbic acid in acerola were associated with the highest levels of N, P and K in leaves. Leao *et al.* (1999) studied the effect of saline stress (0.0, 42.79, 85.56, 128.34 and 171.11 mM NaCl) in acerola in pot for 5 months. Leaf was reduced by 93 per cent in plants subjected to salinity compared to control. Plant N and P contents were not significantly affected by treatment but K content was highest in plants without salt and Na content was highest under highest salt stress. In a greenhouse trial with acerola cv. Okeinawa seedling in Brazil, Brasil *et al.* (1999) applied 4 levels of N (0, 80, 160 and 240 mg/kg soil) as urea; 4 levels of K (0, 100, 200 and 300 mg/kg soil) as potassium chloride; and 2 levels of limestone (absent or present). They observed quadratic response with N application for plant height increment; stem diameter and number of lateral branches, whereas total dry matter showed a positive linear response. Potassium and limestone application, however, showed little effect on seedling development.

Applications of lime bringing the soil pH from 4.5 to 5.5 and 6.5 increased growth, yield and the ascorbic acid content of the fruit (Hernández medina and Vélez-santiago, 1960). Total acidity and ascorbic acid content were significantly correlated and reached a peak in fully developed green fruits, decreasing with increasing maturity. Foliar sprays of Mg and minor elements had no effect on yield or ascorbic acid content (Hernández medina and Vélez-santiago, 1960).

In the majority of Amazonian low fertility soils, the nutrients deficiency resource limits the productivity of the plants, affecting economically the majority of small farmers. This research aimed to evaluate acerola plants macro and micronutrients contents in response to fertilization on yellow oxisol located in the Brasileirinho Agricultural Community, Manaus. The experimental design was randomized blocks with a 3x9 split plot scheme with three collect times, nine treatments and five repetitions, with chemical, foliar and organic fertilization. There were significant variations in the acerola foliar contents in different treatments. The plants that received the treatments with chemical fertilization showed higher Ca, Mg, K, P and N leaf contents. The plants that received foliar fertilization showed the highest Fe, Mn and Zn leaf contents.

10.0 Flowering, Fruit Growth and Development

10.1.0 Flowering

Miyashita *et al.* (1964) studied the details of reproductive morphology of acerola. They observed that floral bud differentiation on primary branches in active vegetative growth occurred 8 to 10 days before the first flower buds emerged. However, in few instances flower buds were also produced in the leaf axils of the node bearing the youngest pair of fully expanded leaves. Although flowering occurred in the leaf axils of actively growing primary branches, maximum flowering usually occurs in leaf axils of numerous secondary branches. After initiation of flowering, the apical meristem of the axillary shoot ceased to produce leaf primordia and gradually flattened out at the apex. The phenomenon was found to be accompanied by cellular enlargement and increased vacuolation. The speals, petals, stamens and carpels differentiated in acropetal sequence.

Normally, a single ovule differentiated in each locule by the time the flower bud emerged from the leaf axils, 6-7 days before anthesis. The inner integument was observed as a short ring of tissue extending to about half the length of the nucellus. The tip of the integument directed towards the receptacle end of the flower at this stage. The outer integument differentiated as an outer ring of tissue before curvature of the developing ovule. Unequal growth of the funiculus caused the ovule to run away from the receptacle end of the bud, becoming semi-anatropous 4 or 5 days prior to anthesis. Wide variations and abnormality in ovule development, mega gametogenesis and embryo-sac development were observed in acerola. Parthenocarpy in acerola seemed to be an inherent character and was evidenced by the normal types of development.

10.2.0 Pollination and Fruit Set

Pollination and fruit set studies showed that low fruit set of acerola in Hawaii was primarily due to absence of effective pollinating agents (Yamane and Nakasone, 1961). The honey bees do not seem to be attracted to the acerola blossoms. Wind was found to be ineffective as an agent for pollination.

Like other Neotropical Malpighiaceae, Acerola produces floral oil in elaiophores, attracting bees of the tribe Centridini (Hymenoptera: Apidae) as pollinators. Among its pollinators, *Centris* (*Heterocentris*) *analis* (F.) stands out due to its abundance at flowers and prompt acceptance of trap-nests. For the first time, Oliveira and Schlindwein (2009) proposed the commercial use of *Centris* bees as orchard pollinators. To develop protocols for rearing and management of these bees, we analyzed trap-nest acceptance, brood-cell construction, and larval diet in Acerola orchards. Although *Centris* species, in general, use numerous pollen host plants, females of *C.analis* showed remarkable flower fidelity to Acerola for pollen supply when nesting in the orchard. Such fidelity was previously expected only for floral oil collection. The ease of acceptance of trap-nests by females of *C. analis*, their prolonged yearly activity period, multivoltine life history, and high pollinator efficiency characterize *C. analis* as an excellent potentially manageable pollinator in Acerola orchards. In another related study, Magalhães and Freitas (2013) tested the feasibility and effectiveness of introducing trap nests inhabited by the oil collecting *bee Centris* (*Heterocentris*) *analis* on increasing productivity of organic orchards of acerola (*Malpighia emarginata*). Trap nest blocks containing 242 nests of *C. analis* were placed on the border of 22 orchards of four acerola varieties and monitored over the blooming season. Results were compared to other 22 orchards without bee introduction and showed an increment of 62 per cent in the number of trap nests inhabited by *C. analis* and acerola yield was significantly higher in the orchards where bees were introduced, independently of variety. The increase in productivity was 1,798 kg/ha. It was concluded that the introduction of inhabited nests of *C. analis* to acerola orchards allows multiplying populations of *C. analis*; to increase the number of bees and to reduce pollination deficits in acerola orchards; and can provide real gains in productivity of acerola commercial plantations.

It was very interesting to read the brief review of Freitas *et al.* (1999) on insect pollination of Acerola. According to their review, trees introduced to Hawaii in the 1950s showed extremely low fruit set (1±3±11±5 per cent) despite profuse flowering, and studies demonstrated that absence of pollination was a contributing factor to its poor fruit set (Yamane and Nakasone, 1961; Miyashita *et al.*, 1964). However, placing honeybee colonies in orchards did not help to increase fruit set (Yamane and Nakasone, 1961). Pollination was therefore another puzzling aspect about the West Indian cherry; little was understood of its pollination requirements (McGregor, 1976 cited by Freitas *et al.*, 1999). Only after Raw (1979) has shown that trees in Jamaica set many fruits after visits of an anthophorid bee, *Centris dirrhoda*, to its flower, has pollination in West Indian cherry been clarified. Apparently, instead of nectar, flowers of the West Indian cherry produce oils in glands of their calyx which are collected by females of *Centris* spp. using specialized clumps of spatulate hairs situated in their forelegs (Raw, 1979).

MALPIGHIALES. Malpighiaceae.

A–N.*Stigmaphyllon ciliatum*. A. Vine, with axillary inflorescences. B. Leaf margin. C. Junction of petiole with blade, showing paired leaf glands. D. Inflorescence base, showing jointed pedicels and bract glands. E. Malpighian trichome. F. Flower, face view, showing characteristic petals. G. Close-up of flower center. Note incurved sepals. H. Flower center, side view. I. Flower base from below, showing paired sepal glands. J. Dissected style/stigma, fertile stamen, and staminodes. K. Ovary cross-section, with one ovule per carpel (appearing as two by curvature of anatropous ovule). L. Ovary longitudinal-section, showing apical-axile placen-tation. M–Q.*Malpighia coccigera*. M. Flower, face view. N. Flower, view from below, showing paired sepal glands. O. Flower, side view. P. Stamens. Note connective gland. Q. Fruit, a drupe, with persistent sepal glands (https://www.sciencedirect.com/topics/agricultural-and-biologicalsciences/malpighiaceae.

It was shown that the West Indian cherry flower has a short lifespan, thus requiring pollination on the day of anthesis, when both cross- and self-pollen grains set fruit. The flower can potentially be pollinated by an array of insect visitors, but only oil-collecting bees of the genus *Centris* and it very attractive. *Centris tarsata* appeared as the main pollinator of West Indian cherry in the area studied, and farmers are advised to encourage its presence in orchards. Despite high levels of natural pollination, only 30 per cent of flowers set fruits. It is suggested that a high proportion of abnormal ovule development may be responsible for low fruit set where pollination is satisfactory, and that West Indian cherry varieties showing a lower percentage of such abnormalities should be selected for (Freitas *et al.*, 1999).

Machado *et al.* (2013) evaluated the physico-chemical in acerola fruit visited by insects in Brazil. Plants were evaluated during two periods: 1. period-October to November, and 2. period-February to March, using the genotypes (CMF 017 and CMF 060). The fruits of each plant were collected in the mature stage to perform the chemical analysis (pH, ascorbic acid, total acidity and total soluble solids) and physical (longitudinal diameter, transverse diameter and weight). The parameters were statistically different in weight, total acidity and ascorbic acid. The visitation of insects in the flowers influence the quality of acerola fruit with less acidity and fruit with higher vitamin C content (Machado *et al.*, 2013).

Vilhena *et al.* (2011) presented parameters to support the conservation and management of Acerola pollinators after monitoring pollinator visits to this fruit in a Brazilian Savanna and identified the pollinator's other food resources. Twenty-three species of pollinators were sampled. Although there were temporal variations in visitor frequency, satisfactory levels of fruit set were obtained. This temporal variation promoted a compensatory effect on plant reproduction as the pollinator species were replaced during the course of flowering. The analysis of pollen loads showed grains of 16 plant species and four pollen types belonging to seven families that formed a network with a nested pattern. Such a network suggests that generalist plants must be available in the crop vicinity. The results suggest that the high species richness of Centridini bees and the stability of the pollination network may have a compensatory effect on Acerola. The data presented suggest the natural viability of cultivating Acerola in the savannas of Central Brazil and the importance of natural lands for providing nesting resources and to enhance the bee diversity in agroecosystems (Vilhena *et al.*, 2011).

Self and cross pollination studies showed significant increase in fruit set with cross pollination being more effective. These studies also showed the existence of some degree of self and cross incompatibility. Self incompatibility appears to be greater than cross incompatibility. Clonal differences in the degree of incompatibility were also noted. In a study conducted in India, Kumar *et al.* (1981) reported incompatibility (genetic-physiological mechanism that makes fertilization impossible or extremely difficult).

Pollen germination studies indicated that pollen is viable only on the day of anthesis. The pollen tube length when grown on artificial media was found to be extremely short in relation to stylar length. In Hawaii, the clone Maunawili, when used as a pollen source, caused higher fruit set of clone No. 21 than the reciprocal cross. This information is important in orchard planning where insects and winds are the main pollinating agents. Pollination and fruit set of acerola showed that flowers are produced in great abundance, but actual fruit set is low, the percentage of fruit set ranging from 1.3 to 11.5 per cent. Where winds or insects are found to be ineffective as pollinating agents, artificial stimulation of fruit setting through the use of growth regulators may be the logical solution to the problem of low fruit set.

10.3.0 Effect of Growth Regulators

Spraying with IBA at 100 ppm or PCA at 50 ppm markedly increased fruit set and subsequent yield (Yamane and Nakasone, 1961). Srinivasan *et al.* (1973) working with West Indian cherry (*M. punicifolia*) observed that post-bloom spraying with GA at 50 to 200 ppm significantly increased the ascorbic acid content, total soluble solids and titratable acidity content of fruit but reduced the total and reducing sugars (Table 2). Michelini and Chinnery (1988) reported increased flowering by spraying uniconizole at 0.25 to 1.0 g a.i. In another trial during 1989 they evaluated the effect of paclobutrazol as soil drenching to container-grown plants and observed that 0.5 g a.i., treatment induced earlier and sustained fruiting as a result of chemical pruning. The final fruit-yield was 192 per cent greater than the control.

11.0 Pests and Diseases

There are no serious pests of acerola. Southern stink bug (*Nezanaviridula*) and leaf-footed plant bug (*Leptoglossus phyllopus*) were reported to damage fruits in Florida (Ledin, 1958).

Cercospora leaf spot (*Cercospora bunchosiae*) on acerola was first reported from Florida in 1960 (Burnett and McFarlin, 1960), later on it was also observed in Hawaii. The fungus caused near-circular spots, which occur on both leaf surfaces, are slightly sunken, brown with grey centres, surrounded by a yellow hallo, and vary in size from a pin point to 10 mm in diameter. In the field under conducive conditions, severe defoliation may take place. Leaf abscission may occur when a leaf is infected near the petiole or when the majority of a leaf surface is covered with spots. It is suggested to maintain fugicidal cover only on young flushes during periods of high rainfall. There is no real need for continued protection, and an occasional spray application should be adequate for controlling leaf spot.

12.0 Harvesting and Yield

12.1.0 Maturity Indices

Fruit skin colour is considered as the main maturity index of acerola fruit. However, guidelines on harvest date need to take into account of the destination of the fruit. In case of freezing or processing into pulp or juice, fruit must be red in colour but sufficiently firm to withstand handling. Fruit quality at this stage of maturation is at a maximum, *i.e.*, sugar content is high and acidity is low. Fruits that are deeply red in colour contain less vitamin C than those at earlier stages of maturation (Alves *et al.*, 1999). Fruits may be harvested at the beginning of maturation (green, yellowish green or with some red pigmentation) for use in the manufacture of powdered products such as pharmaceuticals or concentrates for food enrichment, where vitamin C content is the most important characteristic.

Table 2: Effect of GA on Chemical Constituents in West Indian Cherry Fruits (on fresh weight basis)*

GA (ppm)	*Fruit Weight (g)*	*Ascorbic Acid (mg/100 g)*	*Total Soluble Solids (per cent)*	*Titratable Acidity as Anhydrous Citric Acid (per cent)*	*Sugar (per cent)*		
					Total	*Reducing*	*Non-reducing*
Control							
	5.4	1590	4.8	0.98	4.28	2.62	1.50
Pre bloom							
50	5.2	1580	5.0	1.10	4.32	2.96	1.37
100	5.0	1875	5.0	1.15	4.31	2.56	1.75
200	5.3	1960	5.2	1.18	4.31	2.32	1.99
At bloom							
50	5.4	1925	4.9	1.19	4.16	2.53	1.53
100	5.6	1885	5.0	1.26	4.15	2.45	1.70
200	5.2	2250	5.6	1.41	3.92	2.03	1.89
Post-bloom							
50	5.6	1805	5.1	1.21	4.23	2.38	1.85
100	5.3	2125	5.6	1.32	3.82	2.12	1.70
200	5.2	2625	5.6	1.58	3.19	1.97	1.22

* Srinivasan *et al.* (1973).

The effect of stage of maturity (3 stages) on the chemical composition of acerola harvested during March was investigated by Vendrameni and Trugo (2000). Stage of maturation was characterized by different indicators such as colour, vitamin C, soluble solids, protein, ash, moisture, titratable acidity, pH and sugars. Titratable acidity, sugars and soluble solids increased, and vitamin C and protein decreased with the progress of maturation. It was possible to identify 31 volatile compounds in the mature (red) fruits, 23 in the intermediate maturity (yellow) fruits and 14 in the immature (green) fruits. According to Aparecida De Assis *et al.* (2008) acerola fruit presents fast metabolic activity and its maturation occurs rapidly; there is no need to use any activator to get the acerola ready to consume. On the other hand, when commercialised in ambient conditions, it requires fast transportation or the use of refrigerated containers to retard its respiration and metabolism partially. This feature may make the trade of acerola to distant areas difficult. After the cropping, metabolic reactions continue in the fruit and the physiological process is maintained for a considerable time. In acerola, as well as in other tropical fruits, a series of alterations occurs during the process of maturation, ripening and senescence, the main ones being the degradation of chlorophyll, the presence of carotenoids decrease in acidity, increase in reductor sugars and, mainly, the acute decrease in ascorbic acid during these stages.

Ferreira *et al.* (2009) conducted studies to determine the ideal time of harvest of acerola (*Malpighia glabra*) for industrial production of pulp. The fruits were collected in November and grouped in six t maturity stages: 1 green intense, 2 clearly green, 3(25 per cent red, 4 75 per cent red, 5. red and 6 intense red. The chemical analyses available were soluble solids (SS, °Brix), pH, total acidity (AT, g/100 g), ascorbic acid (AA, mg/100 g) and total soluble sugars (AST, g/100 g). It was found, through pH, SS, AT, AA and AST, that the fruits in six maturation stages could be used for pulp processing. However, for the pulp colour criterion, only acerolas in 4, 5 and 6 maturation stages could be commercialized for processing (Ferreira *et al.*, 2009).

12.2.0 Ripening

Oliveira *et al.* (2012) described the major changes in the antioxidant properties during development of acerola from five different clones. Ripening improved fruit physicochemical quality parameters; however, total vitamin C and total soluble phenols (TSP) contents declined during development, which resulted in a lower total antioxidant activity (TAA). Despite the decline in TSP, at ripening, the anthocyanin and yellow flavonoid content increased and was mainly constituted of cyanidin 3-rhamnoside and quercetin 3-rhamnoside, respectively. The activities of oxygen-scavenging enzymes also decreased with ripening; furthermore, the reduction in vitamin C was inversely correlated to membrane lipid peroxidation, indicating that acerola ripening is characterized by a progressive oxidative stress. Among the studied clones, II47/1, BRS 237, and BRS 236 presented outstanding results for vitamin C, phenols, and antioxidant enzyme activity. If antioxidants were to be used in the food supplement industry, immature green would be the most suitable harvest stage; for the consumer's market, fruit should be eaten ripe (Oliveira *et al.*, 2012).

Chemical composition and antioxidant activity of juice from immature and mature acerola and of concentrated juice from immature acerola were determined (Righetto *et al.*, 2005). Tartaric, malic and citric acids and a high content of ascorbic acid were found in all the juices. Vitamin C contents were 4.80, 1.90 and 0.97 g/100 g for the concentrated immature, the immature, and the mature acerola juices respectively. The total phenol contents decreased during ripening, from 3.8 mg of catechin/g for immature acerola juice to 1.4 mg of catechin/g for mature acerola juice. The concentrated immature juice had a content of 9.2 mg of catechin/g of juice. Catechin, gallic acid, coumaric acid, syringic acid, caffeic acid and ferrulic acid were detected in immature acerola juice by HPLC analysis whereas mature acerola juice showed only one predominant peak with a retention time similar to that of ferrulic acid. The concentrated juice from immature acerola reduced the oxidation of methyl linoleate by 57.2 per cent while the juice from immature acerola reduced the oxidation by 28.1 per cent. These results stated that the antioxidant potential of the acerola juice depended on its content of phenolic compounds and the vitamin C (Righetto *et al.*, 2005).

Acerola (*Malpighia emarginata* DC.) has been known as a fruit rich in vitamin C. However, changes in composition of constituents in relation to cultivar, growing region and maturity have not been studied. Hanamura *et al.* (2008) investigated the composition for five types of acerola fruits: three different cultivars ('BOK', 'Flor Branca' and 'NRA309') growing in Brazil, and one cultivar each in Vietnam ('Vietnam') and in Japan ('Okinawa'). They also investigated the influence of fruit maturity on the compositional profiles. 'BOK' showed the highest vitamin C and malic acid contents whereas 'Vietnam' had the lowest vitamin C content. On the other hand, 'NRA 309' showed the highest level of total polyphenols. In the study of compositional changes during maturation, it was found that at immature stages proanthocyanidin was the major flavonoid, with high vitamin C content, but the levels of these constituents decreased as ripening progressed. At mature stages, anthocyanins appeared to be the major polyphenol in the ripe acerola fruit. This study showed that composition of acerola fruit differed significantly according to cultivar, growing region and maturity. The information provided in this study may help consumers to make an appropriate selection of functional food (Hanamura *et al.*, 2008).

12.3.0 Harvesting

When the time comes to do the fruit harvest, there is a great concern about the correct time, frequency, procedures and forms of storage. Acerola fruit has the maturation process divided into different stages, varying from immature up to ripe mature. Fruit produced for markets needs to be harvested at its optimal maturation stage. Due to the fragility of the fruits, this process is delicate and makes up most of the cost in farming of acerola. It is necessary to avoid injuring fruit, which can accelerate its deterioration (Aparecida De Assis *et al.*, 2008).

Acerola fruits require harvesting by hand, and this needs to be done daily or atleast on alternate days. Complete ripe fruits are extremely soft and easily damaged by impacts. Harvesting should be carried out in the early hours of the morning individually by hand. It is suggested that the pedicel should be cut while holding the fruit with the fingertips (Alves *et al.*, 1999). The fruit should be placed immediately into the harvest box. Harvest boxes should not be large or deep. Plastic boxes that allow no more than 15 cm depth of fruit are recommended. When mature fruits are to be sold in distant markets, they need to be packed in boxes and piled up in low layers, because the weight of high layers can cause injuries to the fruits in lower layers. Whenever possible, transportation should be done in refrigerated trucks in relatively high humid conditions (Aparecida De Assis *et al.*, 2008).

12.4.0 Yield

The seed propagated plants start producing after second year of planting while the asexual propagated plants start bearing fruits in the very first year of planting, and fruiting occurs three to four times a year (Alves *et al.*, 1999). In some regions of north east Brazil, where there is high light intensity and high temperature, irrigated plants begin to fruit in less than a year and production lasts almost round-the-year (Bleinroth *et al.*, 1996). Nakasone *et al.* (1968) recorded yield as high as 23-32 kg/

tree/year from some clones in Hawaii. An orchard with 200 trees produced 2722-4082 kg yield per year (Moscoso, 1956).

13.0 Post-harvest Management

After harvest, fruits should be kept in shade to minimize water loss. Direct exposure of fruits to the sun's radiation causes significant loss of vitamin C (Nakasone *et al.*, 1966). Protection from the sun can be achieved by placing the fruit under the canopy of the plants until transportation. Alves *et al.* (1999) recommended the following steps of post-harvest handling of fruit:

- ✰ Handle and transport carefully to avoid mechanical damage.
- ✰ Cover the vehicles with canvas, preferably light coloured, leaving free space for ventilation between the cover and the product.
- ✰ Arrange the boxes inside the vehicle in a way that does not impair ventilation.
- ✰ Ensure that the bottom of a box is not in contact with fruit inside the one underneath.
- ✰ Transport in the morning or at the end of afternoon, when ambient temperature is low.
- ✰ Reduce as much as possible the time in the field or in transit.

Osmotic dehydration is a process that promotes small changes in the sensory and nutritional properties of the fresh product and is generally used as a pre-treatment for drying, freezing, frying or minimally processing foods. Silva *et al.* (2010) studied of the mass transfer of osmotically dehydrated acerola. The experiments were performed by immersing the fruits in a sucrose solution at 65°Brix for 12 h at ambient temperature (27°C), using three different fruit:solution ratios (1:4; 1:10 and 1:15 (w:w). The kinetics of water loss, solids gain and weight reduction was determined. Effective mass diffusivity was calculated by the inverse method, using the Levenberg-Marquardt minimisation algorithm. The mathematical model used to describe the physical phenomenon of osmotic dehydration was based on Fick's Second Law, considering the fruits as geometrically perfect spheres. The influence of the fruit:solution ratio was not significant in the range of this study. Diffusivity ranged from 1.558×10^{10} to 1.760×10^{10} $m^2 s^{-1}$.

Acerola fruits have a short shelf life. Being highly perishable and acidic, the fruit is by and large consumed after being processed, in the form of pulps and juices. The fruit is commercially processed into puree, juice or juice concentrates and is perfect for the preparation of jams, jellies, fruit juices and supplements (Prakash and Baskaran, 2018). The fruit can also be used to prepare a range of other products like ice cream, gelatin, juice, soft drinks, nectar, gum, fruit conserve, nutraceutics, yogurts and sodas. It is also used in the fortification of infant foods and for the production of nutritional and pharmacological products (Badejo *et al.*, 2008). Of late, many new and diversified products have appeared in the Brazilian market like blends of acerola and cashew, acerola and orange, and blends with guarana, powdered refreshments and concentrated juices (Matta *et al.*, 2004).

Fruit Bearing, Staking, Harvesting and Marketing of Acerola

Acerola Tree with Fruit Bearing

Staking of Dwarf Acerola Tree (Left) and Hand Picking of Fruits (Right)

Acerola is Ready for Marketing

Selling of Acerola at Local Market

Freeze-drying is a technique that results in high-quality dehydrated products due to the absence of liquid water and the low temperatures required in the processSeveral quality parameters, such as water activity (*a*w), glass transition temperature (*T*g), Vitamin C content, shrinkage and rehydration capacity were investigated for the freeze-drying of acerola fruits (Marques *et al.*, 2007). The variation of temperature with time at different positions of the samples was measured during the freezing of samples, performed prior to the freeze-drying. Drying kinetic curves were obtained for different types of samples. The extent of shrinkage after freeze-drying was investigated and related to the glass transition temperature, *T*g. The variation of water activity (*a*w) along the drying and sorption isotherms was obtained for different freezing techniques. It was observed that freeze-dried acerola fruits can be easily reconstituted, and important nutritional parameters are well preserved after the process. They may be considered as a good source of Vitamin C, whose content was best preserved for freeze-drying of fruits at an intermediary stage of ripening.

In Japan, acerola is used in the production of juice, soft drinks, gums and liqueurs. The USA and Europe are great potential markets due to the scarcity of similar products. In Europe, mainly in Germany, France, Belgium and Hungary, acerola extracts are used to enrich pear or apple juices. In the USA, its main use has been in the pharmaceutical industry. Latin America is also a promising market, mainly in countries such as Argentina, Chile and Uruguay (Aparecida De Assis *et al.*, 2008).

13.1.0 Storage and Transportation

The marketing of fresh acerola fruits is limited to the areas in the close proximity to the producing regions due to its high perishability at ambient temperatures (two to three days). Even under refrigeration its life is limited to about 8 days. For long transportation freezing of acerola fruit at temperatures below –18°C has been used. However, pre-cooling (forced air or hydro-cooling) is strongly recommended to reduce the time of freezing in order to avoid problems such as yellowing of fruit. Storage of frozen fruit must be below a temperature of –20°C. Frozen acerola fruit for international trade is transported by sea. Boxes are palletised and put in containers with temperatures not above that of storage, *i.e.* –18°C. Keeping the cold chain during transit is essential for the maintenance of product quality.

Acerola fruits can be stored at 8°C and 85-90 per cent RH with PVC wrapping for one week compared with 2-3 days at room temperature (Alves *et al.*, 1995) (Table 3).

Chai *et al.* (2014) got a patent for the invention relates to an acerola cherry powder and a method for producing the same, wherein the powder has a high content of juice Solids, a good fluidity, and is easy for processing and preservation. The said powder comprises at least 51-60 mass percent of acerola cherry juice solids and 40-49 mass percent of oxidized starch. The manufacturing method comprises: preparing a concentrate of acerola cherry juice, adding oxidized starch to the concentrate and spray-drying.

Table 3: Characteristics of Acerola Fruits Harvested at Pink Stage and Stored at under Refrigeration, 8°C and 85-90 per cent Relative Humidity with (Ambient atmosphere - A.A.) and Without (modified atmosphere - M.A.) PVC Wrapping*

Characteristics	*Days after harvest*					
	1	*4*	*7*	*8*	*11*	
Weight loss	-A.A.	0.000A1	8.723B	13.537B	16.193B	27.704B
	-M.A.	0.000A	1.768A	1.970A	5.549A	5.828A
Juice yield	-A.A.	61.333A	54.016B	48.867B	46.769B	39.735B
	-M.A.	61.333A	58.321A	59.439A	58.486A	52.201A
Soluble solids	-A.A.	6.875	6.663	6.941	6.538	5.499
	-M.A.	6.875	6.950	7.400	6.752	6.028
Soluble sugar	-A.A.	4.305	4.286	4.332	4.288	3.745
	-M.A.	4.305	4.286	4.288	3.745	2.842
Reducing sugar	-A.A.	3.952	4.024	4.113	3.875	3.258
	-M.A.	3.952	3.943	4.089	3.438	2.538
Acidity	-A.A.	1.206	1.259	1.148	1.182	0.871
	-M.A.	1.206	1.283	1.182	1.190	0.841
pH	-A.A.	3.500	4.100	4.200	4.175	4.500
	-M.A.	3.500	4.375	4.450	4.350	4.650
SS/TA ratio	-A.A.	5.702	5.313	6.041	5.819	6.362
	-M.A.	5.702	5.422	6.272	5.699	7.208
Vitamin C	-A.A.	1.773	1.070	0.988	0.869	0.591
	-M.A.	1.773	0.942	0.866	0.818	0.683

1 Means in the same column, followed by the same letter are not significantly different (P>0.05).

* Alves *et al.* (1995)

The acerola pulp and residue extracts were microencapsulated by spray and freeze-drying, using gum arabic and maltodextrin as encapsulating agents (Rezende *et al.*, 2018). Total anthocyanins, carotenoids (CA), ascorbic acid (AA), phenolic compounds (PC), total flavonoids, antioxidant activity, color, moisture, water activity, solubility, hygroscopicity and microstructure of the powders were analyzed. The acerola residue extract had higher concentrations of bioactive compounds (except for AA) and lower antioxidant activity. The microencapsulated powders by spray and freeze-drying showed good physico-chemical properties. Spray-dried powder has lower microencapsulation efficiency for CA, AA and PC and in general, it presented the best characteristics due to its higher concentration of bioactive compounds (except AA), and antioxidant activity by FRAP and ORAC assays. The spray-dried particles were of spherical shape while the freeze-dried products were of irregular structures. Overall, these results demonstrate the better utility of agro-industrial acerola residue in the form of microparticles of bioactive compounds, retaining good antioxidant activity (Rezende *et al.*, 2018).

14.0 Biotechnology

14.1.0 Micropropagation

Alloufa *et al.* (1999) cultured leaf explants from 3- and 5-month-old *M. glabra* seedlings. Callus formation was best on MS medium supplemented with 5 µM kinetin + 0.5 µM 2,4-D or 10 µM kinetin + 0.5 µM 2,4-D. De Melo *et al.* (1999) cultured nodal segments of *M. emarginata* cultivars in MS, DKW or WPM media. The number of shoots produced did not differ between media, but WPM and DKW produced more large shoots than MS.

Culture conditions for shoot tips excised from cutting plants of acerola propagated through cuttings were investigated to develop a shoot-tip culture system (Yamamoto *et al.*, 2017). Shoot tips of acerola "Sanmikei (Hosoba)" were grown on a medium, consisting of MS basal components with 3 per cent sucrose. The addition of 0.1 ppm 6-benzylaminopurin (BA), 0.05 ppm BA + 0.05 ppm zeatin or 0.1 ppm BA + 0.1 ppm zeatin was effective for the proliferation of shoots. Those that had grown two to three cm from the subculture, were examined further for rooting, which had been stimulated when the shoots were immersed in 250 ppm indole-3-butyric acid (IBA) for 15 seconds. Ten to eleven days of dark treatment enhanced this. All plantlets were recovered after acclimatization (Yamamoto *et al.*, 2017).

14.2.0 Molecular Markers

14.2.1 Diversity Analysis

Chowdhury *et al.* (2005) used morphometric descriptors and random amplified polymorphic DNA (RAPD) markers wer for the classification of acerola cultivars 'Okinawa Sweet' and 'Kagoshima Sweet' grown in Okinawa and Kagoshima Prefecture, respectively by comparison with seven cultivars introduced from Hawaii into the germplasm bank of Japan International Research Center for Agricultural Sciences (JIRCAS). Even though cultivar identification could not be achieved by using morphometric descriptors, all the genotypes were classified according to RAPDbased DNA analysis using 57 RAPD markers amplified by 24 primers. Genetic similarities between all the pairwise combinations of cultivars were calculated using Jaccard's coefficient, which ranged between 0.00 to 1.00 with a mean value of 0.304, and 0.077 to 0.967 with a mean value of 0.406 for the morphometric descriptors and RAPD markers, respectively. RAPD similarity coefficients were correlated with morphometric similarity coefficients (r=0.578). The relative position of different cultivars in a RAPD-based dendrogram was partly similar to that in a dendrogram between morphometric descriptors. The RAPD-based dendrograms revealed that the 9 cultivars could be classified into two groups. Four cultivars namely, 'Hawaiian Queen', 'Manoa Sweet', ' Kagoshima Sweet', and 'Okinawa Sweet', were clustered very closely, indicating the presence of a narrow genetic variability among them, based on both marker types.

Event thought it is a tropical fruit originated from American continent; there are very few varieties of acerola available to Brazilian producers, limiting the crop's potential. Moraes Filho *et al.* (2013) evaluated the genetic diversity of 42 accessions from the Acerola Active Germplasm Bank of Universidade Federal de Pernambuco. Using 15 RAPD primers, 182 markers were obtained, of which 166 were polymorphic and 16 were monomorphic. They found high genetic variability among the accessions ($E = 0.29), with no redundancy. Considering the accessions from the states of Pernambuco, Bahia and Pará as distinct groups, there was greater diversity in accessions from Bahia than from the other two states. In Brazilian orchards, there was high variability among cultivated genotypes. On the order hand, high variability allows the identification of superior genotypes for cropping industry. This study aimed to evaluate the genetic variability among 56 genotypes using ISSR (Inter Simple Sequence Repeats) primers. Leaf samples were collected in Pacajus-CE and taken to the laboratory of Molecular Biology post-harvest, in Fortaleza. Altogether, 20 primers were used which yielded 148 polymorphic bands (79.57 per cent), enabling the differentiation within the population study. As a result, this information may be used in future studies on breeding programs, such as choosing best combinations for parental crossings (Lima *et al.*, 2015).

Twenty-four acerola accesses, of the Active Germplasm Bank were analyzed using molecular markers obtained with random amplified polymorphic DNA (RAPD) and simple sequence repeated (SSRs) primers (Salla *et al.*, 2002). A total of 164 and 73 markers were obtained with RAPD and SSR primers, respectively. Polymorfic markers were scored as present or absent and analyzed using the UPGMA cluster analysis. The results presented reveal high levels of polymorphism in the studied collection. Comparative analysis of the phenograms, generated with both RAPD and SSR primers, revealed that while some accesses clustered in different groups other accesses presented the same association. However, there was large RAPD variation among the accesses. The associations observed with molecular markers were, for many accesses, the same of those determined on the basis of morphological characters (Salla *et al.*, 2002).

In Brazilian orchards, there was high variability among cultivated genotypes which allows the identification of superior genotypes for cropping industry (Lima *et al.*, 2015). They studied the genetic variability among 56 genotypes using ISSR (Inter Simple Sequence Repeats) primers. Leaf samples were collected in Pacajus-CE and taken to the laboratory of Molecular Biology post-harvest, in Fortaleza. Altogether, 20 primers were used which yielded 148 polymorphic bands (79.57 per cent), enabling the differentiation within the population study. As a result, this information may be used in future studies on breeding programs, such as choosing best combinations for parental crossing (Lima *et al.*, 2015).

The genetic diversity in four acerola cultivars were evaluated by molecular markers ISSR, and the pollen viability of five genotypes were estimated (Nascimento *et al.*, 2017). The selected ISSR primers amplified 561 bands. Of this total, 247 bands were polymorphic, representing an average of 48 per cent. In the experiment of pollen viability, the Cabocla, Costa Rica and Flor Branca cultivars had ahigh pollen viability potential, standing out the Cabocla cultivar with the highest value (95.9 per cent) among the cultivars. The ISSR markers allowed to estimate the genetic diversity and cultivars that showed high pollen viability, can they recommend as donor genotypes of pollen in future studies of genetic improvement (Nascimento *et al.*, 2017).

14.2.2 Genetic Fidelity

Sequence-related amplified polymorphism (SRAP) was conducted to validate the genetic homogeneity of the regenerated plants from shoot tip culture. In SRAP analysis, using 13 primer combinations, no polymorphisms among the regenerated and original plants were detected (Yamamoto *et al.*, 2017).

15.0 References

Adeodato, T., Duran, J.A.R., Soares, A. and Diniz, J. (2011) *Proc. the Third Scientific Conf. Int. Soc. Organic Agril. Res.earch* (*ISOFAR*), 28th September - 1. October 2011 in Namyangju, Korea Republic.

Alloufa, M.A.I. and Campos, M.de A. (1999) *Rev. Bras. Frutic.*, **21**: 284-287.

Alves R.E. and Menezes J.B. (1995) Botânica da aceroleira. In: São José AR and Alves R.E. (eds.) *Acerola no Brasil: produção e mercado.* DFZ/UESB, Vitória da Conquista, p. 7-14.

Alves, R.E., Chitarra, A.B. and Chitarra, M.I.F. (1995) *Acta Hort.*, **370**: 228-229.

Alves, R.E., Filgueiras, H.A.C., Mosca, J.L. and Menezes, J.B. (1999) *Acta Hort.*, **385**: 31-36.

Anonymous (1996) https: / /www.crfg.org/pubs/ff/acerola.html

Aparecida De Assis, S., F.P. Fernandes, A. B. G. Martins, De Faria Oliveira, O. M. M. (2008) *Fruits*, **63**: 93–101.

Arostegui, F., Asenjo, C.F., Muniz, A.I. and Alemany, L. (1954) *Proc. Florida Sta. Hort. Soc.*, **67**: 250-255.

B.M. Freitas, J. E. Alves, G. F. Branda And. Araujo, Z. B. (1999) *J. Agric. Sci., Cambridge*, **133:** 303-311.

Badejo, A.A., Tanaka, N., Nobukazu Esaka, M. (2008) *Plant Cell Physiol.*, **49**(1): 126–132.

Bleinroth, E.W., Alves, R.E. and Menezes, J.B. (1996) *EMBRAPA-SPL. FRLIPEX.*, pp. 13-21.

Brasil, E.C., Silva, A.M.B., Muller, C.H. and Silva, G.R. Da (1999) *Rev. Bras. Frutic.*, **21**: 52-56.

Burnett, H.C. and McFarlin, J.R. (1960) *Pl. Dis. Reptr.*, **44**: 505.

Carpentieri-Pípolo V., Destro D., Prete C.E.C., Gonzáles M.G.N., Popper I.O., Zanata S., Silva F.A. (2000) *Pesquisa Agropecuária Brasileira*, **35**(8): 1613-1619.

Carrington CMS and King RAG (2002) *Sci. Hortic.*, **92:** 1-7.

Cavalcante, Í.H.L., Beckmann, M.Z., Martinsa, A.B.G. and Campos, M. C. C. (2007) *Fruits*, **62:** 27–34

Chai,Y.,H.Weng and Xue L. (2014) United States (12) Patent application Publication (10) Pub. No.: US 2014/01994.64 A1. Pub. Date: Jul. 17, 2014.

Chowdhury, A.K., Y.Yonemoto, H. Kato and Macha, M.M. (2005) *Jpn. J. Trop. Agr.*, **49**(4): 255 – 263.

Costa, L. C., M.C.M.D. Pavani, F.V. Moro and Perecin, D. (2003) *Rev. Bras. Frutic.*, **25**(3): 532-534.

Dantas, K. A., da Figueiredo, T.C., de Mesquita, E.F., da Silva Sá F.V. and Ferreira, N.M. (2014) *Revista Verde de Agroecologia e Desenvolvimento Sustentável*, **9** (1): 157-162.

De Freitas, N.S.A. de., Burity, H.A., Bezerra, J.E.F. and Silva, M.V.da (1995) *Pesquisa Agropecuariac Brasileira*, **30**: 1453-1457.

De Melo, N.F.de., Okasaki, W.Y., Leite, C.B., Fari, M., De Melo, N.F. (1999) *Ciencia e Agrotecnologia*, **23**: 102-107.

Dey, K., Ghosh, A., Bauri, F.K., Bhowmick, N., Dey, A.N., Mishra, D.K. and Das, B. C. (2018) *Agric. Rev.*, **39**(2): 144-150.

Esashika, T., Oliveira, L. A. de; Moreira, F. W. (2013) *Revista de Ciências Agrárias* (Portugal), **36:** 399-410.

Ferreira, R. M. de A.; Aroucha, E. M. M.; Souza, P. A. de; Queiroz, R. F. de; Pontes Filho, F. S. T. (2009) *Revista Verde de Agroecologia e Desenvolvimento Sustentável* **4** (2): 13-16.

Gomes J.E., Perecin, D., Martins, A.B.G. and Almeida, E.J. (2000) *Pesqui. Agropecu. Bras.*, **35:** 2205–2211.

Gomes, J.E., Perecin, D., Martins, A.B.G. and Ferraudo, A.S. (2000) *Rev. Bras. Frutic.*, **22:** 36-39

Gomez, P., Reynes, M., Dornier, M. and Hebert, J.P. (1999) *Fruits (Paris)*, **54**: 247-260.

Hanamura,T., Uchida, E. and Aoki, H. (2008) *J. Sci. Food Agril.*, **88:** https: //doi. org/10.1002/jsfa.3285

Hernández medina, E. and Vélez-santiago, J. (1960) *Proc. Amer. Soc. Hort. Sci.*, **4:** 20-25.

Jackson, G.C. and Pennock, W. (1958) *J. Agric. Univ. Puerto Rico*, **42**: 196-205.

Kumar, R.S., Parthasarathy, V.A. and Sundaram, P.K. (1981) *Prog. Hort.*, **13**(2): 73-76.

Kumar, S., Bhajipale, N.S., Sanghai, D.B., and Padgilwar, S. S. (2017) *Int. J. Res. in AYUSH and Pharma. Sci. (IJRAPS)*, **1:** 69-72.

Landrau, P.Jr. and Hernandez-Medina, E. (1959) *Univ. Puerto Rico, J. Agric.*, **43**: 19-33.

Landrau, P.Jr. and Samuels, G. (1956) *Univ. Puerto Rico J. Agric.*, **40**: 224-234.

Leao, P.C., Nagueira, R.J.M. and Araujo, E.De.L. (1999) *Ihertngla Serie Botanica*, **52**: 3-10.

Ledin, B.R. (1958) *Univ. of Florida Agr. Exp. Sta. Bull.*, 594.

Lima, E. N., *et al.* (2015) *Comunicata Scientiae*, **6:** 174-180

Lima, E.N.M.E., Bezerra de Araujo, C.H.C.M., Bertini, Moura, C.F.H. and Hawerroth, M. C. (2015) *Comunicata Scientiae* **6:** 174-180.

Machado, C. S., C.A.L. Carvalho, de, and Brito, P.O. (2013) *Magistra,* **25:** 24-29.

Magalhães,C. and Freitas, B. (2013) *Apidologie,* **44**: 234-239.

Marques, L.G., M. C. Ferreira and Freire, J.T. (2007) *Chem. Engg Proc.,* **46**: 451–457.

Matta, V.M., Moretti, R.H. and Cabral, L.M.C. (2004) *J. Food Eng.,* **61:** 477–482.

Mezadri, T., Pérez-Gálvez, A. and Hornero-Méndez, D. (2005) *Eur. Food Res. Technol.,***220:** 63–69.

Prakash, A. and Baskaran, R. (2018) *J. Food Sci Technol.,* **55:** 3373–3384.

Michelini, S. and Chinnery, L.E. (1988) *Proc. Interamerican Soc. Trop. Hort.,* **32**: 65-73.

Michelini, S. and Chinnery, L.E. (1989) *Acta Hort.,* **239**: 281-284.

Miller, C.D., Wenkam, N.S. and Fitting, K.O. (1961) *Hawaii Ag. Exp. Sta. Univ. Circular,* No. 59.

Miyashita, R.K., Nakasone, H.Y. and Lamoureux, C.H. (1964) In: Reproductive morphology of acerola (*Malpighia glabra* L.). Honolulu, Hawaii, *Hawaii Agricultural Experimental Station, Technical Bulletin* No. 63.

Miyashita, R.K., Nakasone, H.Y. and Lamoureux, C.H. (1964) *Hawaii Ag. Exp. Sta. Tech. Bull.,* No. 63.

Mondin, M., C.A.De Oliveira and Vieira, M.L.C. (2010) *Rev. Bras. Frutic., Jaboticabal - SP,* **32:** 369-374

Moraes Filho, R.M., Martins, L.S.S., Musser, R.S., Montarroyos, A.V.V. and Silva, E.F. (2013) *Gen. Mol. Res.,* **12** (4): 5145-5151.

Morton, J. (1987) In: *Fruits of warm climates* (Ed. Julia F. Morton), Miami, FL. pp. 204-207.

Moscoso, C.G. (1956) *Econ. Bot.,* **10**: 280-294.

Nakasone, H.Y., Miyashita, R.K. and Yamane, G.M. (1966) *Proc. Amer. Soc. Hort. Sci.,* **89**: 161-166.

Nakasone, H.Y., Yamane, G.M. and Miyashita, R.K. (1968) *Hawaii Ag. Exp. Sta. Circular,* No. 65.

Nascimento, P.H.D., S.S. Souza F. de F. Souza and de Melo, N. F. (2017) In: *Jornada de Integracao da Pos-Graduacao da Embrapa Semiarido (II: 2017: Petrolina, 2017):* Anais da II Jornada de Integracao da Pos-Graduacao da Embrapa Semiarido, Petrolina: Embrapa Semiarido, 2017. pp. 153-159.

Oliveira' L.D.S., Moura, C.F.H., De Brito, E.S., Mamede, R.V.S. and De Miranda, M.R.A. (2012) *J. Agric. Food Chem.,* **60** (32): 7957–7964.

Oliveira, R. and Schlindwein, C. (2009) *J. Econ. Entomol.,* **102**(1): 265-273.

Paiva, J.R., Alves, R.E., Barros, L.M., Crisóstomo, J.R., Moura, C.F.H., Almeida, A.S. and Norões, N.P. (2003b) In: *Embrapa Agroindústria Tropical, Fortaleza,* p. 3 (Comunicado Técnico, 87).

Paiva, J.R., Alves, R.E., Melo, F.I.O., Cordeiro, E.R. and Almeida, A.S. (2002) *Crop Breed. Applied Biotechnol.*, **2:** 299-306.

Paiva, J.R., Alves, R.E., Santos, F.J.S., Barros, L.M., Almeida, A.S., Moura, C.F.H., Cacau, C.F.H. and Norões, N.P. (2003a) *Ciência Agrotécnica* **27:** 1038-1044.

Paiva, J.R., Cordeiro, E.R., Corrêa, M.C.M. and Resende, M.D.V. (2007) *Crop Breed. Appl. Biotechnol.*, **7:** 125-132.

Raw, A. (1979) *Revista de Biologia Tropical,* **27:** 203-205.

Rezende, Y.R.R.S., Nogueira J.P. and Narain, N. (2018) *Food Chem.,* **254:** 281-291.

Righetto, A.M. Netto F. M. and Carraro, C. (2005) *Food Sci. Technol. Internatl.,* **11:** https: //doi.org/10.1177/1082013205056785.

Ritzinger, R., Ritzinger, C.H.S.P., Fonseca, N.F. and Machado, C.F. (2018) *Rev. Bras. Frutic., Jaboticabal,* **40** (3): DOI: http: //dx.doi.org /10.1590/0100-2945201892.

Salla, M.F.S., Ruas, C.D.F., Ruas, P.M. and Valéria, C.P. (2002) *Rev. Bras. Frutic.,* **24**(1): 15-22.

Silva de, P. N., Costa, L., E., Ferreir A.F.A., Silva, A. C. da R. and Gomes, A. V. do (2010) *Revista Agrarian,* **3:** 126-132.

Silva, M.A.C., Corre^a, J.L. G. and da Silva, Z.E. (2010) *Int. J. Food Sci. Technol.,* **45:** 2477–2484.

Srinivasan, C., Pappiah, C.M. and Doraipandian, A. (1973) *Indian J. Exp. Biol.,* **11:** 469-470.

Vendramini, A.L. and Trugo, L.C. (2000) *Food Chem.,* **71:** 195-198.

Vichiato, M. and de Vichiato, M.R.M. (2008) *Ecossistema,* **33:** 59-64.

Vilhena, A., Rabelo, L., Bastos, E. and Augusto, S. (2011) *Apidologie,* **43** (1): 51-62.

Yamamoto, M., Ito, A. , Sogabe, A., Kajiwara, Y. and Ishihata, K. (2017) *Trop. Agril. Dev.,* **61**(2): 57-61.

Yamane, G.M. and Nakasone, H.Y. (1961) *Proc. Amer. Soc. Hort. Sci.,* **78:** 141-148.

Index

Q

R

S

T

U

V

W

X

Y

Z

www.ingramcontent.com/pod-product-compliance
Ingram Content Group UK Ltd.
Pitfield, Milton Keynes, MK11 3LW, UK
UKHW022218051026
15624UKWH00001B/220

9 789354 616198